COMMENTS ON THE PREVIOUS EDITIONS

'I am very impressed with the book's empirical and theoretical range.'

Mark Bouman, Professor and Department Chair, Chicago State University

'An excellent book, monumental in scope, breathtaking in depth . . . a great achievement.'

Christopher Smith, Professor of Geography and Planning, SUNY at Albany

'This remains the most definitive text and the best introductory one.'

Kevin Ward, Senior Lecturer in Geography, University of Manchester

'It has huge depth and breadth of coverage, with which no other text can (or probably wants to) compete.'

Tim Coles, Lecturer in Human Geography, University of Exeter

'This text is more comprehensive than any text I know in the subject area. It is nearly 700 pages long and is a tour de force.'

Linda Stainer, School of Geographical Sciences, University of Derby

Urban Geography: A Global Perspective offers the most contemporary, comprehensive, and insightful presentation on urban geography, a compelling readable, spectacularly exciting, and pleasingly sophisticated investigation of an extensive range of vital urban issues at the local and global scales. It is a must-read for any urban student and scholar the world over.

James O. Wheeler, The Merle Prunty Jr. Emiritus Professor of Geography, University of Georgia

An impressive overview of the urban process, from the origins of cities to the problems of contemporary urban regions in the developed and developing worlds.

L. S. Bourne, Professor of Geography, University of Toronto

Michael Pacione's *Urban Geography: A Global Perspective* provides a comprehensive global view of urban geography that is chock full of information and lavishly illustrated with maps, photographs, and illustrations. At the same time it is theoretically smart, rich in historical detail, and questioning about urban futures in a global world. It is the ideal text for undergraduate courses in urban geography and urban studies.

Professor Neil Smith, City University of New York

I heartily endorse Pacione's new text. He has set himself the daunting task of introducing students to the full breadth and richness of urban geography and rises magnificently to the occasion.

Professor A. G. Champion, University of Newcastle

Urban Geography

Third edition

Today, for the first time in the history of Humankind urban dwellers outnumber rural residents. Urban places, towns and cities, are of fundamental importance – for the distribution of population within countries; in the organisation of economic production, distribution and exchange; in the structuring of social reproduction and cultural life; and in the allocation and exercise of power. Furthermore, in the course of the present century the number of urban dwellers and level of global urbanisation are destined to increase. Even those living beyond the administrative or functional boundaries of a town or city will have their lifestyle influenced to some degree by a nearby, or even distant, city.

The analysis of towns and cities is a central element of all social sciences including geography, which offers a particular perspective on and insight into the urban condition. The principal goal of this third edition of the book remains that of providing instructors and students of the contemporary city with a comprehensive introduction to the expanding field of urban studies. The structure of the first two editions is maintained, with minor amendments. Each of the thirty chapters has been revised to incorporate recent developments in the field. All of the popular study aids are retained; the glossary has been expanded; and chapter references and notes updated to reflect the latest research. This third edition also provides new and expanded discussion of key themes and debates including detailed consideration of metacities, boomburgs, public space, urban sprawl, balanced communities, urban economic restructuring, poverty and financial exclusion, the right to the city, urban policy, reverse migration, and traffic and transport problems.

The book is divided into six main parts. Part One outlines the field of urban geography and explains the importance of a global perspective. Part Two explores the growth of cities from the earliest times to the present day and examines the urban geography of the major world regions. Part Three considers the dynamics of urban structure and land use change in Western cities. Part Four focuses on economy, society and politics in the Western city. In Part Five attention turns to the urban geography of the Third World, where many of the countries experiencing highest rates of urban growth are least well equipped to respond to the economic, social, political and environmental challenge. Finally Part Six affords a prospective on the future of cities and cities of the future. New to this edition are: further readings based on the latest research; updated data and statistics; an expanded glossary; new key concepts; additional study questions; and a listing of useful websites.

The book provides a comprehensive interpretation of the urban geography of the contemporary world. Written in a clear and readable style, lavishly illustrated with more than 80 photographs, 180 figures, over 100 tables and 190 boxed studies and with a plethora of study aids *Urban Geography: A Global Perspective* represents the ultimate resource for students of urban geography.

Michael Pacione is Professor and Chair of Geography at the University of Strathclyde in Glasgow.

URBAN GEOGRAPHY

A GLOBAL PERSPECTIVE

Third edition

MICHAEL PACIONE

Routledge
Taylor & Francis Group

LONDON AND NEW YORK

First edition published 2001
reprinted 2001, 2003, 2004
Second edition 2005
reprinted 2006, 2007, 2008, 2009
Third edition 2009
by Routledge
2 Park Square, Milton Park, Abingdon, Oxon OX14 4RN

Simultaneously published in the USA and Canada
by Routledge
270 Madison Ave, New York, NY 10016

Routledge is an imprint of the Taylor & Francis Group

Typeset in Bell Gothic and Times New Roman
by RefineCatch Limited, Bungay, Suffolk
Printed and bound in Great Britain by
the MPG Books group

British Library Cataloguing in Publication Data
A catalogue record for this book is available from the
British Library

Library of Congress Cataloging in Publication Data
A catalog record for this book has been requested

ISBN 13: 978–0–415–46201–3 (hbk)
ISBN 13: 978–0–415–46202–0 (pbk)
ISBN 13: 978–0–203–88192–7 (ebk)

ISBN 10: 0–415–46201–0 (hbk)
ISBN 10: 0–415–46202–9 (pbk)
ISBN 10: 0–203–88192–3 (ebk)

TO CHRISTINE, MICHAEL JOHN
AND EMMA VICTORIA

Contents

Preface to the Third Edition

The first and second editions of *Urban Geography: A Global Perspective* published in 2001 and 2005 respectively elicited a gratifyingly favourable response from an international readership of academic professionals and students, and the book has been adopted as a course text in colleges and universities in Britain, North America and elsewhere. This revised edition benefited from constructive feedback from this international readership.

The principal goal of this third edition of the book remains that of providing instructors and students of the contemporary city with a comprehensive introduction to the expanding field of urban studies. The structure of the first two editions is maintained, with minor amendments. Each of the thirty chapters has been revised to incorporate developments in the field since publication of the second edition. All of the popular study aids are retained; the glossary has been expanded; and chapter references and notes updated to reflect the latest research.

Users of the first two editions will recognise and benefit from the structural continuity, as well as the additional features of this third edition. For those who are new to the book – if you are reading this it is likely that you share my interest in the city. Cities are fascinating, dynamic and complex socio-spatial phenomena. As the Prologue to the book illustrates, knowledge of cities and of city living is of real importance both for academic understanding and in negotiating our daily lives in the urban world of the twenty-first century. *Urban Geography: A Global Perspective* is designed to provide an accessible introduction to the study of urban geography in the contemporary world.

Acknowledgements

Few books make the passage from conception to publication without the assistance of a variety of outside agencies, and this book is no exception. Preparation of this third edition of *Urban Geography: A Global Perspective* was facilitated by the professional support provided by the staff at Routledge. Academic colleagues, students and an international panel of reviewers offered valuable comments on the second edition of the book that aided preparation of this revised third edition. In writing the book I enjoyed and benefited from reading the work of other researchers engaged in the study of contemporary urban environments. I am grateful to all the authors and publishers who gave permission to reproduce material in the book. While considerable effort has been made to trace and contact copyright holders prior to publication, the author and publishers apologise for any oversights or omissions and if notified will endeavor to remedy these at the earliest opportunity. A number of friends and colleagues from the international geographic community kindly provided some of the photographs to illustrate different themes in the book. I am grateful to Stuart Aitken (for Plate 4.1), John Cole (Plate 25.2), Bruce Connolly (Plate 23.2a and b), Roman Cybriwsky (Plates 5 and 7.1), Larry Ford (Plate 26.3), Brian Godfrey (Plate 9), Mallik Hossain (Plate 28.1), Hamish Main (Plate 5.2), Arthur Morris (Plate 2.1a), Tony O'Connor (Plate 22.1), Jong-Kie Park (Plate 4), Philip Rushton (Plate 6), Denis Shaw (Plate 8.2), David Simon (Plate 23.1), and Neil Smith (Plate 16.1). All other photographs are taken from my own collection. The maps and diagrams in the book were drawn by Sharon Galleitch, Michael John Pacione and Margaret Dunn.

The greatest debt is owed to my family – my wife, Christine, who continues to manage the demands of career and home in a way that is beyond the ability of most males, my son, Michael, whose uncanny ability to solve computing problems is a continuing source of comfort, and my daughter Emma, who contributed simply by being the person she is. Collectively all three provided a comfortable and supportive environment in which to work and, when required, essential distractions from the task of writing.

Michael Pacione
Milton of Campsie
1 August 2008

Prologue
Daily life in the global village

You awake to the sound of an alarm clock bought in the local hardware store but manufactured in Taiwan. You breakfast on fresh orange juice from Florida, tea or coffee from Sri Lanka or Brazil, bread made from wheat grown on the prairies of North America, jam from Bulgaria, butter from New Zealand, bacon from Denmark and free-range eggs from a local organic farm. Depending on your job, and income, your choice of clothing for the day may have been designed and made in a Milanese fashion house or manufactured under sweatshop conditions by child or female labour in the Far East. As you leave for work in the city you nod a greeting to your neighbour who is teleworking from home, and to her partner who works flexitime to accommodate domestic commitments. You travel into the city in a car constructed in South Korea. In your office the ballpoint pen on your desk was manufactured in Germany. Your notepad is made from timber harvested from a renewable forest plantation in Sweden. You perform your job using a personal computer constructed and sold locally but designed in Silicon Valley, California, and loaded with software produced in Seattle. You may spend much of your working day in cyberspace, using e-mail and teleconferencing facilities to establish real-time communications with business contacts around the globe. All your discussions are in English, the emerging global language.

In your lunch break you visit a McDonald's, an icon of cultural globalisation in which the same menu and decor are available in almost every large city of the world. With time to spare, you stroll past the security officer at the entrance to a new shopping mall and enter a travel agency where you make a reservation for a summer trip to Bali. In the adjacent music store you pick up a copy of a CD that is currently a chart hit in over twenty countries. You might spend some time sitting on a public bench in the town square reading a newspaper that, in addition to local and national news and weather, contains reports of a bank robbery in Bogotá, floods in Dhaka, share prices on the Tokyo stock market, pollution in Lagos and baseball results from the USA. Across the square a small crowd has gathered around a group of street musicians from Peru who are playing traditional folk tunes of the Andes. By the fountain an unemployed man distributes leaflets protesting against the loss of local jobs as a result of downsizing by a foreign-owned TNC. Overhead a Russian jet airliner comes in to land at the city's international airport. At the street intersection enterprising youths rush to clean car windscreens before the traffic lights turn to green. As you rise to return to the office, a party of Japanese tourists ask for directions to the cathedral. You direct them past the redeveloped harbour-side where postmodern pavement cafés and gentrified loft apartments located around a yacht marina are replacing the abandoned warehouses and wharves of an industrial past. On the opposite bank of the river the site of a recent garden festival awaits redevelopment under a public–private partnership scheme to build a concert hall, science museum, I-max theatre and a millennium tower.

After work you meet some friends in an Irish theme pub, before going on to eat dinner in an Indonesian restaurant. As the city centre empties of its daytime work force the space is reoccupied by a night-time population of service workers and entertainment seekers. The homeless search for a sleeping space for the night in the doorways of now silent commercial premises, while in the shadows prostitutes and drug dealers ply their trade.

The journey out to your exurban home takes you past inner suburban areas of social housing where groups of unemployed people congregate on street corners and regard passing strangers with a mix of curiosity and suspicion. Many of the young have never had a job and lack the educational qualifications to enter the labour force, while older men formerly employed in heavy engineering are without the flexible skills needed to compete for a place in the post-Fordist economy. Some inner-city neighbourhoods are occupied by ethnic minorities, while others accommodate lifestyle communities, each contributing to the cultural diversity of the metropolis. Farther out, towards the edge of the city, the proliferation of middle-class gated communities underlines the levels of segregation and social polarisation within the city. Once you are on the urban motorway the lights of the city are soon left behind. Out of sight, out of mind – until tomorrow.

Introduction

The contemporary world is an urban world. This is apparent in the expansion of urban areas and the extension of urban influences across much of the habitable surface of the planet. Today, for the first time in the history of humankind, urban dwellers outnumber rural residents. Urban places – towns and cities – are of fundamental importance: for the distribution of population within countries; in the organisation of economic production, distribution and exchange; in the structuring of social reproduction and cultural life; and in the allocation and exercise of power. Furthermore, in the course of the present century the number of urban dwellers and level of global urbanisation are likely to increase. Even those living beyond the administrative or functional boundaries of a town or city will have their lifestyle influenced to some degree by a nearby, or even distant, city. We inhabit an urban world in which the spread of urban areas and urban influences is a global phenomenon. The outcomes of these processes are manifested in the diversity of urban environments that characterise the contemporary world.

The study of towns and cities is a central element of all social sciences, including geography, which offers a particular perspective on and insight into the urban condition. The scope and content of urban geography are wide-ranging, and include the study of urban places as 'points in space' as well as investigation of the internal structure of urban areas. Within the general field of urban geography specialised sub-areas attract researchers interested in particular aspects of the urban environment (such as population dynamics, the urban economy, politics and governance, urban communities, housing or transport issues). This eclectic coverage, allied to the synthesising power of a geographical perspective, is a major advantage for those seeking to understand the complexity of contemporary urban environments.

Students of urban geography draw on a rich blend of theoretical and empirical information to advance their knowledge of the city. The breadth of urban geography and the volume of published research may appear daunting to someone approaching the field for the first time. This book synthesises this wealth of material to provide a comprehensive introduction to the study of urban geography in the contemporary world. It is intended primarily for undergraduate students of geography and for those taking urban-based courses in cognate social sciences.

STRUCTURE OF THE BOOK

In this introduction I explain the aims and objectives of the book, identify the intended readership and indicate how the material is arranged. I also provide instructors and students with detailed guidance on how to make the best use of the book. Following the Introduction, the book is organised into six main parts. In Part One of the book I lay the foundations for the study of urban geography. In the opening chapter we explore the importance of, and relationship between, global and local factors in the processes of urbanisation and urban change. I highlight the major outcomes of these processes, and identify the main themes and issues of importance in urban geography. In Chapter 2 I introduce a number of key theoretical and conceptual issues, and provide a brief history of the subject in order to establish a framework for analysis in urban geography.

Part Two focuses on the world, regional and national scales. In this part of the book we examine the origins and growth of cities from the earliest times to the present day, establish the global context for the processes of urbanisation and urban growth, identify recent developments in the urban geography of the major world regions, and examine national systems of cities and different types of urbanised region in the world.

Part Three considers urban structure and land use in Western cities. We examine the key agents and processes underlying patterns of urban change and develop an understanding of the construction and reconstruction of urban areas with particular reference to major urban land uses and issues such as post-war suburbanisation, new community development, residential mobility and neighbourhood change, housing problems and policies, retailing and transportation.

Part Four focuses on economy, society and politics in the Western city. Discussion of the urban economy is set within the context of the changing nature of employment in the global economy and the post-Second World War restructuring of metropolitan space-economies. We consider the nature and incidence of poverty and deprivation, and assess national and local responses to urban economic change. The question of social justice in urban service provision is also addressed. We explore the key social processes of congregation and segregation in the city and examine the concept of community and the different bases of residential differentiation. We then employ the notion of urban liveability to consider differential quality of life within cities along a number of major dimensions. Finally, we discuss the role of local government and the distribution and use of power in the city.

In Part Five our attention turns to the urban geography of the Third World, a region that exhibits some of the highest rates of urbanisation and urban growth in the world, as well as the greatest incidences of urban social, economic and environmental problems. I establish the global context for Third World urbanisation before analysing the internal structure of Third World cities. The process of rural–urban migration is identified as a major factor in urbanisation and urban growth. Our attention then focuses on experiences of life in Third World cities, with detailed examination of the urban economy, housing issues, environment,

health, transport, and poverty, power and politics in the Third World city.

Finally, in Part Six we employ a prospective viewpoint to consider the future of cities with particular reference to the concept of sustainable urban development, and critically examine the nature of cities in the twenty-first century.

URBAN GEOGRAPHY IN COLLEGES AND UNIVERSITIES

The importance of urban systems and environments in contemporary society is reflected in the availability of urban-based courses at all levels of the educational system. In the further and higher education sectors, most college and university departments of geography, as well as academic departments in cognate social sciences, offer at least an introductory class in urban geography. In many departments the importance and scope of urban geography is reflected in the provision of several urban classes organised either in terms of a systematic division of the field (with, for example, classes on urban social geography, urban economic geography or urban historical geography) and/or according to a regional specialisation (with a major division between the urban geography of the Western world and study of urban phenomena in the Third World).

This book is designed to be used as a basic resource for those engaged in the teaching and study of urban geography. In the following section I offer some guidance for students and instructors on how to make best use of the book.

HOW TO USE THIS BOOK

INSTRUCTORS

The book provides an introduction to the study of urban geography. Its principal characteristics are:

1. Comprehensive coverage of the major themes and issues of importance in urban geography.
2. A global perspective which examines urban environments in both the developed world and the Third World, and in all the major world regions.
3. An approach to teaching that relates explanation of theories and concepts to revealed processes and patterns, employing examples and case studies from appropriate environments worldwide.

4. Support of the text by a number of user-friendly learning strategies, including numerous boxed studies to amplify key concepts and illustrate particular issues; maps, diagrams and tables of statistics; photographs; an introductory preface for each chapter; a guide to further reading; a list of key concepts, set of study questions, example seminar discussion topics (listed in Table 1.1) and a student project for each chapter, that may also form the basis for a dissertation; a comprehensive list of references; a glossary of key terms; and a work-file approach to encourage active student participation in the learning process. Glossary terms have been highlighted initially in bold type.

For instructors the material presented in the book can be used in a flexible way to construct and organise several different types of class in urban geography the content of which can be varied to fit differing needs in terms of breadth and depth of coverage of the subject. For example, in an introductory urban class you may decide to make less use of Section 1 that deals with theory and methodology; whereas in a more advanced class greater use may be made of the many study aids – references, further readings, notes, glossary, projects and study questions that can be used either for revision or as part of class assessment.

Three basic ways in which the book may be used as a foundation resource for class instruction in urban geography are:

1. In its entirety as a means of providing introductory-level students with a *general overview* of our urban and urbanising world.
2. As a foundation for the study of the urban geography of *particular world regions*.
3. As the basis for more specialised classes dealing with *specific urban themes*.

These options can be combined to provide an integrated course across several years of undergraduate urban geography teaching with, for example, a general first-year introduction to urban geography followed by semester-length classes that develop more detailed, in-depth analyses of particular issues and themes in the second and subsequent years of study.

We can illustrate the potential application of this flexible structure with reference to a selection of the range of urban geography classes currently on offer in colleges and universities in Britain, Europe, North America and Australasia. The shaded areas in Table 1 indicate chapters in *Urban Geography: A Global Perspective* that may be used as a core learning resource for some of the many different possible classes in Urban Geography. In addition, depending on study objectives, level of student knowledge and engagement, and time available, each of the core groups of chapters identified can be supplemented by cross-reference to linked material located elsewhere in the book.

STUDENTS

The book provides a comprehensive introduction to the study of urban geography. It explains the major theories and concepts underlying processes of urban change in the contemporary world and relates these to revealed patterns and outcomes in towns and cities across the globe.

It also incorporates a number of features designed to help your understanding and revision of material learned. Thus, the basic resource of the text is supported by a wealth of additional information provided in a variety of forms. These include:

1. Chapter previews and introductions that outline the main objectives and issues to be covered.
2. Boxed studies that provide supplementary information on key issues and detailed case studies.
3. Maps, diagrams and tables that complement the text with supporting graphical and numerical information.
4. Photographs that provide a visual illustration of urban environments throughout the world.
5. Further reading that identifies key readings to enable you to follow up the themes covered in each chapter.
6. Key concepts that act as a shorthand revision guide by checking your understanding of important issues in each chapter.
7. Study questions designed to test your knowledge and promote critical consideration of the theories, concepts, themes and issues discussed in each chapter.
8. A project that enables you to undertake independent research and develop a detailed understanding of a particular topic.
9. A glossary that defines the key terms in urban geography.
10. A comprehensive list of major references that acts as a guide for further independent study. and
11. The work-file approach, which involves you as an active partner in the production of knowledge on the urban environment.

Book chapter	Course					
	An Introduction to Urban Geography	The Urban Economy	Urban Historical Geography	The Built Environment	Urban Social Geography	Third World Urbanisation
1. Urban geography: from global to local	▓	▓		▓	▓	▓
2. Concepts and theory in urban geography	▓	▓	▓	▓	▓	▓
3. The origins and growth of cities	▓		▓			
4. The global context of urbanisation and urban change	▓		▓			
5. Regional perspectives on urbanisation and urban change	▓		▓			
6. National urban systems	▓					
7. Land use in the city	▓			▓		
8. Urban planning and policy	▓			▓		
9. New towns	▓			▓		
10. Residential mobility and neighbourhood change	▓			▓	▓	
11. Housing problems and housing policy	▓	▓		▓	▓	
12. Urban retailing	▓	▓		▓		
13. Urban transportation	▓	▓		▓		
14. The economy of cities	▓	▓				
15. Poverty and deprivation in the Western city	▓	▓			▓	
16. National and local responses to urban economic change	▓	▓				

| Book chapter | Course | | | | | |
	An Introduction to Urban Geography	The Urban Economy	Urban Historical Geography	The Built Environment	Urban Social Geography	Third World Urbanisation
17. Collective consumption and social justice in the city						
18. Residential differentiation and communities in the city						
19. Urban liveability						
20. Power, politics and urban governance						
21. Third World urbanisation within a global urban system						
22. Internal structure of Third World cities						
23. Rural–urban migration in the Third World						
24. Urban economy and employment in the Third World						
25. Housing the Third World urban poor						
26. Environmental problems in Third World cities						
27. Health in the Third World city						
28. Traffic and transport in the Third World city						
29. Poverty, power and politics in the Third World city						
30. The future of the city – cities of the future						

STUDY TIPS

In addition to the particular guidance offered above on how best to use the book my 'top ten' tips for the successful study of urban geography are as follows:

1. To obtain maximum advantage from the book you should make full use of the in-built study features.
2. Ensure that you understand the key concepts and terms. Re-read the text to refresh your knowledge of any you are unsure about. The glossary can also act as a ready reference source for a definition. In addition, there will be several dictionaries of geography available in your college library.
3. Use the end-of-chapter study questions to test your knowledge and understanding. It is also always helpful to try to relate general urban processes and patterns with conditions in a town or city with which you are familiar (for example, can you identify local incidences of social polarisation, gentrification, ghettoisation, de-industrialisation or suburbanisation?).
4. Take time out to read the boxed studies that accompany each chapter in the book. These not only provide detailed information on particular concepts, issues and situations but can help you relate to the broader themes of the chapter.
5. Study the maps, diagrams and tables. Think about what they are depicting, and identify what trends, patterns and processes they display.
6. Use the photographs as a means of entering into the urban scene. Look at the detail of the environment and think about what is being shown (and not shown). How well does it relate to the theme it represents?
7. The glossary can be used as a quick-fire question-and-answer test for yourself by seeing how well your own definitions match those given.
8. Cultivate the habit of making a regular trip to the college or university library to check recent issues of the main academic journals dealing with urban matters. These include *Urban Studies*, *Urban Geography*, *Annals of the Association of American Geographers*, *Transactions of the Institute of British Geographers*, *Environment and Planning A*, *Environment and Planning D: Society and Space*, *Cities*, *Urban Affairs Quarterly*, *Journal of the American Planning Association*, *Journal of Urbanism*, *Town Planning Review*, *International Development Planning Review* (formerly *Third World Planning Review*), *Habitat International*, *Environment and Urbanisation*.
9. Engage directly in the production of knowledge by creating your own personal work-file of urban materials. These can be obtained from a variety of sources, including:
   - The textbook – make notes, prepare bullet points, annotate diagrams and write short essays on key issues.
   - Additional reading – write summaries of important readings.
   - Lecture notes and handouts provided by the instructor.
   - Class essays, projects or other assignments.
   - Your own maps, sketches or photographs of townscapes with which you are familiar and which are representative of particular themes covered in the class.
   - Government reports and statistics – for example, published census data and urban planning documents.
   - The media – newspaper and magazine cuttings, television reports and documentaries.
   - The World Wide Web – provides a vast array of potentially useful information on urban issues, but must be used with critical appreciation of the sources, which can range from international agencies and government departments, through scholarly research articles published in online journals, to non-refereed statements by individuals. You can use a search engine (such as Yahoo, Lycos, Alta Vista, Google or Copernic), to locate sites of interest or, if known, type in a specific address. For example: for direct access to organisations such as the World Bank, www.worldbank.org; the United Nations, www.undp.org, www.un.org; the United Nations Centre for Human Settlements (Habitat), www.unchs.org; the World Health Organisation, www.who.int/en/; the OECD, www.oecd.org/home; or for access to newspapers, magazines and television such as *National Geographic*, www.nationalgeographic.com; *One World* magazine, www.oneworld.net; *The Times*, www.the-times.co.uk. Web sites relating to themes covered in each chapter are listed extensively at the end of the book.

Ideally, your work-file should be in the form of a loose-leaf binder that allows later material to be inserted in appropriate places. This will also help you organise your thoughts, identify gaps in your knowledge and questions for further investigation, highlight the linkages among different themes, and aid revision.

10. Two final points before we commence our study of urban geography. First of all, bear in mind the fact that knowledge of urban geography will be of both academic and practical importance in the urban world of the twenty-first century. Second, and most important, enjoy the learning experience!

Copyright Acknowledgements

The author and publishers would like to thank the following for granting permission to reproduce images in this work.

ILLUSTRATIONS

Alexandrine Press, Marcham, for figure 6 from *Built Environment* by P. Gaubatz, in 'Understanding Chinese urban form', Vol. 24(4), 1998, p. 265 (Figure 22.11).

American Geographical Society, New York, for the figure of 'The Latin American city' from the *Geographical Review*, by L. Ford, in 'A new and improved model of Latin American city structure', Vol. 86(3), 1996, pp. 437–40 (Figure 22.1).

American Geographical Society for figure 3 from the *Geographical Review*, by L. Ford, in 'A model of Indonesian city structure', Vol. 83(2), 1993, pp. 374–96 (Figure 22.10).

Association of American Geographers, Washington DC, for figure 11.2 from the *Annals of the Association of American Geographers*, by S. Rowe and J. Wolch, in 'Social networks in home and space', Vol. 80, 1990, p. 191 (Figure 17.3).

Association of American Geographers for a figure from *Resource Paper for College Geography 75–2*, by P. Muller, in 'The outer city', 1976, p. 19 (Figure 18.12).

Association of Pacific Coast Geographers, Sacramento CA, for figure 8 from the *Yearbook of the Association of Pacific Coast Geographers* in 'Order and disorder – a model of Latin American urban land use', Vol. 57, 1995, p. 28 (Figure 22.2).

Blackwell Publishers, Oxford, for figure 3.11 from *Geografiska Annaler Series B* by E. Bylund in 'Theoretical considerations regarding the distribution of settlement in inner north Sweden', Vol. 42, 1960, pp. 225–31, © Swedish Society for Anthropology and Geography, 1960 (Figure 6.4).

Blackwell Publishers and the joint editors for figure 21.1 from the *International Journal of Urban and Regional Research* by D. Harvey in 'The urban process under capitalism', Vol. 2, 1978, pp. 101–31 (Figure 7.9).

Blackwell Publishers for the figure of a 'Stress model of the residential decision process' from *Geografiska Annaler Series B* by L. Brown and E. Moore in 'The intra-urban migration process', Vol. 52, 1970, pp. 1–13, © Swedish Society for Anthropology and Geography, 1970 (Figure 10.3).

Blackwell Publishers for figures 2.2, 3.1 and 4.1 by J. Friedmann in *Empowerment*, 1992, pp. 27, 50 and 67 (Figures 24.3, 29.1 and 29.2).

Centre for Urban Policy Research, New Brunswick NJ, for figure 11.1 by G. Galster and E. Hill in *The Metropolis in Black and White: Place, Power and Polarization*, CUPR, 1992 (Figure 15.2).

Elsevier Science, Amsterdam, for figure 1.1 from *Progress in Planning* by D. Bennison and R. Davies in 'The impact of town centre shopping schemes in Britain', Vol. 14(1), 1980, pp. 1–104 (Figure 12.7), copyright Elsevier 1980.

Elsevier Science for figure 2 from *Habitat International* by D. Shefer and L. Steinvortz in 'Rural-to-urban and urban-to-urban migration patterns in Colombia', Vol. 17(1), 1993, pp. 133–50 (Figure 23.2), copyright Elsevier 1993.

Guilford Press, New York, for figure 12.5 by R. Golledge and R. Stimson in *Spatial Behaviours*, p. 456, 1997 (Figure 10.2).

International Books, Baltimore MD, for the figure from *Carfree Cities* by J. Crawford, 2000, p. 130 (Figure 30.6).

International Institute for Environment and Development, London, for figure 1 from *Environment and Urbanization* by C. Hunt, in 'Child waste pickers in India', Vol. 8(2), 1996, p. 115 (Figure 26.1).

James & James Publishers, London, for figure 6.1 from J. Hardoy *et al.*, *Environmental Problems in Third World Cities*, Earthscan, 1992, p. 181 (Figure 30.1).

Liverpool University Press, Liverpool, for figure 5 from *Third World Planning Review* by S. Angel and S. Boonyabancha in 'Land sharing as an alternative to eviction', Vol. 10(2), 1988, pp. 107–27 (Figure 25.4), and for figure 3 from *International Development Planning Review* by P. Barter in 'Transport, urban structure and lock-in in the Kuala Lumpur metropolitan area', Vol. 26(1), 2004, p. 18 (Figure 28.3).

MIT Press, Cambridge MA, and the author for figure 3 by K. Lynch in *The Image of the City*, 1960, p. 19 (Figure 19.2).

Orion Publishing Group, London, for figures 1, 9, 10, 14 and 15 by M. Thomson in *Great Cities and their Traffic*, Penguin, 1977 (Figures 13.2 and 13.4), first published by Victor Gollancz, an imprint of the Orion Publishing Group.

Oxford University Press, Oxford, for figure 10.1 by J. Gugler in *The Urbanization of the Third World*, 1988, p. 167 (Figure 24.2), by permission.

Pearson Education, Harlow, for figure 3.3 by E. Morris in *History of Urban Form*, Longman, 1994, p. 52 (Figure 3.3).

Pearson Education for figure 2.5 by E. Morris in *British Town Planning and Urban Design*, Longman, 1997, p. 29 (Figure 3.7).

Pearson Education for figure 2.2 by J. Dawson in *Shopping Centre Development*, Longman, 1983, p. 25 (Figure 12.4).

Pearson Education for figure 7.1b by R. Potter and S. Lloyd-Evans in *The City in the Developing World*, Longman, 1998, p. 142 (Figure 25.2).

Pion Ltd, London, and A. R. Veness for figure 4 from *Environment and Planning D: Society and Space* by A. R. Veness in 'Home and homelessness in the United States', Vol. 10, 1992, pp. 445–68 (Figure 11.2).

Russell Sage Foundation, New York, for figure 7.1 by M. White in *American Neighborhoods and Residential Differentiation*, Russell Sage Foundation, 1987, p. 237 (Figure 7.8).

Taylor & Francis, Abingdon, for the figure of 'Mann's model of a typical medium-sized British city' by P. Mann in *An Approach to Urban Sociology*, Routledge, 1965 (Figure 7.5).

Taylor & Francis, Abingdon, for the figure 'The concept of the rent gap' by N. Smith in *The New Urban Frontier*, Routledge, 1996 (Figure 10.5).

Taylor & Francis for the figure 'A general model of economic activity within a developed economy' by P. Ekins and M. Max-Neef in *Real Life Economics*, Routledge, 1992 (Figure 16.4).

Taylor & Francis for the figure 'Franks model of dependent development', by J. Dickensen *et al.* in *A Geography of the Third World*, Routledge, 1996 (Figure 21.1).

Taylor & Francis for figure 3.2 by L. Brown in *Place, Migration and Development in the Third World*, 1991 (Figure 23.1).

Taylor & Francis for figure 6.6 by D. Hilling in *Transport and Developing Countries*, Routledge, 1996 (Figure 28.2).

John Wiley & Sons, Chichester, for figure 2.1 by F. Boal in D. Herbert and R. Johnston (eds) *Social Areas in Cities*, Vol. 1, 1980, p. 397 (Figure 18.8), reproduced with permission.

TABLES

Elsevier Science for table 3 from *Progress in Planning* by L. Bourne in 'Reinventing the suburbs', Vol. 46(3), 1996, pp. 163–84, (Table 4.13), copyright Elsevier 1996.

Elsevier Science for table 3.1 from *Progress in Planning* by D. Benison and R. Davies in 'The impact of town centre shopping schemes in Britain', Vol. 14(1), 1980, pp. 1–104 (Table 12.3), copyright Elsevier 1980.

Elsevier Science for table 3 from *Progress in Planning* by P. Heywood in 'The emerging social metropolis', Vol. 47(3), 1997, pp. 159–250 (Table 20.1), copyright Elsevier 1997.

International Thomson Business Press, London, for the table 'Public housing completions 1939–91, USA' by E. Howenstine in G. Hallett (ed.) *The New Housing Shortage*, Routledge, 1993 (Table 11.3).

ITBP for the table 'The hierarchy of planned shopping centres' by K. Jones and J. Simmons in *The Retail Environment*, Routledge (Table 12.1).

James & James Publishers for tables 3.1, 3.2 and 4.1, pp. 68, 20 and 103, by J. Hardoy and D. Satterthwaite in *Squatter Citizen*, Earthscan, 1989 (Tables 25.2, 25.3 and 25.6).

James & James Publishers for tables 3.2 and 5.1, pp. 77 and 152, by J. Hardoy *et al.* in *Environmental Problems in Third World Cities*, Earthscan, 1992 (Tables 26.2 and 27.1).

Oxford University Press, New York, for tables 1.3 and 2.1 by R. J. Waste in *Independent Cities: Rethinking US Urban Policy*, copyright 1998 by Oxford University Press Inc. (Tables 15.3, 15.6).

Taylor & Francis for the table 'The main goals of planning' by A. Thornley in *Urban Planning under Thatcherism*, Routledge, 1991 (Table 8.1).

Taylor & Francis for the table 'Pressures on the welfare state' by S. Pinch in *Worlds of Welfare*, Routledge, 1997 (Table 17.1).

Taylor & Francis for table 2.1 by D. Drakakis-Smith in *The Third World City*, Methuen, 1987 (Table 21.1).

Taylor & Francis for the table 'Principal reasons for migration from rural north-east Thailand', by M. Parnwell, in *Population Movements and the Third World*, 1993 (Table 23.1).

Taylor & Francis for the table 'Principal policy responses to rural–urban migration in the Third World', by M. Parnwell in *Population Movements and the Third World*, Routledge, 1993 (Table 23.4).

BOXES

Pearson Education, Harlow, for figure 1.7 by A. Morris in *History of Urban Form*, Longman, 1994, p. 7 (Box 3.3).

Pearson Education for figure 3.45 by A. Morris in *History of Urban Form*, Longman, 1994, p. 84 (Box 3.5).

Taylor & Francis for material used in box 14.4 by A. King in *Global Cities*, Routledge, 1990 (Box 14.5).

Taylor & Francis for material used in box 16.7 by C. Wood, 'Renewal and regeneration' in D. Chapman (ed.) *Creating Neighbourhoods and Places*, E. & F.N. Spon, 1996, pp. 194–221 (Box 16.7).

Taylor & Francis for the material in boxes 21.4 and 24.4 by D. Drakakis-Smith in *Pacific Asia*, Routledge, 1993 (Boxes 21.4 and 24.4).

Every effort has been made to contact copyright holders for permission to reprint material in this book. The publishers would be grateful to hear from any copyright holder who is not here acknowledged and will undertake to rectify any errors or omissions in future editions of the book.

PART ONE

The Study of Urban Geography

1

Urban Geography

from global to local

Preview: the study of urban geography; global triggers of urban change – economy, technology, demography, politics, society, culture, environment; globalisation; localisation of the global; the meaning of geographic scale; local and historical contingency; processes of urban change; urban outcomes, why study urban geography?

INTRODUCTION

Urban geography seeks to explain the distribution of towns and cities and the socio-spatial similarities and contrasts that exist between and within them. If all cities were unique, this would be an impossible task. However, while every town and **city** has an individual character, urban places also exhibit common features that vary only in degree of incidence or importance within the particular urban fabric. All cities contain areas of residential space, transportation lines, economic activities, service infrastructure, commercial areas and public buildings. In different world regions the historical process of urban evolution may have followed a similar trajectory. Increasingly, similar processes, such as those of suburbanisation, gentrification and socio-spatial segregation, are operating within cities in the developed world, in former communist states and in countries of the Third World to effect a degree of convergence in the nature of urban landscapes. Cities also exhibit common problems to varying degrees, including inadequate housing, economic decline, poverty, ill health, social polarisation, traffic congestion and environmental pollution. In brief, many characteristics and concerns are shared by urban places. These shared characteristics and concerns represent the foundations for the study of urban geography. Most fundamentally, the character of urban environments throughout the world is the outcome of interactions among a host of environmental, economic, technological, social, demographic, cultural and political forces operating at a variety of geographic scales ranging from the global to the local.

In this book we approach the study of urban geography from a *global perspective*. This acknowledges the importance of macro-scale structural factors in urban development, but also recognises the reciprocal relationship between global forces and locally contingent factors in creating and recreating the geography of towns and cities. A global perspective also demonstrates the interdependence of urban places in the contemporary world, and facilitates comparative urban analysis by revealing common and contrasting features of cities in different cultural regions. This global perspective resonates with recent transnational[1] and postcolonial[2] approaches to urban study. The global perspective eschews the analytical value of simplistic distinctions between cities based on, for example, level of development; rejects prioritisation of knowledge produced in one world region or city over another; and seeks to understand cities by drawing on the richness and diversity of urban experience that characterises our contemporary urban world.

While the organisation of this book into sections and chapters is a necessary pedagogic device to bring order to the complex and diverse subject matter of urban geography, the global perspective underlines the interconnections among the different sections, chapters and themes presented. In addition, the global perspective not only promotes integrated insight into our urban world but also encourages readers to 'think outside the box' and seek to relate experiences across different urban and cultural realms.

The framework we shall employ in our study of urban geography is depicted in Figure 1.1. In reading from left to right, we move from the global to the local scale. This schematic diagram highlights the major 'trigger factors' and processes underlying contemporary urban change (Fig 1.1a), as well as the principal urban outcomes of these processes (Fig 1.1b). In this chapter we examine the nature of these factors and processes, and the relationship between global and local forces in the construction of contemporary urban environments.

GLOBAL TRIGGER FACTORS

THE ECONOMY

Economic forces are regarded as the dominant influence on urban change. Since its emergence in the sixteenth century, the capitalist economy has entered three main phases. The first phase, from the late sixteenth century until the late nineteenth century, was an era of *competitive* **capitalism** characterised by free-market competition between locally oriented businesses and *laissez-faire* economic (and urban) development largely unconstrained by government regulation. In the course of the nineteenth century the scale of business increased, consumer markets expanded to become national and international, labour markets became more organised as wage-rate norms spread and government intervention in the economy grew in response to the need for regulation of public affairs. By the turn of the century these accumulated trends had culminated in the advent of *organised capitalism*. The dynamism of the economic system – the basis of profitability – was enhanced in the early decades of the twentieth century by the introduction of **Fordism**. This economic philosophy was founded on the principles of mass production using assembly-line techniques and 'scientific' management (known as Taylorism), together with mass consumption fuelled by higher wages and high-pressure marketing techniques. Fordism also involved a generally mutually beneficial working relationship between **capital** (business) and labour (trade unions), mediated by government when necessary to maintain the health of the national economy.

The third and current phase of capitalism developed in the period following the Second World War. It was marked by a shift away from industrial production towards services (particularly financial services) as the basis of profitability. Paradoxically, the explanation for this shift lay in part with the very success of Fordism. As mass markets became saturated, and profits from

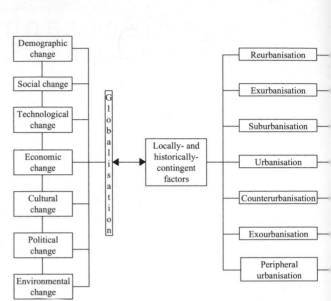

TRIGGERS PROCESSES

Figure 1.1a   Triggers, processes and outcomes in urban geography

mass production declined, many enterprises turned to serve specialised 'niche markets'. Instead of standardised production, specialisation required flexible production systems. This current phase of capitalism is referred to as *advanced capitalism* or *disorganised capitalism* (to distinguish it from the organised nature of the Fordist era). The transition to advanced capitalism was accompanied by an increasing **globalisation** of the economy in which **transnational corporations** (TNCs) operated often beyond the control of national governments or labour unions (see Chapter 14).

The evolution of the capitalist economy is of fundamental significance for urban geography, since each new phase of capitalism involved changes in what was produced, how it was produced and *where* it was produced. This meant that 'new industrial spaces' (based, for example, on semiconductor production rather than shipbuilding) and new forms of city (such as a **technopole** instead of a heavy industrial centre) were required.[3]

TECHNOLOGY

Technological changes, which are integral to economic change, also influence the pattern of urban

OUTCOMES

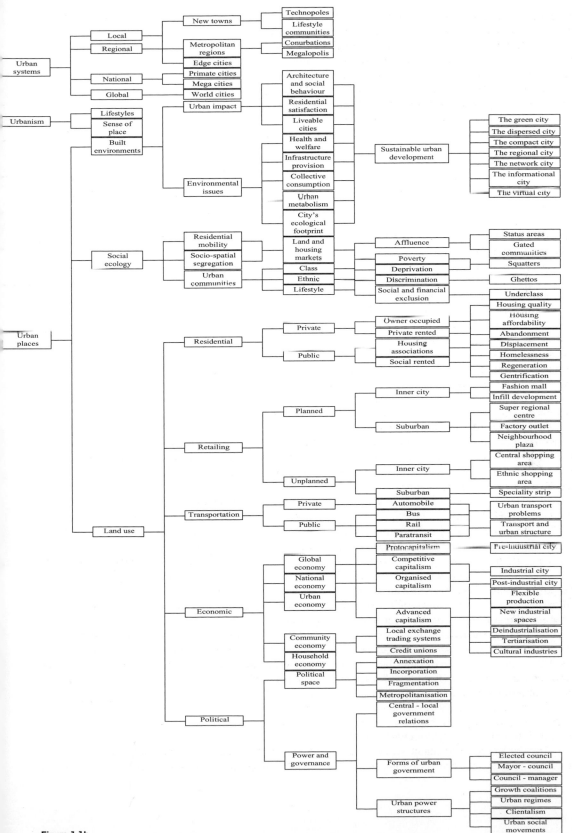

Figure 1.1b

growth and change. Innovations such as the advent of global telecommunications have had a marked impact on the structure and functioning of the global economy. This is illustrated by the **new international division of labour** in which production is separated geographically from research and development (R&D) and higher-level management operations, while almost instantaneous contact is maintained between all the units in the manufacturing process, no matter where in the world each is located.

The effects of macro-level technological change are encapsulated in the concept of economic long waves or cycles of expansion and contraction in the rate of economic development (see Chapter 14). The first of these cycles of innovation (referred to as **Kondratieff cycles**) was based on early mechanisation by means of water power and steam engines, while the most recent (and still incomplete) cycle is based on micro-electronics, digital telecommunications, robotics and biotechnology. The different technology eras represented by Kondratieff cycles shape not only the economy but also the pace and character of **urbanisation** and urban change. Modification of the urban environment occurs most vigorously during up cycles of economic growth. Technological changes that directly affect urban form also occur at the local level. Prominent examples include the manner in which advances in transportation technology promoted suburbanisation (see Chapter 7) or how the invention of the high-speed elevator facilitated the development of skyscrapers in cities such as New York.

DEMOGRAPHY

Demographic changes are among the most direct influences on urbanisation and urban change. Movements of people, into and out from cities, shape the size, configuration and social composition of cities. In Third World countries expectations of improved living standards draw millions of migrants into cities (see Chapter 23), while for many urban dwellers in the West the 'good life' is realised through suburbanisation or exurbanisation. The condition of the urban environment can also affect the demographic structure of cities by influencing the balance between rates of fertility and mortality. Intra-urban variations in health are pronounced in Third World cities where well above-average mortality rates are recorded in overcrowded squatter areas (see Chapter 27), but socio-spatial differences in health status are also characteristic of many Western cities. Demographic changes are related to other 'trigger

factors' such as economic growth or decline, and political change. For example, population growth in a Third World country may induce political attempts to restrict **migration**, whether within the country to control 'overurbanisation' or between countries, as along the Mexico–USA border.

POLITICS

Cities reflect the political ideology of their society. The urban impact of political change is demonstrated by the case of Eastern Europe and the former Soviet Union (see Chapter 5). During the middle fifty years of the twentieth century the development of **new towns** and reorganisation of existing cities reflected the imperatives of a **command economy** and centralised political apparatus. The planned **socialist city** was intended to promote national economic development and to foster social and spatial **equity** in **collective consumption**. Accordingly, high priority was afforded to urban industrial development and the construction of large estates of public housing. With the collapse of communism and relaxation of strict urban land-use controls, capitalist tendencies such as suburbanisation and social differentiation in housing are becoming increasingly evident. An example of how local–regional changes can have a global impact is the way in which the end of the Cold War influenced the cities of the US **sunbelt**/gunbelt, whose economies had been dependent on defence-related industries.[4]

Within Western societies, changes in political ideology and subsequent modifications of economic and urban policy have had major impacts on city development. The rise of the **New Right** governments of Ronald Reagan in the USA and Margaret Thatcher in the UK led to reductions in public expenditure and increased dependence upon the private sector in urban development. This is evident in the rise of agencies such as **urban development corporations** and **enterprise zones**, public–private partnership schemes, property-led urban regeneration, and strategies such as the private finance initiative in the UK (see Chapter 16). Politics and economics exist in a reciprocal relationship, the outcomes of which can have a major impact on urban change. On the one hand, the formulation of urban policy may be influenced by political forces such as the opposition of middle-class suburban voters to increased taxation to pay for inner-city services. On the other hand, a political decision by central government not to provide a financial incentive package to attract inward investment by a foreign-owned TNC can affect the future economic prosperity of a city and its residents.

SOCIETY

Macro-scale social changes can have a significant impact on the character of towns and cities. For example, social attitudes towards abortion or use of artificial methods of birth control may influence the demographic composition of a society and its cities. Popular attitudes towards ethnic or lifestyle minorities can determine migration flows between countries and cities, as well as underlying patterns of residential **seg-regation** within cities. This is evident in recent trends towards increased suburbanisation of the African-American population in the USA, which has been facilitated not only by the economic advancement of individuals and households but by changing social attitudes to the education, employment and residential status of minorities (see Chapter 18). In a similar manner, the attitude of society towards other minority groups, such as single-parent households, the unemployed, disabled people and elderly people, and towards women, conditions their status and location in the city (see Chapter 19).

CULTURE

One of the most significant of cultural changes in Western society in the post-Second World War era, particularly from the late 1970s onwards, has been the rise of materialism. This is displayed in conspicuous consumption by those who can afford it. At the urban scale this is manifested in the appearance of a 'cappuccino society' characterised by stores selling designer clothes, wine bars, pavement cafés, **gentrification**, yuppies (young upwardly mobile professionals), marbles (married and responsible but loaded executives), and bumper stickers proclaiming 'Dear Santa, I want it all.' It is also evident in an increasing gap between rich and poor in cities (see Chapter 15). While some observers claim to have identified signs of an emerging 'post-consumerism', materialism remains a dominant cultural influence in urban society.

The effect of cultural change on cities is encapsulated in the concept of postmodernity or **postmodernism** (see Chapter 2). This embraces social difference and celebrates variation in urban environments, whether expressed in architectural or social terms. Youth cultures have flourished in some urban settings (for example, in inner-city ethnic areas energised by hip-hop and rap music), as have alternative lifestyle communities such as the gay districts of West Hollywood in Los Angeles or Paddington in Sydney. Postmodern **urbanism** is also evident in the growth of **cultural industries** (related to media and the arts) and in the regeneration (and place marketing) of historic urban districts such as Gastown in Vancouver, Covent Garden in London or the Merchant City in Glasgow.

THE ENVIRONMENT

The impacts of environmental change on patterns of urbanisation and urban change are seen at a number of geographic scales. At the planetary scale, global warming due to the greenhouse effect (caused in part by the waste outputs of urban civilisation) may require the construction of coastal defences to protect cities such as Bangkok, Jakarta, Venice and London from the danger of inundation. Equally, any significant deflection of the North Atlantic Drift current (the Gulf Stream) which carries warm water from the Caribbean to the coasts of Britain and northern Europe would affect local weather patterns and require a costly response from cities in terms of urban infrastructure such as improved winter heating systems, public transport and road maintenance. At the local scale, 'natural' phenomena such as earthquakes and landslides can force the abandonment of settlements (for example, Third World **squatter settlements** constructed on marginal lands) or require major works of reconstruction, as in Mexico City after the 1985 earthquake.

For reasons of clarity we have examined each of these important 'trigger factors' independently. In practice, of course, they are interrelated and operate simultaneously along with other local-scale factors to influence urban change. These trigger factors are integral elements in the processes of *globalisation*.

GLOBALISATION

Globalisation is a term used to describe a complex of related processes that has served to increase the interconnectedness of social life in the (post)modern world[5] (Box 1.1). For Robertson (1992 p. 8), the concept refers 'both to the compression of the world and the intensification of consciousness of the world as a whole'.[6] Globalisation is evident in three forms:

1. *Economic globalisation*, seen in arrangements for the production, exchange, distribution and consumption of goods and services (such as the rise of TNCs, the new international division of labour, increases in foreign direct investment, flexible forms of production and a global financial system).[7]
2. *Political globalisation*, seen in arrangements for the concentration and application of **power** (such as the growth of multi-state political-economic groupings, and consideration of local issues within global context).[8]
3. *Cultural globalisation*, seen in arrangements for the production, exchange and expression of symbols that represent facts, meanings, beliefs, preferences, tastes and values (such as the global distribution of images and information, and the emergent cosmopolitanism of urban life).[9]

Significantly, there is a reflexive relationship between the global and the local. While global forces lead to change in the city, cities modify and embed globalisation within a local context. Within the global–local nexus, global forces are generally held to be most powerful and their control more spatially extensive. Local forces are seen to be relatively weaker and geographically limited in effect. However, it is important not to allow the discourse of globalisation to obscure the fact that urban change is not effected by global forces alone. As we shall see in the course of the book, national, regional and local agents are also of significance. For example, national taxation policies, regional trade-union power and local planning regulations all have an impact on urban development and change. Local actions may also have global consequences, as when urban-based protests against socially regressive economic activities (such as the exploitation of child labour) result in changes in patterns of trade and consumption at the world scale.

Globalisation should not be regarded as leading automatically to the disintegration of local life. Individuals can either disembed themselves from a

BOX 1.1

Principal characteristics of globalisation

1. Globalisation is not a new phenomenon. The processes of globalisation have been ongoing throughout human history, but the rate of progress and its effects have accelerated since the 'early modern' period of the late sixteenth century, and more especially over the past few decades in tandem with the transition towards postmodernity.
2. Globalisation involves both an intensification of worldwide social relations through time–space compression of the globe, and local transformations involving enhancement of local identity as well as of local consciousness of the world as a whole.
3. In the global–local nexus, global forces are generally held to be most powerful and their control more spatially extensive. Local forces are seen to be relatively weaker and geographically limited in effect, although certain local actions can have global consequences.
4. Globalisation takes place in cities, but the relationship between the city and global processes is dialectic. While global forces lead to changes in the city, cities modify and embed globalisation within local context.
5. Global forces are mediated by locally and historically contingent forces as they penetrate downwards, coming to ground in particular places.
6. A number of 'trigger forces' underlie globalisation but the dominant force is generally regarded as economic.
7. Globalisation reduces the influence of nation-states and political boundaries at the same time that states are organising the legal and financial infrastructures that enable capitalism to operate globally.
8. Globalisation operates unevenly, bypassing certain institutions, people and places. This is evident at the global scale in the disparities between booming and declining regions and at the urban scale by social polarisation within cities.
9. The differing interests of actors mean that global forces are sometimes embraced, resisted or exploited at lower levels.
10. The mobility of capital diminishes the significance of particular places, although it may also strengthen local identity by engendering a defensive response by local actors.
11. The location of agents that command, control and finance the global economy defines a global city. These are the bases from which transnational corporations (TNCs) launch offensives around the world. The extent to which cities achieve global status is a major determinant of their prosperity.

Plate 1.1 An example of economic and cultural globalisation: the British high-street retailer Marks & Spencer's store in Hong Kong

locality by operating within a global milieu or embed themselves by attachment to a particular place. Neither condition is exclusive. The reflexive nature of the global–local relationship is evident in the world of international finance, where the disembedded electronic space of the international financial system actually compels embedded social relations in specific locations (such as the City of London), to facilitate discussion of new financial products and engaging in interpersonal exchanges of information. More generally, in the place-bound daily lives of most people, particularly those outwith the mainstream of advanced capitalism, globalisation may promote a search for local identity in a mobilised world.[10]

Globalisation is a highly uneven set of processes whose impact varies over space, through time, and between social groups. Global forces bypass many people and places. Many towns in the Third World, as well as in rural areas of Western society, produce mainly for local consumption using local techniques. Even within global cities, certain **neighbourhoods** where poverty and disadvantage prevail are peripheral to the working of the global economy. It should be recognised, of course, that the current situation of social and economic disadvantage in such areas may have been triggered by macro-scale forces such as the investment decision of an executive in a TNC based in a city on the other side of the globe. The unevenness of globalisation is apparent at all levels of society. At the world scale it is seen in the disparities between booming and declining regions, and at the urban scale in the social polarisation between affluent and marginalised citizens.

The uneven penetration of globalisation is a question not simply of which institutions, industries, people and places are affected but also of how they are affected.[11] Local people and places may be overwhelmed and exploited by the forces of globalisation or they may seek to resist, adapt or turn globally induced change into an opportunity. Resistance is a common response in which global forces are mediated at lower spatial scales. During the 1980s, mass protests were held in a number of Third World cities in an effort to resist the imposition of economic austerity measures by the International Monetary Fund (as a response to the burgeoning Third World debt crisis). More recently, the chamber of commerce in the predominantly Mexican Pilsen neighbourhood of Chicago opposed the siting of a Starbucks coffee shop in an effort to preserve local ethnic diversity from the homogenising cultural influence of a global corporation. Frequently, however, people and organisations must acquiesce and adapt to global forces. The proliferation of clothing sweatshops in Los Angeles and New York City during the 1980s was an attempt by local manufacturers to maintain their competitiveness in world markets. Global connections may also be pursued actively by local agents, particularly if they create new opportunities for local economic growth. In world cities such as London and New York foreign investment in property development, especially during the boom years of the 1980s, boosted commercial property markets by raising rental income, increasing property values and generally expanding business opportunities.[12] Provincial cities less attractive to finance capital are usually less able to attain similar benefits.

GLOCALISATION: THE LOCALISATION OF THE GLOBAL

As we have seen, there is a dialectical or reflexive relationship between global and local processes in constructing contemporary urban environments. The term **glocalisation** has been used to describe the simultaneous operation of processes of:

1. De-localisation or de-territorialisation evident, for example, in the instantaneity of e-mail communication across the globe.
2. Re-localisation or re-territorialisation, whereby global influences interact with and are transformed within local context as, for example, in the creation of historic heritage districts in cities.

The effects of 'glocalisation' are generally apparent. Persky and Weiwel (1994) have demonstrated how in many US cities a significant share of economic activity continues to serve local markets.[13] In labour market terms globalisation is of relevance only for a small minority of workers with the skills necessary to compete in international labour markets. Equally, firms must strike a balance between being sufficiently mobile to take advantage of opportunities that may arise elsewhere, and being sufficiently embedded in a local context in order to develop links with local suppliers and labour forces. In social terms, globalisation has positive and negative consequences for different peoples and places. On one hand, it may promote the growth of vibrant multicultural urban communities; on the other it can lead to socio-spatial concentrations of disadvantage. Politically, glocalisation focuses attention on cities as both strategic sites for global interests seeking to maximise profit and spaces where local grassroots and civil society assert their right to liveable places. By combining the de-territorialisation tendency of globalisation and the re-territorialisation of localisation the concept of glocalisation underscores the dialectic nature of the relationship between global and local forces in the construction and reconstruction of contemporary urban environments.

THE QUESTION OF SPACE AND SCALE

The complexity of the global–local nexus forces us to think about the meaning of spatial scale itself. As we shall see in Chapter 2, space is a social construct. The nature of urban (and other) spaces is the product of social relations among actors with different interests. When we refer to the global and the local as being related dialectically, we do not mean to imply that scale or geographic space *per se* should be given theoretical or political priority in determining the social processes underlying urban change. When we speak of

global and local scales we are, in fact, focusing on the *processes* operating at these (and other intermediate) scales, not on space or scale *per se*.[14] Further, in the real world the significance of different geographic spaces and scales is in a constant state of flux, with certain spaces/scales declining in social importance and others increasing. Changes in the relative importance of geographic spaces/scales are reflected in changes in the distribution of power among social groups. This may result in a strengthened position for some (for example, a spatially mobile TNC) and a weakened position for others (for example, a place-bound minority **community**). We shall develop the concept of social power later in the book (see Chapters 20 and 29). Here it is sufficient to recognise that the notion of discrete spatial scales, global and local, is a simplification of a reality in which different actors have interests, and operate simultaneously, at multiple spatial scales. A TNC makes decisions on the local–regional scale in relation to its branch plant operations, negotiates tax advantages with national governments, secures capital in the international financial markets and adapts its planning strategy to satisfy institutional shareholders based in a variety of countries. Equally, local actors opposed to the closure of a factory providing local employment may use direct action through a 'work-in' at the plant, liaise with regional trade union organisers, lobby national government to exert pressure on the parent company or employ international legal experts to pursue the option of an employee buyout of the plant.[15]

LOCAL AND HISTORICAL CONTINGENCY

As was suggested above, in addition to the operation of macro-scale trigger factors, the processes of urban change are also influenced by a host of locally and historically contingent factors that interact with and mediate more general forces of change. There are as many examples of such forces as there are cities in the world. They range from the physical impact of a constricted site on urban development, as in the high-rise architecture of Hong Kong island, to socio-demographic influences such as the concentration of retirees in seaside towns of southern England, or the particular cultural economy of Las Vegas.

More generally, global forces come to ground in cities. As Sassen (1996 p. 630) observes, cities are the locations where much of 'the work of globalisation gets done'.[16] Local context, such as the urban economic base, social structure, political organisation, tax regulations, institutions and competing interest groups, exerts a powerful influence on urban change. The central importance of cities within the global economy and society is affirmed also in the 'hollowing out of the state' thesis. This contends that globalisation has led to the supranational and local scales becoming more important loci of economic and political power than the nation-state, which, in contrast to its central role during the Fordist era, has assumed the role of enabler rather than that of regulator of economic activity.[17] The importance of local action within a globalising world is also a direct consequence of the post-Fordist drive for economic flexibility and product differentiation. The economic advantages that may accrue to cities that can provide attractive production environments (for example, an educated and skilled labour force, good climate and high-quality life space) stimulate innovation and encourage local communities to emphasise the particular advantages of their city as a place to live and work.[18] The concept of a dialectic relationship between processes operating at global, local and other intermediate scales provides an essential backcloth to our study of urban geography.

PROCESSES OF URBAN CHANGE

The interaction of global 'trigger forces' and locally contingent factors results in a number of different processes of urban change. These are summarised in Figure 1.1a and will be discussed in detail later (see Chapters 4 and 21).

Urbanisation occurs when cities grow at the cost of their surrounding countryside, *suburbanisation* and *exurbanisation* when the inner ring or commuter belt grows at the expense of the urban core, *counterurbanisation* when the population loss of the urban core exceeds the population gain of the ring, resulting in the agglomeration losing population overall, and *reurbanisation* when either the rate of population loss of the core tapers off or the core starts to regain population while the ring still loses population. These processes of urban change are visible to varying degrees in **metropolitan areas** of both the developed world and the Third World. The phenomena of **peripheral urbanisation** and **exourbanisation** are characteristic of cities in the Third World. The concept of peripheral urbanisation reflects the expansion of capitalism into Third World regions and employs a **political economy** perspective (see Chapter 2) to describe the impact of global capitalism on national **urban systems** in the Third World. Exo-

urbanisation is promoted by foreign direct investment in Third World countries leading to a pattern of urban growth based on labour-intensive and assembly manufacturing types of export-oriented industrialisation, as in the Pearl River delta region of China.

The varying processes of urban change underline the value of examining global forces within local context in order to understand the complexity of contemporary urban forms. In practice, the reciprocal interaction between global and local forces leads to a plethora of urban outcomes.

URBAN OUTCOMES

The outcomes of the processes of globalisation and urban change are depicted in Figure 1.1b. Three principal outcomes refer to:

1. Changes in *urban systems* at local, regional, national and global scales.
2. The spread of *urbanism*.
3. Changes in the socio-spatial construction of *urban places*.

Each of these principal outcomes subsumes a host of related processes and outcomes. As we progress through Figure 1.1, from left to right, we obtain more detailed insight into the major themes and issues of relevance to the study of urban geography. In essence, Figure 1.1 provides a set of 'signposts' to the subject matter of urban geography. All these topics will be discussed later in the course of the book.

At this point, however, it is useful to think about the kinds of question we need to ask to develop an understanding of urban geography. Appropriate questions include where, when and why did the first cities arise? What are the major urban land uses? Further questions focused on the urban economic base, on the social composition of residential environments, urban political structure, urban policy and planning, and the concept of sustainable urban development, also shed light on the dynamics of urban change. The list of questions that could offer insight into the form and functions of cities is almost limitless. (You can think of some of your own.) Some key questions that reflect the nature of urban geography, and the content of this book, are shown in Table 1.1. Examination of Figure 1.1 and Table 1.1 provides prospective students with a useful overview of the scope and content of urban geography.

A second method of gaining an understanding of the nature of urban geography is to engage in field observation in the tradition of nineteenth-century social

TABLE 1.1 THEMES AND QUESTIONS IN URBAN GEOGRAPHY

Theme	Question
Urban geography: from global to local	Why study urban geography? What is the nature of the relationship between global and local forces in the production of urban environments? How has globalisation influenced the form and function of cities?
Concepts and theory in urban geography	What do we mean by urban? What contribution have the different major theoretical perspectives made to understanding in urban geography?
The origins and growth of cities	When, where and why did the earliest cities arise? What are the social and spatial characteristics of pre-industrial, industrial and post-industrial cities?
The global context of urbanisation and urban change	Are socio-cultural differences between cities declining as a result of globalisation? How important is your own city in local, regional, national or global terms?
Regional perspectives on urbanisation and urban change	Why are cities in the Third World growing faster than cities in the developed world? How did the urban settlement pattern of North America develop? Why are many of the greatest centres of urban life in Europe?
National urban systems	What functions do cities perform? Does any single city or set of cities dominate the national urban system? How extensive is the hinterland of your city?
Land use in the city	What are the major urban land uses? What factors and agents underlie land-use change in the city? How do different populations use urban space?
Urban planning and policy	Why does the impact of urban planning and policy vary between countries? How does public planning affect urban development in the capitalist city? What effect has the collapse of communism had on the form and function of the socialist city?
New towns	What is the difference between an 'old town' and a 'new town'? Should the growing demand for housing be met by building on greenfield land or on brownfield sites in existing cities?
Residential mobility and neighbourhood change	Why do households move? How do urban housing markets work? What factors lead to housing abandonment, gentrification and neighbourhood regeneration?

TABLE 1.1 CONTINUED

Theme	Question
Housing problems and housing policy	What is the relative importance of different housing tenures in the city? Is there an adequate supply of decent and affordable housing? Why are people homeless? What can governments do to improve the quality and availability of housing?
Urban retailing	What forms of retail facility are available in the city? What impact has suburbanisation had on the structure of retail provision? What difficulties confront disadvantaged consumers?
Urban transportation	What are the dominant modes of travel in the metropolitan area? What is the relationship between transport policy and urban structure? Is the car-oriented city sustainable?
The economy of cities	What is the economic base of the city? How is the urban economy related to the global economy? To what extent has your city been affected by industrialisation, deindustrialisation or tertiarisation in recent decades? What role does the informal sector play within the urban economy?
Poverty and deprivation in the Western city	What aspects of multiple deprivation are evident in the city? Is there an inner-city problem? Do particular neighbourhoods suffer from the effects of long-term decline? What can be done to alleviate the problems of poverty and deprivation?
National and local responses to urban economic change	What strategies have been employed to address the problems of urban economic decline? Is there a role for local economic strategies or for the community economy in urban economic regeneration? What has been the impact of property-led regeneration on the socio-spatial structure of the city? Are cultural industries important?
Collective consumption and social justice in the city	Is there any evidence of social or spatial inequity in access to public services? To what extent have reductions in public expenditure impacted on the provision of welfare services in the city?
Residential differentiation and communities in the city	Does the city, or particular neighbourhoods within the city, generate a sense of place for residents? Why does residential segregation occur? What is the extent and incidence of social polarisation in your city? How has it changed over recent years?

TABLE 1.1 CONTINUED

Theme	Question
Urban liveability	How would you define a liveable city? What are the main hazards of urban living? Is there a relationship between urban design and social behaviour? How do different populations, such as women, elderly people, children, members of minorities and disabled people, cope with the pressures of urban life?
Power, politics and urban governance	What form of government controls the city? Does the political geography of the metropolitan area facilitate or constrain good government? Do citizens participate in urban government? What is the relationship between urban government and central government?
Third World urbanisation within a global urban system	How are the processes of urbanisation and urban change in the Third World related to those of the developed world? What evidence is there of the influence of colonialism on urban development? What is the likely role of Third World cities within the global urban system of the twenty-first century?
The internal structure of Third World cities	What are the main structural differences between cities of the Third World and those of the developed world? How has the abolition of apartheid affected the form of cities in South Africa? Is the concept of the Islamic city valid?
Rural–urban migration in the Third World	What are the relative contributions of migration and natural population change to urban population dynamics? Why do people migrate to cities and what are the consequences for the urban geography of the Third World? What is the response of governments to rural–urban migration?
Urban economy and employment in the Third World	What are the major components of the urban economy? What is the role of women and children in the urban economy? How important is the household economy in the coping strategies of the urban poor?
Housing the Third World urban poor	What are the main sources of shelter for the Third World urban poor? When, where and why do squatter settlements arise? What response have Third World governments made to the problem of low-income housing provision?

TABLE 1.1 CONTINUED

Theme	Question
Environmental problems in Third World cities	What are the major environmental hazards faced by poor urban residents in the Third World? How do Third World cities manage the wastes produced by their burgeoning populations? What is the ecological impact of the city on the surrounding area?
Health in the Third World city	What are the chief causes of morbidity and mortality in the Third World city? Are the health problems in Third World cities different from those experienced in cities in the developed world? What can be done to address the health problems of Third World cities?
Traffic and transport in the Third World city	What are the main forms of urban transport? What role is played by paratransit? What are the major traffic and transport problems?
Poverty, power and politics in the Third World city	How do the poor respond to their disadvantaged position within the urban power structure? How important is clientelism? What is the significance of urban social movements in the Third World city?
The future of the city – cities of the future	What is sustainable urban development? What is the nature of the relationship between energy consumption and urban form? What will be the form of the city of the future?

commentators such as Charles Booth in London, Friedrich Engels in Manchester and Jacob Riis in New York. The central area of most cities may be explored easily on foot at a pace that enables the 'participant observer' to absorb the structure of the city and engage with the ebb and flow of urban life. In Glasgow a stroll of less than a mile from the central railway station through the old city to the cathedral represents a journey back in time, over a millennium of urban history. In cities with extensive public transportation systems the urban explorer can cover large areas. In London, for example, the Underground rail system affords access to a variety of urban milieux: the financial heart of the City of London (Bank station), the political hub (Westminster), an ethnic Bangladeshi community (Shoreditch), Georgian residential squares and terraces (Leicester Square), an environment of the super-rich (Kensington), disadvantaged council estates (Hackney) and areas of waterfront regeneration (Tower station). In many cities, especially the low-density metropolitan areas of North America, private transport is a prerequisite for exploration beyond the urban core.

All towns and cities can be 'read' by an observer trained in the basic principles and concepts of urban geography. In the prologue to the book I present a 'virtual field trip' that illustrates the daily rhythms of life in an advanced capitalist city. You can undertake a similar exercise for your own city.

WHY STUDY URBAN GEOGRAPHY?

You should now be in a good position to answer this question. Urban geography provides an understanding of the living environments of a majority of the world's population. Knowledge of urban geography is of importance for both students and citizens of the contemporary world. Urban places are complex phenomena. Urban geography untangles this complexity by explaining the distribution of towns and cities, and the socio-spatial similarities and differences that exist between and within urban places.

This book adopts a global perspective for the study of urban geography. The perspective acknowledges:

1. The interconnectedness of global urban society as a result of globalisation.
2. The dialectic relationship between global and local processes in the construction and reconstruction of urban environments.
3. The importance of local and regional variations in the nature of urbanism within the overarching concept of a global economy and society.

The concepts, themes and issues introduced in this opening chapter illustrate the complexity of urban phenomena and the explanatory power of urban geography. Concepts and theory provide the essential framework for explanation in any academic subject. Accordingly, proper understanding of urban geography

must be based on a combination of theoretical insight and empirical analysis. For the author of a textbook, deciding on how and when to present students with the (sometimes difficult) theoretical content is always a tricky question. This book favours a dual approach. Throughout the text, relevant theory and concepts are integrated with empirical evidence to illuminate particular themes and issues under investigation. In addition, however, since students should be familiar with the key philosophical perspectives and debates at an early stage, these are introduced in the following chapter. Together, the two chapters that comprise Part One of the book provide a foundation for our subsequent discussion of the wide range of themes and issues that constitute the subject matter of urban geography.

FURTHER READING

BOOKS

J. Benyon and D. Dunkerley (2000) *Globalization: The Reader* London: Athlone Press

L. Budd and S. Whimster (1992) *Global Finance and Urban Living* London: Routledge

K. Cox (1997) *Spaces of Globalization: Reasserting the Power of the Local* New York: Guilford Press

P. Dicken (1998) *Global Shift: Transforming the World Economy* New York: Guilford Press

J. Eade (1997) *Living the Global City: Globalisation as Local Process* London: Routledge

P. Knox and P. Taylor (1995) *World Cities in a World System* Cambridge: Cambridge University Press

S. Sassen (1994) *Cities in a Global Economy* Thousand Oaks CA: Pine Forge Press

J. Short and Y.-H. Kim (1999) *Globalization and the City* London: Longman

JOURNAL ARTICLES

N. Brenner (1999) Globalisation as reterritorialisation: the re-scaling of urban governance in the European Union *Urban Studies* 36(3), 431–51

K. Cox (1995) Globalization, competition and the politics of local economy development *Urban Studies* 32(2), 213–24

D. Graham and N. Spence (1995) Contemporary deindustrialisation and tertiarisation in the London economy *Urban Studies* 32(6), 213–24

C. Hamnett (1994) Social polarisation in global cities: theory and evidence *Urban Studies* 31(3), 401–24

J. May (1996) Globalization and the politics of place: place and identity in an inner London neighbourhood *Transactions of the Institute of British Geographers* 21, 194–215

G. Mohan (2000) Dislocating globalisation: power, politics and global change *Geography* 85(2), 121–33

S. Sassen (1996) Cities and communities in the global economy: rethinking our concepts *American Behavioral Scientist* 39, 629–39

D. Walker (1996) Another round of globalization in San Francisco *Urban Geography* 17(1), 60–94

KEY CONCEPTS

- Urbanisation
- Urban growth
- Urbanism
- Global perspective
- Trigger factors
- Globalisation
- Glocalisation
- Global–local nexus
- Spatial scale
- Local contingency
- Hollowing out of the state
- Global village

STUDY QUESTIONS

1. Why study urban geography?
2. Examine the influence of global-scale 'trigger factors' in urban change.
3. With reference to particular examples, illustrate how locally or historically contingent factors can mediate the urban impact of globalisation.
4. In his play *Coriolanus* the English dramatist William Shakespeare asked, 'What is the city but the people?' What did he mean? Write an account of a city as seen through the eyes of a member of a particular social group, such as a yuppie, homeless person, public official, property speculator or member of an ethnic or lifestyle minority.

PROJECT

Most towns and cities can be 'read' by an observer with a knowledge of urban geography. Using the example of the 'virtual field trip', presented in the *Prologue* to this book, as a guide, prepare an analytical description of the pattern of daily activity in any city with which you are familiar. In your account you should:

1. Attempt both to describe and to interpret the observed patterns of daily life in your city.
2. Consider how the different population groups and areas fit into the life of the city.
3. Relate your observations of urban life to the key themes and issues in urban geography identified in this chapter. Which of the general themes and issues are of particular relevance to your city? Does your city exhibit specific local characteristics that influence its form and function?
4. Illustrate your account, if you wish, with relevant supporting material such as field sketches, photographs, newspaper cuttings, published statistics or government reports.

2
Concepts and Theory in Urban Geography

Preview: the scope of urban geography; the concept of urban; the urban as an entity; urban as a quality; the importance of a spatial perspective; the value of the urban dimension; a history of urban geography; environmentalism; positivism; behaviouralism; humanism; structuralism; managerialism; postmodernism; transnationalism; postcolonialism; moral philosophy; levels of urban analysis

INTRODUCTION

As we saw in Chapter 1, in approaching a subject for the first time it is useful to begin by obtaining an overview of the main conceptual approaches, themes and issues that comprise the field. In this chapter we describe the scope of urban geography and its links with other branches of geography. We define the concept of urban and explain the value of an urban geographical perspective for an understanding of contemporary towns and cities. We establish the academic context for the study of urban geography by providing a brief history of the subject. In this discussion we relate work in urban geography to the major theoretical developments in the discipline of geography. Finally, we employ the concept of levels of analysis to illustrate the kind of research undertaken by urban geographers from the global to the local scale.

THE SCOPE OF URBAN GEOGRAPHY

Urban geographers are concerned to identify and explain the distribution of towns and cities and the socio-spatial similarities and contrasts that exist within and between them. There are thus two basic approaches to urban geography:

1. The first refers to the spatial distribution of towns and cities and the linkages between them: *the study of systems of cities*.
2. The second refers to the internal structure of urban places: *the study of the city as a system*.

In essence, urban geography may be defined as the study of cities as systems within a system of cities.[1] Figure 2.1 indicates the scope of urban geography as well as the subdiscipline's links with other branches of geography. The diagram also indicates the power of urban geography to synthesise many different perspectives so as to advance our understanding of urban phenomena. This eclectic approach to the analysis of urban places extends beyond geography to incorporate research findings and knowledge across traditional disciplinary boundaries. The integrative power of urban geography is a key characteristic of the subdiscipline.

A second principal characteristic of geographical analysis of the city is the centrality of a spatial perspective. This distinguishes urban geography from cognate areas of urban study such as urban economics, urban sociology or urban politics. We shall see later that there is a long-standing debate among social scientists over the relative importance of spatial and social forces for the explanation of urban phenomena. However, as we saw in Chapter 1, it is important to be clear about the place of space in urban geography. By acknowledging the importance of spatial location we are not implying that space *per se* is the key explanatory variable underlying patterns of human activity in the city. The significance of space varies with context. For example, spatial location is of no real significance in the electronic hyperspace occupied by flows of finance between cities in the global economy but may be of fundamental importance for the spread of infectious diseases in a Third World squatter settlement. The spatial perspective of urban geography is of real

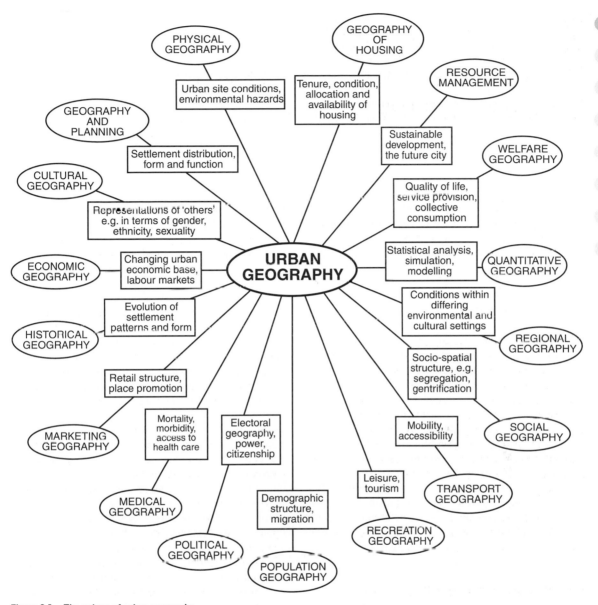

Figure 2.1 The nature of urban geography

analytical value because, as Massey (1985)[2] observed, the world does not exist on the head of a pin.

DEFINING THE URBAN

In approaching the concept of urban it is useful to draw a distinction between the question of what is an urban place, and what is urban. This is more than an exercise in semantics. The distinction between the urban as a physical entity and the urban as a quality helps us to understand the complexity of urban life, and illuminates different approaches to the study of cities.

THE URBAN AS AN ENTITY

Four principal methods are employed to identify urban places:

1. *Population size.* Since urban places are generally larger than rural places, at some point along the population–size scale it should be possible to decide when a village becomes a town. In practice, this urban population threshold varies over time and space. In Sweden any settlement with more than 200 inhabitants is classed as urban in the national census, whereas in the USA the population minimum for urban status is 2,500; in Switzerland it is 10,000, rising to 30,000 in Japan. Such diversity reflects social context. Given the sparse distribution of settlement in many areas of Sweden, a threshold of 200 may be appropriate, whereas in a densely settled country such as Japan virtually all settlements would exceed such a low urban threshold population. If not made explicit, these differences may complicate international comparison.
2. *Economic base.* In some countries population size is combined with other diagnostic criteria to define an urban place. In India, for example, a settlement must have more than 75 per cent of the adult male population engaged in non-agricultural work to be classified as urban.
3. *Administrative criteria.* The majority of towns and cities in the world are defined according to legal or administrative criteria. The definition of urban places by national governments leads to great diversity, which creates difficulties for comparative research that can be overcome only by urban analysts constructing their own definitions and applying them uniformly across the globe. A second problem with administrative definitions is that these may have little correspondence with the actual physical extent of the urban area. A frequent problem is underbounding, where the built-up area of the city extends beyond the urban administrative boundary. This may lead to major fiscal difficulties for the central city deprived of taxes from commuters resident beyond the legal boundaries of the city.
4. *Functional definitions.* To address problems such as underbounding (and its converse, overbounding), urban researchers devised 'functional urban regions' which reflect the real extent of urban influence. The concept of the extended urban area was first introduced by the US Bureau of the Census in 1910 and later developed into the Standard Metropolitan Statistical Area (SMSA) in 1960 and, since 1983, the Metropolitan Statistical Area (MSA). Consolidated Metropolitan Statistical Areas (CMSAs) are formed by two or more contiguous MSAs.[3]

In 2000 a review of standards for defining US metropolitan statistical areas retained the two main principles established in 1960:

1. Settlement form (based on the population size of a central core city).
2. Functional integration between central and outlying counties (reflected in journeys to work, with this criterion raised from 15 per cent to 25 per cent).

Other criteria for inclusion within a metropolitan area, (for example, percentage of the labour force that is non-agricultural), were dropped. The 2000 standards identify two main types of core based statistical areas (CBSAs). These are:

1. *Metropolitan* statistical areas, defined around at least one Census Bureau-defined urbanised area of 50,000 or more population.
2. *Micropolitan* statistical areas, defined around at least one urban cluster of at least 10,000 and less than 50,000 population.

Adjacent CBSAs that have sufficient employment interchange (measured using journey-to-work data) are grouped to form larger 'combined statistical areas'.

Another significant development was the replacement of a 'central cities' classification with one of 'principal cities' (defined on the basis of a variety of population and employment data). While this will capture many of the previous central cities it will also reflect recent changes in the US urban landscape by identifying newer outlying employment centres as principal cities.

Within metropolitan statistical areas the 2000 standards identify two types of counties as a basis for metropolitan divisions. These are:

1. *Main* counties, with 65 per cent or more of employed residents who remain in the county to work, and with a jobs-to-resident-workers ratio of 0.75 or greater;
2. *Secondary* counties, with a high jobs-to-resident-workers ratio (0.75 or greater), but a lower percentage of employed residents working within the county (50–64.9 per cent).

Main counties can stand alone as a metropolitan division or can provide the organising basis for a metropolitan division. Secondary counties must combine with another secondary county or with a main county to form the basis of a metropolitan division. The remaining counties of an MSA are assigned to the main and secondary counties with which they have the highest commuting interchange. Metropolitan divisions, if present in an MSA, will account for all of its territory.

Comparison of the 2000 metropolitan standards with the 1990 standards revealed that the new definitions accounted for 90 per cent of the US population (compared with 80 per cent for the 1990 definitions). Clearly, the application of the new urban standards will result in a change of status for many US counties. In 2003 the US Office of Management and Budget introduced a new category of places called micropolitan areas (Box 2.1).

Geographers in the UK have sought to define a similar set of **daily urban systems**. A first attempt identified Standard Metropolitan Labour Areas comprising a core plus metropolitan ring that together formed the daily urban system. A development of this system added an outer ring consisting of all local authorities that send more commuters to the core in question than to any other core, the whole being designated a Local Labour Market Area (LLMA). A conceptually similar scheme is that based on Functional Urban Regions, which have been used to compare changing patterns of urbanisation in Western Europe.[4]

THE URBAN AS A QUALITY

In contrast to definitions of the city as a physical entity, the concept of the urban as a quality is related more to the meaning of urban places and the effect of the urban milieu on people's lifestyles (and vice versa). Clearly, although cities exist as physical objects, it is by no means certain that they are perceived by their inhabitants in the same way that they are objectively structured. It is reasonable, therefore, to think of a city as having both an objective physical structure and a subjective or cognitive structure.

We each have our own conception of what a city is and of our local town or city. The same urban space can be seen in different ways by residents, tourists, workers, elderly people, unemployed people, women and children. For the homeless person the city may be a cold, anonymous and inhospitable place; for the elderly a spatially restricted world; for the wealthy a cornucopia of opportunity and well-being. Understanding these subjective interpretations of the urban is important, because meanings inform us not only about the places to which they refer but also about the people who articulate them and the social context in which they live.

Urban geographers and others have sought to identify urban meaning through two main approaches.

1. *Cognitive mapping.* Geographers, planners and environmental psychologists have employed **mental maps** or cognitive mapping techniques to explore the subjective world of urban places, with a view to both obtaining a better understanding of human behaviour in the urban environment and improving the quality of urban life.[5] Whereas

BOX 2.1

Micropolitan America

A place is defined as metropolitan or micropolitan based on the size of its centre rather than total population. This is relevant to the definition of 'urban'. The traditional view holds that a large core city anchors subsequent suburbanisation and creates a metropolitan area. But micropolitan areas reverse this standard pattern by growing to metropolitan scale without a large central city.

Some of the largest micropolitan areas can be classed as 'decentralised cities' yet are differentiated from the edge city phenomenon (see Chapter 4) because growth is taking place at a distance from a 'central city'.

Some micropolitan statistical areas have larger populations than the smallest metropolitan statistical areas but are classified as the former because their core urban areas have fewer than 50,000 people.

Collectively micropolitan areas account for 690 of the 3,141 US counties; over 28.3 million people (one in ten Americans) live in micropolitan areas. Overall, population growth in micropolitan areas lags behind that of the USA as a whole. Most fast-growing micropolitan areas are in the shadow of booming metropolitan areas like Las Vegas NV, Jacksonville FL and Dallas TX. Almost all are in the fastest-growing parts of the country, with none in the northeast and only one in the Mid-west.

The micropolitan area classification permits a more sensitive differentiation between urban and rural America. The previous Census Bureau designation of non-metropolitan was too broad to be synonymous with rural areas. Now the remaining non-metropolitan counties that fall below the micropolitan level can be classified as truly rural.

traditional means of cognitive mapping provide subjective spatial representations of urban environments, more recently postmodern approaches seek to 'map' the meanings of the city for different 'textual communities' who share a common understanding of the '**text**' and organise their lives accordingly as, for example, in the creation of a 'suburban mentality'.[6]

2. *Urbanism as a way of life.* Early efforts to identify urban places in terms of a distinct lifestyle were based on Wirth's (1938) concept of a **rural–urban** continuum.[7] This argued that as the size, density and heterogeneity of places increased, so did the level of economic and social disorganisation. Wirth, a member of the Chicago school of **human ecology**, regarded urbanisation as a process leading to the erosion of the moral order of society due to the concomitant decline of community. He saw the urban as a separate spatial realm with its own environmental influences on individuals, and he contrasted the social disorganisation of urban life (in which much social interaction is of a transitory and superficial nature with 'unknown others') with the strong extended family links and communities in small settlements and rural areas. More recent perspectives that acknowledge the interpenetration of social realms have rejected the crude dualism of bipolar concepts such as urban–rural or public–private.

Accordingly, although cities do exert a particular influence on their inhabitants the concept of a rural–urban continuum has been criticised:

1. For its Western ethnocentrism (which assumed that the rural–urban change process is universally applicable);
2. By studies which reveal the presence of 'village communities' in cities (including Wirth's own work on the ghetto);[8] and
3. For failure to locate the process of urbanisation within the political economy of capitalism (not least the impact of wider social, economic and political changes in rural areas, as demonstrated by the presence of 'urban' societies in supposedly rural areas).[9]

In seeking to reinterpret the meaning of 'urban', Harvey (1973, 1985) and Castells (1977) dispensed with the notion of a separate urban realm and concluded that while urbanism (as a way of life associated with residence in an urban area) has a distinctive structure and character, it exists within a larger framework created by the forces of capitalism.[10] This means that 'urban lifestyles' can spread beyond the physical limits of the city.

The quintessential diversity of urban life is central to postmodern representations of the city. Informed by processes of globalisation, social polarisation, cultural fragmentation and advances in information and communications technology, these focus on the rise of new cultural groupings and urban spaces, such as those defined by lifestyle communities.[11] Postmodern readings of the city as 'text' employing urban metaphors, such as the city as jungle, bazaar, organism and machine,[12] produce a multitude of representations of cities from the perspectives of different populations. In order to understand the truth of the city we need to acknowledge the 'reality' of the city as a concrete construction (thing) and as an abstract representation (idea), and examine how each influences the form of the other.

THE SIGNIFICANCE OF SPACE AND PLACE

Place is part of but different from space. Place is a unique and special location in space notable for the fact that the regular activities of human beings occur there. Moreover, because it is a site of such activities and all that they entail, place may furnish the basis of our sense of identity as human beings, as well as for our sense of community with others. In short, places are special sites in space where people live and work and where, therefore, they are likely to form intimate and enduring connections. As we see in Chapter 18, even in a globalising world a sense of place is of real importance in people's daily lives.

Paradoxically, the advent of cyberspace has refocused attention on the importance of places in urban life. There is growing recognition among urban scholars that place is a central concept in the analysis of how urban areas are constructed and come to have meaning for their residents. Furthermore, as the constraints of geographical distance become less important, the specific features of particular **locales** are becoming more important in the locational decisions of businesses and households. The 'construction' of place is also a characteristic of the restructuring of many contemporary cities from being centres of production (for example, the steeltowns of yesteryear) to being centres of consumption (for example, Las Vegas of today), in the sense that they provide the context in which goods and services are compared, evaluated, purchased and used.

Places such as London's Covent Garden or Fisherman's Wharf in San Francisco obtain a distinctive character that not only reinforces the place's sense of identity but transforms the locality into an 'item of consumption', a process often boosted by city marketing strategies (see Chapter 16).

In striving to redefine 'urban' in other than empirical terms, and to explain the meaning of urban life, both Harvey and Castells rightly rejected notions of 'spatial fetishism' (which assigned causal power to space *per se* in determining human action) and emphasised the need to examine the role of urban places in capitalist society. However, it is important not to dismiss completely the power of space. Space is more than a medium in which social, economic and political processes operate. The dimensions of space – size, density, distance, direction, territory and location – exert powerful influences on urban development and on human interaction. Distance and direction have a direct effect on social networks, journeys to work and physical access to place-bound facilities. The size and density of residential developments have been shown to influence the incidence of deviant behaviour and social pathologies. The concept of **territoriality** contributes to the formation of discrete sub-areas within cities, often segregated on ethnic or class lines. The partitioning of space by local government boundaries has implications for the social composition and fiscal health of different areas and the **quality of life** of residents. Competition between local administrations to attract desirable developments, such as shopping centres, and exclude undesirable facilities, such as mental hospitals, also has a fundamental spatial basis. There is a reciprocal relationship between society and space. As Massey (1984 p. 52) observed, 'just as there are no purely spatial processes, neither are there any non-spatial social processes'.[13]

A more recent challenge to the significance of geographical space has accompanied the advent of cyberspaces (such as the Internet), which for some authors has undermined the relevance – conceptual and actual – of 'real' space. It is argued that as a result of 'time–space compression', cyberspace negates the effects of physical separation and produces new 'spaceless, placeless social spaces' in which people can interact.[14] While cyberspace undoubtedly has spatial implications, the suggestion that it heralds the 'end of geography' is excessive, not least because even in advanced societies connections to cyberspace are unevenly distributed.[15] This is illustrated clearly by the contrast between elite, and often physically enclosed, communities with extensive links to the global information superhighway and spatially proximate 'off-line'

urban spaces occupied by the poor, where time and space constraints are profoundly real.[16]

THE VALUE OF THE URBAN DIMENSION

The Wirthian association of particular lifestyles with different settlement sizes is now regarded as unacceptably simplistic. Some have extended this argument to claim that the concept of urban has become redundant on the grounds that, in Western societies at least, we are all 'urban' no matter where we live. This has led some to conclude that it is neither fruitful nor appropriate to study the city in its own right.[17]

Leaving aside the fact that this represents a Western view of contemporary society which is not yet applicable to all parts of the world, the suggestion that the city cannot be a significant unit of study confuses two different questions. These are:

1. Whether cities can be objects of analysis.
2. Whether explanations of 'urban phenomena' can be restricted to the urban level.

As Agnew *et al.* (1984)[18] point out, a negative answer to the second question does not require a negative answer to the first. We can reject the idea of 'urban explanation' while accepting the urban as a valid object of analysis. Cities are places – or, more accurately, conglomerations of overlapping and interrelated places – and, as we have seen, places matter! Social relations occur in, and help to constitute, places. Individuals, households, communities, companies and public agencies exist and operate in particular places. People's social, economic and political relations are embedded in the particularity of specific places. The results of global-level structural forces come to ground in particular places. To maintain that cities are merely elements in the capitalist **mode of production** ignores significant variations between and within urban places, as well as the complex suite of other social and cultural factors shown to be involved in the construction and reconstruction of urban environments.[19]

More recent arguments against the urban as a frame of reference reflect those that dismiss the value of space, and contend that advances in telecommunications technologies will lead to the effective dissolution of the city on the grounds that, once time has become instantaneous, spatial congregation becomes unnecessary.[20] Such technologically deterministic views ignore the relational links between new information technologies and urban form which have been apparent ever since the introduction of the telegraph, wireless and telephone in the nineteenth century.[21] New technologies contribute to the reconstruction of urban space but do not render it redundant. Visions of the 'spaceless city' greatly overestimate the degree to which virtual reality can substitute for place-based face-to-face interaction.[22] As Amin and Graham (1997) conclude, the contemporary city, while housing vast arrays of telematic entry points into the burgeoning worlds of electronic space, is a cauldron of emotional and personal worlds and attachments, an engine of reflexivity, trust and reciprocity.[23] While many cities and citizens are linked into an electronic 'non-place urban realm',[24] place-based relational networks that rely on propinquity and physical interaction – the key characteristics of urban places – remain central to the experience of human social, economic and cultural life.

The fundamental importance of towns and cities as focal points in contemporary society, in both the developed and the developing realms, emphasises the significance of the urban dimension and underlines the value of the study of urban geography.

A BRIEF HISTORY OF URBAN GEOGRAPHY

Urban geography is an established branch of geography that attracts researchers and students in significant numbers, and produces a large and expanding volume of published work to aid our understanding of the city. The advance of urban geography to a central position within the discipline has occurred over the past half-century. As Herbert and Johnston (1978 p. 1) noted:

> whereas in the early 1950s a separate course on urban geography at an English-speaking university was quite exceptional, today the absence of such a course would be equally remarkable; indeed, in many institutions students can opt for a group of courses treating different aspects of the urban environment.[25]

Urban geography is a dynamic subdiscipline that comprises a combination of past ideas and approaches, current concepts and issues that are still being worked out (Table 2.1). It may be likened to:

> a city with districts of different ages and vitalities. There are some long-established districts dating back to a century ago and sometimes in need of repair; and there are areas which were once fashionable but are so no longer, while others are being rehabilitated. Other districts have expanded recently and rapidly; some are well built, others rather gimcrack.[26]

Since the late 1970s the scope of urban geography has expanded rapidly. For some commentators the increased diversity is a source of potential weakness that may lead ultimately to its disintegration. For others, including the present writer, the breadth of perspective strengthens urban geography's position as an integrative focus for research on the city.

Urban geographers have approached the study of the city from a number of philosophical perspectives. While the significance of each for the practice of urban geography has changed over time, none of the main approaches has been abandoned completely, and research informed by all perspectives continues to be undertaken under the umbrella of urban geography. It is important for students of urban geography to be familiar with the different philosophies of science which underlie the subject. To aid this understanding

Plate 2.1   The variety of urban environments is shown by examples of 'waterfront living' in (*upper*) a shanty town built on stilts over a coastal marsh in the port city of Manta, Ecuador, and (*lower*) the gentrified St Katharine's Dock in central London

TABLE 2.1 THE EXPANDING SCOPE OF URBAN GEOGRAPHY

	Systems of cities	Cities as systems
1900	Urban origins and growth	Site and situation of settlements
	Regional patterns of settlement	Urban morphology Townscape analysis Urban ecology
	Central place theory Settlement classification	Social area analysis Factorial ecology Delimitation of the central business district
	Population movements Migration decisions Suburbanisation	Residential mobility Retailing and consumer behaviour Urban imagery
	Urban and regional planning	Power and politics Territorial justice Differential access to services
	The role of cities in the national political economy	Urban problems in structural context
	Edge cities Counter-urbanisation	Economic restructuring Poverty and deprivation The inner-city problem
	Rural–urban migration in the Third World	Housing markets and gentrification The urban property market Traffic and transport problems The urban physical environment Housing, health and economy in Third World cities
	Globalisation of culture and society The global economy The global urban system World cities and global cities Megacities	The urban impact of globalisation Social construction of urban space Cultural diversity in cities Social justice Urban liveability Sustainable cities
2000	Technopoles	Future urban form

we shall structure our discussion of the changing nature of urban geography with reference to the main epistemological developments in the discipline.

ENVIRONMENTALISM

During the first half of the twentieth century the major concerns of urban geography reflected the more general geographical interest in the relationship between people and environment, and in regional description. Early work on urban site and situation, and on the origins and growth of towns, was largely descriptive. One of the first English-language texts in urban geography comprised a classification and analysis of over 200 towns in terms of their site, situation, relief and climate.[27] Despite their simplicity, these investigations provided a foundation for the more conceptually refined practice of urban morphological analysis which continues to illuminate patterns of urban growth and change to the present day.[28]

At the inter-urban scale, studies of regional patterns of settlement focused attention on the importance of transportation systems.[29] Together with the work of land economists,[30] this shifted attention from environmental factors towards the economics of location. The economic analysis of cities as points in space was developed most fully in **central place theory**.[31] At the intra-urban scale, work on **urban ecology** also moved attention from the environment to human behaviour, and introduced a social dimension to urban geography. Nevertheless, at mid-twentieth century the focus of urban geography was primarily on land use and related issues. The first major 'paradigm shift' to affect urban geography reflected a desire to make geographical investigation more scientific. This led to the introduction of the philosophy of **positivism**.

POSITIVISM

Positivism is characterised by adherence to the 'scientific method' of investigation based on hypothesis testing, statistical inference and theory construction (Box 2.2). Although evident in the work of Christaller (1933) and Losch (1954) on the spatial patterning of settlements,[32] positivism blossomed in urban geography in the late 1950s with the development of the spatial analysis school. The redefinition of urban geography as the science of spatial relationships[33] was accompanied by a shift in emphasis away from exceptionalism (the study of the unique and particular) towards a nomothetic approach (aimed at a search for abstract or universal laws). This was aided by the emergence of quantitative analytical techniques fuelled by the 'quantitative revolution' in geography of the 1950s and 1960s.[34]

Box 2.2

The assumptions of positivism

1. Events that occur within a society or which involve human decision-making have a determinate cause that is identifiable and verifiable.
2. Decision-making is the result of a set of laws to which individuals conform.
3. There is an objective world comprising individual behaviour and the results of that behaviour which can be observed and recorded in an objective manner on universally agreed criteria.
4. The researcher is a disinterested observer.
5. As in the study of inanimate matter, there is a structure to human society (an organic whole) that changes in determinate ways according to observable laws.
6. Application of the laws and theories of positivist social science can be used to alter societies in determinate ways (social engineering).

Source: adapted from R. Johnston (1983) *Philosophy and Human Geography* London: Arnold

---

The new approach led to multivariate classifications of settlement types,[35] investigations of the rank–size rule for the populations of urban places,[36] and analyses of spatial variations in urban population densities.[37] Stimulated by translations of Christaller's work, urban geographers devoted attention to modelling settlement patterns and the flows of goods and people between places.[38] Concepts such as **distance decay** (the attenuation of a pattern or process over distance) were also introduced in the study of urban phenomena such as consumer behaviour, and trip generation and travel patterns.[39] The subsequent expansion of computing power and development of geographical information systems has ensured that modelling and simulation remain a vibrant, if minority, area of urban geography.[40]

The new methods of spatial science were also applied to analysis of the internal structure of cities. The urban land-use models of the Chicago school of human ecology reflected the positivist philosophy in their proponents' belief that human behaviour was determined by ecological principles or 'natural laws' which stated that the most powerful group would obtain the most advantageous position (e.g. the best residential location) in a given space. During the 1970s the development of a range of multivariate statistical techniques extended the social area approach of the ecologists in the form of *factorial ecologies* designed to reveal the bases of residential differentiation within the city.[41] The positivist spatial science approach was also central to the models of urban structure introduced into geography from **neoclassical economics**. These were founded on the assumption of *Homo economicus* or the economic rationality of human behaviour. This stated that individual decisions were based on the goal of utility maximisation – that is, the aim of minimising the costs involved (usually in terms of time and money) and maximising the benefits.

For two decades, spatial science was the dominant paradigm in urban geography. However, during the 1970s a growing awareness among geographers of alternative philosophies of science led to strong criticism of positivism:

1. The adequacy of an approach which focused on spatial form to the neglect of underlying causal processes was questioned. It was argued that, since spatial form was largely the outcome of the prevailing *social* forces, the focus of urban research should switch from the study of spatial relations to the study of social relations.
2. Particular criticism was directed at positivism's mechanistic view of the role of humankind, and its failure to recognise and account for the

idiosyncratic and subjective values that motivate much human behaviour.

3. The science of spatial relations affords no insight into the meaning of urban places to their inhabitants. To explore this **sense of place** requires an approach that focuses on the daily activities and perceptions of urban residents.

One response to a growing dissatisfaction with positivist science and the poor predictive ability of many spatial models was a move towards direct observation of human behaviour and decision-making. This led to the development of **behaviouralism** in urban geography.[42]

BEHAVIOURALISM

The behavioural approach sought to overcome the shortcomings of spatial analysis by highlighting the role of cognitive processes and decision-making in mediating the relationship between the urban environment and people's spatial behaviour. Urban geographers employed cognitive mapping techniques to examine a host of issues, including migration,[43] consumer behaviour,[44] residential **mobility**,[45] residential preferences, perceived neighbourhood areas and images of the city.[46]

The behavioural approach introduced greater realism into urban studies, as the emphasis on empirical investigation of human behaviour countered the abstract nature of 'spatial' theory. But behaviouralism did not break away wholly from the positivist tradition. Much of the methodology of positivism was retained, and although focused on exposing the values, goals and motivations of human behaviour, it was still concerned with seeking law-like generalisations. As a consequence, behaviouralism has attracted much of the same criticism that has been levelled at positivism, in particular its failure to recognise and account for the 'untidiness', ambiguity and dynamism of everyday life.

HUMANISM

The **humanistic** approach views the individual as a purposeful agent of change in the city rather than a passive respondent to external stimuli. Although it is acknowledged that people do not act free of constraints, the humanist philosophy accords central importance to human awareness, agency, consciousness and creativity. The aim of a humanistic approach is to understand human social behaviour using methodologies that explore people's subjective experience of the world. In

practice, this means a change from the positivist principles of *statistical* inference based on representative random samples of the population to the principle of *logical* inference based on unique case studies using methods such as ethnography and analysis of literary texts to demonstrate 'the social construction of urban space'.[47]

The humanistic perspective has been criticised for placing excessive emphasis on the power of individuals to determine their own behaviour in the city, and affording insufficient attention to the *constraints* on human decision-making. This critique was advanced most forcefully by proponents of structuralism (see below), who regarded the focus on the individual in society as a distortion of a reality in which people's behaviour is conditioned by forces over which they have little control.

STRUCTURALISM

Structuralism is a generic term for a set of principles and procedures designed to expose the underlying causes of revealed patterns of human behaviour. In practice this means that explanations for observed phenomena cannot be found through empirical study of the phenomena alone but must be sought by examination of prevailing social, economic and political structures.

Structural analysis in urban geography has been based primarily on the work of Marx.[48] According to the **Marxian** or political economy approach, every society is built upon a mode of production – a set of institutional practices by which the society organises its productive activities, provides for its material needs, and reproduces the socio-economic structure. Capitalism is a specific mode of production (others being slavery, **feudalism**, socialism and communism). Cities are viewed as an integral part of the capitalist mode of production by providing an environment favourable to the fundamental capitalist goal of **accumulation**. This is the process by which the value of capital is increased through the continual reinvestment of profits from earlier investments. The effect of this expansionary dynamic is most visible in the changing urban land market and, as we shall see later, in processes such as urban redevelopment, gentrification and suburbanisation.

The political economy approach entered urban geography in the early 1970s in response to the continuing social problems of urban areas, highlighted in the USA by the civil rights movement. In seeking to uncover the structural forces underlying observed social problems located in the dynamics of the capitalist system, it was argued that:

1. Capitalist society is characterised by conflict between socio-economic groups over the distribution of resources. A key resource is power, most of which is held by an elite who are able to manipulate the majority.
2. Since quantitative spatial analysis describes patterns but fails to reveal underlying causes, any proposals or policies based on this analysis will be supportive of the status quo and unable to lead to progressive social change.

Much attention has been directed to the analysis of urban property and housing markets, and studies of residential patterns.[49] Notwithstanding the exercise of 'constrained choice' by individuals, the political economy approach interpreted urban residential segregation primarily as a result of decisions by those with power in the property market, including building society managers,[50] estate agents[51] and local authority housing managers.[52] Harvey (1976)[53] offered an incisive exposition of the relationship between urban residential patterns and the dominant political economy of monopoly capitalism.

The dominance assigned to social structure over human agency in the structuralist perspective was rejected by humanistic geographers. Other critics have attacked the emphasis attached to class divisions in society to the neglect of other lines of cleavage such as gender, **ethnicity** and sexuality, all of which cut across class boundaries and which exert a significant influence on urban lifestyle and the processes of urban restructuring.[54] Nevertheless, the political economy approach has had a major impact in urban geography and has provided real insight into the economic and political forces underlying urban change.

MANAGERIALISM

Doubts over the analytical value of class in modern societies led some writers to abandon Marx's class-based analysis in favour of Weber's concept of 'social closure' – a process by which social groups seek to maximise their benefits by restricting access to resources and opportunities to a limited circle of 'eligibles'. (The practice of 'exclusionary zoning' of land use in the USA city provides a good example.) This perspective on power and conflict in society is closely related to the concept of urban managerialism, which focuses attention on the power of **urban managers** –

professionals and bureaucrats – to influence the socio-spatial structure of cities through their control of, for example, access to public housing, or the allocation of mortgage finance.[55] Structuralists are dismissive of **managerialism**'s focus on intermediate-level decision-makers within a social formation. However, at the interface between consumers and allocators of scarce resources, managerialism introduces a humanistic perspective that can help to expose the operation and rationalities of the distributive process in cities.

POSTMODERNISM

Postmodern theory began to exert an influence on urban geography in the late 1980s and 1990s. The postmodern perspective is characterised by the rejection of grand theory and an emphasis on human difference. This distances postmodernism from both positivism, with its search for general laws and models, and structuralism, with its base in grand theory relating to the capitalist mode of production. The most visible impact of postmodern thinking on the city is in its architecture, where the 'concrete functionalism' of the modern era is replaced by a diversity of styles. In terms of the social geography of the city, the most important contribution of a postmodern perspective is how its focus on difference, uniqueness and individuality sensitises us to the needs and situations of all members of a society. This emphasis on the need to study urban phenomena from the multiple viewpoints of diverse individuals and groups was an integral part of the 'cultural turn' in urban geography (Box 2.3), and has been reflected in studies of gender differences in urban labour markets,[56] as well as of the 'spaces of exclusion' occupied by minority groups defined by class,[57] marital status,[58] sexuality,[59] race,[60] age[61] and disability.[62]

In addition to fundamental societal divisions based on class and ethnicity postmodernism has helped focus particular attention on social cleavages based on gender and sexuality as presented in feminist geography and in queer theory. Men and women not only use cityscapes in different ways but experience and perceive them differently. City environments both create and reflect gender roles in society. McDowall (1995) describes how female merchant bankers in the City of London felt obliged to conform to working practices of an aggressive male dominated environment.[63] Clearly, gender roles in the city reflect not only sexual differences but the system of power relations. Under a system of patriarchy the dominant social arrangements

Box 2.3

The cultural turn in urban geography

The cultural turn of the 1980s and 1990s high-lighted the study of culture in cities. Culture, in this context, is best regarded as 'ways of life' – it is a process of social significance wherein meaning is fluid and contested as it emerges from the shared discourse (sets of meanings) of different human groups. Culture viewed as a way of life comprises the interrelated elements of:

Values, i.e. people's ideals and aspirations, e.g. personal freedom.
Norms, i.e. rules and principles that govern people's lives, e.g. parking restrictions.
Objects, i.e. material things that people use, e.g. automobiles.

The cultural values of a society may be 'read' from the 'text' of the cityscape. Compare, for example, an urban expression of sprawling low-density suburbia with that of a compact, highly regulated form of development. A city's dominant value systems are contested – for example, a multi-storey office building may represent both an icon of financial power and a symbol of capitalist oppression.

A key impact of the cultural turn in urban geography was greater recognition of the voices of diverse 'others' and incorporation of the concerns of those previously excluded from urban geographical study. Conversely critics point to over-emphasis on representation, ideology and meaning and extensions into literary and psychoanalytical theory as under-playing key issues of power, inequality and material welfare.

and institutional structures promote domination of men over women and of masculinity over femininity. Cities can also provide supportive environments for formation of feminist associations to oppose patri-archical social systems – as evident in the increasing representation of women in the spheres of business and politics.

A postmodern perspective, and in particular queer theory, emphasises the socially constructed character of sexual identities. Cities can be places of repression and sites of liberation for those whose sexuality trans-gresses conventional boundaries. Gay and lesbian spaces – comprising clubs, retail outlets and areas of residential gentrification – have contributed to the establishment of alternative sexual identities, as well as to creation of a 'pink economy' and communities in many cities.[64] Postmodern perspectives on gender and sexuality in the city also illustrate that planes of division are cross-cutting and overlapping. For exam-ple, those who do not qualify on class, ethnic, age or gender are as likely to be excluded from a gay enclave as from any other neighbourhood where their identity marks them as an 'other'.

A major criticism directed at the postmodern approach to the city is its apparently unlimited relativism. Because it privileges the views of all indi-viduals, there appears to be no limit to the range of possible interpretations of any situation – there is, in effect, no 'real world'.[65] This has drawn particular criticism from 'socially concerned' urban geographers who decry postmodernism's inability to address the 'real' problems of disadvantaged urban residents.

TRANSNATIONALISM

Transnational urbanism refers to contemporary forms of urbanism resulting from the forces of globalisation. Particular attention is focused on transnational flows of migration and cultural practices that link residents in 'sending' and 'receiving' localities in a form of 'transnational social formation'.

According to Smith (2001 p 4), 'social networks of transnational migrants comprise one of the key cir-cuits of communicative action connecting localities beyond borders and constituting translocal ties across the globe'.[66] Transnational urbanism is evident in many ways. Examples range from the economic and cultural links between small rural villages in Mexico and large US cities (as in the flow of remittances and communications between places); to political activities of transnational networks of urban pressure groups, such as Slum/Shack Dwellers International, a network concerned with mutual learning through shared experience that has federations in several countries (see Chapter 29). The perspective of transnational urbanism helps illuminate the commonality within the diversity of our urban world.

POSTCOLONIALISM

Although the colonial era is largely past the attributes of the period may persist in Western representations of non-Western societies – both in the 'Third World' and within cities in the West. The **postcolonial** stance of critical engagement with the after-effects of colonialism attempts to expose the **ethnocentricism** of the dominant culture.[67] In the urban context, postcolonial theory contributes to understanding of cities in both 'colonising' and 'colonised' states. This is evident, for example, in the imprint of colonialism in cities of former colonial powers, as in the cultural and ethnic hybridity introduced by Algerian migrants in Paris, Puerto Ricans in New York City, and Jamaicans in London. Equally a postcolonial perspective can highlight the construction and reconstruction of cities in former colonies in practices ranging from promotion of 'heritage conservation' in Singapore to creation of a new capital city (Lilongwe) in Malawi as a conscious break with the colonial past.

The particular value of a postcolonial approach to urban study is its sensitivity to the diversity of urban experience and its advocacy of a global perspective that views different forms of urbanism as integral to an understanding of the contemporary world.

MORAL PHILOSOPHY

An approach based on moral philosophy or ethics represents an emergent perspective in urban geography. This seeks to examine critically the moral bases of society. Central to the ethical perspective is the concept of normative judgement that focuses on what *should be* rather than what is.[68] This involves critical evaluation of actual situations against **normative** conditions as defined by ethical principles. In urban geography, researchers are confronted constantly by ethical issues. Questions addressed include the extent to which there is equity in the distribution of welfare services, employment opportunities and decent housing for different social groups in the city; how to interpret the causes of inner-city poverty (including the relative weight attached to personal deficiencies of the population, or structurally induced constraints on behaviour); and the social acceptability of existing urban conditions – for example, what is an acceptable level of air pollution or of infant mortality?

Although moral perspectives on the city formed an important part of social science in the nineteenth century,[69] the foundations of the current ethical perspective in urban geography lie in the humanistic approach, in the consideration afforded to issues of social justice within the political economy perspective of the early 1970s,[70] and in more recent critiques of the ethics of market-oriented individualism.[71] There is also a degree of commonality with the postmodernist emphasis on the importance of difference (seen, for example, in feminist critiques of mále-centred interpretations of what constitutes a liveable urban environment).[72] Most fundamentally, however, the ethical perspective rejects the postmodernist denial of the possible existence of generally applicable moral bases for societal behaviour. Further, an ethical perspective contends that in a society not all manifestations of 'otherness' should be fostered; some (racial discrimination or child prostitution, for example) should be constrained.[73] As Smith (1994 p. 294) observed, while acknowledging the importance of 'difference' and 'otherness', 'we should not allow uncritical deference to other people's views and cultures to deny the possibility that some kinds of behaviour, ways of life and even moral codes are wrong'.[74]

IN SEARCH OF COMMON GROUND

Each of the major philosophical perspectives considered can claim to illuminate some part of the complex dynamics and structure of the city. But no single approach provides a full explanation of urban phenomena. The question of whether an accommodation is possible among the different approaches has been polarised between those who accept a pluralist stance – 'agreeing to differ' on the grounds that there is no single way to gain knowledge – and those who insist on the need to make a unitary choice of theoretical framework due to the perceived superiority of a particular theory of knowledge. Others have sought to combine approaches in different ways.[75] The latter route, which incorporates a search for a middle ground between the *generalisation* of positivism and the *exceptionalism* of postmodern theory, is the approach favoured here.

The analytical value of employing different theoretical perspectives is illustrated in Table 2.2, with reference to the question of urban residential structure.

To fully understand urban phenomena in the contemporary world requires consideration of both the *general* structural processes related to the 'mode of production' and an empirically informed appreciation of the *particular* social formations that emerge from the interaction of structural forces and local context. The importance of employing a combined multi-layered 'realist'

TABLE 2.2 ANALYTICAL VALUE OF DIFFERENT THEORETICAL PERSPECTIVES IN URBAN GEOGRAPHY: THE EXAMPLE OF URBAN RESIDENTIAL STRUCTURE

Theoretical perspective	Interpretative insight
Environmentalism	Although the notion of environmental determinism is now discredited, the influence of environmental factors on residential location can be seen in the problems of building in hazardous zones, and in the effects of architectural design on social behaviour (see examples in Chapter 19)
Positivism	Uses statistical analysis of objective social, economic and demographic data (e.g. via factorial ecology) to reveal areas in the city that display similar residential characteristics (see examples in Chapter 18)
Behaviouralism	Addresses the key question of why people and households relocate by examining the motives and strategies underlying the intra-urban migration of different social groups (see examples in Chapter 10)
Humanism	Explains how different individuals and social groups interact with their perceived environments, as in the differential use of public and private spaces within a city or residential neighbourhood (see examples in Chapter 19)
Managerialism	Illustrates how urban residential structure is affected by the ability of professional and bureaucratic gatekeepers to control access to resources, such as social housing or mortgage finance (see examples in Chapter 11)
Structuralism	Examines the ways in which political and economic forces and actors (e.g. financial institutions, property speculators and estate agents) influence the residential structure of a city through their activities in urban land and housing markets (see examples in Chapter 7)
Postmodernism	Explores the place of different social groups in the residential mosaic of the city by focusing on the particular lifestyles and residential experiences of various populations, such as ethnic minorities, affluent groups, gays, the elderly, disabled, and the poor (see examples in Chapter 18)
Transnationalism	Emphasises the interrelationships between cultural and residential environments across the globe as a consequence of globalisation, as evidenced in links between Third World rural villages and minority ethnic communities in Western cities (see Chapter 21)
Postcolonialism	Illuminates the effects of the colonial era on contemporary urban environments in both former colonising and colonised states as, for example, in the continuing influence of Western planning regulations on the form of urban development in Third World cities (see Chapter 22)
Moral philosophy	Critically evaluates the ethical underpinnings of issues such as homelessness or the incidence of slums and squatter settlements (see examples in Chapter 25)

perspective[76] that encompasses the global and local scales, social structure and human agency, and theory and empirical investigation in seeking to interpret the city informs the organisation and content of this book.

LEVELS OF ANALYSIS IN URBAN GEOGRAPHY

As we have seen, the character of urban environments is the outcome of the interplay of a host of private and public interests operating at a variety of geographical scales. In order to understand the geography of towns and cities, therefore, it is necessary to look both within and beyond the settlement, and to examine the complex of factors involved in urban change at all levels of the global–local continuum.

Although the factors and processes involved in urban development are not confined to any discrete level of the global–local spectrum, the concept of 'levels of analysis' offers a useful organising framework which simplifies the complexity of the real world and illustrates some of the issues of concern to urban geography at different spatial scales. We can identify five main levels of analysis.

THE NEIGHBOURHOOD

The neighbourhood is the area immediately around one's home; it usually displays some homogeneity in terms of housing type, ethnicity or socio-cultural values. Neighbourhoods may offer a locus for the formation of shared interests and development of community solidarity. Issues of relevance to the urban geographer at this level include the processes of local economic decline or revitalisation, residential segregation, levels of service provision and the use of neighbourhood political organisations as part of the popular struggle to control urban space.

THE CITY

Cities are centres of economic production and consumption, arenas of social networks and cultural activities, and the seat of government and administration. Urban geographers examine the role of a city in the regional, national and international economy, and how the city's socio-spatial form is conditioned by its role (for example, as a financial centre or manufacturing base). Study of the distribution of power in the city

would focus on the behaviour and biases of formal organisations as well as the informal arrangements by which public and private interests operate to influence government decisions. The differential socio-spatial distribution of benefits and disbenefits in the city is also an important area of investigation in urban geography.

THE REGION

The spread of urban influences into surrounding rural areas and, in particular, the spatial expansion of cities have introduced concepts such as urban region, metropolis, metroplex, **conurbation** and **megalopolis** into urban geography. Issues appropriate to this level of analysis include the **ecological footprint** of the city, land-use conflict on the **urban fringe**, growth management strategies and forms of metropolitan governance.

THE NATIONAL SYSTEM OF CITIES

Cities are affected by nationally defined goals established in pursuit of objectives that extend beyond urban concerns. Successive New Right governments in the UK (under Margaret Thatcher and John Major) and the USA (under Ronald Reagan and George Bush) followed an economic policy that focused on *national* economic development largely irrespective of its consequences for the growth or decline of individual urban areas. Cities were encouraged to become more competitive to attract inward investment. National-level policy guidelines, incentives in the shape of competitive grants, and financial and other controls over the actions of local government have a direct influence on urban decision-making and management. In order to comprehend processes and patterns of urban change, geographers need to have an understanding of national policy and the ways in which it affects the inter- and intra-urban geography of the state.

THE WORLD SYSTEM OF CITIES

The concept of a world system of cities reflects the growing interdependence of nations and cities within the global political economy. In this urban system, 'world cities' occupy a distinctive niche owing to their role as political and financial control centres. This status is evident in the concentration of advanced **producer services** such as education, R&D, banking and insurance, accounting, legal services, advertising and real estate services. Drawing on a 'world cities' perspective enables the urban geographer to reframe many urban questions previously defined in the context of the city or regional boundaries. This is illustrated graphically by the way in which investment decisions by managers in a Japanese-owned multinational company with headquarters in New York can lead to job losses and deprivation at the neighbourhood level in Liverpool or Lagos.

In studying the contemporary city, urban geographers must remain aware of the relationship between global and local forces in the production and re-production of urban environments. The need for such a perspective is reinforced by the process of globalisation, which, as we have seen, emphasises the linkages between different 'levels of analysis'. In particular, the global and the local must be regarded not as analytical opposites but as two sides of the same coin. The process of globalisation underscores the need for urban geographers to employ a multi-level interdisciplinary perspective in the search for urban explanation.

FURTHER READING

BOOKS

B. Berry and J. Wheeler (2005) *Urban Geography in America 1950–2000* New York: Routledge

K. Cox and R. Golledge (1981) *Behavioural Problems in Geography Revisited* London: Methuen

D. Gregory and J. Urry (1985) *Social Relations and Spatial Structures* London: Macmillan.

D. Harvey (1973) *Social Justice and the City* London: Arnold

D. Harvey (1985) *The Urbanisation of Capital* Oxford: Blackwell

D. Ley and M. Samuels (1978) *Humanistic Geography* London: Croom Helm

S.A. Marston, B. Towers, M. Cadwallader and A. Kirby (1989) The urban problematic, in G. Gaile and C. Willmott (eds) *Geography in America* Columbus OH: Merrill, 651–72

R. Peet (1998) *Modern Geographical Thought* Oxford: Blackwell

JOURNAL ARTICLES

A. Amin and S. Graham (1997) The ordinary city *Transactions of the Institute of British Geographers* 22, 411–29

B. Berry (1964) Cities as systems within systems of cities *Papers and Proceedings of the Regional Science Association* 13, 147–63

M. Dear and S. Flusty (1998) Postmodern urbanism *Annals of the Association of American Geographers* 88(1), 50–72

M. Gottdeiner (2000) Lefebvre and the bias of academic urbanism *City* 4(1), 93–100

F. Schaefer (1953) Exceptionalism in geography: a methodological examination *Annals of the Association of American Geographers* 43, 226–49

D. Walmsley (2000) Community, place and cyberspace *Australian Geographer* 31(1), 5–19

L. Wirth (1938) Urbanism as a way of life *American Journal of Sociology* 44, 1–24

KEY CONCEPTS

- Meaning of urban
- Urbanisation
- Urbanism
- Rural–urban continuum
- Space and place
- Cyberspace
- Social processes
- Spatial processes
- Environmentalism

- Positivism
- Behaviouralism
- Humanism
- Structuralism
- Managerialism
- Postmodernism
- Moral philosophy
- Levels of analysis

STUDY QUESTIONS

1. What do you understand by the term 'urban'?
2. Consider the value of an urban geographical perspective for understanding contemporary towns and cities.
3. With the aid of relevant examples, identify the concerns of the urban geographer at different levels of the global–local spectrum.
4. Dr Johnson (1709–84) said, 'When a man is tired of London he is tired of life; for there is in London all that life can afford,' yet Shelley (1792–1822) thought that 'Hell is a city much like London.' Make a list of the positive and negative features of urban life.
5. With reference to appropriate examples examine the view that large cities concentrate diversity.
6. Select any one of the major theoretical perspectives (such as positivism, structuralism or postmodernism) and write a reasoned critique of its value for the study of urban geography.

PROJECT

Select a major academic journal dealing with urban issues (e.g. *Urban Studies* or *Urban Geography*) and undertake a content analysis over a period of time to illustrate the changing focus of the subject. For each of the journal issues sampled you should classify the papers by their main theme. This information can then be tabulated and/or graphed to indicate the percentage coverage of different themes and hence the changing emphasis of urban geography over time. You should expect to find some enduring issues, some that decline in importance, as well as new entries to the field.

PART TWO

An Urbanising World

3

The Origins and Growth of Cities

Preview: pre-industrial cities; theories of urban origin; hydraulic theory; economic theory; military theories; religious theories; early urban hearths; the spread of urbanism; the Dark Ages and urban revival in Western Europe; the medieval town; early modern urbanism; industrial urbanism; the form of the industrial city; residential segregation; the origins of urban USA; the westward progress of urbanism in the USA; post-industrial urbanism; the post-industrial/postmodern city

INTRODUCTION

Three major transformations have altered the course of human life. The first was the revolution that led to the development of agriculture around 7000 BC and the growth of Neolithic farming settlements such as Jarmo in Iraq and Jericho in modern Israel. The second was the pre-industrial revolution that brought cities into being. The third was the **industrial revolution** of the eighteenth and nineteenth centuries that created the urban industrial forerunners of our present cities. This chapter begins by focusing on the pre-industrial revolution to examine the origins and development of the earliest urban settlements, before going on to consider the character of industrial and post-industrial cities.

PRECONDITIONS FOR URBAN GROWTH

In Chapter 2 we saw how problematic the definition of a city can be. With reference to the **pre-industrial city**, Wheatley (1971 p. xviii)[1] defined urbanism as 'that particular set of functionally integrated institutions which were first devised some 5,000 years ago to mediate the transformation of relatively egalitarian, ascriptive, kin-structured groups into socially stratified politically organised, territorially based societies'. This emphasis on institutional change

relates the growth of cities to a major socio-political restructuring of society, which he regards as a key element in the development of civilisation. In similar vein, Childe (1950)[2] provides a listing of ten characteristics of an urban civilisation. These may be separated into five primary characteristics referring to fundamental changes in the organisation of society and five secondary features indicative of the presence of the primary factors (Box 3.1). Thus, for example, a community that was capable of building monumental public works probably had not only the craft specialists to undertake the task but also sufficient surplus to support the work. It is important to recognise that these advances in social organisation occurred in particular environmental settings. The combination of environmental and social forces underlying early urban development is encapsulated in Duncan's (1961)[3] ecological perspective, which emphasises how external (environmental) stimuli and internal (social) interrelationships operated together to promote the growth of cities in the pre-industrial era (Box 3.2).

With these preliminary observations in mind, we can consider a number of different theories advanced to explain the origins of urban society. In doing so we should note that, as in much social science, it is not a question of whether one hypothesis is correct and the others wrong. Rather, each contributes some insight on the rise of cities.

BOX 3.1

Childe's ten characteristics of an urban civilisation

Primary characteristics

1. *Size and density of cities.* The great enlarge-ment of an organised population meant a much wider level of social integration.
2. Full-time *specialisation of labour.* Speciali-sation of production among workers was institutionalised, as were systems of distribu-tion and exchange.
3. *Concentration of surplus.* There were social means for the collection and management of the surplus production of farmers and artisans.
4. *Class-structured society.* A privileged ruling class of religious, political and military func-tionaries organised and directed the society.
5. *State organisation.* There was a well struc-tured political organisation with membership based on residence. This replaced political identification based on kinship.

Secondary characteristics

6. *Monumental public works.* There were col-lective enterprises in the form of temples, palaces, storehouses and irrigation systems.
7. *Long-distance trade.* Specialisation and exchange were expanded beyond the city in the development of trade.
8. *Standardised, monumental artwork.* Highly developed art forms gave expression to sym-bolic identification and aesthetic enjoyment.
9. *Writing.* The art of writing facilitated the processes of social organisation and man-agement.
10. *Arithmetic geometry and astronomy.* Exact, predictive science and engineering were initiated.

BOX 3.2

Preconditions for pre-industrial urban growth

Population
The presence of a population of a certain size residing permanently in one place is a fundamental requirement. The environment, level of technology and social organisation all set limits on how large such a population would grow. Particularly important was the extent to which the agricultural base created a food surplus to sustain an urban population. The earliest cities were relatively small in modern terms, with few exceeding 25,000 inhabitants.

Environment
The key influence of the environment, including topography, climate, social conditions and natural resources on early urban growth, is illustrated by the location of the earliest Middle Eastern cities on the rivers Tigris and Euphrates, which provided a water supply, fish and fertile soil that could be cultivated with simple technology.

Technology
In addition to the development of agricultural skills, a major challenge for the early urban soci-eties of the Middle East was to develop a technol-ogy for river management to exploit the benefits of water and minimise the risk of flooding.

Social organisation
The growth of population and trade demanded a more complex organisational structure, including a political, economic and social infrastructure, a bureaucracy and leadership, accompanied by social stratification.

Source: adapted from O. Duncan (1961) From social system to ecosystem *Sociological Inquiry* 31, 140–9

THEORIES OF URBAN ORIGINS

HYDRAULIC THEORY

The importance of irrigation for urban development, especially in the semi-arid climates of the Middle East where the **agricultural revolution** took place, was identified by Wittfogel (1957),[4] who argued that the need for large-scale water management required cen-tralised co-ordination and direction, which in turn required concentrated settlement. The principal char-acteristics of a 'hydraulic society' are that it:

1. Permits an intensification of agriculture.
2. Involves a particular **division of labour**.
3. Necessitates co-operation on a large scale.

As Box 3.2 indicates, these are reflected in Duncan's (1961) preconditions for urban growth. Agricultural intensification allows a concentration of population, while co-operation leads to a need for managers and bureaucrats. Those who control the water resources (whether a temple elite or secular state) exert power over others (e.g. farmers). The division of labour, centralisation of power and administrative structure all promote concentrated settlement, and hence the emergence of a town. As well as in the Middle East, evidence of a relationship between the adoption of irrigation and rapid population growth, nucleation, monument construction, intense social stratification and expansionist warfare has also been found at Teotihuacán in Mexico, which was the site of a pre-industrial city of 70,000–125,000 inhabitants.

There is now little doubt that irrigation was a key factor in the growth of pre-industrial cities in the ancient world. The problem lies in disentangling cause and effect. This is particularly difficult if one is seeking to support the belief that urbanisation *followed upon* the development of irrigation. A more likely scenario is that the institutions of centralised urban government and large-scale irrigation grew side by side. At first, small-scale irrigation schemes would have required a certain amount of administration, which would have expanded the irrigation system. This in turn would have required greater administration and so on, eventually leading to large-scale irrigation works and an urban political organisation with a monopoly of power.

ECONOMIC THEORY

Several theorists have suggested that the development of complex large-scale trading networks stimulated the growth of urban society.[5] Certainly, the fact that southern Mesopotamia did not have many raw materials such as metallic ores, timber, building stone or stone for tools made trade essential. This required an administrative organisation to control the procurement, production and distribution of goods. Such an organisation would have been a powerful agent in the community, and its power may well have extended beyond trade into other aspects of society. The need to increase production for trade purposes as well as to feed an expanding population would have led to continued specialisation and intensification, and the growing sedentary population would itself constitute a market for local produce and trade goods. Once again, however, it is unclear whether trade was a cause of city growth or the product of an already existing urban administrative elite.

MILITARY THEORIES

Some theorists suggest that the origin of cities lay in the need for people to gather together for protection against an external threat, the initial agglomeration leading to subsequent urban expansion. The excavation of a massive defensive wall built on bedrock would appear to indicate the defensive origins of Jericho, but not all early towns have such defences. Wheatley (1971)[6] believed that warfare may have contributed to the *intensification* of urban development in some places by inducing a concentration of population for defensive purposes and by stimulating craft specialisation.

RELIGIOUS THEORIES

Religious theories focus on the importance of a well-developed power structure for the formation and perpetuation of urban places and, in particular, how power was appropriated into the hands of a religious elite who controlled the disposal of surplus produce provided as offerings.[7] There is clear evidence of shrines and temples in ancient urban sites and there can be little doubt that religion played a significant part in the process of social transformation that created cities. However, it is unlikely to have been the sole factor.

THEORETICAL CONSENSUS?

It is doubtful if a single autonomous, causative factor will ever be identified in the nexus of social, economic and political transformations that resulted in the emergence of urban forms of living. A more realistic interpretation is generated if the concept of an 'urban revolution' is replaced by the idea of an *urban transformation* involving a host of factors operating over a long period of time. As Redman (1978 p. 229)[8] explained:

urbanisation was not a linear arrangement in which one factor caused a change in a second factor, which then caused a change in a third, and so on. Rather, the rise of civilisation should be conceptualised as a series of interacting incremental processes that were triggered by favourable ecological and cultural conditions and that continued to develop through mutually reinforcing interactions.

This approach represents a significant departure from monocausal views on the origin of cities. Its value lies in exposing the key roles of social stratification and individual and group decision-making underlying the complex reality of the transformation from nomadic life to settled urban life in the ancient world. For urban geographers, this poses the particular question of *where* such developments occurred.

EARLY URBAN HEARTHS

As Figure 3.1 illustrates, there is evidence of early city growth in four areas of the Old World and one area of the New World:

1. *Mesopotamia*. The first cities are thought to have begun around 3500 BC in lower Mesopotamia (Sumer) around the Tigris and Euphrates rivers (Figure 3.2). One of the earliest cities was Ur, which from 2300 BC to 2180 BC was the capital city of the Sumerian Empire, which extended north along the Fertile Crescent, possibly as far as the Mediterranean. In 1885 BC Ur and the other southern cities were conquered by the Babylonians[9] (Box 3.3).
2. *Egypt*. There is a long-standing debate in archaeology over theories of urban diffusion or independent invention but it is most probable that agricultural and other technologies, possibly including city-building, spread along the Fertile

Crescent, then south-west into the Nile valley. By 3500 BC a number of the Neolithic farm hamlets along the lower Nile had risen to 'overgrown village' status and were clustered into several politically independent units, each containing large co-operative irrigation projects. The transition from settled agricultural communities to cities occurred around 3300 BC when the lower Nile was unified under the first pharaoh, Menes.

The early Egyptian cities were not as large or as densely settled as those of Mesopotamia because:

- The early dynastic practice of changing the site of the capital, normally the largest settlement, with the ascendancy of a new pharaoh limited the growth opportunity of any single city.
- The security provided by extensive desert on both sides of the Nile meant that once the valley was unified politically, Egyptian cities, unlike those of Mesopotamia, did not require elaborate fortifications and garrisoned troops for protection.

3. *The Indus valley*. The Harappa civilisation appeared around 2500 BC in the Indus valley in what is now Pakistan. It was distinguished by twin capital cities, a northern one of Harappa in the Punjab and Mohenjo-daro, 350 miles downriver.

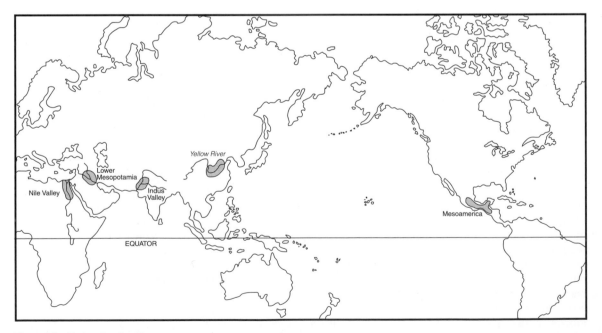

Figure 3.1   Early urban hearths

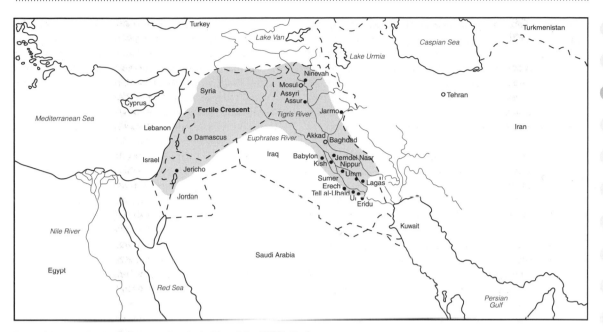

Figure 3.2   The Fertile Crescent and ancient cities of the Middle East

BOX 3.3

Ur of the Chaldees

The city of 2000 BC was surrounded by a wall 26 ft (8 m) in height which enclosed an area of eighty-nine acres (36 ha) and a maximum population of 35,000. The walled city was an irregular oval shape, about three-quarters of a mile (1.2 km) long by half a mile (0.8 km) wide. It stood on a mound formed by the ruins of previous buildings, with the river Euphrates flowing along the western side and a navigable canal to the north and east.

The harbours to the north and west provided protected anchorages (A and B in the plan). The *temenos* or religious precinct occupied much of the north-western quarter of the city (C in the plan). The remainder of the city within the walls was densely built up as residential quarters (D in the plan), with two-storey houses of burnt brick below and mud brick above, with plaster and whitewash concealing the change in material.

House size and ground plan varied according to the space available and the owner's means, but generally rooms were arranged around a central paved courtyard that gave light and air to the house. A processional avenue (X–X in the plan) possibly ran straight through to the *temenos*. Including those resident outside the walls, a figure of 250,000 has been estimated for the total population of the city-state.

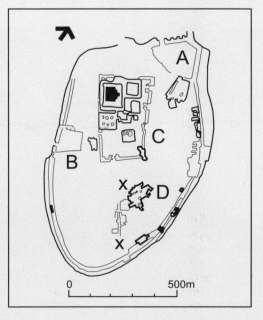

Source: A. Morris (1994) *History of Urban Form* Harlow: Longman

The planned layout of each city was in marked contrast to the organic growth of Mesopotamian cities such as Ur. Both cities were laid out on a gridiron pattern with wide, straight streets forming 1,200 ft × 800 ft (370 m × 240 m) rectangular blocks. Socio-spatial segregation was common, with blocks or precincts occupied by a specific group such as potters, weavers, metalworkers and the elite. Each city covered approximately one square mile in area (640 acres/250 ha) and accommodated 20,000 people.

The Harappa kingdom was ruled from the twin capitals by a single 'priest-king' who wielded absolute power. There is some evidence of trade with the Sumerian city-states by 2000 BC but the unchanging material culture and still undeciphered written language suggests that, in contrast to the cities of the Nile valley, the Harappa culture and cities emerged independently. Following a thousand years of stability, the Harappa civilisation was destroyed by invaders around 1500 BC.

4. *The Yellow River*. The valley of the Huangho (Yellow River) was the birthplace of the Shang civilisation that arose around 1800 BC. The most significant feature is that individual cities, such as An-Yang, were linked into a network of agricultural villages; a town wall did not separate an urban **subculture** from a rural one.[10] This form of 'urban region' is without precedent in the early civilisations of Mesopotamia, the Nile and the Indus.

5. *Mesoamerica*. The earliest cities in the New World appeared around 200 BC – in southern Mexico (Yucatán), Guatemala, Belize and Honduras. Thus Mesoamerican peoples were entering a stage of development equivalent to the Neolithic of the Old World at a time when Mesopotamian cities had been in existence for 2,000 years. Of the several civilisations that evolved in Mesoamerica, the Mayan, which flourished between AD 300 and AD 1000, was the most culturally advanced. Cities such as Tikal, Vaxactum and Mayapán were centres of small states ruled by a leader drawn from a

Plate 3.1   The pyramid at Chichen Itza in Mexico functioned as a solar calendar for the Mayan civilisation that flourished in Meso-America and the Yucatan between 300 BC and AD 1300

priesthood and organised into a loose confedera-tion. Society was highly stratified, with the elite occupying central city land around the palaces and temples, and the lower classes the urban periphery.

Although the debate over urban diffusion or independent origin remains unresolved, identification of a number of common features among early urban societies encour-aged Sjoberg (1960) to propose a model of the socio-spatial structure of the pre-industrial city[11] (Box 3.4).

THE SPREAD OF URBANISM

By 800 BC cities such as Athens, Sparta and Megara had arisen on the Greek mainland. Greek cities subse-quently spread to other parts of the Mediterranean and along the Black Sea coast. The Greek urban diaspora was a direct response to population pressure and the poor agricultural base available to the mainland cities. Individual cities equipped expeditions to establish new cities. A first wave beginning around 750 BC led to

BOX 3.4

A model of the pre-industrial city

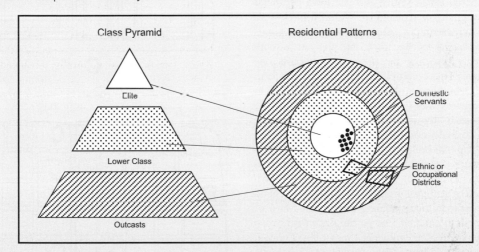

Sjoberg's model attempts to identify some common cross-cultural features of the pre-industrial city. Like all models claiming a wide range of application, this model of a generic pre-industrial city has been criticised for oversimplification and overextension. Sjoberg's model is most appropriate in cities where domination by an elite occurred. He identifies three social groups with social boundaries rigidly defined and often formally codified. The class pyramid was translated into a distinct spatial pattern. The elite occupied the central district around the ceremonial and symbolic institutions, including the religious, edu-cational and political structures. Interspersed within the inner core were the servants of the elite, but the major concentration of lower-class residences was in a zone outside the core. The outcasts were relegated to the periphery, completing the spatial gradation of social status. Within each of the zones further dif-ferentiation occurred by occupation or ethnicity;

however, there was limited segregation of distinct land-use functions. This generalisation has been ques-tioned by Vance,[12] who gives greater emphasis to the pattern of occupational subdistricts and downplays the extent of a zone of lower-class labourers, placing them instead as lodgers scattered within the various subdistricts. Notwithstanding differences of detail, both Sjoberg and Vance depict a walking-scale city with a general social degradation from core to periphery.

Despite its simple structure, at a general level the Sjoberg model continues to serve as a useful standard against which individual cities may be assessed. For example, while Langton[13] found limited correspon-dence between the model and the social geography of seventeenth-century Newcastle upon Tyne, Radford[14] revealed a close relationship between the model's description of pre-industrial urban society and the situation in the antebellum city of Charleston SC.

Source: after G. Sjoberg (1960) *The Pre-industrial City* New York: Free Press

settlements on the coast of the Ionian Sea, in Sicily and in southern Italy (e.g. Ephesus, Syracuse and Naples), with a second wave spreading east to reach the Black Sea by 650 BC. A significant feature of Greek urban colonisation was that in contrast to the organic layout of mainland cities like Athens, the new independent cities frequently adopted a gridiron plan, irrespective of the topography of a site.

The expansion of the Roman Empire, following the defeat of Carthage in the Punic Wars (264–146 BC), carried the practice of city-building, and in particular the gridiron plan, into Western Europe. In Britain, prior to the Roman conquest in 55 BC, systematic town-planning was unknown. Figure 3.3 shows the typical form of a Roman imperial urban settlement. The town perimeter was usually square or rectangular. Within it there were two main cross-streets, the east–west *decumanus* through the centre of the town, and the north–south *cardo*, usually bisecting the *decumanus* towards one end. Secondary streets completed the grid layout to form the *insulae* or building blocks. The *forum* (equivalent to the Greek *agora*) was typically located in one of the angles formed by the intersection of the two principal streets and normally comprised a colonnaded courtyard with a meeting hall built across one end. The main temple, the theatre and the public baths were also located near the forum. Venta Silurum, a Romano-British town on the western military frontier in Wales, is a good example of Roman town planning[15] (Box 3.5). Many of the great modern cities of Europe can trace their urban beginnings to Roman influence, including London, Brussels, Seville, Cologne, Paris, Vienna and Belgrade.

With the fall of the Roman Empire in the fourth century AD, Muslim control of the Mediterranean trade routes during the seventh and eighth centuries and increasing Viking raids from the north in the ninth century, much of Europe entered a Dark Age of economic and cultural stagnation. Under such unsettled conditions long-distance trade of any significance was impossible, towns became isolated and inward-looking, and urban life in Western Europe declined to its nadir by the end of the ninth century.

URBAN REVIVAL IN WESTERN EUROPE

During the tenth century, the reopening of long-distance trade routes and the presence of urban nuclei represented by fortified settlements and ecclesiastical centres stimulated the beginnings of a commercial revival. But the rebirth of urbanism was hindered by

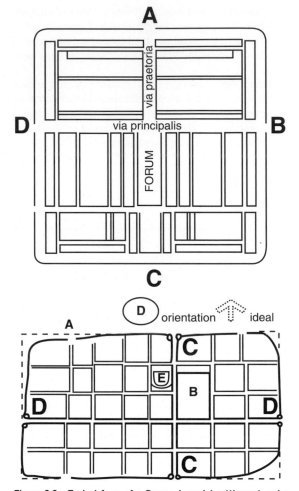

Figure 3.3 Typical form of a Roman imperial settlement and a generalised example from Roman Britain
Source: A. Morris (1994) *History of Urban Form* Harlow: Longman

the social and economic constraints inherent in the dominant social system of feudalism. Feudalism was a form of social organisation (or mode of production) characterised by the interrelationship of two social groups. Direct producers (peasants working the land) were subject to politico-legal domination by social superiors (feudal lords) who formed a status hierarchy headed by the monarch. Social relations under feudalism are not defined primarily through market interactions (as under capitalism) but are legally defined, and the whole society was organised in terms of a formal and rigid hierarchy of power and dominance. A key feature of the manorial (estate-based) system of economic organisation prevalent in medieval feudalism was the focus on self-sufficiency and local markets.

BOX 3.5

Plan of Venta Silurum (Caerwent)

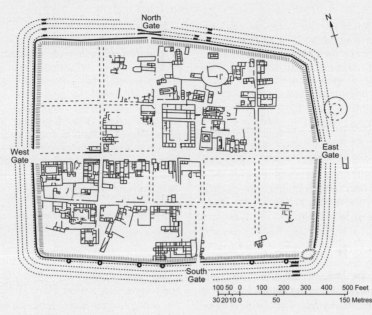

The town dates from AD 75 and was founded to replace a tribal hill fort situated on high ground a mile to the north-west. The settlement covered an area of forty-four acres (18 ha) and typically was divided in two by the main east–west street that was lined with shops and ran between the two main gates. Two other streets running east–west and four running north–south divided the town into twenty blocks. In a north–south direction there is not one dominant street line but two, each leading to a gate in the defences in a manner typical of a military settlement.

This restricted innovation and commercial expansion. In addition, since the amount of **surplus value** extracted by feudal lords in rents and taxes was dependent less on market forces than on their expenditure requirements (which increased as they competed for political status through conspicuous consumption), there was insufficient reinvestment in the productive capacity of agriculture and industry.[16] The regrowth of towns in Europe coincided with the decline of the feudal system and was related to the increasing economic and political power of an emerging middle class of merchants opposed to the economic regulation of their activities by feudal lords (Box 3.6). The urban bourgeoisie of merchants, manufacturers and financiers, often organised into guilds, obtained certain freedoms, notably the right to supervise their own markets. Successful towns obtained charters from the king and, in England, were known as boroughs. In the more urbanised parts of Europe many towns such as Milan,

Florence, Cologne and Bruges were able to become self-governing municipalities (communes) capable of independent political action. In northern Italy a fiercely competitive society of rival city-states had emerged by the thirteenth century (providing a foundation for the Italian Renaissance). The economic and political freedoms obtained by the emerging class of burgesses (free merchants and craftsmen) was of central importance in the rise of European urban society and the corresponding decline of medieval feudalism. As the power of the urban bourgeoisie increased, the goal of the economy became expansion and economic profit the function of city growth.

THE MEDIEVAL TOWN

The power of commerce and trade in medieval urbanism is exemplified in the German city of Lübeck,

BOX 3.6

The merchant class and urban growth in medieval Europe

During the centuries of the Dark Ages an economy based on exchange was replaced by an economy of consumption. But a new merchant class was appearing on the edge of feudal society. Probably originating among landless men, escaped serfs, casual harvest labourers, beggars and outlaws, the bold and resourceful among them – the fast talkers, quick with languages, ready to fight or cheat – became chapmen or pedlars, carrying their wares to remote hamlets. They were paid in pennies and farthings and in portable local products, such as beeswax, rabbit fur, goose quills and the sheepskins for making parchment. If they prospered, they could settle in a centre and hire others to tramp the forest paths.

The merchants made the towns. They needed walls and wall builders, warehouses and guards, artisans to manufacture their trade goods, cask makers, cart builders, smiths, shipwrights and sailors, soldiers and muleteers. They needed farmers and herdsmen outside the walls to feed them; bakers, brewers and butchers within. They bought the privilege of self-government, substituting a money economy for one based on land, and thus were likely to oppose the local lord and become supporters of his distant superior, the king. Towns recruited manpower by offering freedom to any serf who would live within their walls for a year and a day.

Source: **M. Bishop (1971)** *The Penguin Book of the Middle Ages* **Harmondsworth: Penguin**

founded in 1159 by the Duke of Saxony in order to attract merchants from neighbouring lands. Situated at the western edge of the Baltic Sea, Lübeck developed into the nucleus of a German Baltic trading system organised by a confederation of towns known as the Hanseatic League that included Hamburg, Lüneberg, Cologne, Danzig and Riga. Development of regular trade with the Russian city of Novgorod brought furs and forest products, which, along with local salt for preserving fish and meat, were distributed throughout northern Europe in exchange for cloth and other manufactures. The wealth and power acquired by the merchant elite enabled them to govern Lübeck and the other Hanse cities.

Within Lübeck the central importance of trade and commerce was reflected in the urban structure (Figure 3.4). The town had a medieval cathedral and monasteries, but it was dominated by the central market place. This consisted of the shops of a range of specialist merchants and craftsmen concentrated together in their own areas (e.g. a street of bakers, spice merchants or tailors), with living quarters located above the shop. A mint, the new urban engine of the medieval money economy, was located close to the market place.

Despite the central importance of commerce, the medieval town did not sever links with the surrounding area but enjoyed a symbiotic relationship with the countryside. In particular, medieval towns, with their poor hygiene and higher mortality, grew only because of the influx of rural immigrants seeking employment

and economic opportunity. Many cities, such as Venice and Nuremberg, annexed large areas of their surrounding lands to guarantee food supplies (a rural–urban symbiosis that remains common in modern China).

In addition to the commercial function, the main characteristic of the European medieval city that distinguished it as a separate entity was its legal status, which afforded citizens privileges denied to rural dwellers. The dictum that 'town air makes free' was a powerful attraction for the rural peasantry and a major force for urban growth. The same privileges did not extend to medieval towns elsewhere. The city of Novgorod, despite its prosperity, was not politically autonomous. Neither were urban freedoms possessed by other important non-European medieval towns such as Baghdad, Delhi or Beijing. Only in the medieval West did cities acquire this special legal status.[17]

More generally, the new force of *capitalism* pushed aside the vestiges of feudalism, ushered in the industrial revolution, and paved the way for the emergence of the industrial city.

THE PRECONDITIONS FOR INDUSTRIAL URBANISM

The complex series of innovations commonly referred to as the industrial revolution emerged in Britain from the mid-eighteenth century onwards, spreading from there to the European mainland and ultimately to other

(a)

(b)

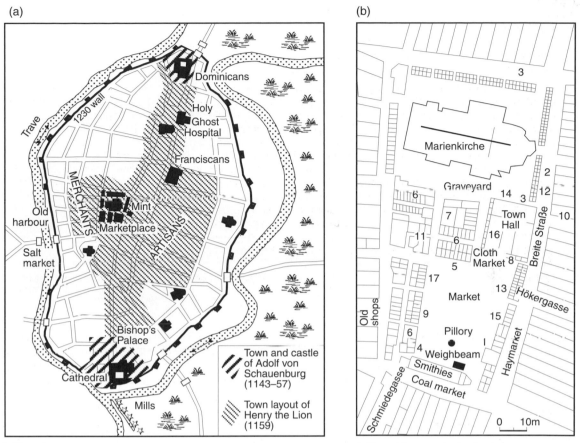

Lübeck

Lübeck's market place

KEY
1: armourers; 2: butchers; 3: bakers; 4: beltmakers; 5: money-changers; 6: shoemakers; 7: spice merchants;
8: needlemakers; 9: fellers; 10: herring merchants; 11: grocers; 12: minters; 13: goldsmiths; 14: cookshops;
15: saddlers; 16: tailors; 17: tanners.

Figure 3.4   The urban structure of medieval Lübeck

parts of the world during the nineteenth and twentieth centuries.[18] The industrial revolution is typically thought of in terms of technology – the invention of complex machines and use of inanimate energy sources that greatly increased worker productivity – but important cultural, social and population changes were also involved. It is important to understand these and their effects on the form of the city.

For Weber (1958),[19] a fundamental change in Western cultural values was a prerequisite for the industrial revolution. Central to this was the concept of profit. In medieval Europe, profit was an alien concept among artisans: the price of a product was fixed by adding to the cost of materials the fair value of one's labour. In the absence of profit (defined as the excess

of the selling price of goods over their cost) there could be no large capital accumulation to invest in the development of industrial society. This change in social outlook was stimulated by the Protestant Reformation, which fostered a new set of values that stressed rationality in interpersonal relations, including trade, hard work and the right to the material rewards of one's labour. This 'work ethic', which originated in northern Europe in the sixteenth century, spread throughout Western society and in Weber's view represented a prime precondition for the industrial revolution, the development of capitalist society and the emergence of the **industrial city**, the foundations of which were laid in the early modern period.

Plate 3.2 The oval-shaped Renaissance-period Piazza San Pietro in Rome, constructed by Bernini in 1656 to symbolise the all-embracing power of the Church, is a potent expression of the link between architecture and ideology

EARLY MODERN URBANISM

In the early modern (post-medieval, pre-industrial) period, from 1500 to 1800, the economic and political powers of the European medieval city were usurped by the expansion of nation-states. Despite this relative decline in urban autonomy, however, the period witnessed increasing levels of urbanisation and the gradual emergence of an urban system in Europe.[20] During the early modern period urban development in Europe was characterised by cycles of growth and decline. Three main phases may be identified:

1. The 'long sixteenth century', from 1500 to 1650, was a period of general prosperity during which urban growth was widely distributed.
2. From 1650 to 1750 population growth slowed in response to war, plague and famines combined with cyclical downturn in the economy as a result of rising labour costs, falling rent incomes from property and technological stagnation. Urban growth occurred differentially. Cities that were

centres of state government (such as Madrid) and port cities engaged in Atlantic trade (such as Amsterdam) experienced growth. By contrast the majority of inland trading centres, ecclesiastical seats and industrial towns suffered at least a relative loss of standing. A net result was a more hierarchical size distribution of urban places.

3. The period of 'new urbanisation' between 1750 and 1800 witnessed the growth of smaller cities and the addition of new cities to the urban system. This process was related to the political economy of **proto-industrial** production that formed as urban and rural-based precursors of factory-based manufacturing industry.[21] Widespread urban growth was stimulated by increases in agricultural production and incomes that promoted the growth of regional marketing, administrative and commercial centres. Many small cities with resource-based industries such as metallurgy grew as a consequence of technological innovation, and many rural places that had developed with proto-industrial textile production emerged

to become industrial towns as technological change encouraged factory organisation. By the end of the early modern period the twin processes of proto-industrialisation and urbanisation had paved the way for the emergence of industrial urbanism.

THE FORM OF THE INDUSTRIAL CITY

The nineteenth century witnessed the flowering of *industrial capitalism* in Western Europe and the rapid growth of the European industrial city. A major driving force was the factory system with its **economies of scale**, increased productivity and higher levels of output. The need for a large pool of labour as well as ancillary services and markets for their products encouraged factories to cluster together in towns. Successful towns attracted further economic activity that drew in more migrants in search of work in a cumulative process of growth (Box 3.7). Town populations grew apace. By 1800 London was the largest city in the world, with a population of over 900,000. The population of Birmingham increased by 273 per cent between 1801 and 1851 from 71,000 to 265,000, Manchester grew from 75,000 in 1801 to 338,000 in 1851 (a growth of 351 per cent) and Glasgow from

84,000 to 350,000 over the same period (an increase of 317 per cent). Lawton (1972)[22] estimated that almost all the 27 million increase in the British population between 1801 and 1911 was absorbed by urban areas. In many of the great nineteenth-century industrial cities, wealth found civic expression in monumental public buildings and cultural institutions.

Industrial capitalism also brought a major realignment of social structures with the creation of two main classes:

1. *Capitalists*, who invested in labour with the goal of realising a profit.
2. *Labour*, that sold its skills to the owner of capital in return for a wage.

The unequal division of power between the two main classes had a direct effect on the form of the industrial city, which was developed primarily to fulfil the needs of capital. This was particularly evident in the socio-spatial segregation of the classes (Figure 3.5). The burgeoning populations of the nineteenth-century industrial cities placed an enormous strain on urban services and infrastructure. Public sanitation and water supplies were inadequate and often non-existent in the **slums**. Mid-century London had 200,000 undrained cesspools, and the river Thames was virtually an open

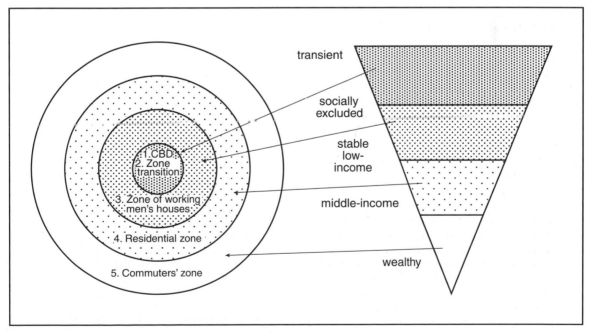

Figure 3.5 Burgess's model of the industrial city

BOX 3.7

The circular and cumulative model of urban growth

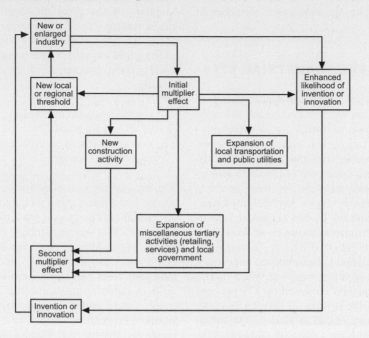

The urban economy may be viewed as being made up of two interrelated sectors:

1. The basic sector, comprising activities that produce goods and services sold outside the city to obtain the means needed to purchase imports (e.g. manufactures).
2. The non-basic sector, comprising activities that provide goods and services for the city itself (e.g. retailing).

The two sectors are functionally interdependent. Thus, if the basic sector expands, workers in that sector will spend more on services available in the city, creating growth in the non-basic sector. The extent of growth will depend on the respective sizes of the two sectors. In a city with a basic : non-basic ratio of 1 : 4 an increase of 1,000 in basic employment will lead to an increase of 5,000 (1,000 + 4,000) in total employment. This is referred to as the urban economic multiplier and is a key concept in the economic explanation of urban growth. In practice, given the variety of economic activity in any city, the mechanisms of growth are highly complex. They are best seen as a circular and cumulative process whereby growth in one sector triggers expansion through secondary multiplier effects elsewhere in the economy.

Consider a situation in which the output of a company manufacturing computer chips expands in response to increased demand. Company profits and employees' wages will rise and expenditure by shareholders and workers will increase, leading to growth in tertiary activities (such as existing and new shopping and retail activities). The company may decide to move to larger premises, which generates growth in the construction industry. It may also take on additional workers, leading to in-migration, and new house-building and urban growth. These changes in the non-basic sector can act as a secondary multiplier effect, lifting the city to a new threshold comprising a larger market that promotes newer forms of manufacturing and service activities. The company research and development team may invent an even faster computer chip, expanding demand for the company's product and leading to a further round of growth.

The interaction between the basic and non-basic sectors of the urban economy and combination of multiplier effects promotes urban growth. Conversely, the opposite situation may also occur, leading to economic stagnation and urban decline.

sewer. Similar conditions existed in other large industrial cities and it is no surprise that cholera and typhoid were rife, especially, though not exclusively, among the poor. A cholera epidemic in 1832 claimed the lives of 5,000 Londoners. In the same year cholera killed 2,800 in Glasgow. The wealthy, with their power to exercise choice, sought to maintain a healthy separation from the masses.

RESIDENTIAL SEGREGATION

Several processes underlay the pattern of residential segregation in the Victorian city. First, in terms of the physical growth of the city and the evolution of housing areas, much of the house building was undertaken by speculative builders, who were rarely interested in building for the lower end of the market. As a result, the demand for cheap housing remained high, especially following the influx of migrants from the Irish famine in the mid-nineteenth century. This demand was met by multiple occupancy, leading to overcrowded conditions in the central slum areas of most cities. The development of new middle- and high-status housing on the urban periphery by speculative builders allowed vacated inner-city properties to filter down to poorer families.

Second, the development of residential segregation was a result of individual locational decisions within the context of a rapidly expanding urban population. Few urban dwellers remained in the same house for long in the nineteenth century, and both migrants and established residents would have been continually re-evaluating the suitability of their residential location. The main economic constraints on residential relocation were those of disposable income, the availability of employment in different areas, and access to accommodation in different sectors of the housing market. Social, demographic and cultural characteristics, as well as knowledge of the city, also influenced residential location.

Third, the process of residential differentiation was also influenced by the development of commercial and industrial areas within the city that imposed constraints on the nature of residential development. These constraints were exerted through pressure on land in certain districts, such as the invasion of inner-city housing areas by commercial development; through the provision of employment in particular areas with associated working-class housing close by; and through industrial pollution that prompted those who could do so to move away. Fourth, although land-use planning did not exist in the Victorian city and institutional forces were weaker than they are today, national and local governments affected residential development in several ways. Local by-laws, together with national health and housing legislation, imposed some control over new housing development, while town improvement schemes aimed at better water supply, sewage disposal, the closure of cellar dwellings and the demolition of slum properties affected existing housing. The rate at which new houses were built, and their tenure and form, were also constrained by private-sector factors such as the problem of land-ownership and the prevailing rate of return on investment. Finally, though based on objective standards, residential differentiation was perpetuated through popular images of different areas. Certain parts of the city could gain a reputation as being, say, of high status, unhealthy or predominantly Irish.[23]

The operation and outcomes of the processes of residential segregation in the Victorian city are apparent in Liverpool where, as in many nineteenth-century industrial cities, immigration was a major factor underlying urban change (Box 3.8). More generally, Pooley (1979)[24] provides a model that summarises the processes of residential sorting by social class in the nineteenth-century British city (Figure 3.6). Intra-urban residential mobility is characterised by three main features – distance, direction and frequency – which, in combination, produce different patterns of movement within the city. These three main features are, in turn, influenced by the personal characteristics of the household (e.g. socio-economic group, or stage in the life cycle) and by general features of city structure (such as the extent to which distinct social and housing areas have developed). The outcome of these factors is seen in the production of social areas characterised by different levels of residential mobility and community stability. The model identifies four types of area, together with their typical locations in the Victorian city:

1. *Type A*. An area that maintains long-term stability owing to low rates of mobility and the replacement of population loss by in-migrants of a similar character. This type occurred most frequently within high-status areas of the city.
2. *Type B*. An area that attains short-term stability through frequent short-distant mobility of a circulatory nature, but which over a longer period succumbs to a general centrifugal flow of population. This was the dominant form of mobility in the Victorian city.

BOX 3.8

Socio-spatial segregation in nineteenth-century Liverpool

By 1871 socio-spatial segregation was a feature of the urban landscape of Liverpool. It was based on three principal factors: (1) socio-economic class; (2) ethnicity (primarily Irish/non-Irish); and (3) stage in lifestyle. Though the result is a complex mix of sectors, zones and nuclei, a series of distinct elements can be identified:

1. The older central residential districts, approximately within the 1851 built-up area, had developed by 1871 into a number of distinct sectors largely constrained by the location of the docks, principal routes radiating from the centre, and surviving islands of high-status property.
2. Outside the 1851 built-up area, broadly zonal characteristics are present: population from the central area had moved out into the newer suburbs, within which there were several areas of a transitional nature. Complementary to the mainly outward direction of 'internal' population movement, the central zones were largely fed by in-migrants, some of whom were to stay permanently in low-status areas, others of whom soon moved out to the newer suburbs. There was also considerable movement between sectors of low-status housing, often caused by displacement due to urban redevelopment, though non-Irish families generally avoided the Irish areas.
3. The high-status population moved sectorally into the third peripheral zone, while, within the suburban ring, lateral movement took socially upward-moving families into the high-status sector to the north.

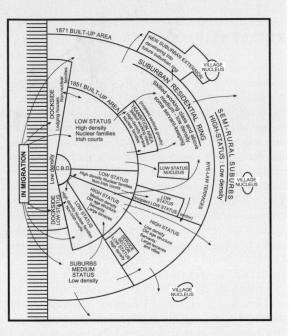

Source: **adapted from R. Lawton and C. Pooley (1976)** *The Social Geography of Merseyside in the Nineteenth Century: Final Report to the Social Science Research Council* **London: SSRC**

3. *Type C.* An area where there are high rates of longer-distance mobility, as from central and industrial areas under pressure for further commercial use. Rapid community disintegration may take place, and such areas are essentially unstable in character, with a largely transient population often in the process of moving towards the urban periphery. Community disintegration may also occur in high-status areas that are under pressure from a lower-status population. As the high-status migrants leave, the area may become unstable, or alternatively develop as a new, lower-status community.

4. *Type D.* An area of rapid in-migration on the urban periphery receiving large numbers of in-migrants from outside the city as well as from intra-urban population flows. Such areas are characterised by strong community formation and long-term stability.

Although the precise form of mobility patterns is dependent on local urban context, the model provides a useful summary of the major causal forces, processes and outcomes of residential mobility in the Victorian city. It is clear that in general the nineteenth-century city formed a dynamic system in which frequent

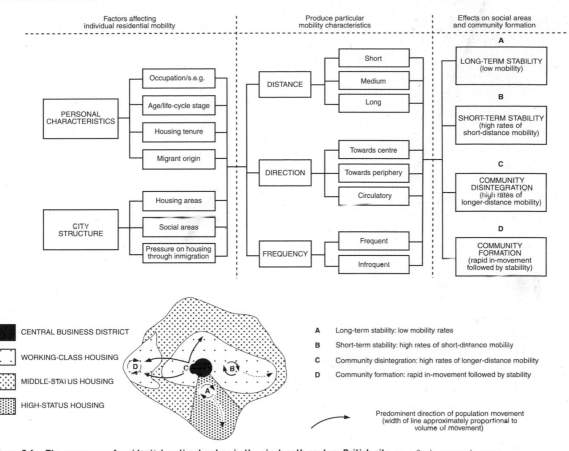

Figure 3.6   The processes of residential sorting by class in the nineteenth-century British city. *s.e.g.* Socio-economic group

Source: C. Pooley (1979) Residential mobility in the British city *Transactions of the Institute of British Geographers* 4, 258–77

individual residential moves were an integral part of a 'socio-spatial sorting mechanism' that influenced the nature of the urban residential mosaic by reinforcing some communities and destroying others.[25]

HOUSING THE POOR

It is important to recognise differences between England, Scotland, North America and continental Europe with regard to housing systems and consequent social patterns in industrial cities. In England, most working-class families rented accommodation from private landlords on weekly terms. Their houses were most likely to be all or part of two- or three-storey terraced houses, whether 'through houses' (with a back door and garden), 'back-to-back' or built around courtyards, perhaps sharing water supply and toilet facilities. In Scotland and continental Europe the working

classes were more likely to live in tenements, managed by a 'factor' on behalf of an absentee landlord and let on yearly terms. In North America, although most poor families rented from private landlords, rates of home-ownership were higher, and there was little involvement of philanthropic bodies and less state intervention in housing provision.[26]

In the industrial towns of nineteenth-century Britain, neither town councils nor employers saw it as their responsibility to build housing for the immigrants. This task was generally left to speculative builders, whose *modus operandi* was to maximise the use of land and minimise expenditure on materials and services. This architectural parsimony reached an apogee in the terraces of back-to-back housing constructed from the 1830s onwards. These dwellings were built in double rows under a single roof, one side facing the street and the other a parallel back street or alley or a courtyard accessible from the street at intervals through narrow

Plate 3.3 Serried rows of nineteenth-century red-brick working-class terraced housing run down the hillside to the now disused shipyards on the river Tyne in the Elswick district of Newcastle upon Tyne

passages or 'entries' (Figure 3.7). At the ends of the street or in the corner of a courtyard a standpipe for water supply and a communal earth closet were provided for the use of about 20 houses, or upwards of 150 people. The density of such developments averaged 60 houses or 200 rooms per acre (150 houses or 500 rooms per hectare), and the overcrowding, poor sanitary arrangements, lack of sunlight and absence of through-ventilation in the houses made back-to-backs the despair of medical officers (Box 3.9). As late as 1913 an inquiry into housing conditions in part of Birmingham found that:

> some 200,000 people were housed in 43,366 dwellings of the back-to-back type already long condemned as injurious to health because of lack of ventilation. For the most part they contained only three rooms, and so were overcrowded. In the six worst wards, from 51 per cent to 76 per cent were back-to-backs. Even more serious was the fact that 43,020 houses had no separate water supply, no sinks, no drains, and 58,028 no separate w.c., the closets being communal and exposed in courts.

This meant that over a quarter of a million people lived in cavernous conditions. The real objection to back-to-back houses lies not so much in their method of construction as in the degrading and disgusting conditions of their outbuildings, which frequently made decency impossible and inevitably tended to undermine the health and morals of the tenants.

> (Bournville Village Trust 1941 p. 16)[27]

Even higher residential densities were created by the tenement form of building in working-class areas of nineteenth-century Glasgow (Box 3.10).

THE OTHER SIDE OF THE COIN

In marked contrast to the slum areas of the nineteenth-century industrial city were the status areas occupied by the elite. Many of Britain's industrial cities were characterised by marked socio-spatial differentiation between the upper-class west end and lower-class east end.

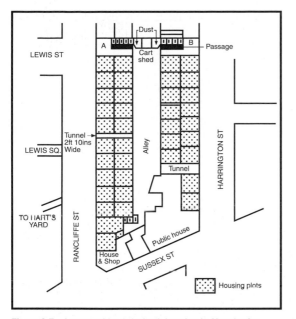

Figure 3.7 Layout of back-to-back housing in Manchester

Source: E. Morris (1997) *British Town Planning and Urban Design*
Harlow: Longman

BOX 3.9

The slum district of Little Ireland in Manchester, 1849

The district of Little Ireland was a typical living environment for 4,000 residents at the bottom of the social order. Engels described how the 'coal-black, stagnant, stinking river' Medlock loops through the area. The rows of back-to-back housing held an average of twenty persons per house, with its two rooms, attic and wet cellar. There was a single privy for six houses. Engels (1958 p. 71) presents a graphic description of the area: 'the cottages are very small, old and dirty, while the streets are uneven, partly unpaved, not properly drained and full of ruts. Heaps of refuse, offal and sickening filth are everywhere interspersed with pools of stagnant liquid. The atmosphere is polluted by the stench and is darkened by the thick smoke of a dozen factory chimneys. A horde of ragged women and children swarm about the streets and they are just as dirty as the pigs which wallow happily on the heaps of garbage and in the pools of filth.'[28]

Source: adapted from T. Freeman (1957) *Pre-Famine Ireland* Manchester: Manchester University Press

BOX 3.10

Living conditions in the tenement slums of mid-nineteenth-century Glasgow

The westward migration of the wealthy from the old town not only introduced socio-spatial polarisation into what had been in the eighteenth century a heterogeneous urban structure but freed land and housing for other uses. Those parts of the old city abandoned by the elite were colonised by a working-class population that was burgeoning in response to the new industries' demand for labour. Residences previously occupied by a single wealthy family were 'made down' to accommodate large numbers of the poor in grossly crowded conditions, characterised most graphically by the wynds and vennels around Glasgow Cross. The end result of decades of making down in such places was addresses such as 'Bridgegate, No. 29, back land, stair first left, three up, right lobby, door facing'.

In 1861 two-thirds of Glasgow's population of 394,864 lived in houses with only one or two rooms, and of these 60 per cent shared a room with at least four others. Overcrowding was intensified by property speculators building on any available open space to produce 'backjams' and 'backlands' tenements, which were either added to existing buildings or erected in the erstwhile gardens of the formerly wealthy burgher residences.

By the mid-nineteenth century the old city was a congeries of poor-quality housing. One street in a dark ravine of a close housed a husband, wife and two children in a 'sort of hole in the wall' measuring six shoe-lengths in breadth, between eight and nine in length from the bed to the fireplace, and of a height that made it difficult to stand upright. Densities of 1,000 persons per acre were commonplace. A notorious example was the rookery in the Drygate, where in a single close 500 people claimed a dwelling.

Source: M. Pacione (1995) *Glasgow: The Socio-spatial Development of the City* Chichester: Wiley

The genesis of this pattern is illustrated by the westward migration of the upper middle classes in Glasgow which began in the mid-eighteenth century with the movement of the 'tobacco lords' away from the old city centre, and culminated in the class-based residential segregation of the Victorian city. The process was driven by two sets of factors. Major push factors were:

1. Commercial encroachment into formerly high-status residential areas adjacent to the central business district.
2. The deteriorating social and physical fabric of the old city centre.

The dangers of crime and vice from slum areas were compounded by the threat of cholera, which, though originating in the slums, was no respecter of class.

In addition to the push factors, the social elite of nineteenth-century Glasgow had a clearly defined set of criteria against which to judge a potential suburb. Most important was the social composition of the neighbourhood, which had to be exclusively upper class. In Glasgow's emerging west end this was ensured by:

1. An informal land-use zoning consensus among owners to reserve the area for high-quality housing despite the presence of exploitable deposits of coal, iron, brick clay, building sand and freestone.
2. Restricted clauses in feu charters, leaseholds and freeholds of houses which laid down what activities were permitted, as well as details of the design of houses, provision of footpaths, type of gas lamps and sewers, and even the nature of the shrubbery. Later feu contracts also often specified a minimum market price for a given house – the whole providing an early example of 'exclusionary zoning'.

The middle classes were also attracted by the picturesque 'healthful and well aired' environment of the west end with its rolling drumlin topography, prevailing westerly winds, and location above and to the west of the industrial areas and insalubrious old town. A further attraction was the proximity of the west end to the central business district, where the majority of Glasgow's leading citizens were engaged.[29]

As well as spatial or territorial social segregation, in cities such as Glasgow, where 'high-rise' tenement building was the norm, vertical segregation was also prevalent. Within the tenement the principal apartments were always on the first floor, away from the noise, smell and dirt of the street but not too far to climb (Box 3.11).

The industrial city first developed in Britain, the cradle of the industrial revolution, but soon spread to Europe and North America.

THE ORIGINS OF URBAN USA

Despite the existence of a Neolithic urban hearth in Mesoamerica, there were no pre-Columbian settlements in the territory of the future USA that could be considered of urban status. The urban origins of the USA are related unequivocally to the process of colonisation by Spanish, French and especially English settlers.

The earliest urban settlements were established during Spanish occupation of the south-west in what are now the states of Texas, Arizona, New Mexico and California. This process also extended north from the Caribbean into Florida. Reps (1965)[30] has identified St Augustine in Florida, founded in 1565, as the first American city. As in other early Spanish settlements, this combined the functions of (1) military post (*presidio*), (2) base for trade and farming (*pueblo*), and (3) religious centre (mission). On the west coast, in California, the Spanish mission settlements formed the pre-urban nuclei for many of the great modern cities from San Diego to San Francisco de Solano.[31] Nevertheless, despite the early beginnings, by the mid-nineteenth century there were no urban settlements in the south-west of comparable size to those of the east and rapidly emerging Midwest.

English colonisation of North America began in 1585 with several unsuccessful attempts by Crown-appointed adventurers to establish a settlement on Roanoke Island in North Carolina. During the seventeenth century the colonial effort passed to corporate enterprises in the shape of royal chartered joint-stock companies. In 1607 the Virginia Company of London founded Jamestown, which became a centre for the tobacco plantations and which by 1619 had attained a population of around 1,000. Reps (1965)[32] has divided these early British settlements into four groups:

1. The earliest tidewater colonies of Virginia and Maryland, including Jamestown and Baltimore.
2. The new towns of New England, the largest of which was Boston.
3. The towns of the middle colonies, dominated by New York and Philadelphia (Box 3.12).
4. The colonial towns of Carolina and Georgia such as Charlestown and Savannah.

Also in the seventeenth century the Dutch established Fort Orange (Albany) in 1624 and New Amsterdam (New York) two years later as their main trading posts. Farther north the French were active, founding Quebec in 1608 and Montreal in 1620, giving them effective control of the St Lawrence river valley. By the end of the seventeenth century, however, English colonisation was dominant, although settlement was still largely confined to a fifty-mile (80 km) fringe of country inland from navigable waterways. In 1800, of a US

BOX 3.11

Tenement life in Glasgow

The tenements of Glasgow have provided accommodation for millions of citizens over the past 150 years. The quality of this form of housing varied across the city, and was signalled by the character of the front entrance. In working-class tenements the entrance to the close (the passage leading to the stairs) was a simple opening in the wall. In better-class tenements the close entrance might be decorated to match the windows on the floor above, or might be adorned by a porch reached by a flight of steps from street level. The close was the communal part of the tenement, and in working-class communities was a place of social interaction where neighbours met as they entered or departed from the building or took their turn to sweep the close or wash the stairs.

At the rear of the tenement was the back court. This enclosed space, or drying area, was the common property of everyone living in the tenement. It sometimes contained a communal wash-house with a cast iron boiler heated by a coal fire, which was available to residents on a rota basis. Strict adherence to the rota was required for a peaceful existence! One could not change one's wash day to suit circumstances, and the whereabouts of the indispensable item, the wash-room key, was a frequent source of friction.

The back court also accommodated the ash pits which replaced the dung heap of earlier generations. Space in the back court was also taken up by the brick w.c. stack, affixed to the back wall of the tenement, and housing a single toilet for communal use on each half-landing. These toilet stacks replaced the dry closet or privy in the back court which served the whole tenement until the Police Act of 1892 made internal sanitation compulsory.

As far as the internal structure of the building is concerned, the prime characteristic of the tenement as a building form was its flexibility. An ingenious architect could provide a reasonable variety of different house sizes and plans within the same solid rectangular framework. The four-storey Victorian tenement could contain five large houses of six or more rooms or, on a building plot of the same size, to meet totally different requirements, twenty-five single-apartment houses. The diagram illustrates the layout of a west-end tenement at the higher end of the market (with two flats of four or five main rooms on the upper floor).

At the other end of the social scale were one-room and two-room apartments, commonly known as a

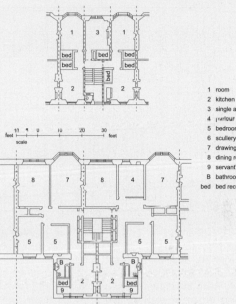

1 room
2 kitchen
3 single apartment
4 parlour
5 bedroom
6 scullery
7 drawing room
8 dining room
9 servant's room
B bathroom
bed bed recess

'single end' and a 'room and kitchen' respectively. In the diagram a 'single end' is grouped between two two-room flats, and this 2 : 1 : 2 formation was probably the most common tenement type in Glasgow. In the 'room and kitchen' arrangement the room or parlour faced the street, as in the better-class tenements. A bed closet was provided on the wall opposite the window, the built-in bed hidden from view behind a door until the practice was forbidden by the Glasgow Buildings Regulation Act of 1900. The parlour would be furnished to the extent of the occupants' means, and, perhaps surprisingly in view of the cramped size of the flat, was often kept for best and used only 'for visitors' or for special occasions such as a wedding or a funeral. The family lived and slept in the kitchen at the rear. In the kitchen the open bed recess contained a built-in bed about 2 ft 6 in. (75 cm) above the floor, below which could be stored a smaller bed that could be wheeled out at night for the children. Toilet facilities were provided by a communal w.c. on the half-landing of the stair. It is salutary to note that at the outbreak of the First World War the great majority of Glasgow families lived in one- or two-room flats, and that as late as 1951 this form of accommodation made up half the city's housing stock.

Source: **M. Pacione (1995)** *Glasgow: The Socio-spatial Development of the City* Chichester: Wiley

BOX 3.12

Philadelphia: a seventeenth-century planned town

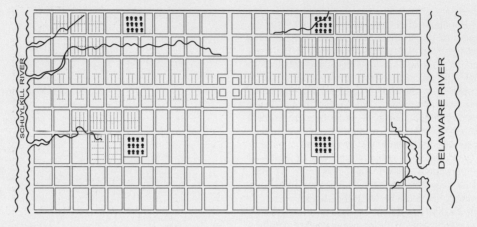

In 1681 William Penn (1644–1718) was granted a charter from Charles II establishing him as governor and proprietor of Pennsylvania (in return for cancellation of a £16,000 royal debt owed to Penn's father). Penn appointed three commissioners to lead the first group of settlers and provided them with a detailed brief for the location and planning of a new town.

The gridiron plan served the immediate purpose of quickly dividing up the urban land into equal-size plots to be allocated to settlers by lottery. The layout also reflected Penn's experience of the Great Plague and the Great Fire of London, and his determination to create a healthy and safe urban environment. Two main avenues 100 ft (30 m) in width crossed in a central square of ten acres (4 ha), with minor squares of eight acres (3 ha) in each quarter of the city. The central square would accommodate a number of public buildings. Penn also instructed that space should be allocated to allow every house to be located at the centre of the plot if the owner desired, thereby providing recreation space and reducing the fire hazard.

After three years Philadelphia had 600 houses, which increased to 2,000 by 1700. Although Penn had envisaged a dispersed pattern of growth between the two rivers, the dominance of the Delaware as a trade route meant that the city expanded outwards from the port area. Not until well into the nineteenth century did the city reach the area of the central square designated in the master plan, a new city hall being completed there in 1890. Subsequently, in the twentieth century, the migration of the central business district (downtown) away from the waterfront led to the gradual decay of the old port area and earliest residential districts, providing opportunities for redevelopment.

population of 4 million, only some 170,000 were settled west of the Allegheny mountains. The nineteenth century saw a shift in the population centre of gravity as the urban frontier moved westward (Box 3.13).

THE WESTWARD PROGRESS OF URBANISM

A major obstacle to westward settlement was the difficult transportation links back to the main markets of the eastern seaboard. The importance of developments in transport technology for the expansion of urbanism is illustrated by the impact of the construction of the Erie Canal between the Hudson river and Lake Erie. When the canal reached the obscure village of Buffalo in 1824 the decimation of freight costs over the 363 miles to the Hudson River led to the construction of 3,400 houses in the following year, with the town reaching a population of 18,000 by 1840. Other settlements on the new waterway such as Rochester, Utica, Syracuse and, principally, New York all grew rapidly into prosperous cities (Box 3.14).

Borchert (1967)[33] has proposed a model of US urban development related to changes in technology.

BOX 3.13

The township and range system of land division

The westward expansion of settlement in the USA was based on a national land policy enshrined in the Land Ordinance of 1785. The law required that the territory to be sold should be laid out in regular sections or townships six miles square. Each township was divided into thirty-six square sections of one square mile or 640 acres (260 ha). Half the land was to be sold as townships, the other half as sections. Section 16 in each township was reserved for the support of schools.

The first seven ranges of townships were surveyed by 1786 immediately west of the Ohio river (numbered 1–7 from east to west). An example township in the seventh range is shown divided into its thirty-six sections. Each section could also be divided into quarter-sections. Section boundaries often provided lines for roads, and villages and towns grew up at junctions or were laid out and promoted by property speculators.

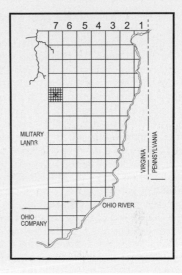

STAGE 1. THE SAILING VESSEL AND HORSE-DRAWN WAGON ERA TO 1830

Throughout the eighteenth and early nineteenth centuries the North American economy was based almost entirely on agriculture and the export of staple products. The main function of the few cities that existed was mercantile, involving the import of manufactured goods from Europe and export of primary produce. The largest cities were on the Atlantic seaboard, with New York, Philadelphia, Boston and Baltimore dominating the trade of the original colonies. The next largest cities were located on the riverine routes into the interior at New Orleans, Quebec and Montreal. Most of the cities beyond the Appalachians were small service centres for surrounding agricultural settlements, the largest in 1830 being Cincinatti with 24,800 inhabitants.

STAGE 2. THE AGE OF STEAM AND THE IRON RAIL, 1830–70

The proportion of the US population living in urban areas rose from 8 per cent to 23 per cent over the period 1830–70. In addition to the continued growth of the cities of the eastern seaboard, which had developed as manufacturing bases and key centres for the commercial exploitation of the developing continent,

the most dramatic trend was the spread of urbanism in the Midwest associated with the construction of canals and railways. The rail network expanded from 9,000 miles of track in 1850 to 53,000 miles by 1870 and, as well as linking all the east-coast cities, had reached Oakland on the west coast. Inland cities in accessible locations boomed. Chicago on Lake Michigan, with waterway links to the east and the Great Lakes and Erie Canal, grew into a city of 300,000 people by 1870. The only large cities beyond the east coast and the Midwest were New Orleans (population 191,000), with a rich agricultural **hinterland**, and San Francisco (population 150,000), which developed rapidly following the gold rush of 1849.

STAGE 3. THE AGE OF STEAM AND STEEL, 1870–1920

Over the period 1870–1920, the urban population of the USA experienced a fivefold increase to 41 million, and urban development was evident across the continent. On the east coast the New York metropolitan area had quadrupled in population to 5.3 million by 1920, while Philadelphia tripled to reach 1.8 million. In the Midwest, Chicago moved from being the fifth-largest city in North America to being the second largest and experienced a ninefold increase in population to

BOX 3.14

New York: the urban explosion

In contrast to Philadelphia, New York began without a master plan, and only after 150 years of organic growth was the first gridiron laid out. When the British captured the city in 1664 the population was around 1,500. Following the opening of the Erie Canal, the city's population mushroomed from 150,000 in 1820 to 300,000 in 1840, 515,000 in 1851 and 942,000 by 1870. To accommodate expansion, city commissioners were authorised to lay out a gridiron on the undeveloped part of Manhattan Island. As the map shows, the plan of 1811 imposed a uniform grid based on twelve 100 ft (30 m) wide north–south avenues and 155 east–west streets 60 ft (18 m) in width. The southern start line of the commissioners' plan is clearly seen against the earlier randomly aligned grids.

By 1850 the infilling of the grid had reached Forty-second Street without any provision for major public open space. In 1858 the culmination of a fourteen-year campaign led to the appointment of Olmsted and Vaux to design Central Park. By 1898 the current five-borough city of New York was constituted with the addition of Brooklyn, Bronx, Queens and Staten Island to Manhattan.

The expansion of New York was facilitated also by advances in structural engineering that allowed the Hudson and East rivers to be bridged and tunnelled under, and that underlay the construction of skyscrapers and high-speed elevators. New York is today a 'world city' and one of the command centres of the global economy.

2.7 million. On the west coast a growing number of cities had populations in excess of 100,000, including Los Angeles. By contrast, in the south, urban development was still at an incipient stage.

The continental urbanisation wave was driven by several forces, including:

1. The increasing integration of the North American transport system due to the standardisation of rail gauges and increase in transcontinental lines.
2. The influx of large numbers of poor immigrants, which provided a continuous supply of low-cost labour.
3. The introduction of the assembly-line factory system, exemplified by the 'Fordist' production system of the Detroit automobile industry, which spread rapidly to all types of manufacturing.
4. The large surpluses being produced by a mechanised agricultural system.

5. The entrepreneurial activity of family-owned corporations. Pittsburgh, for example, became the base of the Carnegie Iron & Steel Corporation, which in 1901 merged with other producers to become US Steel.

Consequently, by the early decades of the twentieth century the basic alignment of the North American urban system had been established.

STAGE 4. THE AGE OF THE AUTOMOBILE AND AIR TRAVEL, 1920–70

Over the period 1920–70 the proportion of the US population living in urban places of 5,000 or more people increased from 47 per cent to 70 per cent. In geographical terms the major changes in the distribution of the urban population were:

1. The emergence of super-metropolitan and megalopolitan urban areas.
2. The spread of metropolitan urban development across most of the continent.
3. The growth of urban areas beyond the old central cities.

These spatial trends were facilitated by transport developments, including the widespread use of the automobile and construction of the interstate highway system.

The highest rates of urban growth occurred in the sunbelt, with slower growth and in some cases negative growth in the older cities of the north-east (Table 3.1). Between 1940 and 1980 Phoenix grew by 1,107 per cent and Albuquerque by 836 per cent. The phenomenal growth rates of the sunbelt cities were related to the exploitation of petrochemicals and the growth of the microelectronics, computer and aerospace industries. The new industries' need for a smaller but highly trained and educated work force freed them from the need to locate close to a large pool of factory labour and laid greater emphasis on other factors, including markets, prevailing wage and tax rates, and a benign climate. Other key factors in the expansion of sunbelt cities included:

1. The growth of federal military expenditure.
2. The migration of large numbers of retirees, often on government pensions, drawn by a warm climate and low cost of living.

3. An improved standard of living and changing lifestyle resulting in more leisure time.
4. The entrepreneurialism and civic boosterism of the political and economic leadership of sunbelt cities, which attracted inward investment.[34]

In terms of morphology, the cities of the sunbelt exhibit a distinctive character: shaped by the automobile, they have no need for a dominant downtown area but sprawl over the landscape in a low-density form of development encapsulated in the phenomenon of the **edge city**.

STAGE 5. THE AGE OF DECONCENTRATION, 1970 TO THE PRESENT

The continuation and extension of these trends to other regions during the latter part of the century allows us to add a fifth stage to Borchert's (1967) model. This is characterised by a decrease in the size of the larger metropolitan areas, a decline in the density of the urban population, and increasing segregation of people in communities according to socio-economic factors such as class, race, age or language. The counterurbanisation trend is driven by industrial decentralisation, the insecurity and discomfort of metropolitan life as perceived by those able to exercise a choice, a search for amenities and retirement loci, and improved transportation and communications networks which enable people to enjoy an urban lifestyle without being permanently resident in a large metropolitan area. These characteristics are exhibited most clearly in the concept of the **post-industrial city**.

TABLE 3.1 THE FASTEST-GROWING US CITIES AND BIGGEST POPULATION LOSERS, 1980–90

The fastest-growing				The big population losers			
City	1980	1990	% change	City	1980	1990	% change
1. Moreno Valley CA	28,309	118,779	319.6	1. Gary IN	151,968	116,646	−23.2
2. Rancho Cucamonga CA	55,250	101,409	83.5	2. Newark NJ	329,248	275,221	−16.4
3. Plano TX	72,331	128,713	77.9	3. Detroit MI	1,203,369	1,027,974	−14.6
4. Irvine CA	62,134	110,330	77.6	4. Pittsburgh PA	423,960	369,879	−12.8
5. Mesa AZ	163,594	288,091	76.1	5. St Louis MO	452,804	396,879	−12.4
6. Oceanside CA	76,698	128,398	67.4	6. Cleveland OH	573,822	505,616	−11.9

Source: adapted from E. Phillips (1996) *City Lights* New York: Oxford University Press

POST-INDUSTRIAL URBANISM

Post-industrialism is a social process that has had a major impact on the city. Its principal characteristics are:

1. Changes in the economy leading to a focus on the service sector rather than on manufacturing.
2. Changes in the social structure that afford greater power and status to professional and technological workers.
3. Changes in the knowledge base, with greater emphasis on R&D.
4. Greater concern for the impact of technological change.
5. The advent of advanced information systems and intellectual technology.

According to Bell (1973) the industrial–post-industrial watershed was reached in 1956 in the USA when, for the first time in the history of industrial civilisation, the number of white-collar workers exceeded the number of blue-collar workers.[35] For Bell (1973) the advent of post-industrial societies meant fundamental changes in the nature of economic organisation and social relations. The resultant post-industrial city is characterised by:

1. An employment profile that reflects the twin processes of deindustrialisation and tertiarisation (see Chapter 14) as part of a restructuring of the economic base from a Fordist mode of industrial production to more flexible (post-Fordist) production systems (e.g. high-tech knowledge-based industries).
2. Greater integration into the global economic system.
3. Restructuring of urban form.
4. Emerging problems of increasing income inequality, social and spatial segregation, privatisation of urban space and growth of defensible spaces.

Post-industrial urbanism is characterised by fragmentation of urban form and its associated economic and social geographies.[36] For some commentators this socio-spatial fragmentation heralds the advent of the postmodern city,[37] and is evident in the 'quartering' of urban space.

THE QUARTERING OF URBAN SPACE

It is possible to identify a number of increasingly separate 'residential cities'. Each has its source in a parallel, (although not always congruent), 'city of business and work'. The five principal formations of parallel residential and business cities are: (1) the luxury city and the controlling city, (2) the gentrified city and the city of advanced services, (3) the suburban city and the city of direct production, (4) the tenement city and the city of unskilled work, (5) the abandoned city and the residual city.

1. *The luxury areas of the city*, while located in clearly defined residential districts, are not tied to any quarter. For the wealthy the city is less important as a residential location than as a location of power and profit. If they reside in the city, it is in a world largely insulated from contact with non-members of their class. The controlling city tends to be located in the high-rise centres of advanced services. The controlling city parallels the luxury areas of the residential city in many ways but not necessarily in physical proximity. The luxury city and the controlling city come together in space in the citadel where those at the top of the economic hierarchy live, work, consume and recreate in protected spaces.
2. *The gentrified city* serves professionals, managers and 'yuppies' who may be doing well for themselves, yet work for others. Their residential areas are chosen for environmental or social amenities or access to white-collar jobs. Gentrified working-class neighbourhoods, older middle-class areas and new developments with modern and well-furnished apartments all serve their needs. The parallel city of advanced services consists of professional offices with many ancillary services internalised in high-rise office towers and is enmeshed in a technologically advanced communications network. The skyscraper centre is the stereotypical form, as at La Défense in Paris or Canary Wharf in London's Docklands.
3. *The suburban city* of the 'traditional' family is sought out by better-paid blue- and white-collar workers, the lower middle class. It provides stability, security and a comfortable world of consumption. The home as a symbol of self, exclusion of those of lower status, physical security against intrusion, political conservatism, comfort and escape from the workaday world (thus often incorporating substantial spatial separation from work) are characteristic. The city of direct production parallels, but is not congruent with, the residential suburban city in either time or space. It includes not only manufacturing but

also the production aspects of advanced services, government offices and the back offices of major firms located in clusters in various locations within the metropolitan area that facilitate easy contact with clients. An example in New York City is the industrial valley for the printing industry between midtown Manhattan and the financial district.

4. *The tenement city* serves as home for lower-paid workers who often have irregular employment, few benefits, limited job security and little chance of advancement. In earlier days, the residents of these 'slum neighbourhoods' were often the victims of clearance and urban renewal programmes. Today they may experience abandonment, displacement, service cuts, deterioration of public facilities and political neglect. Struggles against displacement by urban renewal or gentrification have led to militant social movements in many cities. The city of unskilled work includes the informal economy, small-scale manufacturing, warehousing, sweatshops and unskilled consumer services. These are closely intertwined with the cities of direct production and advanced services

and thus are located near them, but separately in scattered clusters. The economic city of unskilled work parallels the tenement city.

5. *The abandoned city* is the place of the very poor, the excluded, the never-employed and permanently unemployed, and the homeless. In older industrialised countries it will have a crumbling infrastructure and deteriorating housing, typical of slums and ghetto areas of the inner city. In less developed countries the excluded often live in peripheral squatter settlements. The abandoned city is paralleled in economic terms by the residual city – the city of the less legal portions of the informal economy. In the developing countries the two cities overlap. In industrialised countries the residual city is the place where otherwise undesirable land uses are located. Many of the most polluting and environmentally detrimental components of the urban infrastructure, necessary for its economic survival but not tied directly to any economic activity, are located here. These include sewage treatment plants, incinerators, AIDS residences, homeless shelters, juvenile detention centres and jails.

Plate 3.4   The social polarisation of the postmodern city is displayed by the scavenger collecting discarded drinks cans from rubbish bins in the up-market Sixteenth Street mall in Denver CO

THE POST-INDUSTRIAL/ POSTMODERN CITY

Just as industrialism left its imprint on the nineteenth-century city,[38] so post-industrialism/postmodernism promoted changes in the form of the late twentieth-century city. Soja (1995)[39] has characterised these trends in terms of six geographies of restructuring:

1. *The restructuring of the economic base of urbanisation*. This involves a fundamental change in the organisation and technology of industrial production, and the attendant social and spatial division of labour. This is marked by a shift from Fordist to post-Fordist urbanisation, from the tight organisation of mass production and mass consumption around large industrial complexes to more flexible production systems vertically disintegrated but geographically clustered in 'new industrial spaces'.
2. *The formation of a global system of world cities*. This has the effect of expanding both the 'outreach' of particular world cities, bringing most of the globe within their effective hinterland, and their 'inreach', bringing into the global city capital and labour from all major cultural realms. In brief, the local is becoming globalised and the global localised in a process of 'glocalisation'.[40]
3. *The radical restructuring of urban form*. This restructuring has generated a large number of neologisms, including **megacity**, outer city, edge city, metroplex, **technoburb, post-suburbia**, technopolis, heterpolis and exopolis to indicate a process whereby the city is 'simultaneously being turned inside out and outside in'.[41] The spatial organisation of the post-industrial/postmodern city is significantly different from that of the early modern city, exemplified in the neat concentric and sectoral models of the Chicago school (see Chapter 7), or even the more disjointed late modern city with its dominant central business district, **inner city** of poor and **blue-collar** workers, and sprawl of middle-class dormitory suburbs.

   Among the most obvious features of the post-industrial/postmodern city is the urbanisation of suburbia, but change has also affected the older central cities including significant reductions in density and the gentrification of former working-class neighbourhoods. As we shall see in Chapter 4, some of the large metropolitan regions have also experienced renewed growth after several decades of population decline. On a larger scale is the emergence of megacities, such as Mexico City.
4. *The changing social structure of urbanism*. Postmodern urbanism is associated with the development of new patterns of social fragmentation, segregation and **polarisation**, most evident in highly visible lifestyle differences and a growing gap between rich and poor. The socio-economic structure of the postmodern city is increasingly fluid and fragmented in ways that reduce the interpretative value of simple class-based divisions.
5. *The rise of the carceral city*. The complex geographies of postmodern cities have made them increasingly difficult to govern via traditional local-government structures. This has promoted the appearance of 'carceral' areas of walled-in residential estates protected by armed guards, shopping malls made safe by electronic surveillance, 'smart' office buildings impenetrable to 'outsiders', and neighbourhood-watch schemes organised by concerned home-owners.[42] More positively, these developments have refocused attention on 'the politics of place', bringing enhanced local political consciousness over who controls and benefits from the restructuring of urban space.
6. *A radical change in urban imagery*. This refers to our images of the city and how these affect our behaviour and lifestyle in the postmodern city. Hollywood and Disneyland, and their equivalents elsewhere, produced a modernist hyper-reality as entertainment, but in the postmodern city hyper-reality has diffused from such specialised factories into everyday life.[43] The popular media and expanding communications network have helped to promote the effects of hyper-reality on how people eat, work, dress, sleep, vote, enjoy leisure – that is, all the activities that underlie the social construction of urban space.

Los Angeles in the late twentieth century assumed a position with regard to urban theory comparable to that of Chicago in the early twentieth century (Box 3.15). A number of academic and popular accounts have constructed Los Angeles as an archetype of contemporary and future urbanisation.[44] These should not be accepted uncritically, however. In particular, it is important to reflect on how widespread the phenomenon of the postmodern city is. To suggest that the postmodern metropolis of Los Angeles offers a general model for the interpretation of contemporary urbanisation would be incorrect. As with the urban models of the Chicago school, so the greatest contribution of the California school of urban geography lies not in the modelling of urban form but in the exposition of the forces influencing contemporary urban landscapes which have been identified first and most visibly in Los Angeles.

BOX 3.15

Los Angeles: the archetypal postmodern metropolis

Los Angeles has been described as the quintessential postmodern metropolis – not because it is a model for all other cities, but because the processes of postmodernism are displayed with particular clarity in the city. These may be illustrated with reference to six geographies or restructuring processes:

1. *The restructuring of the economic base of urbanisation,* seen in:
   - The growth of new technopoles outside the old industrial zone in what were once dormitory suburbs or agricultural land, exemplified by the clusters of high-technology aerospace and electronics firms and industrial parks in Orange County and the San Fernando valley.
   - The growth of the film and entertainment industry.
   - The presence of low-skill, labour-intensive, fashion-sensitive growth industries such as the apparel, furniture and jewellery industries, mostly around downtown Los Angeles but increasingly dispersed throughout the metropolitan area.
   - The growing finance, insurance and real estate (FIRE) sector providing business services to both domestic and foreign capital.

2. *The formation of a global system of world cities,* seen in:
   - The city's growing importance as a world financial and trade centre.
   - The creation of urban multiculturalism, with more than one-third of the 9 million residents of Los Angeles County foreign-born.

3. *The radical restructuring of urban form,* seen in the complex geography of population and employment growth in the metropolitan area. The city of Moreno Valley, for example, grew rapidly during the 1980s as families moved there in anticipation of jobs. The non-appearance of the promised jobs led to overcrowded schools, poor social services, gridlocked freeways and four-hour daily commuting journeys.

4. *The changing social structure of urbanism.* Inequality is a deeply embedded facet of Los Angeles life, and the disparity between rich and poor continues to widen, with an increasing percentage of low-wage workers and a steady rise in the poverty rate (from 11 per cent in 1969 to 15 per cent in 1989) and in levels of homelessness. Economic inequalities often coincide with racial and ethnic divisions, with African-Americans, Latinos and Asians represented disproportionately at the bottom of the economic ladder, and located in inner-city areas such as Watts. The fluidity of postmodern urban restructuring is evident in the rapid racial transition in areas such as Huntington Park and Maywood, which have moved from being 80 per cent Anglo in 1965 to 90 per cent Latino.

5. *The rise of the carceral city.* The carceral (imprisoning) landscape of Los Angeles has been characterised as 'the built environment of security-obsessed urbanism', seen in the gated communities and voracious consumption of private security services, NIMBY protest movements, increasing suburban separation, exclusionary zoning regulations and the hardening of the city against the undesirable poor, with absent public lavatories, razor wire-protected trash bins, and overhead sprinkler systems that operate randomly through the night to deter sidewalk (pavement) sleepers.

6. *A radical change in the urban imagery.* Los Angeles is the world's leading centre for the production and marketing of hyper-reality. Not only have these image products reached around the globe (who is not familiar with the voyages of the Starship *Enterprise*?) but they also affect the local urban landscape. In the 'theme-parked' environment of Los Angeles one can choose to live in a place that caters for a particular lifestyle – for example, for the elderly, swinging singles or gay/lesbian communities – or occupy a 'dreamscape' replica of a Greek island, Little Tokyo or old New England.

FURTHER READING

BOOKS

C. Chant and D. Goodman (1999) *Pre-industrial Cities and Technology* London: Routledge

M. Dear (2000) *The Postmodern Urban Condition* Oxford: Blackwell

D. Goodman and C. Chant (1999) *European Cities and Technology* London: Routledge

P. Hall (1998) *Cities in Civilisation* London: Weidenfeld & Nicolson

P. Hohenberg and L. Lees (1995) *The Making of Urban Europe 1000–1994* Cambridge, MA: Harvard University Press

M. Pacione (1995) *Glasgow: The Socio-spatial Development of the City* Chichester: Wiley

J.W. Reps (1965) *The Making of Urban America* Princeton, NJ: Princeton University Press

G. Roberts and J. Steadman (1999) *American Cities and Technology* London: Routledge

G. Sjoberg (1960) *The Pre-industrial City: Past and Present* New York: Free Press

E. Soja (2000) *Postmetropolis* Oxford: Blackwell

JOURNAL ARTICLES

J. Borchert (1967) American metropolitan revolution *Geographical Review* 57, 301–22

V.G. Childe (1950) The urban revolution *Town Planning Review* 21, 3–17

J. Curry and M. Kenney (1999) The paradigmatic city: postindustrial illusion and the Los Angeles school *Antipode* 31(1), 1–28

M. Dear (2002) Los Angeles and the Chicago school: invitation to a debate *City and Community* 1(1), 5–32

R. Lawton (1972) An age of great cities *Town Planning Review* 43, 199–224

C. Pooley (1979) Residential mobility in the British city *Transactions of the Institute of British Geographers* 4, 258–77

A. Scott (1999) Los Angeles and the LA school: a response to Curry and Kenney *Antipode* 31(1), 29–36

KEY CONCEPTS

- Theories of urban origin
- Hydraulic theory
- Urban hearths
- Pre-industrial city
- Gridiron plan
- Medieval feudalism
- Medieval urbanism
- Early modern urbanism
- Industrial revolution
- Industrial capitalism
- Socio-spatial segregation
- Tenements
- Basic : non-basic employment
- Post-industrial urbanism
- Postmodern city

STUDY QUESTIONS

1. Where, when and why did the first cities appear?
2. Examine the contribution of ancient Greece and Rome to the spread of urbanism.
3. The nineteenth century witnessed the flowering of industrial capitalism in Western Europe. Identify the major features of the socio-spatial structure of the European industrial city.
4. Trace the origins and development of urbanism in the USA.
5. Identify the principal characteristics of the post-industrial/postmodern city.
6. With the aid of a relevant example, examine the use of the gridiron plan in urban development.

PROJECT

Study the origins and development of a city with which you are familiar. Information may be collected from secondary sources such as maps and books held in your local or institutional library and from primary sources via your own fieldwork, which may involve sketching/photography or land-use mapping. The end product of your research should take the form of an annotated map of the city showing the important growth phases from urban origins to the present day.

4

The Global Context of Urbanisation and Urban Change

Preview: global urbanisation trends; the changing distribution of world urban population; causes of urban growth; the size of the world's settlements; million cities, megacities and metacities; the relationship between urbanisation and economic growth; the urbanisation cycle; stages of urban development; urbanisation; counterurbanisation; reurbanisation; suburbanisation; exurbanisation; types of urbanised regions

INTRODUCTION

The global urban pattern is changing in three main ways as a result of:

1. *Urbanisation:* an increase in the proportion of the total population that lives in urban areas.
2. *Urban growth:* an increase in the population of towns and cities.
3. *Urbanism:* the extension of the social and behavioural characteristics of urban living across society as a whole.

In this chapter we first examine recent patterns of urbanisation and urban growth at the global scale in order to provide a framework for subsequent discussion of regional, national and local patterns and processes of urban change.

THE URBANISATION OF THE GLOBE

As we have seen in Chapter 3, the current high level of urbanisation at the global level is a relatively recent phenomenon. At the end of the nineteenth century the extent of world urbanisation was limited, with only Britain, north-west Europe and the USA more than 25 per cent urban in 1890.[1] With less than 3 per cent of the world's population living in towns and cities, levels of urbanisation elsewhere were insignificant. In the USA, urban development was confined primarily to the cities of the east coast and emerging Midwest.

An indication of the rapid progression of urbanisation across the globe is provided in Figures 4.1–4.4. The spread of urbanisation in Europe, North America and the Middle East is apparent, as are the rising levels of urbanisation in Africa and Asia, which were almost wholly rural in 1950. Figure 4.5 confirms that for most countries of the world urbanisation is a contemporary and ongoing process. Over the course of the past half-century, a world in which most people lived in rural areas has been transformed into a predominantly urban world. This trend has influenced not just the physical location of population but also the organisation and conduct of economic and social life of most people on the planet – both urban and rural dwellers.

Figure 4.5 also reveals the differential incidence of urbanisation in the world. As Table 4.1 indicates, the more developed regions (MDRs) exhibit high levels of urbanisation. More than 75 per cent of the population of Europe, North America, Japan and Australia–New Zealand were urban dwellers in 1994, and by 2025 at least eight out of every ten people in these regions are expected to live in urban areas. Accordingly, in the developed countries the pace of urbanisation has slackened and has in some instances gone into reverse, as is indicated in the **demographic-transition** model (Box 4.1). By contrast, the less developed regions (LDR) are characterised by rapid urbanisation that is expected to continue for decades (Figure 4.5 and Table 4.1). As Table 4.1 shows, in 1970, 25 per cent of the population of LDRs lived in urban areas. By 1994 almost 40 per cent were urban dwellers. Since the

Figure 4.1   The urban world in 1890

Figure 4.2   The urban world in 1950

urban population is growing at an average rate of 3.5 per cent per annum while it's rural counterpart grows at less than 1 per cent per annum, it is estimated that by 2025 almost 60 per cent of the population of LDRs (i.e. 4 billion people) will live in towns and cities.

THE CHANGING DISTRIBUTION OF THE WORLD'S URBAN POPULATION

The rapid growth of the world's population has been accompanied in most countries by the multiplication and growth of urban places. The United Nations

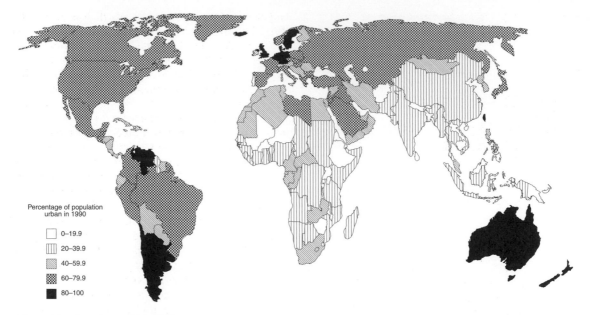

Figure 4.3   The urban world in 1990

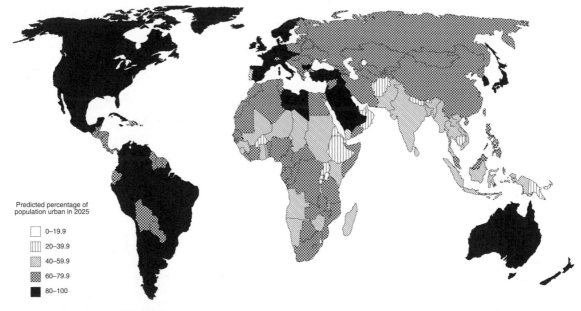

Figure 4.4   The urban world by 2025

estimates that between 1950 and 2025 the number of urban dwellers will increase nearly sevenfold, from 738 million to 5.1 billion. But the world urban population is not distributed evenly among regions. As Table 4.1 reveals, in 1970 the more developed regions and less developed regions had a similar number of urban

dwellers, 677 million and 676 million respectively. The year 1970 represents a 'tipping point' in the global distribution of urban population. Prior to 1970, most urban dwellers lived in the MDR but this dominance has been declining since 1950, when 442 million (60 per cent) of the 737 million urban dwellers worldwide

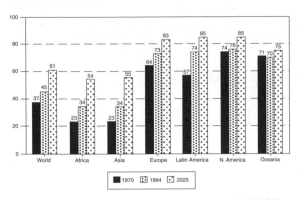

Figure 4.5 The urban population of world regions, 1970, 1994 and 2025 (%)

Source: United Nations (1995) *World Urbanisation Prospects: The 1994 Revision* New York: United Nations

lived in the MDRs. Since 1970 the number of urban dwellers in LDRs has overtaken that of MDRs, and the gap continues to widen. Currently 1.7 billion urban dwellers (60 per cent of the world urban population) live in the LDRs while 868 million live in the MDRs. By 2025 4 billion of the 5 billion urban dwellers are expected to live in the LDRs.

The distribution of urban population is also changing within the LDRs and the MDRs. In the former realm, Asia is a major region of urban growth. Whereas in 1970 Asia was home to 503 million urban dwellers (37 per cent of the world total), by 1994 1.2 billion (46 per cent) of the 2.5 billion global urban dwellers were Asian. It is anticipated that 2.7 billion (more than

half the world's urban dwellers) will live in Asia by 2025. This trend is in marked contrast to the situation in Europe. As Table 4.1 shows, with 423 million urban dwellers in 1970 Europe was second only to Asia. Between 1970 and 1994 Europe added 110 million urban dwellers with a further 65 million expected by 2025. During the period 1994–2025 Asia is expected to add 1.5 billion urban residents – or about 23 new urban Asians for every new European urban resident.

Latin America and the Caribbean are also growing rapidly, with their urban population more than doubling from 163 million in 1970 to 349 million in 1994. The urban population of the region is expected to reach 601 million by 2025, slightly more than the number projected for Europe (598 million). Africa exhibits the fastest urban growth rate of any major world region. From 84 million urban residents in 1970 Africa by 1994 had 240 million, and by 2025 the number of urban dwellers is expected to reach 804 million. All these trends are confirmed by the analysis of urban growth rates (Table 4.2 and Figure 4.6).

THE CAUSES OF URBAN GROWTH

The increasing levels of urbanisation and urban growth identified are the result of a combination of natural increase of the urban population and net in-migration to urban areas. The two major processes reinforce each other, although their relative importance varies. Findley (1993),[2] for example, found that

TABLE 4.1 URBAN POPULATION AND PERCENTAGE URBAN IN MORE DEVELOPED AND LESS DEVELOPED REGIONS, 1970, 1994 AND 2025

Region	Urban population (million)			Urban share (%)		
	1970	1994	2025	1970	1994	2025
More developed regions	**677**	**868**	**1,040**	**67.5**	**74.7**	**84.0**
Australia–New Zealand	13	18	26	84.4	84.9	89.1
Europe	423	532	598	64.4	73.3	83.2
Japan	74	97	103	71.2	77.5	84.9
Northern America	167	221	313	73.8	76.1	84.8
Less developed regions	**676**	**1,653**	**4,025**	**25.1**	**37.0**	**57.0**
Africa	84	240	804	23.0	33.4	53.8
Asia[a]	428	1,062	2,615	21.0	32.4	54.0
Latin America	163	349	601	57.4	73.7	84.7
Oceania[b]	1	2	5	18.0	24.0	40.0

Source: **adapted from United Nations (1995)** *World Urbanization Prospects: The 1994 Revision* **New York: United Nations**

Notes: **[a]Excluding Japan. [b]Excluding Australia–New Zealand**

The demographic transition model

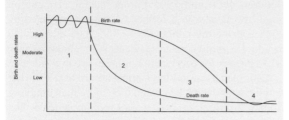

This model describes the main stages of population change that accompany industrialisation and urbanisation:

1. Birth rates and death rates are both high, leading to fluctuating population growth determined more by mortality than by fertility. This is associated with conditions in pre-industrial society and in some developing areas today (e.g. sub-Saharan Africa), where most people live in rural environments at a subsistence level.

2. In the early stage of development, death rates begin to fall, owing to improvements in medicine and public health. Family and religious values can contribute to maintain a high birth rate, leading to rapid population growth. Many of the population migrate from rural to urban areas, feeding city growth and urbanisation. This phase is typical of many Third World countries today.

3. With the move to cities, large families are less of an economic asset than they had been in rural society, and more of an expense on the family budget. Birth rates fall to a level more in line with the death rate, leading to a slow-down in population growth. Europe and North America have passed through this phase.

4. This stage has been reached in highly developed countries, such as the UK and the USA, where both birth and death rates are low and population growth rates are near zero.

for twenty-four developing countries between 1975 and 1990 the average contribution of migration to urban growth was 54 per cent. This underestimates the real situation, however, since many migrants who live and work in the city do not move there permanently (see Chapter 23). Finally, it is important to note that although it is often assumed that the largest migration flows are in the countries of the South, more migration takes place in the USA, where one person in five moves each year, albeit between urban areas, than in all the LDRs. As we shall see in the following chapter, the restructuring of the urban system in the USA during the 1980s was more rapid and fundamental than that taking place in most countries of the LDRs.

SETTLEMENT SIZE

The world's urban population is distributed among settlements of differing sizes along a continuum from small towns with several thoussnd people to giant cities with populations of tens of millions. Most of the urban population live in settlements with fewer than 500,000 inhabitants.[3] Most of these intermediate settlements function as links between town and country where agricultural surpluses are exchanged for manufactured goods and services in accordance with the precepts of central place theory (see Chapter 6).

The distribution of urban population in terms of settlement size is shown in Table 4.3. In 1950 only one city (New York) had a population of 10 million or more, accounting for 1.7 per cent of the world urban population. By 1990 twelve cities had attained this size and shared 7.1 per cent of the world urban population. It is anticipated that by 2015 twenty-seven cities will reach the 10 million mark, accommodating 10.9 per cent of the world urban population. In absolute numerical terms this represents a rise from 12 million people living in a single city in 1950 to 450 million in several giant cities by 2015. Between 1990 and 2015 nearly all the population growth in the largest urban agglomerations is expected to occur in the LDRs. At the other end of the population-size continuum, cities with fewer than 500,000 inhabitants were home to 63.7 per cent of the world urban population in 1950. While their share is expected to decrease slowly in the future, more than 50 per cent of world urban population is still expected to live in such cities in 2015 (Table 4.3).

MILLION CITIES, MEGACITIES AND METACITIES

One of the most striking features of the global urban pattern is the degree to which the urban population lives in giant cities that dominate the global urban and economic systems. Against the background of a general increase in the number of people living in urban places, it is the metropolitan regions that are proliferating and expanding most rapidly.

TABLE 4.2 AVERAGE ANNUAL RATE OF CHANGE OF URBAN POPULATION, 1965–70, 1990–5 AND 2020–5 (%)

Region	1965–70	1990–5	2020–5
World	**2.64**	**2.53**	**1.93**
Less developed regions	**3.58**	**3.51**	**2.33**
Africa	4.64	4.38	3.34
Asia[a]	3.28	3.68	2.31
Latin America	3.97	2.60	1.26
Oceania[b]	2.26	3.13	3.32
More developed regions	**1.74**	**0.75**	**0.45**
Australia–New Zealand	2.35	1.32	1.05
Europe	1.68	0.84	0.24
Japan	2.20	0.37	−0.06
North America	1.63	1.29	0.99

Source: adapted from United Nations (1995) *World Population Prospects: The 1994 Revision* New York: United Nations
Notes: [a]Excluding Japan. [b]Excluding Australia–New Zealand

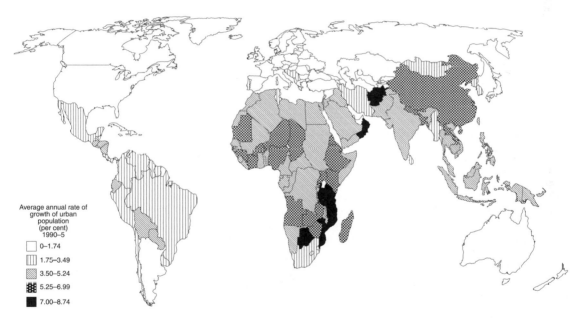

Figure 4.6   Average annual rate of urban population growth, 1990–5

Average annual rate of
growth of urban
population
(per cent)
1990–5

- 0–1.74
- 1.75–3.49
- 3.50–5.24
- 5.25–6.99
- 7.00–8.74

While noting the difficulties of urban definition and cross-national comparison, we can identify a number of significant trends in the geographical distribution of megacities. Table 4.4 lists the fifteen largest urban agglomerations at different points in time, and enables us to map the major changes over the post-Second World War period. Tokyo, with a population of 27.9 million in 2000, has been the largest city in the world since 1970 and is projected to retain that rank (Box 4.2). By contrast, New York is projected to continue to slip down the ranking over the next twenty-five years. Other expected changes include the entry of Lagos, Karachi, Delhi and Dhaka to replace Rio de Janeiro, Osaka, Buenos Aires and Seoul by the year 2015. By then, Lagos is expected to be the third-largest city in the world after Tokyo and Bombay. Comparison of the lists for 1950 and 2000 demonstrates the remarkable shift in the global distribution of largest cities from the MDRs to the LDRs, a trend that will continue for the foreseeable future.

TABLE 4.3 URBAN POPULATION, NUMBER OF CITIES AND PERCENTAGE OF URBAN POPULATION BY CITY-SIZE CLASS: WORLD, MORE DEVELOPED REGIONS AND LESS DEVELOPED REGIONS

Size class	World				More developed regions				Less developed regions			
	1950	1970	1990	2015	1950	1970	1990	2015	1950	1970	1990	2015
10 million or more												
No. of agglomerations	1	3	12	27	1	2	4	4	0	1	8	23
Population	12	44	161	450	12	33	63	71	0	11	98	378
% urban	1.7	3.2	7.1	10.9	2.8	4.8	7.5	7.2	0.0	1.7	6.9	12.0
From 5 million to 10 million												
No. of agglomerations	7	18	21	44	5	8	6	8	2	10	15	36
Population	42	130	154	282	32	61	44	56	10	69	110	226
% urban	5.7	9.6	6.8	6.8	7.2	9.0	5.2	5.7	3.5	10.2	7.7	7.2
From 1 million to 5 million												
No. of agglomerations	75	144	249	472	43	73	98	120	32	71	151	352
Population	140	265	474	941	84	136	191	240	56	129	283	701
% urban	19.0	19.6	20.8	22.7	19.1	20.1	22.7	24.2	19.0	19.0	20.4	22.2
From 500,000 to 999,999												
No. of agglomerations	105	175	295	422	59	85	104	123	46	90	191	299
Population	73	122	203	293	42	61	72	84	31	61	132	209
% urban	9.9	9.0	8.9	7.1	9.5	9.0	8.5	8.5	10.5	9.0	9.2	6.6
Fewer than 500,000												
Population	470	792	1,284	2,178	272	386	472	540	198	406	812	1,638
% urban	63.7	58.5	56.4	52.6	61.5	57.1	56.1	54.5	67.0	60.0	56.6	52.0

Source: **adapted from United Nations (1995)** *World Urbanization Prospects: The 1994 Revision* **New York: United Nations**

Also, the largest cities are becoming larger; the average population of the world's largest cities was over 5 million inhabitants in 1990, compared with 2.1 million in 1950, and less than 200,000 in 1800. The number of megacities (defined by the United Nations as cities with 8 million or more inhabitants) is increasing rapidly, particularly in LDRs. Whereas in 1950 only New York and London had a population of 8 million or more, by 1970 eleven cities had become megacities (Table 4.5). Three were located in Latin America and the Caribbean (São Paulo, Buenos Aires and Rio de Janeiro), two in North America (New York and Los Angeles), two in Europe (London and Paris) and four in Asia (Tokyo, Shanghai, Osaka and Beijing). By contrast, in 1994 sixteen of the twenty-two megacities were in LDRs, and by 2015 it is expected that twenty-seven of the thirty-three megacities will be located in LDRs. The geographical shift in the focus of megacity growth is repeated in the distribution of 'million cities' (Table 4.6) and in the emergence of **metacities** (defined as conurbations of more than 20 million people) in Asia, Latin America and Africa. Many of these metacities, (or 'hypercities'), have

populations greater than some countries – the population of Greater Mumbai is larger than that of Norway and Sweden combined. While Tokyo is the only metacity today, by 2020 it is likely to be joined by others including Delhi, Mumbai, Mexico City, Sao Paulo, New York City, Dhaka, Jakarta and Lagos. Some of these giant cities have emerged as global or 'world cities' (see Chapter 14).

URBANISATION AND ECONOMIC GROWTH

Regional differences in rates of urbanisation throughout the world provide clear evidence of the relationship between urbanisation and industrialisation. As we saw in Chapter 3, prior to the industrial revolution urban development was restricted by the amount and value of surplus product that could be generated and accumulated in a single place. The low level of economic growth imposed a ceiling on the number of people who could be sustained in urban places.

TABLE 4.4 THE FIFTEEN LARGEST URBAN AGGLOMERATIONS, RANKED BY POPULATION SIZE, 1950, 2000 AND 2015

Rank	Agglomeration and country	Population (million)
1950		
1.	New York, USA	12.3
2.	London, UK	8.7
3.	Tokyo, Japan	6.9
4.	Paris, France	5.4
5.	Moscow, Russian Federation	5.4
6.	Shanghai, China	5.3
7.	Essen, Germany	5.3
8.	Buenos Aires, Argentina	5.0
9.	Chicago, USA	4.9
10.	Calcutta, India	4.4
11.	Osaka, Japan	4.1
12.	Los Angeles, USA	4.0
13.	Beijing, China	3.9
14.	Milan, Italy	3.6
15.	Berlin, Germany	3.3
2000		
1.	Tokyo, Japan	27.9
2.	Bombay, India	18.1
3.	São Paulo, Brazil	17.8
4.	Shanghai, China	17.2
5.	New York, USA	16.6
6.	Mexico City, Mexico	16.4
7.	Beijing, China	14.2
8.	Jakarta, Indonesia	14.1
9.	Lagos, Nigeria	13.5
10.	Los Angeles, USA	13.1
11.	Calcutta, India	12.7
12.	Tianjin, China	12.4
13.	Seoul, Republic of Korea	12.3
14.	Karachi, Pakistan	12.1
15.	Delhi, India	11.7
2015		
1.	Tokyo, Japan	28.7
2.	Bombay, India	27.4
3.	Lagos, Nigeria	24.4
4.	Shanghai, China	23.4
5.	Jakarta, Indonesia	21.2
6.	São Paulo, Brazil	20.8
7.	Karachi, Pakistan	20.6
8.	Beijing, China	19.4
9.	Dhaka, Bangladesh	19.0
10.	Mexico City, Mexico	18.8
11.	New York, USA	17.6
12.	Calcutta, India	17.6
13.	Delhi, India	17.6
14.	Tianjin, China	17.0
15.	Metro Manila, Philippines	14.7

Source: **adapted from United Nations (1995)** *World Urbanization Prospects: The 1994 Revision* **New York: United Nations**

BOX 4.2

Tokyo: global metacity

Tokyo is the largest urban agglomeration in the world and is approaching the threshold of a 30 million population super-city. It has long been the dominant urban centre in Japan. As early as 1700 Tokyo had a population of over 1 million, when the population of London was 650,000. Following the Second World War, the city's population grew from 6.9 million in 1950 to 16.5 million in 1970, reaching 26.5 million in 1994. The outer limit of the urbanised area now reaches 100 miles (160 km) from the urban core.

Tokyo meets all the criteria of world-city status. The city dominates the Japanese urban system. It is the major commercial centre, housing the headquarters of all major Japanese companies; a manufacturing and wholesaling centre; the chief financial centre; the seat of government; the cultural capital; the centre for the media and advertising industries; and has the largest concentration of institutions of higher learning in the world. The Tokyo metropolitan region is also the world's largest consumer market.

The unprecedented growth of the city has led to urban problems, including escalating land prices, shortage of office space, high-cost housing, congestion and excessive commuting journeys. In recent years the government has sought to encourage decentralisation through the construction of technopoles, small cities of around 50,000 built for high-technology industries, research institutes and colleges near existing medium-size cities. It has also discussed the relocation of capital functions to other parts of the Tokyo metropolitan region. This ties in with the long-range plan for the region, which has among its goals the development of a multi-core urban structure in which the functions of the city are dispersed to reduce distances between place of employment and place of residence.

The key role that cities play in dynamic and competitive economies and the relationship between the scale of a national economy and the level of urbanisation is illustrated by the fact that most of the world's largest cities are in the world's largest economies.[4] In 1990 the world's twenty-five largest economies had over 70 per cent of the world's 'million cities' and all but one of its twelve agglomerations with 10 million or more inhabitants.

Despite such categorical evidence, the relationship between urbanisation and level of economic development is complex, with many factors at work, and while economic growth may result in increased levels of

TABLE 4.5 NUMBER OF MEGACITIES, 1970, 1994, 2000 AND 2015

Region	1970	1994	2000	2015
World	**11**	**22**	**25**	**33**
Less developed regions	**5**	**16**	**19**	**27**
Africa	0	2	2	3
Asia[a]	2	10	12	19
Latin America	3	4	5	5
More developed regions	**6**	**6**	**6**	**6**
Europe	2	2	2	2
Japan	2	2	2	2
Northern America	2	2	2	2

Source: adapted from United Nations (1995) *World Urbanization Prospects: The 1994 Revision* New York: United Nations

Note: [a]Excluding Japan

TABLE 4.6 REGIONAL DISTRIBUTION OF THE WORLD'S POPULATION IN 'MILLION CITIES' AND THE LOCATION OF THE WORLD'S LARGEST 100 CITIES

	Proportion of the world's:				No. of the world's 100 largest cities in:		
	Urban population		Population in 'million cities'				
	1950	1990	1950	1990	1800	1950	1990
Africa	**4.5**	**8.8**	**1.8**	**7.5**	**4**	**3**	**7**
Eastern Africa	0.5	1.7	—	0.8	—	—	—
Middle Africa	0.5	1.0	—	0.8	0	0	1
Northern Africa	1.8	2.8	1.8	3.2	3	2	5
Southern Africa	0.8	0.9	—	0.8	0	1	0
Western Africa	0.9	2.6	—	2.0	1	0	1
Americas	**23.7**	**23.0**	**30.1**	**27.8**	**3**	**26**	**27**
Central America and the Caribbean	2.8	4.2	2.2	3.5	2	2	3
Northern America	14.4	9.2	21.2	13.1	0	18	13
South America	6.5	9.7	6.7	11.1	1	6	11
Asia	**32.0**	**44.5**	**28.6**	**45.6**	**64**	**33**	**44**
Eastern Asia	15.2	19.7	17.6	22.2	29	18	21
South-East Asia	3.7	5.8	3.4	5.6	5	5	8
South-Central Asia	11.2	14.8	7.0	14.6	24	9	13
Western Asia	1.8	4.1	0.6	3.3	6	1	2
Europe	**38.8**	**22.8**	**38.0**	**17.9**	**29**	**36**	**20**
Eastern Europe	11.8	9.3	7.7	6.3	2	7	4
Northern Europe	7.7	3.4	9.0	2.1	6	6	2
Southern Europe	6.5	4.0	6.7	3.2	12	8	6
Western Europe	12.8	6.2	14.6	6.2	9	15	8
Oceania	**1.1**	**0.8**	**1.6**	**1.3**	**0**	**2**	**2**

Source: adapted from United Nations (1996) *An Urbanising World: Global Report on Human Settlements* Oxford: Oxford University Press

urbanisation, higher levels of urbanisation in turn can stimulate more economic growth. We can illustrate this in the context of the Third World, where a number of factors can mediate the relationship between urbanisation and economic growth. These include the following:

1. The nature of economic activity within each sector of the national economy can affect urbanisation. For example, the type of agricultural enterprise can affect the scale of urban settlement. High-value crops that provide good incomes for farmers and workers within intensive farming systems can support rapid growth of local urban centres to the point where agricultural surpluses support a relatively urbanised population.
2. The nature of land ownership is important in determining whether profits are spent or invested within the country or taken abroad.
3. Cultural preferences for a type of lifestyle can influence the level of urbanisation.
4. Government policies and the activities of state institutions are also important. An increasing share of national income is controlled by the public sector. Since most government employees are urban residents, most public spending flows to towns and cities. Some observers have equated this trend with 'urban bias' in the investment strategies of Third World governments and have argued that this has encouraged the migration of people from rural to urban areas, thereby leading to overurbanisation (Box 4.3).

In general, although there is a close relationship between the level of urbanisation and national levels of economic development, the range of potential influences suggests that different factors may be of importance in different national settings. Furthermore, the relative importance of factors that influence levels of urbanisation may change over time as a country becomes more urbanised.

THE URBANISATION CYCLE

The simplicity of the logistic-growth urbanisation curve (Box 4.4) should not be taken to mean that urbanisation

---

BOX 4.3

Overurbanisation

The concept of overurbanisation in less developed countries implies that economic growth is unable to keep pace with urban population growth, thereby leading to the major social and economic problems evident in most large cities of the Third World. It is also argued that overurbanisation is detrimental to national economic growth, since, with limited investment available to most Third World states, the political influence of cities and resultant high level of investment in urban areas will reduce investment in other productive sectors of the economy.

The concept of overurbanisation is not accepted by those who reject the implication that the form of urbanisation in the North represents a model for the South and that any departure from this pattern represents 'overurbanisation'. Despite the simplicity of the logistic curve (see Box 4.4) urbanisation is not a single process followed by all countries in the process of development, and it would be surprising if urban trends in the South mirrored patterns in the North. The much weaker position of most developing countries in world markets, and the fundamental differences between the world in the nineteenth century and late twentieth century must affect the social economic and political factors that influence levels of urbanisation.

The notion of overurbanisation is useful to the extent that it emphasises the dynamic links between different economic sectors and areas of a country. Given that there are 600,000 agricultural villages in India where the infant mortality rate is still 100 per thousand, literacy is less than 40 per cent, malnutrition is widespread, the supply of drinking water is inadequate, electricity is lacking and half the population live in mud huts,[5] it is little wonder that rural–urban migration is a feature of many Third World nations. In such situations the problem of 'overurbanisation' might be addressed via rural development policy, which could also benefit the national economy in countries such as India where a significant part of GDP comes from the agricultural sector.

Although it is not inappropriate to use the term 'overurbanisation' to describe a situation of high levels of primate city urbanisation, rapid in-migration, saturated urban labour markets and overburdened urban services, in seeking explanations we must look beyond these urban problems to national and international economic and political forces.

BOX 4.4

The urbanisation curve

Urbanisation is a process of population concentration whereby towns and cities grow in relative importance through, first, an increasing proportion of the national population living in urban places and, second, the growing concentration of these people in the larger urban settlements. It has been suspected that all nations pass through this process as they evolve from agrarian to industrial societies. For Davies (1969)[6] the typical course of urbanisation for a nation is represented by a logistic curve.

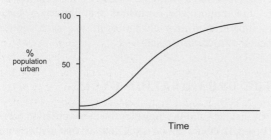

The first section of the curve is associated with very high rates of urbanisation associated with large shifts of population from rural areas to towns and cities in response to the creation of an urban economy. This is followed by a longer period of consistent moderate urbanisation. As the urban percentage reaches above 60 the curve begins to flatten, approaching a ceiling of around 80 per cent. This is the level at which rural and urban populations appear to reach functional equilibrium. At any one time individual countries are at different stages of the urbanisation curve.

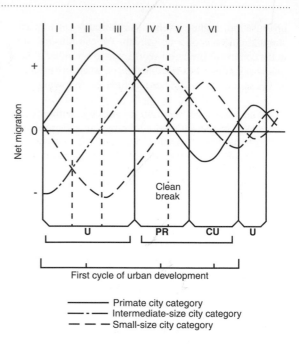

Figure 4.7 Generalised stages of differential urbanisation
Source: H. Geyer and T. Kontuly (1993) A theoretical foundation for the concept of differential urbanisation *International Regional Science Review* 17(2), 157–77

intermediate-size and small cities go through successive periods of fast and slow growth in a cycle of development. The main stages of this urbanisation cycle are summarised in Figure 4.7.

THE PRIMATE CITY PHASE

During the initial phase of urbanisation, the primate city phase, an increasing proportion of economic activity and population in a country concentrates in a limited number of rapidly growing **primate cities** (Box 4.5). This phase can be subdivided into three stages:

1. First is an *early primate city* stage in which a primate city attains spatial dominance of the urban system, attracting a large proportion of net inter-regional migration.
2. An *intermediate primate city* stage follows, in which the rapidly growing primate city is still monocentric in form but with suburbanisation a prominent feature. Suburban nodes, the nuclei of a future multinodal primate city, may emerge. As a result of favourable locational attributes, certain intermediate-size cities develop faster than others.

is a unidirectional (low to high) process. In a large number of advanced countries the level of urbanisation actually decreased between 1965 and 1990. Although this can be a result of statistical underbounding of urban areas (see Chapter 2), it is now generally acknowledged to indicate a process of population redistribution which involves either the relatively faster growth of smaller urban places or the absolute decline of the largest cities.[7] The shift in the incidence of strongest population growth away from the largest cities in the national urban system has been termed 'polarisation reversal'[8] or, more commonly, 'counterurbanisation'.[9] Geyer and Kontuly (1993)[10] have incorporated these concepts into a theory of differential urbanisation which postulates that large,

BOX 4.5

The law of the primate city

The law of the primate city refers to a situation in which a single city accommodates a dispropor-tionately large number of a country's population. In some instances the primate city size distribu-tion is the result of outside or foreign influences on the settlement pattern. In many present-day Third World countries, for example, primate cities developed as a result of the intervention of a colonial power. Bangkok is one such example.

Jefferson (1939)[11] argued that, in the early stages of a country's urban development, the city that emerges as larger than the rest develops an impetus to self-sustaining growth that enables it, over time, to attract economic and political functions to such an extent that it dominates the national urban system. Capital cities such as Paris or Vienna occupy this niche.

In some countries a variety of forces, such as nationalism in Spain and territorial size in the USA, have led to several cities growing to compa-rable size rather than the emergence of a single primate city.

The law of primacy is most relevant to coun-tries that have a relatively simple economy and spatial structure, a small area and population, low incomes, economic dependence upon agriculture and a colonial past.

3. Phase 2 is followed by an *advanced primate city* stage in which the primate city becomes so large that owing to agglomeration diseconomies, such as congestion costs, a monocentric urban structure becomes unwieldy. By means of intra-regional decentralisation the primate city develops a multi-centred or megalopolitan character, and dominates the rest of the urban system economically and spa-tially. Expansion of the urban system as a whole may lead to one or more intermediate-size cities becoming as large as existing cities. At some point in the development history of most countries the growth rate of the primate city slows and a process of spatial deconcentration commences.

THE INTERMEDIATE CITY PHASE

The slowing growth rates of the primate city and spatial deconcentration of urban population are often accompanied by growth of intermediate-size centres close to the primate city. This 'population turnaround' or 'polarisation reversal' has been documented in a number of countries including the USA, the UK, France, Greece, Brazil and Botswana[12] This phase can be subdivided into two stages:

1. An *early intermediate city* stage characterised by the uneven growth of a limited set of intermediate-sized cities that are close to but not contiguous with the primate metropolitan region. Although the primate city is still gaining population in absolute terms, it is starting to lose in relative terms to the intermediate cities. Suburban centres within the primate metropolitan region are now growing faster than the central city.

2. An *advanced intermediate city* stage in which the suburbanisation process that characterised the development of the primate city during the advanced primate city stage is reproduced in the fastest-growing intermediate-size cities but on a smaller scale. In addition, in contrast to the early intermediate city stage, all centres within the primate metropolitan region begin to lose popula-tion in absolute terms, with the central city losing more than the suburban centres.

THE SMALL CITY PHASE

The small city phase represents a continuation of the previous stage during which deconcentration takes place from the primate and intermediate-size cities towards small urban centres which may eventually grow at a faster rate than either the primate city or the inter-mediate-size cities. By the end of this phase the urban system has reached a 'saturation point' where the rural population cannot be reduced much further (see the logistic curve in Box 4.4) and rural–urban migration ceases to be a major contributory factor in the urbanisa-tion cycle. Since population growth through natural increase may also be very low in advanced societies (see the demographic transition model in Box 4.1), urban growth in general may be slow.

However, as the model shows (Figure 4.7), the 'small city phase' marks not only the end of the first cycle of urban development but also the beginning of a new one which follows the sequence of major metro-politan, intermediate-size city, and small city growth. Significantly, the set of major metropolitan areas ex-periencing the highest rates of net in-migration during the urbanisation phase of the second cycle may or may not be the same set of large urban areas from the first cycle (as indicated by differential urban growth rates in the rustbelt and sunbelt areas of the USA). Large urban centres able to retain their dominant position in the national and international urban hierarchy, along with a limited group of rapidly growing intermediate-size urban areas from the first cycle, will constitute the new set of major metropolitan centres.

A 'STAGES OF URBAN DEVELOPMENT' MODEL

The concept of a cycle of urbanisation has also been employed by Klaassen *et al.* (1981) and van den Berg *et al.* (1982) to study the growth patterns *within* individual urban agglomerations.[13] As Figure 4.8 shows, four stages of urban development are envisaged:

1. *Urbanisation*: when certain settlements grow at the cost of their surrounding countryside.
2. *Suburbanisation* or *exurbanisation*: when the urban ring (commuter belt) grows at the cost of the urban core (physically built-up city).
3. *Disurbanisation* or *counterurbanisation*: when the population loss of the urban core exceeds the population gain of the ring, resulting in the agglomeration losing population overall.
4. *Reurbanisation*: when either the rate of population loss of the core tapers off, or the core starts regaining population with the ring still losing population.

As Figure 4.8 indicates, the model is based on changes in the direction and rate of population movement between urban core and urban ring (which together comprise a functionally related daily urban system). The

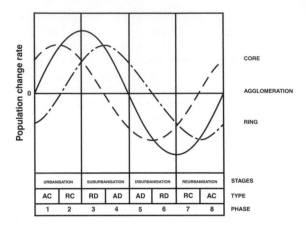

Figure 4.8 The stages of urban development model

Source: A. Champion (2000) Urbanization, suburbanization, counter-urbanization and reurbanization, in R. Paddison and W. Lever (eds) *Handbook of Urban Studies* London: Sage

two types of change are absolute shifts when the directions of population change in the two zones differ, and relative shifts when change occurs in the same direction but at different rates. These trends are summarised in Table 4.7. We have already examined the processes and patterns of urbanisation. Here we focus on population movements associated with the other major dimensions of urban change identified in Table 4.7.

REURBANISATION

The empirical evidence for reurbanisation is mixed. A study of 241 functional urban regions (FURs) in Europe found that between 1981 and 1991 the proportion of urban cores gaining population reached 47 per cent,

TABLE 4.7 STAGES OF DEVELOPMENT OF A DAILY URBAN SYSTEM

Stage of development	Classification type	Population change characteristics			
		Core	Ring	DUS	
I Urbanisation	1. Absolute centralisation	++	−	+	
	2. Relative centralisation	++	+	+++	Total growth
II Suburbanisation/ Exurbanisation	3. Relative decentralisation	+	++	+++	(Concentration)
	4. Absolute decentralisation	−	++	+	
III Disurbanisation/ Counterurbanisation	5. Absolute decentralisation	− −	+	−	
	6. Relative decentralisation	− −	−	− − −	Total growth
IV Reurbanisation	7. Relative centralisation	−	− −	− − −	(deconcentration)
	8. Absolute centralisation	+	− −	−	

compared with only 22 per cent over the period 1975–81.[14] However, it was mainly the smaller FURs (particularly those with ancient cathedrals and universities) that exhibited reurbanisation, not the larger, older urban regions. In the UK reurbanisation occurred in only four of thirty-six FURs (Glasgow, Oxford, Cambridge and Canterbury), with only Glasgow confirming model expectations.

On the other hand, there is a growing body of case-study evidence that indicates a recovery of large cities from the high levels of population loss experienced in the 1970s era of counterurbanisation. As Table 4.8 shows, the rate of population loss for all 280 of Britain's urban areas fell from 4.2 per cent in 1971–81 to 0.1 per cent for 1981–91, while the growth rate for the urban rings declined from 9 per cent to less than 6 per cent. Evidence from the 2001 Census in the UK suggests that, during the 1990s, where repopulation of city centres occurred a large proportion of the new residents were young socially mobile adults seeking proximity to work and leisure facilities and the 'stylishness' of city-centre living – a gentrification trend (see Chapter 10) encouraged by government policies in support of new residential developments in inner urban areas (such as London docklands)[15] In the USA the 1980s witnessed the re-emergence of the larger metropolitan areas as the fastest-growing elements of the urban landscape[16] (Table 4.9). Overall, metropolitan areas with 1 million or more residents grew by 12 per cent in the 1980s compared with 8 per cent in the previous decade. While much of this growth was in the south and west even the north's large metropolitan areas switched from a population decline of 0.9 per cent between 1970 and 1980 to a 2.7 per cent increase in the 1980s.

The population growth that has occurred in the central areas of US cities was fuelled by two principal

TABLE 4.8 CITIES AND DOWNTOWN POPULATION CHANGES IN THE USA, 1990–2000

Zone type	Rate for decade				Deviation from GB rate			
	1951–61	1961–71	1971–81	1981–91	1951–61	1961–71	1971–81	1981–91
Great Britain	4.97	5.25	0.55	2.50	0.00	0.00	0.00	0.00
Core	3.98	0.66	−4.20	−0.09	−0.99	−4.59	−4.75	−2.59
Ring	10.47	17.83	9.11	5.89	5.50	12.58	8.56	3.39
Outer area	1.74	11.25	10.11	8.85	−3.23	6.00	9.56	6.35
Rural area	−0.60	5.35	8.84	7.82	−5.57	0.10	8.29	5.32

Source: adapted from A. Champion (2000) Urbanization, suburbanization, counterurbanization and reurbanization, in R. Paddison and W. Lever (eds) Handbook of Urban Studies Beverly Hills, CA: Sage

TABLE 4.9 TYPES OF POPULATION CHANGE IN METROPOLITAN AREAS OF THE USA, 1980–90 AND 1990–96

Type	1980–90		1990–6	
	No. of cities	%	No. of cities	%
1	39	17.0	15	6.6
2	54	23.5	53	23.3
3	40	17.4	31	13.7
4	60	26.1	85	37.4
5	37	16.1	43	18.9

Source: adapted from J. Mercer (1999) North American cities: the micro-geography, in F. Boal and S. Royle (eds) North America: A Geographical Mosaic London: Arnold, 191–206

Notes: Type 1 cities: central-city decline added to metropolitan decline. Type 2 cities: central-city decline and metropolitan growth. Type 3 cities: central-city stagnation (equal to a less than 5% population change over the period) and metropolitan growth. Type 4 cities: strong central-city growth (between 5.1–19.9% over the period) and metropolitan growth. Type 5 cities: booming central-city growth (20.0% or more over the period) and metropolitan growth

migration streams. First, new migrants, primarily from Latin America and Asia, moved into lower-value areas of cities such as New York and Los Angeles as well as into other metropolitan areas on the west coast (San Diego and San Francisco), in the south-west (Houston) and Florida (Miami) that historically had attracted relatively fewer migrants (see Chapter 5). The second stream comprised a flow of 'baby-boomers' (those born just after the Second World War and during the affluence of the 1950s and early 1960s), investing in high-status residential areas. During the 1980s the strongest magnets for adult 'boomers' were metropolitan areas with expanding high-tech and defence-oriented economies, including coastal cities such as Boston and Seattle and sunbelt locations like Dallas and Atlanta. Australia and Canada also provide evidence of strengthening metropolitan areas and inner-city growth in

the 1980s.[17] Evidence for US cities over the period 1990–2000 indicates that the occurrence and extent of downtown 'population rebound' varies considerably (Table 4.10). While in some cities the downtown contribution to metro growth is small, in others it represented a significant proportion of total population growth, and in others downtown population growth offset city-wide population decline.

In general, these empirical observations suggest that:

1. There are widespread signs of renewed growth or reduced population decline for larger metropolitan areas, as well as a population recovery for urban cores.
2. There is no evidence of suburban-ring areas losing out to core areas, not even in relative terms, let alone in accordance with the absolute change

TABLE 4.10 TYPES OF POPULATION CHANGE IN METROPOLITAN AREAS OF THE USA, 1980–90 AND 1990–96

Cities where downtown population growth contributed to city population growth

	Downtown population			City population			Downtown share of city growth (%)
	1990	2000	Increase	1990	2000	Increase	
Miami FL	15,143	19,927	4,784	358,548	362,470	3,922	122.0
Boston MA	77,253	80,903	3,650	574,283	589,141	14,858	24.6
Atlanta GA	19,763	24,931	5,168	394,017	416,474	22,457	23.0
San Francisco CA	32,906	43,531	10,625	723,959	776,733	52,774	20.1
Chicago IL	56,048	72,843	16,795	2,783,726	2,896,016	112,290	15.0
Seattle WA	12,292	18,983	6,691	516,259	563,374	47,115	14.2
New York	153,927	170,708	16,781	7,322,564	8,008,278	685,714	2.5
Dallas TX	11,858	15,198	3,340	1,006,877	1,188,580	181,703	1.8
Los Angeles CA	34,655	36,630	1,975	3,485,398	3,694,820	209,422	0.9
Houston TX	7,029	7,565	536	1,630,553	1,953,631	323,078	0.2

Cities where downtown population growth reduced city population loss

	Downtown population			City population			Downtown offset of city loss (%)
	1990	2000	Increase	1990	2000	Decrease	
Detroit MI	34,872	35,618	746	1,027,974	951,270	−76,704	1.0
Baltimore MD	28,579	30,067	1,470	736,014	651,154	−84,860	1.7
Norfolk VA	2,390	2,881	491	261,229	234,403	−26,816	1.8
Washington DC	26,597	27,667	1,070	606,900	572,059	−34,841	3.0
Milwaukee WI	15,039	16,359	1,320	628,088	596,674	−31,114	4.1
Philadelphia PA	74,686	78,349	3,663	1,585,577	1,517,550	−68,027	5.1
Cleveland OH	7,261	9,599	2,338	505,616	478,403	−27,213	7.9
New Orleans LA	6,988	8,051	1,063	496,938	484,674	−12,264	8.0
Pittsburgh PA	6,517	10,216	3,699	369,879	334,563	−35,316	9.5
Jackson MS	5,253	6,762	1,509	196,637	184,256	−12,381	10.9

Source: adapted from E. Birch (2002) Having a longer view on downtown living *Journal of the American Planning Association* 68(1), 5–21

associated with the later phase of reurbanisation specified in the 'stages of urban development' model.

Reurbanisation as defined by Klaassen *et al.* (1981)[18] has not yet emerged as a significant feature in the urban systems of advanced economies. There is also considerable disagreement over the extent to which the inner-city revitalisation that took place in the 1980s will be able to continue and lead to a fundamental change in the form of the Western city. The process of decentralisation, on the other hand, is likely to continue as a major feature of post-industrial urbanisation, albeit in a form very different from the dormitory-style suburbanisation of the early post-Second World War period.

We can identify two main forms of population decentralisation. The first, counterurbanisation or urban deconcentration, is characterised by net population movement from metropolitan regions into smaller urban regions and rural areas that lie beyond the primary commuter-sheds of the major cities. The second, suburbanisation, reflects a long-established centrifugal movement of population which progressively has involved a broader range of urban functions than just housing taking place over longer distances as personal mobility has grown and urban centres have expanded to embrace their previously rural hinterlands.

COUNTERURBANISATION

Signs of a population reversal in rural areas were first identified in the USA,[19] but similar trends were soon detected in other advanced nations, including Canada,[20] Australia,[21] Western Europe[22] and Britain.[23] Counterurbanisation in Britain dates from the early 1960s, when for the first time areas situated well away from metropolitan influence began to grow faster than the main conurbations and their dependent regions. Population growth in rural Britain was particularly strong in the late 1960s and early 1970s but has continued over recent decades, with net out-migration from the main metropolitan areas to the rest of the UK averaging around 90,000 people per year, a rate of 0.5 per cent.[24]

The reasons for this reversal of long-established trends are so multifaceted that any attempt to apply a single explanation to the widely diverse changes under way in different regions would be unduly simplistic. Synthesising findings from a range of investigations

provides a useful inventory of contributory factors. These include:

1. Continuing growth of metropolitan centres and their spillover into adjacent non-metropolitan counties.
2. Decentralisation of manufacturing in pursuit of lower land and wage costs.
3. Increased employment in service occupations.
4. Early retirement, coupled with higher retirement incomes not tied to a particular location.
5. Increased *per capita* disposable real income.
6. Increased pursuit of leisure activities at all ages, centred on amenity-rich areas outside the daily range of metropolitan commuting.
7. Increased enrolments in rural colleges and universities in the USA, especially in the late 1960s and early 1970s as a result of the post-war 'baby boom'.
8. Growth of state government in the USA.
9. Levelling off of the loss-of-farm population.
10. Growth of an anti-materialist perspective among the young.
11. Narrowing of the traditional gap between urban and rural lifestyles with the extension of electricity, water and sewage systems, telecommunications and access to modern facilities.
12. More long-distance commuting.
13. Growth associated with energy and extractive industries.
14. Completion of the interstate highway system in the USA.
15. Lower cost of living in rural areas.
16. Growth of anti-urbanism as characterised by increased fear of crime and concern with urban disamenities such as congestion and pollution.
17. Growth in importance of military establishments in some US counties during the 1960s.
18. Residential preference for lower-density rural living.
19. Government decentralisation policies in some countries such as Sweden, France and Britain (in the case of the last-named, via the New Towns programme; see Chapter 9).

The list of factors is diverse and many are interrelated. The extent to which each contributes to the population turnaround will depend on local conditions.

THE SUBURBANISATION WAVE

The suburbanisation process began on a significant scale in the 1920s and accelerated after the Second

World War, especially in North America. The USA is the world's first predominantly suburban nation.[25] As Table 4.11 shows, by the early 1960s the suburbs held 51 per cent of the US urban population, by 1980 they accounted for half of total metropolitan employment, and by 1990 about two-thirds of the metropolitan population and 55 per cent of metropolitan employment. By the last of these dates, US suburbia contained more than half the entire national population, having expanded from 41 million to 115 million since 1950 (an increase of 180 per cent). In 2000 the **suburbs** accommodated 140 million Americans (50 per cent of the total population).

The suburban wave was driven by the following factors:

1. The rapid growth of urban population and rising disposable incomes enabled people to meet both the cost of new housing and the associated transport costs.
2. Widespread diffusion of the automobile enhanced individual mobility. The number of US automobiles rose from under 1 million in 1910 to 27 million by 1930, the latter amounting to one for every five persons.
3. New suburbs started to resist annexation by central cities through legal incorporation, which enabled them (and their residents) to shield themselves from the problems of the central city (such as low-quality housing, rising taxes, congestion, racial tension and crime), and to provide the particular living environment they desired and could pay for.
4. There was a huge pent-up demand for housing.
5. There was a need to generate employment after fifteen years of low investment during the Depression of the 1930s followed by the war years.

6. These goals were promoted by public policies that favoured new house-building over rehabilitation and highway construction over mass transit.

Consequently, in the USA, the 1950s represented 'the largest suburban decade ever'.[26]

In the UK the most ubiquitous form of new rural residential development has been the residential sub-division or housing estate located as an adjunct to an existing rural settlement within commuting distance of an urban workplace. These dormitory settlements, growing almost solely because of out-migration from central cities, have been termed **metropolitan villages**, defined as settlements where more than one in five of the work force is employed in towns or cities.[27] In the USA the suburbanisation process has created transit- and freeway-dependent dormitory subdivisions, infill developments and automobile-dependent dispersed suburbia. The impact of post-war suburban expansion in the USA is demonstrated most vividly in the growth of **boomburgs** (Box 4.6). The functions of suburbs range from undifferentiated residential areas to a more recent mix of specialised retail corridors, high-technology industrial clusters, and high-density office and commercial nodes or 'edge cities'[28] that have contributed to the formation of increasingly polycentric metropolitan regions (Box 4.7). The impact of suburbanisation on central cities has been profound. Most strikingly, in the 1960s several US cities, notably New York, became technically bankrupt and unable to finance their current expenditure on services. Rising local taxes and deteriorating local services merely served to accelerate the flight of better-off residents and footloose firms into the suburbs, leaving behind the less dynamic economic sectors and less wealthy people, notably African-Americans and recent immigrants from overseas. Even

TABLE 4.11 CHANGING POPULATION AND EMPLOYMENT DISTRIBUTIONS IN THE USA, 1950–2000

	Census year					
	1950	1960	1970	1980	1990	2000
Central cities as percentage of:						
Metro population[a]	57	49	43	40	37	38
Metro employment	70	63	55	50	45	37
Suburbs as percentage of:						
Metro population	43	51	57	60	63	62
Metro employment	30	37	45	50	55	63

Source: **US Census Bureau statistics**

Note: [a]**Metro refers to the metropolitan statistical area (MSA) as defined by the US Census at each census date**

BOX 4.6

The Boomburg phenomenon

A 'boomburg' is an incorporated place with more than 100,000 residents that is not the largest city in its metropolitan area and that has maintained a double-digit rate of population growth in recent decades. Boomburgs typically do not resemble traditional central cities or even older satellite cities. While they have many urban elements – such as apartment buildings, retail centres, entertainment venues and office complexes – they usually lack a dense business core. Boomburgs, (such as Henderson NV, Gilbert AZ and Fontana CA), are distinct from traditional cities not so much in terms of their functions as in terms of their low density and loose spatial configuration.

Most (forty-four) of the USA's fifty-three boomburgs as of 2003 were in the south-west. Many are products of master-planned community developments (see Chapter 9). Between 2000 and 2003 five of the fastest-growing cities in the USA were boomburgs, with the four top-growth boomburgs being in the Phoenix and Las Vegas metropolitan areas. Over this three-year period the fastest-growing boomburg (Gilbert AZ) gained 35,301 people (a growth rate of 32.1 per cent).

A possible constraint on the continued growth of boomburgs in the west and south-west is the problem of ensuring a reliable water supply. Even if this obstacle is overcome no place can boom for ever. Todays boomburgs may be tomorrows mature cities. But a new phase of boomburg is already evident – in Arizona's Central Valley as growth slows in Tempe and Mesa it is increasing in places such as Apache Junction and Buckeye.

Source: adapted from R. Lang (2006) Are the boomburgs still booming? in A. Berube, B. Katz and R. Lang (eds) *Redefining Urban and Suburban America: Evidence from Census 2000* III, Washington DC: Brookings Institution Press, 83–91

BOX 4.7

Edge City, USA: Tyson's Corner VA

The post-Second World War movement of housing, industry and commerce to the outskirts of urban areas has created perimeter cities that are functionally independent of the urban core. In contrast to the residential or industrial suburbs of the past, these new cities contain along their superhighways all the specialised functions of a great metropolis: commerce, shopping malls, hospitals, universities, cultural centres and parks.

This new peripheral urban form is referred to by various names, including technoburb, post-suburbia, cyberbia, stealth city or edge city. Driving time, not space, determines its fluid boundaries. Garreau (1988)[29] defined an edge city as a place that has 5 million ft^2 (460,000 m^2) or more of leasable office space, 600,000 ft^2 (56,000 m^2) or more of leasable retail space, more jobs than bedrooms, is perceived by the population as one place, and that grew from practically nothing in the early 1960s.

Tyson's Corner, just beyond the beltway around Washington DC, is the archetypal edge city. In the mid-1960s Tyson's Corner was a rural corner of northern Virginia marked only by the intersection of Interstate 66, the Washington Beltway and the access road to Dulles International Airport. Administratively it is still rural, an unincorporated 6,000 acre (2,400 ha) area that contains 30,000 residents and over 75,000 jobs, all under the jurisdiction of Fairfax County but split between three different county supervisory districts and three county planning districts. Tyson's Corner does not exist as a postal address; residents' mail must go to either McLean or Vienna. Within this framework in 1990 was the ninth largest concentration of commercial space in the USA, including more than 20 million ft^2 (1.9 million m^2) of office space, 3,000 hotel rooms and parking for more than 80,000 cars. The area was also the largest east-coast retail concentration outwith Manhattan. Yet it had little of the apparatus of urban governance or civic affairs.

Source: J. Garreau (1991) *Edge City* New York: Random House

in the more stable 1980s, when New York City's population grew by 3.5 per cent, its white non-Hispanic population fell by 11.5 per cent and the proportion of its total residents accounted for by the 'minority population' rose to over 60 per cent in 1990. As Table 4.12 indicates, similar patterns of 'white flight' were recorded by other major cities of the USA. While the exodus of white-flight population from New York and Chicago slowed during the 1990s in other major cities, the rate of white out-migration from the central core increased (Box 4.8).

In Detroit the loss of almost half the white population of the central city during the 1990s matched the rate of the previous decade, with the result that nearly nine out of ten residents of central Detroit are from minority ethnic groups (Table 4.12).

TABLE 4.12 POPULATION CHANGE IN US CENTRAL CITIES, 1960–2000

City	2000 population (000)	Population change (%)				White population change (%)		Minority share of total population (%)	
		1960–70	1970–80	1980–90	1990–2000	1980–90	1990–2000	1990	2000
New York	8,008	1.4	−10.4	3.5	−9.4	−11.5	−6.6	60.5	55.3
Chicago IL	2,896	−4.7	−10.7	−7.4	−4.0	−17.2	−3.8	62.7	58.0
Philadelphia PA	1,518	−3.1	−13.5	−6.1	3.2	−12.3	−19.5	48.4	55.0
Detroit MI	951	−8.5	−19.2	−14.6	7.5	−47.7	−47.6	79.7	89.5
Baltimore MD	651	−2.8	−12.5	−6.4	11.5	−15.8	−28.4	61.6	69.4
Cleveland OH	478	−14.3	−23.6	11.9	5.5	−19.0	−20.7	52.5	58.5
Pittsburgh PA	335	−14.1	−18.5	−12.8	9.5	−14.9	−15.2	28.5	32.4
Cincinnati OH	231	−9.8	−15.0	−5.5	9.1	−11.2	−20.3	39.9	47.0

Source: US Census Bureau statistics

Notes: Data refer to central cities only. Cities are ranked by 2000 population size. Hispanics are included in the minority population

BOX 4.8

The changing ethnic composition of US central cities

A combination of 'white flight' and an influx of Hispanic populations has left US whites as a minority in nearly half the 100 largest cities. According to the 2000 census non-Hispanic whites are now a majority in only fifty-two of the largest 100 cities, compared with seventy a decade earlier. Whites now account for 44 per cent of the 58,441,915 people in the largest 100 cities (compared with 52 per cent in 1990). Some 2.3 million whites or 8.5 per cent of the white urban population left the largest cities between 1990 and 2000. Birmingham AL lost 40 per cent of its white population. Other cities where whites have become a minority include Anaheim CA and Riverside CA, where immigration from Mexico is particularly strong. Cities experiencing economic difficulties such as Rochester NY saw 'white flight' to the suburbs. The Hispanic population of the 100 largest cities grew from 17.2 per cent to 22.5 per cent, an increase of 3.8 million. The Asian population grew by 1 million, from 5.3 per cent to 6.6 per cent. The black population increased by 876,000 but as a proportion of the urban population fell slightly from 24.6 per cent to 24.0 per cent.

Although attempts have been made to classify suburbs, they are better viewed as dynamic entities with a diversity that reflects their role in the postmodern city.

Hanlon *et al.* (2006) illustrate the diversity of suburban places in the USA, identifying poor suburbs, manufacturing suburbs, immigrant suburbs and black suburbs in addition to the stereotypical white middle-class suburb.[30] The diversity of the suburbanisation phenomenon is encapsulated by Bourne (1996)[31] in a list of ten differing interpretations (Table 4.13). The first and more traditional interpretation views suburban development as a 'natural' process of accommodating growth by extension of the urban margin, and is characterised by the classical ecological models of the city.[32] The second perspective sees the suburbs as an escape route from the social and environmental problems of cities either via individual decisions or through centralised planning initiatives.[33] The third and fourth views are based on a structural or political economy perspective which interprets suburbanisation as a tool of government macro-economic policy and a means of generating employment and promoting capital accumulation for the land development, building and financial sectors.[34] The fifth characterises suburbanisation as social engineering, as a means of rescuing the poor from themselves and perhaps as an indirect means of inculcating an assumed superior moral order of the past.[35] The sixth and seventh explanations are market-driven and derive from micro-economic theory and the capitalist logic of individual utility maximisation. One views suburban sprawl as the expected outcome of rational individuals, households and firms seeking more efficient and less costly environments; the other emphasises the dominant role of consumer preferences for more space, new housing, privacy and private consumption.[36] The eighth sees suburban development as a socio-political strategy of exclusion designed to satisfy

TABLE 4.13 ALTERNATIVE INTERPRETATIONS OF THE SUBURBANISATION PROCESS

1. Suburbs as *natural ecological extensions*: suburbanisation as a natural process of organic and evolutionary growth; expansion takes place from the inside outward to the fringe, but is still tied to the urbanised core for jobs and services.

2. Suburbs as a means of *escapism*: as a means of escape from the health, housing and environmental problems of the industrial inner cities.

3. Suburbs as *macro-economic policy tools*: suburbs as Keynesian policy instruments of macro-economic management and regulation, and for generating local employment multipliers.

4. Suburbs as vehicles for *capital accumulation*: as a means for landowners, the financial sector and the property industry to capture the social surplus, deriving from the profits from the development of newly built suburban environments on the fringe.

5. Suburbs as a means of *social engineering*: as a means of rescuing the poor and the disadvantaged from themselves, and of re-establishing a traditional and presumed superior moral order of earlier times and communities.

6. Suburbs as the logical outcomes of *rational locators*, reflecting the rational decisions of firms and households seeking lower-cost locations and more efficient and less regulated landscapes, within a competitive urban environment.

7. Suburbs as maps of *consumer preferences and choices*, emphasising the dominant role in suburbanisation of the preferences of individual consumers for more space, new housing, social homogeneity and certain public goods.

8. Suburbs as *socio-political strategies*: Strategies building on manipulation of the political fragmentation of the metropolis, entrenched local autonomy and the demands for social exclusiveness.

9. Suburbs as *asylums*: as defensive strategies, driven by fear of others, of the inner city and by uncertainty over property values, and stressing security and exclusion.

10. Suburbs as *rural nostalgia*, reflecting a desire to return to the countryside and rural roots, but without also severing

Source: **L. Bourne (1996) Reinventing the suburbs: old myths and new realities** *Progress in Planning* **46(3), 163–84**

demands for local autonomy, social homogeneity and differential consumption of collective goods and services.[37] The final two perspectives return to the view of suburbs as a defensive strategy that is driven either by fear of 'others' who are different and who may pose a threat to a preferred lifestyle, or by a desire to recapture an assumed simple rural way of life but without losing the advantages of urban living.[38] Suburbs are open to all these interpretations, with the applicability of each ranging from place to place, and over time. Nowhere else is the postmodern message of difference, and the difficulty of generalisation, more relevant than in suburbia.

EXURBANISATION

The suburbanisation wave reaches its greatest extent in the phenomenon of extended suburbanisation or exurbanisation. Nelson (1992)[39] identified four principal factors to explain exurbanisation:

1. Continued deconcentration of employment and the rise of exurban industrialisation.
2. The latent anti-urban and rural location preferences of US households.
3. Improved technology that makes exurban living possible.
4. An apparent policy bias favouring exurban development over compact development.

These developments on the margins of suburbia represent a transition state between urban and rural life akin to the second-home phenomenon. Exurbia tends to be dominated by middle-class residents, many of whom commute long distances to work in the city or in the newer suburbs, but other groups are also present, including retirees and young households seeking social status, more land and new housing at a lower cost than is available in the suburbs. In the USA the **exurbs** have captured as much as one-quarter of recent national population growth and 60 per cent of recent manufacturing investment.[40] While the national population grew by 113.2 per cent between 1990 and 2000 population growth in exurbia averaged 31.4 per cent. In 2000 11 million people lived in the exurbs of larger metropolitan areas, with greatest numbers of exurbanites in the south (47 per cent of the nation's exurban population) and mid-west (24 per cent).[41] For some this heralds a 'post-suburban' era characterised by inner suburban population loss and relative income decline, an increase in suburban employment, a reduction in suburban out-commuting, an increase in exurban population and income, and increased farmland conversion to urban use.[42]

The acid test for any model is how well it corresponds with reality. The first three stages of population change indicated by the model shown in Figure 4.8 accord well with the pattern of urban development in North America and Western Europe. Urbanisation followed by central city decline and suburban growth have been characteristic features of the US city for several decades, and national urban systems in Europe also appear to have followed the model sequence.[43] There is, however, less evidence for the final stage of the model. This casts doubt on the hypothesised progression through all stages. Despite many examples of gentrification and 'urban renaissance' in cities of the MDRs, the weight of demographic evidence seems to indicate the continuing dominance of centrifugal trends within urban regions rather than a general shift into a reurbanisation stage.[44] In contrast, as we have seen, urbanisation remains the dominant process in the LDRs.

TYPES OF URBANISED REGIONS

The increasing scale of urbanisation, urban growth and development of national urban systems has given rise to a number of different forms of urbanised regions. An important distinction is between monocentric forms of urban development (e.g. the city region) in which there is a single dominant centre, and polycentric urban areas that comprise a regional system of cities with complementary functions connected by a network of transport and communication links:

1. *The city-region.* This is an area focused on the major employment centre in a region and encompassing the surrounding areas, for which it acts as the primary high-order service centre. The functional relationship between a city and its region was a key feature of central place theory (see Chapter 6). The city-region remains an appropriate description of monocentred urban areas of up to a million people found in the less densely populated parts of even the most highly urbanised countries. Variants employed for statistical purposes include functional urban regions (FURs) and standard metropolitan statistical areas (SMSAs) (see Chapter 2).

2. *Conurbation.* This is the term coined in 1915 by Geddes to describe a built-up area created by the coalescence of once-separate urban settlements.[45] With improvements in transportation and communications the functional influence of the conurbation has spread beyond the limits of the built-up area, so the term is now widely used in the UK and elsewhere to describe multi-nodal functional urban units. The functional relationships within a conurbation differ from those of a city-region; in essence, while there is a degree of dominance by the largest city, the other urban places also have their own functional linkages.

3. *The urban field.* This is a unit, similar to the conurbation, used in the USA. An urban field is generally regarded as a core urban area and hinterland of population at least 300,000, with an outer limit of two hours' driving time. Defined in this manner, urban fields range in population size from 500,000 to 20 million and cover one-third of the USA and 90 per cent of the national population. Urban fields are more spatially extensive than European conurbations, since they are based on higher levels of personal mobility. The southern California 'urban field' extends 150 miles from north to south

Plate 4.1 The exploding postmodern metropolis of Los Angeles CA with its restricted high-rise central business district surrounded by low-density suburbia reaching into the mountains

and includes Tijuana in Mexico (in the process creating a transnational city in which the largest 'Mexican' city is Los Angeles). The concept may become increasingly relevant for understanding the functional reality of urbanised regions outwith the USA as similar levels of mobility are achieved through improvements in transport and communications. The urban field is one form of polycentric urban region.[46] A second is the polynucleated metropolitan region or megalopolis.

4. *Megalopolis.* This is the term introduced by Gottmann in 1961 to describe the urbanised areas of the north-eastern seaboard of the USA encompassing a population of 40 million oriented around the major cities of Boston, New York, Philadelphia, Baltimore and Washington DC.[47] Gottmann subsequently defined a megalopolitan urban system as an urban unit with a minimum population of 25 million. The central importance of transactional activities (in terms of international trade, technology

and culture) would indicate a location at a major international 'breakpoint' (such as a port city). A megalopolis would typically have a polynuclear form but with sufficient internal physical distinctness for each constituent city to be considered an urban system in its own right. The cohesiveness of the megalopolitan system depends on the existence of high-quality communications and transportation facilities.[48] This megalopolitan phenomenon was identified initially in six zones: the archetype model of the north-eastern USA, the Great Lakes area extending from Chicago to Detroit, the Tokaido area of Japan centred on Tokyo–Yokohama and extending west to include Osaka–Kobe, the central belt of England running from London to Merseyside, the north-west European megalopolis focused on Amsterdam–Paris–Ruhr, and the area around Shanghai. Since then, twenty-six growth areas of the USA have exhibited megalopolitan patterns (Figure 4.9),

1 Metropolitan Belt	6 Southern Piedmont	13 Willamette Valley	20 Central Illinois
1A Atlantic Seaboard	7 North Georgia–	14 Central Oklahoma–	21 Nashville Region
1B Lower Great Lakes	South East Tennessee	Arkansas Valley	22 East Tennessee
2 California Region	8 Puget Sound	15 Missouri–Kaw Valley	23 Memphis
3 Florida Peninsula	9 Twin Cities Region	16 North Alabama	24 El Paso–
4 Gulf Coast	10 Colorado Piedmont	17 Blue Grass	Ciudad Juárez
5 East Central Texas–	11 St Louis	18 Southern Coastal Plain	
Red River	12 Metropolitan Arizona	19 Salt Lake Valley	

Figure 4.9 Megalopolises of the USA

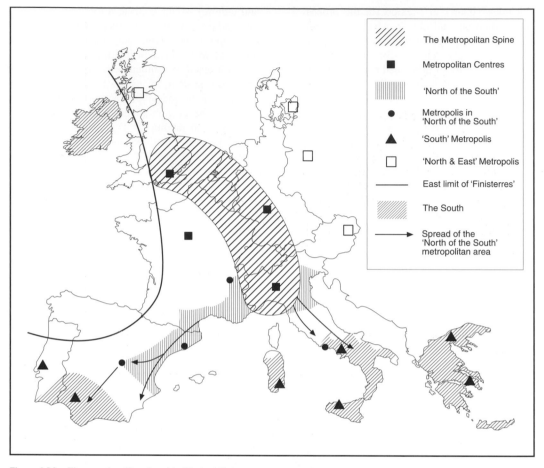

Figure 4.10 The megalopolitan trend in Western Europe

while similar trends are evident in Brazil (between Rio de Janeiro and São Paulo), in China[49] and in Europe[50] (Figure 4.10). At the international scale the notion of connectivity (though not necessarily physical contacts) among different cities has given rise to the concept of the trans-national sub-regional urban corridor – evident, for example, in the 1,500 km BESETO urban belt stretching from Beijing to Tokyo via Pyongyang and Seoul and containing 100 million people and 112 cities of over 200,000 inhabitants (Figure 4.11). An even larger international urban system is envisaged based on flows of goods and services, investments, information and people between major mainly coastal metropolitan cities in the Asian-Pacific region. This international regional city system contains a number of smaller scale urban corri-

dors such as BESETO, the Pearl River delta, and JABOTABEK.[51]

5. *Ecumenopolis.* This is the term employed by Doxiades in 1968 to describe a projected urbanised world or universal city by the end of the twenty-first century[52] (Figure 4.12). Although highly speculative, the ecumenopolis concept does focus attention on the potential consequences of unrestrained urban growth and underlines the importance that is currently being attached to the concept of sustainable urban development (see Chapter 30).

In the next chapter we switch our scale of analysis to provide a detailed examination of recent processes and patterns of urban change within the major regions of the world.

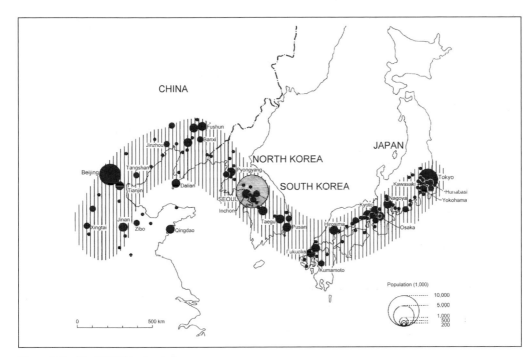

Figure 4.11　The BESETO urban belt

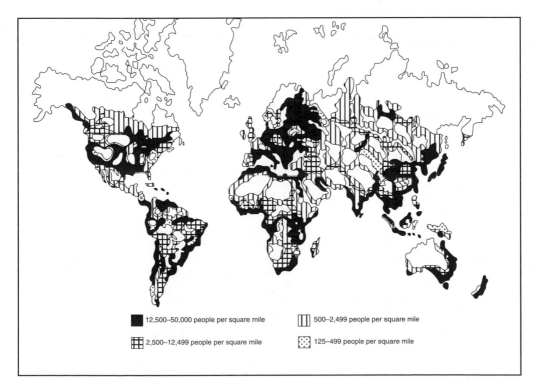

■ 12,500–50,000 people per square mile
▦ 2,500–12,499 people per square mile
▥ 500–2,499 people per square mile
⸬ 125–499 people per square mile

Figure 4.12　Doxiadis's concept of ecumenopolis: an urbanised world at the end of the twenty-first century
Source: C. Doxiadis (1968) *Ekistics* London: Hutchinson

FURTHER READING

BOOKS

A. Berube, B. Katz and R. Lang (2006) *Redefining Urban and Suburban America* Washington DC: Brookings Institution Press

A. Champion (2000) Urbanization, suburbanization, counterurbanization and reurbanization, in R. Paddison and W. Lever (eds) *Handbook of Urban Studies* Beverly Hills CA: Sage

W. Frey and A. Speare (1988) *Regional and Metropolitan Growth and Decline in the United States* New York: Russell Sage Foundation

J. Garreau (1988) *Edge City* New York: Doubleday

J. Gottmann (1961) *Megalopolis: The Urbanized Northeastern Seaboard of the United States* Cambridge MA: MIT Press

R. Harris and P. Larkham (1999) *Changing Suburbs: Foundation, Form and Function* London: Spon

United Nations (2006) *World Urbanization Prospects: The 2005 Revision* New York: United Nations

JOURNAL ARTICLES

M. Bontje and J. Burdack (2005) Edge cities European style: examples from Paris and the Randstad *Cities* 22(4), 317–30

L. Bourne (1996) Reinventing the suburbs: old myths and new realities *Progress in Planning* 46(3), 163–84

J. Elliot (1997) Cycles within the system: metropolitanisation and internal migration in the US, 1965–1990 *Urban Studies* 34(1), 21–41

A. Fielding (1989) Migration and urbanisation in Western Europe since 1950 *Geographical Journal* 155, 60–9

H. Geyer and T. Kontuly (1993) A theoretical foundation for the concept of differential urbanisation *International Regional Science Review* 15(12), 157–77

T. Gweba (2006) Towards a theoretical exploration of the differential urbanisation model in sub-Saharan Africa: the Botswana case *Tijdschrift voor Economische en Sociale Geografie* 97(4), 418–33

B. Hanlon, T. Vicino and J. Short (2006) The new metropolitan reality in the US *Urban Studies* 43(12), 2129–43

W. Lucy and D. Phillips (1997) The post-suburban era comes to Richmond: city decline, suburban transition, and exurban growth *Landscape and Urban Planning* 36, 259–75

A. Nucci and L. Long (1996) Spatial and demographic dynamics of metropolitan and non-metropolitan territory in the United States *International Journal of Population Geography* 1(2), 165–81

M. Pacione (1980) Differential quality of life in a metropolitan village *Transactions of the Institute of British Geographers* 5(2), 185–206

KEY CONCEPTS

- Urbanisation
- Urban growth
- Urbanism
- Megacities
- Million cities
- Metacities
- Urban bias
- Logistic-growth urbanisation curve
- Urbanisation cycle
- Primate city
- Polarisation reversal
- Demographic transition
- Differential urbanisation
- Suburbanisation
- Exurbanisation
- Counterurbanisation
- Reurbanisation
- Metropolitan village
- Edge city
- Overurbanisation
- City region
- Conurbation
- Urban field
- Megalopolis
- Ecumenopolis

STUDY QUESTIONS

1. Over the course of the second half of the twentieth century a world in which most people lived in rural areas was transformed into a predominantly urban world. Examine the changing distribution of the world's urban population over that period.
2. One of the most striking features of the global urban pattern is the degree to which the urban population live in giant cities. Identify and explain the distribution of the world's megacities and 'million cities'.
3. With reference to the theory of differential urbanisation, outline the main stages of the urbanisation cycle.
4. The concept of a cycle of urbanisation has also been employed to study growth patterns within individual urban agglomerations. Explain the four main stages in this model.
5. With the aid of particular examples, examine the evidence for urbanisation, counterurbanisation, suburbanisation or reurbanisation in any country with which you are familiar.
6. Identify the principal characteristics and functions of suburbs. How do these correspond to suburban areas in your locality?
7. Identify the locations of boomburgs in the USA. Describe their characteristics and current and likely future patterns of development.
8. Examine the character of the edge city form of urban development in either the USA or UK–Europe.

PROJECT

Use census data to plot the population of your city over time. Identify and explain the significant trends. Use disaggregated data to determine whether certain parts of the metropolitan area have lost population, gained population or remained static since 1950. Explain any changes revealed by your analysis.

5

Regional Perspectives on Urbanisation and Urban Change

Preview: urbanisation and urban change in world regions; North America; Latin America and the Caribbean; Western Europe; East and Central Europe; Asia and the Pacific; Africa

INTRODUCTION

This chapter will examine the diverse processes of urbanisation and urban change within the world's major regions. In our discussion we shall examine the dynamics of recent urban change in six global regions. In each region we shall identify the main social, economic and demographic forces underlying urban change and the effects of these on the pattern of urban development at present and in the immediate future.

NORTH AMERICA

Between 1950 and 1990 the population of the USA increased by 65 per cent to reach 248.7 million; in Canada it increased by 95 per cent to reach 27.3 million. The proportion of the population living in urban centres grew from 64 per cent to 82 per cent in the USA and from 61 per cent to 77 per cent in Canada. Over the period, major changes occurred in the distribution of population, and in the relative importance of different cities. The dominance of the larger industrial cities and older eastern gateway ports diminished and complex new urban hierarchies emerged. Economic growth was accompanied by changes in methods of production, communication and transportation, modification of traditional occupational structures and of the spatial division of labour, and an increasingly uneven distribution of wealth, all of which influenced the structure and relative importance of regional and urban economies.

DEMOGRAPHIC CHANGE

The demographic basis of the post-Second World War urban transformation in North America is the outcome of two principal components:

1. A series of socio-demographic shifts and internal migration movements.
2. Immigration.

Both the USA and Canada experienced a 'marriage and baby boom' between 1948 and 1963. The resulting population expansion contributed to the rapid growth of the consumer economy, to increased housing demand and new residential construction, and to the evolution of new urban forms through widespread suburbanisation. Parallel to these demographic changes were equally dramatic changes in household and family composition, especially the proliferation and diversification of household types. Average household size declined from 4.0 persons to 2.7 persons between 1950 and 1970, and is even lower in metropolitan areas. In the USA over 20 per cent of all households have only one person and 13 per cent are female-headed single-parent households. Fewer than half of all households are now family households in the sense of having members related by blood or marriage. The impact of these shifts in living arrangements on metropolitan development include a substantially greater consumption of housing space, urban land and public resources *per capita* and per household than would otherwise be the case.[1]

As rates of natural increase in population declined after 1963 they also became more uniform across the continent. As a result, internal migration flows assumed greater importance as determinants of urban growth and population redistribution, and as a source of social change within localities. In any given year 18 per cent of North Americans change their place of residence, and after five years over 50 per cent have moved. With such high mobility levels, the potential for redistributing population and economic activity is high – and this contributes to a degree of uncertainty regarding future settlement patterns.

Another consequence of the declining rate of natural increase in population has been the growing importance of immigration as a factor in population change. During the 1970s and 1980s immigration flows represented 20 per cent of the USA's and 30 per cent of Canada's total population growth. By the early 1990s the USA was admitting 1.1 million official immigrants, and Canada 200,000 each year. The origins of immigrants have shifted away from long-established sources in Europe to countries in the South, especially Mexico and Latin America (for the USA) and Asia. In 2000, of the 28.4 million foreign-born people living in the USA (representing 10.4 per cent of the total population), 51 per cent were born in Latin America, 25.5 per cent in Asia, 15.3 per cent in Europe and the remaining 8.1 per cent in other regions of the world. The foreign-born population from Central America (including Mexico) accounted for nearly two-thirds of the total foreign-born population. As we saw in Chapter 4, nearly one in two of the foreign-born live in a central city in a metropolitan area (compared with 27.5 per cent of the native-born population). The focused residential geography of Latino and Asian peoples (for example in cities such as Miami and Los Angeles), along with the long-standing urban concentration of blacks and redistribution of whites from high-immigration metropolitan areas indicates an increasing 'demographic balkanisation' of the US population in particular regions and cities. Significantly for the social composition of cities, many of the new immigrant populations are visibly and culturally distinct from earlier European immigrants.

Plate 5.1   The Banco Popular in the Harlem district of New York testifies to the growing multicultural identity of US cities

REGIONAL AND URBAN CHANGE

As we saw in Chapter 3, the system of metropolitan areas in the USA has expanded and become more complex in the post-war period. By 1990 there were 284 metropolitan areas with 193 million residents, or 77.5 per cent of the total US population. Half the US population now lives in large metropolises, with consolidated metropolises such as the New York–New Jersey–Connecticut CMSA being of megalopolitan proportions (Table 5.1). Figure 5.1 depicts the largest metropolitan areas in Canada and the USA.

The impact of population mobility on urban patterns since the Second World War has exhibited regional variation, with certain regions and urban areas being prime destinations for inflows, whereas others have lost population. Most significantly, since migrants, on average, tend to be younger (excluding retirement migrations) and better educated than non-migrants, these population flows also transfer considerable wealth, market power and labour skills to the receiving places. They also shift the locations of future generations of population growth. During the 1980s the contribution of net migration (combining immigration and internal migration) to metropolitan population growth in the USA varied widely

from a low of –8.9 per cent in Pittsburgh to +27.8 per cent in Phoenix (one of the world's most rapidly growing cities in the decade). For internal migration only, the contributions ranged from –11.1 per cent for Chicago to +25.2 per cent in Tampa. In terms of international immigration, most migrants are attracted by job opportunities and follow earlier migration paths established by friends and kin. In Canada three-quarters of all recent immigrants have settled in the three largest metropolitan areas, namely Toronto, Montreal and Vancouver, while in the US gateway cities such as New York, Los Angeles, San Francisco and the border cities of Texas and the south-west have attracted most of these flows. This concentration has transformed the social structure and ethno-cultural character of the recipient cities; in some cases, such as Los Angeles and Miami, the 'new minority' constitute a majority of the central city population.

The characterisation of these changing urban growth patterns since 1950 in terms of polar extremes (in the USA between the sunbelt and frostbelt or between the sun and gunbelt of the southern states and the northern rustbelt) is generally useful but also potentially misleading, since growth has been uneven within both of these broad regions. We can shed more light on these trends by grouping metropolitan areas according

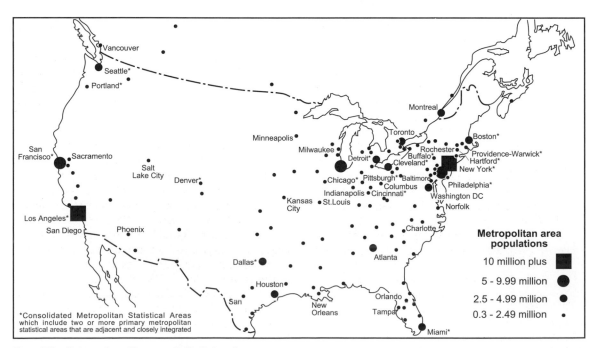

Figure 5.1   Major metropolitan areas in North America

TABLE 5.1 RANKING OF US METROPOLITAN AREAS BY POPULATION, 2000

Rank	Metropolitan area	Population (000)	% change	
			1980–90	1990–2000
1.	New York CMSA	21,200	3.4	9.6
2.	Los Angeles CMSA	16,374	26.3	12.7
3.	Chicago CMSA	9,158	1.5	11.1
4.	Washington–Baltimore CMSA	7,608	16.1	13.1
5.	San Francisco–Oakland–San José	7,039	15.4	12.6
6.	CMSA	6,189	4.3	5.0
7.	Philadelphia CMSA	5,819	–2.0	10.9
8.	Boston CMSA	5,456	6.5	0.0
9.	Detroit CMSA	5,222	32.5	29.4
10.	Dallas–Fort Worth CMSA	4,670	19.6	25.2
11.	Houston CMSA	4,112	33.0	38.9
12.	Atlanta MSA	3,876	20.7	21.4
13.	Miami CMSA	3,555	23.2	19.7
14.	Seattle MSA	3,252	40.0	45.3
15.	Phoenix MSA	2,946	–2.6	3.0
16.	Cleveland–Akron CMSA	2,969	15.5	16.9
17.	Minneapolis–St Paul MSA	2,814	34.1	12.6
18.	San Diego MSA	2,604	3.2	4.4
19.	St Louis MSA	2,396	28.1	15.9
20.	Tampa–St Petersburg MSA Pittsburgh MSA	2,358	–6.8	–1.5

Source: **US Census Bureau statistics**

Notes: **A consolidated metropolitan statistical area (CMSA) includes two or more primary metropolitan statistical areas (i.e. two or more large cities), while a metropolitan statistical area (MSA) is centred on one large city**

to their main economic bases. Doing so identifies a number of factors underlying urban growth:

1. The fastest-growing urban areas in the USA in the 1980s and early 1990s were, typically, smaller metropolitan areas with economies based on retirement and recreational pursuits (e.g. Fort Pierce FL). The population of these communities grew by 46.9 per cent in the 1980s, compared with an average of 11.9 per cent for all metropolitan areas. Most of these places are in Florida, Arizona or Nevada.[2]
2. A second group, typically larger places, included the finance and service centres acting as 'control and management' centres for the national economy. US metropolitan areas in this category (e.g. New York) grew by 14.5 per cent during the 1980s.
3. Third, cities involved primarily in public administration as national, state or provincial capitals, or as military centres showed an average growth rate of 17.5 per cent in the 1980s.
4. Manufacturing-based cities grew by only 1.5 per cent on average while mining and resource communities declined by 2 per cent on average over the decade.
5. Other cities have grown as regional service centres (e.g. Dallas and Atlanta) or as hosts for expanding high-technology industries (e.g. Silicon Valley in San José, Software Valley near Salt Lake City, and Kanata near Ottawa).

Analysis of the 2000 census also suggests that metropolitan areas in the USA may be distinguished by the degree to which they attracted or lost international and domestic migrants during the 1990s. The largest metropolitan areas gained the greatest number of migrants from abroad over the period 1995–2000 but lost most domestic migrants. As Table 5.2 shows, nine of the ten largest metropolitan areas adhered to this pattern. The New York metropolitan region gained almost a million migrants from abroad but at the same time lost 874,000 residents. The five-county Los Angeles region gained nearly 700,000 immigrants but lost 550,000 domestic migrants. For some cities, the pattern of large immigration gains and significant domestic migration losses is not new. Over several decades New

TABLE 5.2 NET DOMESTIC MIGRATION AND MOVERS FROM ABROAD, 1995–2000, FOR LARGEST US METROPOLITAN AREAS

Metropolitan area	Total population 2000	Net domestic migration [a]		Movers from abroad [b]	
		No.	Rate [c]	To central cities	To suburbs
New York–Northern New Jersey–Long Island, NY–NJ–CT–PA CMSA	21,199,865	−874,028	−44.4	614,057	389,602
Los Angeles–Riverside–Orange County, CA CMSA	16,373,645	−549,951	−36.8	324,013	375,560
Chicago–Gary–Kenosha, IL–IN–WI CMSA	9,157,540	−318,649	−37.6	172,597	150,422
Washington–Baltimore, DC–MD–VA–WV CMSA	7,608,070	−58,849	−8.6	65,837	234,429
San Francisco-Oakland–San José, CA CMSA	7,039,362	−206,670	−32.3	194,220	179,649
Philadelphia–Wilmington–Atlantic City, PA–NJ–DE–MD CMSA	6,188,463	−83,539	−14.5	58,131	69,790
Boston–Worcester–Lawrence, MA–NH–ME–CT CMSA	5,810,100	−44,973	−8.5	99,790	93,708
Detroit–Ann Arbor–Flint, MI CMSA	5,456,428	−123,009	−24.2	36,179	72,796
Dallas–Fort Worth, TX CMSA	5,221,801	148,644	33.6	151,679	79,815
Houston–Galveston–Brazoria, TX CMSA	4,669,571	−14,377	−3.5	138,826	75,442
Atlanta, GA MSA	4,112,198	233,303	68.4	15,975	146,997
Miami–Fort Lauderdale, FL CMSA	3,876,380	−93,774	−27.4	60,493	239,412
Seattle–Tacoma–Bremerton, WA CMSA	3,554,760	39,945	12.6	47,001	75,765
Phoenix–Mesa, AZ MSA	3,251,876	245,159	93.6	104,609	30,408
Minneapolis–St Paul, MN–WI MSA	2,968,806	34,207	12.9	31,145	34,975
Cleveland–Akron, OH CMSA	2,945,831	−65,914	−23.7	13,969	22,288
San Diego, CA MSA	2,813,833	−6,108	−2.4	63,695	45,127
St. Louis, MO–IL MSA	2,603,607	−43,614	−17.9	13,915	21,432
Denver–Boulder–Greeley, CO CMSA	2,581,506	93,586	42.3	44,472	49,498
Tampa–St Petersburg–Clearwater, FL MSA	2,395,997	103,375	49.5	24,728	42,936

Source: Adapted from W. Frey (2003) *Metropolitan Magnets for International and Domestic Migrants* Washington, DC: Brookings Institution

Notes: [a] A negative value for net migration or the net migration rate is indicative of net out-migration, meaning that more migrants left an area than entered it. Positive numbers reflect net in-migration to an area.

[b] This category includes movers from foreign countries, as well as from Puerto Rico, US Island Areas and US minor outlying islands.

[c] The net migration rate is based on an approximated 1995 population, which is the sum of people who reported living in the area in both 1995 and 2000 and those who reported living in the area in 1995 but now live elsewhere. The net migration rate divides net migration, in-migration minus out-migration, by the approximated 1995 population and multiples the result by 1,000.

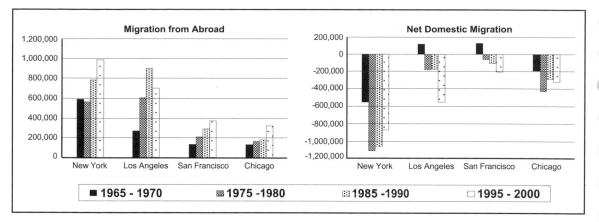

Figure 5.2   Migration flows for selected US immigrant magnet metropolitan areas
Source: W. Frey (2003) *Metropolitan Magnets for International and Domestic Migrants* Washington DC: Brookings Institution

York, an established immigrant port-of-entry city, has gained migrants from abroad even as it has lost significant numbers of residents to other parts of the USA. In fact, the domestic migration losses sustained by New York from 1995 to 2000 were less severe than the city experienced in the late 1970s and late 1980s (Figure 5.2). In other cities, including Los Angeles and San Francisco, net domestic out-migration accelerated during the 1990s. This trend established these two west-coast immigrant ports of entry as 'redistributors' of population to fast-growing metropolitan areas in the interior USA; a role played by New York and Chicago in earlier decades.

Two main explanations have been proposed for strong domestic out-migration from the largest immigration magnet metropolitan areas. One views the process as an emerging national version of the 'white flight' phenomenon that characterised local city-to-suburb movements, especially in the 1950s and 1960s. While this remains a factor underlying metropolitan population movements the ethnic composition of out-migration also reflects the overall ethnic composition of the metropolitan area. Out-migrants from New York and Chicago are more likely to be white than those leaving Los Angeles or San Francisco (Figure 5.2). A second hypothesis relates out-migration to displacement of lower-skilled workers from immigrant magnet areas by new immigrants who compete disproportionately in lower-skilled labour markets. Evidence from cities, such as Los Angeles and San Francisco, suggests that highest rates of domestic out-migration involved adults without a college degree, while conversely in-migrants tended to be college graduates.[3]

The inflow of immigrants has also pushed up house prices to the benefit of black homeowners in south central Los Angeles. One result has been that between 2000 and 2006 the black population of towns such as Victorville, two hours drive from central Los Angeles, rose from 11,900 to 24,500. Other Los Angeles suburbs such as Palmdale and Lancaster reveal an incipient process of 'black flight' from inner city ghettos such as Compton and Crenshaw.

Geographically, at the national scale the metropolitan areas that gained most in-migrants from elsewhere in the USA over the period 1995–2000 are located in the south-east and non-California west. Movement of new residents into these areas followed and fuelled the growth of new industries and expanding urban and suburban developments in cities like Phoenix, Atlanta and Las Vegas. These domestic migration magnets also attracted new immigrant populations, in response to demands for low-skilled labour (Box 5.1).

Within metropolitan areas immigrants have boosted urban core populations in cities such as New York, San Francisco, Washington DC and Boston (Table 5.2), offsetting their losses of domestic migrants to the suburbs and elsewhere, and contributing to a degree of reurbanisation (see Chapter 4). At the same time, core counties in the Midwest and rustbelt continue to sustain some of the greatest domestic out-migration. The city of St Louis lost 105,000 domestic migrants over the period 1995–2000 but received fewer than 12,000 migrants from abroad. Some cities have considered strategies to attract in-migrants to reinvigorate declining urban neighbourhoods.

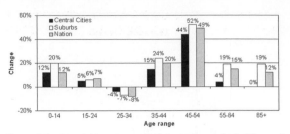

Figure 5.3   Population change by age group, central cities and suburbs of large metropolitan areas, 1990–2000

Source: Adapted from W. Frey (2003) *Boomers and Seniors in the Suburbs* Washington DC: Brookings Institution

While not all migrants from abroad head primarily for core urban counties, in general population gains in peripheral counties occur almost entirely from domestic migration. The disparity between growing suburban and exurban counties, where domestic migration dominates, and inner metropolitan counties dependent on migration from abroad to offset population decline is characteristic of both domestic migrant magnet metropolises, such as Atlanta and Denver, and immigrant magnet metropolises, such as New York and Washington DC.

Analysis of the 2000 census also revealed the changing age composition of US metropolitan areas, and highlighted the 'middle-aging' and 'greying' of many suburbs.[4] In 2000, for the first time, more than half of the US population (50.5 per cent) was aged 35 years or over (cf with 46 per cent in 1990). The 1990–2000 suburban growth rate for the 35 and over population was 28 per cent compared to 15 per cent for the central cities (Figure 5.3). It would appear that, in contrast to the traditional image of suburbia as a haven for young families, most US suburbs are now 'middle-aging' due to the pressure of post-war baby-boomers (now 35–54 years old); while other suburbs are 'greying' as the parents of the boomers, who

settled here several decades ago, age in situ. Such demographic trends clearly impact on local demand for goods and services. Suburbs with affluent boomers now enriching the tax base and younger populations continuing to enter may expect to 'middle age' more comfortably and enjoy high levels of services and amenities. Conversely, suburbs with less affluent aging populations and declining or at best moderately growing younger populations face a major challenge in meeting senior demands for medical and social services (Table 5.3).

Age differentiation among US suburbs is also related to ethnic difference, particularly in metropolitan areas such as New York and Los Angeles where there is a greater percentage of ethnic minority females of child-bearing age compared to the US population in general. One result is the emergence of a 'racial generation gap' in many suburbs where younger populations are largely and increasingly minority and older populations largely white. This reflects a phenomenon that has been evident in many US central cities. The overlay of ethnic and age differences and divergence of interests of young adult parents of largely minority child populations (with concerns about schools, parks and public safety) and of a predominantly white middle-aged or senior population (who prefer lower property taxes, elder care services and facilities for the disabled) adds another dimension to the social geography of suburban America.

It is likely that future patterns of urban growth and change in North America will continue to reflect the high levels of population and economic mobility. Extrapolation of trends evident in the 1990s suggests that among the most significant factors underlying the future metropolitan pattern of North America are immigration, an ageing population, lifestyle changes, the uncertain growth in income, revisions in government welfare policies, technological innovations, trade liberalisation and increased global competition.

TABLE 5.3 SUBURBS WITH GREATEST GROWTH IN 35-AND-OVER POPULATION, AND DECLINE IN UNDER-35 POPULATION, 1990–2000

Rank	Suburbs	% change 35-and-over population	% change under-35 population
Suburbs with greatest 35-and-over growth			
1	Las Vegas, NV–AZ MSA	89.8	75.4
2	El Paso TX MSA	83.2	39.5
3	Austin TX MSA	74.8	42.4
4	Phoenix–Mesa AZ MSA	70.9	47.5
5	Colorado Springs CO MSA	63.4	17.7
6	McAllen–Edinburg–Mission TX MSA	57.7	50.7
7	Dallas TX PMSA	56.0	28.2
8	Atlanta GA MSA	55.7	35.5
9	Jacksonville FL MSA	53.9	16.2
10	Raleigh–Durham NC MSA	53.1	32.5
Suburbs with greatest under-35 declines			
1	Syracuse NY MSA	17.3	−12.9
2	Pittsburgh PA MSA	9.7	−11.2
3	Scranton–Hazleton PA MSA	7.2	−11.0
4	Charleston–North Charleston SC MSA	26.8	−10.9
5	Buffalo NY MSA	12.8	−9.6
6	Springfield MA NECMA	14.9	−9.0
7	Youngstown–Warren OH MSA	12.9	−8.7
8	Albany–Schenectady–Troy NYMSA	17.3	−8.3
9	Dayton–Springfield OH MSA	13.9	−7.5
10	Hartford CT NECMA	16.1	−7.4

Source: adapted from W. Frey (2003) *Boomers and Seniors in the Suburbs: Aging Patterns in Census 2000*
Washington DC: Brookings Institution

LATIN AMERICA AND THE CARIBBEAN

In 1990 the population of Latin America and the Caribbean totalled 440 million, having doubled in size since 1960. Over this period the region shifted from being predominantly rural to predominantly urban. With 71.4 per cent living in urban areas in 1990, this represented a level of urbanisation comparable to that of Europe. There is also a heavy concentration of population in large cities, with, in 1990, more people living in 'million cities' in the region than in rural areas.[5] By 1990 most countries with more than a million inhabitants had more than half their population in urban areas. By 2001 these urbanisation trends had strengthened, with more than three-quarters of the region's population living in cities (Table 5.4). The rapidity of urbanisation in the region is related to the scale of economic growth, with Brazil, Mexico, Colombia and the Dominican Republic experiencing both rapid economic growth and high levels of urbanisation.

In terms of the level of urbanisation we can identify three groups of nations in the region:

1. The most urbanised, with over 80 per cent of their population in urban areas, includes Venezuela and the three Southern Cone nations of Argentina, Chile and Uruguay with a long tradition of urban development based on rapid immigration from Europe in the late nineteenth and early twentieth centuries, and a land ownership structure which provided little opportunity for immigrants to acquire farmland, thereby ensuring their settlement in the cities.
2. The second, with between 50 per cent and 80 per cent in urban areas, includes most of the countries that experienced rapid urban and industrial development between 1950 and 1990 – Brazil, Mexico, Colombia, Ecuador, and the Dominican Republic – as well as others such as Bolivia, Peru, and Trinidad and Tobago (Table 5.4).
3. The third, with less than 50 per cent of the population in urban areas, comprises the less populous countries such as Paraguay, Haiti and Costa Rica.

By 1990 the region had 300 million urban inhabitants and thirty-six 'million cities', including three with

TABLE 5.4 LATIN AMERICA: URBAN POPULATIONS FOR 1990 AND 2001 AND URBAN CHANGE SINCE 1950

Country	Urban population (000)		% urban			Change % urban	
	1990	2001	1950	1990	2001	1950–90	1990–2001
Caribbean							
Cuba	7,801	8,482	49.4	73.6	75.5	24.2	1.9
Dominican Republic	4,293	5,615	23.7	60.4	66.0	36.7	5.6
Haiti	1,855	3,004	12.2	28.6	36.3	16.4	8.7
Jamaica	1,217	1,470	26.8	51.5	56.6	24.7	5.1
Puerto Rico	2,518	2,987	40.6	71.3	75.6	30.7	4.3
Trinidad & Tobago	854	969	63.9	69.1	74.5	5.2	5.4
Central America							
Costa Rica	1,429	2,448	33.5	47.1	47.8	13.6	59.5
El Salvador	2,269	3,935	36.5	43.9	61.5	7.4	17.6
Guatemala	3,628	4,688	29.5	39.4	39.9	9.9	0.5
Honduras	1,985	3,351	17.6	40.7	53.7	23.1	13.0
Mexico	61,335	74,846	42.7	72.6	74.6	29.9	2.0
Nicaragua	2,197	2,943	34.9	59.8	56.5	24.9	−3.3
Panama	1,240	1,639	35.8	51.7	56.5	15.9	4.8
South America							
Argentina	28,158	33,119	65.3	86.5	88.3	21.2	1.8
Bolivia	3,665	5,358	37.8	55.8	62.9	18.0	7.1
Brazil	110,789	141,041	36.0	74.6	81.7	38.6	7.1
Chile	10,954	13,254	58.4	83.3	86.1	24.9	2.8
Colombia	22,604	32,319	37.1	70.0	75.5	32.9	5.5
Ecuador	5,625	8,171	28.3	54.8	63.4	26.5	8.6
Paraguay	2,109	3,194	34.6	48.9	56.7	14.3	7.8
Peru	15,068	19,084	35.5	69.8	73.1	34.3	3.3
Uruguay	2,751	3,097	78.0	88.9	92.1	10.9	3.3
Venezuela	17,636	21,475	53.2	90.4	87.2	37.2	3.2
Latin America and the Caribbean (total)	314,161	399,269	41.6	71.4	75.8	29.8	4.4

Source: adapted from United Nations (2002) *World Urbanization Prospects: The 2001 Revision* New York: United Nations

more than 10 million (Figure 5.4). The region also had two of the world's five largest cities (São Paulo and Mexico City) and in total eight of the world's fifty largest cities. São Paulo is the largest city, reflecting its dominant economic role within the region's largest economy, with Mexico City second largest. While the region's urban population increased more than twentyfold between 1900 and 1990, most of the major urban centres of today were founded by the Spanish and Portuguese, with some, such as Mexico City and Quito, being even older pre-Colombian cities.[6]

However, in Brazil and Mexico there was a reordering of the urban system in the twentieth century during which many small cities emerged as prominent urban nodes. In Brazil, Brasilia was a political creation in 1958, while Porto Alegre was an unimportant town in 1800. Belo Horizonte was created as a new city at the end of the nineteenth century, while neither Fortaleza nor Curitiba was an important town before they became state capitals in the nineteenth century. In Mexico, too, the restructuring of the urban hierarchy has created new cities away from long-established centres of economic and political power (Box 5.2). This reordering of the urban system is more comparable to that in the USA than to what has happened in other countries in the region. Restructuring may also affect the internal arrangement of cities.

In general, all major metropolitan areas experience a decentralisation of population and of production as they grow. As we have seen, this process has gone

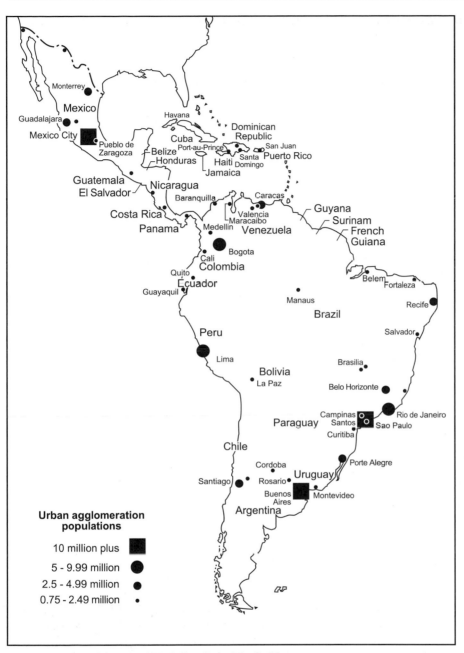

Figure 5.4   Major urban centres in Latin America and the Caribbean

furthest in the USA, where decentralisation within a core region is commonplace, but it is also evident in some parts of Latin America. In the Mexico City region mega-urbanisation is linked to a process of territorial restructuring that is creating a polycentric extended metropolitan region. The expanded urban system and integrated road network have stimulated flows of people, commodities and capital among the constituent urban centres, and have tended to break down the differences between rural and urban activities.[7]

BOX 5.2

Urban restructuring and the effect of the Mexico–USA border

The distribution of population and of urban population in Mexico has been changed significantly by the economic interaction between settlements in its north and the USA. The increasing population concentration in the north of Mexico is strongly associated with the development of the *maquila* industries there. This increase in population was strongly concentrated close to the border; the population in the thirty-six municipalities that adjoin the USA grew fourteenfold between 1930 and 1990, from 0.28 million inhabitants in 1930 to 2.9 million in 1980 and close to 4 million by 1990. These industries originated in the mid-1960s, when the government of Mexico began a programme to promote industrial development in the border area. This permitted Mexican and foreign-owned factories within the border area to import machinery, materials and components without paying tariffs as long as the goods produced were re-exported. Most of the *maquila* industries that developed were owned by US companies, and they could also take advantage of the US tariff regulations. In 1967 there were fewer than 100 plants with around 4,000 workers; by the early

1990s there were 2,000, employing more than half a million people. The devaluation of the Mexican currency against the US dollar in the 1980s also boosted industrial and agricultural exports. Since the early 1970s *maquila* industries have been allowed to set up in the interior and since 1989 to sell products in the domestic market. During the 1980s some cities away from the frontier were attracting major investments – for instance, new motor vehicle export plants were set up in Chihuahua, Hermosillo and Saltillo. The table lists the border towns that to date have attracted most *maquila* industries and the 'twin city' in the United States with which they are connected. Of these, Ciudad Juárez and Tijuana concentrated most *maquila* industries in 1990. By 2000 the border cities accommodated almost 1,500 *maquiladora* assembly plants (40 per cent of the national total), employing over 600,000 people (50 per cent of national *maquiladora* employment), the greatest concentration being in Cuidad Juarez (with 308 plants employing 250,000, or 10 per cent of city employees) and Tijuana (with 779 plants employing 185,000, or 14 per cent of city employment).

Urban centre	Population (000)			Compound growth rates (%)						Partner city in USA
	c. 1900	1950	1990	1900–30	1930–50	1950–60	1960–70	1970–80	1980–90	
Ciudad Juárez	8.2	124.0	807.0	5.4	5.9	7.8	5.3	2.2	4.0	El Paso
Tijuana	0.2	60.0	755.0	12.6	10.3	9.8	8.3	2.4	5.8	San Diego
Nuevo Laredo	6.5	57.7		4.1	5.0	4.9	5.1	2.8		Laredo (and to San Antonio, Houston)
Reynosa	1.9	34.1	296.0	3.1	10.3	8.1	6.6	3.3	4.3	McAllen
Matamoros	8.3	45.8	319.0	0.5	8.1	7.3	4.3	3.0	5.4	Brownsville
Mexicali	0.0	65.7	637.0		7.7	10.6	4.4	2.1	6.4	Calexico

Sources: United Nations Centre for Human Settlements (1996) *An Urbanising World* Oxford: Oxford University Press; S. Pena (2005) Recent developments in urban marginality along Mexico's northern border *Habitat International* 29, 285–301

SMALL AND INTERMEDIATE URBAN CENTRES

Notwithstanding the number of 'million cities' in most Latin American countries, a considerable proportion of the urban population live in urban centres other than the large cities. The 1991 census in Argentina showed that 46.5 per cent of the national population lived in urban centres with fewer than 1 million inhabitants, including 18 per cent in urban places with fewer than 100,000.

There are great contrasts in the size, growth rate and economic base among the urban centres with fewer than 1 million inhabitants. These include thousands of small market towns and service centres with a few thousand residents, as well as some of the region's most prosperous and rapidly growing cities over the past twenty years. The latter include:

1. Cities that served agricultural areas producing high-value crops for export (e.g. Zamora in Mexico). The pattern of land ownership is a major

influence on the extent to which high-value crop production stimulates local urban development. Whereas a large number of small but prosperous farmers with intensive production can promote local urban development, large land holdings or plantations divert much of the economic stimulus to more distant cities, including those overseas.

2. Cities that attract substantial numbers of tourists. For example, Cuautla in Mexico grew from being a small market town with 18,000 inhabitants in 1940 to a city of 120,000 in 1991 as a result of the growing importance of tourism.

It is difficult to predict the scale and nature of urban change in Latin America given that it is so dependent on economic performance. For countries that sustain rapid economic growth, urbanisation is likely to continue. Whatever the economic performance, however, the dominance of the region's largest metropolitan areas is likely to decline as a result of the emergence of new cities with a comparative advantage. These include important centres of tourism, and cities well located to attract new investment in export-oriented manufacturing or to benefit from forward and backward linkages from high-value export agriculture. In many of the higher-income countries in Latin America the factors that stimulated urban decentralisation in North America and Europe, including the development of good-quality inter-regional transport and communications systems, will also exert an influence on urban structure.

WESTERN EUROPE

Western Europe is highly urbanised, with over 75 per cent of the population resident in urban areas (Table 5.5). Consequently, between 1980 and 1995 the region exhibited one of the smallest increases in the level of urbanisation of all major world regions. As in North America, changes in household size and composition and in age structure have had an important influence on housing markets and settlement patterns.

TABLE 5.5 URBAN POPULATIONS AND LEVELS OF URBANISATION FOR ALL WEST EUROPEAN COUNTRIES WITH 1 MILLION OR MORE INHABITANTS

Country	Urban population (000)		% Urban	
	1990	2001	1990	2001
Northern Europe				
Denmark	4,357	4,538	84.8	85.1
Estonia	1,131	955	71.8	69.4
Finland	3,063	3,031	61.4	58.5
Ireland	1,993	2,276	56.9	59.3
Latvia	1,902	1,437	71.2	59.8
Lithuania	2,552	2,532	68.8	68.6
Norway	3,068	3,365	69.3	75.0
Sweden	7,112	7,358	83.1	83.3
United Kingdom	51,136	53,313	89.1	89.5
Southern Europe				
Greece	6,413	6,408	62.6	60.3
Italy	38,050	38,565	66.7	67.1
Portugal	3,303	6,601	33.5	65.8
Spain	29,592	31,073	75.4	77.8
Western Europe				
Austria	4,267	5,444	55.4	67.4
Belgium	9,606	9,997	96.5	97.4
France	41,218	44,903	72.7	75.5
Germany	67,699	71,948	85.3	87.7
The Netherlands	13,262	14,272	88.7	89.6
Switzerland	4,070	4,826	59.5	67.3
Total (including all countries)	**94,889**	**14,161**	**76.6**	**71.7**

Source: adapted from United Nations (2002) *World Urbanization Prospects: The 2001 Revision* New York: United Nations

Among the most significant economic changes were the rapid growth in the financial-services sector, which boosted the economy of certain key cities, and the continued decline in most cities that had traditionally specialised in heavy industry and port-based activities.

DEMOGRAPHIC CHANGES

Northern, southern and western Europe have long had the slowest population growth rates of any of the world's regions, and there has been a steady decline in annual average population growth rates over recent decades. The twelve countries that made up the EU in 1994 saw their annual average growth rate of 0.96 per cent in 1960–5 fall to 0.71 per cent, 0.61 per cent, 0.37 per cent and 0.27 per cent in the next four five-year periods. The slight increase to 0.34 per cent for the period 1985–90 was due entirely to net immigration, which rose from 0.5 per thousand in 1985–90 to 1.6 per thousand in 1985–90.[8] The very low population growth rates help explain why it is common for cities to experience a net loss of population, since a relatively low level of net out-migration can still exceed the rate of natural increase.[9]

Important changes have also occurred in relation to household size and composition, in divorce and cohabitation, and in the level and incidence of childbearing. In much of Europe the average household size is now below 3.0 persons, and more women are starting their family later in life. In addition, over a quarter of households contain only one person, at least 10 per cent of families are headed by a lone parent, and around one in three marriages end in divorce. These trends indicate an increase in the number of households if not in population, with significant implications for housing markets, migration and settlement structures.

The influence of demographic changes on cities and urban systems is seen clearly in the way the 'baby boom' of the late 1940s and early 1950s was associated with strong suburbanisation and deconcentration pressures as couples reaching the family-building stage sought affordable homes with gardens and a safe and pleasant environment within commuting distance of work. Similarly, as the population ages, so retirement migration becomes increasingly important. This generally takes the form of population movement away from larger metropolitan centres down the urban hierarchy, taking advantage of the lower house prices and more congenial living environments of smaller towns and more rural areas. However, the preferred location for such migration has changed from seaside and spa towns in the 1950s to countryside areas in the 1960s and 1970s, and to the Mediterranean sunbelt in the 1980s.

EUROPE'S CITIES

London and Paris are the largest cities in the region, after which the size ranking becomes more ambiguous depending on the boundaries used (Figure 5.5). Comparing cities by their dominant characteristics provides greater insight into current urban processes. Table 5.6 presents eleven types of cities and their dominant characteristics, with examples of cities that represent each functional type. This highlights the major dichotomy between the growing high-tech/services cities, which include many cities that are not among the largest in Europe, and the declining industrial and port cities, many of which still are among the largest cities although most have declining populations.

There is little agreement about the likely future pattern of urban growth in Western Europe, not least because of lack of knowledge of how the factors that affect population distribution will themselves develop. The advent of a post-Fordist economy embracing 'flexible specialisation' has induced major structural change in urban economies, while political changes associated with the EU and removal of the Iron Curtain have the potential for large-scale economic restructuring. A major debate is whether these changes will lead to the growth of a European core region at the expense of the peripheral zones, with economic activity concentrated in a belt stretching from south-east England through Benelux, south-west Germany and Switzerland to Lombardy and north-west Italy. However, not all places within a favoured zone prosper and not all places outside a favoured zone are necessarily at a disadvantage. Research at the 'locality' level suggests that the economic fortunes of different places have more to do with their inherent characteristics and the way in which they position themselves with respect to the global economy and international capital than with their particular geographic location.[10] This perspective sees a Europe of the future made up of a large number of individual, largely urban-centred regions that compete for jobs, people and capital investment, sometimes in co-operation with their neighbours but usually in competition.

EAST AND CENTRAL EUROPE

Until the lifting of the Iron Curtain in 1989 the countries of Eastern and Central Europe and the former

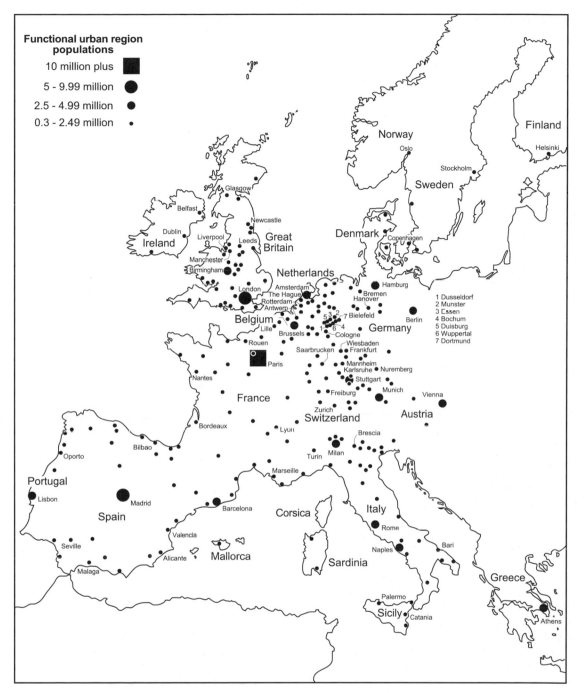

Figure 5.5   Major urban centres in Western Europe

TABLE 5.6 FUNCTIONAL TYPES OF CITIES IN WESTERN EUROPE

City type	Characteristics	Examples
Global	Accumulation of financial, economic, political and cultural headquarters of global importance	London, Paris
Growing high-tech/ services	Modern industrial base, national centre of R&D, production-oriented services of international importance	Bristol, Reading, Munich
Declining industrial	Traditional (monostructured) industrial base, obsolete physical infrastructure, structural unemployment	Metz, Oberhausen, Mons, Sheffield
Port	Declining shipbuilding and ship-repair industries, environmental legacies, in the South burdened by additional gateway functions	Liverpool, Genoa, Marseille
Growing without modern industrialisation	Large informal economy and marginal underclass, uncontrolled development and deteriorating environment	Palermo, Thessaloniki, Naples
Company towns	Local economy depending to a high degree on a single corporation	Leverkusen, Eindhoven
New towns	New self-contained cities with overspill population in the hinterland of large urban agglomerations	Milton Keynes, Evry
Monofunctional satellites	New urban schemes within large agglomerations with focus on one function only (e.g. technopole, airport)	Sophia-Antipolis, Roissy
Small towns, rural centres, rurban belts	Smaller cities and semi-urban areas in rural regions, along coasts or transport corridors with weak economic potential	All over Europe
Tourism and culture	Local economic base depending on international tourism and cultural events of European importance	Salzburg, Venice
Border and gateway	Hinterland divided by national border; gateways for economic migrants and political refugees	Aachen, Basel

Source: **K. Kunzmann and M. Wegener (1991)** *The Pattern of Urbanization in Western Europe 1960–1990*
Dortmund: Institute of Town Planning

Soviet Union were united by economic and political characteristics, distinct from those of Western Europe, that affected their settlement system and the form of their cities.[11] These included the fact that government direction rather than market forces determined the nature and location of most productive investment. In general, priority was given to industry over services, and in many instances industries were located outside the major cities in places that differed from those locations a market economy would have produced. Also, industries were kept in operation long after they would have been deemed unprofitable in the West, so that until the economic and political changes of the late 1980s most industrial centres had avoided the negative impacts of **deindustrialisation**. The abolition of an urban land market under communism, the limited role permitted for a private housing market and private enterprises, and the large-scale public housing estates produced a very different logic to the form and spatial distribution of urban areas (see Chapter 8). Since 1989, however, there has been a transformation in the economies of Eastern Europe towards more market-oriented growth. This is likely to bring major changes to settlement systems through changes in the scale, nature and spatial distribution of economic activities.[12]

INTERNAL MIGRATION AND URBANISATION IN EAST AND CENTRAL EUROPE

Internal migration has generally been from poorer to richer regions and from rural to urban areas.[13] In the period up to 1989 a significant part of this migration was attributed to official government policies that promoted the growth of large towns and cities to provide the necessary industrial labour force. At the same time, communist planners also sought to develop urban centres lower down the size hierarchy by means of new town development, often in relation to a particular economic function (such as coal-mining), and by imposition of controls on the growth of the largest cities (e.g. Budapest). As Table 5.7 and Figure 5.6 indicate, many of the region's main population concentrations are around national capitals and major industrial centres.

ECONOMIC AND URBAN CHANGE IN THE FORMER SOVIET REPUBLICS

Throughout the Soviet era there was a strong relationship between industrial and urban growth, and the central government's industrial development policies had a major impact on population and urban change. After 1953 the post-Stalin leadership continued the emphasis on heavy industry and the military sector. The priority attached to industrial modernisation and to the introduction of plastics and chemical industries led to investment in oil and natural gas production, which explains why several of the region's rapidly growing cities in the period 1959–89 were in the Volga–Urals area. These include cities such as Novocheboksarsk that had not existed in 1959.[14] Although the exploitation of oil and gas stimulated urban and industrial

developments in peripheral producing regions, such as Siberia and Kazakhstan, the growing network of oil and gas pipelines and electricity grids also permitted the continuing expansion of the major urban industrial centres in the western parts of the country. The concentration of urban development in European Russia was also promoted by the numerous environmental and social advantages over the east and the higher level of infrastructure provision. An additional factor was the increasing importance of trade links with Eastern Europe (through COMECON) and Western Europe, especially after 1970 (Table 5.8).

An increasing proportion of the urban population was concentrated in large cities. By 1989 61 per cent of the region's urban dwellers lived in cities with 100,000 or more inhabitants (compared with 49 per cent in 1959), with 22 per cent living in 'million cities' (compared with 9 per cent in 1959). This illustrates the failure of government attempts to restrict the growth of the largest cities by means of controls on internal migration. Large cities played a key role in the command economy for two reasons. First, they had the highest-quality infrastructure, skilled labour forces, and relatively low wages. Unlike in a market economy, in the Soviet command economy wage rates, taxes, land costs, rent and infrastructure costs were not necessarily higher in the major cities, as they would be normally in Western Europe or North America. Second, in a

TABLE 5.7 URBAN POPULATION IN EASTERN EUROPE IN 1992 AND 2001, AND URBAN CHANGE SINCE THE 1930s

Country	Urban population (million)		% population urban		
	1992	2001	Pre-war	1992	2001
Albania	1.2	1.4	15.4[a]	35.0	42.9
Bulgaria	5.9	5.3	21.4[b]	66.0	67.4
Czech Republic	7.8	7.7		75.2	74.5
Slovakia	3.7	3.1		69.2	57.6
Hungary	6.1	6.4	33.2[c]	59.0	64.8
Poland	23.1	24.1	37.3[d]	60.9	62.5
Romania	12.3	12.4	21.4[c]	54.4	55.2
Bosnia/Herzegovina	1.5	1.8		34.2	43.4
Croatia	2.4	2.7		50.8	58.1
Macedonia	1.1	1.2		53.9	59.4
Slovenia	1.0	0.9		48.9	49.1
Montenegro	0.3			50.7	
Serbia	4.6			46.5	
Yugoslavia	4.9	5.5		48.6	52.0
Eastern Europe	**71.1**	**72.5**		**56.4**	**60.0**

Source: adapted from United Nations (2002) *World Urbanization Prospects: The 2001 Revision* New York: United Nations
Notes: [a]1938. [b]1934. [c]1930. [d]1939

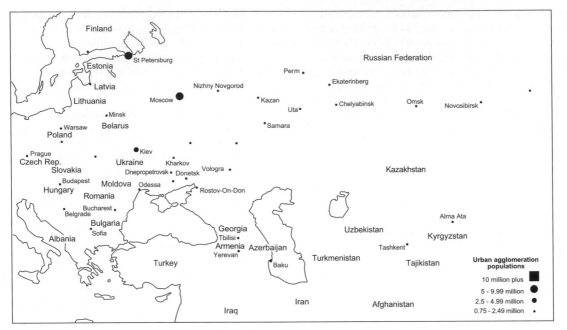

Figure 5.6   Major urban centres in Eastern and Central Europe

TABLE 5.8 LEVELS OF URBANISATION OF THE FORMER SOVIET REPUBLICS 1959, 1989 AND 2001 (%)

Republic	Level of urbanisation			Change	
	1959	1989	2001	1959–89	1989–2001
Baltic Republics					
Estonia	56.5	71.6	69.4	15	−2.2
Latvia	56.1	71.7	59.8	15	−11.9
Lithuania	38.6	68.0	68.6	29	0.6
Republics bordering Europe					
Belarus	30.8	65.5	69.6	35	4.1
Ukraine	45.7	66.4	68.0	21	1.6
Moldova	22.3	47.0	41.4	25	−5.6
Russian Federation	52.4	73.6	72.9	21	−0.7
Caucasus					
Georgia	42.4	55.8	56.5	13	0.7
Azerbaijan	47.8	53.9	51.8	6	−2.1
Armenia	50.0	67.8	67.2	18	−0.6
Kazakhstan	43.7	57.2	55.8	14	−1.4
Central Asia					
Uzbekistan	33.6	40.7	36.6	7	−4.1
Kyrgyzstan	33.7	38.3	34.3	5	−4.0
Tajikistan	32.6	32.6	27.7	0	−4.9
Turkmenistan	46.2	45.4	44.9	−1	−0.5
Total (former USSR)	**47.9**	**65.9**	**55.0**	**18**	**−10.9**

Source: adapted from United Nations (2002) *World Urbanization Prospects: The 2001 Revision* New York: United Nations

command economy the fact that success depended as much on access to political and administrative decision-makers and government officials as economic performance enhanced the attraction of major administrative centres. The transition to a market economy has been accompanied by less central direction of urban development and reduced levels of urbanisation as a consequence of emigration to the West and the beginnings of a suburbanisation process in some cities.[15]

Looking ahead, the fundamental political and economic changes heralded by the break-up of the Soviet Union are likely to have far-reaching implications for population and urban change in the region. Urban growth will be promoted by the emergence of capital cities and national urban systems in the new republics. Cities in general may benefit from an expansion of service activity restrained by Soviet economic-development policy. On the negative side, cities associated closely with the heavy industrialisation of the Soviet era (e.g. Donetsk in Ukraine) may experience the deindustrialisation and out-migration that have affected similar cities and regions in North America and Western Europe. **Smokestack** industries were so widespread within the command economy that deindustrialisation represents a major challenge for the new independent states. Similarly, policies to reduce military expenditure and restructure the military-industrial complex will also impact upon the economy of established industrial regions and lead to a restructuring of the urban system.

ASIA AND THE PACIFIC

In 1990 Asia contained three-fifths (3,186 million) of the world's population and an increasing share (32 per cent in 1990) of its urban population. In 2001 Asia had 58 per cent of the world's population and 46 per cent of its urban population (Table 5.9). It also contains many of the world's largest and fastest-growing cities, reflecting the presence of most of the nations with the highest economic growth rates since 1980. However, few generalisations are valid for the region, given the variety of countries, which range from the richest in the world to the poorest, and from the largest and most populous to the smallest and least populous (Figure 5.7). With almost two-thirds of the population still living in rural areas, Asia remains a predominantly rural continent. Significantly, however, this proportion of urban to rural population is partly explained by the criteria employed to define urban places. In Asia, if India and China were to alter their definitions of 'urban centres' to those commonly used in many European or Latin American

nations the level of urbanisation would increase to 50–60 per cent as hundreds of millions of those now classified as rural dwellers became urban.

We can identify three main groups of countries in the Asia-Pacific region according to their levels of urbanisation:

1. The most urbanised group includes Australia, New Zealand, Japan, Hong Kong, Singapore and Korea. In these countries agriculture plays a minor role in the economy and more than half of GDP comes from the service sector.
2. The second group comprises Thailand, Indonesia, Malaysia, the Philippines, Fiji and Pakistan, where agriculture contributed less than one-third of GDP and where (apart from Thailand) between 30 per cent and 50 per cent of the population lives in urban areas.
3. The third group contains China and all South Asian countries except Pakistan. These remain predominantly rural, with agriculture having a greater importance in their GDP.

Such classifications are purely indicative, however, and are susceptible to change over time as, for example, the speed of economic growth in China propels the country into the second group. In addition, parts of India and China currently exhibit the economic and urban characteristics of group 2 or even group 1 nations (Box 5.3).

METROPOLITAN AREAS AND EXTENDED METROPOLITAN REGIONS

As we have seen, a process of population deconcentration tends to affect the core areas of most metropolitan regions, with relatively slow population growth rates or even population decline within the central city and much higher growth rates in the outer areas. In Asia this process is marked in Jakarta, where the population of the metropolitan region is deconcentrating rapidly.[16] During the 1980s central Jakarta grew at a rate of 3.1 per cent per year, while the growth rates of the urban population in the three neighbouring districts that constitute the Jakarta greater metropolitan area were 11.7 per cent, 20.9 per cent and 19.8 per cent. In the main cities within the more successful economies there is also a major restructuring of the central areas of cities, with development of office buildings, convention centres, hotels and a diverse range of supporting service enterprises as well as modern transport facilities. All these trends tend to involve the removal of people from central cities. In Australian cities such as Sydney,

TABLE 5.9 POPULATION AND URBAN CHANGE, 1950–2001, IN ASIAN COUNTRIES WITH 10 MILLION OR MORE INHABITANTS

Country	Urban population, 2001 (000)	% of population in urban areas (level of urbanisation)			Change in level of urbanisation (%)	
		1950	1990	2001	1950–90	1990–2001
Eastern Asia						
China	471,927	11.0	26.2	36.7	15.2	11.5
Democratic People's						
Republic of Korea	13,571	31.0	59.8	60.5	28.8	0.7
Japan	100,469	50.3	77.2	78.9	26.9	1.7
Republic of Korea	38,830	21.4	73.8	82.5	52.5	8.9
South-Central Asia						
Afghanistan	5,019	5.8	18.2	22.3	12.4	4.1
Bangladesh	35,896	4.2	15.7	25.6	11.4	9.9
India	285,608	17.3	25.5	27.9	8.3	2.4
Iran	46,204	27.0	56.3	64.7	29.3	8.4
Nepal	2,874	2.3	10.9	12.2	8.6	1.3
Pakistan	48,425	17.5	32.0	33.4	14.5	1.4
Sri Lanka	4,409	14.4	21.4	23.1	7.0	1.7
South-East Asia						
Indonesia	90,356	12.4	30.6	42.1	18.2	11.5
Malaysia	13,154	20.4	49.8	58.1	29.4	8.3
Myanmar	13,606	16.2	24.8	28.1	8.6	3.3
Philippines	45,812	27.1	48.8	59.4	21.7	10.6
Thailand	12,709	10.5	18.7	20.0	8.2	1.3
Vietnam	19,395	11.6	19.9	24.5	8.2	5.2
Western Asia						
Iraq	15,907	35.1	71.8	67.4	36.7	−4.4
Saudi Arabia	18,229	15.9	77.3	86.7	61.4	9.4
Syria	8,596	30.6	50.2	51.8	19.6	1.6
Turkey	44,755	21.3	60.9	66.2	39.6	5.3
Yemen	4,778	5.8	28.9	25.0	23.1	−3.9
Total	**340,529**	**18.6**	**31.2**	**37.4**	**12.6**	**6.2**

Source: adapted from United Nations (2002) *World Urbanization Prospects: The 2001 Revision* New York: United Nations

Melbourne, Brisbane, Adelaide and Perth similar trends are evident in a growing centralisation of wealth combined with the selective dispersal of poverty to suburbs vulnerable to structural-change processes such as deindustrialisation and unemployment (Box 5.4). This form of urban restructuring contrasts with the 'doughnut' stereotype that has been employed to characterise US cities, following several decades of white flight from the urban core. Some of the largest relocations of people, to make way for new urban developments, have taken place in the major cities of Asia.

In many cities, such as Tokyo, Jakarta and Bangkok, population growth has extended outwards beyond the boundary of the metropolitan area, a trend which is often acknowledged by governments in defining new planning regions (for example, the **extended metropolitan region** around Bangkok stretches some sixty miles (100 km) from the central core). Some Asian urban agglomerations cross existing (or former) national boundaries. Hong Kong, for example, is the centre of the Hong Kong–Zhujiang delta region, and a large part of Hong Kong's manufacturing production has been relocated in southern Guangdong, where some 3 million workers are employed in factories financed and managed by Hong Kong entrepreneurs. At the same time Hong Kong has emerged as a centre of manufacturing-related producer services within the emerging transnational polycentric metropolitan region.[17] In a

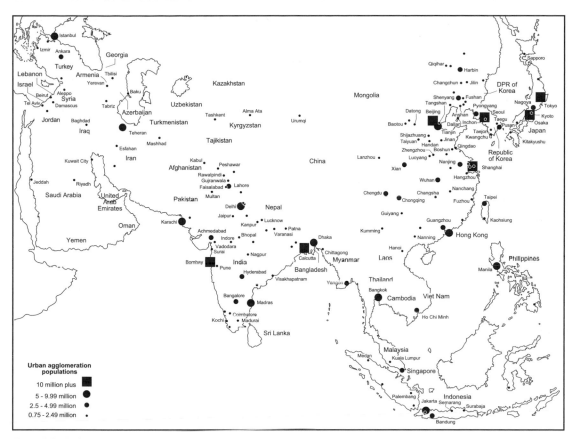

Figure 5.7   Major urban centres in Asia

number of Asian countries mega-cities have expanded to the extent that they form 'systems of cities' linked together functionally in networks of settlements encompassing huge tracts of highly urbanised as well as rural areas. The generic term of mega-urban region (see Chapter 4) may be used to describe these agglomerations but in practice we can identify four types:

1. Mega-city-centred extended metropolitan regions, such as Bangkok, Metro Manila and Jakarta, where development emanates from a dominant urban core to envelope adjacent settlements.
2. Extended metropolitan regions, such as the Shanghai–Nanjing–Hangzhou–Suzhou region and the Beijing–Tianjin–Tangshan national capital region, where a number of urban nodes form a regional network.
3. Polynucleated metropolitan regions where no one city-region dominates but a number of highly urbanised settlements form a system of cities,

such as in the Pearl River delta region made up of Guangzhou, Shenzhen, Hong Kong, Macao and Zhuhai.
4. True megalopolitan regions such as the Tokyo–Nagoya–Osaka bullet train corridor where several large mega-cities with their own extended metropolitan regions encompass an extensive highly urbanised area.

In general Asian mega-urban regions tend to be fragmented administratively and politically and are often characterised by an absence of unified or coordinated governance structures. The daunting challenge in Asia mega-urban regions, therefore, is to manage development to ensure that these agglomerations can continue to be economically productive, meet ever-rising levels of demand for urban infrastructure and basic urban services, foster civic involvement, promote social justice, and ensure the sustained liveability of the urban environment.[18]

BOX 5.3

The economic and urban development of Shanghai

In 1949 Shanghai was the country's pre-eminent city, with over half of China's modern industry. But after 1949, throughout the first phase of heavy industrialisation, it was deprived of significant investment funds because of its location, which was vulnerable to attack from the sea. Its population remained remarkably low, given that it had been the largest industrial centre in the world's most populous nation. In 1982 the city's urban core districts contained only 6.27 million inhabitants, with the rest of its 11.8 million inhabitants distributed over what was then over 5,000 km² (1,900 square miles) of mainly farming and small-town communities in neighbouring counties. It was not until the early 1980s that it was again given priority – when it was designated one of the fourteen 'open cities', and the Shanghai Economic Zone was established in 1983 – but it was still compared unfavourably with Guangzhou. In the early 1990s a combination of strong local pressure and a concentration of ex-Shanghai leaders in the higher echelons of national politics led to a major programme of redevelopment, especially in its eastern hinterland, a large area of agricultural and marginal land generally referred to as Pudong (literally 'east of the Huangpu river'). Here a new high-technology metropolis is being developed, modelled on Singapore.

By 1992 Greater Shanghai's population had risen to 12.87 million, but the urban core districts now contained 7.86 million inhabitants. There was also a large floating population that was not included in this figure – for instance, in 1993 it had a *registered* temporary population of over 600,000. Migrants into Shanghai generally find employment in the suburban districts and the six counties annexed to the city, and in almost every *zhen* and smaller town within Shanghai's orbit there are established communities of migrants. Most are engaged in new, smaller plants and off-farm enterprises.

Shanghai's official boundaries encompass more than 2,300 square miles (6,000 km²), and many of the remaining pockets of untouched farmland are being developed as housing estates, relocated factories and local enterprises, especially in ribbon developments along each highway. The development of rural industrialisation in the city's attached counties has also introduced a new sense of urbanness into much of the peri-urban core. Within a mega-urban region of 76 million inhabiting the whole Yangtze river delta Shanghai may be designated as the 'head of the dragon', with other cities like Hangzhou, Nanjing, Suzhou and Wuxi providing much of the economic and social energy for the region's rapid growth.

FACTORS UNDERLYING URBAN CHANGE IN ASIA

Urban change in Asia demonstrates how the growth or decline of cities must be understood in terms of the effects of the globalisation of the **world economy** on the one hand, and economic, social or political changes that are specific to that city or region on the other. For example, the size and rapid growth of Delhi owe more to its role as capital of India than to a position within the global economy. Similarly, Karachi's population growth over recent decades to become one of the world's largest cities has been increased by immigration of refugees, including 600,000 from India after Partition, then from Bangladesh during the 1970s, and Afghanistan and Iran in the late 1970s and 1980s. By contrast, urban dynamics in Singapore are shaped more by the city-state's role within the global economic system than by its acting as a political or administrative centre. Most of the other large cities in Asia come between these two extremes. Tokyo is the world's largest urban agglomeration because it is both the national capital of the world's second-largest

economy and one of the three pre-eminent global cities (see Chapter 14). Tokyo's role within the global economy has also helped to counter a tendency towards decentralisation from the central city. During the 1980s the city attracted many head offices of major national and international companies, while the deregulation and internationalisation of Japan's financial markets helped to create a concentration of service activities in the city.

Urban trends in all Asian nations are also influenced strongly by government actions. A significant policy change during the 1980s was the relaxation of government controls on urban growth in various countries and the downscaling of programmes directing new investment into peripheral regions. The most dramatic change was in China. During the Maoist era (1947–77) urbanisation had three main characteristics. First, the level of urbanisation was low due to government migration controls and a system of food-rationing and household regulation (*hukou*). Second, industrial and urban centres were developed in inland areas at the expense of coastal areas, in pursuit of spatial balance and for national security considerations. Third, since the Chinese economy

BOX 5.4

Australasian cities

The most noticeable feature of the physical form of Australasian cities is their low density. The average density of fourteen persons per hectare (5.6 per acre) is similar to that of the USA and compares with average densities of 54 p.p.h. and 157 p.p.h. for European and Asian cities respectively.

Most cities in Australia and New Zealand were developed at a time when suburban living in detached housing was practicable. High wages, high employment, cheap land and good public transport encouraged low-density suburbanisation in the late nineteenth and early twentieth centuries. As in the USA, the suburbanisation process was accelerated by a rapid rise in car ownership after the Second World War. The development of residential suburbs was aided by government provision of public housing estates (as in the UK) and was accompanied by sub-urbanisation of manufacturing, retailing and other services (as in the USA). Subsequently, with a few exceptions, such as Perth, urban transport planners have used the low-density urban structure to justify reductions in public transport services and increased provision for the automobile.

In keeping with a neoliberal response to globalisation, cities in Australia and New Zealand must compete for investment within the international market place. Waterfront redevelopments, as in Wellington's Lambton Harbour and Darling Harbour in Sydney, reflect similar developments in North American cities (see Chapter 16). Inner-city redevelopment has been accompanied by the growing centralisation of wealth and increasing suburban poverty. Although there are inner-city areas with high proportions of low-income households (mostly single households living in private rented accommodation, as in the UK), the greatest concentrations of low-income households are in the outer suburbs (mostly families with children). This pattern is in contrast to the model of the typical US city.

In Australasian cities, as in the UK and the USA, polarisation of income has increased over recent decades and the geography of poverty and affluence is much sharper. There is also an ethnic dimension to socio-spatial variations in income. This is evident in a concentration of Maori in public housing in many of Auckland's outer suburbs, and of overseas immigrants (most recently from South East Asia and Latin America) in the middle and outer suburbs of cities such as Sydney, Melbourne and Adelaide. The extent of ethnic segregation, however, is not as extreme as in the ethnic ghettoes of US cities (see Chapter 18).

Sources: **G. Robinson, R. Loughran and R. Tranter (2000)** *Australia and New Zealand: Economy, Society and Environment* **London: Arnold; S. Hamnett and R. Freestone (2000)** *The Australian Metropolis* **London: Spon**

was isolated from the rest of the world its urbanisation was unaffected by external forces, such as foreign capital. Following the introduction of economic reforms and the 'open door' policy in 1978 three main changes occurred in the pattern of Chinese urbanisation. First, the rate of urbanisation accelerated (Table 5.9), fuelled by rapid rural–urban migration in the 1980s and 1990s. This was made possible by changes in the household regulation system that during the 1960s and 1970s had controlled migration, and by the growing private food and housing markets and employment opportunities that allowed people to find a livelihood outside the official system. Second, coastal areas, including Shanghai and the Pearl River (Zhujiang) delta, benefited from preferential policies, including fiscal incentives, administrative autonomy and, most important, designation as special economic zones and open development areas. Third, external factors, especially foreign direct investment, played an increasingly important role in shaping urbanisation and urban growth.[19] Much of this urban expansion has been concentrated in large cities.[20] The

post-Maoist urbanisation process transformed a pre-reform situation of **underurbanisation**, marked by achievement of high industrial growth without a parallel growth of urban population (Box 5.5).

AFRICA

Most of the nations with the fastest-growing populations are now in Africa (Table 5.10 and Figure 5.8). In the period since the early 1960s, when most countries gained formal independence, African cities have changed in four main ways:

1. Most have grown in size. There are two main trends. First, while the largest cities have continued to expand, rates of population growth have slackened since the spectacular increases of the 1960s and 1970s. Also, whereas the principal component of large city growth was rural–urban migration in the earlier post-independence period,

BOX 5.5

Underurbanisation in China

China's path to urbanisation does not mirror that of developed economies in which the process of urbanisation was linked closely with the level of economic development; nor does it reflect the condition of overurbanisation that characterises many developing economies where increases in urban population outpace economic development (see Chapter 4). China's current urbanisation status has emerged from a situation of underurbanisation prior to 1978. Underurbanisation was a typical phenomenon of socialist economies, including pre-reform China. This arose as a result of several factors, including state emphasis on industrialisation and a related view of urbanisation as primarily a cost incurred in pursuit of industrialisation. This led to efforts to minimise the costs of urbanisation by, for example, the use of capital-intensive production technologies in manufacturing and labour-intensive modes of production in agriculture, and policy restrictions on rural–urban migration. Such strategies were embedded in an ideological commitment to the ideal of income equality, full employment and a more uniform level of economic and social development across regions.

Source: **L. Zhang and S. Zhao (2003) Reinterpretation of China's under-urbanization** *Habitat International* **27(3), 459–83**

natural increase is now the major growth element in many cities. Second, in many countries medium-sized cities are now growing at least as quickly as the largest cities, possibly owing to the deteriorating condition of infrastructure and public services in the major cities.

2. The deterioration of services and infrastructure is the result of the mismatch between economic and urban growth. As national and urban economies stagnate in absolute terms and urban populations continue to grow (at around 4.5 per cent per year for the region as a whole), the resources needed for roads, sewers, water systems, schools, housing and hospitals cannot keep up with demand (Box 5.6). The impact of these adverse living conditions is distributed differentially between the small elite of upper-level managers, foreign diplomats, senior politicians and successful businessmen on the one hand and the growing number of low-income urban dwellers on the other.[21]

BOX 5.6

The deteriorating infrastructure of African cities

As African cities continued to increase in size during the 1980s and 1990s their declining economic situation led to a precipitous deterioration of basic infrastructure and urban services. In many cities most refuse is uncollected and piles of decaying waste are allowed to rot in the streets and on vacant lots. Schools are becoming so overcrowded that many students have only minimal contact with their teachers. A declining proportion of urban roads are tarmacked and drained, and many turn into quagmires during the rainy season.

Basic medical drugs and professional health care are extremely difficult to obtain, except for the rich. Public transport systems are seriously overburdened, and an increasing number of people are obliged to live in informal housing on unserviced plots where clean drinking water must be purchased from water sellers at a prohibitive cost, and where telephones and electrical connections are scarcely available.

The lack of investment in urban infrastructure and services also inhibits economic expansion. For example, in Lagos unreliable infrastructure services impose heavy costs on manufacturing enterprises. Virtually every manufacturing firm in the city has its own electric power generator to cope with the unreliable public power supply. These firms invest 10–35 per cent of their capital in power generation alone and incur additional capital and operating expenses to substitute for other unreliable public services.

Generally, in Nigeria, and many other low-income countries, manufacturers' high costs of operation prevent innovation and the adoption of new technology, and make it difficult for them to compete in international markets.

Source: **United Nations Centre for Human Settlements (1996)** *An Urbanising World* **Oxford: Oxford University Press**

TABLE 5.10 URBAN POPULATION CHANGE IN AFRICAN COUNTRIES WITH 1 MILLION OR MORE INHABITANTS, 1950, 1990 AND 2001

Country	Urban population, 2001 (000)	(%) of population in urban areas (level of urbanisation)		
		1950	1990	2001
Eastern Africa				
Burundi	603	1.7	6.3	9.3
Eritrea	730	5.9	15.8	19.1
Ethiopia	10,222	4.6	12.3	15.9
Kenya	10,751	5.6	23.6	34.4
Madagascar	4,952	7.8	23.8	30.1
Malawi	1,745	3.5	11.8	15.1
Mauritius	486	28.8	40.5	41.6
Mozambique	6,208	2.4	26.8	33.3
Rwanda	497	1.8	5.6	6.3
Somalia	2,557	12.7	24.2	27.9
Tanzania	11,982	3.8	20.8	33.3
Uganda	3,486	3.1	11.2	14.5
Zambia	4,237	8.9	42.0	39.8
Zimbabwe	4,630	10.6	28.5	36.0
Middle Africa				
Angola	4,715	7.6	28.3	34.9
Cameroon	7,558	9.8	40.3	49.7
Central African Rep.	1,575	16.0	37.5	41.7
Chad	1,964	3.9	20.5	24.1
Congo, Republic of the	2,056	30.9	53.5	66.1
Gabon	1,038	11.4	45.7	82.3
Zaire (now Dem. Rep. of Congo)	16,120	19.1	28.1	30.7
North Africa				
Algeria	17,801	22.3	51.7	57.7
Egypt	29,475	31.9	43.9	42.7
Libya	4,757	18.6	82.4	88.0
Morocco	17,082	26.2	46.1	56.1
Sudan	11,790	6.3	22.5	37.1
Tunisia	6,329	31.2	54.9	66.2
Southern Africa				
Botswana	768	0.3	23.1	49.4
Lesotho	592	0.3	23.1	28.8
Namibia	561	9.4	31.9	31.4
South Africa	25,260	43.1	49.2	57.7
West Africa				
Benin	2,774	5.3	29.0	43.0
Burkina Faso	1,999	3.8	17.9	16.9
Côte d'Ivoire	7,197	13.2	40.4	44.0
Ghana	7,177	14.5	34.0	36.4
Guinea	2,312	5.5	25.8	27.9
Liberia	2,414	13.0	42.1	45.5
Mali	3,606	8.5	23.8	30.9
Mauritania	1,624	2.3	46.8	59.1
Niger	2,366	4.9	15.2	21.1
Nigeria	52,539	10.1	35.2	44.9
Senegal	4,653	30.5	39.8	48.2
Sierra Leone	1,714	9.2	32.2	37.3
Togo	1,579	7.2	28.5	33.9
Total	**303,481**	**11.8**	**31.7**	**37.7**

Source: adapted from United Nations (2002) *World Population Prospects: The 2001 Revision* New York: United Nations

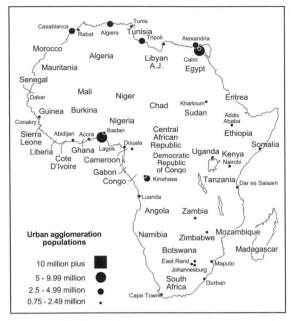

Figure 5.8 Major urban centres in Africa.

3. Although most sub-Saharan states aimed to industrialise in the years after independence, most made only limited progress, and the **import substitution model** that was adopted proved costly and of only limited success. Marginalisation in the world economy, deteriorating terms of trade for primary products, limited availability of domestic capital, failure to attract foreign direct investment, increasing indebtedness, wars and

natural disasters, limited government capacity and economic mismanagement have contributed to low or negative rates of economic growth for most countries, for most of the time since the 1960s. As a result, many African towns and cities, have economies that cannot support their growing populations.[22] Also, the urban labour market has changed since the 1960s. In the post-independence decade the educational system was expanded and African graduates had little difficulty in finding good jobs in either the public sector or the large-scale private sector. Subsequent contraction and **privatisation** of the public sector have reduced these opportunities. Most African cities also have a burgeoning informal economy which has arisen in direct response to the needs of the poor (Box 5.7) (see Chapter 24).

4. All these changes have had a major effect on city form. Where once the colonial central business district (CBD) was the focus of urban life, now the centre of gravity has shifted as more of the population have moved to the urban periphery where land is cheaper and more accessible, shelter can be constructed economically using locally available materials, and official planning regulations are rarely enforced (see Chapter 25). Urban encroachment on rural areas on the edge of cities leads to conflict. The peri-urban zone is an area of economic and social change characterised by pressure on natural resources, changing employment opportunities and constraints and changing patterns of land use. While rising numbers of urban inhabitants seeking jobs, housing and waste-disposal sites have a clear

---

BOX 5.7

Informal employment in African cities

Beginning in the 1970s, and gaining strength during the economic downturn of the 1980s and early 1990s, the urban informal sector has become a powerful force for employment creation in virtually all African cities. The main informal activities include carpentry and furniture production, tailoring, trade, vehicle and other repairs, metal goods fabrication, footwear, and miscellaneous services. Estimates made in the mid-1970s suggested that in the typical African country the informal sector employs 60 per cent of the urban labour force. Figures for individual cities included Abidjan 44 per cent, Kumasi 65 per cent, Lagos 50 per cent and Lomé 50 per cent. The

proportion of the urban labour force in informal activities has almost certainly risen since then.

A trend of the 1980s and 1990s was the supplementation by informal activities of formal-sector jobs on the part of the public-sector employees. Although many informal-sector workers are poor, this is not always the case, as such activities as market trading, the operation of public transport vehicles or renting out of housing in informal neighbourhoods may involve considerable income. Nevertheless, in general there is a marked gap between formal wages and informal earnings.

Source: **United Nations Centre for Human Settlements (1996)** *An Urbanising World* **Oxford: Oxford University Press**

Plate 5.2   In Kano, Nigeria, the line of the now demolished old city wall separates the high-density ancient Islamic city from the more spacious twentieth-century new town

interest in the expansion of cities into peri-urban areas, the residents of these areas are often some of the poorest in the city region, dependent on natural resources for food, fuel, water and building materials. The horizontal expansion of the African city attenuates infrastructure such as piped water, electricity, sewerage and roads beyond system capacity and adds significantly to the costs of education, health and other social services. The combination of peripheral expansion of cities and declining public resources to service them represents the major challenge for the planning and management of African cities in the twenty-first century.

Having examined urban patterns and processes at the global and major world regional scales, in the next chapter we change the level of analysis to examine urban systems at the national scale.

FURTHER READING

BOOKS

A. Bagnasco and P. Le Gates (2000) *Cities in Contemporary Europe* Cambridge: Cambridge University Press

G. Chapman, A. Dutt and R. Bradnock (1999) *Urban Growth and Development in Asia* Aldershot: Ashgate

A. Gilbert (1996) *The Mega-city in Latin America* New York: United Nations University Press

J. Gugler (2004) (ed) *World Cities Beyond the West* Cambridge: Cambridge University Press

F. Lo and Y. Yeung (1996) *Emerging World Cities in Pacific Asia* New York: United Nations University Press

P. Marcuse and R. van Kempen (2000) *Globalizing Cities: A New Spatial Order* Oxford: Blackwell

J. McDonald (2008) *Urban America: Growth, Crisis and Rebirth* London: M. E. Sharp

B. Rakodi (1997) *The Urban Challenge in Africa* New York: United Nations University Press

S. Yusuf and W. Wu (1997) *The Dynamics of Urban Growth in Three Chinese Cities* Oxford: Oxford University Press

JOURNAL ARTICLES

A. Aguilar (2002) Megaurbanization and industrial location in Mexico's central region *Urban Geography* 7, 649–73

L. Brown, T. Mott and E. Malecki (2007) Immigrant profiles of US urban areas and agents of resettlement *Professional Geographer* 59(1), 56–73

P. Coprani (1999) Rapid urbanization in the Gambia *Third World Planning Review* 21(2), 155–75

T. Firman and G. Dharmapatri (1995) The emergence of extended metropolitan regions in Indonesia *Review of Urban and Regional Development Studies* 7, 167–88

J. Klink (1999) The future is coming: economic restructuring in the São Paulo fringe: the case of Diadema *Habitat International* 23(3), 325–38

G. Lintz, B. Muller and K. Schmude (2007) The future of industrial cities and regions in central and eastern Europe *Geoforum* 38, 512–19

C. Pannell (2002) China's continuing urban transition *Environment and Planning A* 34, 1571–89

S. Osada (2003) The Japanese urban system 1970–1990 *Progress in Planning* 59, 125–231

Y. Zhu (1998) Formal and informal urbanisation in China *Third World Planning Review* 20(3), 267–84

KEY CONCEPTS

- Internal migration
- Immigration
- Port-of-entry cities
- Retirement migration
- Sunbelt/gunbelt
- Frostbelt/rustbelt
- Baby boom
- Greying suburbs
- Racial generation gap
- Flexible specialisation
- Command economy
- Smokestack industries
- Deindustrialisation
- Extended metropolitan region
- Underurbanisation
- Informal economy

STUDY QUESTIONS

1. Identify the main factors underlying the future metropolitan pattern of North America and consider the likely urban outcome of these trends.
2. Discuss the major functional types of cities in Western Europe.
3. Consider the likely changes to the settlement system of Eastern Europe following the collapse of communism.
4. Asia contains many of the world's largest and fastest-growing cities. Examine the process of urban change in any one city.
5. Consider the impact of independence on the nature of urban development in Africa.
6. Examine the changing character of urbanisation in China from 1947 until the present day.

PROJECT

Using published data (for example, that produced by UN agencies), examine the relationship between key variables such as level of urbanisation and level of industrialisation, or urbanisation and quality of life (measured in a variety of ways). Draw graphs of these relationships and interpret the revealed patterns. Calculation of a correlation coefficient and, if appropriate, insertion of a regression line with confidence limits can help by identifying statistically significant outliers from the general trend. Remember, however, that a statistically significant relationship does not necessarily indicate a causal relationship between any two variables.

6

National Urban Systems

Preview: national urban systems; classifications of cities; principles of central place theory; assessment and applications of central place theory; network theory; diffusion theories of settlement location; a mercantile model of settlement evolution

INTRODUCTION

At the national level, cities are part of a complex system of interrelated urban places, and are key elements in the economic, social and political organisation of regions and nations. The interdependence among towns and cities makes it important to view a country as a system of urban places rather than as a series of independent settlements. The concept of an urban system refers to a set of towns and cities that are linked together in such a way that any major change in the population, economic vitality, employment or service provision of one will have repercussions for other places[1] (Box 6.1). In this chapter we examine urban functions and the differing types of urban places within a national territory, and review the main theories to describe and explain the development of national urban systems.

NATIONAL URBAN SYSTEMS AND THE OUTSIDE WORLD

Although the principal focus of this chapter is on the national level, we must remember that national urban systems vary in terms of their degree of closure or openness to outside influences. Pred (1977)[2] has suggested a fourfold classification of national city systems based on their degree of openness/closure and level of internal interdependence (Table 6.1):

1. Countries where no real systems of cities exist and which are characterised by little economic and

---

BOX 6.1

Urban systems at the intra-national scale

From the eighteenth century onwards the urban structure of Western nations became more complex. Traditional cities became more functionally integrated into regional and national urban systems, while larger cities expanded and developed their economic and social systems. Three levels of urban system may be identified in Western industrial countries:

1. A national system dominated by metropolitan centres and characterised by a step-like population-size hierarchy, with the number of places at each level increasing in a regular manner with decreasing size of place.
2. Nested within the national system are regional sub-systems of cities, displaying a similar but less clearly differentiated arrangement, usually organised about a single metropolitan centre.
3. Contained within regional sub-systems are local sub-systems or daily urban systems representing the life space of urban residents.

---

social exchange between settlements and with the outside world (i.e. low interdependence and high closure). There are few present-day examples of this category since even Third World countries generally have some form of urban system, a degree of internal exchange and communication

TABLE 6.1 CLASSIFICATION OF CITY SYSTEMS BASED ON LEVELS OF CLOSURE AND INTERDEPENDENCE

	Level of closure	
	Low closure	*High closure*
Low interdependence	Little exchange between a nation's settlements but considerable external influence. No national system of cities. Example: earlier colonial states (with port cities)	Little exchange between settlements or with the outside world. No national system of cities. Example: medieval Europe
Level of interdependence *High interdependence*	Strong internal (within nation) and external interdependencies. A national system of cities with a high level of 'openness'. Example: the city systems of most present-day European countries	Strong internal interdependencies but *relatively* low external links (though these links will be large in absolute terms). Example: USA

and some external links. The best examples are therefore historical. Many countries of medieval Europe were characterised by a few independent urban centres acting as market places for their surrounding areas, but with limited exchange between centres (see Chapter 3).

2. Countries where little exchange occurs between urban centres, but rather than being independent, each centre has strong external ties (i.e. low interdependence and low closure). Here no national system of cities exists but instead individual centres might form part of an international system of exchange. The port cities of eighteenth- and nineteenth-century colonial states exhibited such features. Even today this form of externally focused urban structure persists in some Third World states owing to inertia, the continued export focus of the economy and the locational preferences of multinational organisations.

3. Countries characterised by towns and cities with a high level of interdependence but where the whole system is also subject to strong exogenous influences (i.e. high interdependence and low closure). This includes most countries of Western Europe, which are generally highly developed and relatively small in population or territorial terms. High levels of development are linked to specialisation, exchange and interdependence within the national city system; while small size leads to a high level of international trade (in relation to domestic production) with strong transport and information links to other cities throughout the world.

4. Large developed countries such as the USA and international groupings such as the EU. In these countries or groups of countries the inter-city movement of goods is stimulated by the absence of tariff and quota barriers on trade, whereas the presence of such barriers to external trade results in a relatively high degree of closure.

TYPES OF PLACES IN THE NATIONAL URBAN SYSTEM

Most cities accommodate a combination of functions, with the relative importance of each varying between places. Researchers have sought to impose some order on this functional diversity by classifying settlements in order to gain insight into their characteristics. City classification schemes range from simple general description[3] to those based on multivariate statistical techniques.[4]

One of the earliest examples of city classification is Harris's (1943) analysis of 605 US cities.[5] Having recognised that all large cities are multifunctional, Harris sought to identify 'critical levels' of employment in certain occupations that would separate cities into clearly defined functional types. Harris constructed the classification framework by examining the employment profiles of the cities and then *intuitively* setting minimum levels of employment for each of nine functional types. For example, 'manufacturing cities' were those in which manufacturing accounted for at least 74 per cent of total employment in manufacturing, retailing and wholesale (the largest city in this group being Detroit).

Harris's typology provided a benchmark for subsequent city classifications which, as computational techniques became more powerful, included larger numbers of variables. Criticism of the subjective identification of classes by Harris also led to approaches in which classes were derived statistically from the raw data. A prime example is the multivariate statistical classification of British towns by Moser and Scott (1961), which employed a total of fifty-seven measures of population size and structure, population change, households and housing, economic character, social class, voting characteristics, health and education.[6] This represented a broader study of urban difference than the earlier, narrower functional analyses. Application of principal components analysis to the data set produced fourteen types of towns (with two, London and Huyton, excluded, since they were 'too different from other towns to be included in any group') (Table 6.2).

TABLE 6.2 CLASSIFICATION OF BRITISH TOWNS

Mainly resorts, administrative and commercial
1. Mainly seaside towns
2. Mainly spas, professional and administrative centres
3. Mainly commercial centres with some industry

Mainly industrial
4. Including most of the traditional railway centres
5. Including many of the large ports as well as two Black Country towns
6. Mainly textile centres in Yorkshire and Lancashire
7. Including the industrial towns of the north-east seaboard and mining towns of Wales
8. Including the more recent manufacturing towns

Suburbs and suburban-type
9. Mainly 'exclusive' residential suburbs
10. Mainly older mixed residential suburbs
11. Mainly newer mixed residential suburbs
12. Including light-industry suburbs, national-defence centres and towns within the sphere of large conurbations
13. Mainly working-class and industrial suburbs
14. Mainly newer industrial suburbs

Source: C. Moser and W. Scott (1961) British Towns:
A Statistical Study of their Social and Economic Differences
Edinburgh: Oliver & Boyd

Despite increasing technical sophistication, city classifications were not recognised as making a major contribution to the theoretical development of urban geography. Many considered the classifications to be essentially descriptive (and therefore outmoded), providing illustrative distribution maps rather than explanations of location. Most fundamentally, by the early 1970s critics were able to argue that the development of different classifications had become an end in itself rather than a basis for understanding urban function or for addressing urban problems.[7] In response to this critique a number of studies employed city classifications as an initial step in the analysis of urban processes and problems. Typical of these is Noyelle and Stanback's (1983) examination of the differential growth of sunbelt and snowbelt cities.[8] In order to test the view that sunbelt cities were accruing population and employment at the expense of snowbelt cities, they first grouped 140 US cities using a 'location quotient' which measured the share of a city's population in a given industry as a percentage of the share of employment in the same industry within the US economy in total. Four major types of city were identified (Table 6.3).

The classification was then used as a basis for an analysis of the flow of people and economic activity between cities in different regions. It was found that although the sunbelt was growing at a faster rate than the snowbelt as a whole, study of particular types of

TABLE 6.3 CLASSIFICATION OF US METROPOLITAN AREAS BY TYPE AND SIZE

Type of centre	Example
I **Diversified service centres**	
National nodal	New York NY
Regional nodal	Philadelphia PA
Sub-regional nodal	Memphis TN
II **Specialised service centres**	
Functional nodal	Detroit MI
Government–education	Washington DC
Education–manufacturing	New Haven CT
III **Production centres**	
Manufacturing	Buffalo NY
Industrial–military	San Diego CA
Mining–industrial	Tucson AZ
IV **Consumer-oriented centres**	
Residential	Nassau FL
Resort–retirement	Tampa FL

Source: T. Noyelle and J. Stanback (1983) The Economic
Transformation of American Cities
Totowa NJ: Rowman & Allenheld

city uncovered significant variations on this general trend. For example, while assembly activities had relocated to suburban and sunbelt locations, high-level producer services remained in the larger cities, most of which were in the snowbelt. This kind of research shows the potential utility for city classification (in this case for illuminating the processes of economic restructuring and population redistribution).

City-classification research has also proved of value in comparing different types of city in relation to measures of need, hardship and fiscal stress. Logan and Molotch (1987) related the classification of city types to particular consequences (advantages or disadvantages) for individuals and social groups (in terms of, for exam-

ple, local rents, wages and wealth, taxes and services, and daily life)[9] (Table 6.4). Thus, affluent residents of 'innovation centres' have the best of all worlds; Hispanic migrant populations will continue to increase in the 'entrepôt' cities, leading to further ethnic segregation; 'module production centres' and most 'retirement centres' are unlikely to provide income redistribution through taxes; growing 'headquarters' cities have the highest rates of housing-price inflation; and 'leisure playgrounds' are characterised by a two-tier economy comprising many low-paid service workers and relatively few highly paid managers. Clearly, research on urban function has moved a long way from the early descriptive urban taxonomies to a position where city

TABLE 6.4 US CITY TYPES

Type	Examples	Key functions
Headquarters	New York City NY; Los Angeles CA	Corporate centres: dominance in cultural production, transport and communication networks; corporate control and co-ordinating functions of many large transnational corporations and international banks
Innovation centres	Silicon Valley towns, including Santa Clara CA; Austin TX; and Research Triangle NC	Research and development of aerospace, electronics and instruments; some are (or were) so involved in military contracting that they are 'war preparation centres'
Module production places	Alameda CA (military base); Hanford WA (nuclear waste); Omaha NE (the '800' phone exchange centre); Detroit and Flint MI (cars)	Sites for routine economic tasks (e.g. assembly of autos, processing of magazine subscriptions or credit card bills), some located near a natural resource (e.g. mining centre) or government function (e.g. Social Security main office in Baltimore MD)
Third World entrepôts (warehouses)	Border cities such as San Diego CA; Tijuana, Mexico; Miami FL	Trade and financial centres for importing, marketing and distributing imported goods, including illegal goods such as drugs and pirated music; major labour centres because of their large numbers of low-paid workers in sweatshop manufacturing and tourist-oriented jobs such as hotel maids
Retirement centres	Tampa FL; Sun City AZ	Home to growing numbers of ageing Americans. Range: affluent towns that maximise services to less affluent cities dependent on pensions, social security and other public programmes to support the local economy
Leisure-tourist playgrounds	Tahoe City CA; Las Vegas NV; Atlantic City NJ; Disney World FL; Williamsburg VA	Range: theme parks, sport resorts, spas to gambling meccas, historical places, and cultural capitals

Sources: adapted from J. Logan and H. Molotch (1987) *Urban Fortunes* Berkeley CA: University of California Press;
E. Phillips (1996) *City Lights* New York: Oxford University Press

classifications can contribute to an understanding of urban processes, and the formulation and monitoring of urban policy.

THEORIES OF THE URBAN SYSTEM

Settlements exist because certain activities can be carried on more efficiently if they are clustered together rather than dispersed. Inspection of a map showing the pattern of settlement in a region can usually be disaggregated into three component elements that reflect different urban functions:

1. A linear pattern consisting of transport centres performing 'break of bulk' or 'break in transportation' services (e.g. transfer of cargo from ship to rail), and for which location is related to the network of transport routes.
2. A cluster pattern consisting of places performing specialised activities such as mining, and for which location is related to the localisation of resources.
3. A uniform pattern consisting of places whose prime function is to provide a range of goods and services to the surrounding area, and which require to be accessible to the population that uses them.

Most towns, even if based initially upon specialised functions, develop a service function for the surrounding area. Settlements that interact with and provide goods and services to an adjacent hinterland (as well as to their resident population) have been termed *central places*. By definition, the location of central places is closely connected with the general distribution of population. If the population of an area is evenly spread, then the settlements that serve it will also be evenly distributed. If the distribution of population is uneven, central places will be concentrated in the most accessible locations. Some central places in favourable locations cater for more people and can offer more specialised services; these centres tend to grow progressively larger. Such differential growth produces various grades of central places characterised by different population sizes and zones of influence. The fact that this arrangement of large and small places has often suggested a degree of regularity encouraged efforts to form generalisations about the size and distribution of central places. The most fully developed theory to explain the spatial organisation of settlements derives from Christaller's (1933) work in southern Germany.[10] Following Christaller's original statement in

the 1930s there have been numerous attempts to test his propositions[11] and to modify and refine his ideas, with a major reformulation proposed by Losch (1943).[12]

CENTRAL PLACE THEORY

Christaller's spatial-equilibrium theory is fundamentally economic in approach and sets out to predict how, through competition for space, an optimal pattern of settlement will emerge. Like all models, central place theory represents a simplification of reality and is predicated on a number of assumptions (Box 6.2).

BOX 6.2

Assumptions underlying Christaller's central place theory

1. There is an unbounded uniform plain on which there is equal ease of transport in all directions. Transport costs are proportional to distance and there is only one type of transport.
2. Population is evenly distributed over the plain.
3. Central places are located on the plain to provide their hinterlands with goods, services and administrative functions.
4. Consumers minimise the distance to be travelled by visiting the nearest central place that provides the function that they demand.
5. The suppliers of these functions act as economically rational human beings, that is, they attempt to maximise their profits by locating on the plain to obtain the largest possible market. Since people visit the nearest centre, suppliers will locate as far away from one another as possible so as to maximise their market areas.
6. They will do so only to the extent that no one on the plain is farther from a function than he or she is prepared to travel to obtain it. Central places offering many functions are called higher-order centres; others, providing fewer functions, are lower-order centres.
7. Higher-order centres supply certain functions that are not offered by lower-order centres. They also provide all the functions that are provided in lower-order centres.
8. All consumers have the same income and the same demand for goods and services.

Economic principles and geometry

Christaller's theory applied to those settlements that are predominantly concerned with serving the needs of the surrounding area. The significance of this service role cannot be measured simply by the population of the place. While population may be a measure of absolute importance it is not a measure of a settlement's *centrality*. Centrality is the degree to which a place serves its surrounding area, and this can be gauged only in terms of the goods and services offered. Clearly, there are different orders of goods and services: some are costly, bought infrequently, and need a large population to support them (e.g. furniture, jewellery); others are everyday needs and require a small population (e.g. groceries). From this two concepts emerge:

1. *The threshold population.* The threshold is defined as the minimum population required for a good or service to be provided – that is, the minimum demand to make the good or service viable.
2. *The range of a good.* This is the maximum distance which people will travel to purchase a good or service. At some range from the central place, the inconvenience of travel as measured in time,

cost and effort will outweigh the value of or need for the good.

From these two concepts an upper and a lower limit can be identified for each good or service. The lower limit is determined by the threshold, the upper limit by the range. Ideally each central place would have a circular trade area. It is obvious, however, that if three or more tangent circles are placed in an area, unserved spaces will exist. In order to eliminate any unserved areas the circular market areas must overlap and, since people in these overlap zones will choose to visit their nearest centre in keeping with the assumption of minimum movement, the final market areas must be hexagonal (Figure 6.1). The resulting hexagonal pattern is the most efficient way of packing market areas on to the plain to ensure that every resident is served.

Christaller started by identifying typical settlements of different sizes in southern Germany. He then measured their average population, distance apart, and extent of their hexagonal tributary areas. Christaller also stated that the number of central places at each level of the settlement hierarchy follows a fixed ratio (the K value) from the largest *Landeshauptstadt* (regional capital) to the smallest *Marktort* (hamlet) (Table 6.5). In its

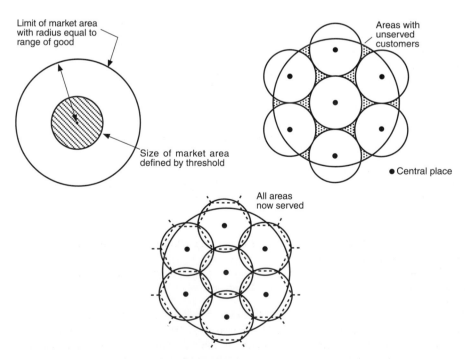

Figure 6.1　Deriving the hexagonal pattern of market areas for central places

TABLE 6.5 CHARACTERISTICS OF CENTRAL PLACES IN SOUTHERN GERMANY

Place	No. of places	Distance apart (km)	No. of complementary regions	Range of region (km²)	Area of region (km²)	No. of types of goods offered	Typical population Place	Typical population Region
Marktort	486	7	729	4.0	44	40	1,000	3,500
Amtsort	62	12	243	6.9	133	90	2,000	11,000
Kreisstadt	54	21	81	12.0	400	180	4,000	35,000
Bezirksstadt	18	36	27	20.7	1,200	330	10,000	100,000
Gaustadt	6	62	9	36.0	3,600	600	30,000	350,000
Provinzhauptstadt	2	108	3	62.1	10,800	1,000	100,000	1,000,000
Landeshauptstadt	1	186	1	108.0	32,400	2,000	500,000	3,500,000

Source: W. Christaller (1933) *Die zentralen Orte Suddeutschenland* Jena: Fischer, translated by C. W. Baskin (1966)
Central Places in Southern Germany Englewood Cliffs NJ: Prentice-Hall

simplest terms, therefore, Christaller's model proposed that settlements with the lowest level of specialisation (*Marktort*/hamlet) would be equally spaced and surrounded by hexagonally shaped hinterlands. For every six hamlets there would be a larger, more specialised central place (*Amtsort*/township centre) which would be located equidistant from other township centres. The *Amtsort* would have a larger market area for specialised services not available in the hamlet. Further up the hierarchy even more specialised settlements would also have their own hinterlands and would be located an equal distance from each other.

In the basic model the smallest centres would be spaced 7 km apart. The next higher centres would serve three times the area (and therefore three times the population) of the lower-order centres, and would be located $\sqrt{3} \times 7 = 12$ km apart. Similarly, the trade area of centres at the next higher level of specialisation would again be three times larger (Table 6.5). This kind of arrangement is called a K-3 hierarchy; in it the number of central places in the settlement hierarchy follows a geometric progression. 1, 3, 9, 27, etc. Thus lower-order centres, in order to be provided with higher-order goods and services, nest within the tributary areas of higher-order places according to a definite rule (the K value).

A settlement pattern with these features exhibits what Christaller has called the *marketing principle*. In this, the major factor influencing settlement distribution is the need for central places to be as near as possible to the population they serve. Thus the K-3 hierarchy and nesting pattern produce the maximum number of central places in accordance with the notion of movement minimisation (Figure 6.2).

By applying the economic principles in conjunction with the geometric properties of the theory plus the simplifying assumption of an isotropic surface,

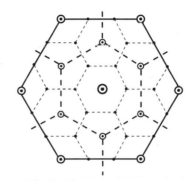

———	Higher-order market area
– – –	Middle-order market area
------	Lower-order market area
⊙	Higher-order centre
⊙	Middle-order centre
⊙	Lower-order centre
•	Lowest-order centre

Figure 6.2 Hierarchical and spatial arrangement of central places

Christaller derived his general model of the location, size and spacing of settlements.

An assessment of central place theory

The main criticisms directed at central place theory are the following:

1. The theory is not applicable to all settlements. Being limited to service centres, it does not include some of the functions, such as manufacturing industry, that create employment and population.

2. The **economic determinism** of the theory takes no account of random historical factors that can influence the settlement pattern.

3. The theory makes unrealistic assumptions about the information levels and mental acumen required to achieve rational economic decisions, even if profit maximisation were the only goal of human behaviour.

4. The notion of a homogeneous population ignores the variety of individual circumstances.

5. Christaller's model assumed relatively little governmental influence on business locational decisions, whereas today national and local governments play a major role in influencing business locations by, for example, offering grants to attract electronics firms into Scotland's 'Silicon Glen' or the lobbying by US sunbelt city mayors to attract investment to their cities.

6. Central place theory is a static formulation that relates to the distribution of service centres under assumed conditions at one point in time. A particular level of mobility is implied by the assumption that consumers look to their nearest central place to satisfy their needs. Levels of personal mobility have increased greatly since the model was proposed. Consumers do not always visit their nearest store, and multi-purpose shopping trips often result in low-order centres being by-passed for low-order goods, thus leading to their decline. Even in the classic field-study area of Iowa, economic restructuring and improved transportation infrastructure have undermined the 'nearest neighbour' travel patterns of earlier generations. Also, in many advanced countries, today telecommunications and 'tele-shopping' have further eroded the 'frictional effect' of distance on consumer behaviour.

Christaller (1966 p. 84) was not unaware of the temporal limitations of his theory, pointing out that:

> the stationary state is only fiction whereas motion is reality. Every factor which adds to the importance of the central place – regional population, supply and demand of central goods, prices of the goods, transportation conditions, sizes of the central places and competition between central and dispersed production of a good – is subject to continuous change.[13]

Unfortunately, he did not translate these qualifications into a dynamic model of the functional and spatial dimensions of the urban system. Accordingly, the relevance of central place theory in explaining current settlement patterns is limited. However, recognition of the limitations of central-place theory is not the same as rejecting it. Even as an ideal, the theory is useful. Study of where theory and reality diverge can lead to explanation. Kolars and Nystuen (1974 p. 73) suggest that the main contributions of both Christaller and Losch 'have been as much to stimulate further geographical thought as to give us any absolute explanations of the real world'.[14] While there is little evidence of a complete central place settlement structure emerging in the real world, the theory has stimulated much work in relation to retailing and consumer behaviour, and, as we see below, in the fields of physical and social planning.

Applications of central place theory

Central place ideas have been employed widely in regional-planning schemes in the USA, Canada, Africa, India, Europe and the Middle East.[15] For example, an Israeli settlement on the Laklish plains to the east of the Gaza Strip was based on a three-level hierarchy of:

1. 'A' settlements of various types (including protective border kibbutzim) housing immigrant settlers and serving as agricultural centres containing facilities used daily.

2. 'B' settlements (rural community centres), each planned to serve four to six 'A' settlements and to supply facilities and buildings used by them once or twice a week.

3. 'C' settlements (regional centres), towns roughly at the geographical centre of their region, providing administrative, educational, medical and cultural facilities, and with factories for crop-processing.

The most clearly articulated application of central place principles has occurred in the Dutch polderlands.[16] The increasing importance attached to a planned settlement pattern can be seen by tracing the development of the Wieringermeer, drained in 1930, and the north-east and east Flevoland polders, drained in 1942 and 1957 respectively.

In the first polder the location of the villages (service centres) was not successful. The settlement pattern did not conform to any model distribution, with the planners expecting that a spontaneous process of settling would lead to certain clusters at road intersections. As a result, the three regional villages of Slootdorp, Middenmeer and Wieringerwerf were clustered in the middle of the polder, which meant they had overlapping trade areas and that people living well away from the villages were inconvenienced by long journeys. The lower than expected population growth on the polder exacerbated the problems, with small villages incapable of providing a satisfactory level of service.

In the second, north-east, polder a settlement pattern was carefully planned in an attempt to avoid the mistakes of the Wieringermeer. Since it was one of the few places in the world where no historical or physical obstacles frustrated the realisation of a theoretical spatial model, Christaller's hierarchical system was applied with some modifications. In the middle of the area a regional centre, Emmeloord, was founded (with a target population of 10,000), with ten surrounding villages as local service centres each with target populations of 1,000–2,000. Despite this careful planning, however, the settlement pattern quickly demonstrated a number of shortcomings. Because of agricultural mechanisation and the reduced demand for labour, the populations of most villages did not reach the target (threshold) figure

and this made it difficult to keep the services feasible and the community viable. Paradoxically, Emmeloord grew more rapidly than anticipated. This was due to the increased **accessibility** to the rural population of the varied services in the regional centre, largely as a result of the general increase in mobility engendered by the spread of the motor car in the 1960s.

Experience gained in the first two polders was applied to the settlement of East Flevoland. The initial settlement plan had been similar to that of the north-east polder, with ten 'A' centres having a local service function surrounding a single 'B' or district centre, Dronten. A 'C' centre, Lelystad, the capital of the polder's province, was planned at the junction of the four polders but in the western corner of East Flevoland (Figure 6.3). Because of the diminishing importance of farm employment and the increasing affluence, aspirations and mobility of the population, this pattern was reduced to only four villages in 1959 and eventually to two by 1965 (Figure 6.3). The

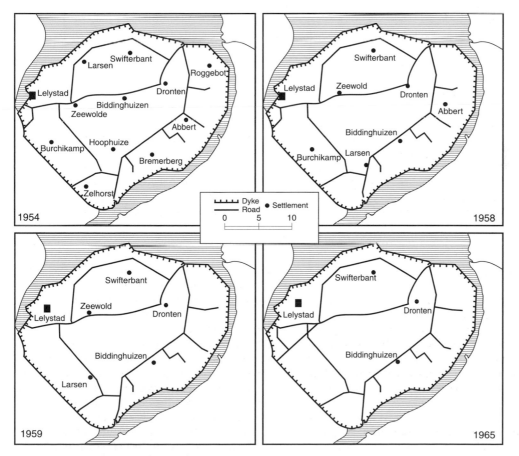

Figure 6.3   The changing settlement pattern of East Flevoland

declining significance of agricultural employment over the course of the development of the first three polders also influenced the location and population composition of settlement in the later southern Flevoland and Markerwaard polders, with less emphasis on placing villages to serve the needs of farming, and the introduction of commuters and other non-agricultural workers from Randstad Holland. The fact that in southern Flevoland, an area within the sphere of influence of Randstad Holland (see Box 8.3), no thought was given to a system of service centres set up according to a hierarchically arranged pattern of villages suggests that in such pressured 'rurban' areas classical central place theory is of limited relevance.

Reconstructing central place theory

The highest central place in Christaller's model is the regional capital (*Landeshauptstadt*) of 500,000 people and regional population of 3.5 million (Table 6.5). The omission of higher levels in the settlement hierarchy reflects the model's basis in southern Germany. Since the time of its inception, globalisation has increased the importance of world or global cities (see Chapter 14), meaning that a modern reconceptualisation of Christaller's model would incorporate at the apex of the hierarchy:

1. *Global cities*, typically with 5 million or more people within their administrative boundaries and up to 20 million in their hinterlands (for example, New York or London).
2. *Subglobal cities*, typically with 1 million to 5 million people and up to 10 million in their hinterlands (for example, national capitals as well as commercial capitals that are not global cities, such as Milan or Barcelona).

As a consequence of increased personal mobility, places at the lower levels of Christaller's hierarchy have declined in significance as central places having lost any service functions (such as a village store) to become mainly residential villages. Only the *Bezirkstadt*, with a population of 10,000 and service hinterland of 100,000, retains a significant service function (with, for example, a superstore and limited range of national chain stores). Some of the most significant changes have affected settlements at the next two higher levels, typically county market towns found across much of southern England, southern Germany and most of France. Many of these have grown because in depopulating regions

they have attracted population outflow from surrounding rural areas, and in more prosperous regions have attracted much of the out-migration from major cities.

NETWORK THEORY

Criticism of the contemporary relevance of central place theory led some researchers to suggest an alternative 'network model' of urban settlement; based on the concept of a 'dispersed city'[17] defined as a group of similar-size politically discrete cities separated by open land, functioning economically as a single urban unit. Such an arrangement of places deviates from the basic nested hierarchical ordering of central place theory in which horizontal relationships between places of similar size was unnecessary (since all provided the same amenities and services).

In the polycentric/polynuclear[18] or network city[19] complementarity (based on two-way flows between places replaces hierarchical relationships (characterised by vertical flows between places). For some, the network model is superior to central place models in describing the spatial patterning of urban places in a polycentric metropolitan region. However, rather than viewing the network model as a replacement for central place theory they may be seen as complementary. Whereas the central place model seems most suited to patterns of urban development in industrial economies the network model appears to be more applicable to economies that have become more service-sector dominated.[20] The Randstad provides an example of one such polycentric urban region.

DIFFUSION THEORIES

As we have seen, one of the most serious disadvantages of Christaller's theory is its static nature, which does not enable it to respond easily to changing social and economic conditions. This has led some writers to suggest that attempts to understand 'natural' as opposed to planned settlement patterns in terms of spatial equilibrium theory are of limited value. An alternative, which explicitly acknowledges the importance of the time dimension and historical perspective, is to examine the processes by which settlement spreads across a region from the initial point of colonisation. A number of models have been devised.

Bylund (1960) has proposed, within a deterministic framework, six hypothetical models of settlement diffusion based upon his study of early colonisation in

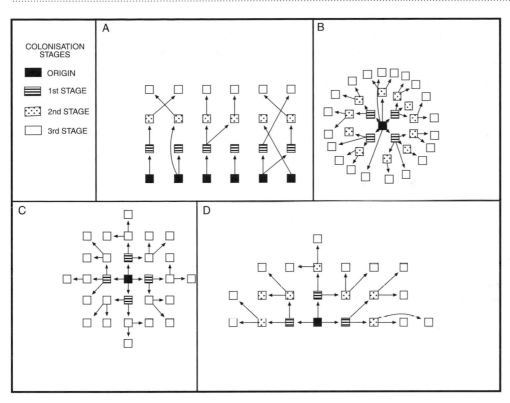

Figure 6.4   Bylund's model of settlement diffusion

Source: E. Bylund (1960) Theoretical considerations regarding the distribution of settlement in inner north Sweden *Geografiska Annaler* Series B 42, 225–31

central Lapland.[21] His four basic models are shown in Figure 6.4; each differs in the number and location of 'mother settlements'. The process of clone-colonisation (model B) appears to imitate the actual pattern of colonisation most closely, and this concept is developed in two further models to replicate the known settlement history of the area. The principles underlying these models have also been developed by Morrill (1962) in a probabilistic simulation of central place patterns over time.[22] The idea behind the model is that the behaviour of persons, as seen in the founding and growth of settlements and transport lines, occurs gradually over time and may be described as random within certain limiting conditions. The aim of Morrill's approach is to account for the general pattern of settlement, not the exact location of places. The model employed is of the Monte Carlo type (a stochastic growth process model) in which a set of probabilities related to both human and physical conditions governs choice of behaviour.[23] The mechanics of the method have been well illustrated by Abler *et al.* (1972).[24]

Morrill (1963) examined the spread of settlement in Sweden using this historical-predictive approach.[25] He begins by acknowledging that as the number, size and location of settlements in any region are the result of a long and complex interplay of forces, any study which proposes to explain the origins of such patterns must take into account four major factors:

1. The economic and social conditions that permit and/or encourage concentration of economic activities in towns.
2. The spatial or geographical conditions that influence the size and distribution of towns.
3. The fact that such development takes place gradually over time.
4. Recognition that there is an element of uncertainty or indeterminacy in all behaviour.

Whereas the first two factors are also explicit considerations in classical spatial equilibrium theories, the latter two are uniquely central to diffusion theory. The historical dimension is crucial not least because, as study of the Dutch polderlands showed, changes in social, economic and technological conditions over time can have a radical effect on the efficient functioning of a settlement pattern. It is also clear that locational decisions made on the basis of incomplete information are subject to error, and consequently real settlement patterns are the outcome of many less-than-perfect decisions. Morrill (1963 p. 9) attempted to simulate the development of the settlement pattern in southern Sweden over the period 1860–1960 in order to discover the major locational forces channelling urban development and migration. He was able to conclude that:

> in sum, the model results can be considered realistic from the point of view of distribution, that is similarity in spatial structure; and from the point of view of process, as a reasonable recognition and treatment of the pertinent forces. Deviations resulted in the main from oversimplified assumptions rather than from a mistaken approach.

The construction of a theory to explain the spread of settlement within a territory also formed the basis of Hudson's (1969) work in Iowa, in which he attempted to integrate diffusion theory with central-place theory.[26] Drawing on the work of plant and animal ecologists he identified three phases of settlement diffusion:

1. *Colonisation*, which involves the dispersal of settlement into new territory.
2. *Spread*, in which increasing population density creates settlement clusters and eventually pressure on the physical and social environment.
3. *Competition*, which produces regularity in the settlement pattern in the way suggested by central place theory.

Empirical testing of these hypotheses using settlement data from six areas of Eastern Iowa, at three different times between 1870 and 1960, found that the suggested increase in settlement regularly over time did occur.

Vance (1970)[27] also employed an historical diffusion perspective to devise a mercantile model of settlement evolution within a colonial setting (Figure 6.5). This envisaged five main stages in the development of settlement systems, in both core (home country) and periphery (colony):

1. *Exploration*. This involves the search for economic information by the prospective colonising power.

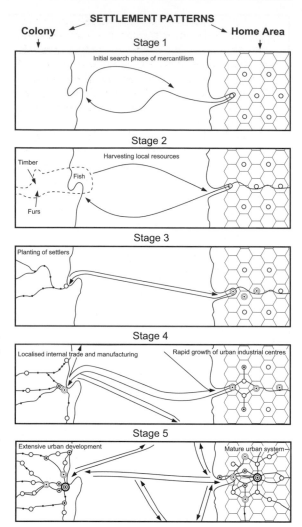

Figure 6.5   Vance's mercantile model of urban development
Source: J. Vance (1970) *The Merchant's World* Englewood Cliffs NJ: Prentice-Hall

2. *Harvesting of natural resources*. This involves the periodic harvesting of staple products, such as fish and timber, with limited permanent settlement.
3. *Emergence of farm-based staple production*. Increased permanent urban settlement permits export of agricultural commodities to the mother country that supplies the colony with manufactured goods. The seaports in the colony act as 'points of attachment' with the home country.
4. *Establishment of interior depot centres*. Settlement penetrates inland along favourable long-distance routes that facilitate the movement

of staple products from the interior to coastal points of attachment which also begin to develop manufacturing. Towns established at strategic locations along these routes serve as 'depots of staple collection'. This period also witnesses rapid growth of urban industrial centres in the home country to supply both the overseas and the domestic markets.

5. *Economic maturity and central place infilling.* The growth of a manufacturing sector leads to economic maturity accompanied by the emergence of a central place form of settlement pattern in which the depots of staple collection take on a service function and develop as regional centres. In time smaller central places are founded to serve local needs. Overall, the spatial structure of the colonial urban network is exogenic, being determined largely by external forces (Box 6.3).

There are close similarities between the patterns of urban development in the USA modelled by Vance (1970) and those identified in West Africa and Brazil by Taffe *et al.* (1963).[34]

Notwithstanding the analytical value of these theories, the great variety of settlement forms and distribution would appear to lend support to Bunce's contention (1982 p. 96) that 'settlement patterns are the product of the area which they occupy'[35] and Grossman's view (1971 p. 192) that 'general laws are meaningless outside the specific cultural and technological context'.[36] Despite the pedagogic value of the seminal models considered above, it is difficult to

---

BOX 6.3

The missing middle in the Third World urban hierarchy

A characteristic of many national urban systems in the Third World is the underdevelopment of the middle tier of the settlement hierarchy. Johnson (1970)[28] showed how the nations of Europe had almost ten times as many central places (towns of over 2,500 people) per village as Middle Eastern states. The resulting argument, that a more even pattern of development in the Third World requires promotion of middle-order settlements, has been taken up by subsequent writers such as Rondinelli (1983).[29] This proposal is also related to the thesis of urban bias,[30] which maintains that major cities of Third World countries have received a disproportionate amount of national resources and that a greater degree of territorial justice needs to be incorporated into national economic and urban development policy. This approach has been employed in Third World socialist states such as China,[31] Cuba[32] and Tanzania.[33]

---

avoid the conclusion that a *general* theory to explain the location, size and spacing of settlements is unattainable. Urban geographers now acknowledge the need to take into account the political and social forces operating at a variety of spatial scales in interpreting the changing spatial and functional dimensions of national urban systems.

FURTHER READING

BOOKS

K. Beavon (1977) *Central Place Theory: A Reinterpretation* Harlow: Longman

W. Christaller (1966) *Central Places in Southern Germany* Englewood Cliffs NJ: Prentice-Hall

E. Keeler and W. Rogers (1973) *A Classification of Large American Urban Areas* Santa Monica CA: Rand Corporation

C. Moser and W. Scott (1961) *British Towns: A Statistical Study of their Social and Economic Differences* Edinburgh: Oliver & Boyd

J. Vance (1970) *The Merchant's World* Englewood Cliffs NJ: Prentice-Hall

C. van der Wal (1997) *In Praise of Common Sense. Planning the Ordinary. A Physical Planning History of New Towns in the Ijsselmeerpolders* Rotterdam: 010 Publishers

JOURNAL ARTICLES

E. Bylund (1960) Theoretical considerations regarding the distribution of settlement in inner north Sweden *Geografiska Annaler* 42, 225–31

W. Coffey (1998) Urban systems research: an overview *Canadian Journal of Regional Science* 21, 327–64

M. van Hulton (1969) Plan and reality in the IJsselmeerpolders *Tijdschrift voor Economische en Sociale Geografie* 60, 67–76

R. Morrill (1962) Simulation of central place patterns over time *Lund Studies in Geography* Series B 24, 109–20

KEY CONCEPTS

- Openness/closure
- Location quotient
- Urban taxonomies
- Break of bulk
- Central places
- Centrality

- Range of goods
- Marketing principle
- Frictional effect of distance
- Network theory
- Diffusion theories
- Mercantile model

STUDY QUESTIONS

1. Prepare a list of the largest twenty-five or fifty settlements in your country. Using one of the city classification schemes described in the chapter, allocate each settlement to an appropriate category. Describe and explain the results.
2. Explain the economic principles and geometry underlying Christaller's central place theory.
3. Since its inception, central place theory has been subjected to critical analysis. Examine the main criticisms of the theory.
4. With reference to particular examples illustrate the ways in which central place theory has been applied in the real world.
5. Explain the network city model of urban settlement and consider how well it describes contemporary patterns of urban development.
6. Explain Vance's mercantile model of settlement evolution.

PROJECT

Determine the sphere of influence of either a single settlement or a group of settlements by selecting a range of services (e.g. banking, education, different forms of retailing, hospitals, newspaper circulation areas) and interviewing the service providers to ascertain where people come from to obtain the services. Each service will have its own **catchment area**. Plot these on a map and examine the pattern shown. Explain any differences in the catchment areas. Is there a clearly defined sphere of influence for the settlement? If you are studying more than a single settlement, do spheres of influence overlap? Does any one settlement dominate the region in terms of service provision?

PART THREE

Urban Structure and Land Use in the Western City

Plate 1 The ancient city: Thebes

Plate 2 The fantasy city: Las Vegas

Plate 3 The post-industrial city: Glasgow

Plate 4 The modern industrial city: Pusan, South Korea

Plate 5 The global city: Tokyo

Plate 6 The mobile city: San Diego

Plate 7 The informal city: Mumbai

Plate 8 The congested city: Hong Kong

Land Use in the City

Preview: urban morphogenesis; town plan analysis; ecological models of the city; Burgess's concentric zone model; Hoyt's sector model; the multiple-nuclei model of Harris and Ullman; Mann's model of a British town; Kearsley's modified Burgess model; Vance's urban realms model; White's model of the twenty-first-century city; urban political economy; major actors in the production of the built environment; growth coalitions; the central business district; urban architecture; public space; architecture and urban meaning; the social construction of the urban landscape

INTRODUCTION

Although most towns and cities have occupied the same location for centuries, the buildings and other physical infrastructure which comprise the built environment are not fixed but affected continuously by dynamic forces of change initiated by public and private interests. This modification of the urban environment occurs at a variety of scales ranging from the residential relocation decisions of individual households to large-scale projects including public road-building programmes and private house-building schemes. In addition, to differing degrees in different countries, the operation of these 'market forces' is influenced (enhanced or constrained) by national and local planning. The net effect of these socio-spatial processes is revealed most clearly in the land-use structure of the city. In this chapter we examine the principal models and theories of urban land use. For analytical convenience these are arranged into four broad types based on the principles of:

1. Morphogenesis.
2. Human ecology.
3. Political economy.
4. Postmodernism.

URBAN MORPHOGENESIS

The study of urban **morphogenesis** or town-plan analysis has a long history in urban geography. Since its high-point in the 1960s the approach has been sidelined, despite the fact that in its more recent formulations it has sought to advance from description and classification of urban forms[1] to analysis of the causal forces underlying changes in the pattern of urban land.[2]

Much current research in the morphogenetic tradition stems from the seminal work of Conzen (1960),[3] who divided the urban landscape into three main elements of town plan, building forms and land use, and demonstrated how each reacted at a different rate to the forces of change:

1. *Land use* is most susceptible to change.
2. Since *buildings* represent capital investments and are adaptable to alternative uses without being physically replaced, change occurs at a slower rate than with land use.
3. The *town plan* or street layout is most resistant to change.

Conzen also introduced the concepts of the **fringe belt** and **burgage** cycle to aid analysis of urban change. The existence of a fringe belt and associated fixation line reflects the fact that urban growth is cyclical rather than continuous, with periods of outward extension alternating with periods of standstill (marked by a fixation line) due to a downturn in the building cycle. A succession of fringe belts can be identified around most towns, related to phases of active growth (Figure 7.1). The burgage cycle indicates the way in which land use on a single plot develops over time.

These concepts have been developed by Whitehand (1991)[4] into an approach that seeks to identify the decision-making behaviour underlying land-use change. This is based on the premise that the town plan at any one time is the outcome of the perceptions, principles and policies of individuals (e.g. landowners) or agencies (e.g. local planning departments) which exercise the necessary power. The westward extension of the city of Glasgow in the eighteenth century illustrates both the economic power of landowners and the influence of the burgage-plot pattern of land-holding on urban form (Box 7.1). More recent evidence of the influence of landowners, developers and planners on urban structure is provided by Whitehand's (1992, 2006) studies of residential infilling in Amersham in Berkshire and of the Edwardian fringe belt of Birmingham in

which he explores the decision-making processes underlying urban change, focusing on negotiations between developers and the local planning authority.[5] In similar vein, Moudon (1992) has studied the evolving residential morphology of the North American city.[6]

These attempts to explore the backgrounds, motivations and actions of the major agents in the creation of **townscapes** at the local level represent a major advance on the earlier descriptive classifications of town plans. However, the difficulty of undertaking such detailed investigations increases as one looks further into the urban past.

ECOLOGICAL MODELS OF THE CITY

According to the ecological perspective developed by the Chicago school of human ecology, the significant processes underlying the spatial configuration of the growing American industrial city were analogous to those found in nature. Hence, *competition* among land uses for space resulted in the *invasion* of the most desired parts of a city and eventually the *succession* of existing land uses by a more *dominant* activity (as in the expansion of the **central business district** (CBD) into the surrounding transition zone). Under free-market conditions, certain parts of the city would be occupied by the function that could maximise use of the site, and in due course **natural areas** would evolve, distinguished by their homogeneous social or ethnic character (such as a slum or ghetto). On the basis of this tendency for ecological processes to sort similar households, Burgess (1925) derived his general concentric zone model of residential differentiation.[7] Figure 7.2 shows Burgess's interpretation of the land-use structure of Chicago (on the left of the diagram) and the general model arising from it (on the right of the diagram). The characteristics of the five zones are described in Box 7.2. Burgess maintained that the city tends to grow outwards in annular fashion from Zone I to Zone V.

It is important to recognise that the concentric zone model was proposed as an ideal type, not as a representation of reality. Based on the study of one city (Chicago) at one point in time (the 1920s) it offers a description of urban development as this would occur if only one factor (radial expansion from the city centre) determined the pattern of urban growth. Burgess was able to point to many examples of invasion and succession underlying the changing occupancy pattern of different zones in Chicago in the early twentieth

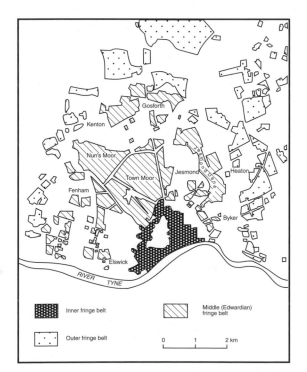

Figure 7.1 The fringe belts of Newcastle upon Tyne

Source: J. Whitehand (1967) Fringe belts: a neglected concept of urban geography *Transactions of the Institute of British Geographers* 41, 223–33

BOX 7.1

Land ownership and the development of the street pattern in eighteenth-century Glasgow

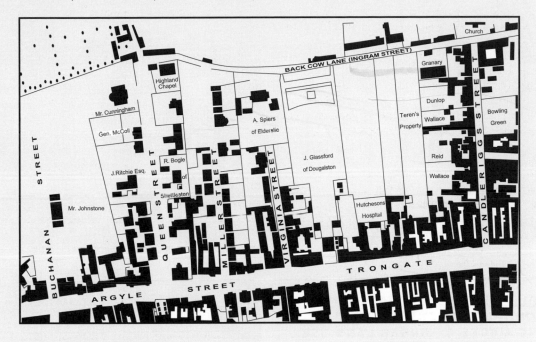

Between 1710 and 1780, eight new streets were developed as the town spread west from Glasgow Cross. The influence of the medieval pattern of land ownership based on burgage plots exerted a controlling influence on these developments. As the street plan shows, whereas a single plot or rig provided sufficient space to form a narrow wynd or vennel, the wider and longer streets were formed by the purchase and amalgamation of several plots. Miller Street (constructed by John Miller, a maltster and town bailie, or magistrate) required eight plots. The large profits to be gained by capitalising on the appreciating land values are confirmed by the fact that the cutting of the street obliged Miller to demolish half his newly built mansion.

The first of the new streets, Virginia Street, was opened in 1793 through two acres (0.8 ha) of cabbage plots. Virginia Street was a furrow long (furlong, about 200 m). Although the plot was evolved for the convenience of tillage, its size and shape were well suited to the speculative builder's goal of maximising the number of properties fronting on to the new streets. These developments on what had been the old burghal tillage lands set in motion a shift in the focal point of the city west from the Cross. It also signalled the emergence of marked socio-spatial segregation as the upper classes moved away from the crowded conditions of the old town.

Source: M. Pacione (1995) *Glasgow: The Socio-spatial Development of the City* **Chichester: Wiley**

century as successive waves of immigrants worked their way from their initial quarters in the zone of transition out to more salubrious neighbourhoods. In the model (Figure 7.2) this is shown by how some of the early immigrant groups (e.g. Germans) have 'made it' to the superior accommodation of Zone III, replacing second-generation American families who had moved

out to settle the outer residential zone, Zone IV. Burgess was not unaware of the many other factors that influence city growth. (For example, in a less well-known model he postulated a relationship between residential status and altitude in 'hill cities'.)[8] Although Burgess maintained that his model would apply to the then-contemporary American city, he did

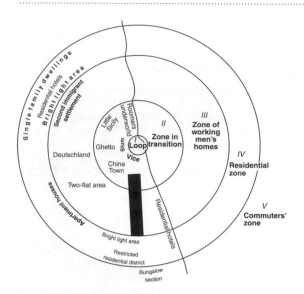

Figure 7.2 Burgess's concentric-zone model of urban land use, applied to Chicago

Source: R. Park and E. Burgess (Eds) (1925) *The City* Chicago: University of Chicago Press

TABLE 7.1   ASSUMPTIONS AND PRINCIPLES OF THE CONCENTRIC-ZONE MODEL OF URBAN LAND USE

1. Cultural and social heterogeneity of the population
2. Commercial–industrial base to the economy of the city
3. Private ownership of property and economic competition for space
4. Expanding area and population of the city
5. Transport is equally easy, rapid and cheap in every direction within the city
6. The city centre is the main centre for employment and near this centre space is limited; competition for this space is high, and therefore it is most valuable. The opposite is true of peripheral areas
7. No districts are more attractive because of differences in terrain
8. No concentrations of heavy industry
9. No historic survival of an earlier land-use pattern in any district

Plate 7.1   The high-rise central business district of San Francisco CA forms a backdrop to the low-rise high-status neighbourhood of Nob Hill

not expect any one city to be a perfect example of the theory.

Subsequent attempts to apply the model have often been less than successful, partly because they failed to recognise its limiting assumptions (Table 7.1). For example, the model is based on the concept of a city with a large population undergoing rapid expansion, with much of the population increase assumed to be due to the arrival of ethnically diverse immigrants from overseas. Both assumptions were met in Chicago, which from 1860 to 1910 increased its population almost twentyfold. The Burgess model was also formulated on the basis of a particular set of economic and political circumstances. In particular, the model assumed private ownership of property and the absence of any city planning constraints on the use of private property. Under these circumstances, property owners were free to develop their land as they wished. It also meant that only the wealthy could afford to live in the better locations away from inner-city

BOX 7.2

Burgess's concentric zone model of urban land use

Zone I

The first and smallest zone is the *central business district* (CBD). This is the focus of the commercial, social and cultural life of the city, and the area where land values are highest. Only activities where profits are high enough to meet the rent demanded can locate in the area. The heart of the zone is the downtown shopping area, with large department stores and the most exclusive shops. The area also contains the main offices of financial institutions, the headquarters of civic and political organisations, the main theatres and cinemas, and the more expensive hotels.

The CBD is the most accessible area in the city. It has the greatest number of people moving into and out of it each day, and the main transport termini are therefore located there. Forming the outer ring of the central area is a wholesale business district with warehouses, light industries and, perhaps, a market. The CBD contains the original nucleus of the settlement, but only scattered pockets of residences remain.

Zone II

Immediately adjacent to the CBD is the *zone in transition*. Early in the history of the city this formed a suburban fringe that housed many of the merchants and well-to-do citizens. With the growth of the city, however, industries encroached into this zone from the inner zone, and the quality of the residential environment deteriorated. The inner margins of the zone in transition are industrial and its outer ring is composed of declining neighbourhoods. The once fashionable town houses have been converted into flats, furnished rooms and even small industries. The population of the zone is heterogeneous and includes first generation immigrants as well as older residents. It is also an area frequented by vagrants and criminals, and rates of crime and mental illness are the highest in the city.

Those who own property in the zone are interested only in the long-term profit to be made from selling out to businesses expanding from the central area, and in the shortterm profits that accrue from packing in as many tenants as possible. As a result, property is run-down. The zone is characterised by a highly mobile population. Not surprisingly, as people prosper they tend to move out into Zone III, leaving behind the elderly, the isolated and the helpless.

Zone III

This is termed the *zone of independent working men's homes*. The population consists of the families of factory and shop workers who have managed to prosper sufficiently to escape the zone in transition but who still need cheap and easy access to their workplaces. The zone is focused on factories, and its population forms the bulk of what may be termed the respectable working class. Unlike in the 'childless' zone in transition, all age groups are represented.

Zone IV

This is an area of *better residences*, a zone of private housing or good apartment blocks. It is the home of the middle class. At strategic locations, subsidiary shopping centres have developed as mini versions of the downtown shopping area.

Zone V

Still farther out from the inner city is the *commuter belt* within thirty to sixty minutes' journey time of the CBD. This is essentially a suburban dormitory zone characterised by single-family dwellings.

Beyond these five main zones Burgess sometimes recognised two additional areas comprising:

Zone VI

The surrounding *agricultural district*.

Zone VII

The wider *hinterland* of the city.

---

slums. These conditions were again met by the Chicago of the 1920s. But in most Western societies today, government has intervened in the housing and property market (see Chapter 8). As a result, the slum housing of the model's zone in transition has been replaced, in many cases, by public redevelopment schemes. Further, around every major British city there are large estates of council housing provided by the government for people who would be unable to compete for such locations in the open market envisaged by Burgess. The general applicability of the model is further reduced by the gentrification of some inner-city slums, and by the continued association between high social status and inner-city residence in many European cities.[9] The value of the concentric-zone model is therefore limited historically and culturally. The model cannot be applied universally, and even within the USA it has become dated. Nevertheless, while the explanatory power of the model is limited in today's world, some of the constituent land-use zones can still be recognised, and it remains a useful pedagogic device against which to test real-world cities.

The earliest constructive criticism of Burgess's model emerged from an analysis of the internal residential structure of 142 American cities by Hoyt (1939).[10] By mapping the average residential rent values for every block in each city, Hoyt concluded that the general spatial arrangement was characterised better by sectors than by concentric zones (Figure 7.3). The resultant model of urban land use starts with the assumption that a mix of land uses will develop around the city centre, then, as the city expands, each will extend outwards in a sector. In this manner the high-rent neighbourhoods of the wealthy follow a definite path along communication lines, on high ground free from flood danger, towards open country, or along lake or river fronts not used by industry. Conversely, low-income groups with limited housing choice consume the obsolete housing of the wealthy, now converted into apartments, or occupy less desirable zones. The sectors undergo growth and change over time but according to the model, outward change occurs only

within sectors. The whole sector may not be geographically or socially similar at any one time, with, for example, better-quality housing moving towards the periphery, leaving decaying housing nearer the centre. A major contrast between the models of Burgess and Hoyt is that whereas residential change is stimulated on the *demand side* in Burgess's model, with immigrants competing for inner-city housing, Hoyt stresses *supply-side* mechanisms, with the construction of new housing for the middle classes on the urban periphery (and subsequent **filtering** of vacated dwellings) being the catalyst for socio-spatial change. Hoyt's model does not replace the concentric-zone scheme but extends it by adding the concept of direction to that of distance from the city centre. A major weakness of the theory is that it largely ignores land uses other than residential, and it places undue emphasis on the economic characteristics of areas, ignoring other important factors, such as race and ethnicity, which may underlie urban land-use change.

The excessive simplicity of the concentric ring and sector models of the city was addressed by Harris and Ullman (1945), who observed that most large cities do not grow around a single CBD but are formed by the progressive integration of a number of separate

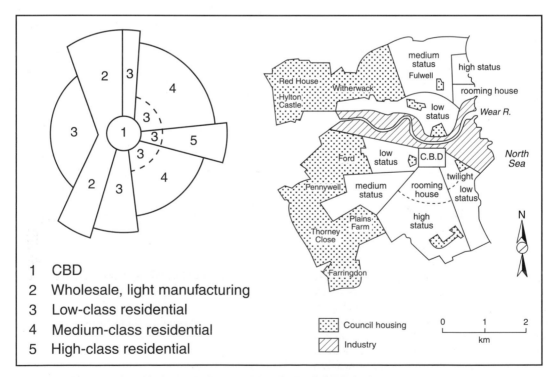

1 CBD
2 Wholesale, light manufacturing
3 Low-class residential
4 Medium-class residential
5 High-class residential

Figure 7.3 Hoyt's sector model of urban land use, and its application in Sunderland

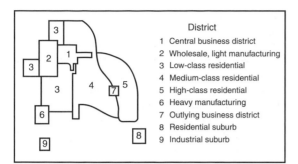

District

1 Central business district
2 Wholesale, light manufacturing
3 Low-class residential
4 Medium-class residential
5 High-class residential
6 Heavy manufacturing
7 Outlying business district
8 Residential suburb
9 Industrial suburb

Figure 7.4 Harris and Ullman's multiple-nuclei model of urban land use

nuclei[11] (Figure 7.4). The location and growth of these multiple nuclei are determined by a number of controlling factors:

1. Certain activities require specialised facilities and congregate where these are available. Industry, for example, requires transport facilities and is often located close to railway lines, major roads or port facilities.
2. Similar activities group together to profit from external economies of association, leading to the emergence of specialised legal districts or financial quarters.
3. Some activities repel each other owing to negative externality effects, as seen in the separation of high-income residences from industry.
4. Some activities which could benefit from a central location in or near the CBD, but which cannot afford the high rents demanded, must locate elsewhere. Warehousing or grocery wholesaling are examples of activities that require large structures and would benefit from a central location but are forced to 'trade off' space for accessibility (Box 7.3).

The value of the Harris and Ullman model lies in its explicit recognition of the multinodal nature of urban growth. Furthermore, they argue that land uses cannot always be predicted since industrial, cultural and socio-economic values will have different impacts on different cities. While the Burgess zonal pattern and, to a lesser extent, the Hoyt sectoral pattern suggest inevitable pre-determining patterns of location, Harris and Ullman suggest that land-use patterns vary depending on local context. Hence the multiple-nuclei model may be closer to reality. In practice, elements of all these models may be identified in many large Western cities. In London,

for example, the annular rings of growth reflect Burgess, and a clear distinction can be drawn between an older and poorer inner city and more affluent and modern outer suburbs. Superimposed on this is a pattern of sector development with a zone of local authority and workers' dwellings from the latter part of the industrial revolution extending from the East End to Dagenham and beyond. To the north and west an affluent residential sector extends from Mayfair to St John's Wood into Hampstead and on into the 'stockbroker belt' of the Chiltern Hills. Finally, multiple nuclei can be found at various scales, the most evident being the financial centre of the City or the concentration of medical services around Harley Street.

One of the most severe criticisms of the 'classical' models of urban land use referred to their economic bias and consequent neglect of cultural influences on urban land-use patterns. In an early study, Firey (1947) demonstrated that neither the concentric zone nor a sector theory was adequate in explaining land-use patterns in Boston MA, where non-economic considerations, centred on 'sentiment and symbolism', lay behind the spatial juxtaposition of the fashionable residential area of Beacon Hill and an area populated by low-income immigrants and their descendants.[12] Firey's work was significant in illustrating how social values could override economic competition as the basis for socio-spatial organisation. Firey recommended a 'cultural ecology' approach instead of urban ecology in order to take into account specific cultural and historical factors influencing a city's land-use patterns. In this he anticipated many of the arguments of postmodernism.

Subsequent refinements of the ecological approach have set aside the crude biotic analogy but have retained useful concepts such as natural areas, albeit reformulated as 'social areas' or 'neighbourhood types' (see Chapter 18). Other work on ecological patterns in cities has sought to reform the traditional models to provide concepts of more direct relevance to contemporary urban society. Four of these merit further consideration.

MODIFICATIONS OF THE CLASSICAL URBAN MODELS

MANN'S MODEL OF A TYPICAL BRITISH TOWN

One of the limitations of the classical ecological models was their specific focus on the US city.

BOX 7.3

The trade-off model of urban land use

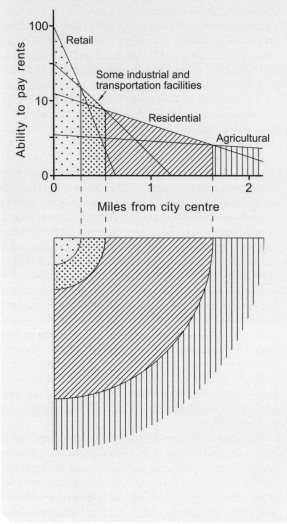

The mainspring of the concentric-zone model of urban land use is the expansion of the inner zone outwards. This movement is triggered by excessive demand for central city land. The neoclassical economics 'trade-off' model employs the concept of bid-rent curves to explain why demand for land, and therefore land-use patterns, vary across the urban area. The basis of the model is the relationship between accessibility and land rent. The more accessible a location the greater the demand for it, which is reflected in the distribution of land values. In the model the city centre is assumed to be the most accessible and therefore most valuable location. Since some land uses place greater importance on accessibility, they are prepared to pay higher rents for central locations.

As the model shows, under normal conditions a department store will always outbid a housing developer for a central city site because the store's success is more dependent on its being accessible to as large a population as possible. Land uses that need to be accessible but which cannot afford to pay the rents demanded for sites in the CBD tend to locate in the surrounding zone in transition. In this way the CBD expands outwards and so sets the pattern of city growth in motion. Different land uses evaluate sites in terms of attributes such as size and proximity to desired facilities (such as clients, workplace or open country). Since each is limited in terms of what it can afford to pay for any site, consumers may have to trade off space against accessibility. (For example, a household may occupy a small city-centre apartment to gain accessibility to workplace or leisure facilities, trading these off against a more spacious single-family dwelling in a suburban commuter location.) As a result of such trade-offs, different land uses occupy different locations in the city in a process that produces the urban land-use pattern.

Mann (1965) combined elements of the Burgess and Hoyt models in his model of a typical medium-size British city.[13] The model also incorporated a climatic consideration relevant to the UK by assuming a prevailing wind from the west. As Figure 7.5 shows, in the model:

1. The best residential area (A) is located on the western fringe of the city, upwind of and on the opposite side of town from the industrial sector (D).

2. The areas of the working class and the main council estates (C) are located close to the industrial zone.

3. The lower middle-class housing (B) borders on each side of the best residential area.

4. The model also identifies a CBD, a transition zone, a zone of small terraced houses in sectors C and D, larger housing in sector B, large old houses in sector A, post-1918 residential areas with post-1945 housing added on the periphery, and dormitory settlements at commuting distance from the city.

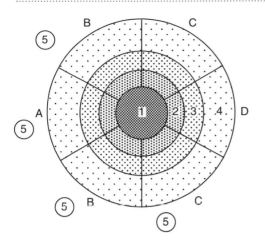

1 City centre

2 Transitional zone

3 Zone of small terrace houses
in sectors C and D, bye-law
houses in sector B, large old
house in sector A

4 Post-1918 residential areas
with post-1945 development
mainly on periphery

5 Commuting distance villages

A Middle-class sector

B Lower-middle-class sector

C Working-class sector (and main
municipal housing areas)

D Industry and lowest
working-class areas

Figure 7.5 Mann's model of a typical medium-size British city
Source: P. Mann (1965) *An Approach to Urban Sociology* London:
Routledge

KEARSLEY'S MODIFIED BURGESS MODEL

Kearsley's model was an attempt to extend Mann's model of urban structure to take into account contemporary dimensions of urbanisation such as the level of governmental involvement in urban development in Britain, slum clearance, suburbanisation, decentralisation of economic activities, gentrification and ghettoisation[14] (Figure 7.6). Manipulation of the model's various elements – such as the extension of inner-city blight, minimisation of local and central government housing, and expansion of recent low-density suburbs – offers a North American variant of the basic model.

VANCE'S URBAN REALMS MODEL

By extending the principles of the multiple-nuclei model, Vance (1964) proposed the urban realms model.[15] The key element is the emergence of large self-sufficient urban areas each focused on a downtown independent of the traditional downtown and central city. The extent, character and internal structure of each 'urban realm' is shaped by five criteria:

1. The terrain, especially topographical and water barriers.
2. The overall size of the metropolis.
3. The amount of economic activity within each realm.
4. The internal accessibility of each realm in relation to its dominant economic core.
5. Interaccessibility among suburban realms. Particularly important here are circumferential links and direct airport connections that no longer require them to interact with the central realm in order to reach other outlying realms and distant metropolises (Figure 7.7).

Though conceived on the basis of work on the San Francisco Bay area, the model has subsequently been applied to describe the general land-use structure of other US cities.[16]

WHITE'S MODEL OF THE TWENTY-FIRST-CENTURY CITY

Since publication of the three classical models of urban land use many new forces have come to influence urban growth. These reflect societal changes such as deindustrialisation of the urban economy, the emergence of a service economy, the dominance of the automobile, a decrease in family size, suburban residential developments, decentralisation of business and industry, and increased intervention by government in the process of urban growth. White (1987) proposed a revision of the Burgess model that incorporates these trends in order to guide our understanding of the twenty-first-century city[17] (Figure 7.8).

1. *Core.* The CBD remains the focus of the metropolis. Its functions may have changed over the years but it still houses the major banks and financial institutions, government buildings and corporate headquarters as well as the region's main cultural and entertainment facilities. A few large

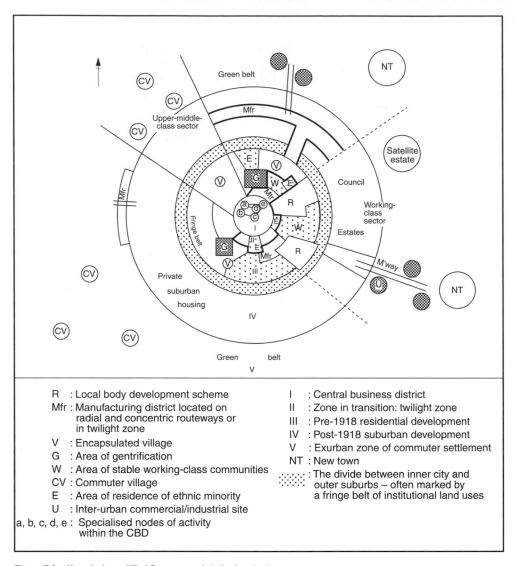

Figure 7.6   Kearsley's modified Burgess model of urban land use
Source: G. Kearsley (1983) Teaching urban geography: the Burgess model *New Zealand Journal of Geography* 12, 10–13

department stores retain flagship establishments downtown, but most retailing has moved with the affluent population to the suburbs, with many remaining outlets being speciality stores catering for daytime commuters.

2. *Zone of stagnation*. While Burgess expected investors from the CBD to expand into the zone in transition, White depicts the area as a zone of stagnation. He argues that rather than extending outwards spatially, the CBD expands vertically. Lack of investment in the zone was compounded by the

effects of slum clearance, highway construction, and the relocation of warehousing and transport activities to suburban areas. Although some older US industrial cities (such as Cleveland OH) have sought to revitalise the zone through conversion of buildings into entertainment, shopping and residential areas, younger cities (such as Dallas TX) have abandoned the zone altogether.

3. *Pockets of poverty and minorities*. These comprise highly segregated groups living on the fringes of society, including the homeless,

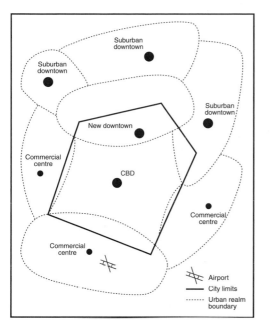

Figure 7.7   Vance's urban realms model, as applied to Atlanta

Source: J. Vance (1964) *Geography and Urban Evolution in the San Francisco Bay Area* Berkeley CA: Institute of Local Government Studies, University of California; T. Hartshorn and P. Muller (1989) Suburban downtowns and the transformation of metropolitan Atlanta's landscape *Urban Geography* 10(4), 375–395

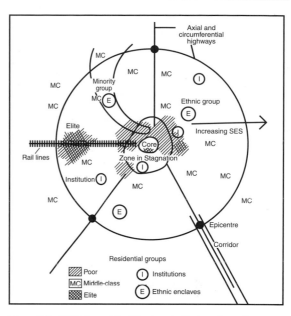

Figure 7.8   White's model of the twenty-first-century city

Source: M. White (1987) *American Neighborhoods and Residential Differentiation* New York: Russell Sage Foundation

addicts, dysfunctional families, the **underclass**, and members of minorities. The surroundings reflect their status, being dominated by deteriorating housing in blighted neighbourhoods. These slum areas are found mostly in the inner city skirting the zone of stagnation, but a few are also located in older suburbs.

4. *Elite enclaves.* The wealthy have the greatest choice of housing environment and are able to insulate themselves from the problems of the metropolis. Most live in neighbourhoods on the urban periphery in expensive houses on spacious lots. 'Gilded' neighbourhoods also remain in the central areas of older large metropolises.

5. *The diffused middle class.* These areas occupy the largest area of the metropolis and are spatially concentrated between the outer edge of the central city and the metropolitan fringe. This suburban zone is characterised by social diversity:
   ■ In the interior sections are older settled neighbourhoods which are now in transition as the original settlers have raised their families and are moving to other dwellings.

Some of these neighbourhoods, often adjacent to the central city, are attracting the black middle class. Although large numbers of African-Americans have moved to the suburbs in recent decades they remain highly segregated.
   ■ Farther out there are the archetypal suburban communities comprising married couples with small children living in single-family detached homes built on spacious lots. The suburbanisation of business and industry means that nucleations of other social groups are also present, with working-class families living in more modest neighbourhoods, the elderly in garden apartments and retirement communities, singles in apartment complexes and ethnic minorities in their own enclaves.

6. *Industrial anchors and public sector control.* Industrial parks, universities, R&D centres, hospitals, business and office centres, corporate headquarters and other large institutional property holders can exert a major influence on patterns of land use and residential development. Institutional actors and other members of a local growth coalition (see below) can pressure city government to modify zoning, lower taxes and construct infrastructure. The location of such activity (e.g. the

siting of a large shopping mall) is of considerable importance in shaping the urban structure.

7. *Epicentres and corridors*. A distinguishing feature of the evolving twenty-first-century metropolis is the emergence of peripheral epicentres located at the convergence of an outer beltway and axial superhighway and providing a range of services to rival those of the CBD. Corridor developments, as along Route 128 near Boston or the Johnson Freeway in Dallas, can also act as a focus for intensive economic activity.

The classical models together with more recent modifications provide a powerful insight into the changing structure of the Western city. A major deficiency, however, is that only limited explicit consideration is given to the processes underlying the revealed patterns of land use. This criticism underlay development of the political economy interpretation of urban change.

A POLITICAL ECONOMY PERSPECTIVE

Despite some success in describing general patterns of urban land use the traditional ecological models, and in particular their positivist basis in neoclassical economics, were criticised in the early 1970s as:

1. *Mechanistic*, viewing humans as rational decision-makers operating in an abstract environment.
2. *Ideological*, retaining the myth of value-free research while legitimising market capitalism and retention of the socio-economic status quo.
3. *Devoid of ethical content*, since questions of equity and fairness of social conditions and resource allocation were excluded.

Energised by the development of a policy-oriented and relevance perspective in human geography, urban geographers sought interpretations of urban change that revealed the structural forces underlying observed land-use patterns. This led researchers such as Harvey (1975) and Castells (1977) to focus explicitly on the place of the city in the capitalist mode of production.[18] This neo-Marxist perspective was based on the premise that:

if the city is considered to start with as a market where labour, power, capital and products are exchanged, it must be equally accepted that the geographical configuration of this market is not the result of chance; it is governed by the laws of capital accumulation.

(Lamarche 1976 p. 86)[19]

The city in advanced capitalist societies is regarded as a particular built form commensurate with the fundamental capitalist goal of accumulation (the process by which capital is reproduced at an ever-increasing scale through continued reinvestment of profits). Thus, as well as concentrating the means of production through agglomeration, cities also develop an infrastructure that facilitates the geographical transfer of profits in search of optimum investment opportunities. Harvey (1985) refers to this process as the 'circulation of capital' and sees it as a key factor in urban development.[20]

As Figure 7.9 shows, Harvey envisaged three *circuits of capital*:

1. The *primary* circuit refers to the structure of relations in the production process (e.g. the manufacture of goods for sale). Surplus value (profits) created in the production process is either reinvested in the primary circuit with a view to generating further profit or, in the event of overproduction (or **underconsumption**), may be channelled via the capital market into the secondary or tertiary circuits.
2. The *secondary* circuit involves investments in fixed capital, such as the built environment (e.g. property development), in the expectation of realising profits either:
   - ■ In the form of rental income from the **use value** of the building; or
   - ■ From the enhanced future **exchange value** (sale price) of the building.
3. The *tertiary* circuit involves investment in science and technology that leads ultimately to increases in productivity, or investment in improving labour capability through education or health expenditure. Much of this investment is made collectively by the state, since individual capitalists are unlikely to take a long-term view of the potential advantages.

Significantly, there is a limit to the process of capital transfer from the primary to secondary circuits (since the market can absorb only so many office buildings or leisure centres). When this point is reached, investments become unproductive and the exchange value of capital put into the built environment is reduced or in some instances lost completely (resulting in bankruptcy for some investors and redundancy for the work

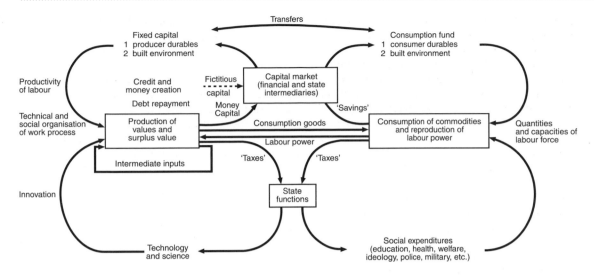

Figure 7.9 Harvey's model of the circulation of capital
Source: D. Harvey (1978) The urban process under capitalism *International Journal of Urban and Regional Research* 2(1), 101–31

force). Since these events occur in specific locations, the outcome is an uneven geography of development, with some areas (e.g. the CBD) benefiting from an inflow of capital investment and other areas (e.g. 'off-centre' commercial areas) tending to decline owing to lack of capital investment.

However, the devaluation of *exchange value* does not necessarily destroy the *use value* of a building or site that can be used as the basis of further development. Since growth (the accumulation of profits) is a constant requirement of the capitalist mode of production, devaluations of fixed capital (e.g. an existing building) represent one of the main ways in which capitalism can speed the accumulation of new capital value (e.g. by redevelopment). This process is illustrated graphically by Feagin (1987), who describes how a property developer in Houston demolished a structurally sound twenty-storey office block to make way for an eighty-two-storey supertower.[21] There is, therefore, a major contradiction in the capitalist city between the capitalist dynamic of accumulation (provoking urban growth and change) and the inertia of the built environment (which resists urban change). As Harvey (1981 p. 113) explained:

> under capitalism there is then a perpetual struggle in which capital builds a physical landscape appropriate to its own condition at a particular moment in time, only to destroy it, usually in the course of a crisis, at a subsequent point in time.[22]

As we shall see later (Chapter 14), the transfer of capital from the primary to secondary circuit in the form of speculative investment in the built environment underlay the property boom experienced by the large cities of Europe, North America and Australia during the late 1970s and early 1980s in the wake of the oil price shock of 1973.

The transfer of capital from the primary to the secondary circuit of Harvey's model has also been postulated to be a major cause of post-Second World War suburbanisation in the USA.[23] Examination of the relationship between the circulation of capital and post-war suburbanisation in the USA enables us to relate theory to the real world and illustrates how the restructuring of capital also involves a restructuring of space. According to this thesis, as a result of the inability of the domestic market to absorb the industrial surpluses that built up as the American war machine returned to peace-time production, other labour- and capital-absorbing activities were promoted by successive governments in the 1950s and 1960s, including suburban capital formation. By engineering a shift of investment into the secondary circuit, the state and specialist financial institutions avoided a crisis of over-accumulation in the primary circuit and simultaneously stimulated new demand for industrial goods in the housing and transportation sectors (Box 7.4). The role of government in promoting post-war low-density suburbanisation by subsidising the costs of land development has also been explored for metropolitan Toronto,

BOX 7.4

The over-accumulation crisis and post-war suburbanisation in the USA

The scale of post-Second World War peripheral expansion in the USA was unprecedented. Between 1950 and 1955, as the US population grew by 11.6 million, 79 per cent of the growth was in the suburbs, and the total volume of new suburban residential construction was three times that in the central cities.[24] This process was fostered by federal government housing and transport policy.

While the 1949 and 1954 Housing Acts laid out a strategy for slum clearance and the rebuilding of inner-city neighbourhoods, in practice federal subsidies flowed overwhelmingly to suburban housing development. In the period 1950–4 the number of low-rent public housing starts fell from 75,000 units to 20,000 units annually, while funding for private suburban housing was liberalised, with increased upper limits for Federal Housing Administration (FHA) home mortgage insurance and relaxed FHA loan terms for single-family dwellings in suburban areas. The electoral significance of the tax deductions associated with owner-occupied dwelling also acquired increasing significance as the level of home ownership grew. As Badcock (1984 p. 135) observed, 'this powerful orientation in the USA towards home owners at the expense of tenants, new construction in the suburbs in preference to the rehabilitation or selective redevelopment of inner city housing, and the largest builders in the industry, had far-reaching consequences for the metropolitan built environment'.[25]

The federal highway programme was also of major importance to post-war suburbanisation. Between 1944 and 1961 the federal government allocated its entire transport budget of US$156 billion to roads.[26] The major commitment to highway construction was signalled in the 1956 Federal-Aid Highway Act, which provided 90 per cent federal support for the construction of the 65,156 km (40,462 mile) Interstate Highways and Defense System. The urban expressway system, constructed as part of the interstate highway programme, served to connect dormitory suburbs with downtown office jobs. Equally important, the 1956 Federal-Aid Highway Act required that central city bypass routes must be built into the urban highway network, and by the early 1970s more than eighty of these circumferential beltways had been constructed in the outer areas of US metropolitan areas.[27] The enhanced accessibility of suburban highway-oriented locations attracted regional shopping centres, industrial and office parks and apartment estates to the interchanges along the beltways. Berry and Kasarda (1977) estimated that between 1960 and 1970 the growth of blue-collar employment in the suburbs was 29 per cent, compared with a loss of 13 per cent for the central cities, while for white-collar jobs the suburbs recorded an increase of 67 per cent over the period, compared with 7 per cent for the central cities.[28]

The over-accumulation/under-consumption thesis aids our understanding of the suburbanisation process in the USA but is of doubtful relevance to the situation in Britain, where successive post-war governments have sought to control the outward spread of cities through the operation of a centrally directed planning system (see Chapter 8). In consequence the physical containment of British cities differentiates them from the leapfrog subdivisions and suburban sprawl of US metropolitan centres.

where, between 1955 and 1964, discriminatory public funding of highway and service infrastructure supported expansion of the outer suburbs.[29]

The political economy approach affords valuable insight into the key processes and agents responsible for the production of the built environment of the capitalist city. This perspective has exposed the roles of and relationship among various factions of capital in influencing urban change. It also highlights the impact that economic and political processes located outside the territory of any particular city have on its internal structure and development. The political economy perspective has been criticised, however, for its apparent reification of the market, and for giving insufficient attention to *human* geography. This critique echoes the view of Form (1954), who argued that the *social* organisation of the land market was more relevant than economic models that depicted the city as a free market in which individuals competed impersonally.[30] In a study of Lansing MI, Form identified the relationships between the dominant agents in the urban land market, and the ways in which each influenced land-use patterns. This viewpoint developed into the concept of urban managerialism.[31] It also informed the humanistic critique of political economy which refocused research attention on the role of human agency in the production of the built environment (see Chapter 2).

MAJOR ACTORS IN THE PRODUCTION OF THE BUILT ENVIRONMENT

The land development industry comprises a variety of builders, subcontractors, architects, marketing agents, developers and speculators, together with their legal and financial consultants. During the conversion of rural land into occupied housing a plot may pass through the ownership of at least five different actors: a rural producer, a speculator, a developer, a builder and finally a household. Assisting with land transfers at each stage is a set of facilitators, including real-estate agents and financiers. Finally, at local and central government level, planners and officials oversee the development process to varying degrees according to the prevailing social formation (Figure 7.10). The motives and methods of these participants vary considerably (Box 7.5). While bearing in mind the interre-

lated nature of the land-development process, we can, in the interests of clarity, examine the operation of several of these individual actors in detail.

SPECULATORS

Property speculators – either individual entrepreneurs or corporations – purchase land with the hope of profiting from subsequent increases in property values. Speculative activity is a characteristic of capitalist urban development that occurs throughout the urban arena. As we saw in our discussion of urban political economy, speculation in the central city can contribute to the creation of slums prior to revitalisation of a neighbourhood either through private-sector upgrading or gentrification or through publicly financed rehabilitation (see Chapter 10). The effects of speculative

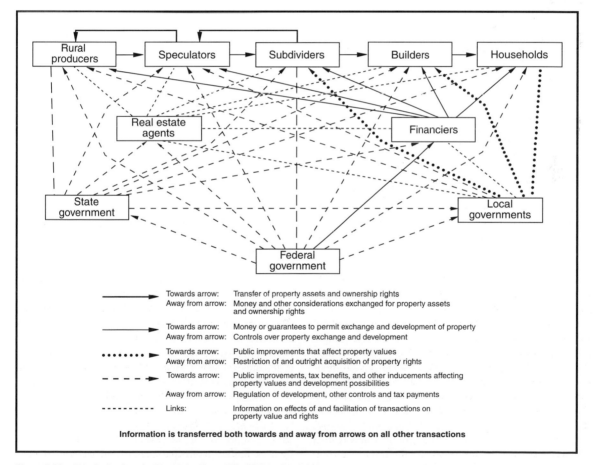

Figure 7.10   Principal actors in the production of the built environment

BOX 7.5

Major agents and motives in the capitalist land development process

The motives and methods of different participants in the land development process vary considerably. Participants include the following groups:

1. *Rural producers*. Landowners, who are primarily concerned with the productive capability of their land, the most obvious group being farmers.
2. *Speculators*. These may own land that is still in productive use, but their basic interest lies in its appreciating value. (They hope to buy low and sell high.) Their decisions are based on financial considerations (such as depreciation rates, and the comparative viability of alternative investment opportunities). Speculators assemble land for subsequent development but can also withhold land from the market, thereby forcing prices up.
3. *Developers*. These can be categorised as subdividers and builders. In the USA the former install basic infrastructure and facilities while the latter construct and sell houses on the prepared lots. In Britain the role of subdivider and builder is usually combined.
4. *Households*. Those who purchase or lease the units are motivated by the functional utility of the house as a place to live (use value) and the land development process improvement, or at least maintenance, of the financial investment represented by their property (exchange value).
5. *Real-estate agents*. These purvey information and act as intermediaries between buyer and seller. Their rewards come from commission charged on each land transaction completed, so agents have a vested interest in promoting property transfers. Some estate agents also engage in speculation.
6. *Financiers*. These provide the funds necessary to the development process. Their decisions are based on a combined desire to obtain the highest possible rate of return on an investment or loan and to minimise or avoid risk.
7. *Other facilitators*. Other professionals involved include lawyers who represent clients in disputes that may arise during the development process, as well as consultants who advise other agents.
8. *Government*. All governments influence the process of urban development, although the level of involvement varies. In Britain a centrally directed planning system determines the framework within which local authorities operate. In North America local governments are less constrained by national legislation.

activity on the residents of the existing built-up area can be pronounced, often leading to **displacement** and the destruction of communities. The impact of speculative development on land use is seen most starkly around the peripheries of cities. In the USA, the area under urban use is over 10 million acres (4 million ha), but 'twice as much land is withdrawn from other uses because of the leapfrogging which characterises much suburban growth'.[32] Leapfrog development usually occurs as builder-developers try to avoid land that is tied up in complex legal arrangements or is being held by speculators in anticipation of very high profits. Some indication of the level of profits possible is provided by the fact that in Los Angeles the price of residential land increased at a rate of 40 per cent per year in the late 1970s, in part owing to speculative activity. In areas around expanding cities in Japan the multiplier between land values for agricultural use and urban use is over two hundredfold,[33] while around some British towns such as Reading in Berkshire the price of £2,000 per acre for agricultural land can rise to £1 million if

planning permission for residential development exists.[34] Such inflationary costs are, of course, built into the final purchase price for suburban housing.

On the urban fringe, the effects of land speculation constitute one of the most important impacts on land use. For agriculture, speculative land-holding can have both positive and negative effects. Beneficial interaction can occur through the rental back to farmers of farmland that has been purchased by non-farm interests. Provided that the lease conditions are not onerous, from a purely economic perspective it may be attractive for a farmer to rent land rather than to purchase it and encumber the business with a heavy mortgage, thus releasing more of the farmer's capital for improvements. On the other hand, rising land prices and land speculation make farm enlargement costly, and where tax is levied on *land values*, as in the USA, sales of surrounding land can push up property taxes. Tax pressures coupled with other urban shadow effects such as pollution, trespass, theft and vandalism may eventually force the suburban agriculturalist to sell out to speculators. The impact of *potential* urban development also affects land husbandry practices on the fringe. Where urban pressures are strong, farmers may become active speculators, disinvesting in their farms while anticipating a large capital gain from the sale of their land in the near future. Some farmers may 'farm to quit' or attempt to 'mine' the soil's fertility, while others may 'idle' their farmland. In the absence of effective land-market regulation, leaving land idle may be a perfectly rational land-use response to the economic incentives created by the urban-fringe property market. Berry and Plaut (1978), for example, estimated that for every acre converted to urban uses in the north-eastern USA another was idled owing to urban pressures. In Japan, one-third of paddy fields, amounting to 750,000 ha (2,900 square miles) are left idle by speculators in the hope of conversion to urban use.[35] Farmers living under less intense urban pressures may be involved in a more passive form of land speculation, watching the appreciation of land values with a view to selling for a large profit on retirement.[36]

REAL-ESTATE AGENTS

Although the principal role of real-estate agents is as middlemen between buyers and sellers of property, some adopt a broader remit and may operate in the assembly of small land parcels for development or as speculators in the urban land market.[37] Direct involvement in the land market can be particularly profitable during periods of inflation and can lead to manipulation of the land market. Gutstein (1975) found that in a resurgent inner-city neighbourhood in Vancouver in the early 1970s some properties were 'sold' several times in a single year between holding companies with the same owner in order to force up rent levels in the local market.[38]

Similar instances of land speculation have been recorded in areas of racial transition where real estate agents also often play an active role in **blockbusting** – a process whereby members of minority (black) groups obtain entry to residential areas previously reserved exclusively for the majority (white) population[39] (see Chapter 10). A real-estate speculator who has bought a property in a block is able to resell to a black family for a substantial profit when demand from the minority group reaches a critical level.[40] Frequently this initial sale stimulates existing homeowners to sell up (owing to fears about falling property prices and/or racial bias) and reduces demand from other majority-group buyers.[41] It can be argued

that the blockbuster is providing much-needed accommodation to groups that are discriminated against in the open housing market. However, in some cases the blockbuster is guilty of exploiting the anxieties of residents (possibly forcing them to sell below the market price) and the housing poverty of minority groups (who are willing to pay above market prices) to realise a significant capital gain. Foreman (1971) estimated that in Chicago during the 1950s and 1960s a small minority of real-estate agents triggered the blockbusting of 70,000 white families over a decade.[42] Similarly, Orser (1990) has described how the west Baltimore neighbourhood of Edmondson Village experienced real-estate blockbusting and massive 'white flight' in response to racial transition between the late 1950s and the mid-1960s. Over a period of less than a decade almost total racial change took place within one of the two **census tracts**, a new population of 11,000 replacing a previous population of 9,000.[43]

The social costs of blockbusting in terms of racial disharmony, resentment and harassment are less easy to calculate. Racial and ethnic steering by real-estate agents is a common practice that contributes to patterns of residential segregation[44] (see Chapter 18). More generally, real-estate agents can influence the social composition of neighbourhoods by directing people to particular housing areas on the basis of their perception of the market, their partial knowledge of the city, and the normal operating territory of their company.[45]

FINANCIAL INSTITUTIONS

Financial institutions have grown in importance in both Britain and North America with the decline in private rented housing and the growth of home ownership. In common with developers, financial institutions seek to maximise profits and minimise risks. Since these opportunities are differentiated over the urban area, financiers adopt spatially discriminating lending practices, a fact that will have a significant impact on the location of new construction as well as on maintenance and improvements to existing structures. The practice of **red-lining** areas by mortgage lenders is well documented. Lambert (1976 p. 33) found in a study of Birmingham that 'the building societies deliberately steered clear of certain of the older areas . . . although they usually emphasised that there was no written policy which precluded houses in these areas'.[46] A contemporary study of Leeds reported that an applicant for a building-society mortgage on a property in

Headingley was informed that 'it was doubtful if a mortgage would be available . . . owing to the nearness of the *blue zone*' – that is, an area with a high proportion of students and immigrants (Weir 1976).[47] A survey of mortgage finance managers in Bristol found that

> all the managers had heard of red-lining . . . and said they did not practise it. One manager stated categorically 'we do not red-line'. Later in the discussion he pointed to the St Paul's area on the map and said that 'there are certain areas in the city in which we won't lend'.
>
> (Bassett and Short 1981 p. 286)[48]

Discrimination has also been practised on racial grounds. In the USA, especially prior to the civil rights legislation of 1968, bank lending to African-Americans was based on the twin criteria of residence in an established black neighbourhood and in a 'good area'.[49] More recently, Stegman (1995) reported that African-Americans and Hispanics were 60 per cent more likely to be rejected for mortgage loans than whites even when financial and employment status and neighbourhood characteristics were controlled for.[50] Such practices may be stimulated by economics as well as prejudice, since banks, concerned to protect property values in areas where they have invested, may be uncertain about the long-term effect on values of racial transition. This inherent need to minimise risk continues to influence the spatial pattern of lending practices. Harvey (1975) has demonstrated how various financial agencies in Baltimore segmented the city into a series of sub-markets on the basis of the individual lending preferences[51] (Box 7.6). The practice of red-lining areas perceived as poor risks has also been followed by the insurance industry.[52]

Despite the passage of the US Mortgage Disclosure Act 1975, which requires lending agencies to disclose the geographic pattern of mortgage awards, discriminatory practices may continue largely through covert means such as discouraging would-be borrowers with higher interest rates, high down payments, lower loan to value rates and shorter loan maturity terms for property in red-lined areas.[53] In Atlanta GA analysis of mortgage-loan decisions in the 1980s and 1990s revealed that even after controlling for differences in borrower characteristics 'prime conventional home-purchase capital avoids a broad swath of Atlanta's inner-city neighbourhoods, as well as suburbs perceived at risk for race and class transition' (Wyly 2002 p. 95).[54] Despite the US Home Mortgage Disclosure Act 1995, and identification of expanding home-

ownership as a national priority in the American Homeownership and Economic Opportunity Act of 2000, red-lining of certain neighbourhoods remains a feature of the geography of housing investment in the city. Given the need for financial institutions to operate within a free-market environment, it is unrealistic to expect them not to engage in what they perceive to be 'economically rational' discrimination.[55] Nevertheless, the practice of red-lining ensures that property values will decline, and generally leads to neighbourhood deterioration. It can also engineer a flow of investment away from inner-city areas towards more affluent suburban home-owners. In a study of Boston, for example, it was discovered that levels of reinvestment of savings deposits in inner-city communities were between 3 per cent and 33 per cent, compared with levels of between 108 per cent and 543 per cent in the outer suburbs.[56] This flow of capital to the suburbs reflects the wider interests and operations of mortgage finance institutions, many of which are involved in financing and controlling the suburban activities of large construction companies.[57]

GROWTH COALITIONS

While we have so far considered the major agents of urban change independently, in practice **growth coalitions** operate to foster urban development. These networks, or **growth machines**,[58] exhibit several key characteristics:

1. The critical actors in growth coalitions tend to be those fractions of capital which are most place-bound, such as *rentiers*, who rely on an intensified use of land or buildings in a particular area for enhanced profits.
2. Growth-machine activists are concerned particularly with the exchange value of a place and tend to oppose government intervention that might regulate growth.
3. Growth networks are often combined private–public coalitions that support themselves through local pro-growth bureaucracies. In the fragmented local government system of the USA (see Chapter 20), local growth elites can exert an influential role in electing local officials, 'watchdogging' their activities and scrutinising administrative detail. However, since local-government bureaucracies are also sensitive to citizen demands (for reasons of political legitimation), the pro-growth stance can be modified by popular pressure.

BOX 7.6

Housing sub-markets in Baltimore MD

Harvey (1975) demonstrated how various financial agencies in Baltimore segmented the city into a series of sub-markets on the basis of their individual lending preferences.

	Under $7,000	$7,000– $9,999	$10,000– $11,999	$12,000– $14,999	Over $15,000
Private	39	16	13	7	7
State S & Ls	42	33	21	21	20
Federal S & Ls	10	22	30	31	35
Mortgage banks	7	24	29	23	12
Savings banks	–	3	5	15	19
Commercial banks	1	1	2	3	7
Percentage of city's transactions in category	21	19	15	20	24

The first table shows the specialised niche in the housing price structure occupied by different lending institutions. Thus, for the lowest-grade housing, 81 per cent of mortgages derived from either state savings and loan (S&L) companies or from private arrangements. The higher perceived risk attached to this sub-market made it unattractive to the larger banks and federal savings and loan firms. The spatial expression of these lending practices is shown in the second table.

District	Total houses sold	Sales per 100 properties	Cash	Private	Federal S&Ls	State S&Ls	Mort-gage bank	Com-mercial bank	Savings bank	Other	FHA	VA	Average sale price ($)
Inner city	1,199	1.86	65.7	15.0	3.0	12.0	2.2	0.5	0.2	1.7	2.9	1.1	3,498
1. East	646	2.33	64.7	15.0	2.2	14.3	2.2	0.5	0.1	1.2	3.4	1.4	3,437
2. West	553	1.51	67.0	15.1	4.0	9.2	2.3	0.4	0.4	2.2	2.3	0.6	3,568
Ethnic	760	3.34	39.9	5.5	6.1	43.2	2.0	0.8	0.9	2.2	2.6	0.7	6,372
1. E. Baltimore	579	3.40	39.7	4.8	5.5	43.7	2.4	1.0	1.2	2.2	3.2	0.7	6,769
2. S. Baltimore	181	3.20	40.3	7.7	7.7	41.4	0.6	–	–	2.2	0.6	0.6	5,102
Hampden	99	2.40	40.4	8.1	18.2	26.3	4.0	–	3.0	–	14.1	2.0	7,059
W. Baltimore	497	2.32	30.6	12.5	12.1	11.7	22.3	1.6	3.1	6.0	25.8	4.2	8,664
S. Baltimore	322	3.16	28.3	7.4	22.7	13.4	13.4	1.9	4.0	9.0	22.7	10.6	8,751
High turnover	2,072	5.28	19.1	6.1	13.6	14.9	32.8	1.2	5.7	6.2	38.2	9.5	9,902
1. North-west	1,071	5.42	20.0	7.2	9.7	13.8	40.9	1.1	2.9	4.5	46.8	7.4	9,312
2. North-east	693	5.07	20.6	6.4	14.4	16.5	29.0	1.4	5.6	5.9	34.5	10.2	9,779
3. North	308	5.35	12.7	1.4	25.3	18.1	13.1	0.7	15.9	12.7	31.5	15.5	12,330
Middle income	1,077	3.15	20.8	4.4	29.8	17.0	8.6	1.9	8.7	9.0	17.7	11.1	12,760
1. South-west	212	3.46	17.0	6.6	29.2	8.5	15.1	1.0	10.8	11.7	30.2	17.0	12,848
2. North-east	865	3.09	21.7	3.8	30.0	19.2	7.0	2.0	8.2	8.2	14.7	9.7	12,751
Upper income	361	3.84	19.4	6.9	23.5	10.5	8.6	7.2	21.1	2.8	11.9	3.6	27,413

(Transactions by source of funds (%); Sales insured (%))

Note: VA stands for Veterans Administration.

Thus the inner city was heavily dependent on cash transactions and money raised from private sources. In the traditional ethnic districts of east and south Baltimore small-scale community-based savings and loan associations (S&Ls) dominated housing finance. Middle-income white areas of north-east and south-east Baltimore were basically served by federal S&Ls. The affluent areas drew on the financial resources of savings banks and commercial banks as well as federal S&Ls, while areas of high turnover and racial change were characterised by mortgage-bank finance combined with Federal Housing Administration (FHA) mortgage insurance. The black community of west Baltimore, historically discriminated against, also relied heavily on mortgage bank finance supported by FHA insurance even though in income terms it was broadly similar to those white areas served by federal S&Ls.

Source: **D. Harvey (1975) The political economy of urbanisation in advanced capitalist societies, in G. Gappert and H. Rose (eds)** *The Social Economy of Cities* **Beverly Hills CA: Sage, 119–62**

4. The power of pro-growth business coalitions over local political strategy is generally less in the UK than in the USA, in part because of a more centralised political and planning system, and greater central funding of local services. Nevertheless, over the period 1979–97 successive New Right governments in the UK encouraged greater private-sector involvement in local economic planning (see Chapter 16) and enhanced the power of business interests in the process of urban development.[59]

5. Despite a common commitment to growth, the composition of growth networks can vary from place to place and may extend beyond the capitalist class, with, in some cases, membership embracing labour (usually the construction trade unions) or minority groups.

6. Within the **pro-growth coalition**, fractions of capital can pursue strategies that impose negative externality effects on each other, as when the production of a new shopping centre leads to the decay of shopping facilities in older areas.

The main ideological battle is between pro-growth and anti-growth factions. The pro-growth ideology proclaims that more development results in increased population, greater aggregate sales, more local tax revenues and more local jobs, owing to increased local spending. Benefits therefore flow to those who need employment and those in favour of lower taxes. The anti-growth ideology emphasises the other side of the coin: development brings more people to an area than local institutions can service, thus any downward trend in taxation is overcome in the later stages of growth by the need for greater fiscal expenditure. In addition, development produces pollution, traffic congestion and social pathologies such as higher crime rates. There is also no guarantee that new jobs will go to locals.[60]

In some cases, growth networks operate in a corrupt manner to exploit rapid development, as for example, when political leaders raise campaign funds by favouring business elites, or public officials receive payments in return for planning and taxation easements.[61] The ability of power groups to manage urban growth led Gale and Moore (1975) to formulate the hypothesis of the **manipulated city**, which contends that urban form is the outcome of conscious manipulation by an alliance of elite interests with social power.[62] The extent of the influence of elite coalitions is difficult to assess, however. In contrast to the more regulated process of urban development in Britain and Europe (see Chapter 8), in the 'free-market' North

American city, growth coalitions of businessmen and municipal politicians have existed since the nineteenth century, supported by an ideology that considers urban growth and commercial growth to be partners. In Canada, in the early 1970s, half the aldermen on the city councils of Toronto, Winnipeg and Vancouver held occupational connections with development interests, and pro-growth parties held control of City Hall in each case.[63] In instances of financially weak cities the business/financial community can exercise close to monopoly control, translating their economic power into political power. The financial crisis of New York City in the mid-1970s is a classic example of how a financial coalition can benefit at the public expense. By 1977 20 per cent of New York's budget was committed to interest payments.[64] Similarly, in Cleveland in 1978 the oligopoly of business interests which formed the Cleveland Trust Company (which had interlocking directorates with major local steel, coal, utility and banking companies) was able to exact significant economic concessions from the city in return for arranging a loan. Globalisation has promoted a common 'growth discourse' among growth coalitions in many US cities based on the belief that entrepreneur-led effort is required for them to compete successfully for capital and jobs within a hypermobile global economy. The power of growth coalitions to influence urban development remains formidable. In Cleveland during the 1980s the growth coalition aggressively promoted gentrifying neighbourhoods along Lake Erie and the construction of a new downtown sports complex as catalysts to improve the local economic investment climate. In the process consideration of the effects of uneven development on disadvantaged populations was muted.[65] It is not surprising that Harvey (1971) characterised much of the political activity of the city 'as a matter of jostling for and bargaining over the use and control of the hidden mechanisms for redistribution'.[66] The land-use changes that result from this activity involve an uneven distribution of the net costs and benefits of urban development (Box 7.7).

THE CENTRAL BUSINESS DISTRICT

The central business district or downtown is a principal element of all major models of urban land use. Although some CBDs experience fierce competition from business nuclei elsewhere in the city, most cities retain a strong downtown. Even the archetypal polynucleated metropolis of Los Angeles has a CBD.

BOX 7.7

Externalities

Cities would not have appeared if the collective effect of land-use decisions were not beneficial. However, many of the effects of land-use decisions are negative in that they impose disbenefits on other land users. These negative externalities can take a variety of forms. They may, for example, arise from commercial invasion of residential areas or from traffic problems caused by the continued intensification of city-centre land use, or from the process of urban sprawl. As well as being unpriced, externalities are unplanned (in the sense that their extent and impact are not predetermined by a public agency). They are not always unintended, however, since the pollution of atmosphere or an urban river by a factory may be a deliberate attempt to transfer part of the real cost of production from the producer to society at large. Once the externality problem reaches a critical level it is likely to provoke a public response in the form of planning intervention, the aim of which is to influence the market mechanism so that either prices are adjusted to reflect full social costs or outputs are controlled to a socially optimal level. Frequently local governments have to carry the costs generated by negative spillovers.

Another land-use problem related to the question of externalities is the free-rider issue. A free rider is defined as an individual who refrains from any active initiatives but who benefits from the activities of others. For example, once a neighbourhood begins to decline in status, individual property owners will frequently, and quite rationally in individual cost–benefit terms, choose to do nothing in the face of the deterioration of their property, waiting instead for someone else to undertake improvements from which they, the property owners, will obtain positive externalities. In most cases the normal effect is that private renewal is postponed indefinitely as individual property owners await the windfall gains, which in practice have little likelihood of appearing unless government rehabilitation programmes are enacted.

The key characteristic of the city centre or CBD is its accessibility. Accessibility is a major factor in the locational decisions of central-city land users. Activities requiring an accessible location for their economic viability or functional efficiency tend to gravitate to the CBD. As neo-classical economic theory explains, different activities have different levels of demand for accessibility. The differential need for, and willingness to pay for, accessible sites determines the internal land-use pattern of the central city, with land uses placing greatest value on accessibility outbidding others for central city sites (see Box 7.3). The demand for central locations is translated into high land values, which in turn produce a high intensity of land use, expressed most visibly in the high-rise physical structure of the downtown created as developers seek to maximise use of costly sites. The CBD/downtown is typically:

1. The main commercial centre of the city.
2. A centre of retailing.
3. An area where manufacturing once concentrated and where light industry may still exist.
4. A locus for service industries, business offices and financial institutions.
5. A zone with only limited residential land uses.

Whereas early geographical studies of the CBD focused on the spatial delimitation of the area,[67] more recent analyses have concentrated on the changing nature of the zone. Since the 1950s the character and land-use structure of the CBDs of cities in the UK and the USA have been transformed by several major economic and social processes. These include decentralisation of population (see Chapter 4) and of retail activities (see Chapter 12), deindustrialisation (see Chapter 14), increased socio-spatial polarisation and segregation (see Chapter 18), and reductions in the traditional accessibility of the central city associated with increased levels of car ownership (see Chapter 13).

Retail suburbanisation has meant that in many cities of the USA the downtown has become less of a retail and commercial district than an office–commercial and cultural and entertainment complex. However, many city centres retain significant advantages for specialist shopping activities, being the point of maximum accessibility to the whole metropolitan area in comparison with the sectoral accessibility of suburban regional centres. Government activities and entertainment and cultural facilities that attract significant numbers of people on a regular basis are also concentrated in the city centre. Many city centres have also developed a strong tourist and convention trade function.[68] Furthermore, as we have seen in Chapter 4, in some larger cities a degree of reurbanisation is taking place. In several US cities, in addition to rehabilitation of decaying properties for residential use (see Chapter 11), revitalisation of the CBD has been aided by development of a major facility such as a sports stadium or performing-arts centre, that in turn attracts complementary service functions such as restaurants, hotels and retail outlets. The CBD also remains a major focus of office employment, notwithstanding the growth of 'suburban downtowns' or edge cities. Companies often locate their top management in the CBD even if they relocate clerical functions to the suburbs. Even in the electronic age of teleconferencing, corporate managers required to respond to rapidly changing market conditions find value in physical proximity to clients, consultants, bankers, government agencies and their competitors' headquarters. For some large companies (for example in law, banking or management consultancy) a prestigious downtown address (and distinctive building) assumes symbolic as well as practical significance. More generally, the skyscrapers of the CBD are symbols of both corporate power and the high land values embedded in the capitalist organisation of urban space. The attacks on the World Trade Center in New York on 11 September 2001 was a malign reflection of the symbolic power of architecture. The terrorists were attacking not only a landmark building but a symbol of American global capitalism.

URBAN ARCHITECTURE

Architecture and urban design are linked into the dynamics of urban change. As elements in the political economy of urbanisation, architecture and urban design:

1. Stimulate consumption by providing products for different market segments, from new office towers to festival shopping malls.
2. Promote the circulation of capital through the creation of a steady supply of new fashions in domestic architecture.
3. Aid the legitimation of existing economic and social relations by using the 'aura' of urban architecture to suggest the stability, permanence and 'naturalness' of the current urban environment.

Changes in urban form over time, from pre-industrial to post-industrial/postmodern cities, have been accompanied by change in the dominant form of architecture. In the nineteenth-century city, away from the mass of working-class vernacular construction, architects responded to the processes of industrialisation, modernisation and urbanisation with designs for new public buildings, factories, office buildings and mansion houses that rejected these contemporary processes of urban change and embraced classical features. The results are seen in the impressive edifices of Victorian

railway stations and town halls that still dominate the centres of many cities.

A second major influence on urban architecture was the modernist movement, which has been characterised as a reaction against the extravagance of bourgeois Victorian society. Promoted by the ideas of William Morris in England and by Frank Lloyd Wright in the USA, **modernism** was boosted by the ideas of the German Bauhaus school (1919–32), which sought to employ industrial production methods, modern materials and functional designs to promote inexpensive architecture available to all citizens. The international impact of modernism on urban design was heightened by the Swiss architect Le Corbusier, who, in response to the challenge of the automobile age, conceptualised the city as a 'machine for living'. The most visible impacts of his grand designs on contemporary cities were seen in comprehensive redevelopment plans, high-rise building and urban motorways of the 1960s.

In contrast to Le Corbusier's high-density lifestyle, in the USA Wright's response to the automobile age was his low-density Broadacre City, designed on the basis of two new technologies: the automobile and mass-produced building using high-pressure concrete, plywood and plastics. The use of new technology and materials was especially evident in the downtowns of post-war cities, where glass-fronted façades that reflected their surroundings became popular for corporate buildings in the 1960s and 1970s. This 'late modern' period also produced buildings with dramatic geometrical forms and buildings that emphasised technology with exposed pipes, ducts and elevators.

The demise of modernist architecture was prefigured by criticism of the uniformity and 'dullness' of Corbusian tower blocks,[69] and of the insufficient attention given to the needs of people for sociable living spaces.[70] For some commentators the demolition of the prize-winning Pruitt-Igoe public housing project in St Louis in 1972 tolled the death knell of modernist architecture. In contrast to the abstract formalism of modernist architecture, postmodern buildings were designed to be decorative and replete with symbolism. Postmodern architecture is eclectic, often combining ('double-coding') modernist styles or materials with historical or vernacular motifs – as in 'neo-traditional' urban designs for upmarket resort communities such as Seaside in Florida, or in kitsch design as seen in 'authentic' Tudor inns.

In terms of the broader context of urbanisation and urban change, postmodern architecture (as well as postmodern culture and philosophy) emerged along

Plate 7.2 The glass facade of an office building in Phoenix AZ reflects the architectural style of the late modern period

with the development of global capitalism. As Harvey (1989) observed, although this transition and critique of modernism had been under way for some time, it was not until the international economic crisis of 1973 that the relationship between art and society was disturbed sufficiently to allow postmodernism to become the 'cultural clothing' of advanced capitalism.[71] The property development industry was quick to adopt postmodern design, to promote product differentiation and to both create and supply demand from an increasingly 'consumerist' society. Postmodern design has also stimulated the preservation of historic buildings and urban areas, often linked with the growth of cultural industries and festival shopping developments. The dystopian side of postmodernism is seen most clearly in the proliferation of security and surveillance systems and the creation of 'fortress' architecture designed to exclude 'undesirables' such as rough-sleepers or panhandlers/beggars from parts of the city,[72] including 'public space'.

PUBLIC SPACE IN THE POSTMODERN CITY

Debate over the meaning and role of public space in cities has grown as a result of (1) the increasing privatisation of urban public space, and (2) growing attention to categories of 'difference' and their associated geographies (e.g. gay villages). An important distinction is between urban spaces as 'representations of space' (the way in which cities are portrayed in, for example, planning documents or city marketing literature), often referred to as 'real space' or 'conceived space'; and 'spaces of representation' (the personal feelings people have towards the spaces they inhabit everyday), often referred to as perceived space'. Significantly, public space often originates as a representation of space as in a city plaza, but as people use the space it becomes also a representational space. The counterpoint of public space – the private spaces of home, neighbourhood and privately owned commercial spaces such as shopping malls, offices and leisure centres – is in many instances increasingly segregated from the public city.

Starting from a normative (ideal) perspective that city life entails the coming together of strangers[73] it is often argued that public space should be democratic and open, offering a site where difference can be acknowledged and celebrated. Zukin (1995) contends that public space is the primary site for 'public culture' where people may socialise in such a way that a shared citizenship is created (a sense of identity transcending class, gender and ethnic boundaries).[74] Others, such as Mitchell (1995, 2003) have emphasised the political role of public spaces as democratic sites of critical public discourse and, on occasion, of protest and resistance.[75] Replacement of open public spaces by closed 'sanitised' spaces from which groups and individuals regarded as 'out of place' are excluded has been characterised as dystopian urbanism by Davis (1992).[76]

In contrast, an alternative 'reading' of urban public space portrays it as undesirable and unsafe. From this perspective fear of the 'other' (especially among the white middle class in contemporary Western societies) leads to avoidance of 'unsafe' public spaces and a withdrawal from the public life of the city. For these groups encounters with difference are not viewed as pleasurable and part of the vitality of the city but as potentially threatening and dangerous.

Some writers have suggested that physical forms of public space have been superseded by electronic means of participation in a new form of 'e-public space'. Clearly, as in the case of material public space, access to the 'city of bits'[77] is not freely available to all. Nor indeed would the disembodied electronic world, in itself, be generally seen as an adequate space for representation of public opinion.

City governments have responded to the dominant discourse on public space by introducing measures to halt the perceived degeneration of public space – including electronic surveillance in everyday places, and urban renewal schemes that involve privatising and commodifying 'public space'. While predictions of the 'end of public space' are premature it is important to recognise the changing nature and contested role of public space in contemporary cities. Debate over the definition and delimitation of public space and on who and what is, or is not, acceptable within it has been ongoing since the time of the Greek *agora* – a space that fulfilled a role both as a place of citizenship, commerce, pleasure and public affairs; and a place of exclusion (of women, slaves and foreigners).

ARCHITECTURE AND URBAN MEANING

Early attempts to read urban meaning from architectural forms and styles regarded urban architecture as symptomatic of the social and cultural values of its age. Thus Mumford (1938) attributed the development of the avenue in the baroque city of the sixteenth century to the growing militarisation of society and consequent need for troop mobility.[78] The difficulty of establishing unequivocally the social values underlying architecture led more recent researchers to analyse urban architecture as a product of specific dominant social groups rather than as a symbol of an historical epoch.[79] While Mumford's attempt to read the city as a text is retained, it is related to particular groups, not to society as a whole. However, there is still a large range of possible interpretations of urban meaning to be derived from analysis of architectural form. The text may be read in many ways – for example, from the viewpoint of class conflict seen in the privatisation of public space, or from that of gender relations, as seen in the characterisation of domestic residential space as primarily a female domain. Further, while architecture can reveal the power of certain groups to influence urban structure, architectural analysis alone is insufficient to explain urban meaning.

The diversity of urban imagery also underlines the fact that urban meaning is created and can be manipulated. As we have seen, many actors, from estate agents to local authorities, have a vested interest in presenting places in their most favourable light. Urban meaning may also be contested politically by disadvantaged groups who seek to define urban space in ways that best fit their needs. This is seen in the diverse actions of urban **social movements** (see Chapters 20 and 29). In short, urban space is socially constructed.

THE SOCIAL CONSTRUCTION OF THE URBAN LANDSCAPE

The diversity of urban landscapes is a key feature of the postmodern perspective on urban development, although the explanation of how the variety of urban environments is produced runs counter to the economic emphasis of the political economy approach. Postmodernism regards all landscapes as symbolic expressions of the values, social behaviour and individual actions of people marked on a particular locality over time. The built environment of the city is seen as the product of a dialectic interaction between society and space.[80] Postmodernism attempts to interpret the nature of this relationship. As we have seen, the approach has been likened to reading the city as a text written by a plethora of different authors which has a series of meanings embedded in it. Postmodern analysis aims to read this text from the viewpoint of the different actors involved in its production.[81]

The method may be illustrated by considering the concept of patterns of consumption, which is one of the most powerful processes in the urban **sociospatial dialectic**. The different patterns of consumption enjoyed by different groups in society are reflected in the polarisation of the retailing landscape as a result of the segmentation of the market into different niches catering for different tastes, preferences and lifestyles. Patterns of consumption influence, and are influenced by, the nature of the physical landscape. The different urban landscapes are the result of this dialectic relationship between social practices and the physical environment. The same urban space can therefore have a different meaning for different social groups, a situation that can lead to conflict over the appropriate use of land. This is evident in debates over public and private space[82] (Box 7.8), and is clearly illustrated in the concept of the *grade-separated* city in which two different environments are interlaced within the architecture of the CBD.[83] On the outside – both physically and, often, socially – are streets, pavements (sidewalks), plazas and parks, while inside, 'eligible' citizens encounter skywalks, tunnels, concourses and atria. In many North American cities these two environments share the same geographic location in the CBD while functioning as separate entities, one under the purview of the local public sector and the other controlled by a loose collective of private-sector interests. The existence of grade-separated space redefines patterns of interaction between activities and between segments of the downtown population. As Byers (1998 p. 200) observed, 'in a sense, the downtown environment is turned inside out, or, perhaps more appropriately, outside in. Groups that once shared the same city streets are now spending their days in environments that rarely intersect with one another.'[84] In the grade-separated city, distinctions in human activity patterns between day and night, work week and weekend, inside and outside, all help to reconfigure the way space in different parts of the CBD is used by different sections of the general public.

As we noted in Chapter 2, a major difficulty inherent in the postmodern approach is that the diversity of meanings attached to the urban landscape makes generalisation difficult if not impossible. Fundamentally, just as it is necessary to avoid reification of the economy as the sole force for urban change, so it would be equally short-sighted to view urban space as the product of the actions of voluntaristic agents free of 'structural' constraints. The urban landscape is the product of both culture and economy, and a proper understanding of urban environments must be based on explicit acknowledgement of this complexity.

In this chapter we have examined the major forces underlying urban structure and land use. The one actor in the urban development process not yet considered is, in some settings, the most influential. The state – central and local government – exerts both a direct (via planning regulations) and an indirect (via, for example, taxation policy) effect on urban form and structure. We consider this factor in the next chapter.

BOX 7.8

BIDs to privatise public space in New York City

The difference between the urban landscapes of rich and poor is illustrated by the growth of business improvement districts (BIDs) in New York City. A BID can be incorporated in any commercial area, and following the decline of city services in the wake of the fiscal crisis of 1975 a growing number of business and property owners created BIDs to exert control over the quality of public space in their vicinity.

The Grand Central Partnership, a fifty-three-block BID established in 1988, employs uniformed street cleaners and security guards, runs a tourist information booth, has closed a street in front of Grand Central Station to create a new outdoor eating space, issues its own bonds and hires lobbyists to press the state legislature for supplemental funding.

Although the architects of this new urban space are apparent, the question of who can occupy the space is less clear. The city government agencies have approved the BID plans, but the local community board representing a variety of local business interests has objected to street closures, and has questioned the effectiveness of the BID services for the homeless and the brusqueness of their removal by the BID security guards. In 1995 an advocacy group, the Coalition for the Homeless, sued the BID for hiring out the homeless as security guards at below the minimum wage.

The Times Square BID promised to improve the run-down image of the area and promote a new public culture by establishing a 'community court' in a disused theatre, thereby enhancing community control over quality of life. When set up in 1994 the court dispensed community service sentences for minor offences such as shoplifting and prostitution.

BIDs can be equated with an attempt to reclaim public space from the sense of menace that drives shoppers and others to the suburbs, but it is also the case that BIDs reconstruct public space in their own image and nurture visible social stratification by catering for a particular lifestyle.

FURTHER READING

BOOKS

E. Burgess (1925) The growth of the city, in R. Park and E. Burgess (eds) *The City* Chicago IL: University of Chicago Press, 47–62

D. Harvey (1981) The urban process under capitalism, in M. Dear and A. Scott (eds) *Urbanisation and Urban Planning in Capitalist Society* London: Methuen, 91–121

H. Hoyt (1939) *The Structure and Growth of Residential Neighborhoods in American Cities* Washington DC: Federal Housing Administration

A. Jonas and D. Wilson (1999) *The Urban Growth Machine: Critical Perspectives, Two Decades Later* Albany NY: State University of New York Press

M. White (1987) *American Neighborhoods and Residential Differentiation* New York: Russell Sage Foundation

J. Whitehand and P. Larkham (1992) *Urban Landscapes: International Perspectives* London: Routledge, 170–206

S. Zukin (1991) *Landscapes of Power: From Detroit to Disneyworld* Berkeley CA: University of California Press

JOURNAL ARTICLES

T. Banerjee (2001) The future of public space *Journal of the American Planning Association* 67(1), 9–24

J. Byers (1998) The privatization of public space: the emerging grade-separated city in North America *Journal of Planning Education and Research* 17, 189–205

R. Checkoway (1980) Large builders, federal housing programmes and post-war suburbanisation *International Journal of Urban and Regional Research* 4(1), 21–45

L. Feagin (1987) The secondary circuit of capital: office construction in Houston, Texas *International Journal of Urban and Regional Research* 11(2), 172–91

A. Harding (1991) The rise of urban growth coalitions UK-style? *Environment and Planning C* 9, 295–317

C. Harris and E. Ullman (1945) The nature of cities *Annals of the American Academy of Political Science* 242, 7–17

T. Hartshorn and P. Muller (1989) Suburban downtowns and the transformation of Atlanta's business landscapes *Urban Geography* 10, 375–95

J. Rust (2001) Manufacturing industrial decline: the politics of economic change in Chicago 1955–1998 *Journal of Urban Affairs* 23(2), 175–90

G. Squires (2003) Racial profiling, insurance style: insurance redlining and the uneven development of metropolitan areas *Journal of Urban Affairs* 25(4), 391–410

C. Teixiera (1995) Ethnicity, housing search and the role of the real estate agent: a study of Portuguese and non-Portuguese real estate agents in Toronto *Professional Geographer* 47(2), 176–83

KEY CONCEPTS

- Urban morphogenesis
- Town-plan analysis
- Fringe belt
- Burgage cycle
- Natural areas
- Concentric-zone model
- Sector model
- Multiple-nuclei model
- Urban ecology
- Urban realms model
- Political economy
- Circulation of capital
- Use value
- Exchange value
- Human agency
- Blockbusting
- Red-lining
- Growth machines
- Manipulated city
- Postmodernism
- Public space
- Grade-separated city
- Overaccumulation
- Socio-spatial dialectic

STUDY QUESTIONS

1. How can the study of town plans help explain the changing pattern of urban land use?
2. Critically evaluate the contribution of Burgess's concentric-zone model to our understanding of urban structure.
3. Explain the political-economy interpretation of urban land-use change. Can you identify any real-world examples of these processes at work?
4. Select one of the major actors involved in the production of the built environment and, with reference to specific examples, write an essay to illustrate their role in urban change.
5. With reference to the concept of externalities, prepare a list of positive and negative externality effects of urban development and relate these to events in a city of your choice.
6. Explain what is meant by the privatisation of public space and think of some examples of where it has occurred.

PROJECT

On a map of your local city draw a transect line (e.g. along a major routeway) from the city centre to the metropolitan edge. In the field, follow this line, mapping the major land uses on either side of the route. Back in the laboratory produce a general land-use map of your urban transect. Explain the variations in land use with distance from the city centre. Compare your empirical findings with the theoretical patterns described by the major models of urban land use discussed in the chapter.

8

Urban Planning and Policy

Preview: the roots of urban planning; post-war urban planning in the UK; urban policy in the UK; the redevelopment phase; the social-welfare phase; the entrepreneurial phase; the competitive phase; the 'Third Way'; urban planning in the USA; urban sprawl; growth management; smart growth; planning the socialist city; socialist urban form; planning for sustainable urban development

INTRODUCTION

The development of most urban areas is influenced, to some degree, by the processes of urban policy and urban planning. In this chapter we examine the nature and operation of urban policy and planning in the UK, Europe and the USA from its origins in the nineteenth century until the present day. Urban planning and urban policy are concerned with the management of urban change. They are state activities that seek to influence the distribution and operation of investment and consumption processes in cities for the 'common good'. However, it is important to recognise that urban policy is not confined to activity at the urban scale. National and international economic and social policies are as much urban policy, if defined by their urban impacts, as is land-use planning or urban redevelopment. In effect, urban policy is often made under another name. Urban policy and planning are dynamic activities whose formulation and interpretation are a continuing process. Measures introduced cause changes that may resolve some problems but create others, for which further policy and planning are required. Furthermore, only rarely is there a simple, optimum solution to an urban problem. More usually a range of policy and planning options exists from which an informed choice must be made[1] (Box 8.1).

Planning is carried out within the broad framework of government policy-making and has its general objectives set out in legislation. A primary objective of the UK planning system is 'to regulate the development and use of land in the public interest' (Department of the Environment 1999 p. 2).[2] Planning can be undertaken for a variety of reasons (Table 8.1). The aims of urban planning may be contradictory and reveal differing attitudes to the roles of the market mechanism and the state. Central to this debate is the question 'planning for whom?' While most of the purposes of planning listed in Table 8.1 assume that benefits should accrue to the 'public as a whole' or, in relation to redistribution aims, to the poorer and less vocal sections of society, the question of the validity of these goals and the extent to which they are met has produced polarised views on the value of planning (Box 8.2).

THE ROOTS OF URBAN PLANNING

THE UK

With the exception of various forms of 'planted settlements' (e.g. Greek colonial towns or medieval planned new towns, *bastides*), the forces underlying urban growth have operated largely free of any form of public accountability, so that for much of the historical period urban development proceeded in an unregulated organic fashion. Today a powerful system of planning exists in the UK and Europe, and to a lesser extent in North America, which aims to circumscribe urban development and direct it towards socially beneficial goals.

For urban planning to exist, there must be a political consensus that the problems affecting cities can best be tackled through government intervention. This, in turn,

BOX 8.1

The nature of urban policy

Urban policy is the product of the power relation between the different interest groups that constitute a particular society. Foremost among these agents are government (both local and national) and capital in its various fractions. Capital and government pursue specific goals that may be either complementary or contradictory. For capital the prime directive is profit maximisation. Government, on the other hand, in addition to facilitating the process of accumulation, must also satisfy the goal of legitimation. These political and economic imperatives have a direct influence on the nature of urban policy. Urban policy is also conditioned by external forces operating within the global system, as well as by locally specific factors and agents.

The form of urban policy employed depends on the problem to be addressed and, most fundamentally, on the ideological position of the state. Adherents of market capitalism view the production of unevenly developed cities as the inevitable outcome of technological change in an economic system that readily adapts to innovation. The negative socio-spatial effects of this restructuring that impinge on disadvantaged people and places are regarded as unavoidable consequences of a process that is of benefit to society as a whole. For those on the political right, market forces are the most efficient allocators of capital and labour, and state intervention is considered unnecessary. Policies involving social welfare expenditure and government financial aid to declining cities are regarded as harmful because they anchor low-wage workers to sites of low employment opportunity, discourage labour-force participation and inhibit labour mobility. Welfare state liberals, on the other hand, while accepting the central role of the market, acknowledge that the institutional and cyclical 'market imperfections' that have left certain people and places in prolonged economic distress must be rectified by compensatory government policies.

TABLE 8.1 THE MAIN GOALS OF PLANNING

1. To improve the information available to the market for making its locational choices
2. To minimise the adverse 'neighbourhood effects' created by a market in land and development
3. To ensure the provision of any 'public goods', including infrastructure or actions that create a positive 'neighbourhood effect', which the market will not generate because such activity cannot be rewarded through the market
4. To ensure that short-term advantage does not jeopardise long-term community interest
5. To contribute to the co-ordination of resources and development in the interest of overall efficiency of land use
6. To balance competing interests in the use of land to ensure an overall outcome that is in the public interest
7. To create a good environment, for example in terms of landscape, layout or aesthetics of buildings, that would not result from market processes
8. To foster the creation of 'good' communities in terms of social composition, scale or mix of development, and a range of services and facilities available
9. To ensure that the views of all groups are included in the decision-making processes regarding land and development
10. To ensure that development and land use are determined by people's needs, not means
11. To influence locational decisions regarding land use and development in order to contribute to the redistribution of wealth in society

Source: A. Thornley (1991) *Urban Planning under Thatcherism* London: Routledge

development is more restricted, with zoning being the major mechanism for land-use control.

Urban planning emerged as a response to the manifest problems of the nineteenth-century industrial metropolis (see Chapter 3). Two types of reaction were evident. The first, represented in the work of Marx and Engels, was revolutionary and advocated the overthrow of the social and political system responsible for creating the polarised social conditions that characterised urban Britain. The conservative alternative involved basic acceptance of the urban industrial system but the use of state intervention to ameliorate its worst excesses. It was the latter argument, articulated

requires willingness by individuals to relinquish some of the rights to property which they enjoy in a free market and to accept the principle that land use should be centrally controlled for the public good.[6] In general, while electorates in the UK and Europe have accepted the implications of a comprehensive system of urban planning, in the US government intervention in urban

BOX 8.2

The value of planning

Urban planning is a ubiquitous activity in the modern world. Nevertheless, the principles and practice of planning have come under attack from both the left and the right of the political spectrum. In contrast to the positive goals identified in Table 8.1, the value of planning has been dismissed by the far left, which regards it as a state apparatus attuned to the needs of capital and designed to maintain the unequal distribution of power in society. Reade (1987), for example, concludes that the legitimacy of the property development industry and its associated financial institutions is maintained by having a 'planning system' that provides a pretence of government intervention in the 'public interest'.[3]

For some critics on the right, planning is seen as a major *cause* of inner-city decline and social unrest through its policies of clearance and decentralisation, rigid land-use zoning and imposition of standards. According to Steen (1981), because of planning, resources are fruitlessly channelled into deprived areas and wasted rather than encouraging wealth creation.[4] For others the chief problems of planning lie in its interference with the market and in the fact that, contrary to the goals of planning, a dynamic and prosperous urban economy requires *inefficiency* in its structure and land use in order to permit innovation and experimentation.[5] The New Right attack on the redistributive goals of urban planning in the UK was articulated most forcibly by the Conservatives under Margaret Thatcher during the 1980s.

in the UK by the factory and sanitary reformers, and reinforced by the success of a number of early housing and new town schemes, that paved the way for the emergence of modern urban planning.[7]

The belief that designing new communities offered a means of escape from the problems of the nineteenth-century industrial city was central to the ideas of the utopian socialists. Robert Owen (1813) proposed the creation of 'agricultural and manufacturing villages of unity and mutual co-operation' to house between 1,000 and 1,500 persons and cater for all the social, educational and employment needs of the community. Owen's model industrial complex at New Lanark in Scotland promised superior working and living conditions, cheap subsidised shops based on bulk-buying of all household necessities, and a community school or 'Institution for the Formation of Character', which operated as a day school for children and a night school for adults. Owen's example led to the development of a number of other model towns.[8]

The social reformer and architect James Silk Buckingham (1849) proposed a utopian temperance community of 10,000, to be named Victoria.[9] The plan envisaged segregation of land uses, with manufacturing trades and noxious land uses near the periphery, and housing and offices in the inner areas. All dwellings were to have flushing toilets and there would be a variety of house sizes to accommodate different households. Public baths were to be provided in each quarter of the town. A **green belt** of 4,000 ha (10,000 acres) of agricultural land would surround the settlement. All land was to be owned by the development company and buildings occupied for rent. Although Victoria was never built, many of the ideas were taken up by later urban reformers, including Ebenezer Howard in his designs for **garden cities** (see below).

Several smaller utopian communities were started (Table 8.2). Among the most significant were Saltaire and Bournville. Saltaire was built by Titus Salt to replace his woollen mills in Bradford by a single large factory and a new town for the work force. He selected a greenfield site crossed by a canal and railway, and between 1851 and 1871 completed a model industrial town of 820 dwellings and a population of 4,389. The town was endowed with a variety of community buildings and parks, and although the residential density of thirty-two houses (170 persons) per acre (eighty houses or 420 persons per hectare) appears crowded by today's standards, it represented a marked improvement on working-class living conditions of the time. The model town of Bournville built by George Cadbury also involved the transfer of factory production from an inner-city to a greenfield site but was the result of a wider concept than Saltaire. From the outset Bournville was intended to be not only a company town but also a general example to society of how it was possible to provide decent living conditions without endangering company profits. With its cottage houses, large gardens, open space and community facilities built around the chocolate factory, Bournville set new standards of working-class living (Figure 8.1).

In quantitative terms the nineteenth-century model towns contributed little to the problems of the factory slums in the large cities, but they did stimulate reformers to question the morality and economic necessity of such living conditions. The idealism of the utopian socialists was mirrored at the turn of the century in the ideas of Ebenezer Howard (1898).[10]

TABLE 8.2 PLANNED SETTLEMENTS IN EIGHTEENTH- AND NINETEENTH-CENTURY BRITAIN

Settlement name	Company/family	Location	Industry	Date started
Cromford	Arkwright	Derbyshire	Textiles	1772
Styal	Greg	Cheshire	Cotton textiles	1784
New Lanark	Arkwright/Dale/Owen	Lanarkshire	Cotton textiles	1785
Nent Head	London Lead Co.	N. Yorkshire	Lead	1825
Turton	Ashworth	Bolton	Textiles	1825
Barrow Bridge	Gardener/Bazley	Lancashire	Textiles	1830
Street	Clark	Somerset	Footwear	1833
Meltham Mills	Brooks	Huddersfield	Cotton thread	1836
Copley	Akroyd	Halifax	Textiles	1844
Wilshaw	Hirst	Huddersfield	Textiles	1849
Saltaire	Salt	Bradford	Textiles	1851
Bromborough Pool	Price	Wirral	Candles	1853
Akroydon	Akroyd	Halifax	Textiles	1861
West Hill Park	Crossley	Halifax	Carpets	1864
Bournville	Cadbury	Birmingham	Cocoa	1878
Port Sunlight	Lever	Wirral	Soap	1888
Foyers	British Aluminium	Loch Ness	Aluminium	1895

Howard, with London in mind, was strongly critical of living conditions in the large towns. His alternative was to design a garden city based on the following principles:

1. Each garden city would be limited in size to 32,000 population.
2. It would have sufficient jobs to make it self-supporting.
3. It would have a diversity of activities, including social institutions.
4. Its layout would be spacious.
5. It would have a green belt to provide agricultural produce, recreation space and a limit to physical growth.
6. The land would be owned by the municipality and leased to private concerns, thereby reserving any increases in land value to the community as a whole.
7. Growth would occur by colonisation.

Howard did not envisage building isolated towns but advocated a cluster arrangement of six interdependent garden cities, connected by a rapid transport route around a central city of 58,000 population, the whole comprising a 'social city' of 250,000 people (Figure 8.2). Howard's ideal of a 'town–country' lifestyle led to the founding, in 1899, of the Garden City Association, which built two garden cities, at Letchworth (1901) and Welwyn (1920), and was a major stimulus to the formation in 1914 of the Town Planning Institute.

A second, parallel stimulus to urban planning in the UK was the sanitary-reform movement. This was stimulated by concern over the health of the urban population as a succession of epidemics ravaged the densely packed inner areas of the major British cities (see Chapter 3). The deficiency of public facilities, such as a clean water supply and adequate sewerage system, reinforced arguments for government intervention, and under pressure from enlightened politicians such as Shaftesbury, Torrens, Cross and Chadwick legislation was passed to establish basic levels of sanitary provision and building standards. The 1875 Public Health Act consolidated previous measures and introduced a set of codes in respect of the construction of new streets, and the structure of houses and their sanitary facilities, and also gave local authorities the power to close down dwellings that were unfit for human habitation. These measures did much to reduce levels of morbidity and early mortality in the nineteenth-century city.

THE USA

The roots of urban planning in the USA can be traced to the ideas of the Progressive intellectuals of the late nineteenth century. In examining the shortcomings of American capitalism this loose-knit group of

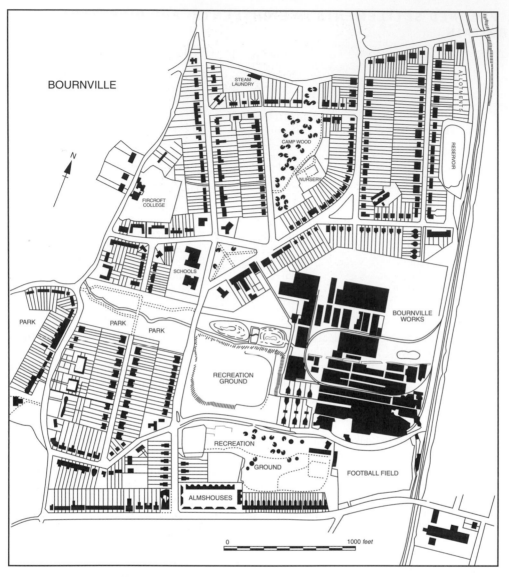

Figure 8.1   Plan of the model settlement of Bournville
Source: S. Ward (1994) *Planning and Urban Change* London: Chapman

sociologists, economists and political scientists, including White, Dewey, Cooley and Park, recognised a need for public intervention in, through not control of, the economy.[11] Among suggestions for government regulation of business, employment and politics in cities they advocated the appointment of specially trained experts to manage cities. Designers sought to create more humane living environments. Landscape architects such as Olmstead, Davis and Vaux designed residential areas as 'cities in a garden', which led to the

building of Riverside IL, Llewellyn Park NJ and Brookline MA as America's first planned 'romantic suburbs'. More generally, the City Beautiful movement, which emerged following the Chicago World Columbia Exposition of 1893, argued for the planned unity of the city as a work of art supported by a master plan for land use and by comprehensive zoning ordinances to maintain that plan. By 1913 over forty cities had prepared master plans, and more than 200 were engaged in some form of major civic improvement.[12]

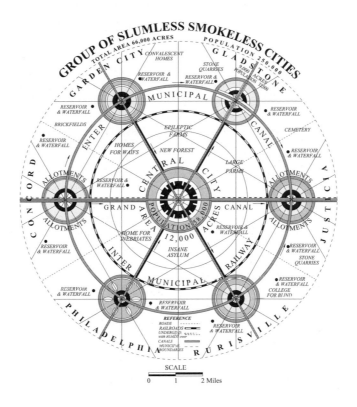

GROUP OF SLUMLESS SMOKELESS CITIES

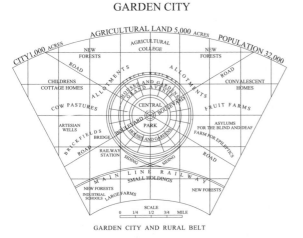

GARDEN CITY

GARDEN CITY AND RURAL BELT

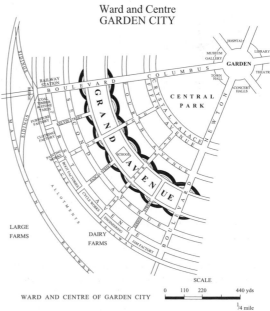

Ward and Centre
GARDEN CITY

WARD AND CENTRE OF GARDEN CITY

Figure 8.2 **Structure of Howard's garden city**

In 1917 a new professional organisation, the American City Planning Institute, was established.[13] In practice, however, the forces of privatisation were too strong to be contained by public officials, and 'urban planners seldom did more than follow residential and commercial developers with transportation and sewerage systems'.[14] The strength of private enterprise within the US political economy has been a major force in shaping the form of US planning and the structure of American urban areas.

WESTERN EUROPE

In parallel with the development of utopian idealism in Britain, equally ambitious alternative urban forms were being advanced in Europe in the ideas of the Italian Futurist movement, launched in a manifesto by Marinetti in 1909. The concept of a new, comprehensively planned city was a key idea in futuristic urban designs, which included high-rise building, elevated roadways, land-use segregation and the use of mass-production techniques and new materials such as glass and concrete. The Swiss architect Charles Jeanneret (Le Corbusier)[15] proposed a city for 3 million people based on four main principles:

1. As a result of increasing size and central area congestion the traditional form of city had become obsolete.
2. Pressure on the central business district could be reduced by spreading the density of development more evenly.
3. Congestion could be alleviated by building at higher density (1,000 persons to the net residential acre, or about 2,500 to the hectare) at local points, with a high proportion (95 per cent) of intervening open space.
4. An efficient urban transportation system incorporating railway lines and segregated elevated motorways would link all parts of the city.

Although his plans were not implemented in their entirety, Le Corbusier's ideas exerted a profound effect on urban planning and the form of cities. The concept of high-rise, high-density building was translated into practice in most large cities during the 1950s and 1960s, although in many instances less attention was given to the quality of space surrounding the tower blocks.

Notwithstanding differing national circumstances, we can identify three broad phases of urban planning in Western Europe:

1. In the immediate aftermath of the Second World War the focus of attention was on reconstruction and satisfying the backlog of housing and basic infrastructure.
2. By the late 1950s, increasing affluence and the growth of centralised planning systems led to comprehensive slum clearance, city-centre redevelopment schemes, and the construction of urban motorways and large-scale public housing projects.
3. From the late 1970s, growing awareness of the social disruption caused by the large-scale remodelling of cities led to greater attention to public participation in planning, and the replacement of redevelopment by rehabilitation.

A significant ongoing problem for all large European cities is that of growth management. Here the experience of the multicentred metropolitan region of Randstad may offer lessons for polycentric urban regions elsewhere (Box 8.3).

POST-WAR URBAN PLANNING IN THE UK

Although urban planning in the UK was inspired by reformist reactions to the nineteenth-century city, its cause was advanced in the 1930s by the growing focus on the distribution of population, and associated land-use issues. The increasing concentration of population in an axial belt stretching from London to Merseyside, high unemployment levels outside this zone, and the threat of urban sprawl induced governmental concern. A number of reports on these and other issues (Table 8.3) contributed to the passage of the 1947 Town and Country Planning Act, which established the structure of modern urban planning in the UK.[16]

The basic principle enshrined in the 1947 Act is that of private ownership of land but public accountability in its use so that landowners seeking to undertake development first had to obtain permission from the local planning authority. The local authority was also given the power to acquire land for public works by compulsory purchase on payment of compensation to the landowner. Local planning authorities were required to prepare and submit quinquennial development plans to the Ministry of Town and Country Planning indicating how land in their area was to be used. A second principle embodied in the 1947 Act was that of community gain, rather than individual gain, from land **betterment**. This meant that when land was developed, the increase in its

BOX 8.3

Urban growth management: the Dutch approach

The Netherlands is a small country with a population density of over 1,030 persons per square mile (400 per square kilometre). The Randstad city-region contains 36 per cent of the national population on 5 per cent of the land area and incorporates the three large cities of Amsterdam, Rotterdam and The Hague as well as a number of smaller urban centres. Pressure on open space is considerable.

Dutch planners seek to maintain the green heart of Randstad by directing growth outwards towards the peripheral regions and, within the urban region, by encouraging higher-density development in rehabilitated inner-city areas. Strong planning controls to protect the centre of Randstad are supported by large-scale public ownership of urban land, with 70 per cent of Amsterdam owned by the city authority. The fact that the polycentric structure of Randstad is functional as well as physical, with different cities specialising in broadly different functions (e.g. heavy industry in Rotterdam, finance in Amsterdam and government in The Hague), also reduces journey times and traffic congestion. Continued planned regulation of urban growth will be needed to prevent the coalescence of individual urban areas in Randstad and enable the 'green heart metropolis' to maintain its open structure.

TABLE 8.3 MAJOR REPORTS AND LEGISLATION LEADING TO THE 1947 PLANNING SYSTEM IN THE UK

Date	Report or new legislation
1940	Report of the Royal Commission on the Distribution of the Industrial Population (Barlow Report, Cmd 6153)
1942	Report of the Expert Committee on Compensation and Betterment (Uthwatt Report, Cmd 6386)
1942	Report of the Committee on Land Utilisation in Rural Areas (Scott Report, Cmd 6378)
1943	Ministry of Town and Country Planning Act
1943	Town and Country Planning (Interim Development) Act
1944	Town and Country Planning Act
1945	Distribution of Industry Act
1945	Abercrombie's Greater London Development Plan
1946	New Towns Act
1947	Town and Country Planning Act

value that resulted from the granting of planning permission was reserved for the community by the imposition of a 100 per cent land development tax. Political opposition ensured that this provision was removed in 1952. Today, compulsory exaction of betterment values from land developers has been replaced by a system of negotiated agreements over **planning gain**, which represents the benefits that a local authority may require of a developer as a condition for planning permission.[17] We can note the similarity, in principle, with the mechanism of incentive zoning in the USA (Box 8.4).

The primary objectives of the 1947 planning system at the city-region scale were urban containment, protection of the countryside and the creation of self-contained balanced communities (e.g. new towns). These goals were advanced by local authorities using the development plan and development control process, and by central government through the New Towns programme, supplemented by the Expanding Towns scheme (which enabled cities with problems of overcrowding to arrange overspill schemes with other local authorities; see Chapter 9).

The first major changes to the 1947 planning system were contained in the 1968 Town and Country Planning Act. This sought to introduce greater responsiveness and flexibility to the plan-making process. The single development plan with its specific land-use focus and five-year life expectancy was replaced by a two-tier system of structure plans and local plans. Structure plans are comprehensive strategic statements designed to translate national and regional economic and social policies into the specific areal context of the local authority. Local plans apply the structure-plan strategy to particular areas and issues, and make detailed provision for development control.[18]

Most commentators would agree that the 1947 planning system has achieved its objectives: post-war suburban growth has been contained to the extent that coalescence of adjacent cities has been prevented and good-quality agricultural land protected. The major mechanism of green belts around the large cities has prevented peripheral sprawl, although they have contributed to inflation in land and housing prices in

BOX 8.4

Incentive zoning in Seattle

Incentive zoning offers bonuses to developers in return for the provision of public benefits. The Washington Mutual Tower gained twenty-eight of its fifty-five storeys as a result of amenities offered by the developer.

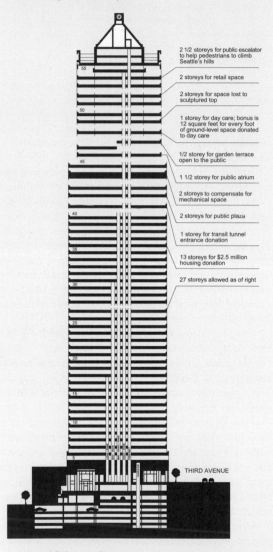

2 1/2 storeys for public escalator to help pedestrians to climb Seattle's hills

2 storeys for retail space

2 storeys for space lost to sculptured top

1 storey for day care; bonus is 12 square feet for every foot of ground-level space donated to day care

1/2 storey for garden terrace open to the public

1 1/2 storey for public atrium

2 storeys to compensate for mechanical space

2 storeys for public plaza

1 storey for transit tunnel entrance donation

13 storeys for $2.5 million housing donation

27 storeys allowed as of right

THIRD AVENUE

In addition to the twenty-seven storeys allowed as of right, the developer gained thirteen storeys for a $2.5 million housing donation and between one and two and a half storeys for various other public benefits.

existing settlements within the green belt and provoked an ongoing conflict between developers and planners over the availability of land for housing[19] (Box 8.5). Also, the green belt has been powerless in the face of increases in transport technology and in the length of acceptable commuting journey, which have seen some residential development leapfrog into settlements beyond the green belt. But, in general, the UK planning system has prevented the kind of scattered urban development in evidence around cities in North America.

URBAN POLICY IN THE UK

Urban planning takes place within the framework of national urban policy, the priorities of which reflect the ideology of the state. We can identify four major phases in British post-war urban policy (Table 8.4):

1. Physical redevelopment.
2. Social welfare.
3. Entrepreneurial.
4. Competitive.

THE PHYSICAL REDEVELOPMENT PHASE

From the end of the Second World War until the late 1960s urban problems were seen largely in physical terms. The policy response to issues of housing quality and supply, transport, and industrial restructuring focused on slum clearance and comprehensive redevelopment strategies, and the planned decentralisation of urban population via regional policy and new town development.

THE SOCIAL WELFARE PHASE

In the early 1970s empirical research highlighted the incidence of poverty within Britain's cities as the long post-war boom faltered (see Chapter 15). Emphasis was placed on supplementing existing social programmes to improve the welfare of disadvantaged individuals and communities. Influenced by US initiatives, such as the Head Start project developed as part of the Model Cities programme, a range of area-based experimental schemes were introduced during the 1970s. These 'Urban Programme' initiatives, which included educational priority areas and community

BOX 8.5

Land-use conflict in the green belt

Various agents and interests are involved in determining whether a parcel of land is transferred from rural to urban use (see diagram). In the UK, where a

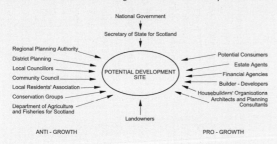

nationally co-ordinated planning system attempts to effect an equitable balance between private profit and public interest, conflict between developers and planners over land for new housing is a key factor determining the character of the green belt. The debate over land availability has intensified since the early 1970s, with, in general, builder-developers arguing that the planning system restricts their ability to obtain a basic factor of production and that development controls inflate the price of land and therefore of houses. The purpose of the planning system, on the other hand, is to ensure the orderly release of building sites within an approved policy framework.

The commuter village of Torrance in the green belt to the north of Glasgow is a typical location for the ongoing battle between house builders and planners around Britain's cities.[20] The issues are complex: in seeking permission to build houses on a disputed

site (Tower Farm) the developers cited a number of arguments, including the need to satisfy demand for housing in attractive locations, and their willingness to provide, via 'planning gain', additional infrastructure and community facilities. The planners argued that demand for new housing could be accommodated on brownfield sites outwith the green belt, and that the local authority would have to meet the ongoing running costs of any facilities.

Failure to resolve the dispute at a local level resulted in a public inquiry that concluded in favour of the planning authority. The developers exercised their right of appeal to the Secretary of State. Their appeal was dismissed. Thus, in the case of the Tower Farm site, the land remained part of the green belt. However, while the battle of Tower Farm may be over, both developers and planners live to fight another day in a continuing struggle between private profit and public interest in the production of the built environment.

Source: **M. Pacione (1990) Private profit and public interest in the residential development process**
***Journal of Rural Studies* 6(2), 103–16**

development projects,[21] operated from a 'culture of poverty' perspective (under which poverty is held to be self-reproducing) and aimed to give deprived communities the capacity to solve their own problems. The widespread increase in unemployment in the mid-1970s, in the wake of the Arab oil embargo and world recession, made it clear, however, that the scale of urban deprivation could not simply be the result of the inadequacies of the poor. An alternative 'structural' explanation of poverty was enshrined in the final report of the Community Development Project. This rejected a social pathological view of deprivation in the conclusion that 'there might certainly be in those areas a higher proportion of the sick and the

elderly for whom a better co-ordination of services would undoubtedly be helpful, but the vast majority were ordinary, working-class men and women who, through forces outside their control, happened to be living in areas where bad housing conditions, redundancies, lay-offs and low wages were commonplace'[22] (Community Development Project 1977 p. 4).

It followed that to tackle the root causes of urban decline would require more than marginal adjustments of existing social policies. This was acknowledged by the 1977 White Paper on policy for the inner cities, which signalled a more broadly based approach to urban problems, combining economic, social and environmental programmes and involving new partnership

TABLE 8.4 MAIN PHASES IN BRITISH POST-WAR URBAN POLICY

	Physical redevelopment, 1945–69	Area-based social welfare projects (the inner-city problem), 1969–79	Entrepreneuralism, 1979–91	Competitive policy, 1991–7
Representational regime	Construction of corporatist consensus around reconstruction programme. Central–local government partnership in council-housing redevelopment programme. Close links with construction industry (system-build, high-density, high-rise)	Area-based projects run by local government gave way in 1978 to 'partnerships' involving central government and designated local authorities, other statutory bodies (such as area health authorities), and local voluntary organisations and local industry (but not labour organisations). 'Programme areas' also designated. Local government seen as the natural agent of regeneration. Local government dominated by urban managerialism	Greater emphasis placed on the role of the private sector in urban policy (privatism). Creation of business elites and growth coalitions. 'Privatised' partnerships and business representation on a range of national and local institutions	Patterns of interest representation shifted, requiring new patterns of local leadership with private sector, community representatives and voluntary sector organisations, alongside city councillors in boards and companies at arm's length from the local authority. Consolidation of trends towards urban governance to prevent municipalisation of policy
Internal structures of the state	Monolithic state characterised by corporate planning, bureaucratic paternalism, functionalism, uniformity and inflexibilty. Constructive partnership between central and local government. Department of Economic Affairs created (1964) together with a set of official regions and regional planning councils and regional planning boards	Home Office and the newly created Department of the Environment were key departments during this period. Urban Deprivation Unit created in the Home Office. Some attempt to co-ordinate policy across a number of areas but thwarted by inter-department rivalry. Attempts to adopt corporate working and the bending of main programmes to address the urban problem. Co-operation between central and local government	Centralisation of power. Shift in local government towards urban governance with private sector having greater role. Confrontation approach to local government (rate capping, cutbacks, abolition, and quangos bypassing local authorities). Urban entrepreneuralism promoted involving leaner and flatter managerial structures, generic roles, team working and flexibility	Purported 'new localism' developed through the establishment of integrated government offices for the regions and the Challenge Fund. Cabinet committee established to oversee Challenge Fund. In reality, 'remote control' exercised via the contract culture. Central–local government relations characterised by an 'authoritarian decentralism' or 'centralist localism'. Local partnerships involved in a process of centrally controlled local regulatory undercutting. New public management (to promote privatisation and competition) creating the enabling authority

Patterns of state intervention

The long post-war boom period was dominated by policies to achieve national unity through regional balance, containment of urban growth and the reconstruction of urban areas, involving slum clearance and comprehensive redevelopment. Instruments included establishment of the development-plan system, New Towns, industrial development certificates, office development permits and development areas. National Plan published in 1965 advocating 'growth through planning'

Area-based experimental social-welfare projects attempting to respond to economic, social and environmental problems resulting from structural decline of the economy. Inner-cities policy, e.g.:

- Urban Programme, 1969
- General Improvement Areas, 1969
- Educational Priority Areas
- Community Development Projects, 1969
- Inner Area Studies
- Housing Action Areas, 1974
- Comprehensive Community Programmes, 1974
- Enhanced Urban Programme, 1978
- Inner Urban Areas Act (1978) acknowledged the economic nature of the urban crisis. Keynesian demand-management techniques used to counter the emerging crisis ('stagflation') of the 1970s

Neo-liberal philosophy pursued involving deregulation, liberalisation and privatisation. Social needs subordinated to the needs of business. Emphasis given to property-led initiatives and the creation of an entrepreneurial culture, e.g.:

- Enterprise Zones, 1979
- Urban Development Corporations, 1979
- Urban Development Grants, 1982
- Derelict Land Grant, 1983
- City Action Teams, 1985
- Estate Action, 1985
- Urban Regeneration Grants, 1987
- City Grant, 1988

Competition for funds and competitiveness of business and localities the leading priorities for regeneration policy. Initiatives to improve the competitive advantage of localities, e.g.:

- City Challenge, 1991
- Urban Partnership, 1993
- City Pride, 1993
- Single Regeneration Budget, 1994
- Rural Challenge, 1994
- Regional Challenge, 1994
- Capital Challenge, 1996
- Local Challenge, 1996
- Sector Challenge, 1996

Source: **N. Oatley (1998)** *Cities, Economic Competition and Urban Policy* London: Chapman

arrangements between central and local government to provide a more co-ordinated response. The emphasis on improving the economic environment of cities was promoted in several ways, including a shift of attention from new towns to urban regeneration and increased powers to enable local authorities to aid and attract industrial developments. The major vehicle for these measures was the expanded Urban Programme.[23] In practice, however, a concerted attack on the urban problem was undermined by the prerogatives of national economic policy, which led to reductions in local-authority finance. The Conservative government elected in 1979 continued the concept of partnership but stressed the involvement of the private sector.

THE ENTREPRENEURIAL PHASE

The advent of the Thatcher government in 1979 'witnessed the fracture of three main pillars upon which post-Second World War social democratic politics was constructed – Fordism, welfarism and Keynesianism'[24] (Gafficken and Warf 1993 p. 71). The reorientation of urban policy by the New Right Conservative government was part of a wider agenda to restructure Britain economically, socially, spatially and ideologically around a new consensus of free-market individualism and unequivocal rejection of the social-democratic consensus of the post-war Keynesian **welfare state**. As Martin (1988 p. 221) observed, the thrust of state policy shifted from welfare to enterprise:

> the aim has been to reverse the post-war drift towards collectivism and creeping corporatism, to redefine the role and extent of state intervention in the economy, to curb the power of organised labour, and to release the natural, self-generative power of competitive market forces in order to revive private capitalism, economic growth and accumulation.[25]

The Keynesian commitment to the macroeconomic goal of full employment was replaced by the objective of controlling inflation by means of restrictive monetary measures, and supply-side flexibilisation. From its inception 'Thatcherism' was a doctrine for modernising Britain's economy by exposing its industries, its cities and its people to the rigours of international competition in the belief that this would promote the shift of resources out of inefficient 'lame duck' traditional industries and processes into new, more flexible and competitive high-technology sectors, production

methods and work practices.[26] The principal mechanisms for achieving this transformation centred on tax cuts and deficit spending, deregulation and privatisation, all of which had geographically uneven impacts. At the urban level these three macro-economic strategies were combined most strikingly in the concepts of the enterprise zone (EZ) and the urban development corporation (UDC) (see Chapter 16).

As part of the broad political and economic agenda, urban policy was also used to restructure central government–local government relations. Five processes characterised the changes:

1. Displacement involving the transfer of powers to non-elected agencies (e.g. UDCs), thereby bypassing the perceived bureaucracy and obstructiveness of local authorities.
2. Deregulation involving a reduction in local authorities' planning controls to encourage property-led regeneration (e.g. in EZs).
3. The encouragement of bilateral partnerships between central government and the private sector.
4. Privatisation, incorporating the **contracting out** of selected local government services, housing tenure diversification and provision for schools to 'opt out' of local education authority control.
5. Centralisation of powers through a range of quangos[27] (quasi non-governmental organisations).

THE COMPETITIVE PHASE

By the early 1990s it was evident that the approach to urban policy pursued since 1979 had failed to reverse urban decline. The limitations of a property-based approach to regeneration had been exposed by a slump in the demand for property in the recession of 1989–91. As Turok (1992 p. 376) observed, although property development has potentially important economic consequences, it is 'no panacea for economic regeneration and is deficient as the main focus of urban policy'.[28] Property development lacks the scope, powers and resources to provide the holistic approach required to tackle urban decline. Its limited perspective and goals are unable to guarantee a rise in overall economic activity in a locality and ignore important 'human issues' such as affordable housing, education and training, social exclusion and investment in basic infrastructure.

The government's response was to reconstruct urban policy around the 'initial catalyst' of competition. This era of competitive urban policy was heralded by the

Plate 8.1 The Canary Wharf development in London symbolises the changing economic structure of the former docklands

City Challenge initiative, which introduced competitive bidding among local authorities for urban regeneration funds.[29] Successful schemes are managed by a multi-agency partnership involving the local authority along with the private, voluntary and community sectors. While local authorities appeared to have a greater role under this arrangement, in practice local autonomy is constrained by the underlying entrepreneurial ethos of the partnership organisations and by the need for successful bids to conform to central government guidelines.

THE 'THIRD WAY'

We can add a fifth phase to this chronology of British urban policy commencing in 1997 with the election of New Labour. This marked a move away from the neoliberal era of policy under the Conservatives towards a situation in which greater attention is paid to the social

consequences of economic policy. In policy terms the key priorities include strengthening local and regional economies, increasing economic opportunities for deprived areas, rebuilding neighbourhoods and promoting **sustainable development**. Pursuit of these goals is based on three key principles:

1. A strategic approach is to be taken which involves the integration of national policies and programmes with EU, regional and local initiatives to ensure a comprehensive attack on the multifaceted problems of social disadvantage.
2. Local authorities are to be given a stronger role in urban regeneration.
3. The 'sweat equity' of local people will be encouraged to promote community economic development.

Specifically, New Labour's approach to urban policy is based on CORA:

1. *Community* involvement, with greater public participation.
2. *Opportunity* to work or to obtain training and education.
3. *Responsibility* in the obligation of citizens who can work to do so.
4. *Accountability* of governments to their publics.

We discuss the particular policies and outcomes of this Third Way in relation to urban regeneration in Chapter 16.

URBAN PLANNING IN THE USA

Although urban planning in the USA shares the same reformist roots as in the UK, its evolution and contemporary structure are very different. By contrast with the situation in Britain, there is no national system of planning in the sense of a common framework with a clearly defined set of physical, social and economic objectives. Planning is not obligatory, and together with the fragmented structure of local government – in addition to the federal government and fifty states, there are about 8,000 counties, 18,000 municipalities and 17,000 townships each with the power to plan or regulate land use, i.e. an average of 760 per state[30] – this means that the content of planning is both local and variable from place to place. In principle a range of techniques for controlling urban growth and land use are available, but in practice the major tool employed is land-use zoning. Specific urban problems, such as the provision of low-income housing, are addressed through federal policy.

The first comprehensive zoning ordinance was passed in New York in 1916. The judgement of the US Supreme Court in 1926 that zoning did not infringe the Fourteenth Amendment to the Constitution (which protects against property being taken without due process of law) led to widespread adoption of the technique. Under this procedure, the effective control of land use was transferred from the state to the municipalities and townships, which were thereafter permitted to limit the types of development on land within their boundaries,[31] including control over the height, bulk and area of buildings constructed after the enactment of zoning regulations. The purposes of such controls were to minimise problems of congestion, fire hazard, shading by high buildings; to control population density; to ensure the provision of urban services; and to promote the general welfare of the public. In practice there are many forms of zoning[32] (Box 8.6). This variety, and the underlying presumption in favour of development, means that

controls over market-induced physical growth and change are much weaker in the USA than in the UK.

Critics of zoning maintain that:

1. It is unnecessary, since market forces will produce a fair segregation of land uses.
2. It is open to corruption, particularly in respect of variances (permitted modifications or adjustments to the zoning regulations).
3. It can lead to premature use of land resources by owners who fear an unfavourable zoning change (down-zoning).
4. It is unequal in its effect, since a piece of property zoned for commercial use provides its owner with windfall profits at the expense of neighbours who must bear the costs of increased traffic noise and congestion.

The most vociferous criticism is reserved for the practice of **exclusionary zoning**. This refers to the adoption by suburban municipalities of legal regulations designed to preserve their jurisdiction against intrusion of less desired land uses. Regulations requiring large lots, excessive floor space, three or more bedrooms, or excluding multiple-unit dwellings, high-density development or mobile homes, all serve to maintain high-cost housing and effectively exclude lower-income population.

Supporters of zoning argue that it is a flexible tool and an effective means of allowing local residents to determine part of the character of their neighbourhood. Certainly its wholesale use during much of the twentieth century has helped to determine the land-use structure of metropolitan America.

URBAN SPRAWL

As we saw in Chapter 4, growth of suburbs has been one of the paradigmatic characteristics of US cities since the early 1950s. Policy analysts generally agree that some degree of suburbanisation was a reasonable and inevitable response to problems of overcrowding and housing shortages in central cities in mid-twentieth century America. Between 1950 and 2000 US metropolitan areas grew by over 141 million people, (an increase of more than 166 per cent), and the pattern of suburbanisation pursued allowed the nation to respond quickly to the needs of rapidly expanding metropolitan populations.

The suburbanisation of America has provoked ongoing debate over the benefits and disbenefits of **sprawl**

BOX 8.6

Common forms of land-use zoning in the USA

Zoning is the division of an area into zones within which uses are permitted as set out in the zoning ordinance. If, however, owing to special circumstances, literal enforcement of the ordinance will result in unnecessary hardship for the landowner, the board of adjustment is empowered to issue a variance or relaxation of zoning conditions. The zoning system is, therefore, characterised by a number of forms of zoning, including:

1. *Cluster zoning and planned unit development.* This involves the clustering of development on part of a site, leaving the remainder for open space, recreation, amenity or preservation. The planned unit development is an extension of cluster zoning, in which developers are given freedom to design developments to meet market demand but within a negotiated set of criteria relating to pollution, traffic congestion, etc.
2. *Special district zoning.* This is designed to maintain the special land-use character of a place, such as the Special Garment Center District in New York City, designated to deflect market forces and maintain the garment industry against pressure to convert manufacturing space into offices and apartments.
3. *Downzoning.* Downzoning is the rezoning of an area to a lower-density use and is often the result of neighbourhood pressure to avoid the development of intrusive land use. Since downzoning is likely to reduce the value of undeveloped land, it is likely to be objected to by the landowners.
4. *Large lot zoning.* This has the ostensible purpose of safeguarding public welfare by ensuring that there is good access for emergency-service vehicles, roads are not too congested and there is ample open space. It can also be employed to exclude undesirable residential development and maintain the social exclusivity of a neighbourhood.
5. *Incentive zoning.* This is basically a means of obtaining private-sector provision of public amenities by offering zoning bonuses in return for private finance of specific infrastructure. It is similar in principle to the UK concept of 'planning gain'.

(Table 8.5). Charges against sprawl focus on a loss of land and destruction of natural habitats; road congestion; excessive energy usage, fuel costs and pollution; accentuation of social and racial divisions; and a decline in community involvement.[33] Proponents contend that sprawl satisfies many of the key goals of US society including the freedom to hold land, to live and travel wherever one wishes, to accumulate wealth, and to participate in democratic government at both local and national levels.[34]

Some evidence suggests that concern over the impacts of sprawl is mostly a 'quality of life' issue for the majority of the voting public and, as such, only emerges in to political prominence during periods of economic prosperity. Two fundamental reasons are suggested for this ebb and surge. First, since growth and construction creates employment the majority of the voting middle class will be less concerned about sprawl during a recession when they are more concerned about their basic material well-being. Conversely, a prosperous economy can generate protests from existing residents over the possible negative impacts of sprawl on their quality of life. The cyclical nature of the issue exacerbates the difficulty of reaching a consensus over the costs and benefits of sprawl.

GROWTH MANAGEMENT IN THE USA

A wide range of growth management strategies has been employed by cities and states in an attempt to moderate the negative effects of urban sprawl[35] (Table 8.6). Many of the techniques in use seek to link residential development to infrastructure provision. The town of Ramapo NY, thirty-five miles from downtown Manhattan, introduced a timed growth plan in the late 1960s to ensure that residential development proceeded in phase with provision of municipal services. To this end, developers were required to obtain a permit for suburban residential development. Where the required municipal services were available the permit was granted but elsewhere development could not proceed until the programmed services had reached the location or were provided by the developer. The town of Petaluma CA, forty miles north of San Francisco, introduced, in 1972, an annual development quota of 500 dwellings. Applications to build were assessed against criteria that included access to existing services with spare capacity, design quality, open-space provision, the inclusion of low-cost housing and the provision of public services. The city of Napa CA introduced an

TABLE 8.5 PERSPECTIVES ON SPRAWL

Salient Issues	Sprawl	
	Anti	Pro
Land	Consumes valuable, limited land resources, including farmland	There is more than enough land and farmland to develop
Open space	Reduces access to and views of open space	Can be protected by public acquisition of development rights
Endangered habitat	Fragments habitats, threatening endangered species	Wildlife is increasing, not decreasing in suburban areas
Travel	Requires longer journeys to work	Journeys to work are shorter when jobs decentralize along with residences
Traffic congestion	Auto-dependent sprawl causes congestion	An urban not suburban problem that can be reduced by road pricing
Energy consumption	Auto-dependent sprawl consumes unsustainable amounts of energy	Auto technology is changing and oil reserves remain adequate
Air pollution	Increased auto travel caused by sprawl contributes to global warming and air pollution	Global warming is still unproven; and air pollution is an urban not suburban problem
Mobility	Careless about the carless; makes public transit less attractive and less efficient	Most suburban households have access to at least one car; public transport is more suited to central cities
Water pollution	Destroys wetlands and contributes to water pollution from increased runoff	Environmental restrictions increase housing costs and are unfair to landowners
Public health	Contributes to obesity and increased stress levels	Risks to public health are unproven
Community	Suburbia is destroying community life and character in America	Suburbs provide many opportunities for community involvement
Economic	Costs more than compact development, requiring longer extensions to public infrastructure; also imposes additional tax costs on existing residents	Cities are more expensive than suburbs; externalities associated with new development can be internalised by imposing appropriate taxes
Social	Geographically divides races and classes	Suburbs are becoming more diverse and more equal
Central cities	Directs investment away from central cities in need of redevelopment; leaves cities problem ridden	A back to the city trend is evident in some cities with some suburbanites seeking an urban lifestyle; cities are responsible for their own problems

Source: Adapted from S. Angel, S. Sheppard and D. Circo (2005) *The Dynamics of Global Urban Expansion* Transport and Urban Development Department World Bank Washington DC, p72

urban-limit line intended to cap the population at 75,000 by the year 2000. Beyond the boundary essential public services would not be provided (the actual population in 2000 was 72,585). Several Californian cities, including San Diego and San José, have passed laws requiring voter approval for proposed developments. While 'voter requirements' have not prevented new development in the long run, they have affected the power balance between 'pro-growth' developers and 'slow growth' community interest groups, and have provided current

TABLE 8.6 MAIN ADVANTAGES AND DISADVANTAGES OF SELECTED GROWTH-MANAGEMENT TECHNIQUES

Technique	Purpose	Advantages	Disadvantages
Urban growth boundaries	To set a limit on the extension of urban services	Encourage more compact and cheaper to service development. Limit sprawl	Need for restrictive zoning outside boundaries, and policies on phased growth within boundaries. Can be changed like zoning. If boundaries are too tight can raise cost of land
Concurrency or adequate facilities policies	To ensure that adequate infrastructure is in place before a development is built. To create phased development	Slow the pace of development. Minimise leapfrog patterns. Put more of infrastructure cost on the developer	Developers may fear they will lead to a moratorium on infrastructure and delay their projects
Growth-rate caps	To set a limit on the percentage rate of annual growth	Promote phased, fiscally affordable development	Can raise the cost of land and housing. Difficult to monitor
Transfer of development rights	To transfer development potential from protected lands to designated growth areas	Private developers pay the costs. Concentrates growth	Difficult to agree on sending and receiving areas
Impact fees	One-time payments from a developer to cover the cost of new services for new development	Place more of servicing costs on to developer	Raise cost of development. Setting fees is an imprecise process

residents with some compensation for the negative aspects of growth.[36]

In contrast to the UK, where a national planning system governs land development, restricting an individual's right to develop their land is a politically contentious issue in the USA. This difficulty has been overcome by separating the development value of the land from its existing use value and permitting the transfer of development value to another site. Under this system of transfer of development rights (TDR) owners of land can sell their development rights to developers in designated receiving areas in which they are permitted to build at an increased density that reflects the value of the transferred rights. While this mechanism has been employed in several areas, such as Montgomery County in Maryland, overall relatively few TDR programmes have been initiated.

SMART GROWTH

The smart thing about the **smart growth** movement that emerged in the USA in the mid-1990s is that while advocates share many of the goals of previous anti-sprawl efforts their language and methods are more pragmatic and inclusive. Rather than appealing narrowly to environmental sensibilities the focus is on broader quality of life issues based on a more comprehensive view of the urban development process.[37] Many of the principles of smart growth (Box 8.7) overlap with those of **new urbanism** (see Chapter 9). Smart growth is concerned to protect land from (premature) development and promote development in desired directions. In 1997 the state of Maryland passed the Neighbourhood Conservation and Smart Growth Act in an effort to curb urban sprawl, encourage investment

BOX 8.7

Major principles of smart growth

1. Mix land uses.
2. Take advantage of compact building design.
3. Create a range of housing opportunities and choices.
4. Foster distinctive, attractive communities with a strong sense of place.
5. Preserve open space, farmland, natural beauty and critical environmental areas.
6. Strengthen and direct development towards existing communities.
7. Provide a variety of transport choices.
8. Make development decisions predictable, fair and cost-effective.
9. Encourage community and stakeholder collaboration in development decisions.

BOX 8.8

Marxist-Leninist principles underlying the Soviet socialist city

The application of Marxist-Leninist principles differentiated the Soviet socialist city from its Western counterpart in the following ways:

1. Nationalisation of all resources (including land).
2. Planned rather than market-determined land use.
3. The substitution of collectivism for privatism, most apparent in terms of the absence of residential segregation, the dominant role of public transport, and the conscious limitation and dispersal of retail functions.
4. Planned industrialisation as the major factor in city growth.
5. The perceived role of the city as the agent of directed social and economic change in backward and frontier regions alike.
6. Cradle-to-grave security in return for some restrictions on personal choice of place of residence and freedom to migrate.
7. Directed urbanisation and the planned development of cities according to principles of equality and hygiene rather than ability to pay.

Source: **J. Bater (1980)** *The Soviet City* **London: Arnold**

within growth centres, and protect countryside. The Act required counties and incorporated towns and cities to identify growth boundaries. Within these 'priority funding areas' state funds are available for infrastructure development; development outside the growth areas must be funded by the counties or the private developers. State office buildings, economic development funds, housing loans and industrial development financing are targeted within growth areas. The smart-growth policy does not prohibit developments outside priority funding areas and major developers that do not require public sector infrastructure investment can proceed unimpeded. But, more generally, state infrastructure dollars can help curb sprawl by influencing the location and amount of urban development. A key challenge for smart-growth strategies is to achieve acceptability across the diverse groups with an interest in urban development. As Cullingworth and Caves (2003 p. 186) acknowledged, 'without the necessary co-operation the system will not work'.[38]

PLANNING THE SOCIALIST CITY

If market capitalism represents the governing philosophy for urban growth and change in the USA, the other extreme of the ideological spectrum is represented by urban planning in the socialist city.[39] Following the Bolshevik revolution in 1917, debate over the ideal form of the Soviet socialist city was influenced strongly by the doctrines of Marx and Lenin (Box 8.8).

The general principles for planning the socialist city were laid out in the 1935 plan for Moscow:

1. *Limited city size.* Reflecting the ideas of Ebenezer Howard, the optimum city size was generally considered to be in the range of 50,000–60,000 (to allow both the economic provision of items of collective consumption and the forging of a community ethos). In the case of Moscow, with a population of 3.5 million, an upper limit of 5 million was envisaged. An internal passport system was introduced in 1932 to control population movements and hence city size.
2. *State control of housing.* Regulations to control the allocation of housing space were considered essential on grounds of equity and public health.
3. *Planned development of residential areas.* This was based on the superblock concept in which all daily facilities would be available within walking distance. Groups of superblock neighbourhoods would

constitute a *micro-region* of 8,000–12,000 population which would offer higher-order services.

4. *Spatial equality in collective consumption*. The distribution of consumer and cultural services was to be based on the general principle of equal accessibility.

5. *Limited journey to work*. Norms were established to govern the time spent travelling to work (in large cities a forty-minute trip was deemed a maximum), and public transport was to be the dominant mode.

6. *Stringent land-use zoning*. To reduce journeys to work, housing and employment foci could not be too far apart, but as industry was to be a major urban employer, strict zoning and the use of green buffers were essential to separate residential areas from noxious industry.

7. *Rationalised traffic flows*. Heavy traffic flows in cities were to be handled by designated streets to minimise pollution externalities and congestion.

8. *Extensive green space*. Parks and green belts were an integral part of urban design, with public ownership of all land facilitating its conversion to open space.

9. *Symbolism and the central city*. The city centre was to be the symbolic heart of the city and locus for public demonstrations. This role was emphasised by grand architecture and the decentralisation of administrative and distributive services.

10. *Town planning as an integral part of national planning*. Urban planning was subservient to national economic planning.

The overriding dominance of industry and its controlling ministries largely frustrated the ideals of Soviet socialist urban planning. By the death of Stalin in 1953, the urban situation in the Soviet Union could be compared with Pittsburgh or Sheffield in the mid-nineteenth century. As French (1995) observed, 'heavy industry dominated and polluted, living conditions were appalling, the workers had inadequate rationed access to food and all daily requirements'.[40] During the post-Stalin era, priority in urban planning was necessarily given to housing construction, and the fringes of most cities were built up with micro-regions of apartment blocks in which the planned norms of service provision were achieved rarely, or reached only years after the housing was occupied (Box 8.9). In the latter days of the Soviet system, and with the introduction of Gorbachev's *perestroika* programme, the mechanisms of urban planning changed, to enable public participation and to incorporate social planning

into purely physical planning for cities. Reality replaced the idealism of the founders of Soviet socialism. The goal of a truly socialist city was never achieved.[41]

SOCIALIST URBAN FORM

Urban form evolves in response to complex interactions among public and private forces. As we have

---

BOX 8.9

The Soviet micro-region

The micro-region was the basic spatial unit of planning in the Soviet socialist city. Comprising a number of apartment blocks (*kvartaly*) the micro-region of population 5,000–12,000 was designed to provide basic services for residents. Theoretically, any flat in a micro-region should be within 100–300 m (90–270 yards) of the nearest shops with 80 per cent of the population not over 1,000 m (0.6 mile) away. As French (1995) describes, in reality the vast majority of micro-regions do not meet the norms of service provision as a result of the pressing need to provide housing. Where shops have been provided they are not in readily accessible precincts but scattered through the micro-region on the ground floors of the apartment blocks, thereby entailing much time and effort in shopping. In particular, any given micro-region is unlikely to have the full range of shops selling daily and weekly necessities. Thus journeys to neighbouring microregions are often necessary.

At a higher level in the urban structure, plans group several micro-regions to form a 'living region' (*zhiloy rayon*) of 30,000–50,000 people with higher levels of service provision. In the largest cities, such as Moscow, plans envisaged groupings of 'living regions' into a 'town region' (*gorodskoy rayon*) with a major service centre which would obviate the need for journeys to the city centre except on special occasions. The 1971 Moscow Genplan contained eight such 'town regions', one of which was the old central district, the overall city focus. In practice the other seven surrounding 'town regions' failed to develop their foci; indeed, such regional centres have not emerged effectively anywhere as yet.

Source: **R. French (1995) *Plans, Pragmatism and People* London: UCL Press**

seen, in socialist cities planned investment and state regulations were the defining instruments that created distinctive patterns of land use. The spatial character of the (formerly) socialist cities typically exhibited a relatively high density of residential settlement in the urban core; an adjacent band of low-density settlement where industry was located, often in large and polluting factories that would have been considered unacceptable urban investments in Western societies; and a swath of increasing residential density at the periphery, commonly in high-rise apartment blocks built to a standard design. Figure 8.3 portrays this archetypal rising density profile or 'camel back' for St Petersburg, where residential density drops off sharply at a radius of only 4 km and then rises consistently to 14 km from the centre. In Budapest residential density also declines at around 4 km from the centre but here the absence of a 'camel back' reflects the local authorities' more liberal policies (compared with the Soviet model) regarding private-housing investment and a lower reliance on high-rise public housing estates. Sofia shows more of a mix of high and low densities at all distances from the city centre. By contrast cities in the long-standing market economies, such as London, generally exhibit a smooth and declining density gradient, although the steepness of the drop and extension of relatively high residential densities over considerable distances from the city centre vary as a function of regulation, tax and investment policies affecting the land market.

Under state socialism urban land and housing markets did not exist. Before the transition to capitalism almost all housing was state-owned and city governments restricted residential mobility as a way to tackle housing shortages. People lived in the same place for long periods and neighbourhood change was slow. Planners located stores and services according to the demographic profiles of neighbourhoods and, once located, urban functions remained in place for decades.[42] In Moscow because large-scale housing construction did not begin until the 1960s most working- and middle-class families lived in old downtown buildings in subdivided, overcrowded apartments shared by several extended households (known as 'communal apartments'). Blue-collar workers from outside Moscow who filled the least desirable jobs in exchange for a Moscow residence permit were given vacant rooms in communal apartments whose former residents had moved to new single-family units in high-rise housing projects on the city's periphery. In marked contrast to the capitalist city people from different classes, professions, ages and ethnic groups lived in the same neighbourhood, the same building and even the same communal apartment.[43] When an old building became unliveable all inhabitants were resettled to new housing and the upgraded building was converted to government offices or commercial space. (Thus, between 1960 and 1988, the total supply of housing in central Moscow dropped by 50 per cent.) In general, the better

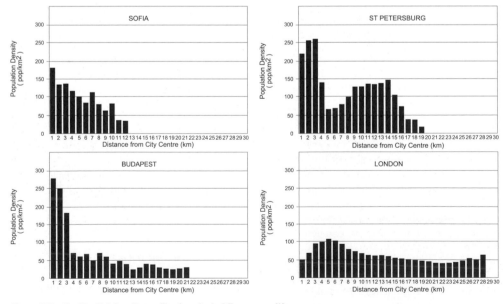

Figure 8.3   Residential density profiles in selected European cities

infrastructure, prestige and accessibility offered by a downtown location contributed to the concentration of all types of economic activities there, including manufacturing, services and state offices.

The uneven impact of post-socialist privatisation has reinforced Soviet-era privileging of the city centre in Moscow. With 7 per cent of the urban population in 1997 the central district (CBD) of Moscow employed 25 per cent of city jobs and contained 27 per cent of all institutions and enterprises. Such spatially focused growth stimulated demand for new office and commercial space in the central city which has prompted Moscow city government, like many entrepreneurial cities in the USA, to undertake large-scale redevelopment of the downtown in partnership with private capital.

In the transition period since 1989 privatisation and the replacement of **central planning** with the free market, together with post-industrial restructuring, have catapulted cities such as Moscow from industrial socialism to post-industrial capitalism in a short period. In the post-Soviet city capitalist tendencies such as suburbanisation and counterurbanisation are in evidence, and social differentiation in housing and quality of life is increasingly apparent. In Moscow between 1990 and 1994 the level of private ownership of housing increased from 9.3 per cent to 49.1 per cent. This tenure change and the reintroduction of market mechanisms have been accompanied by increased residential mobility that has enabled some individual citizens to improve their living conditions. As well as suburbanisation, gentrification has appeared in central city redevelopment schemes[44] while, more generally, there is a trend towards a 'European' urban residential structure characterised by an affluent inner core and a poor periphery.[45] Nevertheless, while individual choice has been enhanced for those able to participate in the housing market, state planning control of zoning regulations and real-estate taxes (coupled with a limited private mortgage system and declining real incomes for the majority) ensure that, in the short to medium term at least, the social geography of Russian cities will continue to be influenced strongly by local government.

As in the former Soviet Union, in China the declining role of state enterprises in the economy, the introduction of land and housing markets and the opening up of cities to foreign investment have meant that the state and

Plate 8.2   High-rise apartment blocks in Moscow are typical of public housing provision in the socialist city

the centrally planned economy have a much reduced influence on urban development. As we have seen (Chapter 5), as a result of the dual effects of internal reforms and globalisation processes, Chinese cities have experienced significant transformation in their socio-spatial structure in the reform era. Just as in the post-Soviet city, 'Western' processes of urban sprawl, spatial segregation and social polarisation are being reproduced in Chinese cities. The transition is evident in Beijing, where the creation of a modern CBD is under way, a real-estate market has emerged and the renewal of inner-city slums is in progress. The rapid growth of the urban economy and population has also resulted in peripheral expansion, with the built-up area of the city expanding by almost 50 per cent, from 335 km² to 488 km² between 1978 and 1998. The introduction of marked social and spatial polarisation in Chinese cities, as well as in other 'transitional societies' such as Vietnam,[46] represents a new problem and challenge for urban authorities.

TOWARDS PLANNING FOR SUSTAINABLE URBAN DEVELOPMENT

Increasingly, urban policy and planning are required to adopt a long-term prospective role. This is particularly so when society, in managing urban change, seeks to strike a balance between economic priorities on the one hand and social and environmental priorities on the other. This issue is central to the question of sustainable urban development.

Economic development is fundamental to human well-being, but growth which fails to recognise the limits of natural resources and the finite capacity of global ecosystems to absorb waste is a basis for long-term decline in the quality of life. Sustainable development aims to meet 'the needs of the present without compromising the ability of future generations to meet their own needs'.[47] Cities are major agents in depleting the quality of the environment for future generations. They are significant generators of gases, such as carbon dioxide, which cause global warming and of nitrogen oxides and sulphur dioxide, which contribute to acid rain. The greenhouse effect and ozone depletion are consequences of processes of urbanisation and industrialisation which consume raw materials and energy, and produce environmentally damaging waste products.

It is important, however, not to allow the rhetoric of sustainable development to obscure the fact that cities will continue to be net consumers of resources and producers of waste products simply because of the relative intensity of social and economic activity in urban

places. Neither should idealism be permitted to cloud the fact that most people will not relinquish voluntarily a cherished lifestyle (such as an urban workplace and rural residence). Furthermore, the goal of sustainability is not an integral element of market capitalism and will encounter opposition from entrenched interests. Most fundamentally, it is unrealistic to expect calls for restraints on economic growth to protect the future environment to be heeded generally in a world where millions of impoverished people face a daily struggle to survive. The importance attached to sustainability in the trade-off between environmental considerations and social, economic and cultural aspirations is clearly a function of general levels of well-being in a society.

Thus the concept of sustainable urban development embraces more than environmental issues, and cannot be achieved merely by imposing pollution taxes or by promoting technical developments to reduce the energy consumption of cars and production processes. Sustainability must also address the key question of people's lifestyle. In essence there are two broad approaches to sustainable urban development:

1. An environmental protection approach with a focus on a municipal programme to reduce the consumption of resources and minimise the environmental impact of development.
2. A holistic approach including an ecological component (stressing the importance of environmentally sound policies), economic aspects (development activities and fiscal issues) and social-equity issues (a fair distribution of resources and the distributional impact of policies).

From this broader perspective the ideal sustainable community is characterised by:

1. *Environmental integrity*: clean air, soil and water, a variety of species and habitats maintained through practices that ensure long-term sustainability. recognition that the manner in which natural resources are used and the impact of the individual, corporate and societal actions on the natural processes directly influence the quality of life.
2. *Economic vitality*: a broadly based competitive economy responsive to changing circumstances and able to attract new investment and provide employment opportunities in the short and long term.
3. *Social well-being*: safety, health, equitable access to housing, community services and recreational activities, with full allowance for cultural and spiritual needs.

The realisation of a sustainable and liveable city requires both an integrated planning and decision-making framework and a fundamental shift in traditional values and perspectives. There needs to be a change in focus from curative measures such as pollution reduction to measures based on prevention, from consumption to conservation, and from managing the environment to managing the demands on the environment. This will require change at the individual, community, business and urban levels.

Some tentative steps along this road may be identified in the European Sustainable Cities project, which frames the search for sustainable urban environments as a challenge 'to solve both the problems experienced within cities and the problems caused by cities [while] recognising that cities themselves provide many potential solutions'.[48] At the local government level,

Agenda 21 of the 1991 Rio Earth Summit agreed a number of proposals relating to waste management, energy conservation, the integration of land-use and transport planning, and the protection of natural habitats. The summit proposed that by 1996 most local authorities should have carried out a consultative process with local residents to develop a Local Agenda 21, including targets and timetables.

In view of the increasing number of people living in urban settlements, strategies designed to substitute sustainable development for unsustainable growth can have a major influence on future urban policy and planning, and on the form and function of cities in the twenty-first century (see Chapter 30). For some analysts, the way towards an improved urban future lies in the creation of new settlements. We examine this theme in the following chapter.

FURTHER READING

BOOKS

K. Axenov, I. Brade and E. Bondarchuk (2006) *The Transformation of Urban Space in post-Soviet Russia* London: Routledge

J. Cullingworth and R. Caves (2008) *Planning in the US* London: Routledge

J. Cullingworth and V. Nadin (2006) *Town and Country Planning in the United Kingdom* London: Routledge

N. Gallent, J. Andersson and M. Bianconi (2006) *Planning on the Edge: The Context for Planning at the Rural–Urban Fringe* London: Routledge

O. Gilham (2002) *The Limitless City: A Primer on the Urban Sprawl Debate* Washington DC: Island Press

R. Jackson (1981) *Land Use in America* London: Arnold

P. Newman and A. Thornley (1996) *Urban Planning in Europe* London: Routledge

M. Pacione (2003) Urban policy, in M. Hawkesworth and M. Kogan (eds) *Encyclopaedia of Government and Politics* London: Routledge, 795–808

K. Pallagst (2007) *Growth Management in the US: Between Theory and Practice* Aldershot: Ashgate

J. Reps (1965) *The Making of Urban America* Princeton NJ: Princeton University Press

JOURNAL ARTICLES

W. Ackerman (1999) Growth control versus the growth machine in Redlands, California *Urban Geography* 20(2), 146–67

J. Claydon and B. Smith (1997) Negotiating planning gains through the British development control system *Urban Studies* 34(12), 2003–22

F. Gaffiken and B. Warf (1993) Urban policy and the post-Keynesian state in the United Kingdom and the United States *International Journal of Urban and Regional Research* 17(1), 67–84

K. Geurs and B. van Wee (2006) Ex-post evaluation of thirty years of compact urban development in the Netherlands *Urban Studies* 43(1), 139–60

M. Pacione (1990) Private profit and public interest in the residential development process *Journal of Rural Studies* 6(2), 103–16

R. Peiser (2001) Decomposing urban sprawl *Town Planning Review* 72(3), 275–98

R. Rudolph and I. Brade (2005) Moscow: processes of restructuring in the post-Soviet metropolitan periphery *Cities* 22(2), 135–50

E. Wentz and C. Redman (2007) The spatial structure of land use from 1970 to 2000 in the Phoenix, Arizona, metropolitan area *Professional Geographer* 59(1), 131–47

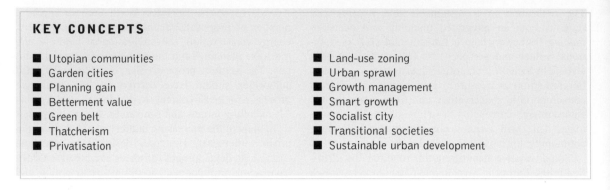

KEY CONCEPTS

- Utopian communities
- Garden cities
- Planning gain
- Betterment value
- Green belt
- Thatcherism
- Privatisation

- Land-use zoning
- Urban sprawl
- Growth management
- Smart growth
- Socialist city
- Transitional societies
- Sustainable urban development

STUDY QUESTIONS

1. Planning seeks to influence the process of urban development for 'the common good'. Consider what is meant by this concept. Make a list of examples of planning actions that have a positive impact on the common good, and a list that you feel have a negative impact on the common good.
2. Identify the origins and development of the current system of urban planning in the UK.
3. Urban planning takes place within the framework of national urban policy. Examine the post-Second World War development of urban policy in the USA or UK and explain its effects on urban planning.
4. Examine the arguments for and against land-use zoning in the USA.
5. Explain what is meant by urban sprawl and examine the benefits and disbenefits of the phenomenon.
6. Identify the characteristics of the socialist city and consider how they may change in response to the demise of the command economies in Eastern Europe.
7. What do you understand by the concept of sustainable urban development?

PROJECT

Conflict over use of land is an integral part of capitalist urban development. Review back issues of your local newspapers (available in your local library) to identify examples of land-use conflicts. Analyse the reasons for the dispute and examine the motives of those involved. How was the dispute resolved and what effect did the decision have on the urban fabric? Who were the winners and losers in the matter?

9

New Towns

Preview: the concept of new towns; the UK new towns; the neighbourhood unit; private-sector new towns in the UK; pressures for urban development; new towns in Europe; new towns in the Third World; new communities in the USA; federal involvement; the profit motive; master-planned communities

INTRODUCTION

The attraction of exchanging a decaying and over-crowded urban environment for life in a new, planned community has a long history that ranges from the Greek colonies in the eastern Mediterranean of the fifth century BC to the present day (see Chapter 3). Modern new towns primarily stem from post-Second World War developments in the UK, where, over a period of three decades, twenty-eight new towns were constructed as part of a government strategy to allevi-ate the social, economic and physical problems of older urban areas. In this chapter we trace the roots of modern new-town development in the UK and its subsequent manifestation and reformulation in the USA and elsewhere. We then consider the future of the new town concept and the role of public and private agencies in new community building.

THE BRITISH NEW TOWNS

The British concept of new towns was inspired by reformist reaction to the evils of the nineteenth-century industrial city and was promoted in the early twentieth century by the 'garden city' ideal. The post-war British new towns were conceived in response to the Barlow report (1940) and to Abercrombie's Greater London Plan (1945).[1] The recommendations of the Barlow report on the geographical distribution of the industrial population identified a need for satellite towns to help reorganise congested urban areas. The Abercrombie plan applied the general principles of decentralisation, urban contain-ment, redevelopment and regional balance of opportu-nity to the particular problems of London. Following the Barlow report recommendations, the Greater London Plan proposed a ban on additional industrial develop-ment within the plan area, with planned redevelopment in the inner ring, and an enlarged green belt. The most radical proposal was the planned decentralisation of over a million people from the inner ring to the outer ring. This was to be achieved by the expansion of existing towns and the construction of eight new towns to accom-modate 400,000 people, with full local employment and community facilities (Figure 9.1). The legal power to implement these proposals was contained in the New Towns Act of 1946, which was the basis for the construction of twenty-eight new towns throughout the UK[2] (Figure 9.2).

The general characteristics of the new towns were defined by the Reith committee (1946),[3] which recom-mended that they should:

1. Be sited around large, densely built-up urban areas to help reduce their populations.
2. Not be located less than 40 km from London or less than 20 km from the regional cities, to allow the development of independent communities.
3. Have populations of between 20,000 and 60,000, the upper limit reflecting one of Howard's prin-ciples and the lower limit based on the concept of a threshold population needed to ensure a satis-factory level of service provision as well as a mixed class structure.

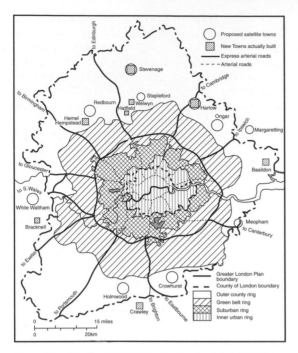

Figure 9.1   The Abercrombie Plan for Greater London, 1944

4. Normally be built on greenfield sites to give planners full scope and avoid disruption of existing
communities (although in practice most new towns
were built around an existing settlement nucleus).
5. Avoid building on the best agricultural land.

The twenty-eight new towns created under the 1946
Act were designated in two main phases (Figure 9.2).
Of the first fourteen (Mark I) new towns, the eight in
the Outer London ring plus East Kilbride outside
Glasgow were intended to advance the objectives
advocated by Howard, Barlow and Abercrombie, of
dispersing population from overcrowded urban areas.
Included in this group was Welwyn Garden City,
which was redesignated as a new town. The remaining
five new towns were designed to aid regional development, with, for example, Corby built to house workers
for a new steel plant, and Glenrothes to accommodate
miners in the East Fife coalfield.

The determination of the post-war new town planners to achieve socially balanced communities led to
the widespread adoption of the **neighbourhood unit**
concept as a basis for physical planning. In Crawley
New Town this attempt at 'social engineering' was
evident in proposals to reflect the national class profile
by integrating 20 per cent middle-class households

	Date of designation		Area in hectares	Pop. at designation	Pop. in 1976	Pop. target in 1976
First-generation (Mark I) new towns						
Stevenage	Nov.	1946	2,532	6,700	74,000	105,000
Crawley	Jan.	1947	2,396	9,100	75,000	85,000
Hemel Hempstead	Feb.	1947	2,391	21,000	78,000	85,000
Harlow	March	1947	2,588	4,500	81,000	90,000
Aycliffe	April	1947	1,254	60	26,000	45,000
East Kilbride	May	1947	4,148	2,400	76,200	90,000
Peterlee	March	1948	1,133	200	27,500	30,000
Hatfield	May	1948	947	8,500	26,000	29,000
Welwyn	May	1948	1,747	18,500	41,000	50,000
Glenrothes	June	1948	2,333	1,100	33,700	70,000
Basildon	Jan.	1949	3,165	25,000	91,890	130,000
Bracknell	June	1949	1,337	5,149	45,000	60,000
Cwmbran	Nov.	1949	1,278	12,000	45,000	55,000
Corby	April	1950	1,791	15,700	53,500	70,000
Second-generation (Mark II) new towns						
Cumberland	Dec.	1955	3,152	3,000	45,000	100,000
Skelmersdale	Oct.	1961	1,669	10,00	41,000	80,000
Livingston	April	1962	2,708	2,000	29,000	100,000
Redditch	April	1964	2,906	32,000	53,200	90,000
Runcorn	April	1964	2,930	28,500	54,600	100,000
Washington	July	1964	2,271	20,000	46,000	80,000
Irvine	Nov.	1966	5,022	34,600	52,305	120,000
Milton Keynes	Jan.	1967	8,900	40,000	77,000	250,000
Peterborough	July	1967	6,453	81,000	109,000	180,000
Newtown	Dec.	1967	606	5,500	7,700	13,000
Northampton	Feb.	1968	8,080	133,000	158,000	230,000
Warrington	April	1968	7,535	122,300	135,400	201,500
Telford	Dec.	1968	7,790	70,000	99,700	220,000
Central Lancashire New Town	March	1970	14,267	234,500	248,000	420,000

Figure 9.2   The British new towns

in each neighbourhood unit as well as in the town as a whole. Such attempts to determine social structure and activity by physical design ignore the 'disorganised complexity' of urban life. In the 1950s growing recognition of the limited social relevance of the neighbourhood unit led to its role being reduced to one of ensuring that essential facilities were located in reasonable proximity to housing.

The Conservative government brought the first phase of New Town designations to a halt in 1952, largely on the grounds of cost. The problems of urban overcrowding were to be tackled instead by overspill schemes arranged between local authorities with the financial support of central government under the provisions of the Town Development Act 1952. London made arrangements with thirty-two towns, including Basingstoke and Swindon, to build a total of 55,000 dwellings for overspill Londoners. Significantly, since a Town Development Act for Scotland was not passed until 1957, a second Scottish new town was started at Cumbernauld in 1956. The New Town concept was not abandoned entirely by the government standstill, since several local authorities built their own new towns, as at Killingworth and Cramlington in Northumberland, and Thamesmead, built by London County Council.

By 1961 several factors led to a decision to recommence the New Town programme:

1. The first generation of new towns were showing signs of becoming profitable.
2. The Cumbernauld plan, with its Radburn-inspired separation of people and traffic, was attracting international acclaim.
3. There was continued criticism of living conditions in many cities, and increasing displacement of population as a result of comprehensive urban redevelopment schemes.
4. There was growing pressure on local councils to release building land to accommodate overspill from the large cities.

Most of the Mark II new towns were designed to accommodate conurbation overspill, with, for example, Skelmersdale and Runcorn sited close to Merseyside, Redditch and Dawley (renamed Telford) for Birmingham, and Washington for Tyneside. The overspill objective was reflected in higher population densities and increased target populations that reached 250,000 in Milton Keynes and 500,000 for Central Lancashire New Town. The desire for flexibility to accommodate subsequent growth was evident in the open-ended plans for later new towns. Runcorn was laid out with three spines radiating from a centre separated by green wedges, while Warrington was built on a linear pattern that could be lengthened if necessary.

The increasing role of the automobile in daily life also exerted a strong influence on New Town design, with vehicle–pedestrian separation an integral feature of Cumbernauld, Washington and Milton Keynes.[4] In Runcorn emphasis was placed on providing a reliable public transport system operating on a separate roadway. In addition to having an overspill function, the new towns were also conceived as regional growth points, with, for example, the Central Lancashire New Town intended to act as a focus for the economic regeneration of the old textile area between Manchester and Preston.

THE FUTURE OF THE BRITISH NEW TOWNS

There is little doubt that the UK new town programme was a success. In the first twenty-five years, fifteen towns were nearing completion and a dozen more started; they provided new housing and a congenial living environment for 1.4 million people, employment for 650,000, as well as a range of public facilities. Critics, however, point to the fact that the new towns contributed less than 5 per cent of the national house-building programme and, in view of the continuing problems of cities, may be regarded as 'too few and too late'.

One option was to build on post-war plans for a larger number of new towns, and during the 1960s several regional studies advocated expansion of the New Town programme. The South East Study identified sites for new towns of 150,000 to house over 500,000 people displaced from London.[5] These included proposals for new towns at Newbury (Berkshire), Stansted (Essex), Ashford (Kent) and Bletchley (Buckinghamshire), the last of which was developed as Milton Keynes (Box 9.1). However, growing concern over the cost of new towns (the development at Milton Keynes had cost £700 million) and over the potential competitive impact of the new towns on declining inner cities led to the curtailment of the New Town programme in 1976. The first casualty of the decision was the newly designated sixth new town for Scotland at Stonehouse in Lanarkshire, which was abandoned. The expertise of the project team was redirected into the Glasgow Eastern Areas Renewal project. The Conservative government of

Margaret Thatcher confirmed this switch of policy emphasis in favour of the inner cities, and set in train the privatisation of the new towns. Under the New Towns and Urban Development Corporations Act of 1985 the Commission for New Towns was to be

---

BOX 9.1

Milton Keynes New Town

The *South East Study* (1964) proposed a major new city at Bletchley in Buckinghamshire, and in 1967 the Milton Keynes Development Corporation was established to create a new town on an area of 21,900 acres (8,860 ha) with an existing population of 40,000, a planned intake of 150,000 and an expected population of 250,000 by the end of the century.

Reflecting the economic and social confidence of the time, the idea was to provide a New Town that could cater for the 'new affluence' of the 1970s. A key goal in the plan was to provide a wide choice of living environments for future residents. The plan emphasised:

1. The desire to give maximum employment opportunities, with the emphasis on the service sector (education, communications, technical and office employment) rather than manufacturing.
2. The need for a variety of housing types, with a target of 50 per cent private housing.
3. Low-density development, with net residential densities of six to ten dwellings per acre (two and a half to four per hectare), to attract the professional and managerial classes.
4. An efficient transport system based primarily on a grid pattern of town roads.
5. Differing degrees of design control, ranging from maximum control for specific areas (such as the city centre), through intermediate design areas to encourage good design from outside developers, to do-it-yourself areas (in which design controls cover only safety, sanitation and access) conducive to experimental housing.

Milton Keynes has attracted criticism for its low-density housing layout, over-elaborate road pattern and poor public transport system, but it continues to grow and has attained many of its objectives. Its particular goal of becoming a 'city of learning' was enhanced by its selection as the location for the Open University.

---

disbanded and the commercial assets of existing new towns were to be progressively sold off, with private developers assuming an increasing role in the completion of the unfinished new towns. Certainly the British new towns represent a policy that was always intended to terminate once the post-war goal of developing a network of new, improved residential and living environments for former inner-city residents had been achieved.[6] Whether this marks the end of new town building in Britain or a temporary standstill, as in 1952, only time will tell.

PRIVATE-SECTOR NEW TOWNS IN THE UK

Although the overspill of population generated by large-scale post-war inner-city redevelopment has come to an end, the process of suburbanisation, combined with *in situ* growth in population and employment, has ensured continuing pressure on 'outer city' land for residential development.[7] The most recent (1997) government projections estimated a need for 4.4 million new houses in England and Wales by 2016. The continuing problem is how to accommodate this growth.

An early indication of a shift in government policy in favour of the development of private new towns came at Lower Earley, adjoining the town of Reading in the M4 growth corridor west of London,[8] where over 6,000 dwellings were built by a consortium of private builders between 1977 and the early 1990s. The planning gain negotiated amounted to 8 per cent of the selling prices of the houses and was used to finance roads, school sites, leisure facilities and open space.

The commercial success of this scheme led to the formation of Consortium Developments Ltd (CDL) in 1983, a group of nine major private house-builders, with the aim of creating new country towns of at least 5,000 dwellings with associated employment and social amenities. By 1989 CDL had lodged specific proposals to build four new towns – in Essex (Tillingham Hall), Hampshire (Foxley Wood), Oxfordshire (Stone Bassett) and Cambridgeshire (Westmere) (Figure 9.3). This presented the Conservative government with the political dilemma of whether to support the development plans of the private house-building industry (in line with the 'enterprise culture') or to heed the landscape protection goals of traditional Conservative voters in the rural areas ('**Nimbyism**'). The first of the CDL proposals at Foxley Wood was approved provisionally in 1989, only for the decision to be reversed within a few months by a

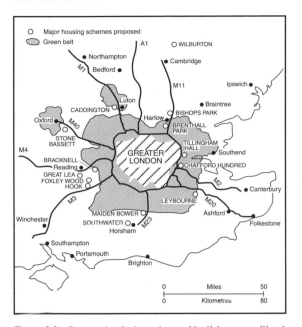

Figure 9.3 Proposed private-sector residential communities in south-east England

will include commercial and industrial employment, schools, churches, shops and parkland. The involvement of the local council (Thamesdown) is limited to the supervision of housing design, leaving the private sector to provide the housing and infrastructure required to accommodate the projected need for 10,000–15,000 additional houses (over 27,000 people) in the area by 2005. Further indication of the emergence of the private sector as a major driving force for new town construction is seen in the responses of local councils to the government's projected requirement for new housing. The Gloucestershire County Council structure plan proposed a new town of 6,000 homes on 670 acres (270 ha) of council-owned land in Berkeley Vale, to the west of Stroud, plus a second new town to the north of Gloucester. The greatest pressure on land is in south-east England (Box 9.2). In November 1997 the newly elected Labour government confirmed the previous government's decision to build 50 per cent of new housing on open countryside, thus fuelling the greenfield–brownfield debate and provoking arguments over sustainable urban development.

new Secretary of State for the Environment, who also rejected the plans for Stone Bassett and Westmere as the policy pendulum swung in favour of 'green issues'. In 1992 CDL disbanded, having spent nine years unsuccessfully attempting to create private new towns. This did not, however, herald the end of private-sector interest in new community construction. As we saw in Chapter 8, the battle over development in the countryside around the major British cities is an ongoing conflict.[9]

In 1993 planning permission was granted for the development of a privately financed township (Hampton) of 5,200 houses for 13,000 people on a 2,000 acre (800 ha) area of abandoned brick and clay pits to the south of Peterborough. When completed, it will comprise four self-contained residential neighbourhoods including low-cost rental, special-needs and sheltered housing; and a 175 acre (70 ha) business park, hotel and leisure complex. Europe's largest private-housing project of 10,000 homes is being developed by a consortium of fourteen contractors on a 1,000 acre (400 ha) site to the north of Swindon at a total cost of £900 million. The development is in the form of several 'new villages'; the first of these (Abbey Meads) opened in 1996 with 200 families, with another 150 homes sold and 400 scheduled for completion within two years. The completed development

NEW TOWNS IN EUROPE

Somewhat paradoxically, in view of the British government's withdrawal from new town development, many countries in Europe and elsewhere have followed Britain's lead, adopting the New Town principle and adapting the concept to their particular needs.[10] In the Netherlands, new towns have been employed in the settlement of the polder lands (see Chapter 6). Central government control is also strong in Israel, where the goals of new-town policy have been to colonise inland territory, strengthen border defences and decongest the three major urban areas of Tel Aviv, Jerusalem and Haifa. In France the principal aim of the new-town programme has been to relieve population pressure on Paris, with five new towns constructed outside the capital (Box 9.3). In Sweden four of the eighteen new communities planned to house 250,000 people from Stockholm were developed by the municipal authority, which has purchased land since 1904 in anticipation of its eventual use for urban development. In Finland the new town of Tapiola was planned and built by a non-profit housing association created by six large private firms and labour unions. In the former Soviet Union many new towns were specialised in function, such as science cities, and most were intended to colonise the country's vast territory.

BOX 9.2

Where will the new houses go?

The garden of England may become Britain's biggest building site, with at least 116,000 homes expected to be built in Kent by 2011, an analysis of county councils' plans reveals.

The second worst affected county is Essex, where granted. 106,000 new houses will be built, followed by Devon, where 99,000 homes are planned. The counties that will escape the bulk of new housebuilding include Durham, where only 22,400 homes are expected, and Cumbria, with 27,500 new homes expected by 2006. In general the survey found that the greatest pressure for homes will be in the South East, South West, East Midlands and Eastern regions, with the North under the least pressure. Clive Aslet, editor of *Country Life*, which published the survey today, said: 'These figures are a uniformly dismal tally. Anyone who keeps their eyes open, travelling around the shire counties, will realise the damage that has already been wrought in recent decades. These plans show the blueprint for development in the future. Enough is enough.'

The survey highlights the pressure on green belt land and Areas of Outstanding Natural Beauty. Government figures show that there will be 4.4 million new households by 2016 because of the growing number of single divorced and elderly people. The councils' structure plans give developers an idea of where planning permission is likely to be. Green-belt land is threatened in Hertfordshire, villages on the Chiltern Downs in Buckinghamshire may be lost in urban sprawl, housing is expected to multiply near Stansted Airport in Essex, and there are fears for green-belt land in Northumberland.

New houses for commuters are planned in Durham, new towns are planned for East Sussex, and Devon villages may become large towns, the survey warns. The Painswick Valley in Gloucestershire is under threat from Stroud District Council, which proposes to build 1,500 new houses. The valley is an Area of Outstanding Natural Beauty immortalised by Laurie Lee, and villages such as Dursley, Cam and Painswick are expected to bear the brunt of Gloucestershire County Council's structure plan. There are plans to develop green-belt land surrounding Slough for 1,000 new homes. Several Berkshire greenfield sites have been earmarked, including Sandelford, near Newbury, where 1,250 dwellings are expected by 2006.

Source: The Times, **13 November 1997**

NEW TOWNS IN THE THIRD WORLD

Osborn and Whittick (1977) list over sixty countries in which planned community development has taken place, as distinct from suburban growth.[11] A large number of new towns have been built in countries of the developing world. In Brazil the urbanisation of the interior was aided by the development of the new capital city of Brasilia, begun in 1956. The new city has functioned as a regional growth pole, but the persistence of problems linked with poverty, including the development of shanty towns and an informal economy, testify to the difficulty of implementing central planning in a developing country with high levels of income polarisation and poor basic services. In Asia several states have created new towns. Construction of Islamabad as the capital city of Pakistan began in 1961, and by 2000 it had a population in excess of 1 million. The Indian new town of Chandigarh was designed by Le Corbusier and built during the 1950s and early 1960s on a neighbourhood plan of fifty-three self-contained sectors. Intended to accommodate 500,000 people, by 2000 the population of Chandigarh has risen to 642,000. Both Islamabad and Chandigarh were built at considerable cost, and in view of the region's poverty the widespread construction of new towns cannot be viewed as a general solution to the housing and infrastructural problems facing Asia's cities. One Asian state that has made extensive use of new towns is Singapore.[12] Following independence in 1959, the government assumed a dominant role in the provision of housing, much of which has been located in sixteen new towns. Singapore provides a good example of how the new-town concept may be adapted to fit local socio-political circumstances. In addition to the goal of overcoming a housing shortage the relocation of population into public housing was linked with a number of other objectives. These included the residential integration of different ethnic groups, physical redevelopment of central city squatter areas, and the release of land for commercial development in line with the creation of an export-oriented modern urban economy. Since 1989 the Singapore government has been engaged in a programme of upgrading earlier new towns, both as a boost to the construction industry and to improve residential environments. In South

BOX 9.3

The new towns of the Paris region

The 1965 *Schème d'aménagement et d'urbanisme* (Urban Management Plan) for Paris reversed the previous objective of controlling the growth of the city and sought instead to accommodate the expected growth (from 8.5 million in 1965 to 14 million by 2000) in an orderly fashion. Eight new towns were initially planned to house a total of 3 million people, but this was reduced to five new towns with a total population of 650,000 by 1990. Unlike the UK New Towns, these were not intended to be independent self-contained settlements but were linked to the agglomeration by rapid transport links. The Parisian new towns were also intended to accommodate future population growth rather than redistribute population from the inner city, as in Britain. In terms of quality of living environment the Parisian new towns represent a major improvement over the high-density *grandes ensembles* (social housing estates), and are now entering a phase of maturity. This was reflected in the revised region plan adopted in 1994, which recommends a shift of policy emphasis from the New Towns and outer suburbs to the inner areas affected by industrial decline and attendant social problems, a trend reminiscent of that enacted in Britain two decades earlier.

Source: D. Noin and R. White (1997) *Paris* London: Fulton

Plate 9.1   Tin Shui Wai new town in Hong Kong

Korea five new towns were constructed around Seoul in the1990s. Located within 25 km of the capital city they are heavily dependent on Seoul as a workplace destination, although in terms of non-work trips a degree of self-containment is emerging as some of the new towns develop as suburban retail centres within the Seoul metropolitan region.[13] In India the new town of Navi Mumbai was viewed on it inception in 1976 as a key component of a regional strategy to decongest central Mumbai. A policy reversal in the early 1990s in favour of reconstructing Mumbai as a 'world city' hindered development of Navi Mumbai as an independent self-contained city.[14] In Malaysia the two 'intelligent new cities' of Cyberjaya (a high-tech R&D core) and Putrajaya (an electronic federal administrative centre) are intended to boost the country's transition from a 'developing' to a 'developed' nation. The experience in Asian cities suggests that new towns and satellite cities can be used as instruments for developing urban peripheral areas. However, the effectiveness of the strategy depends on three main factors. First, new towns have to be built on a large enough scale to be self-contained settlements, rather than merely dormitory communities for workers in the central city. This also negates plans to restrict new towns to specific functions (e.g. an industrial complex). Secondly, the location of a new town in relation to the central city is of crucial importance. Though new settlements should be linked to the central city by appropriate transport modes, if they are too close to the city they are likely to be absorbed by urban sprawl. Third, new towns should be designed to link with their environs, in order to stimulate local-regional development.

NEW COMMUNITIES IN THE USA

The origins of the new-community concept in the USA can be traced to the first settlements. As we saw in Chapter 3, many early colonial communities were built according to carefully prepared plans. These were followed in the nineteenth century by a number of company towns, including Lowell MA (1822) and Pullman IL (1880), and in the early twentieth century by several privately developed suburban planned communities such as Forest Hill Gardens, New York (1911), and Palos Verdes Estates near Los Angeles (1923) (Figure 9.4).

The first attempt to create a garden city was made at Radburn NJ by the New York-based Regional Planning Association of America, which functioned between 1923 and 1933. Although Radburn never honoured three of the basic garden-city principles (the green belt, the provision of industry, municipal ownership and control), it remains one of the most significant US experiments in comprehensive urban planning.[15] Among the innovative elements of the Radburn design were:

1. The use of the superblock as a building unit instead of the typical narrow rectangular block.
2. The separation of pedestrian and vehicular traffic through the use of overpasses and underpasses.
3. Roads constructed for different functions.
4. A continuous park linking the superblocks.

The Radburn development failed, owing to the onset of the Depression. Arguably the most significant contribution of the Radburn experiment for new-town development was the conclusion that even under normal conditions, a garden city could not be completed without federal assistance for land assembly, infrastructure financing and provision of low-income housing.[16]

The first large-scale federal involvement in new-community building took place during the 1930s, when the Suburban Resettlement Program sought to create alternatives to the urban slum. The resulting three green-belt towns (Greenbelt MD, Greenhills OH and Greendale WI) were each encircled by a green belt to limit expansion but did not include sites for industries and were not self-contained new towns in Howard's sense of the term. Rather, they were small, well-planned suburbs that housed only 2,100 families in total when the programme ended in 1938. By 1957 Greenbelt in Maryland had a population of 7,500, at a gross density of two persons per acre (five per hectare).[17] The symbolic importance of the green-belt towns was much greater than their achievement in housing provision. Nevertheless, they were remarkable for:

1. The degree and quality of physical planning.
2. The inclusion of suburban housing for low- and moderate-income groups.
3. Social innovations, such as the introduction of consumer co-operatives for retail services, reminiscent of Robert Owen's initiative at New Lanark.

Congressional sensitivity to popular criticism of the green-belt towns resulted in government withdrawal from the towns in 1952. Two main objections were raised by opponents. First, they felt that the towns were socialist ventures and part of a trend towards

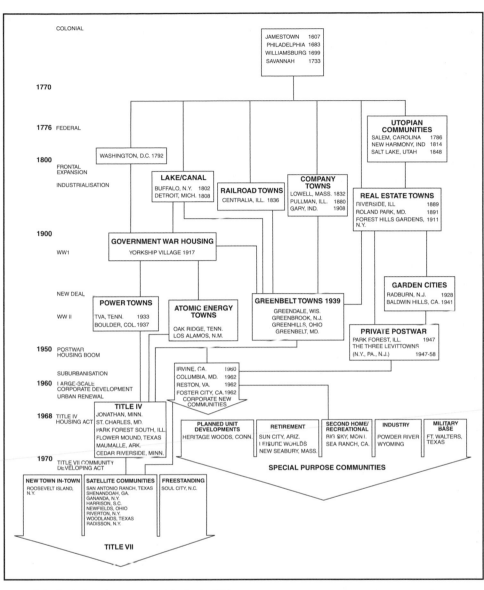

Figure 9.4   Evolution of new communities in the USA

Source: G. Golany (1978) *International Urban Growth Policies* New York: Wiley

'socialistic regimentation disguised as co-operative planning'. Second, the towns placed the government in unnecessary competition with private enterprise, which, it was argued, could provide housing more effectively and more cheaply. The sale of the three projects to private parties brought the federal government about half the original cost of the towns and left the local governments without sufficient tax bases or effective control over undeveloped areas.[18]

Federal interest in new communities was revived during the 1960s by the Johnson administration, which passed the 1968 Housing and Urban Development Act, Title IV of which enabled the Department of Housing and Urban Development 'to guarantee, and enter into commitments to guarantee the bonds, debentures, notes and other obligations issued by new town developers'. A ceiling of $250 million was put on the programme, with a maximum of $50 million for any one developer. The effects of the 1968 Act on new-community development were small, however. The economic situation in

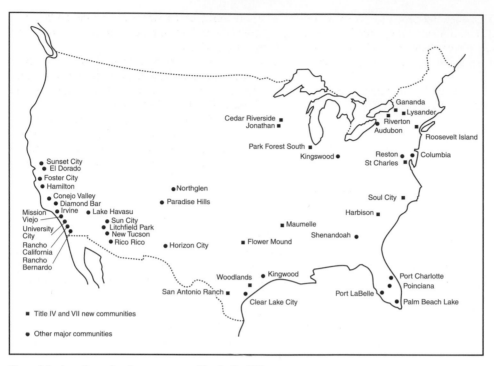

Figure 9.5 Locations of major new communities in the USA

1960–70 was such that private developers had difficulty in locating finance ('patient money') even with a federal guarantee, and only one new community (Jonathan MN) was guaranteed under this legislation.

The 1970 Housing and Urban Development Act was more successful. This doubled the ceiling for loan guarantees to $500 million, extended the guarantee cover to public as well as private developers, and established a New Community Development Corporation within the Department of Housing and Urban Development to manage the new towns programme and to undertake new community demonstration projects on federal land. Fifteen new communities stemmed from the 1970 Act (Figure 9.5). However, the impact of the energy crisis of 1973 and a depressed housing market created financial difficulties for most of the new towns which the Ford administration, committed to cutbacks in public spending and withdrawal from public housing and urban-renewal programmes, was unwilling to alleviate. With the exception of Woodlands TX, north of Houston (which survived because of revenue from the developer's substantial natural-gas holdings), all the towns were foreclosed by the federal government, which sold the assets at a total loss of $570 million.

More generally, the principle of federal involvement in new-town construction was undermined by a lack of

ongoing political support (with the advent of the Nixon administration), and opposition to the concept of new towns from various interest groups, including:

1. Mortgage bankers opposed to federal competition in the financial marketplace.
2. Small builders opposed to a perceived bias in favour of large-volume builders.
3. Big-city mayors who feared loss of business, population and tax revenue. This led the Johnson administration to promote 'new towns in town' at Cedar-Riverside MN and Roosevelt Island NY.

Underpinning this lack of commitment to a publicly supported new-town programme were national reservations about what is properly a public responsibility and what should be reserved to the private sector. These reservations were encapsulated by Eichler and Norwitch (1970 p. 308),[19] who stated that

> If new towns are an anti-megalopolic concept . . . we in America are all for it. But if in order to have new towns we must grant to the state, especially the federal government, the power to control, own, and develop land, and the regulatory devices to channel growth to that land, we resist. . . . If private

ownership of land and freedom of movement are sacrosanct and if new towns are needed, there is only one answer: private industry, with little or no aid or control from government, must build them.

Such attitudes signposted the way ahead for new-community development in the USA.

PRIVATE-SECTOR NEW COMMUNITIES IN THE USA

In contrast to the social-welfare motives underlying public-sector new towns, the profit potential has been the principal force behind private-sector involvement in new community development. Two of the earliest large-scale new communities of the post-Second World War era were Park Forest IL and Levittown NY.[20] Undertaken by builders who had been involved in federal housing programmes during the war, these developments capitalised on the pent-up demand for housing, the easy finance provided by the Veterans Administration and Federal Housing Administration, and highway construction, which opened up peripheral areas for residential development. These and other new communities such as North Palm Beach FL were built for the middle class, with little recourse to the broader planning principles pioneered in Radburn, the green-belt towns or British garden cities.

During the 1960s house-builders were joined in new-community construction by entrepreneurs from a variety of backgrounds, including:

1. Large landowners with properties ripe for development such as the 88,000 acre (35,000 ha) Irvine ranch in California,[21] and the 50,000 acre (20,000 ha) O'Neill ranch, which became the site of Mission Viejo.
2. Large corporations with cash available for diversification into real-estate projects, including Gulf Oil in Reston and the Humble Oil and Refining Company's interest in Clear Lake City.
3. Banks and insurance companies, with, for example, Chase Manhattan Bank, Teachers' Insurance and Annuity Association, and Connecticut General Life Insurance Company underwriting the development of Columbia.[22]

In contrast to the UK, where new town locations were determined by central government, new communities in the USA are sited on the basis of individual developers' perception of the economic situation in an area. Underlying factors which may influence this decision include climate (with California, Florida and Arizona favoured retirement areas), favourable legislative mechanisms for developers (such as the opportunity to create **special districts** to help finance public services) and the availability of large areas of undeveloped land (especially if it is under single ownership, to simplify acquisition). As Figure 9.5 shows, most new communities are located in a few states, including California, Florida and New York. These private-sector developments cater for a variety of residential and lifestyle preferences ranging from retirement communities (such as Sun City AR) to 'postmodern' neo-traditional planned developments (e.g. Seaside FL) that seek to simulate a particular version of small-town America for residents.[23] Many of the new communities are designed in accordance with the principles of the new urbanism[24] (Box 9.4) that have much in common

---

BOX 9.4

New urbanism: principles of neighbourhood development

1. Neighbourhoods should be compact, pedestrian-friendly and mixed-use.
2. Many activities of daily living should occur within walking distance, allowing independence to those who do not drive, especially the elderly and the young. Interconnected networks of streets should be designed to encourage walking, reduce the number and length of automobile trips, and conserve energy.
3. Within neighbourhoods, a broad range of housing types and price levels can bring people of diverse ages, races and incomes into daily interaction, strengthening the personal and civic bonds essential to an authentic community.
4. Appropriate building densities and land uses should be within walking distance of transit stops, permitting public transport to become a viable alternative to the automobile.
5. Concentrations of civic, institutional and commercial activity should be embedded in neighbourhoods and districts, not isolated in remote, single-use complexes.

Plate 9.2 In private retirement communities, such as Sun City Center in Florida, residents are furnished with a lifestyle and facilities that suit their particular requirements

BOX 9.5

Columbia MD

Begun in 1967 and located twenty miles (32 km) from Washington DC, Columbia covers 15,600 acres (6,300 ha) and will eventually accommodate 110,000 people. In 1995 it had 79,000 residents living in nine 'villages'. Each village comprised several neighbourhoods of 700–1,200 dwellings which supported a neighbourhood centre with an elementary school, day-care centre, and local shopping and community facilities. Four neighbourhoods support a village centre with higher-order facilities. There is also a larger shopping centre for the whole community in the downtown city centre which contains a seventy-acre (28 ha) shopping mall, entertainment facilities and office buildings. The road system also adheres to a hierarchical arrangement, from culs-de sac, pedestrian and cycle paths, neighbourhood streets and collector streets to the main US29, which bisects the town. Around 1,800 acres (730 ha) were set aside for industrial parks. One-fifth of the site is reserved for parkland, and a continuous open-space system runs through the town.

Socially, Columbia has made a conscious effort to be a racially integrated community, one-third of residents being black. Subsidised housing units are provided and dispersed over five different sites to avoid the creation of a low-income ghetto. Residents belong to a village association that provides one member to the city-wide Columbia Association. This 'new community' of Columbia represents a $2.5 billion market investment in a particular lifestyle.

with the smart growth strategies discussed earlier (see Chapter 8).

Despite efforts to include subsidised housing for lower-income groups in some new communities, such as Columbia MD (Box 9.5), the market principle ensures that in general those who occupy the new **master-planned communities** are of upper-middle income, of above-average education, white and, with the exception of retirement communities, young to middle-aged. The USA's new communities are not, therefore, new towns in the UK sense of being balanced communities with a range of employment opportunities and housing choice; but they are home to an increasing number of Americans, for whom they provide a particular form of Ebenezer Howard's 'town–country' lifestyle.

FURTHER READING

BOOKS

C. Corden (1977) *Planned Cities* Beverly Hills CA: Sage

G. Golany (1978) *International Urban Growth Policies: New Town Contributions* Chichester: Wiley

N. Griffin (1974) *Irvine: The Genesis of a New Community* Washington DC: Urban Land Institute

P. Hall and C. Ward (1998) *Sociable Cities* Chichester: Wiley

A. Jacquemin (1999) *Urban Development and New Towns in the Third World: Lessons from the New Bombay Experience* Aldershot: Ashgate

F. Osborn and A. Whittick (1977) *New Towns: Their Origins, Achievements and Progress* London: International Textbook Company

E. Talen (2005) *New Urbanism and American Planning* New York: Routledge

JOURNAL ARTICLES

K. Al-Hindi and C. Staddon (1997) The hidden histories and geographies of neo-traditional town planning: the case of Seaside, Florida *Environment and Planning D* 15, 349–72

C. Ellis (2002) The new urbanism: critiques and rebuttals *Journal of Urban Design* 7(3), 261–91

K. Parsons (1990) Clarence Stein and the greenbelt towns *Journal of the American Planning Association* 56(2), 161–83

M. Pacione (2006) Proprietary residential communities in the United States *The Geographical Review* 96(4), 543–66

A. Stockdale and G. Lloyd (1998) Forgotten needs? The demographic and socio-economic impact of free-standing new settlements *Housing Studies* 13(1), 43–58

KEY CONCEPTS

- Neighbourhood unit
- Nimbyism
- Greenfield–brownfield debate
- Green-belt towns
- Neo-traditional planned developments
- Master-planned communities

STUDY QUESTIONS

1. Explain the origins and growth of the new-town concept.
2. Examine the new-town development programme in any country with which you are familiar. Consider how successful it has been in relation to its primary objectives.
3. There is an ongoing debate on whether demand for new housing is best satisfied on greenfield or brownfield land. Make a list of the arguments for and against each form of development.
4. A major contrast between Britain and the USA in new-town development is that whereas in Britain the programme has been driven by the public sector, in the USA the private sector has dominated new community construction. Explain why this is so. Consider the advantages and disadvantages of each approach.
5. Explain what is meant by 'new urbanism'. With the aid of relevant examples illustrate the impact of the principles of new urbanism on urban development.

PROJECT

Select a new town or new community with which you are familiar and write a report explaining its rationale, origins, growth and development to the present day. Consider the future prospects for the new town/community. How does it compare with other new towns, and with the classical models of UK new towns?

10

Residential Mobility and Neighbourhood Change

Preview: residential mobility; why people move; the residential-location decision process; housing search; housing markets; neighbourhood change; housing abandonment; gentrification; displacement

INTRODUCTION

Housing is the largest user of space in the city, and exerts a profound influence on the structure of metropolitan regions. We have examined already the key private-sector (Chapter 7) and public-sector (Chapter 8) agents involved in the land development process. Here we focus on the households that acquire, occupy and exchange the dwelling units produced. We begin by examining the nature of residential mobility in the city, then consider how the aggregate of decisions by individual agents affects the process of neighbourhood change.

WHY PEOPLE MOVE

Intra-urban mobility constitutes the vast majority of moves made by individuals and households within advanced Western nations. Residential relocation can be voluntary or involuntary (Figure 10.1). Although forced relocations, due to property demolition or eviction, can be significant in particular parts of a city, most individuals and households move by choice. Nevertheless, the stimulus for a voluntary move may be externally induced. It is generally agreed that the most important reasons for relocation refer to the characteristics of the housing unit. In Rossi's (1955) classic study of residential mobility in Philadelphia,[1] more than half the movers cited too little or too much living space as a contributory factor, with 44 per cent identifying it as the primary reason behind their desire to move. Adjustment moves accounted for 52 per cent of

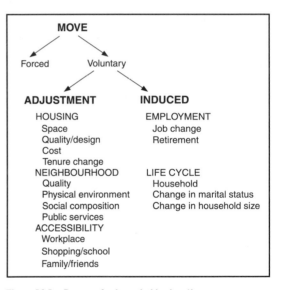

Figure 10.1 Reasons for household relocation

reasons for moving home in a study by Clark and Onaka (1983),[2] with housing characteristics including space, quality and design of the unit, and a desire to shift from renting to owner-occupation of major importance. Neighbourhood characteristics were of lesser importance, as were accessibility considerations, with households generally prepared to trade off a longer journey to work to obtain housing with more amenities at less cost. As we saw in Chapter 4, in larger metropolitan areas the 'indifference zone' for travel to work can extend to over two hours in parts of

TABLE 10.1 TEN LIFE-CYCLE EVENTS RELATED TO RESIDENTIAL ADJUSTMENT OR RELOCATION

1. Completion of secondary education
2. Completion of tertiary education
3. Completion of occupational training
4. Marriage
5. Birth of the first child
6. Birth of the last child
7. First child reaches secondary-school age
8. Last child leaves home
9. Retirement
10. Death of a spouse

Source: **D. Rowland (1982) Living arrangements and the later family life cycle in Australia *Australian Journal of Ageing* 1, 3–6**

southern California. Similar extensive 'commuter-sheds' surround London and other large cities.

Induced moves are related to employment and life-cycle factors (Figure 10.1). Traditionally the life-cycle concept refers to the explicit material needs of families as they move through the various stages of the reproductive cycle. A number of major life-cycle events have been related to residential adjustment or relocation[3] (Table 10.1). More recently the life-cycle concept has been reformulated as the less deterministic concept of the life course. This avoids age stereotypes and acknowledges that economic and cultural factors can induce households at the same life-cycle stage to adopt different residential behaviours.[4]

Fundamentally, people move in the expectation of achieving a superior living environment. This premise underlies the 'value expectancy' model (Figure 10.2), in which migration behaviour is seen as the result of:

1. Individual and household characteristics (e.g. life cycle, household density).
2. Societal and cultural norms (e.g. community *mores*).
3. Personal traits (e.g. attitude to risk).
4. Opportunity structure (e.g. areal differentials in economic opportunity).
5. Information (e.g. volume and accuracy).

The first four factors combine to produce a set of migration goals, while the addition of the fifth factor influences the expectation of attaining these goals. Interaction between migration goals and expectancy of achievement results in migration intention that can be either to make an *in situ* adjustment or to relocate.

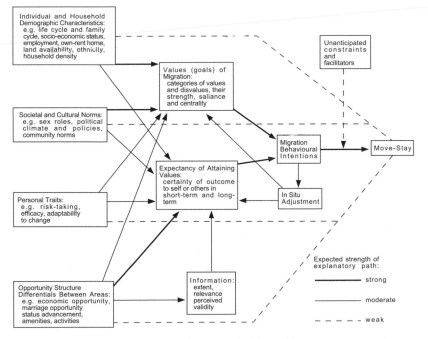

Figure 10.2   A value expectancy model of migration decision-making
Source: R. Golledge and R. Stimson (1997) *Spatial Behaviors* New York: Guilford Press

Plate 10.1   Moving home. In the USA one person in five moves home every year. In some instances the home also moves

Significantly, migration decisions are also affected by unanticipated events (which may be either constraints or facilitators). These include changes in family structure, the financial cost of moving, or distance. Enforced mobility may also create incongruities between migration intentions and actual behaviour. The model also acknowledges the causal significance of cultural signals or *mores* in motivating a desire to move or stay. How people 'read' these signals (which is a central concern of the postmodern perspective on urban change) may not necessarily be rational in a positivist sense, as for example when people have an affinity for a particular neighbourhood.

THE DECISION TO MOVE

The residential-location decision process may be conceptualised as the product of stress created by discordance between a household's needs, expectations and aspirations and its actual living environment (Figure 10.3). As we have seen, the sources of stress may be internal (such as a change in family size)

or external (e.g. the expiry of a dwelling lease). Furthermore, they may be dwelling- specific (such as a need for an extra bedroom) or location-specific (e.g. an extended journey to work due to a job change). Once it appears, a stressor affects the relationship between a decision-maker's achievement level (what he or she already has) and aspirations. The degree to which this occurs varies between individuals according to factors such as socio-economic status, stage in the life cycle, experience and personality. Also, each individual will have a threshold aspiration level that adjusts on the basis of these factors, giving rise to different levels of tolerance of stress. When tolerable stress becomes intolerable strain, the household must decide on one of three courses of action:

1. Remain at the present location and either:
   ■ Improve the environment, or
   ■ Lower expectations.
2. Relocate.

Environmental improvement embraces a range of activities depending on the nature of the stressors

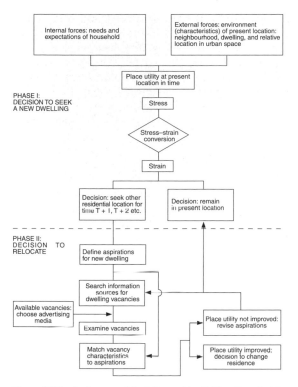

Figure 10.3   A stress model of the residential-location decision process

Source: L. Brown and E. Moore (1970) The intra-urban migration process *Geografiska Annaler* Series B 52(1), 1–13

involved. A small dwelling can be enlarged with an extension, while dilapidated dwellings can be rewired and redecorated. Neighbourhood stressors such as crime or noise pollution can be tackled by approaching the local authorities or through residents' action groups. Households constrained to remain *in situ* may have to lower their aspirations as a means of coming to terms with the situation. They could change their lifestyle or modify their plans (e.g. delay starting a family). More often it is a psychological matter of 'dissonance reduction', or learning to become indifferent to what one cannot attain. A large number of households, however, are able to realise their desire to relocate, and this process involves the search for and selection of a new residence (Figure 10.3).

THE SEARCH FOR A NEW HOME

Whether the decision to move house is voluntary or forced, all relocating households must:

1. Specify an 'aspiration set' of criteria for evaluating new dwellings and living environments.
2. Undertake a search for dwellings that satisfy these criteria.
3. Select a specific dwelling unit.

New dwellings may be evaluated in terms of site characteristics (attributes of the dwelling itself) and situational characteristics (the physical and social environment of the neighbourhood). The lower limits of the household aspiration set are defined by the characteristics of the dwelling currently occupied, while the upper limits are set by standards to which the household can reasonably aspire. In most instances these standards will be determined by income constraints but, as the value expectancy model suggests, other factors may be involved, including a desire to avoid certain areas that do not conform to a particular lifestyle.

On the basis of their aspiration set, individuals initiate a search procedure to locate a suitable new dwelling. This search has a spatial bias.[5] This may be illustrated by conceptualising the city as comprising four types of space:

1. *Action space* is the most extensive and refers to those parts of the city with which the individual is familiar, and includes a subjective evaluation of places. Generally, the longer an individual lives in a city the larger the action space and the greater the differentiation. New areas are assimilated into the action space as travel and information spread.
2. *Activity space* is the territory within which daily movement takes place and is normally organised around often-used nodes, including home, workplace, friends' houses and shopping centres.
3. *Awareness space* indicates the extent of the territory on which a household has information concerning housing opportunities. This area is conditioned by the household's action and activity spaces.
4. *Search space* is a subset of awareness space within which possible new residential locations are evaluated.

The geography of these spaces affects the outcome of the residential relocation decision. In addition to information gained from active experience of the city, households also obtain information on new housing opportunities from secondary sources, including newspapers and, as we have seen in Chapter 7, real-estate agents.[6] The relative importance of different information sources varies for different households. Access to information is also affected by search barriers that can:

1. Raise the costs of gathering information (e.g. due to lack of transport, or time constraints on women with young children).
2. Limit the choice of housing units and locations available (e.g. due to financial constraints, or discrimination in the housing market).[7]

As Figure 10.3 indicates, the eventual choice of new dwelling is based on the increase in satisfaction (place utility) produced by a move. It is important to recognise, however, that in all cities there are many households where residential location is constrained to the point where behavioural models are of limited significance.[8] These subgroups include poor people, elderly people, unemployed people, transient people and special-needs groups such as lone-parent families and former inmates of institutions, as well as homeless people whose living space is the street. Focusing on the constraints involved in residential relocation underlines the fundamental importance of housing-market structure in conditioning residential mobility.

HOUSING MARKETS

In market societies private housing is a commodity produced and exchanged for profit. Outside the social- or public housing sector, where housing may be allocated on the basis of some criterion of need or at a below-market price, the residential choices available to urban households depend primarily on their ability to compete in the housing market, which comprises housing providers, facilitators and consumers (see Chapter 7). While the wealthy can compete in the market for the most attractive and expensive housing in the best areas, those with low incomes or the long-term unemployed are unlikely to be able to gain access to the private housing market except at the lowest levels. Those unable to buy must rent either from a private landlord or in the public social rented sector. The variety of households served by the housing market is demonstrated by the concept of housing class (Box 10.1). This identifies different sections of society with varying access to the housing market, and indicates that within a city there exist a number of distinctive sub-markets catering for different social groups. These sub-markets tend to be localised and are, therefore, reflected in the residential mosaic of the city. In Table 10.2, which refers to the London borough of Southwark, six main housing groups are identified according to income, socio-economic status and national origin. Each group tends to seek accommodation of a different type (distinguished chiefly by tenure) within varying geographic limits. Each housing group is supplied with housing opportunities by different combinations of agencies offering different degrees of security of tenure.

In British cities generally, a major distinction is that between the housing markets for owner-occupied dwellings and for public social housing. As Harrison (1983 p. 180) observed, 'British cities are segregated by class ... No laws of apartheid are needed to enforce this separation: it occurs as naturally as oil divides itself from water, by the play of unequal incomes and savings in the housing market.'[10] In the US city, ethnicity may be substituted for tenure class,

---

BOX 10.1

The concept of housing class

The concept of housing class was introduced by J. Rex (1971)[9] in an attempt to understand the position of immigrant groups in the operation of the British housing market. It denotes different groups of people characterised by their access to particular types of housing, usually defined by tenure. Rex identified seven housing classes: (1) outright owners, (2) mortgagees, (3) tenants in purpose-built public housing, (4) tenants in publicly acquired slum housing awaiting demolition, (5) tenants of whole properties belonging to private owners, (6) house owners who must sub-let parts of their property to cover the repayments and (7) lodgers who occupy one or more rooms in a dwelling shared with other households.

The concept has been criticised subsequently for its assumption of a common system underlying the desirability of certain types of housing and housing tenure, and for being simply an inductive generalisation from a particular case (Birmingham). However, it does illuminate the important role of constraints as well as choice in urban housing markets.

TABLE 10.2 HOUSING GROUPS IN THE LONDON BOROUGH OF SOUTHWARK

Housing group	Socio-economic group	Housing opportunities in terms of tenure	Location
1. Upper-income	Primarily professional managerial and self-employed	Permanent: owner occupation (high-cost unfurnished) Temporary: furnished preceding ownership	Degree of flexibility in location, throughout London
2. Indigenous, middle-income	Skilled and semi-skilled	Permanent: owner occupation unfurnished, local authority	Owner-occupied throughout London, tenants mainly inner London and locally
3. Indigenous, lower-income	Semi-skilled and unskilled	Permanent: unfurnished, local authority, long-standing owner occupation Temporary: furnished	Mainly inner London, emphasis on Southwark
4. Immigrant, middle-income, non-white (West Indian, Pakistani, Greek, etc.)	Skilled and semi-skilled	Permanent: furnished, owner occupation Temporary: furnished	London area, often in immigration community, may be temporary UK residents
5. Immigrant, middle-income, white (mainly Irish)	Skilled, semi-skilled and unskilled	Permanent: unfurnished, local authority furnished Temporary: furnished	Mainly inner London for jobs, may be temporary UK residents
6. Immigrant, lower-income, white and non-white	Semi-skilled and unskilled	Permanent: furnished, local authority Temporary: furnished	Mainly inner London, many with their own communities

Source: **P. Knox (1995)** *Urban Social Geography: An Introduction* **London: Longman**

with the housing market for African-Americans commonly representing a separate submarket with its own distinct spatial arrangement (Box 10.2). In addition to, and often overlapping with, class- and ethnicity-based housing markets, sub-markets also operate within cities to accommodate the needs and preferences of particular groups differentiated on the basis of age,[11] religion[12] or lifestyle.[13]

NEIGHBOURHOOD CHANGE

The aggregate of individual household residential decisions is reflected in the changing socio-spatial structure of city neighbourhoods. Neighbourhoods are in a constant state of flux. Some analysts have sought to understand neighbourhood dynamics by identifying stages in the process of neighbourhood change[14] (Table 10.3). It

is important to recognise, however, that the life-cycle analogy inherent in many models can be misleading, since, unlike biological species, neighbourhoods do not follow a predetermined course of growth or decline. At any stage in a neighbourhood's decline it may reverse direction (owing to inward investment) and begin a period of revitalisation. Similarly, upwardly mobile neighbourhoods may have their progress halted owing, for example, to negative externalities (such as the construction nearby of a noxious facility).

Downs (1981) identified a five-stage neighbourhood change continuum in which neighbourhoods at any stage can be stable, improving or declining[15] (Figure 10.4). Factors that may increase a neighbourhood's susceptibility to decline or revitalisation are shown in Table 10.4. In the remainder of this chapter we will focus on the extremes of housing abandonment and gentrification.

BOX 10.2

The arbitrage model of neighbourhood racial transition

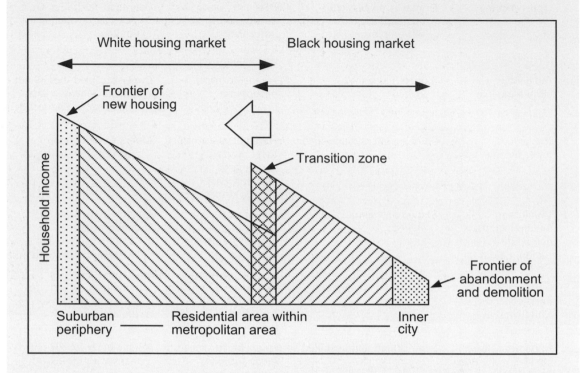

Racially segregated housing markets are characteristic of US metropolitan areas. This means that when a neighbourhood changes from 'white' to 'black' occupancy it actually shifts from one housing market to another.

This process is explained by the arbitrage model. In the model city, two housing groups occupy segregated zones, spanned by a transition zone accommodating both black and white households. Within each interior zone, levels of poverty increase towards the inner city, which means that the most affluent black neighbourhoods and the poorest white neighbourhoods lie adjacent to the transition zone. Average housing prices also decline from the outer city to the inner city. Neighbourhood racial change occurs when the transition zone moves into the white interior. The transfer of houses in the transition zone in response to the differences in prices that white and minority households are willing to pay for the same unit is termed *arbitrage*.

The arbitrage process can be stimulated by an inflow of poor minority households to the black interior that drives up prices for housing there. Households in the black interior are then willing to pay more for housing in the transition zone than white households are willing to pay for the same units. White home owners who move out of the transition zone are more likely to sell to black households, who are prepared to pay higher prices.

As was noted in Chapter 7, speculators may hasten change by buying homes from white households and selling them at higher prices to minority households in a process known as 'blockbusting'. As more housing in the transition zone shifts to minority ownership the area may become less desirable for the remaining whites owing to concern over the exchange value of their property and/or racial prejudice. These households may sell in a depressed market, allowing the transition zone to move further into the white interior, thereby facilitating neighbourhood change.

Stage	Physical changes: dwelling type (predominant additions)	Social changes: Level of construction	Population density	Family structure	Social status (income)	Migration mobility	Other changes
Suburbanisation (new growth), 'homogeneity'	Single-family (low-density multiple)	High	Low (but increasing)	Young families, small children, large households	High (increasing)	High net in-migration, high mobility turnover	Initial development stage; cluster development; large-scale projects, usually on virgin land
In-filling (on vacant land)	Multi-family	Low, decreasing	Medium (increasing slowly or stable)	Ageing families, older children, more mixing	High (stable)	Low net in-migration, low mobility turnover	First transition stage: less homogeneity in age, class, housing first apartments; some replacements
Downgrading (stability and decline)	Conversions of existing dwellings to multi-family	Very low	Medium (increasing slowly), population total down	Older families, fewer children	Medium (declining)	Low net out-migration, high turnover	Long period of depreciation and stagnation; some non-residential succession
Thinning out	Non-residential construction – demolitions of existing units	Low	Declining (net densities may be increasing)	Older families, few children, non-family households	Declining	Higher net out-migration, high turnover	Selective non-residential succession
Renewal	Public housing	High	Increasing (net)	Young families, mainly children	Declining	High net in-migration, high turnover	The second transition stage: may take either of two forms, depending on conditions
	Luxury high-rise apartments	Medium	Increasing (net)	Mixed	Increasing	Medium	
	Town house conversions	Low	Decreasing (net)	Few children	Increasing	Low	

Source: L. Bourne (1976) Housing supply and housing market behaviour in residential development, in D. T. Herbert and R. J. Johnston (eds) *Social Areas in Cities* volume 1 Chichester: Wiley, 111–58

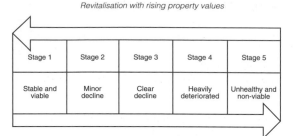

Stage 1: *Stable and viable*: healthy neighbourhoods that are relatively new and thriving or relatively old and stable, with no symptoms of decline and rising property values. Some neighbourhoods remain in this stage for decades.

Stage 2: *Minor decline*: generally older areas with some functional obsolescence and minor physical deficiencies in housing units. Accommodate many younger families at higher densities than when the neighbourhood was first developed. Property values are stable or increasing slightly. The level of public services and the social status of the neighbourhood are below those of stage 1 areas.

Stage 3: *Clear decline*: rented housing is dominant, with poor tenant–landlord relations owing to high absentee ownership. Minor physical deficiencies are evident and many structures have been converted to higher-density uses than those for which they were designed. Overall confidence in the area's future is weak and there may be some abandoned housing.

Stage 4: *Heavily deteriorated*: most housing requires major repairs, and properties are marketable only to the lowest socio-economic groups. The profitability of rental units is poor, and cash-flows are low or negative. Subsistence-level households are numerous; pessimism about the future is widespread, as is housing abandonment.

Stage 5: *Unhealthy and non-viable*: characterised by wholesale abandonment; expectations about the area's future are nil. Residents are those with the lowest social status and lowest incomes in the city, and the neighbourhood is regarded as one to escape from rather than move to.

Figure 10.4   Downs's continuum of neighbourhood change
Source: A. Downs (1981) *Neighborhoods and Urban Development* Washington DC: Brookings Institution

HOUSING ABANDONMENT

Housing abandonment has been a major problem in inner-city neighbourhoods of most cities in the north-east USA since the 1960s. In New York City between 1965 and 1968 the number of dwelling units abandoned was sufficient to house 300,000 persons. By 1978 tax arrears on abandoned property amounted to $1.5 billion, substantially contributing to the city's **fiscal crisis**.[16] The principal reasons behind large-scale housing abandonment were:

1. 'White flight' from the city to suburban residential areas.
2. Physical deterioration.
3. Lack of demand for housing in neighbourhoods marked by social pathologies.
4. Declining rent incomes, owing to increasing municipal real-estate taxes and the limited rent-paying capability of low-income tenants.

5. Withdrawal of investment due to red-lining by banks and mortgage lenders.
6. The imposition of stringent rent controls, which prompt landlords to reduce spending on repair and maintenance.

If a situation is reached in which the housing unit becomes a liability to the owner, bringing in no rental income but generating taxes, the owner may take the economically rational step of halting payment of taxes due and in effect abandoning the building to the city (Box 10.3). The widespread abandonment of inner-city housing in the major American cities had slowed by the mid-1980s as:

1. The worst housing had been withdrawn from the market over the preceding twenty years.

BOX 10.3

Housing abandonment in St Louis MO

In the early 1970s St Louis MO had one of the highest housing abandonment rates in the USA, with an average 4 per cent of the city's housing stock abandoned. In the Montgomery neighbourhood up to 20 per cent of units were abandoned in 1971, and a domino, or negative externality, effect operated to add more territory to the impacted area.

Montgomery was an area of low-quality housing that experienced a transition from being a white to becoming a black neighbourhood during the 1960s. This process culminated in the late 1960s in a sharp increase in the number of abandoned units. In 1970 nearly 80 per cent of housing was rented; crime and vandalism were rife, market values plummeted to zero and property owners ceased making mortgage payments. Slum landlords moved in, paying a token sum for units, then renting them out cheaply, providing little maintenance. The high foreclosure rate eventually slowed as the low value of the property and the lack of potential resale market made the procedure meaningless. Code enforcement programmes were also self-defeating as these too encouraged abandonment. Similarly, rehabilitation was a futile exercise in the face of terminal blight. As soon as units became vacant, fixtures were stripped, frequently by the owner, leaving demolition by the city as the only viable alternative.

TABLE 10.4 FACTORS UNDERLYING NEIGHBOURHOOD DECLINE AND REVITALISATION

Revitalisation factors	Decline factors
1. High-income households	Low-income households
2. New buildings with good design or old buildings with good design or historical interest	Old buildings of poor design and no historical interest
3. Distant from very low-income neighbourhoods	Close to very low-income neighbourhoods or to those shifting to low-income occupancy
4. Inner city gaining (or not losing) population	Inner city rapidly losing population
5. High owner-occupancy	Low owner-occupancy
6. Small rental units with owners living on premises	Large rental apartments with absentee owners
7. Close to strong institutions or desirable amenities, such as a university, a lakefront, or downtown	Far from strong institutions and desirable amenities
8. Strong, active community organisations	No strong community organisations
9. Low vacancy rates in homes and rental apartments	High vacancy rates in homes and rental apartments
10. Low turnover and transience among residents	High turnover and transience among residents
11. Little vehicle traffic, especially trucks, on residential streets	Heavy vehicle traffic, especially trucks, on residential streets
12. Low crime and vandalism	High crime and vandalism

Source: **Adapted from R. Downs (1981)** *Neighborhoods and Urban Development* **Washington, DC: Brookings Institution**

2. Continued in-migration of minority groups, particularly Hispanics and Asians, maintained a demand for inner-city housing.
3. Federal low-interest loans and rehabilitation programmes, combined with action by local neighbourhood groups, prevented the decline into abandonment in many neighbourhoods.

Nevertheless, the problem of abandoned housing remains severe.[17] In Philadelphia in 1995 29,000 residential structures lay abandoned, of which 19,000 were so seriously dilapidated that rehabilitation would cost upwards of $110,000 per unit at a total cost of $2 billion[18] (Figure 10.5). In some extreme cases of market failure, investor confidence can be restored only by public-sector involvement to pump-prime the regeneration of distressed neighbourhoods.

GENTRIFICATION

Gentrification is a process of socio-spatial change where the rehabilitation of residential property in a working-class neighbourhood by relatively affluent incomers leads to the displacement of former residents unable to

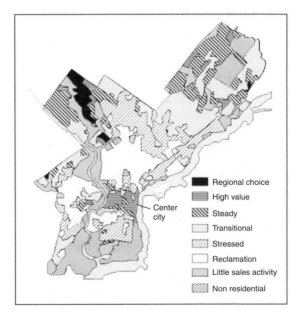

Figure 10.5   Housing market clusters in Philadelphia

afford the increased costs of housing that accompany regeneration. Gentrification commonly involves residential relocation by people already living in the city, and as such is not a 'back to the city' movement by suburbanites. Gentrification should also be distinguished from both the intensification of existing high-status areas and the process of neighbourhood revitalisation involving 'incumbent upgrading' where no spatial mobility is involved in an area's social transformation.[19]

The classic example of gentrification occurs in a residential area that is initially middle-class. Typically the original residents move out from the central city as they establish families and their incomes rise, to be replaced by households of successively lower income. Eventually the cost of maintenance and reinvestment in the housing exceeds the financial ability of occupants and the area undergoes significant deterioration. The result is further in-migration of low-income households, overcrowding, subdivision of large houses into rental units to provide a rent income acceptable to landlords, and the eventual transition from owner-occupation to rented tenure. Landlords may decide to act as 'free-riders' and take no steps to invest in the maintenance or rehabilitation of their property (possibly in response to declining real profits owing to inflation and/or rent controls, and the presence of alternative investment opportunities). If so, the process of deterioration accelerates, possibly leading to abandonment (including 'torching' for insurance purposes). As Beauregard (1986 p. 48) observes, 'the housing stock in this area is now inexpensive, and may attract gentrification'.[20] A second form of gentrification refers to 'loft conversions', or the creation of improved housing units in mixed-use districts, as has occurred in areas of disused industrial and waterfront premises in many older cities.[21] In addition to these manifestations of gentrification produced by individual households, large-scale developers and speculators can also purchase multi-family housing units and derelict industrial spaces for conversion into luxury condominiums and co-operative apartments. We can also identify a gentrification process in which local government takes the initiative through a major urban renewal project (see Chapter 11).

Many cities began to experience the gentrification of select central and inner-city neighbourhoods during the 1950s, most notably London and New York. Today gentrification is a feature of the urban geography of the majority of larger cities in the advanced capitalist world including London,[22] Paris,[23] New York,[24] Washington DC,[25] Vancouver,[26] Adelaide,[27] Amsterdam[28] and Madrid[29] (Box 10.4). The phenomenon has also been identified in parts of Eastern Europe[30] and the Third World.[31] By the 1970s gentrification was widely regarded as an integral residential component of a larger process of urban restructuring linked with wider economic and social trends in capitalist society. Gentrification was seen as a facet of globalisation, the changing economic base from manufacturing industry to producer services, and the boom in office developments in locations such as Canary Wharf and Battery Park City.[32] Gentrification is a circular process. Since the mid-1990s in some global cities, notably London and New York, the growth of a financial services economy has led to 're-gentrification' of areas that were originally gentrified twenty-five years previously. In certain neighbourhoods, such as Brooklyn Heights in New York and Battersea in London, the '**rent gap**' between old gentrified and newly gentrified property is as great as the difference between ungentrified and gentrified property was in the 1970s.[33] This price difference is driving a new wave of 'super-gentrification' in favoured locations where high incomes from the financial services industry underwrite high-standard renovations of previously gentrified properties.[34] Recent research in London has identified gentrification occurring not only in the original working class inner suburbs but elsewhere in inter-war suburbs and further into the wider metropolitan hinterland.[35] This suggests a series of gentrification processes – involving the key features of class change and displacement (but no longer only of 'working class') – occurring in specific social and spatial contexts. Some researchers have associated this trend with emergence of new variants of the middle-class neighbourhood, characterised not only by class but also through **habitus** – (i.e. the construction of meaning in everyday **lifeworlds**) – based on shared aesthetics and tastes, as manifested in the **symbolic capital** of dress codes, use of language and patterns of material consumption. Webber (2007) identified six forms of metropolitan habitus in London, two of which ('city adventurers' and 'new urban colonists') are representative of middle-class gentrification.[36] Striking examples are found in the growing phenomenon of gated communities (see chapter 18) and super-gentrification, as in Barnsbury, London.[37]

THEORIES OF GENTRIFICATION

Theories of gentrification have focused on two alternative types of explanation:

1. In *consumption-side* explanations, neighbourhood change is accounted for primarily in terms of who moves in and out.[38] Recent postmodern interpretations have emphasised the role of

BOX 10.4

Geography and gentrification in London: a tale of two postcodes

West Hampstead has for long been the poor relation of Hampstead in north-west London. Anyone living in NW6, rather than NW3, would pay much less for their home and have fewer facilities to enjoy.

Now, though, things are changing. The Burford Group is building a 220,000 ft² (20,000 m2) development off the Finchley Road, to be called Gourmand, which will be a massive leisure and entertainment complex. It will include a leisure centre, an eight-screen cinema, restaurants, a health and fitness centre, and a Sainsbury's superstore that should attract people in droves.

Prices may still be 25 per cent lower than in Hampstead proper, but James Armstrong, an agent for Foxton's, reckons that it will not be long before prices start to catch up. 'As Hampstead gets more expensive and crowded, so people are happy to move ten minutes further away to a quieter street where lots of pubs and restaurants are opening,' Armstrong said.

Lots of people purchasing in the area have rented in Hampstead for a while but found it too expensive to buy in. However, in West Hampstead a one-bedroom flat will cost about £100,000 as against £145,000 in Hampstead. And West Hampstead is so different. The streets are wider, the houses more spread out and there are fewer red-brick mansion blocks. There are also none of the large multimillion-pound detached houses which are a trade mark of Hampstead. The most expensive houses in West Hampstead are to be found in Fawley Road and Crediton Hill, where an Edwardian six-bedroom property will sell for between £600,000 and £800,000. In Narcissus Road there are pretty two-storey three-bedroom properties, reminiscent of Fulham terraced houses, selling for £330,000. In Pandora Road you could buy a five-bedroom house for about £320,000.

Unlike Hampstead, the area still has a lot of property ripe for modernisation. Although some local developers are hard at work converting and rebuilding, it is surprising that some of the bigger names have not yet surfaced. 'A lot of people are buying now for rental investment,' said Armstrong, 'because they know that in a year's time they will have made their money back as prices rise. I see West Hampstead as a Notting Hill without the market or carnival.' West Hampstead's only drawback is parking. There are no restrictions on any of the streets, so people dump their cars and head for Finchley Road Tube station.

Source: Sunday Times 1 June 1997

cultural factors in gentrification,[39] although it is necessary to avoid more extreme postmodern formulations in which culture supplants rather than complements economics in the process of gentrification.[40] As we saw in our discussion of the value expectancy model (Figure 10.2), culture is an important but not predominant factor in residential location.

2. *Production-side* explanations place emphasis on the role of the state in encouraging gentrification[41] and the importance of financial institutions in selectively providing the capital for rehabilitation.[42] According to Smith (1989), before gentrification occurs in a neighbourhood there must exist a rent gap, which is the difference between the potential ground rent and the actual ground rent under the present land use[43] (Figure 10.6). This arises as a result of cyclical patterns of disinvestment and reinvestment in the built environment. In urban neighbourhoods where physical deterioration and economic devalorisation have reached a critical level, the rent gap becomes sufficiently wide to attract investors seeking to purchase structures cheaply, undertake renovation, then either occupy the unit or resell or rent it to realise a capital gain. Empirical evidence of a causal link between a rent gap and gentrification is inconclusive, however, and there is ongoing debate over the explanatory value of the concept.[44] In Hamnett's (1991) view, although rent-gap theory can help to explain where gentrification may occur, it fails to explain why and when it occurs – in some neighbourhoods and not in others.[45] In contrast to production- or supply-side theories, consumption-side explanations focus on the actors involved in the process.

THE AGENTS OF GENTRIFICATION

All the main agents of change identified in the urban land development process (Chapter 7) are involved

Plate 10.2. Gentrification has transformed abandoned riverside warehouses on Butler's Wharf, London, into luxury apartments and stimulated the growth of a 'cappuccino society'

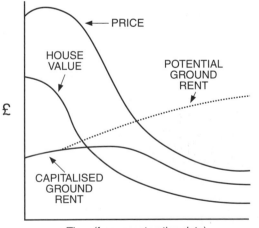

Figure 10.6   The concept of the rent gap

Source: N. Smith (1996) *The New Urban Frontier* London: Routledge

to varying degrees in effecting gentrification, each pursuing a particular set of objectives. Government involvement is both direct and indirect. Central government policies to promote home ownership, including taxation policy (e.g. relief on mortgage interest repayments) and grants for home improvement, can facilitate gentrification. In North America more direct influence may be exerted by local governments, which stand to benefit from the replacement of low-income groups, with their attendant demands for social welfare programmes, by middle-class consumers whose incomes boost the local economy and whose investments enrich the tax base.

Among strategies employed by local governments in the USA to promote desirable neighbourhood change are: (1) advertising certain neighbourhoods judged to have gentrification potential; (2) providing tax abatements for rehabilitation (e.g. the J-51 programme in New York City); (3) using community development funds to rehabilitate and improve public services in selected neighbourhoods; (4) employing code enforcement to make owners rehabilitate or sell their property; (5) designating 'historic' neighbourhoods; (6) reducing publicservice provision in some neighbourhoods to encourage decline prior to facilitating reinvestment; and (7) re-zoning a mixed-use district or failing to enforce existing zoning statutes to facilitate gentrification. For gentrification to occur, such enabling strategies by government have to be accompanied by financial and property agencies becoming interested in the redevelopment potential of a neighbourhood. Landlord developers and real-estate agents have an important part to play in guiding potential gentrifiers to a neighbourhood, buying property and speculating, and displacing residents (e.g. by raising rents). Property interests, in turn, require the co-operation of financial institutions able to lend capital for investment in the built environment. When all these agents come together in a particular spatial context, gentrification can occur. Potential locations are generally characterised by substandard but structurally sound housing 'with potential', clustered to allow a contagion effect to occur, often with a unique spatial amenity such as a view, proximity to good transport links with the central business district (CBD), and the presence of local commercial activities (shops, restaurants) attractive to gentrifiers. Clearly, a satisfactory theory of the gentrification phenomenon must involve both supply-side and demand-oriented explanations. The major demand-side and supply-side factors conducive to neighbourhood revitalisation are listed in Table 10.5.

TABLE 10.5 DEMAND-SIDE AND SUPPLY-SIDE FACTORS UNDERLYING NEIGHBOURHOOD REVITALISATION

City and metropolitan		Neighbourhood	
Factor	Operation	Factor	Operation
Demand side			
Strong downtown business district with growing employment	Creates demand for housing close to downtown jobs	Proximity to amenity such as lakefront, seafront, park or downtown	Enhances long-term value
Rising real incomes	Increasing households' ability to rehabilitate housing	Good public transport	Enhances convenience, especially for households with more than one worker
Formation of many small, childless households	Increases households that need less space, are orientated to urban amenities, and do not need public schools	Access to high-quality public or private schools	Enhances attractiveness to households with school-age children
Rapid in-migration of non-poor households	Increases demand for good-quality housing	No nearby public housing with school-age children who would dominate the public schools	Enhances attractiveness to households with school-age children and incomes high enough to support renovation
No in-migration of poor households	Permits older neighbourhoods to stabilise	Perceived in community as safe. Proximity to revitalised neighbourhoods	Enhances attractiveness as place to live. Creates expectation that revitalisation will work here as well
Supply side			
Long commuting times to downtown business district	Make living near downtown more desirable	Single family housing	Simpler to rehabilitate than multi-family housing; fewer management problems
Strong restrictions on suburban development	Limit suburban housing and jobs, enhancing city housing and jobs	Housing with interesting architectural features such as high ceilings, fireplaces, carved woodwork	Attractive to young households, which are most likely to rehabilitate
Rapid increases of suburban housing	Make city housing more attractive	Brick housing	Easier to rehabilitate than frame housing, easier to care for, lasts longer
Loose housing market	Enables poor households displaced by revitalisation to find adequate housing, possibly reducing resistance to revitalisation	Multi-family housing suitable for condominium ownership	Owner-occupied property better maintained and residency more stable than rented property
Rents not controlled	Encourages property maintenance and investment in new rental units	Financial institution willing to provide mortgages and home-ownership loans	Makes ownership and rehabilitation easier
Easy condominium conversion	Increases owner-occupancy	Commitment by local government to upgrade infrastructure and public services	Reassures private investors of long-term value of homes
		Strong neighbourhood organisation dominated by home owners	Creates pressure on local government to enforce housing codes and improve public services
		Housing and other structures in relatively good condition	Encourages private investment by owners and lenders

WINNERS AND LOSERS

As with all processes of urban change, the benefits and costs of gentrification are distributed unevenly. Among the main beneficiaries are the following:

1. *Local government* gains increased property tax revenue, and attracts higher-income residents with their associated spending power. On the other hand, middle- and upper-income newcomers may demand costly services as, for example, in Philadelphia's Queen Village,[46] where residents requested more police protection, better garbage collection, an improved school curriculum, cobblestone streets, buried electricity wires and new landscaping.
2. *Initial owners of property* benefit from rising property values in the neighbourhood which they capitalise through rental income or the sale of the asset.
3. *Incoming owner-occupiers* also gain. Early arrivals have the advantages of low property prices and higher potential capital gain but run the risk that no gain will accrue, and must usually invest more in property improvement. Later arrivals assume a smaller risk but gain less from rising property values.
4. *Developers* derive profits from buying run-down units and rehabilitating them for rent or resale to households who can afford higher occupancy costs than the previous residents.
5. *Speculators* realise a profit by buying properties, holding them without improvement, then selling at a higher price as the market improves.
6. *Citizens* may gain generally from a higher tax base and lower crime rates in a revitalised area, which must be offset against increased demand for city services.

The major costs of neighbourhood revitalisation fall principally on displaced households who would not otherwise have moved but are forced to do so by rising occupancy costs that they cannot afford to pay.[47]

DISPLACEMENT

The displacement of former residents is a major social consequence of both public clearance and rehabilitation programmes on the one hand, and 'free-market' gentrification on the other. The displacement of working-class families reflects their weak position in society in general and the housing market in particular. Paradoxically, whereas they were once concentrated in the inner city because of their limited purchasing power, they are now being displaced from gentrifying inner neighbourhoods for the same reason.[48]

Displacement due to the super-heating of a local housing market by gentrification can occur in four ways:

1. The eviction of low-income tenants from buildings scheduled for rehabilitation as upper-income residences.
2. The involuntary departure of long-term families or elderly residents on fixed or limited incomes because of their inability to pay sharply escalating property taxes.
3. The inability of newly married children of existing residents to afford housing within the area they have traditionally regarded as their community.
4. Reluctant migration of residents from an area because of the loss of friends or supportive social, religious or economic institutions.

Property abandonment and gentrification are related processes in a cycle of neighbourhood change. Gentrification creates social conflicts but it does upgrade the physical environment. Continued neighbourhood decline, on the other hand, eventually leads to the development of a slum, an area characterised by an impoverished social and physical environment. The question of how best to address this, and other issues related to the quality and quantity of urban housing, are addressed in the next chapter.

FURTHER READING

BOOKS

R. Atkinson and G. Bridge (2005) *Gentrification in a Global Context: The New Urban Colonialism* London: Routledge

A. Downs (1981) *Neighborhoods and Urban Development* Washington DC: Brookings Institution

C. Hamnett (1984) Gentrification and residential location theory: a review and assessment, in D. Herbert and R. Johnson (eds) *Geography and the Urban Environment* volume 6 Chichester: Wiley, 283–330

L. Lees, T. Slater and E. Wyly (2008) *Gentrification* London: Routledge

P. Rossi (1955) *Why Families Move* New York: Free Press

N. Smith (1996) *The New Urban Frontier: Gentrification and the Revanchist City* London: Routledge

JOURNAL ARTICLES

W. Clark and J. Onaka (1983) Life cycle and housing adjustment as explanations of residential mobility *Urban Studies* 20, 47–57

W. Grigsby, M. Baratz, G. Galster and D. MacLennan (1987) The dynamics of neighbourhood change and decline *Progress in Planning* 28(1), 1–76

C. Hamnett and D. Whitelegg (2007) Loft conversion and gentrification in London *Environment and Planning A* 39, 106–24

L. Lees (2000) A reappraisal of gentrification *Progress in Human Geography* 24(3), 389–408

J. Rex (1971) The concept of housing class and the sociology of race relations *Race* 12, 293–301

K. Temkin and W. Rohr (1996) Neighbourhood change and urban policy *Journal of Planning Education and Research* 15, 159–70

D. Wilson, H. Margulis and J. Ketchum (1994) Spatial aspects of housing abandonment in the 1990s: the Cleveland experience *Housing Studies* 9(4), 493–510

KEY CONCEPTS

- Life-cycle concept
- Value-expectancy model
- Aspiration set
- Action space
- Activity space
- Awareness space
- Search space
- Housing class
- Neighbourhood-change continuum
- Housing abandonment
- Gentrification
- Super-gentrification
- Incumbent upgrading
- Free-rider
- Rent gap
- Displacement

STUDY QUESTIONS

1. Examine the major reasons why people move house.
2. Consider the mechanics underlying the search for a new home and identify any constraints on residential relocation experienced by different groups in society.
3. With reference to a neighbourhood with which you are familiar employ Downs' model of neighbourhood change as a basis for constructing a 'development history' of the selected neighbourhood over either the past twenty-five or fifty years.
4. Explain the phenomenon of housing abandonment.
5. With reference to particular examples, explain the process of gentrification. Who are the principal winners and losers in the process?

PROJECT

Read the property pages of the local newspapers over several months and make a note of the location, price and types of property being advertised. Are some parts of the city more dynamic in terms of property sales than others? Is there any evidence of gentrification? Can you identify different sub-markets in different areas (e.g. for apartments or detached houses)? Do some real-estate agents focus on particular niche markets in terms of spatial coverage, housing type or market price? How do these residential patterns relate to your knowledge of the social geography of the city?

11
Housing Problems and Housing Policy

Preview: trends in housing tenure; the growth of owner occupation; the decline of private renting; public-sector housing; council-house sales; the residualisation of council housing; public housing in the USA; housing affordability; homelessness; housing strategies; filtering; clearance; rehabilitation; alternative housing strategies; balanced communities

INTRODUCTION

The availability of shelter is a basic human need. Western governments have adopted different attitudes towards meeting this need. At one extreme, housing is regarded as a consumer good rather than a social entitlement. In the USA the government provides only 1 per cent of the total stock in the form of social housing, primarily as a complement to private urban-renewal programmes, and this effort is directed mainly at the poor, one-parent households, non-working families with dependent children or the low-income elderly. At the other extreme, in Eastern Europe housing was long considered a universal right and an essential part of the 'social capital', although in practice the inability of state resources to meet housing demand led to the promotion of alternative forms of provision, including housing co-operatives and owner occupation. The intermediate position is illustrated by the states of Western Europe, in which housing is considered to be a limited social right, and where the state has intervened in the housing market to ensure a basic level of shelter for the majority of the population. In the absence of intervention, between one-quarter and one-third of the total population of most Western European countries would be unable to meet the full economic cost of the housing it occupies.[1] In this chapter I identify and explain the main trends in the distribution of housing across the major tenures of owner occupation, private rental and public housing. We examine important issues including the residualisation of council housing in the UK, housing affordability and homelessness, and

I provide a critical analysis of the main policy initiatives designed to improve the quality and availability of housing in Western society.

TRENDS IN HOUSING TENURE IN THE UK

The extent of state support for housing provision and its distribution among tenures varies. In the UK central government has played a key role in determining the quality and quantity of housing available for different social groups, with most assistance generally being directed to the owner-occupied sector (via mortgage-tax relief) and the social rented sector (e.g. via rent subsidies). The private rented sector has remained largely unassisted and subject to a lengthy period of rent control and regulation. The effects of this policy regime are revealed in the changing tenure structure of Britain's housing stock.

Over the course of the twentieth century a major transformation was effected in the tenure balance of housing in the UK. Prior to the First World War almost 90 per cent of households rented their accommodation from private landlords, 2 per cent rented from local authorities and the remainder were owner-occupiers (Table 11.1). By 1991 the position had been largely reversed, with only 7 per cent of households still renting privately compared with 68 per cent owner occupation and 25 per cent who were council tenants. Most of the transformation occurred during the post-Second World War period.

TABLE 11.1 HOUSING TENURE AND DWELLING STOCK IN GREAT BRITAIN, 1914–2001 (MILLION)

Year	Rented from local authority or New Town		Privately rented		Owner-occupied		Total No. of dwellings
	No.	%	No.	%	No.	%	
1914	0.1	1.8	7.5	88.2	0.9	10.0	8.5
1944	1.6	12.4	8.0	62.0	3.3	25.6	12.9
1951	2.5	17.9	7.3	52.5	4.1	29.6	13.9
1961	4.4	26.8	5.0	30.5	7.0	42.7	16.4
1971	5.8	30.6	3.6	18.9	9.6	50.5	19.0
1981	6.6	31.0	2.7	12.6	11.9	56.4	21.2
1991	5.7	24.8	1.7	7.4	15.6	67.7	23.1
2001	3.6	15.4	2.4	10.6	17.1	74.0	24.7

THE GROWTH OF OWNER OCCUPATION

During the 1920s owner occupation was not the preferred housing option of most people. Most households occupied privately rented accommodation, or council houses produced under the 1923 Housing Act (the Chamberlain Act). The latter, because of the high standards of construction and consequently of rent, were reserved primarily for better-off members of the working class. Owner occupation of housing became attractive for the following reasons:

1. Local authority housing became more restricted to families displaced from slum-clearance schemes following the 1930 Housing (Greenwood) Act.
2. Wages increased in real terms.
3. The period of mortgage repayment was lengthened from fifteen years to twenty or twenty-five years.
4. Local authorities started to act as mortgage guarrrantors.
5. Levels of car-ownership increased.
6. Lack of effective planning controls over suburban development meant that cheap land could be used for speculative house-building.
7. An increasing proportion of privately rented housing was diverted into the owner-occupied sector as landlords found it more profitable to sell property than remain in the regulated rented sector.

As Table 11.1 shows, during the post-war period owner occupation grew to become the largest sector of the UK housing stock. This was encouraged by several factors, including the following.

1. Government policy favoured owner occupation via:

   - Mortgage tax relief (although the value of this was reduced progressively and removed finally in 2000).
   - Reduced stamp duty on house sales (although again this is subject to change).
   - Abolition of the tax on imputed net income derived from a dwelling by the owner.
   - Government loans and grants to building societies to fund the purchase of pre-1919 privately rented dwellings and to keep interest rates below market rates.
   - The exemption of homes from capital gains tax.
   - The introduction of the option mortgage scheme to provide cheap loans to first-time buyers from low-income groups.
   - The sale of council houses.

2. Specialist financial institutions such as building societies developed, and banks expanded into the mortgage business.
3. The investment climate has generally been favourable to property development.
4. Households have had increasing difficulty finding alternative accommodation, owing to the decline in the privately rented and local authority sectors.

The growth of owner occupation has been particularly marked since 1979, following the election of a Conservative government committed to the creation of a 'property-owning democracy'. The extent of the government's commitment was indicated by the unequal assistance provided for different tenure types with, in 1990–1, tax relief on mortgages for owner-occupiers amounting to £7,500 million compared with the £900 million support for council housing.[2] The increasing numbers and proportion of owner-occupied housing in the UK is revealed in Table 11.2.

In addition to the stimulus provided by housing policies, the growth of home-ownership in Britain reflected the belief that ownership of a house represents a sound financial investment. This assumes that over the long term house prices will rise at least as fast as and probably faster than general inflation. Accordingly, it is better to pay mortgage interest on a loan of fixed capital value that results in the acquisition of an asset at diminishing real cost than to pay rent, probably at a similar level, but escalating over time and with no asset to show for it. This argument has proved to be generally correct. However, it is also predicated on the continuing ability of the mortgagee to service repayments. In a period of recession, unemployment combined with high debt/equity ratios can result in people's homes being repossessed for non-payment of loan instalments. In the early 1990s in the UK, rising unemployment with the onset of recession (following a government rise in interest rates to control inflation) meant that 1.8 million people who had taken on large mortgages to purchase houses during the boom of the late 1980s found themselves in a position of **negative equity**, where the exchange value of their house had fallen below the amount of the outstanding loan. In 1992 350,000 households were in mortgage arrears of more than six months, and 68,000 houses were repossessed.[3] By 1996 400,000 mortgages were in arrears and more than 3 million households were unable to move because of being in a position of negative equity whereby the market value of their house was insufficient to cover the repayment of the mortgage on the property. Similar problems were evident in the wake of the 'credit crunch' of 2008. In general, however, the advantages of owner occupation have ensured that, since 1971, the majority of the UK housing stock is under this form of tenure.

TABLE 11.2 UK DWELLING COMPLETIONS BY TENURE, 1946–2001

Year	Private	Local authority	Housing association	Year	Private	Local authority	Housing association
1946	31,297	25,013	–	1974	140,865	103,279	9,920
1947	40,980	97,340	–	1975	150,752	129,883	14,693
1948	32,751	190,368	–	1976	152,181	129,202	15,770
1949	25,790	165,946	–	1977	140,820	119,644	25,127
1950	27,358	163,670	–	1978	149,021	96,196	22,771
1951	22,551	162,854	–	1979	140,481	75,573	17,835
1952	34,320	186,920	–	1980	128,406	76,597	21,097
1953	62,921	229,305	–	1981	115,033	54,867	19,291
1954	90,636	223,731	–	1982	125,416	33,244	13,137
1955	113,457	181,331	–	1983	148,067	32,833	16,136
1956	124,161	154,971	–	1984	159,492	31,699	16,613
1957	126,455	154,137	–	1985	156,530	26,115	13,123
1958	128,148	131,614	–	1986	170,565	21,548	12,531
1959	150,708	114,524	–	1987	183,736	18,809	12,571
1960	168,629	116,358	–	1988	199,485	19,002	12,764
1961	177,513	105,529	–	1989	179,593	16,452	13,913
1962	174,800	116,424	–	1990	159,034	15,609	16,779
1963	174,864	112,780	–	1991	151,694	9,645	19,709
1964	218,094	141,132	–	1992	139,962	4,085	25,640
1965	213,799	151,305	–	1993	137,293	1,768	34,492
1966	205,372	161,435	–	1994	147,370	1,982	36,612
1967	200,348	181,467	–	1995	152,841	2,135	38,284
1968	221,993	170,214	–	1996	145,891	868	32,148
1969	181,703	162,910	–	1997	152,841	468	27,597
1970	170,304	157,067	–	1998	145,891	428	23,325
1971	191,612	134,000	–	1999	149,151	165	23,315
1972	196,457	104,553	–	2000	144,250	302	22,910
1973	186,628	88,148	8,852	2001	140,352	462	21,070

THE DECLINE OF PRIVATE RENTING

As Table 11.1 reveals, from a dominant position in the housing stock prior to the First World War the privately rented sector has diminished in importance. Although it accounted for half of UK housing in 1951, by 1991 its share had fallen to less than 10 per cent of the total stock, although there was a small recovery by 2001. The principal factors underlying this decline are:

1. *Slum clearance.* The private rental sector included a high proportion of the oldest and poorest dwellings, many built before 1919. From the 1930s onwards slum-clearance programmes demolished hundreds of thousands of these houses.
2. *Policies to deal with overcrowding.* The Housing Acts of 1961, 1964 and 1969 increased controls over multiple occupation, and thereby reduced the number of private tenants.
3. *Housing rehabilitation.* Rehabilitation aided by improvement grants was often followed by tenant displacement and the sale of properties for owner occupation, particularly in gentrifying neighbourhoods.
4. *Poor investment returns.* The comparatively poor rate of return from investment in private rented housing was apparent by the late nineteenth century. With the extension of the principle of limited liability, the development of the stock exchange and building societies, the expansion of government and municipal stock and increased overseas investment opportunities, capital flowed out of private rented property into these alternative investment opportunities. During the twentieth century, increased public intervention and, in particular, rent control further reduced the attraction of housing investment, while the increasing cost of repair and maintenance also eroded profits.
5. *The ethic of home-ownership.* Although there was some building for private rental in the inter-war period, the bulk of private house construction during the building boom of the 1930s, when interest rates were low and land, materials and labour cheap, was for owner occupation.
6. *Subsidies and tax allowances.* In contrast to the subsidies provided to council tenants and the tax advantages for owner-occupiers, only since 1973 have private-rented tenants received rent allowances. Landlords of privately rented housing were not permitted to set a 'depreciation allowance' against their tax liabilities, and this encouraged them to sell for owner occupation.

An investigation into the decline of the private rented sector by the House of Commons Environmental Committee (1982)[4] concluded that private rented housing fell short on all the criteria for sound investment – the level of risk, liquidity, expected return on capital, and management responsibility – when compared with alternative opportunities. In London the large commercial landlords were withdrawing from the market, selling off their freeholds to insurance companies and friendly societies, which in turn were selling long leaseholds to owner-occupiers. Although there were 50,000 purpose-built flats in 1,300 blocks in central London in 1966, less than one-third were still let under the Rent Acts in 1982, the majority having been sold off for prices ranging from £30,000 to £200,000.[5] Overall, more than 3 million dwellings have been sold by landlords to owner occupation since the Second World War. Since relatively little new property has been built for private renting, the remaining private rented stock is old and, owing to a succession of rent controls (Box 11.1), in poor condition. In the USA (where there is negligible competition from public-sector housing) and in France, Germany, Denmark and the Netherlands (where there are only mild rent controls, coupled with subsidies) private renting is also in decline and confined to mainly older and less attractive housing. It should be noted, however, that the shrinkage of the private rented sector has been socially selective. In the larger cities of Europe, North America and Australia the demand for centrally located luxury flats remains sufficient to maintain investment in this type of property.[6]

THE DEVELOPMENT OF PUBLIC-SECTOR HOUSING

The decline of private rented housing, the traditional source of shelter for those denied access to the owner-occupied sector, led to direct government intervention in the provision of housing. By 1939 1.18 million houses had been constructed by the public sector, amounting to 10.3 per cent of total stock. The growth of municipal housing also stemmed from:

1. The failure of nineteenth-century housing associations to demonstrate that it was possible to provide decent dwellings in sufficient numbers at a rent affordable by the low-paid.
2. Increased working-class militancy over housing conditions, vividly displayed in the Glasgow rent strikes of 1915.
3. The shortage of housing in the post-First World War era.

BOX 11.1

Rent controls

Rent controls represent one of the most contentious pieces of housing legislation ever enacted. Introduced to control profiteering by unscrupulous landlords who sought to take advantage of housing shortages following the First World War, rent controls have tended to persist because of government reluctance to risk electoral unpopularity by their removal. The case for rent control was boosted in the UK between 1957 and 1965 when, following the 1957 Rent Act, which decontrolled the rents of 2.5 million dwellings, the practice of letting furnished accommodation to house-hungry and vulnerable immigrants at extortionate rents ('Rachmanism') came to public attention. Overcrowding, illegal sub-letting and insanitary conditions became intense in areas where immigrants concentrated, leading to spasmodic racial conflict and urban riots from the late 1950s onwards.

Although it was introduced for a commendable social purpose, a major defect of rent control has been its role in reducing the supply of private rented accommodation available by restricting the ability of landlords to gain an adequate profit. This has led to physical deterioration of the housing stock through neglect and, in some cases, to property abandonment. The reintroduction of rent regulation in the 1965 Rent Act contributed to the ongoing decline in the privately rented housing stock.

The Housing Acts of 1980 and 1988 attempted to reinvigorate the private rented sector by introducing a form of decontrol for new lets: (1) assured shorthold lettings for a fixed period at the end of which the landlord had vacant possession, and with rents set by a rent assessment committee if necessary; and (2) assured tenancies, with a high degree of security of tenure and rents agreed between landlord and tenant. Existing tenants of rent-controlled housing would continue to be protected by the Rent Acts. A negative aspect of this dual provision was that some landlords sought to harass tenants in rent-controlled flats to switch to an assured shorthold or assured tenancy arrangement.

Rent control is, therefore, a double-edged sword: while seeking to protect tenants from exploitation, it has also induced a reduction in the stock of available privately rented accommodation.

4. A gradual shift away from the Victorian ideology of self-help and the pervading belief in the efficiency of the market mechanism.

The 1919 Housing (Addison) Act provided subsidies to local authorities to make up losses incurred in providing houses over and above those borne by a one-penny rate.[7] Houses built under this Act were intended for the 'general needs' of the working class, but the high standards inspired by the garden city ideal and high rents placed them beyond reach of ordinary working-class people. In the Mosspark scheme in Glasgow, where the rent for a three-bedroom house was £28 per year, 16.2 per cent of the first tenants were 'professionals', 30.3 per cent 'intermediate class' and 27.5 per cent 'skilled non-manual'. Only 18.0 per cent were manual workers. These garden suburbs remain among the most desirable of the municipal housing stock.[8] The 1924 Housing (Wheatley) Act gave further impetus to council house building with over 500,000 houses constructed, but the poorest of the working class remained in overcrowded slum conditions. The 1930 Housing (Greenwood) Act gave local authorities direct responsibility for rehousing communities displaced by slum clearance, for keeping rents within reach of the poor, and for targeting their efforts at slum dwellers previously excluded from council housing. In total, almost 300,000 slum properties were demolished in the 1930s, 1 million council houses were built under the slum clearance programme, and 4 million people relocated. Although the programme tackled some of the worst physical conditions, the minimum standards employed in the rehousing schemes, the lack of community facilities and limited local employment created a basis for future social problems.

Britain emerged from the Second World War with a housing shortage. During the conflict there had been very little new construction, maintenance had been neglected, over 200,000 dwellings had been destroyed by bombing and more than 250,000 had been made uninhabitable. In addition, because of Britain's early urbanisation (see Chapter 3), much of the housing was old and substandard; and the necessary slum clearance programme added to the shortage of housing. Local authorities were given the main role in providing new houses for rent, and for most of the post-war period there was general agreement across the political spectrum of the need for a council house building programme. By the 1970s the crude housing shortage had been resolved, at least in national terms, although regional shortages remained.

Plate 11.1 The Easterhouse estate on the periphery of Glasgow is typical of the kind of public housing constructed in British cities during the early post-war period

Despite significant progress towards the goal of 'a decent home for every family at a price within their means',[9] by the late 1970s a number of criticisms were directed at the UK housing stock, including the following:

1. The social-housing sector was too large and expensive to support; the subsidised rents were not sufficiently well targeted and benefited many who did not need such a level of support.
2. The below-market level of rents contributed to low levels of maintenance undertaken by many local authorities, which stored up problems for the future.
3. Novel building methods employed in the 1960s and 1970s had given rise to structural problems in some estates (with, for example, flat-roofed housing designed originally for a scheme in Algeria employed in the much wetter climate of Glasgow).
4. Problems of damp penetration, inadequate insulation and costly-to-run heating systems provided

poor living conditions and had adverse effects on health.
5. The size of the council-housing operation made effective management difficult.
6. The large, homogeneous council estates lacked a mix of social classes and were often constructed without social amenities.
7. As unemployment rose, problems relating to poverty and deprivation began to concentrate in many estates (see Chapter 15).

All these factors contributed to a re-examination of national housing policy. The greatest force for change, however, was the ideological conviction of an incoming Conservative government committed to reducing public expenditure. With reference to local authority housing, a major divide opened between the government's view of council housing as a 'welfare service', assisting households unable to afford other types of housing, and the Labour Party's belief that the public sector should supply housing for 'general needs' (i.e. should satisfy the demand from all households who wish to rent rather than buy). The new government's housing policy set out to:

1. Increase owner occupation.
2. Provide a wider choice of landlord in the social rented sector.
3. Revive the private rented sector.
4. Reduce the role of local authorities as direct housing providers.

A key component of the Conservatives' housing policy, introduced in the 1980 Housing Act, was the right of council tenants to buy their homes at a discount price that reflected length of tenancy. The case for the sale of council houses rested on the following arguments:

1. Council housing was rising in cost and the growing gap between cost rents and the 'fair rents' charged under the Rent Acts required increasing government subsidies. Some proponents of the 'right to buy' argued that if long-standing tenants (those of more than thirty years' standing) were given their houses and others sold at favourable terms, the savings on subsidies and revenue raised could either lead to a reduction in taxation or allow funds to be directed into other areas.
2. Over the long term, it was thought, it would be of more economic benefit for tenants to own than to rent their council house. Kilroy (1982) demonstrated that over thirty-five years the net outgoings (at 1982 prices) of an average tenant who bought his £14,000 house (at a 40 per cent discount) would be £12,000 less than those of one who continued to rent. The purchaser would also have a well constructed, appreciating asset which they could pass on to their children.[10]
3. Owner occupation offers a feeling of independence, freedom over the care and condition of the property, and greater locational mobility.

Critics of council house sales countered these arguments on the grounds that:

1. The 'rising costs of council house subsidies' case is often based on extreme examples such as inner London boroughs, and ignores the fact that local authorities are able to distribute these costs over the whole of their stock by using the rent surpluses from older homes to keep down rents on newer houses built at a time of higher capital costs and interest rates.
2. The argument that selling houses can contribute to a major cut in public expenditure was rejected by those who questioned the financial calculations employed. They pointed out, for example, that proponents of council house sales included Supplementary Benefit (welfare) payments in the cost of council housing, yet these would still need to be paid to low-income householders if they converted to owner occupation. Omitting this income support cost meant that in 1977–8 the cost of council house subsidy (at £1,100 million) would have been comparable to the tax relief given to owner-mortgagers.
3. Normally only higher-income tenants could afford to buy, even with a discount, and even some of these might face financial difficulty if they reached retirement age before they had paid off their mortgage, or if mortgages were based on contributions from the income of several family members. Also, if family members contributing to mortgage payments moved away, they would be unable to qualify for a mortgage of their own unless they forced the sale of the mortgaged council house.
4. It is mainly the better-sized, higher-quality dwellings that are sold off, leaving families on lower incomes with a reduced chance of improving their housing position through a transfer within the local authority system.
5. The sale of older rent surplus-yielding housing in attractive locations would reduce the possibility of cross-subsidisation and thereby necessitate higher rents for disadvantaged tenants in inner-city locations.

The re-election of the Conservative government in 1983 was followed by the 1984 Housing and Building Control Act, which introduced higher discounts (up to a maximum of 60 per cent) and a reduction in the eligibility period (from three to two years); and by the 1984 Housing Defects Act, which provided 90 per cent repair grants to remedy defects not apparent at the time of purchase. Between 1980 and 1991 1.46 million sales had been completed and the Treasury had received over £20 billion, making the right to buy (RTB) scheme one of the largest privatisation initiatives of the Thatcher era. The privatisation of council housing was also promoted in the provisions of the 1988 Housing Act, which enabled local authorities to transfer their housing stock to housing associations with the approval of a majority of tenants. By 1994 thirty-two local authorities, mostly under Conservative control, had transferred the whole of their stock, amounting to 149,478 dwellings, and in

England, by 2000, the 'loss' of local authority housing through 'large-scale voluntary stock transfer' (LSVT) had reached 450,000 dwellings.

The Conservative view of local authorities as 'enablers' facilitating action by other bodies, rather than direct providers of social housing, is evident in Table 11.2, which shows the reducing role of council house building and the growing importance of the housing association sector as an alternative source of rented accommodation. Housing associations obtain their loan finance for investment in new-house construction and rehabilitation from the Housing Corporation (which oversees their activities) and, since the 1988 Housing Act, from private-sector institutions. By 1992 there were 2,550 registered housing associations in Britain with a total stock of 750,000 houses (3 per cent of total stock). By 1998 housing associations accounted for 1.2 million dwellings or 5 per cent of total housing stock. Housing associations may cater for general housing needs or for the special needs of groups such as elderly or disabled people. They vary in size from national associations owning in excess of 10,000 houses to a majority which own less than 1,000.[11]

THE RESIDUALISATION OF COUNCIL HOUSING

As a consequence of RTB and estate transfers, the public rented sector in the UK is becoming increasingly residualised. This is evident in several ways:

1. Since 1981, when almost 7 million council houses accounted for 31 per cent of the UK stock, there has been both an absolute and a relative decline in the contribution made by public-rented housing.
2. The quality of the council stock has been declining, with an increased incidence of design and structural faults and difficult-to-let dwellings. Whereas in the inter-war and immediate post-war years council housing was a privileged tenure, it is now widely seen as only for those unable to gain entry to owner occupation.
3. Since the early 1970s the social composition of council tenants has altered, with a relative decline in the population of non-manual/skilled manual workers and an increase in personal service/unskilled manual workers. This has led to less of a social mix within the working class, and a higher degree of socio-spatial polarisation.[12]
4. An increasing and disproportionate number of council tenants (such as lone parents, elderly people, sick people, disabled people and unemployed people) are dependent upon income support. A growing proportion of council tenants belong to an 'under-class' dependent on the state rather than the labour market for most of its income.

It is important not to confuse the symptoms of residualisation with the causes. The residualisation of council housing in the UK is a consequence of broader economic and social processes, including high unemployment resulting from technological innovation and deindustrialisation, the withdrawal of private landlords from the low-rent market, increased demand for owner occupation among the more affluent, and government policies aimed at reducing public expenditure. There is an increasing possibility that the future role of council housing will be to provide shelter of last resort to those unable to go elsewhere (Box 11.2).

PUBLIC HOUSING IN THE USA

The restricted role envisaged for council housing by the political right in Britain finds an echo in the history of US attempts to enact a programme of public housing.[13] Like many other social programmes, public housing in the USA is a product of the New Deal era as enunciated in the Housing Acts of 1937 and 1949. Two basic means of supplying housing to low-income families have been used (Figure 11.1). The first employed a public agency to develop, own and manage housing. By the end of 1974 67 per cent of the total public housing supply (860,000 units) had been developed in this way. Another 230,000 units (turnkey housing) were produced by private interests and sold to a public agency, usually a local housing authority. The second main approach authorised by the 1965 Housing Act was the leasing of privately owned housing, and by the end of 1974, 169,000 units accounted for 14 per cent of the total public stock.

A fundamental reason for the failure of the US public housing initiative was that it did not have widespread political support, being alien to the dominant market philosophy. Even the initial rationale behind the programme was multifaceted: it was seen as a way of simultaneously reducing the high levels of unemployment in the 1930s and assisting the housing industry, as well as a means of eliminating slums and increasing the supply of cheap, decent housing to the poor. Weak federal controls over development costs and housing quality inevitably led to major difficulties.

BOX 11.2

Residualisation and social housing in the UK

Over the last quarter of a century the role of social housing has changed. The sector has become much smaller as a proportion of the total, although nearly 4 million households still live in it. While post-war provision was aimed at households on a range of incomes, since the 1980s provision has become more tightly constrained and new lettings focused on those in greatest need. As a result, the composition of tenants has changed, with tenants much more likely to have low incomes and not to be in employment than in the past or than those in the other tenures. Seventy per cent of social tenants have incomes within the poorest two-fifths of the overall income distribution, and the proportion of social tenant householders in paid employment fell from 47 per cent to 32 per cent between 1991 and 2006. Tenants have high rates of disability, are more likely than others to be lone parents or single people, and to be aged over 60. More than a quarter (27 per cent) of all black or minority ethnic householders are social tenants (including around half of Bangladeshi and 43 per cent of black Caribbean and black African householders), compared with 17 per cent of white householders. Looking at today's social housing stock, 93 per cent of it was already within the sector nine years ago (although 750 000 dwellings were transferred between local authority and housing association ownership). For tenants, there is much less movement between dwellings than in the private rented sector, and more than 80 per cent of those living in social housing today were also in the sector ten years ago (if born by then).

Source: J. Hills (2007) *Ends and Means: The Future Roles of Social Housing in England* LSE Centre for Analysis of Social Exclusion Report 34, London: LSE

Figure 11.1 Public housing policy in the USA

Source: G. Galtner (1995) *Reality and Research* Washington DC: Urban Institute

In particular, local housing authorities received no operating subsidy from federal government, their sole source of income being rental payments. As costs rose, minimum and average rents increased steadily. Prior to the 1974 Housing and Urban Development Act, which introduced a degree of control, rent payments equivalent to half a tenant's gross income were common. The 1974 Act marked a major switch of emphasis away from conventional public housing towards the use of leased housing. However, signs of the increased privatisation of public housing had emerged during the 1960s with the 1965 Housing Act authorisation of turnkey housing, the sale of public housing to tenants, and the contracting out of management services to private firms. By the mid-1970s construction of public housing projects had fallen significantly and some of the most deteriorated schemes, such as Pruitt-Igoe in St Louis, had been demolished.

The switch of emphasis in social-housing policy from supply-side to demand-side intervention was reflected in Section 8 of the 1974 Housing and Urban Development Act, which introduced a housing allowance programme for lower-income households (Figure 11.1). Rather than making payments direct to needy tenants, the programme consisted of long-term contracts between the Department of Housing and Urban Development and private developers, landlords or public housing agencies designating specific apartments for Section 8 subsidies. Approved apartments could be existing units or units to be constructed or rehabilitated. Rental payments for tenants were on a par with those for public housing, amounting to between 15 per cent and 25 per cent of their total income. Housing owners received, from Section 8 funds, the difference between the tenant's contribution and a 'fair market' rent based on prevailing private rents in the locality. Section 8 subsidies have been successful in that tenants have a degree of choice of housing unit, and the scheme helped to reduce the overconcentration of low-income tenants in 'projects', many of which have been razed.

Federal disengagement from public housing was signalled in the belief of the Reagan administration that 'the genius of the market economy, freed of the distortion forced by government housing policies and regulations that swing erratically from loving to hostile, can provide for housing far better than federal programs' (President's Commission on Housing 1982 p. xvii).[14] This resulted in an extension of the demand-side approach via Section 8 certificates and the introduction of vouchers worth a certain amount of rental payment that could be 'spent' on any available apartment, and the continued decline in public housing construction (Table 11.3). In 1985, inspired by the example of the UK, the Department of Housing and Urban Development launched a demonstration project to promote the privatisation of public housing. The residualisation of public housing is well advanced in the USA, with the sector catering for a dependent population characterised by low income, unemployment, reliance on public assistance, and a high concentration of the very old, very young and minorities, often occupying a deteriorating physical environment.

Continued federal disengagement from low-cost housing provision was indicated by the 1997 Housing Opportunity and Responsibility Act. The goal is to transform public housing developments into 'more self-sufficient communities'. This will be achieved by creating lower-density mixed-income communities through a combination of demolition of the worst high-rise apartment blocks, the sale of public housing to tenants and the construction of new low-rise developments to accommodate a wider range of income groups (by easing the current restrictions on public housing to households whose income is less than 80 per cent of the area median income). Displaced public housing tenants will be provided with Section 8 (Housing Choice) vouchers to obtain accommodation in the private sector. The underlying premise, that mixed-income housing will create a sense of community in which

TABLE 11.3 US PUBLIC-HOUSING COMPLETIONS, 1939–91

Year	Units	Year	Units
1939	4,960	1966	31,483
1940	34,308	1967	38,756
1941	61,065	1968	72,638
1942	36,172	1969	78,003
1943	24,296	1970	73,723
1944	3,269	1971	91,593
1945	3,080	1972	58,590
1946	1,925	1973	52,791
1947	466	1974	43,928
1948	1,348	1975	24,514
1949	547	1976	6,862
1950	1,255	1977	6,229
1951	10,246	1978	10,259
1952	58,258	1979	44,019
1953	58,214	1980	15,109
1954	44,293	1981	33,631
1955	20,899	1982	28,529
1956	11,993	1983	27,876
1957	10,513	1984	24,092
1958	15,472	1985	19,267
1959	21,939	1986	15,464
1960	16,401	1987	10,415
1961	20,565	1988	9,146
1962	28,682	1989	5,238
1963	27,327	1990	6,677
1964	24,488	1991	9,174
1965	30,769		

Source: E. Howenstine (1993) The new housing shortage: the problem of housing affordability in the United States, in G. Hallett (ed.) *The New Housing Shortage* London: Routledge, 8–67

working residents will provide role models for the unemployed, is reminiscent of the philosophy behind nineteenth-century utopian communities and the early post-Second World War New Towns in the UK[15] (see Chapter 9).

In 2000 1.3 million units of public housing accounted for 5 per cent of all rental housing and 1 per cent of the total housing stock in the USA. The poor condition of some public housing areas, identified by the National Commission on Severely Distressed Public Housing in 1992, led in 1999 to enactment of the Hope VI programme that aimed to revitalise distressed public housing communities (Box 11.3) In Indianapolis revitalisation of the Near Westside neighbourhood involved the replacement of substandard row houses with lower-density duplex units, similar to those in surrounding neighbourhoods. Units were set aside for households

BOX 11.3

The Hope VI programme

Home-ownership and Opportunity for People Everywhere (HOPE) was developed as a direct result of the report of the National Commission on Severely Distressed Public Housing. Originally called the Urban Revitalisation Demonstration, Hope VI funds can be used for the following objectives:

1. Improving public housing by replacing severely distressed projects, such as high-rise and barrack-style apartments, with town houses or garden-style apartments that blend into surrounding communities.
2. Reducing concentrations of poverty by encouraging a mix of income among public-housing residents and encouraging working families to move into housing in revitalised communities.
3. Providing support services, such as education and training programmes, child-care services and counselling to assist public-housing residents obtain and retain jobs.
4. Establishing and enforcing high standards of personal and community responsibility through explicit lease requirements.
5. Forming partnerships, involving public-housing residents, government officials, the private sector, non-profit groups and the community at large in planning and implementing new communities.

who participated in a family self-sufficiency programme that required them to enter into a formal contract with specified goals and participation requirements, leading towards self-sufficiency and, in many cases, home-ownership. The hope was that through training programmes, access to employment opportunities and a ceiling-rent programme many public housing families would become home-owners. Nationally, by 2003, 45,000 households had been relocated to other public housing, to private market-rate housing, or with Section 8 (Housing Choice) vouchers, and 8,000 original households had returned to their redeveloped areas. In Chicago, by 2000, the Housing Authority had received $160 million in six HOPE VI grants to revitalise some of the nation's worst public housing projects including Robert Taylor Homes to the south of downtown, and Cabrini Green west of North Michigan Avenue. While this represented a large infusion of capital for public

housing, in practice, a significant portion was used to subsidise construction of market rate housing that led to displacement of public housing residents, 'vouchered out' into private rental markets.[16] In general the scale of demolition has raised some concern that, while providing improved housing environments, Hope VI has also reduced the amount of **affordable housing** available to the lowest-income households.

HOUSING AFFORDABILITY

Housing affordability is measured by the proportion of household income spent on obtaining housing. An expenditure of more than 30 per cent of income in the USA and over 20 per cent in the UK has been taken to indicate an affordability problem.[17] Such measures are rather crude, however, since the same proportion of household income has a different impact for those at different ends of the social scale. Neither does it take account of the subjectivity of affordability. Nevertheless, despite definitional problems there is evidence of a growing affordability problem in both rented and owner-occupied housing, particularly in overheated housing markets.[18]

Since the early 1980s several structural forces in the UK have operated to make housing more expensive in relation to incomes:

1. Increases in unemployment during periods of recession have reduced incomes and effectively debarred households from the owner-occupied sector. For existing mortgagers this has led to problems over mortgage repayment and in some cases substantial negative equity.
2. As a result of global economic restructuring and the changing nature of labour markets in advanced capitalist societies, those with low levels of skill and limited education have the greatest difficulty in entering the job market and securing an income sufficient to meet the market cost of housing.
3. The private rented sector has continued to decline, and rents have increased following the easing of rent control.
4. There is reduced access to social rented housing as a result of government policies such as the 'right to buy', and constraints on new public-sector construction.

Evidence of the growing unaffordability of housing for some social groups is found in the 400,000 households in mortgage arrears in 1995 and the repossession of 250,000 houses between 1991 and 1994.[19] In the USA between 1997 and 1999 while inflation rose by 6.1 per cent house rents increased by 9.9 per cent and house prices by 16 per cent. In 1997 5.4 million very low-income families (those with incomes below 50 per cent of the local MSA median) paid more than half their income for housing or lived in severely inadequate housing, a situation classified by the Department of Housing and Urban Development as 'worst-case needs'. This represented an increase of 12 per cent in worst-case needs housing since 1991. A growing proportion of these families are working households. In 2002 in no jurisdiction in the USA would a minimum-wage job (at $5.15 per hour) provide enough income for a household to afford the fair market rent (FMR) for a two-bedroom home. Many low-income earners must work two or three jobs to pay their rent. Between 1991 and 1997 the number of housing units affordable to extremely low-income families (those with incomes below 30 per cent of median MSA income) fell by 5 per cent or 370,000 units. In New York City an extremely low-income household could afford a monthly rent of no more than $471, while the FMR for a two-bedroom unit was $1,031. The financial difficulties for low income households seeking to purchase a home may be compounded by redlining by mortgage lenders (see chapter 7). This practice is also evident in the emergence of a 'sub-prime lending market' in which loans carry high interest rates and fees. A subset of the sub prime lending market, termed 'predatory lending' involves excessive terms, poor underwriting, high prepayment penalties and, in some instances, use of illegal and deceptive practices, such as inflated house appraisals. Predatory loans, and sub prime loans more generally are marketed to elderly, low income and minority groups.[20] In 2007 US financial institutions collectively owned $1 trillion (£480 billion) of sub-prime debt, and one in five US mortgages was in the sub-prime category. Nationally, there were over two million foreclosures as people defaulted on mortgage repayments, while in Cleveland, (the then sub-prime capital of the USA), one in ten homes were vacant and whole neighbourhoods were blighted by foreclosed, vandalised and boarded-up houses.

Housing affordability is a problem for places as well as for people. Many people cannot find housing near their workplace or find work at a reasonable distance from where they can afford to live. In many areas workers crucial to the local economy, such as teachers and police, cannot afford to live in the communities they serve. The problem is particularly severe in cities with overheated housing markets. These include high-tech hot-spots in places like Boston, Denver and San Francisco, where rents increased by more than 20 per cent in the late 1990s, and global cities such as London. In the UK the 'key worker' scheme introduced in 2004 was designed to ensure access to housing in London and south-east England for designating key workers (such as teachers, nurses and police) by subsidising the cost of buying or renting a home. More generally, a government review of housing supply in the UK[21] recommended the introduction of a development levy attached to the granting of planning permission for new housing development. The in-cash payments rather than in-kind tariffs (i.e. provision of infrastructure by developers) would be used by regional planning authorities to subsidise affordable housing. This proposal represents an extension of the practice of planning gain, and reenergises debate over taxation of land betterment (see Chapter 8).

The starkest indication of an affordability crisis, however, is represented by the growing incidence of homelessness, which is most visible in the inner areas of large cities, where the sight of people sleeping in doorways and living in cardboard shelters is redolent of conditions in the Third World city (see Chapter 25).

HOMELESSNESS

Homelessness is an extreme form of social exclusion, caused by a combination of personal and structural factors. Most homeless people leave their home either because parents or friends are no longer willing to accommodate them or because of the breakdown of a relationship. Structural factors underlying homelessness include insufficient construction of affordable housing, gentrification, cutbacks in welfare budgets, stagnating or falling real incomes, and the rise of part-time and insecure employment. Discriminatory practices can also contribute to the problem for some social groups.[22] Those most at risk of homelessness include unemployed people, single mothers, disabled people and frail elderly individuals, runaway youths, battered women and children, immigrants and refugees, substance abusers and deinstitutionalised mental patients (see Chapter 17). While many of the homeless are highly visible (e.g. beggars and rough-sleepers), most are not noticeably different from other citizens. The invisible homeless have to cope with the day-to-day strain of living in temporary accommodation, hostels, bed-and-breakfast hotels, or in cramped conditions with friends and relatives. Figure 11.2 illustrates the types of shelter available to the homeless in post-industrial society.

Plate 11.2   For some disadvantaged people home is a public park in central Tokyo

Definitions of homelessness vary between countries, making estimates of the number of homeless persons difficult. In the UK the number of those officially homeless rose from 53,100 households in 1978 to 146,000 households (460,000 persons) in 1994. These figures exclude people not accepted by the local authority as homeless or who are not felt to be in 'priority need'. The latter category excludes most single people and childless couples who approach local authorities for housing assistance. A 1995 report on single homelessness in London enumerated 47,000 persons living in squats, bed-and-breakfast accommodation or sleeping rough, none of whom had been accepted as being in priority need and therefore officially classified as homeless.[23] In Britain as a whole, at any time during the 1990s, over 500,000 people lacked a permanent roof over their heads. In Canada more than 26,500 different individuals make use of Toronto's emergency shelter system each year; and this does not include the number who sleep rough (Box 11.4). In the USA the national

total of homeless people in 1995 was estimated at between 500,000 and 600,000.[24] In New York half the 80,000 homeless live on the street and the remainder in public or private emergency sheltered accommodation. One in five are parents and children, one-third young adults aged between 16 and 21 years, and the remainder single people, of whom 80 per cent are men. As many as 90 per cent of the residents of sheltered accommodation are from ethnic minority groups even though such groups constitute only 40 per cent of the city population.[25] In Los Angeles, where the city council has employed barrel-shaped 'bum-proof' benches and sprinkler systems to discourage people from sleeping in parks, the homeless established a series of encampments in the heart of the city (Box 11.5). The pressure on the homeless population of Los Angeles' downtown Skid Row was increased in the late 1990s by the prospect of a real estate boom driven by gentrification and the accompanying desire to 'clean up' the area by removing its homeless residents.

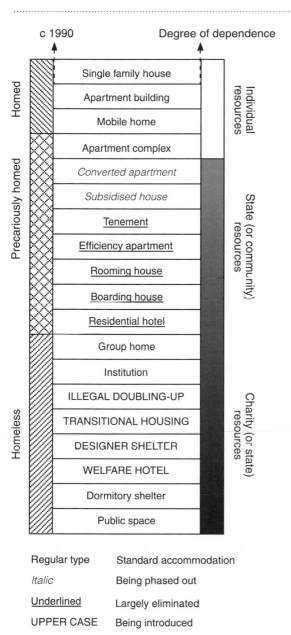

Figure 11.2 Types of accommodation in a post-industrial society
Source: A. R. Veness (1992) Homes and homelessness in the United States *Environment and Planning D: Society and Space* 10, 445–68

STRATEGIES TO IMPROVE HOUSING QUALITY

The process of neighbourhood decline examined in Chapter 10 may culminate in the creation of slum

conditions. Apart from relocating people to new living environments outside the city, the three main approaches to the problem of slums are via the filtering mechanism, by clearance and through rehabilitation.

FILTERING

Filtering requires no direct government involvement. The raising of housing standards is entrusted to the interplay of market forces whereby the provision of improved accommodation for the lower-income members of the community is effected through construction of new properties for upper- and middle-income groups. In principle a chain of sales is initiated by the insertion of new housing at the top end of the market. Properties vacated by those moving to the new housing filter down the social scale and enable individual households to filter up the housing scale, leaving the worst housing vacant and ready for demolition.

Empirical research has questioned the ability of the filtering mechanism to work towards the elimination of slums.[26] A major weakness is that the filtering process at best works slowly and many of the vacancy chains are broken before the poor benefit to a significant degree. Filtering also assumes that poor people are mobile physically and economically, whereas

BOX 11.5

Homeless in Hollywood

Los Angeles skid row covers a seventy-block area east of the central business district (CBD), and by the mid-1980s contained the largest population of homeless people in the USA. In the heart of a major city of advanced capitalist society around 1,000 street dwellers survived in over a dozen encampments. Throughout the CBD other homeless people squatted on pockets of disused land, including that under the parking ramps of the Bonaventure Hotel, a symbol of conspicuous consumption in the city.

The homeless of skid row are engaged in a struggle for *Lebensraum* in the face of attempts by the civic authorities to evict them (e.g. as part of the clean-up of downtown for the 1984 Olympics and in 1987 prior to the Pope's visit to Los Angeles). The conflict over the use of space focused attention on the social status of the homeless. While self-help agencies talked of a sense of community in the camps, the police referred to them as breeding grounds of crime and the narcotics trade. At one end of the spectrum was Love Camp, a co-operatively run encampment on Towne Avenue that pooled resources to rent portable

toilets, shared chores such as cooking, cleaning and firewood search, and ran a newsletter outlining to other homeless groups how to establish a 'homeless block association'. At the other end, across from Jack's Market, was a pool of 'failed robber barons' who extorted money from anyone trying to pass by.

The homeless sought to bring their situation into the political spotlight by, for example, establishing a tent city on state land across from City Hall in the run-up to Christmas 1984. They also confront the new, gentrified users of their turf. On Fourth and Main, where office and government buildings encroach on skid row, clusters of homeless people engage with professionals and office workers on a daily basis. Fashion commentary has become a running part of street repartee, as a way of acknowledging, challenging and reversing the stigma of homelessness: 'Lady, your suit went out of style five years ago. I know because I designed it.' Such strategies invert the relationship between commentator and subject, albeit temporarily, and challenge the meaning of the urban space they both occupy in different ways.

Source: adapted from S. Ruddick (1996) *Young and Homeless in Hollywood* London: Routledge

---

unemployed and underemployed people are often unable to move out of the slum dwellings because they cannot afford to devote more of their income to rent. Continued migration of lower-income and, especially in the USA, ethnic minority groups into the urban housing system also increases demand for the relatively small number of units vacated and can thus maintain price and rent levels beyond the means of many households.

A final criticism of filtering is that, unmitigated, the process will lead to the permanent concentration of poorer households in the worst housing, and such distributional consequences are unacceptable in many countries. The concept of filtering has played an important part in the formulation of housing policy in the USA. Until the early 1930s Britain also relied almost entirely on the filtering process for improving the housing conditions of the working classes. As we have seen, not until the 1930 Housing Act (the Greenwood Act) did the government officially acknowledge that the housing problems of the lower-income groups were unlikely to be resolved through private house-building alone.

CLEARANCE

In 1955 the UK government estimated a total of 850,000 slum dwellings and proposed an annual clearance rate of 75,000. Unofficial estimates that included the continuing obsolescence of Victorian housing stock suggested a replacement rate of 200,000 per year as a more accurate requirement.[27] By 1979 1.5 million dwellings had been demolished or closed and 3.79 million people relocated.

This was a remarkable quantitative achievement, and there is no doubt that the slum clearance and redevelopment programmes provided a superior housing and residential environment for most families. However, the scale of the programme, the timing of the process, the effect on existing communities and the stress imposed on residents have all been criticised. Slum clearance can be a lengthy process and the delay and uncertainty that often surround a programme can cast a shadow of **planning blight** over a neighbourhood and exacerbate the disruptive effect. In Britain it generally took two years for a compulsory purchase order (CPO) to be confirmed and another two years for most of the residents to be rehoused. In the period prior to confirmation of the order, repair and maintenance work practically ceased, and by the time residents were eventually rehoused 'their homes and the state of the surrounding streets are naturally far worse than when the public-health inspector represented the homes as unfit two years or more previously' (Gee 1974 p. 6).[28] Delays caused by national cutbacks in house-building or failure at the local level to tie demolition to the completion of new dwellings intensify the strain on residents.

In Liverpool, for example, during the late 1960s and early 1970s the rate of clearance outpaced the capacity to rebuild in demolished areas. The resultant expansion of vacant land was particularly marked where sites were set aside for non-residential purposes such as public open space, community centres or schools for which finance was in short supply, or for private industrial or commercial development for which demand was low.

Often the policy decision to clear an area would be made five to ten years before a CPO was issued, and although most authorities kept detailed clearance programmes confidential, the probability of an area being affected was advertised by the reduction of public-service investment and by the council buying property and leaving it unoccupied. Empty houses attract the attention of antisocial activities, including those of squatters, fly-tippers, vandals and vermin. Such physical conditions have a demoralising effect on the remaining residents of an area.

The extended nature of the redevelopment process would also result in disruption of friendships and neighbouring patterns, leaving individuals isolated. Feelings of insecurity arising from the physical and social break-up of a community, the loss of friends and relatives, and the disappearance of familiar features in the environment were compounded by the uncertainty over the future caused by a lack of communication between redevelopment authorities and residents. The majority of families relocated from slums moved into council housing, most of which was built on the edge of cities. These overspill estates were remote from the city centre, poorly served by public transport and, initially at least, lacking in amenities.

During the 1960s the need to increase the number of houses to accommodate those displaced by slum clearance and comprehensive redevelopment led to the introduction of high-rise building. Multi-storey flats could be erected speedily at high densities on inner-city slum clearance sites and at a cost that, until 1968, was underwritten by government subsidy. Many tenants welcomed the modern living conditions in the tower blocks, but over time major disadvantages appeared in the form of lift breakdown, poor maintenance of communal areas, lack of facilities for young children and isolation of the elderly. In addition, the land economies proved to be less and the construction costs greater than anticipated. A reduction in the Exchequer's multi-storey subsidy meant that by the mid-1970s most local authorities had retreated from high-rise building.[29]

The clearance and redevelopment strategy was also criticised for disrupting working-class communities. Moreover, there was increasing opposition from the former tenants of condemned properties, who often had to pay more for their council house than they had been paying in tenancies protected by rent controls. Also, residents of clearance areas were often dissatisfied with what they saw as a lack of choice in their rehousing by the local authority, with some municipalities imposing a limit on the number of housing offers they were prepared to make in order to maintain the speed of the clearance programme. Owner-occupiers in designated slum clearance areas also protested about having to accept what many saw as a reduction in housing status, as compensation provision was rarely adequate to enable households to purchase alternative housing. The slum clearance and comprehensive redevelopment era in the UK ended in the mid-1970s in the wake of government concern over costs, growing public pressure for greater participation in planning decisions and the realisation that, with 1.8 million unfit dwellings, 2.3 million lacking one or more basic amenities (such as a sink, washbasin or inside w.c.) and 3.7 million in a state of disrepair, clearance and rebuilding alone were not sufficient to address the problem of inadequate housing within a reasonable time scale.

URBAN RENEWAL IN THE USA

The adverse consequences of the urban renewal process have also been revealed in US cities. The principal complaints have centred on the quality of the new accommodation and the increased costs of replacement housing. Underlying the first issue is the fact that unlike those in the UK, US local authorities are not obliged to provide alternative housing for those displaced by renewal. In a survey of clearance programmes in forty-one cities Anderson (1964) found that 60 per cent of displaced households relocated in other slums, and a general conclusion of US clearance activity is that most households move to neighbourhoods similar to those from which they have been moved, usually on the fringe of the clearance area and often scheduled for clearance in the near future.[30] As Hartman (1964 p. 278) concluded, for many households relocation means 'no more than keeping one step ahead of the bulldozer'.[31] In many US urban neighbourhoods *racial tipping* occurred as displaced black families moved *en masse* into formerly white communities. This aggravated racial tensions and conflicts, and contributed to the racial unrest in US cities.

The cost of urban renewal for displaced residents is partly explained by the strength of opposition to the concept of public housing, which ensured that when urban renewal was introduced under the 1954 Housing Act it was defined broadly to embrace redevelopment for non-residential uses, for which 10 per cent (later raised to 35 per cent) of project grants could be used. With the administrative latitude allowed, the proportion of commercial urban renewal could be pushed up to two-thirds. In effect, public funds 'subsidised the purchase of prime land by private entrepreneurs'.[32] Understandably, private developers co-operated enthusiastically with the urban renewal programme. By the end of 1964 970 different redevelopment projects had been approved, accounting for a total of 36,400 acres (14,700 ha) of inner-city land. A major problem for low-income housing provision was the inability of local governments to direct the nature of the redevelopment process. In practice, many slum areas were cleared to allow the expansion of libraries, colleges, hospitals and commercial luxury housing projects. The West End project in Boston, for example, cleared twenty acres (8 ha) of working-class private housing and replaced it with 2,400 new dwellings designed almost entirely for middle- to upper-income occupancy, with only a 150-unit housing project for elderly people directly aided by federal subsidy.[33] Downs (1970) estimated that households displaced by urban renewal suffered an average uncompensated loss amounting to 20–30 per cent of one year's income.[34] Increased housing costs were imposed largely without any direct consideration of the ability or desire of households to absorb these costs.

The residents of clearance areas are typically among the poorest and least powerful members of society and, in the USA, most clearance zones are located in ethnic-minority neighbourhoods. In such instances, racial discrimination can compound the difficulty of relocatees' obtaining decent affordable housing. The evidence of racial discrimination in the allocation of housing in the USA is unequivocal.[35] The practice has been furthered

by the activities of property developers, estate agents and lending institutions as well as by local government segregation ordinances, restrictive covenants on building occupancy and exclusionary zoning laws. Since the 1968 Civil Rights Act, fair housing legislation has attempted to make discrimination on the grounds of race, colour, religion or national origin illegal in connection with the sale, rental or financing of housing and land offered for residential use. However, the enaction and the enforcement of a law are not the same thing, and housing-market discrimination on racial grounds continues. For the victims this means adverse effects on rents, house prices and access to job opportunities, in addition to the more obvious physical segregation of ethnic groups. In general, African-Americans typically pay between 5 per cent and 20 per cent more than white households for the same housing package, and are less likely to be home-owners and therefore to receive the capital gains ensuing from that tenure status.

REHABILITATION

In the UK the 1969 Housing Act marked a shift away from clearance and redevelopment towards rehabilitation of existing dwellings as the main tool of urban renewal. The 1969 Act extended the system of housing improvement grants and introduced the General Improvement Area (GIA), under which local authorities could undertake environmental improvements financed by the Exchequer. Between 1969 and 1979 1.9 million improvement grants were taken up. However, several criticisms could be levelled at the programme:

1. Although the lack of basic amenities was most serious in the pre-1919 housing stock, the proportion of grants taken up by these dwellings was less than for inter-war houses. Similarly, owner-occupied housing accounted for 70 per cent of grants in 1973 even though private rented property was most in need of improvement.
2. In social terms the effects of the improvement grant boom were largely regressive, with the better-housed and better-off sections of the community deriving most benefit.[36] In London the majority of improvement grants went to landlords and property developers (the proportion varying between 40 per cent and 70 per cent in different boroughs), and in parts of west London the take-up of grants led directly to gentrification and the displacement of working-class residents. Since properties with vacant possession are worth

considerably more to a landlord than their value with sitting tenants (the ratio of 2:5 in West London in 1984 translated into a difference of more than £100,000), the incentive to displace tenants and rehabilitate, to increase value even further, led at times to harassment by neglect until tenants had little option but to vacate. Some property developers made huge profits from the grant-aided conversion of large multi-occupied houses into small self-contained flats that were sold or let to higher-income groups attracted to selected areas.

3. The impact of GIAs was limited. While in council-housing areas public ownership ensured systematic improvement on a block contract basis, private-sector GIAs (which accounted for two-thirds of those designated between 1969 and 1973) made a disappointingly small contribution, with only 18 per cent of houses in need of improvement actually improved by 1973.

4. In contrast to the gentrifying areas of London, the provisions of the 1969 Act had little impact in urban areas where there was no middle-class demand for rehabilitation. In many northern industrial towns local authority intervention was minimal and few GIAs were declared.

Revision of housing-improvement strategy led to the 1974 Housing Act, which introduced Housing Action Areas (HAAs) to tackle small areas (up to 300 houses) of the worst housing conditions within a five- to seven-year time scale, leaving GIAs to deal with better areas. The Act also envisaged a greater role for **housing associations** to combat the abuses of the improvement grant system by private landlords. The effect of the legislation is apparent in the fall in grants taken up by private owners from 238,000 in 1973 to 66,000 in 1979, while grant approvals to housing associations increased from 4,000 to 19,000 over the same period. By the early 1980s 500 HAAs covering 56,000 dwellings had been declared. However, this amounted to only 3 per cent of the pre-1919 housing stock. The scale of the remaining problem was revealed in the 1986 English House Condition Survey, which found 909,000 unfit dwellings (4.8 per cent of total stock), 463,000 which lacked one or more of the basic amenities, and 2.4 million in poor repair.[37] The 1989 Local Government and Housing Act replaced GIAs and HAAs with Urban Renewal Areas. This recognised that poor housing was but one component of disadvantaged urban environments and set out to tackle local economic regeneration as well as substandard housing.

Urban renewal areas (URAs) are larger than GIAs or HAAs and must meet the following conditions:

1. They must have a minimum of 300 properties.
2. Over 75 per cent of dwellings must be privately owned.
3. At least 75 per cent of dwellings must be considered unfit or must qualify for improvement grants.
4. At least 30 per cent of households must be in receipt of specified state benefits.

By 1995 eighty URAs had been designated, comprising 122,000 dwellings in forty-six local authorities. Initially, grants for repairs to housing were mandatory (demand-led) but since 1996, owing to the flood of applications, they are now awarded at the discretion of the local authority.

Poor housing is not confined to the older privately owned or rented stock. From a total public-sector stock of 4,654,000 dwellings in 1985, 3,836,000 (82 per cent) required repair or improvement. The Priority Estates Project attempted to achieve a lasting improvement in conditions on 'difficult to let' estates by directly involving tenants in estate management. This initiative provided the basis for a more broadly based programme of estate action launched in 1985 with the aim of:

1. Bringing empty properties back into use.
2. Reducing levels of crime and 'incivility'.
3. Encouraging tenant participation in decision-making.
4. Diversifying tenure.
5. Attracting private investment.

By 1991 350 schemes were in operation and £2,000 million had been diverted into the programme. An assessment of six early estate-action schemes identified problems in involving residents in management as well as the more general difficulty of employing physical planning in pursuit of social and economic goals.[38] Reflecting the substantial cost of renovating local authority housing stock in the UK (estimated in 2000 at £19 billion over ten years), local authorities have been encouraged to transfer council stocks to 'registered social landlords', a collective term for housing associations and other approved not-for-profit housing organisations.[39] In 2002 the city of Glasgow, following a ballot of tenants, transferred its entire stock of 90,000 dwellings to a housing association, thereby relieving the city of a heavy housing debt burden and repairs backlog.

ALTERNATIVE HOUSING STRATEGIES

An international review of alternative housing strategies reveals a wide range of schemes designed to overcome the shortage of decent and affordable housing. These include:

1. Equity sharing, in which the occupier owns part of the equity in the house and rents the rest from a local authority, with the option to buy the remaining equity as they can afford it.
2. Sale of public land to private developers on condition that it is used to provide low-cost 'starter homes'.
3. Local authority building for sale, or improvement of older houses for sale.
4. Sheltered housing, a common form of accommodation for the elderly in Britain. The Anchor Trust houses 24,000 tenants in 21,000 sheltered units, two-thirds of tenants being dependent on state benefits.
5. Self-build housing, whereby the eventual occupiers purchase a plot and organise construction. It accounts for 80 per cent of detached housing in Australia, 60 per cent in Germany and 50 per cent of all new building in France.[40]
6. Homesteading, in which the 'sweat equity' of people's self-help efforts reduces the amount of cash the homesteader is required to put into the scheme. In practice, the skilled nature of the work involved often requires the use of contractors. Although initially seen as a means of providing low-income homes, the high costs of rehabilitation and the skills required work against this. A further problem is that the selection of homesteaders depends on their ability to meet the costs rather than on their housing needs, with the result that, unless restrictions can be applied, gentrification may take place, in contradiction of the original spirit of the idea.

The notion of housing improvement through a combination of self-help, private investment and limited government intervention is particularly strong in the USA. More than 4,000 community-based non-profit housing organisations (CBHOs) have grown up to undertake housing projects for low- and moderate-income residents considered too small or too great a risk by financial institutions and other for-profit developers. The capital is supplied by special housing trusts established by state and local governments and by foundations such as the Social Initiatives Support Corporation and the Enterprise Foundation.[41] In addition, the Neighborhood Reinvestment Corporation, established in 1978, organises a modest Neighborhood Housing Services (NHS) programme in 137 cities, providing loans and advice to CBHOs undertaking housing rehabilitation.

The criteria for selecting a neighbourhood for an NHS programme are:

1. Housing must be basically sound but showing warning signs of lack of maintenance and deterioration.
2. Mortgages and home loans must be difficult to obtain in the area.
3. At least 50 per cent of the dwellings must be owner-occupied.
4. The area must have distinct boundaries and be large enough to stimulate the imagination of potential participants but small enough to allow for early visible success. On average, NHS areas have 4,900 dwellings, 2,300 buildings and 12,500 people.
5. The median income of the residents must average 80 per cent of that of the city.

We can note the similarity between these criteria and those used to denote URAs in the UK.

The involvement of the private sector is central to the NHS approach. The NHS programme would not be applied to areas that had declined too far to ensure the participation of private investment. The selection of areas with a good chance of success is considered necessary to maintain the credibility of the programme. An attempt to transfer the NHS idea to the UK in the 1980s failed, owing to the designation of areas of very poor housing and the lack of a significant commercial component in redevelopment.[42] Non-profit CBHOs have enjoyed success in a number of US cities including, Portland OR, where the Burnside Community Council used foundation grants to gain control of 40 per cent of the downtown single-room occupancy units (SROs) which were rehabilitated and re-let at reduced rents. For CBHOs to have national significance, further resources and central leadership are required. One option recommended by the National Housing Task Force (1988) is for the federal government to set up a national corporation to promote CBHOs with the emphasis on encouraging benevolent lending (by industry, corporations, Churches and foundations) to finance low-income housing.[43]

BALANCED COMMUNITIES

In the USA a goal of the Hope VI programme is to create mixed-income communities by breaking up concentrations of poor public housing and encouraging middle-income households into redeveloped neighbourhoods. Between 1992 and 2005 more than 120,000 public housing units were demolished but only 40,000 'mixed income' units were built.[44] Despite some notable success in raising medium incomes, labour force participation and property prices,[45] empirical research on Hope VI sites in several cities has highlighted 'the difficulty of attaining an income mix that will lead to meaningful social interaction across social class lines' (Varady *et al.* 2005 p. 163).[46]

In the UK the aim of creating 'balanced communities' is a central plank of government urban policy. This is expressed in official statements that a key requirement of sustainable communities is a 'well integrated mix of decent homes of different types and tenures to support a range of household sizes, ages and incomes'[47] and in the view that 'communities function best when they contain a broad social mix'.[48] As with the US Hope VI experience, research suggests that government attempts to 'socially engineer' balanced communities

have achieved only modest success, with little interaction between the market-price inhabitants and the affordable-housing residents.[49] In practice the ideological and moral argument that mixed-income communities can offer better opportunities for lower-income families, by for example diluting the 'area effect' of concentrated deprivation (see Chapter 15) runs up against the reality of residential sorting based on socio-economic class (see Chapter 18). This suggests that, government policy notwithstanding, socio-spatial segregation will remain a characteristic of the residential landscape of our cities for the foreseeable future.

As our discussion of housing policy and problems indicates, the goal of providing a decent home in a reasonable living environment at an affordable price for all citizens is an ongoing challenge that can be addressed only in part by 'housing policy'. The housing problem is but one element in the complex of social and economic difficulties that confront disadvantaged people and places in urban society.

Plate 11.3   Failed public housing projects, such as Cabrini Green in Chicago, are being cleared under the Hope VI programme

FURTHER READING

BOOKS

M. Anderson (1964) *The Federal Bulldozer* Cambridge MA: MIT Press

G. Daly (1996) *Homeless* London: Routledge

N. Gallent and M. Tewdwr-Jones (2007) *Decent Homes for All* Basingstoke: Palgrave Macmillan

S. Harriot and L. Matthews (1998) *Social Housing: An Introduction* London: Longman

D. Mullins and A. Murie (2006) *Housing Policy in the UK* London: Routledge

A. Ravetz (2001) *Council Housing and Culture* London: Routledge

D. Varady, W. Presier and F. Russell (1998) *New Directions in Urban Public Housing* New Brunswick NY: CUPR Press

JOURNAL ARTICLES

R. Farley, C. Steeh, M. Kryson, T. Jackson and K. Reeves (1994) Stereotypes and segregation: neighborhoods in the Detroit area *American Journal of Sociology* 100(3), 750–80

G. Galster (2008) US housing scholarship, planning, and policy since 1968 *Journal of the American Planning Association* 74(1), 5–16

C. Hartman (1964) The housing of relocated families *Journal of the American Institute of Planners* 30, 266–86

D. MacLennan and A. More (2001) Changing social housing in Great Britain *European Journal of Housing Policy* 1(1), 105–34

P. Malpass (2004) Fifty years of British housing policy *European Journal of Housing Policy* 4(2), 209–227

J. McDonnell (1997) The role of race in the likelihood of city participation in the United States public and Section 8 existing housing programmes *Housing Studies* 12(2), 231–45

J. Smith (1999) Youth homelessness in the UK: a European perspective *Habitat International* 23(1), 63–77

L. Takahashi (1996) A decade of understanding homelessness in the US *Progress in Human Geography* 20(3), 291–310

E. Wyly *et al.* (2008) Subprime mortgage segmentation in the American urban system *Tijdschrift voor Economische en Sociale Geografie* 99(1), 3–23

KEY CONCEPTS

- Rent control
- Owner occupation
- Council housing
- Private-rented sector
- Negative equity
- Slum clearance
- Housing rehabilitation
- Filtering
- Housing associations
- Social housing
- The right to buy
- Residualisation
- Difficult-to-let estates
- Turnkey housing
- Fair rent
- Housing affordability
- Homelessness
- Planning blight
- Urban renewal
- Discrimination
- Community-based housing associations
- Balanced communities

STUDY QUESTIONS

1. Identify and explain the changing tenure distribution of housing in the UK.
2. Illustrate the role of the public sector in the provision of housing in either the UK or the USA.
3. Examine the arguments for and against the sale of council housing. Consider the effect of this policy on the public sector.
4. Map the main areas of owner-occupied, privately rented and public-sector housing in a city with which you are familiar. Comment on the revealed distributions.
5. Critically examine the major strategies employed by governments to improve the quality of the national housing stock.

6. Identify the problem of homelessness in Western cities and consider possible strategies for its alleviation.
7. Critically examine the arguments for and against balanced mixed-income communities. Consider whether it is feasible for governments to create such communities in contemporary urban contexts.
8. Between 2003 and 2025 the population of the USA will grow by almost 60 million. How and where will these people find housing?

PROJECT

The desirability of public-sector housing varies across a city. Plot the main areas of public housing on a map of your city or a city with which you are familiar. Consult the records of the public housing authority to identify the differential quality of housing in the public sector. This can be gauged from statistics on the number of transfer requests received from tenants, vacancy levels, levels of rent arrears, and proportion of abandoned housing. Explain why some areas of public sector housing are more desirable than others.

12
Urban Retailing

Preview: shopertainment; spaces and places of consumption; the spatial switching of retail capital; urban retail forms; the changing structure of urban retailing; suburban shopping centres in North America; retail decentralisation and the central business district in the USA; concentration versus decentralisation in Britain; town-centre shopping schemes; retail decentralisation in Britain; the impact of regional shopping centres; retail recentralisation in the UK; disadvantaged consumers

INTRODUCTION

Retailing land uses constitute a significant part of the urban environment in all developed economies. In the UK there are almost 350,000 retail outlets plus 200,000 outlets devoted to services (such as hairdressers and dry-cleaners) usually found in shopping centres. The amount of floor space devoted to retail and service activities in Britain is over 800 million ft^2 (74 million m^2) and the retail sector employs over 2 million workers, with retail sales contributing 25 per cent of Britain's gross domestic product.[1] In the USA, where retailing and wholesaling activity employs 20 per cent of the national work force, retailing accounts for over $1.5 trillion in sales annually in over 1.5 million establishments,[2] while in the Canadian economy retail sales absorb one-third of disposable income.[3]

The geography of urban retailing has been studied from two main perspectives. From a cultural perspective retailing is viewed as a particular form of consumption, while from an economic perspective attention is focused on the physical construction of retailing environments.

In this chapter we employ both cultural and economic perspectives to illuminate the geography of retailing in the contemporary city. In the first part of the discussion we explain the emergence of 'shopertainment' as a mode of consumption, and identify different spaces of consumption. The focus then turns to the relationship between retailing and urban form.

We examine the effects of spatial switching of retail capital, and identify the different forms of urban retail areas and the changing structure of the urban retailing system. Attention is then devoted to the growth and nature of suburban and out-of-town shopping centres and the impact of this decentralisation dynamic on central shopping areas in British and North American cities. We also consider the role and effectiveness of UK public planning interventions designed to protect the traditional position of the town centre in the urban retailing hierarchy.

SHOPERTAINMENT

For most people shopping is primarily a means to satisfy basic needs. In addition, for many in contemporary urban societies, shopping is, at least in part, a social activity (often referred to as 'retail therapy'). Retailing along with dining, entertainment, and education and culture, is one of the principal consumption activities in the post-industrial/postmodern metropolis. For Hannigan (1998 p. 89) the overlap of these consumer activity systems has given rise to 'three new hybrids which in the lexicon of the retail industry are known as shopertainment, eatertainment and edutainment'.[4] The modern concept of linking shopping and entertainment experiences derives from the post-war suburban malls that have increasingly incorporated leisure facilities. The West Edmonton mall is a world

Plate 12.1 Reconstructing reality: the Rialto bridge, Venice (*upper*) and Las Vegas NV (*lower*)

of shopping, entertainment and social space 'where Spanish galleons sail up Main Street past Marks & Spencer to put in at New Orleans'.[5] Retail and entertainment also merge in festival market-places where shopping and dining experiences constitute the entertainment in a visual environment that projects an aura of historic preservation. Such novel consumption spaces capture the customers' imagination and potential spending-power by permitting them to engage in simulated forms of non-shopping entertainment. In Seattle the retailer Recreational Equipment contains a 65 ft-high free-standing artificial rock for climbing, a glass-enclosed wet stall for testing rain gear, a vented area for testing camp stoves and an outdoor trail for mountain-biking. In Niketown, New York, customers can take part in a range of athletic pursuits as well as watch giant video screens of sports events. As Crawford (1992 p. 16) observed, 'shopping has become intensely entertaining and this in turn encourages more shopping'.[6] This convergence is described as shopertainment.

SPACES AND PLACES OF CONSUMPTION

Retail consumption takes place in a variety of settings, including the home, informal markets, stores, shopping streets and malls.

THE MALL

Shopping malls have been a feature of US retail geography since the world's first fully covered mall opened in Southdale MN in 1956. Promoted as 'tomorrow's main street today' and as 'a whole new shopping world in itself' the centre provided a prototype for most subsequent mall developments. During the peak period of the 'malling of America', from 1960 to 1980, almost 30,000 malls were constructed. This process culminated in the development of super-regional mega-malls in West Edmonton and Bloomington MN (see below). The proliferation of shopping malls has also focused attention on the people who frequent them. As well as those who come to shop malls have become a 'hangout' for adolescent 'mall rats', a 'third place' beyond the home and work/school environments where they can congregate, commune and 'see and be seen'. Many malls open early in the morning to cater for 'mall walkers' who like to practise their keep-fit activities in the safety of the centres.[7] The mega-malls of North America and Britain's regional shopping centres market themselves as 'open to all' and provide a range of activities from tea dances to teenage school-break activities in order to support this image. As discussed in Chapter 7, however, public use of mall space is regulated by surveillance systems designed to exclude 'undesirables'.

The ability of the mall concept to adapt to changing market conditions since the late 1980s is evident in the development of new landscapes of consumption, such as specialty centres, downtown mega-structures and festival market-places. The specialty centre is an 'anchorless' collection of up-market shops with a particular retail and architectural theme. It is prone to 'quaintification' with typical designs in North America including Mediterranean villages (Atrium Court, Newport Beach CA), mining camps (Jack London Village, Oakland CA) and Spanish-American haciendas (the Pruneyard, San José CA). The downtown mega-structure, on the other hand, is a self-contained complex that includes retail functions, hotels, offices, restaurants, entertainment, health centres and luxury apartments (e.g. Water Tower Plaza, Chicago IL). These worlds ensure that the needs of affluent residents, office workers, conference delegates and tourists can be met entirely within a single enclosed space. The festival market-place combines shopping and entertainment with an idealised version of historical urban community and the street market. The market-place is typically decorated with antique signage and props upon which the modern consumer can act out some history. Street entertainers, barrow vendors and costume staff often support the image. As well as examples in North America (e.g. South Street Seaport, New York City; Faneuil Hall Marketplace, Boston MA; Fisherman's Wharf, San Francisco CA), specialty centres and festival market-places have become familiar spaces of consumption in UK cities (e.g. Covent Garden, London; Glasgow's Princes Square and the Corn Exchange in Leeds).

The development of shopping malls in the form of streets acknowledges the traditional role of the street as a locus of retail consumption. Universal Studios' City Walk in Los Angeles and the Fremont Street Experience in Las Vegas both seek to create the diversity of the street in the form of a mall. City Walk, part of MCA's 'Entertainment City', aims to capture the 'real' feel of a Los Angeles street, with boutiqued façades borrowed from Melrose Avenue, 3-D billboards copied from Sunset Strip and a faux Venice Beach, complete with sand and artificially induced waves. The sanitised version of street life is promoted as a safe alternative to traditional Los Angeles streets, especially for family shopping.

THE STREET

The street was the original retail space in cities and remains an important social space. The shopping street can be a destination as well as a thoroughfare. Some streets enjoy a reputation as specialised shopping venues, and retailers are willing to pay high rents to have a presence in such prestigious locations. Bond Street in Mayfair has long been one of the pre-eminent locations for London's luxury shops but in the mid-1990s this dominance was given a renewed emphasis as large numbers of foreign fashion-designer stores were attracted there. Indigenous UK designers with less financial muscle were often relegated to locations in less prominent streets in Covent Garden and South Kensington. The explanation for this trend involves both economic and cultural factors. Physically there is nothing peculiar about Bond Street; it is not a main street, nor is it pedestrianised. It has no particular attractions along its length and is not close to the main

tourist attractions in London. Yet, as in the case of Madison Avenue, New York, Bond Street has become a 'branded street' for international fashion houses who seek to exploit the cultural cachet and market potential of the street. Less exclusive street-based spaces of consumption are represented by traditional street markets (and their modern equivalent of the car-boot sale[8]). Such spaces incorporate both the retailing and the socio-cultural components of the consumption experience. This is particularly evident in ethnic-based spaces of consumption.[9] In many cities such reconstructed and re-imaged ethnic areas play a key role in urban regeneration (see Chapter 16).

HOME

Home shopping began in the USA in the late nineteenth century in the form of mail-order catalogues. (In the 1930s the Sears Roebuck catalogue was even merchandising off-the-shelf houses.) By the early 1990s, however, the market share of traditional 'agency' catalogues had declined to a point where several went out of business. New entrants to this branch of the non-store retail sector (such as Lands' End) have focused on direct marketing aimed at specific lifestyle groups; often those with limited time for traditional modes of consumption, such as professionals or dual-income families. Other forms of home shopping, such as television and Internet sales, are also growing in importance but e-commerce still accounts for a relatively small proportion of total retail sales. In 1998 e-commerce sales in the USA were estimated at $4.6 billion or 0.2 per cent of total retail sales in that year. This was expected to rise to $76 billion or 3 per cent of total retail sales by 2003. In the UK in 2002 e-commerce accounted for £23.3 billion of business representing 1.2 per cent of total sales In the EU e-commerce sales were anticipated to capture $48 billion or 2 per cent of total retail sales by 2002.[10] The convenience of non-store shopping must be set against the loss of the sociality offered by the street, store and mall experience. Though likely to gain an increasing market share, non-place shopping is unlikely to replace more conventional landscapes of consumption.

THE SPATIAL SWITCHING OF RETAIL CAPITAL

As we saw in Chapter 7, 'creative destruction' of the built environment by the spatial switching of capital in search of profit is a leitmotiv of the capitalist city. This 'perpetual struggle in which capital builds a physical landscape appropriate to its own condition at one particular moment in time only to destroy it . . . at a subsequent point in time' (Harvey 1981 p. 113)[11] is illustrated clearly in the changing retail landscape. As we see below, since the mid-twentieth century the process of recurrent innovation has engineered shifts of retail activities from downtowns to suburbs, devalorising retail environments of the inner city, before returning to revitalise some abandoned sites in order to create new spaces of profit extraction. The same profit dynamic has stimulated central city redevelopments, such as Faneuil Hall/Quincy Market in Boston MA (see below), and the reinvention of retail centres in affluent inner-city suburbs as 'urban street' shopping environments. Streetscapes have been rediscovered by corporate US retailers in towns such as Bethesda MD that have attracted 'anti-chain-store chain stores'. These boutiques cater for people who consider themselves too refined and individualistic to shop at the mall or the mass-market big-box stores. Paradoxically, while the boutiques look and feel independently owned they are not.

URBAN RETAIL STRUCTURE

Urban geographers have a particular interest in the spatial and hierarchical organisation of retailing within the city. An early attempt to impose order on the complexity of different types of retail outlet and forms of shopping area identified three main kinds of retail area[12] (Figure 12.1). These are:

1. Nucleated centres that constitute the main shopping areas in the city. A distinction is drawn between the older, unplanned centres of the inner city, including the metropolitan central business district (CBD), and the newer planned suburban centres.

2. Ribbon developments of unplanned retail units that evolve along highway corridors and which are a conspicuous feature of most US cities. These accommodate large space users and affiliated service activities that require good accessibility for passing trade. Typical activities include gasoline stations, automobile dealers, fast-food restaurants and home-supply stores.

3. Specialised function areas comprising retail and service functions that cluster because of mutual attraction or interdependence. These include the theatres, restaurants and bars in an entertainment zone; the automobile row concentration of garages and auto-dealers which benefits from comparison shopping; and the planned or unplanned medical district.

These categories are not mutually exclusive, with many retail functions being found in more than one location, but the classification offers a general description of the major forms of retail location in the city. This early model also provided a basis for subsequent taxonomies, including that proposed by Jones (1991), which differentiates retail areas according to their morphology, location, functional composition, market size and the type of market served[13] (Figure 12.2).

As Figure 12.2 illustrates, inner-city retailing has been dominated historically by the unplanned shopping area, comprising the CBD, speciality product areas, and retail clusters at major route intersections. Planned inner-city shopping areas are a more recent

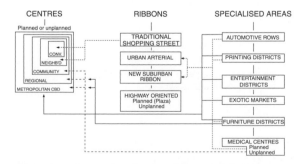

Figure 12.1 Berry's classification of urban retail locations

Source: B. Berry, J. Simmons and R. Tennant (1963) *Commercial Structure and Commercial Blight* Research Paper 85, Chicago: Department of Geography, University of Chicago

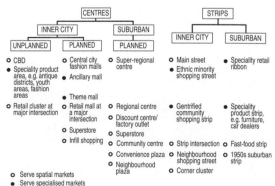

Figure 12.2 A typology of the contemporary urban retail system

Source: T. Bunting and P. Filion (1991) *Canadian Cities in Transition* Toronto: Oxford University Press

phenomenon but have become a common feature in urban North America since the mid-1970s. In the downtown core two types of planned centres have emerged: the central city fashion mall, which has often been the focus of a major urban renewal project; and the ancillary retail complex, which has become the normal ground-floor or underground use in major office, hotel or condominium developments.

Since the 1970s four other forms of planned centre have emerged in the inner city: theme malls, infill shopping centres, retail mall developments at major intersections and superstores or **hypermarkets**. Theme malls are a recent phenomenon and are normally tourist-oriented, occupy waterfront locations and promote a distinctive speciality-product theme and atmosphere. The infill centre is a typical suburban shopping centre that has been developed in the inner city when major retail development companies were confronted with a saturated suburban market and new growth opportunities were restricted to neglected inner-city areas. The development of planned centres at major inner-city intersections represents a modernisation of

traditional retailing at these shopping nodes. The superstore represents the return of the major **supermarket** chains to previously abandoned inner-city locations. Typically these single retail units occupy a minimum of 50,000 ft^2 (4,600 m^2), offer discount prices and a wide product assortment, and rely on extensive trade areas.

As Figure 12.2 indicates, the classification of planned suburban shopping centres is essentially hierarchical, ranging from the neighbourhood shopping plaza to the **regional** and **super-regional shopping centre**. The characteristics of each reflect different numbers of stores, establishment types, total area, selling area, number of parking spaces, customer volumes, trade area size, rental rates, and sales-per-unit area values (Table 12.1). More recent variants of the suburban shopping centre include the mega-mall/recreation complex, and shopping complexes directed to distinct market segments such as the family or young urban professionals.

The retail strip form can be differentiated primarily according to location. In the inner-city main streets, strip intersections, neighbourhood shopping streets

Plate 12.2   This factory-outlet fashion mall outside Las Vegas NV contains over 100 stores, and is linked to a casino and hotel complex to provide a combined retail–recreational experience

TABLE 12.1 HIERARCHY OF PLANNED SHOPPING CENTRES

Measure	Neighbourhood	Community	Regional	Super-regional
Market size (000 population)	20–40	40–200	100–500	500+
Average sales ($ million)[a]	2–5	4–30	10–150	50+
Floor area (gross leasable area) (000 ft^2)	50–100	100–300	300–500	750+
Average sales/ft^{2a}	100	125	200	200
Anchor tenant	Supermarket	Junior dept store	Major dept store	Two or more dept stores
Type of business	Convenience, service	Shopping, convenience	Shopping goods	Shopping goods
Average No. of stores	10–25	20–40	40–100	100+
Average No. of business types	10–20	15–30	20–40	40+
Site area (acres)	5–10	10–30	30–100	100+
Location	Often within a suburban development	Intersection of two arterial roads	Expressway and arterial	Intersection of expressways
Trade area radius	Up to 1.5 miles	1–3 miles	3–7 miles	10+ miles

Source: K. Jones and J. Simmons (1990) *The Retail Environment* London: Routledge

Note: [a]1982 dollars

and corner-store clusters have served essentially the same functions since the early 1900s. In addition, ethnic minority and gentrified community shopping strips have developed specialised retail functions that may serve a metropolitan-wide market, particularly for fashion goods and restaurants. In the suburbs the strip forms of retailing reflect the dominance of the automobile. These include the unplanned 1950s suburban strip shopping malls that were the first stage of retail expansion into the suburbs.[14]

Although other classifications of urban retail forms are available,[15] including those specifically directed to the structure of retailing in UK cities,[16] the classification indicated in Figure 12.2 provides a useful general conceptual framework for understanding the complexity of the urban retail environment. It also underlines the dynamic nature of the urban retail structure as new products, store types, market segments and retail locations emerge as a result of the constant interplay among retailers, consumers, developers and, in the UK, planning authorities.

THE CHANGING STRUCTURE OF URBAN RETAILING

The urban retail pattern is affected by two main forces of change. Changes of the first type are those that have resulted from the effects of market forces and which may be regarded as a 'natural' process of evolution. For most of the post-Second World War period this has been characterised by a general process of decentralisation reflected in the relative decline of retail facilities in the inner city and expansion of trade in the suburbs. Second, sometimes reinforcing these trends and sometimes working against them, there are more specific changes effected by planning. These are most evident in the form of inner-city redevelopment schemes and in the layout of new types of outlying shopping centres.[17]

These changes in urban retail structure have been driven by a range of economic and social forces. The main factors on the *demand side* include:

1. *Changes in residential location.* The single most important factor underlying change in the urban retail structure in the post-war era has been the widespread suburbanisation of population. Despite growth in some inner-city areas related to gentrification, the overall metropolitan population trend has been one of decentralisation (see Chapter 4). In general it has been the younger, richer and more mobile elements of society that have migrated to the suburbs, creating new, large sources of demand in areas where few shopping facilities existed previously. An older, poorer and less mobile population has been left behind in the inner city, where its lower level of purchasing

power has been insufficient to support the surfeit of shopping facilities that remain. These changes have been most pronounced in the US city, where the earlier pattern of retailing has been altered by the growth of new suburban and outlying centres and the decline of central shopping areas. In the UK, by contrast, strict planning controls have sought to restrain the process of retail decentralisation.

2. *Changing consumer attitudes and expectations.* These may be characterised as a general demand for more convenience and comfort in shopping which newer rather than older centres can satisfy more readily by providing one-stop shopping in environmentally friendly malls for car-borne consumers (Box 12.1).

3. *Growth in female employment.* The fact that women account for a higher proportion of the work force than they did a quarter of a century ago has had two major implications for consumer behaviour: first, through the increased purchasing power created by additions to household incomes; and second, through the time constraints imposed on shopping, particularly for women engaged in full-time employment. One result has been an increase in bulk-buying, especially of food, and a reduction in the frequency of shopping trips.

4. *Changing levels of purchasing power.* For three decades after the Second World War most sections of society experienced an increase in purchasing power. Coupled with the growth in population numbers, the resultant rise in levels of consumption stimulated expansion of retail trade. Subsequent downturns in retail expenditure reflecting the condition of the national economy have had a differential impact on retailing. Larger companies have maintained their economic health by increasing their market share at the expense of traditional outlets by offering lower prices and attracting customers from farther afield. As we shall see later, however, although the benefits of cheaper shopping are available to the majority of consumers, there remain large numbers of low-income households, particularly the elderly on limited pensions, single parents and unemployed people, who are unable to buy in bulk or to travel to superstores, hypermarkets and regional shopping centres.

5. *Increased mobility.* The growth of car ownership and car-borne shopping has been another major factor in the trend towards less frequent shopping trips, involving bulk-buying.

BOX 12.1

The Minneapolis skywalk system

The Minneapolis skywalk system is an ingenious device to provide a climate-controlled shopping mall environment in the downtown area. For at least half the year in Minnesota it is unpleasant to be outside: cold, wind and snow fog up your glasses, ruin your shoes and require several pounds of cold-weather gear. Covered shopping malls have a great advantage over unplanned centres, unless a way is found to link together stores and customers. The downtown areas in both St Paul and Minneapolis have developed skywalk systems in order to compete. The skywalk (a tunnel in the sky) links together downtown buildings at the second-storey level, so that shoppers, office workers and visitors can move easily around the city, spending an entire day and visiting several different shopping arcades without going outside. In St Paul the system is supported by the city, which buys a right of way through intermediate buildings. In Minneapolis the skywalk is constructed and managed by individual developers. In both cases the result is a closely integrated and successful downtown area where people can move about easily, ignoring the barriers of traffic or parking garages. In addition, the Minneapolis skywalk system is connected directly to a number of large-scale parking garages on the edge of the downtown that link to the interstate highway leading to the wealthy western suburbs. By integrating several major downtown malls Minneapolis has created a retail attraction that no outlying shopping centre can match. The visitor may get lost temporarily, but it all adds to the sense of exploring an extended bazaar.

Sources: K. Jones and J. Simmons (1990) *The Retail Environment* London: Routledge; J. Byers (1998) The privatisation of downtown public space *Journal of Planning Education and Research* 17, 189–205

On the *supply side* the most significant developments in urban retailing have been:

1. *Structural change.* Most far-reaching has been the expansion of the multiple retailers' share of turnover, largely at the expense of independent shopkeepers. By 1989 the five largest retail firms in the UK accounted for 19 per cent of all retail sales and 41 per cent of food retailing. Independent food retailers increasingly have had to make use of voluntary buying groups (e.g. Mace or Spar) or

cash-and-carry warehouses in order to withstand competition from multiple traders. Declining profit margins on food retailing have also altered operating strategies in retailing. Self-service outlets have mushroomed and the pursuit of economies of scale has promoted the development of larger units. Taken together, these trends have led to a reduction in the total number of retail establishments. In North America the greatest decline in the number of small independent shops occurred in the 1950s in response to the spread of supermarkets and suburban shopping centres. The total number of establishments in the USA fell from 537,000 in 1948 to 319,000 in 1963, while over the same period the number of supermarkets increased from 2,313 to 14,518. Correspondingly, in Britain between 1961 and 1971 the number of shops fell from 542,301 to 504,781, with a further decline to 354,131 by 1980.

2. *New technology*. The use of bar-coding and electronic point of sale (EPOS) terminals linked with in-store computers has a number of advantages for the efficiency of the retail operation, including speeding the sale of goods through a checkout (and checking the work speed of operators), providing an itemised customer receipt and maintaining stock control. Because of the capital costs involved, the benefits of new technology are most likely to accrue to large retailers, further extending their competitive advantage. The potential impact of new technology on urban retailing patterns is considerable, although the rate of adoption of innovations such as teleshopping remains unclear.

In considering the cumulative effects of these demand- and supply-side forces some important differences between the experience in North America and that in the UK must be acknowledged. In particular, the pressures for retail decentralisation have been much greater and more effective in North America than elsewhere. Against a background of a relatively free market economy and the absence of a strong body of state and local government planning laws, a large number of new suburban shopping developments have taken place. At the same time, in many US cities there has been a rapid deterioration of shopping facilities in the inner city and considerable erosion of the former status of the central area. In the UK, in contrast, much stricter and nationally co-ordinated planning policies (see Chapter 8) have attempted to curtail the negative effects of decentralisation. Instead of there being large-scale suburban developments, emphasis has been placed on the redevelopment of the central area and other inner-city shopping facilities. Since the mid-1990s government has assessed proposals for new retail developments against a 'sequential test' whereby first preference is accorded to town-centre sites where suitable sites or buildings for conversion are available, followed by edge-of-centre sites, and only then by out-of-centre sites in locations that are, or can be made, accessible by a choice of means of transport.[18] Further, it is government policy that out-of-centre developments should not be of a scale that would undermine the vitality and viability of town centres that would otherwise serve the community well. As a safeguard, the Secretary of State must be notified of retail proposals in excess of 20,000 m^2 gross floor space (reduced from 40,000 m^2), and of proposals objected to by another local authority from where significant retail spending power would be drawn by the new development. Retail developments in other countries of Western Europe have followed a course between the extremes of the USA and the UK.[19]

THE GROWTH OF SUBURBAN SHOPPING CENTRES IN NORTH AMERICA

The planned shopping centre has been the dominant element in the process of retail change affecting North American cities. The origins of the planned centre can be traced to Roland Park, built in Baltimore in 1907, while the first out-of-town planned centre was the Country Club Plaza built on the outskirts of Kansas City in 1926. However, the main period of growth was in the post-Second World War era. Five development phases, each reflecting a different development philosophy, may be identified:

1. The 1950s were characterised by a 'consequent' development strategy in which the shopping mall was built after the housing in an area (i.e. when the market was established and known). Most plazas were small, and developed independently to serve the convenience of a single residential community. At first the major department stores, with large property investments in the CBD, were reluctant to move into the suburbs. In 1954, however, construction of the Northland Center in suburban Detroit by Hudson's department store company signalled the start of the competitive attack on downtown USA.

2. The 1960s saw a shift in strategy towards 'simultaneous' development, in which the shopping plaza and housing stock were built at the same time, with the planned centre designed to be the focus of the community. The large regional shopping plaza flourished as downtown department stores began to develop suburban branches. During this period, shopping centre development became an 'industry' with established policies and common procedures that led to a 'pattern book' approach to the production of a series of homogeneous shopping environments (Box 12.2).

3. The 1970s witnessed the emergence of the 'catalytic' shopping centre whereby the large mall was seen as a growth pole that would stimulate future residential construction. Typically, a super-regional shopping centre, built on a greenfield site at the intersection of two major expressways, would precede residential development by three to five years. In some cities, such as Detroit, the commercial revitalisation of central areas provided a second major focus for shopping-centre development during this period. By the end of the 1970s the North American shopping centre industry appeared to have reached a point of market saturation, except in the high-growth cities of the sunbelt. Developers pursued a number of alternative growth strategies:

■ A number of first-generation regional centres built in the early 1960s were rejuvenated by enclosing and remixing the outlets.

BOX 12.2

The anatomy of the shopping mall

In addition to the fundamental importance of location, the profitability of the planned shopping mall is dependent to a large extent on the careful selection of tenants and a layout designed to manipulate customer flows and maximise sales. Most large shopping malls incorporate a number of common design principles:

1. The large anchor (generative) tenants are situated at the ends of an internal mall so that customers will visit both ends and in so doing pass by the smaller stores.

2. Access is strictly controlled to minimise the number of customer exit routes at intermediate locations.

3. The mall 'street' is often curved or zigzagged in order to extend its length, increase the number of store fronts and reduce the perceived length of walking distance.

4. Clusters of closely related or competitive activities (e.g. shoe stores) may provide sub-foci within the mall. Food outlets often form a 'gourmet court' with communal tables and seating facilities. Retailers serving different age or income groups are kept apart; stores catering for teenagers, for example, are segregated from up-market retailers. In the largest malls the fashion stores and mass-market retailers may be located on different floors or in different wings.

Typical early mall layouts are shown in the diagrams for Toronto and Montreal.

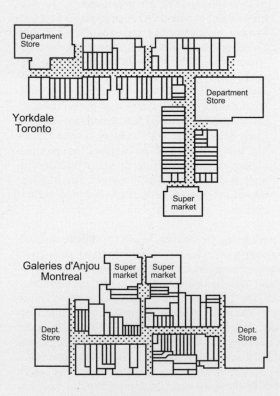

Source: R. Davies (1980) *Marketing Geography* Corbridge: Retail Planning Associates

- Through a strategy of infilling, a number of smaller cities were targeted for enclosed regional or community malls on the edge of town, which resulted in a consequent decline in their downtown cores.
- Developers began to compete with unplanned shopping areas by building strip malls along arterial roads and renting space to low-rent retailers and services.

4. The 1980s saw the emergence of two new retail forms. The first was the power centre, an agglomeration of big-box retailers that often includes **category killers** (e.g. Home Depot), discounters (e.g. Wal-Mart), warehouse clubs (e.g. Costco) and other value-oriented retailers, mostly operating in large industrial-style buildings. The first power centre (the 280 Metro Centre) opened in 1986 in Colma CA between San Francisco and San Mateo. By 1998 there were 313 power centres in the USA with a combined GLA of 266 million ft^2 accounting for over 5 per cent of national shopping-centre sales[20] (Box 12.3). The second significant development saw the emergence of the shopping mall as entertainment centre or tourist attraction, the most ambitious examples being the 3.8 million ft^2 (350,000 m^2) West Edmonton mall, developed in

Alberta in 1985 (Box 12.4), and the 3.5 million ft^2 (325,000 m^2) Mall of America, built in 1992 at Bloomington MN, eight miles (13 km) south of the city centre of Minneapolis. The latter development provides over 400 shops, including four major department stores, on three shopping levels with a seven-acre (3 ha) amusement park at the heart of the mall (Figure 12.3). Super-regional developments of this scale and diversity can challenge the attractions of any downtown area, and reflect the polycentric structure of post-industrial metropolitan America. As early as 1982 there were fourteen metropolitan areas in which a total of twenty-five major suburban retail concentrations exceeded the retail sales levels of the CBD, the trend being

BOX 12.3

Power centres in the USA

A power centre in the USA is a type of super community shopping centre that includes:

1. More than 250,000 ft^2 of gross lettable area (GLA). (The average power centre has a GLA of 375,000 ft^2 and the largest more than 1 million ft^2.)
2. At least one super anchor store with at least 100,000 ft^2 of GLA.
3. At least four smaller anchors with a GLA of 20,000–25,000 ft^2 each.
4. Only a small number of smaller shops with a GLA of less than 110,000 ft^2.
5. Generally an open-air centre.
6. A trading area similar to a regional shopping centre.
7. A unified shopping-centre management.

Most power centres are located in California, followed by Florida, Illinois, New York and Pennsylvania.

BOX 12.4

The West Edmonton Mall, Edmonton, Alberta

With 600 stores and 3.5 million ft^2 (325,000 m^2) of retail space, WEM is the largest shopping centre in North America. It has almost twice the retail floor area of downtown Edmonton and is climate-controlled. (Edmonton is Canada's northernmost metropolitan area.) In size alone, the mall dominates the city's retail environment, but the developer (Triple Five Corporation) has provided further attractions. Recreational facilities (1.4 million ft^2, 130,000 m^2) attract tourists from all over western Canada. Visitors come by bus and plane from as far away as Saskatchewan and British Columbia to see Fantasyland (the amusement park), the Deep Sea Adventure (aquarium), the National Hockey League scale ice rink (where the Oilers practise on occasion) and the five-acre (2 ha) indoor Water Park. Many of them stay in the on-site hotel.

Needless to say, the mall does a lot of business: the Edmonton Planning Department estimates that it accounts for one-third of the city's retail sales and generates 15,000 jobs. The mall cost almost $1 billion and pays $11 million per year in municipal taxes. Central Edmonton, aside from the downtown planned centres, looks devastated. Other major plazas in the city are feeling the pinch as the mall lures away their tenants, and all over western Canada merchants complain that local shoppers are taking their money out of town to the mall.

Sources: **D. Herbert and C. Thomas (1997)** *Cities in Space, City as Place* **London: Fulton; K. Jones and J. Simmons (1990)** *The Retail Environment* **London: Routledge**

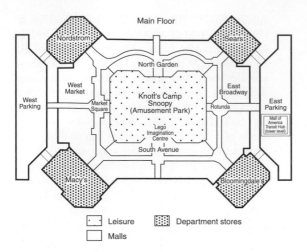

Figure 12.3   Plan of the Mall of America, Bloomington MN

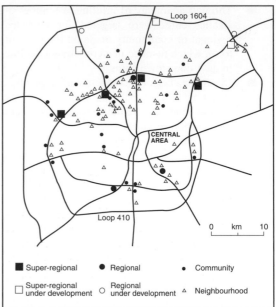

Figure 12.4   Retail centres in San Antonio TX
Source: J. Dawson (1983) *Shopping Centre Development* Harlow: Longman

particularly marked in Atlanta GA with three and Indianapolis IN with six centres. By 1991 there were 1,800 regional shopping centres of over 400,000 ft² (37,000 m²) in the USA, almost all of them in suburban locations. Figure 12.4 provides an example for one city.

5. During the 1990s many markets became saturated owing to the extensive shopping centre building activity of the 1960s, 1970s and early 1980s (Figure 12.5). In addition, escalating construction costs made new shopping centres expensive to build. Owing to these difficulties, and because many of the earlier shopping centres were outdated, the shopping centre industry concentrated on renovation of existing centres rather than on new construction. In the early 1990s three times more centres were under renovation than under construction in the USA. The number of shopping centres grew by 7.5 per cent each year in the 1970s, by 5.2 per cent in the 1980 and by only 2.3 per cent in the period 1990–7. The US shopping centre industry continues to grow, adding over 1 per cent of new retail selling space per year. However, in the 1990s retail capital in the USA entered a new flexible phase, reflected in the reduced production of large malls, and the growth of smaller and more diffuse shopping centres designed to attract a wide range of consumers (see above). The USA retail landscape now comprises a variety of formats, ranging from super-regional centres, mega-malls and power centres to street-based stores in *ersatz* cultural and historical settings.

THE IMPACT OF RETAIL DECENTRALISATION ON THE US CBD

The adverse impact of retail suburbanisation on the CBD was characterised by Berry *et al.* (1963) as 'commercial blight'[21] (Box 12.5). The impact of decentralisation on cities of the US Midwest between 1958 and 1963 is depicted in Figure 12.6. This indicates the more general finding that the greatest competitive impact of retail suburbanisation was felt by the larger metropolitan areas in which the central area lost the bulk of its 'external' market and became increasingly dependent on an 'internal' market comprising the poor, elderly and non-white residents. More recently a study of Charlotte NC found that between 1972 and 1991 the amount of retail space in the central area declined to less than one-sixth of its former size under the impact of three super-regional shopping centres and other suburban retail development.[22]

Several factors indicate that the decline of downtown retailing is not necessarily an inexorable process, however, since:

1. Many city centres retain significant advantages for the provision of specialist shopping opportunities.

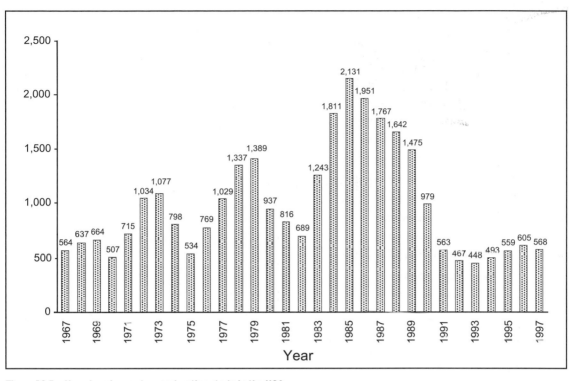

Figure 12.5   New shopping centre construction starts in the USA

BOX 12.5

The nature of commercial blight

Berry *et al.* (1963) suggested that four main kinds of blight can affect the older and weaker elements of the retail system:

1. *Economic blight,* which involves the closure of large numbers of businesses as a result of a reduction in the amount of trade area or purchasing power support.
2. *Physical blight,* which refers to the structural deterioration of buildings, due primarily to age but also because of an inability or unwillingness of owners to undertake maintenance.

3. *Functional blight,* which refers mainly to the obsolescence of small businesses due to the impact of mass merchandising techniques and the growth of automobile shopping.
4. *Frictional blight,* which encompasses a wide range of negative environmental effects created by problems such as traffic congestion, litter accumulation and vandalism of vacant properties.

The combined result of these conditions is the formation of the commercial slum.

Source: **B. Berry, J. Simmons and R. Tennant (1963)** *Commercial Structure and Commercial Blight*
Research Paper 85, Chicago: Department of Geography, University of Chicago

2. The CBD accommodates concentrations of office employment, government functions, entertainment and cultural facilities which bring significant numbers of people into the area.
3. Many city centres have developed an important tourist and convention trade function.

4. Urban renewal has led to the gentrification of formerly declining districts and the creation of retail demand by high-income apartment dwellers. Similarly, the deindustrialisation of the inner city has provided development opportunities on waterfronts for festival shopping districts.

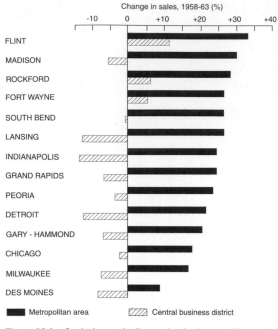

Figure 12.6   Central-area decline and suburban retail growth in the US Midwest, 1958–63

Source: R. Davies (1976) *Marketing Geography* Corbridge: Retail Planning Association

The declining centres of a number of cities in the USA have experienced at least partial revival. A number of mixed office, hotel, convention centre and shopping complexes, exemplified by the Renaissance Center in Detroit MI, have added a distinctive but separate commercial precinct to traditional central city retail areas. Speciality, theme or festival centres have also been developed in a number of city centres, including Quincy Market in Boston MA and Ghirardelli Square on the San Francisco CA waterfront.[23] Although the central shopping areas (CSAs) of many cities are reclaiming some of the trade share lost to suburban centres, this does not signal a return to a former pattern but rather indicates a segregation of the CSA into two distinct parts comprising a growth area of new, specialised shops catering for fashion demands and a deteriorating area made up of older remnants of the retail trade. As we saw in Chapter 7, the recommercialisation of run-down central city areas by private-sector investment can also promote the 'privatisation of public space', as evident in Los Angeles CA[24] and New Orleans LA.[25]

As the suburbanisation wave moves outwards, the need for redevelopment is often more pressing for the older, outlying shopping centres of the metropolitan area than for the central area itself. Table 12.2 shows the changing land values in six different shopping centres in the Chicago metropolitan area over a forty-year period. It is instructive to consider the position of the outlying centre at Sixty-third and Halsted Streets. In the 1950s this was the largest retail centre in Chicago outside the Loop. Located eight miles (13 km) from the downtown area at a point where the elevated transit line connected with a network of bus and trolley (tram) routes, and surrounded by blue-collar and middle-class neighbourhoods, the centre contained several major department stores and hundreds of smaller outlets. In 1952 total retail sales were higher than in any of the new suburban plazas. In the 1960s a period of racial transition commenced in which African-Americans replaced whites, household incomes declined and retail activity reduced sharply. The major department store (Sears) moved out in the mid-1970s. Although 45,000 households still live in the trade area of the centre, the average household income is less than $10,000; 40 per cent of households earn less than $7,500 and 45 per cent of the population are under 25 years of age. Households with cars shop at Evergreen Plaza, a successful mall used by African-Americans six miles (10 km) farther out. The Sixty-third and Halsted centre survives as a cluster of 100 stores, many locally owned, surrounded by abandoned properties and vacant lots. The cycle of decline is in marked contrast to the growth cycle enjoyed by North Michigan Avenue (Box 12.6).

CONCENTRATION VERSUS DECENTRALISATION IN BRITAIN

There is a strong contrast between post-war changes in the urban retailing pattern of Britain and those described for North America. A fundamental difference is the limited amount of suburban development that has been permitted in Britain and the relative protection afforded to the central area and other inner-city shopping areas. There are several factors that explain the application of strict planning controls over retail suburbanisation in Britain:

1. In general, the central shopping area has always held a more dominant position over the rest of the retail system than its counterpart in North America.
2. There has also been a richer historical legacy of buildings to be preserved, as well as a greater mix of non-retailing land uses to be controlled.

TABLE 12.2 CHANGES IN SHOPPING-CENTRE LAND VALUES IN THE CHICAGO METROPOLITAN AREA

Year	Traditional centres (peak land values per front foot, $)		
	State Street, W. Side Madison–Washington (central area)	63rd and Halsted Street (outlying centre)	Church–Orrington Evanston (outlying centre)
1929	22,000	10,000	4,500
1933	14,000	6,500	4,000
1940	10,000	10,000	3,000
1952	17,000	9,000	3,500
1960	17,500	5,500	3,500
1967	22,000	3,000	3,200
	New planned centres (land values per acre, $)		
	Old Orchard	Evergreen Plaza	Ford City
1929	4,000	5,000	3,000
1933	2,250	2,250	1,000
1940	800	1,500	600
1952	2,000	3,000	3,000
1960	50,000	50,000	26,000
1967	152,000	130,680	109,000

Source: R. Davies (1976) *Marketing Geography* Corbridge: Retail Planning Associates

3. More specifically, during the years of rapid post-Second World War suburban development in North America many of the central areas and other inner shopping centres of British towns remained damaged from wartime bombing, necessitating reconstruction and new retail investment.
4. There was also less immediate pressure for retailing to move out of the central area since, as late as the mid-1950s, shopping habits in Britain had still not been affected greatly by the impact of the automobile or mass-selling techniques.

From the 1970s onwards, however, increasing pressure for suburban development was expressed in a growing number of applications for planning permission to build hypermarkets, superstores, discount stores and various sizes of shopping centre on predominantly greenfield sites. The ensuing debate over the desirability of concentration versus decentralisation focused on three main issues:

1. *Economic considerations.* Most planners argue that the competitive impact of new suburban shopping facilities will lead to the same relative decline in the status of the central area and other inner-city retail areas that has been experienced in many US cities. They point to an obligation to prevent this, not least because of the vast amount of public investment in post-war renewal schemes. Against this, many sections of the business community contend that high rents and congestion costs in the traditional retail locations mean that planners are supporting inefficient trading practices that ultimately lead to unnecessary price increases for consumers.
2. *Environmental considerations.* Planners are suspicious of any development that might contribute to urban sprawl and are particularly reluctant to allow new suburban shopping facilities to erode the rural character of the green belt. Opponents argue that this presumes that green-belt land is uniformly attractive, whereas a new shopping facility on a derelict site could contribute to visual upgrading of the landscape.
3. *Social considerations.* Many planners believe that decentralisation will create two different standards of shopping provision, with most of the benefits of suburban developments accruing to a car-oriented middle-class population, leaving the poorer sections of society dependent upon

BOX 12.6

The polarisation of retail provision in central Chicago

For years the Loop in Chicago has been one of the great retail concentrations in the USA. It is not spatially dispersed like downtown Los Angeles, nor mixed in with other land uses like New York. The Loop is based on an intense concentration of access to a large and prosperous market. Enclosed by the elevated transit track, which is itself bounded by the lakefront land uses and the Chicago river, it is also served by underground transit and commuter lines. Offices for the financial district and government are concentrated in the west of the Loop, and continue to expand.

The retail activity is located to the east. In the 1950s six great department stores were located along State Street, on one block after another. In 1948 the Loop generated $6 out of every $1,000 retail sales in the country, one-eighth of all sales in Chicagoland. However, forty years later both Chicago and the Loop have declined in significance: the $6 has dropped to $1, and the 12 per cent to 3 per cent. Of the six great department stores, only Marshall Field and Carson Pirie Scott remain.

What happened? To the usual list of suburban growth, freeway construction, shopping plazas and inner-city decline we can add the dramatic reduction of purchasing power in the immediate trade area south and west of downtown, as shown in the text. At the same time, gentrification and expensive condominiums flourish on the north side of the city. Upscale retailers have detached themselves from the mass markets of the Loop and created a flashy new retailing environment on North Michigan Avenue. It is only a mile or so away but the atmosphere is worlds apart: different customers (different access patterns), different stores (Neiman-Marcus, Saks) and dramatic success. Michigan Avenue has become the greatest fashion retail location between Fifth Avenue and Rodeo Drive, anchored by Water Tower Place (Marshall Field) and the Atrium (with Bloomingdale's). The malls are designed both to exclude unwanted customers and to attract the well-to-do.

Source: **K. Jones and J. Simmons (1990)** *The Retail Environment* **London: Routledge**

minority needs must be addressed, the majority should not be deprived of greater choice.

TOWN-CENTRE SHOPPING SCHEMES

No country in Western Europe has sought to contain the process of decentralisation to the same extent as Britain. As a result of planning policy, much of the post-Second World War pressure for retail growth has been directed into town-centre developments. Between 1950 and 1980 two distinct stages can be identified in the evolution of town-centre shopping schemes (Figure 12.7). During the first period, from 1950 to 1965, precinct development was fostered by local authorities initially prompted by the need for core-replacement schemes in war-damaged cities such as Coventry, Hull and Plymouth. Town-centre developments were stimulated by:

1. The abolition in 1954 of the wartime system of licences for new buildings.
2. Rising personal incomes, which were reflected in a demand for new retail floor-space.
3. Growth in vehicular traffic, leading to local congestion in cities.
4. An increase in the amount of private finance capital available, and a shift in investment focus from residential to commercial property.

The last of these factors is reflected in the greater involvement of the private sector from the early 1960s (Figure 12.7). Most of the new central-area shopping

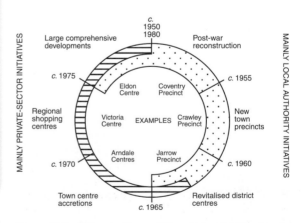

Figure 12.7 Stages in the evolution of town-centre shopping schemes in the UK
Source: D. Bennison and R. Davies (1980) The impact of town-centre shopping schemes in Britain *Progress in Planning* 14(1), 1–104

traditional shopping areas which are likely to run down. A general argument against this is that different standards of retail provision exist already within the urban retail system, and that while

centres were the result of collaboration between a city planning department and a property-development company. The planning authority maintained overall control of the scheme in terms of location, its relation to the existing retail pattern and its integration with other parts of the urban renewal programme, while the centre design and shop layout were left to the property developers, who also managed the precinct as an integrated unit in much the same way as in planned suburban developments in North America. The impact of town-centre shopping schemes is summarised in Table 12.3. The desire of planners to protect existing retail outlets is revealed in the fact that between 1965 and 1980, of 387 new shopping centres built, 315 (81 per cent) were located in town centres.[26]

RETAIL DECENTRALISATION IN BRITAIN

The concentration on the redevelopment of existing retail areas did not eliminate pressure from developers for out-of-town shopping centres, although initially this pressure achieved little result. In 1964 a major planning application for a 1 million ft² (90,000 m²) centre at Haydock, between Manchester and Liverpool, was refused because of the potential impact on existing town centres in the region. During the remainder of the 1960s and 1970s government opposition meant that few proposals for out-of-town regional shopping centres were submitted. In addition, a property-market slump in 1973–4 (Figure 12.8) reduced the amount of capital available for investment in new shopping centres. Only where a suburban shopping development formed the basis of a new district centre in an area of former underprovision or substantial population growth did local planning authorities relax their opposition. The 760,000 ft² (70,000 m²) Brent Cross regional shopping centre was developed in 1977 in suburban

north-west London in an area without major existing facilities, and the first free-standing super-regional centre was built with 1,065,000 ft² (100,000 m²) of retail floor space as the central shopping area for Milton Keynes New Town in 1979.

The 1980s saw a revival of interest in shopping-centre development, with fifty proposals for out-of-town regional shopping centres submitted between 1982 and 1991. This trend was encouraged by an apparent change in central government's philosophy on land-use controls. With the election of the Conservative government under Margaret Thatcher in 1979 there was a general presumption in favour of development, 'enterprise' and employment creation. Government liberalisation of planning controls was taken furthest within enterprise zones (EZs), and in several of these, major retail developments were permitted, as in the 1.63 million ft² (150,000 m²) MetroCentre, built on the site of the ash tip of a disused power station within the Gateshead EZ, and the 1.1 million ft² (102,000 m²) Meadowhall, which replaced a derelict steelworks outside Sheffield (Figure 12.9). In practice, government encouragement

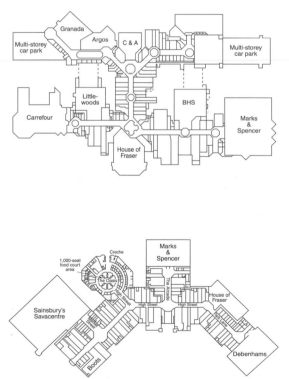

Figure 12.9 The layout of out-of-town regional shopping centres in the UK: (*upper*) the Metro Centre, Gateshead and (*lower*) Meadowhall, Sheffield

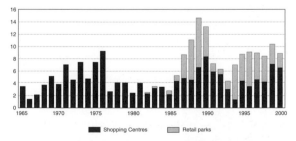

Figure 12.8 British shopping-centre development: total floor space opened each year, 1965–2000 (million ft² gross)

TABLE 12.3 THE IMPACT OF TOWN-CENTRE SHOPPING SCHEMES IN BRITAIN

Economic		Environmental		Social	
Positive	*Negative*	*Positive*	*Negative*	*Positive*	*Negative*
Adds new stock	Reduces old stock	Modernises outworn areas	Changes traditional shopping	Allows for efficient shopping	May favour car-borne shoppers
Accommodates larger, modern stores	Discriminates against small independents	Reduces land-use conflicts	Creates new points of congestion	Provides new shopping opportunities	May limit choice to stereotypes
Increases rates and revenues	Increases monopoly powers	Scope for new design standards	Intrusive effects on older townscapes	Safer	Creates new stress factors from crowds
Creates new employment	Changes structure of employment	Provides weather protection	Creates artificial atmosphere	Provides more comfort and amenities	Attracts delinquents and vandals
Improves trade on adjacent streets	Reduces trades on peripheral streets	Leads to upgrading of some streets	Causes blight on other streets	Concentrates shopping in one area	Breaks up old shopping linkages
Enhances status of central area	Affects status of surrounding centres	Integrates new transport	Causes pressure on existing infrastructures	Potentially greater social interaction	Becomes dead area at night

Source: **D. Bennison and R. Davies (1980) The impact of town centre shopping schemes in Britain** *Progress in Planning* **14(1), 1–104**

TABLE 12.4 MAJOR IN-TOWN SHOPPING CENTRES IN THE UK, 1990–2008

Town	Centre	Year opened	Size (000 ft²)
Leicester	Shires	1991	538
Nottingham	Victoria Centre	1996	820
Reading	The Oracle	1999	700
Southampton	West Quay	2000	800
Romford	The Brewery	2001	510
Solihull	Touchwood	2001	665
Basingstoke	Festival Place	2002	850
Bournemouth	Castlepoint	2003	645
Birmingham	Bullring	2003	1184
Croydon	Centrale	2004	800
Norwich	Chapelfield	2005	676
Doncaster	Frenchgate Centre	2006	720
Glasgow	Silverburn Centre	2007	1017
Leicester	Shires West	2008	772
Motherwell	Ravenscraig	2008	861
Bristol	Broadmead Centre	2008	1196
Liverpool	Paradise Street	2008	1914
London	Westfield	2008	1932

Plate 12.3   The ornate entrance to the 120,000 m² (1.3 million ft²) Trafford regional shopping centre near Manchester

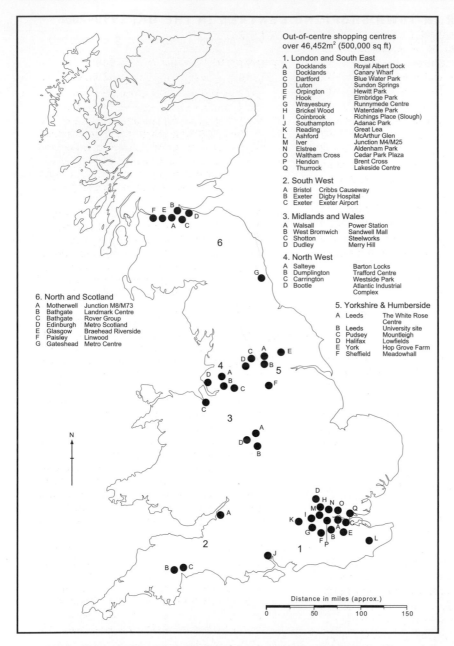

Out-of-centre shopping centres
over 46,452m² (500,000 sq ft)

1. London and South East

A	Docklands	Royal Albert Dock
B	Docklands	Canary Wharf
C	Dartford	Blue Water Park
D	Luton	Sundon Springs
E	Orpington	Hewitt Park
F	Hook	Elmbridge Park
G	Wrayesbury	Runnymede Centre
H	Brickel Wood	Waterdale Park
I	Coinbrook	Richings Place (Slough)
J	Southampton	Adanac Park
K	Reading	Great Lea
L	Ashford	McArthur Glen
M	Iver	Junction M4/M25
N	Elstree	Aldenham Park
O	Waltham Cross	Cedar Park Plaza
P	Hendon	Brent Cross
Q	Thurrock	Lakeside Centre

2. South West

A	Bristol	Cribbs Causeway
B	Exeter	Digby Hospital
C	Exeter	Exeter Airport

3. Midlands and Wales

A	Walsall	Power Station
B	West Bromwich	Sandwell Mall
C	Shotton	Steelworks
D	Dudley	Merry Hill

4. North West

A	Salteye	Barton Locks
B	Dumplington	Trafford Centre
C	Carrington	Westside Park
D	Bootle	Atlantic Industrial Complex

5. Yorkshire & Humberside

A	Leeds	The White Rose Centre
B	Leeds	University site
C	Pudsey	Mountleigh
D	Halifax	Lowfields
E	York	Hop Grove Farm
F	Sheffield	Meadowhall

6. North and Scotland

A	Motherwell	Junction M8/M73
B	Bathgate	Landmark Centre
C	Bathgate	Rover Group
D	Edinburgh	Metro Scotland
E	Glasgow	Braehead Riverside
F	Paisley	Linwood
G	Gateshead	Metro Centre

N

Distance in miles (approx.)

0   50   100   150

Figure 12.10   Major out-of-town shopping centres open or under development in the UK

of competition among retailers did not herald a major relaxation of planning controls over greenfield developments. Between 1985 and 1992 only four super-regional centres opened although several others had received planning permission. By 2000 a further six of the proposed regional centres had opened (Figure 12.10). All these had received planning permission prior to the

change in government policy on retail development in 1996. This new policy goal 'to sustain and enhance the vitality and viability of town centres' was detailed in Planning Policy Guidance Note 6 (PPG6) that introduced a 'sequential approach' to site selection for major new retail developments. This gave priority to development in existing town centres, then to edge-of-town locations, then finally to out-of-town centres sites. In addition, in 1997 an application for expansion of the Merry Hill RSC by 400,000 ft² of retail and leisure space was refused by the Secretary of State. These trends effectively signalled the end of the 'third wave' of retail decentralisation in Britain[27] (see Box 12.7). The fourth and fifth waves of retail decentralisation in the UK are characterised by the rise of retail warehouse parks and factory outlet centres respectively. By 1997 planning consent had been granted for 1,600 **retail parks** (with 230 under construction) and 158 factory outlet centres (with eighty-one under construction).[28] The centres open or likely to open are all within or close to major conurbations and are in areas of industrial or mining dereliction; none of the completed out-of-town regional shopping centres and few of the proposed sites are in the American mould of edge-of-town developments on greenfield sites.

THE IMPACT OF REGIONAL SHOPPING CENTRES IN THE UK

Assessing the impact of a new regional shopping centre is complicated by the difficulty of isolating this effect from that of general economic processes relating to the national or global economy. Nevertheless, several studies have sought to gauge the impact of a new out-of-town development on surrounding retail centres.[29] It is clear that the competitive impact depends greatly on local context.

The MetroCentre, located four miles (6 km) west of Newcastle city centre, is one of the largest enclosed shopping centres in Europe, with 1.63 million ft² (150,000 m²) of retail floor space, an additional 107,000 ft² (9,900 m²) of leisure floor space and 10,000 car-parking spaces. Survey evidence indicates that 12 per cent of shopping trips have been directed to the MetroCentre from Newcastle city centre. This has led to a compaction process in Newcastle, with a refocusing of the CBD on the refurbished enclosed central mall (Eldon Square) and contraction of retailing in peripheral shopping streets. In general, however, large-scale and prescient redevelopment of the city centre retailing area, together with the insertion of an urban metro system in the 1970s, has enabled Newcastle to retain a thriving central retail function in the face of out-of-town competition.

The impact of other regional shopping centres on nearby central areas has been more severe. The central shopping area of Dudley in the West Midlands lost 70 per cent of its market share to the Merry Hill mall,

---

BOX 12.7

Retail decentralisation in the UK

1. *First wave (late 1960s to early 1970s).* The development of out-of-centre and even out-of-town sites of the first free-standing superstores and hypermarkets. By 1993 the existing 868 superstores transacted almost 50 per cent of grocery-shopping trade, and the top five retail companies (Sainsbury's, Tesco, Argyll-Safeway, Asda, Dee-Gateway) controlled 61 per cent of the British grocery market.
2. *Second wave (late 1970s to mid-1980s).* The arrival of retail warehouses selling durable goods (electrical white goods, carpets, furniture, DIY products) and garden equipment. More recently the range has expanded to include clothing, footwear and toys. By 1992 there were over 2,000 retail warehouses.
3. *Third wave (mid-1980s to early 1990s).* The move of high-street chain stores to out-of-town centres in accessible locations, and providing a range of comparison goods shopping that was previously reserved for the town centre. A significant extension of this trend was the development of out-of-town regional shopping centres such as the Gateshead MetroCentre.
4. *Fourth wave (late 1980s to mid-1990s).* The development of the retail warehouse park, of which there were 250 by 1992. Initially, small groups of second-generation purpose-built retail warehouses on edge-of-town sites; later developments exhibit a more co-ordinated layout developed in co-operation with local planning authorities.
5. *Fifth wave (mid-1990s–).* The emergence of factory outlet centres as a further form of retail decentralisation in which groups of leading manufacturers sell products direct to the public. Though these are common throughout the USA, in 1995 there were only four operating in the UK, with planning applications for another twenty.
6. *Sixth wave (2000–).* The ultimate level of decentralisation, home shopping has long been available through catalogue sales but is becoming more widely available via specialised cable television channels and by means of teleshopping in a virtual superstore via the Internet.

with more than half the retailers present in 1986 gone by 1992; some of them moved direct to the new 1,240,000 ft^2 (115,000 m^2) regional shopping centre. In 1997 an application from the developers to expand the centre by the construction of an additional 375,000 ft^2 (35,000 m^2) of gross floor space for retail and leisure uses was refused by the Secretary of State in order to protect the 'vitality and viability' of existing centres. By then, however, the Merry Hill regional shopping centre had effectively replaced Dudley town centre as the main retail focus of the area.[30] In similar fashion Meadowhall has also exerted a significant impact on the CSA in Sheffield. Since the mid-1990s the policy priority in favour of existing retail centres has encouraged developers of 'very large stores' (over 7,000 m^2) to redevelop existing stores (by, for example, converting storage space to sales space and using just-in-time logistics) or to seek to link their retail development to urban regeneration programmes (see below).

RETAIL RECENTRALISATION IN THE UK

Planning Policy Guidance Note 6 has had a significant impact on the pattern of new retail development in the UK. It has restricted the flow of new regional shopping centres, out-of-centre food superstores, and retail parks (see Figure 12.8). In 1990 the amount of new retail floorspace in town centres amounted to 4,419,000 ft^2; in non-town centres 2,448,000 ft^2; and in retail parks 6,057,000 ft^2. By comparison the corresponding figures for 2000 were 3,852,000 ft^2, 1,683,000 ft^2 and 1,953,000 ft^2. In 2000, for the first time since the early 1980s, new shopping floor space in major town centre schemes exceeded that in out-of-town shopping centres and retail parks.

Significantly, PPG6 is intended to apply not solely to retail but to a range of other town centre land uses including leisure and entertainment facilities, offices, and arts, cultural and tourism activities, as well as residential developments in order to promote mixed-use urban regeneration. The policy has been implemented in several large in-town shopping–leisure–entertainment–residential redevelopments, including the Westfield in Shepherd's Bush, London; Shires West, in the heart of Leicester; West Quay, in Southampton;[31] and Broadmead, in Bristol, the last being an effort to counter competition from the out-of-town centre at Cribb's Causeway. The Paradise Street project in central Liverpool was also a response to the loss of retail trade to a nearby out-of-town regional shopping centre, the

Trafford Centre, near Manchester. Undertaken by a public–private partnership (see Chapter 16) the forty-three-acre retail-led mixed-use development will serve to link the city centre with the completed waterfront redevelopment at Albert Dock.

DISADVANTAGED CONSUMERS

The post-Second World War changes in the structure of urban retailing have brought material advantages to a majority of the population through increased choice, comfort and cheapness in shopping. On the other hand, the increasing size and decentralised location of new large stores, disinvestment in smaller supermarkets by the major food retailers who dominate the grocery market, and the decline in the number of small independent shops have contributed to the formation of a class of disadvantaged consumers. This group includes the poor, elderly and mobility-deprived residents of under-served inner-city areas and, in the UK, peripheral public housing estates.[32]

Some local authorities in the UK have sought to address the problems of disadvantaged consumers by seeking to influence the geography of retailing. At the neighbourhood level, in 1980 the borough council of Islington in north London identified a number of 'shopping deficiency areas' within which new retail investment would be encouraged, possibly using the 'planning gain' mechanism (see Chapter 8). In principle this approach could also be employed at the metropolitan level, particularly in situations where retail developers seeking planning permission for new large stores are willing to work with local authorities to tackle social exclusion. As well as bringing employment benefits to areas of long-term unemployment and a low-skilled work force, very large store (VLS) developments can also provide deprived communities with an outlet offering low prices and a wide range of merchandise in areas where the existing retail environment is poor.[33] More generally, however, although local authorities can use planning controls to prevent specific developments, they are less able to direct investment to areas of need. In view of this it is necessary to consider alternative ways of assisting mobility-deprived households.

One possibility is the use of telecommunications and computer facilities to link individual homes or community focal points to a large store or shopping centre some distance away. The first such scheme in the UK, the Gateshead shopping and information service (SIS), was set up in 1980. The scheme aims to improve:

1. The choice of shopping opportunities available to disadvantaged consumers and to provide them with access to stores offering goods at cheaper prices.
2. The access of the relatively housebound to a variety of sources of information on local events, welfare, entertainments, transport timetables and other social services.

The system has 1,000 regular users.[34] Unlike the commercial teleshopping schemes common in the USA, the Gateshead SIS is operated as a joint venture between the local authority and a major retailer. It is aimed at providing a food and convenience-goods shopping service rather than one for non-food luxury items, and is oriented to disadvantaged rather than affluent consumers. Although electronic home shopping (EHS) and the general growth of e-commerce have considerable potential to overcome mobility disadvantage, the cost of providing terminals means that in the short term, most of the residents of inner-city and deprived council estates will continue to rely on conventional retail outlets for their basic shopping needs. The uneven distribution of modern retail facilities within the metropolitan area and the differential ability of consumers to benefit from such developments underline the importance of an efficient urban transportation system. This issue will be considered in Chapter 13.

FURTHER READING

BOOKS

B. Berry, J. Simmons and R. Tennant (1963) *Commercial Structure and Commercial Blight* Research Paper 85, Chicago: Department of Geography, University of Chicago

R. Davies (1995) *Retail Planning Policies in Western Europe* London: Routledge

K. Jones and J. Simmons (1990) *The Retail Environment* London: Routledge

P. McGoldrick and M. Thompson (1992) *Regional Shopping Centres* Aldershot: Avebury

N. Wrigley and M. Lowe (2002) *Reading Retail* London: Arnold

JOURNAL ARTICLES

J. Brooks and A. Young (1993) Revitalising the central business district in the face of decline *Town Planning Review* 64, 251–71

C. Guy (1998) Controlling new retail spaces: the impress of planning policies in Western Europe *Urban Studies* 25(5/6), 953–79

B. Hahn (2000) Power centers: a new retail format in the US *Journal of Retailing and Consumer Services* 7, 223–31

E. Howard (1993) Assessing the impact of shopping centre development: the Meadowhall case *Journal of Property Research* 10, 97–119

A. Loukaitou-Sideris (1997) Inner-city commercial strips *Town Planning Review* 68(1), 1–29

M. Lowe (2005) The regional shopping centre in the inner city: a study of retail-led regeneration *Urban Studies* 42(3), 449–70

K. Robertson (1995) Downtown redevelopment strategies in the United States *Journal of the American Planning Association* 61, 429–37

KEY CONCEPTS

- Shopertainment
- Spaces of consumption
- E-commerce
- Nucleated centres
- Ribbon development
- Specialised function areas
- Power centres
- Theme malls
- Planned shopping areas
- Unplanned shopping areas
- Retail strip
- Regional shopping centres
- Out-of-town centres
- Commercial blight
- Retail suburbanisation
- Town-centre developments
- Super-regional shopping centres
- Retail recentralisation
- Disadvantaged consumers

STUDY QUESTIONS

1. With the aid of relevant examples, explain what is meant by the concept of shopertainment.
2. Identify the major forces influencing the structure of urban retailing in the post-war period.
3. Examine the extent to which planning policies in the UK have constrained the process of retail decentralisation.
4. Examine the effects of retail suburbanisation on inner-city retail areas in either the UK or the USA. Can you provide evidence of commercial blight or the development of polarised levels of retail provision in a city with which you are familiar?
5. Identify the difficulties faced by 'disadvantaged consumers'. Should public authorities intervene in the market process to influence retail developments in favour of such groups and, if so, in what ways?

PROJECT

Map the distribution of shopping areas in a city with which you are familiar. Gather information, from public sources and/or field visits, on the size and retail composition of each centre. Construct a hierarchy of shopping areas. Compare the geographical and hierarchical distribution of the shopping areas with the models discussed in the chapter.

13
Urban Transportation

Preview: patterns of travel demand; urban transport problems; strategic responses to the urban transport problem; private versus public transport; mass transit systems; transport system management; road pricing; auto-restraint measures; non-transportation initiatives; transport and urban structure; transport and sustainable urban development

INTRODUCTION

There is a close relationship between the nature of urban transportation and urban structure (Table 13.1). When most people had to walk to engage in essential daily tasks, cities were necessarily compact. Citizens lived at or close to their workplaces, circumstances that favoured high-density living environments in small, functionally integrated cities that rarely achieved populations of 50,000.[1] Only during the industrial revolution

did vehicles of relatively high capacity and speed allow greater distances to be travelled and larger quantities of goods to be exchanged. This relaxed restrictions on city size and established the interdependence between transport technology and urban form.

During the nineteenth century the development of railways and trams (streetcars) was critical in separating home and workplace, encouraging functional specialisation of land uses in the city, and in promoting the penetration of the surrounding countryside along the more

TABLE 13.1 THE RELATIONSHIP BETWEEN TRANSPORT AND URBAN FORM IN WESTERN CITIES

Stage	Urban functions	Transport technology	Transport system	Urban form
1. Pre-industrial	Defence, marketing, political-symbolic, craft industry	Pedestrian, draught animal	Route convergence, radial	Compact
2. Early industrial	Basic industries, secondary, manufacturing	Electric tram, streetcar, public transport	Radial improvements, incremental additions	High-density suburbanisation, stellate form
3. Industrial	Broadening industry, tertiary service expansion	Motor bus, public transport, early cars	Additional radials, initiation of 'ring' roads (incomplete)	Lower-density suburbanisation, industrial decentralisation
4. Post-industrial	Addition of quaternary activities	Towards universal car-ownership	Integrated radial and circumferential road network	Low-density suburbanisation, widespread functional decentralisation

Source: **D. Herbert and C. Thomas (1997)** *Cities in Space: City as Place* **London: Fulton**

accessible transport corridors. In London between 1860 and 1914 the Underground system and suburban railway lines were formative influences in urban development that permitted extensive suburbanisation and growth of commuter settlements well beyond the urban core. Similar patterns were evident around major cities in the north-eastern USA. In the post-Second World War era widespread car-ownership led to a significant increase in personal mobility and a massive expansion in the built-up areas of cities (Box 13.1).

In this chapter we shall examine the increasing volume and complexity of urban travel, and the problems posed for urban transportation. We shall analyse the major strategic responses to the urban transport problem, examining a range of both transport- and non-transport-based options. Finally, the essential relationship between transport and urban form will be elaborated and the links between transport and sustainable urban development considered.

PATTERNS OF TRAVEL DEMAND

In the USA between 1945 and 1977 the total number of vehicles rose from 29.5 million to 128.6 million (+336 per cent) and by the latter date one-third of the 22,843,000 households had more than one car. By 2000 55 per cent of US households owned more than one car. In 2000 the motor-vehicle population of the USA had reached 235 million for a total population of 281.4 million. The percentage of the US population licensed to drive has also increased, from 72 per cent of those eligible in 1960 to 88 per cent in 1990, and to over 90 per cent by 2000. Those increases have in turn fuelled a rise in vehicle-miles travelled. In addition, the general trend away from walking and public transport to auto use, and lower levels of vehicle occupancy (Table 13.2), have meant that increasingly more cars are being used to serve the same number of travellers.[2]

BOX 13.1

Major eras of metropolitan growth and transport development in the USA

The relationship between transport technology and urban structure may be summarised as a series of stages:

1. *Walking–horsecar era* (1800–90) during which dependence on walking and horsedrawn vehicles for urban transport ensured the retention of a compact, high-density form of city.
2. *Electric streetcar era* (1890–1920) in which the invention of the electric traction motor led to the growth of streetcar suburbs clustered around stops on radial trolley routes.
3. *Recreational automobile era* (1920–45) in which the enhanced personal mobility bestowed by the automobile transformed the 'tracked city' of electric trolleys and trains into a suburban metropolis.
4. *Freeway era* (1945–) during which the automobile became a necessity of modern urban life, and highway developments promoted urban sprawl and the growth of edge cities. This transition to a polycentric metropolis may be seen to have had five growth phases:
   - *Bedroom community* (1945–55) dominated by a post-war residential building boom but only modest expansion of suburban commercial activity.
   - *Independence stage* (1955–65) acceleration of economic growth led by the first

wave of industrial and office parks and, after 1960, by rapid diffusion of regional shopping centres.
   - *Catalytic growth* (1965–80) characterised by further growth of regional shopping malls and associated office, hotel and restaurant facilities at accessible locations in the suburbs.
   - *High-rise/high-technology* (1980–90). Suburban downtowns developed as functional equals of the central business district, characterised by high-rise office buildings and clustering of high-technology R&D facilities as part of the post-industrial service economy.
   - *Mature urban centres* (1990–) in which the suburban downtowns evolve into complexes with diversified land uses and acquire enhanced roles as cultural, entertainment and civic centres, with greater local involvement in resolving problems such as traffic management via, for example, traffic management associations (TMAs) set up by business and local government.

The leading urban transport challenge for the twenty-first century focuses on the efficiencies of moving people about the dispersed, polycentric metropolis.

TABLE 13.2 CHANGING MODE OF JOURNEY TO WORK IN THE USA, 1990–2005

Mode	1990		2005		% change
	No.	%	No.	%	
Drove alone	84,215,298	73.2	102,458,267	77.0	+3.8
In car pools	15,377,634	13.4	14,200,426	10.7	−2.7
Public transport	6,069,589	5.3	6,202,014	4.7	−0.6
Other means	1,512,842	1.3	2,142,757	1.6	+0.3
Walked or worked at home	7,894,911	6.9	8,087,579	6.0	−0.9

Similarly, in the UK the number of cars grew from 2 million in 1950 to 26 million in 2005 (Table 13.3), while the proportion of households owning a car rose from 14 per cent in 1951 to 73 per cent in 2000 (Table 13.4). Between 1955 and 2005 the length of the road network grew from 302,710 km to 388,008 km (an increase of 28.2 per cent). Since the increase in road space available has been much less than the increase in traffic, the 'average daily flow' on all roads increased by nearly half between 1981 and 1993. Much of this increase is the result of the dispersed pattern of urban activities and the increased separation of home and work.

These trends are particularly evident in the US metropolis. People now travel longer distances to work,

TABLE 13.3 NUMBER OF VEHICLES, GREAT BRITAIN, 1950–2005 (000s)

Year	Cars	All vehicles
1950	1,979	3,970
1960	4,900	8,512
1970	9,971	13,548
1980	14,660	19,199
1990	19,742	24,673
2000	23,196	28,898
2005	26,208	32,897

TABLE 13.4 PERCENTAGE OF HOUSEHOLDS WITH REGULAR USE OF A CAR, GREAT BRITAIN, 1951–2004

Year	No. of cars			
	None	1	2	3
1951	86	13	1	–
1960	71	27	2	–
1970	48	45	6	1
1980	41	44	13	2
1990	33	44	19	4
2000	27	45	23	5
2004	25	44	25	5

although travel time has not increased to the same extent. In 1975 the average travel distance to work was nine miles and average travel time twenty minutes. In 1990 the average work trip covered eleven miles and took twenty minutes. In 2001 the average journey to work covered twelve miles and took twenty-four minutes. By 2005 the mean travel time to work in the USA was twenty-five minutes. Clearly, national average figures conceal considerable variation among groups differentiated by age, gender, travel mode and place of residence. Nevertheless, the fairly regular average travel time despite the increase in distance covered reflects improvements in the transportation system and also an 'improvement in individual speeds obtained by shifts to the single-occupant vehicle from car pooling, mass transit and walking' (Pisarski 1992 p. 69).[3]

Accompanying this trend, the spatial pattern of commuting flows has become increasingly complex (Figure 13.1). Reflecting the dispersed structure of the US metropolis, the suburb to central-city trip has not been the dominant work trip since at least 1970.[4] In 2000 the suburb-to-suburb commute accounted for 39 per cent of all metropolitan work trips, with the 'traditional' suburb-to-central-city commute constituting 17.4 per cent and the reverse commute (central city to suburb) 7.6 per cent. In view of the complexity of flows, it is not surprising that the proportion of work trips made on public transport has continued to decline. By 1990 those commuting by private vehicle accounted for 88 per cent of all work trips.[5] Non-work travel has grown at a faster rate than work travel, accounting for 59 per cent of all local travel in the USA in 1969, 74 per cent by 1990 and 85 per cent by 2001.[6] This increase may be attributed to a rise in the number of affluent and dual-income households, which promotes more leisure trips, and a decline in household size and corresponding rise in the number of households.

In combination, these trends of more vehicles on the roads, increased mileage travelled, longer distances covered, increased reliance on and use of a private car and the growing number of households reflect

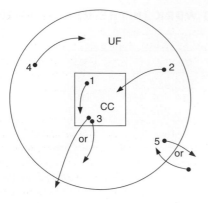

Type 1: 'within central city' movements are trips made by workers who both live and work within the city's legal boundaries.
Type 2: 'inward commuting' encompasses both the traditional commuters from suburbs and metropolitan villages to central cities, plus those workers living in one central city who commute to another.
Type 3: 'reverse commuting' is composed of workers residing in the central city who work anywhere outside that city's boundaries.
Type 4: 'lateral commuting' takes place within the commuter range of the city but both workplace and residence locations are outside the central city.
Type 5: 'cross-commuting' flows are those entering or leaving the central city's commuter zone, meaning that only the workplace or residence is located inside the urban field.

Figure 13.1   A typology of commuting flows
Source: D. Plane (1981) The geography of urban commuting fields *Professional Geographer* 33, 182–8

an unprecedented level of personal mobility, but have also created many of the problems associated with urban transportation.

THE URBAN TRANSPORT PROBLEM

The major facets of the urban transport problem are shown in Figure 13.2. These comprise:

1. *Traffic movement and congestion.* The primary function of urban transport is to provide mobility for people and goods within the city (Box 13.2), but the efficiency with which this is achieved is reduced by congestion. The major cause of urban traffic congestion is the increasing number and use of vehicles on the roads. More specifically, it stems from the concentration of travel flows at certain times during the day, with the principal reason for the typical double-peak distribution of daily trips being the journey to and from work.

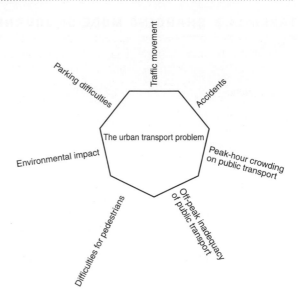

Figure 13.2   Dimensions of the urban transport problem
Source: M. Thomson (1977) *Great Cities and their Traffic* Harmondsworth: Penguin

BOX 13.2

Accessibility and mobility

The concepts of accessibility and mobility are central to understanding urban transportation. 'Accessibility' refers to the number of opportunities or activity sites available within a certain distance or travel time. 'Mobility' refers to the ability to move between different activity sites As the distances between activity sites have become larger, because of lower-density settlement patterns, accessibility has come to depend more and more on mobility, particularly in privately owned vehicles.

Although the need for mobility is a consequence of the spatial separation of different types of land uses in the city, enhanced mobility also contributes to increased separation of urban land uses because improved transportation facilities enable people to travel farther in a given amount of time than they could previously.

Sources: **S. Hanson (1995)** *The Geography of Urban Transportation* **New York: Guilford Press; S. Handy and D. Niemeier (1997) Measuring accessibility** *Environment and Planning A* **29, 1175–94**

Different parts of the city may experience traffic congestion at different times of the day, depending on the mix of traffic, but most large cities experience serious congestion in their central areas during peak hours. In central London average traffic speed fell from 12.9 mph (20.7 km/hr) in 1972 to 10.9 mph (17.6 km/hr) in 1990. This is close to the postulated equilibrium speed of 10 mph (16 km/hr) for peak-hour traffic in the city centre. According to Thomson (1977), motorists will tolerate speeds as low as this before they begin to avoid the area in large enough numbers to create equilibrium at that critical speed.[7]

2. *Crowding on public transport.* In nearly every city the use of public transport is concentrated in the morning and evening rush hours. Whatever the volume of demand, there is invariably insufficient capacity to provide comfortable travel conditions at these times. During conditions of peak-hour loading, passengers are often subjected to lengthy queues at stops, crowding at termini, and excessively long periods of hot and claustrophobic travel in overcrowded vehicles. In Tokyo the metro rail system employs 'pushers' to ensure that passengers are forced into trains to allow the automatic doors to close.

3. *Off-peak inadequacy of public transport.* The difference between peak volume and off-peak usage of public transport means that operators face the financial problem of maintaining sufficient vehicles, plant and labour necessary to provide a peak-hour service which is underused for the rest of the time. A usual response has been to reduce off-peak services, leading to inadequate, unreliable and often non-existent provision of public transport at certain times of the day.

4. *Difficulties for pedestrians.* Paradoxically, although a large number of trips in cities are

Plate 13.1   Rush-hour passengers crowd into a train in Tokyo

made on foot, pedestrians are not often included in urban transportation studies. Pedestrians (and pedal cyclists) encounter two kinds of problem. The first is the problem of accessibility to facilities. The replacement of local outlets (e.g. shops, hospitals, etc.) by larger units serving wider catchment areas puts many urban activities beyond reach of the pedestrian. The second refers to the quality of the pedestrian environment, with footbridges and underpasses often inadequately cleaned and policed, and traffic noise and fumes affecting foot travellers most directly.

5. *Environmental impact*. Transport is a major source of air pollution in cities, with exhaust gases (carbon dioxide, carbon monoxide, nitrogen oxides, hydrocarbons) and other pollutants, such as lead and particulates, contributing to a range of health and environmental problems. Other traffic-induced environmental impacts include noise pollution, visual intrusion, the destruction of natural habitats and segregation of communities by transport routes.

6. *Accidents*. Road traffic accidents constitute a significant social problem, and the majority occur in urban and suburban areas. A minority of the victims are car occupants and a high proportion are pedestrians, cyclists and motor cyclists. Most transportation studies, however, regard road traffic accidents as an unfortunate offshoot of an urban transport system.

7. *Parking difficulties*. In most city centres, finding a place to park a car is difficult. In many cities parking is often seen, by drivers at least, as the major problem of urban transport. As we shall see later, since it is physically impossible for a large city to provide car parking space for all who wish to enter the centre, restrictions apply. These may lead to illegal parking, which can impede the flow of traffic. Parking difficulties are different from other aspects of the urban transport problem, however, as they are often maintained deliberately by local authorities in an attempt to reduce other problems (Figure 13.2).

RESPONSES TO THE URBAN TRANSPORT PROBLEM

Three general approaches may be identified:

1. During the 1950s and 1960s the *supply-fix* approach emphasised the provision of new infra-

structure to increase the capacity of the road system to meet demand. This policy was manifested in large-scale road-building and road-improvement programmes. The emphasis was firmly on the supply side, and transport policy was effectively a *vehicle-oriented* policy. Large increases in highway capacity in many cities tended to exacerbate the problems of traffic congestion and environmental degradation in accordance with Down's law, which states that on urban commuter expressways peak-hour traffic congestion rises to meet maximum capacity.[8] In addition, the increased mobility of car-owners stimulated the development of low-density urban areas with unstructured travel demands, resulting in substantial reductions in the role and effectiveness of public transport.

2. During the late 1960s and early 1970s the character and emphasis of urban transport policy shifted. The policy objective became accessibility rather than simply mobility. Greater emphasis was placed on the exploitation of existing facilities, on minimising the environmental impact of the automobile, and on the equity with which all sections of the urban population were served by transport. The shift to *people-oriented* non-capital-intensive policy options was stimulated by reductions in the growth rate of many large urban areas, lower levels of national economic growth, the rise of a popular movement opposed to the environmental and social costs of major highway projects, and the energy crisis initiated by the Arab oil embargo of 1973. The resultant transport-system management strategies sought to manipulate demand and make more efficient use of existing highway capacity.

3. The third approach is by means of *non-transportation initiatives*, which can range from staggering work hours to designing cities to reduce the need for travel by limiting distances between home and other centres of activity.

We can examine each of these three approaches.

PRIVATE VERSUS PUBLIC TRANSPORT

The major policy debate during the era of the supply-fix approach focused on the relative merits of public and private transport. As Figure 13.3 shows, the emphasis on the provision of additional highway capacity to accommodate suburbanisation and rising levels of car

Plate 13.2 Growing concern with the social and environmental costs of highway construction signalled the end of the road for some urban freeways, such as this project in San Francisco CA

ownership induced a decline in public-transport use. Throughout the industrialised world, declining fare revenues and increasing costs have produced a situation in which few metropolitan mass-transit systems operate without subsidies. The continued provision of transport as a public rather than commercial service depends on the attitude of government. The city of Curitiba in Brazil is often cited as a model for urban transport planners elsewhere (Box 13.3).

In the UK the 1974–9 Labour government supported urban public transport, and several cities, including London, introduced publicly subsidised fare-cutting experiments to increase the attractiveness of public transport. This approach was contrary to the competitive-market philosophy of the incoming Conservative government in 1979, which abolished subsidies to public transport and introduced deregulation to open up the sector to competition. This led to cost savings but a reduction in services outwith peak hours. Similarly, in Toronto a subsidised rapid-transit system in the central city integrated with suburban rail and bus services led to an increase in journeys to work

by public transit, reaching a peak of 22 per cent in 1988. Subsequent withdrawal of subsidies resulted in a sequence of fare increases, declining quality of service and a 15 per cent reduction in passenger traffic by 1993.[9] In some countries, including the USA, Germany and France, travel voucher schemes have been employed to encourage workers to use public transit. The basic model is that the employer issues travel vouchers to be used to pay for travel on public transport. As the vouchers are tax-free there is an incentive for employers and employees to use them instead of a (taxed) wage increase of equivalent value. Transit makes a minor contribution to mobility in American cities, with only 2.5 per cent of all person trips made by transit in 1990 compared with 86 per cent by automobile. With reference to journeys to work, in 2005 only 4.7 per cent were made by public transport compared with 88 per cent by automobile (Table 13.2). In the USA transit use is concentrated in the large cities, with New York accounting for 40 per cent of national transit passenger mileage. Work trips are the mainstay of transit, and in 1990 accounted for

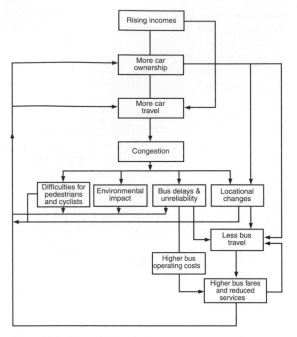

Figure 13.3 The relationship between car ownership and bus ridership

55 per cent of all weekday transit trips.[10] In smaller cities, and for most non-work trips, public transport cannot compete with the private car.

The benefits of most transit systems for the alleviation of peak-hour congestion on work trips to and from the central city underlie efforts to expand its use. To be effective such strategies must enhance the attractiveness of public transport and restrict the use of the private car. A good-quality transit system must:

1. Offer fares low enough for the poor to afford.
2. Operate sufficient vehicles to run a frequent service throughout the day.
3. Have routes which reflect the 'desire lines' of the travelling public, with extensive spatial coverage of the catchment area to ensure that no one is too far from a transit stop.
4. Increase bus speeds relative to those of cars.
5. Provide a co-ordinated multimodal one-ticket system with convenient connections between transport modes.

Many cities in North America and Europe have introduced rapid mass-transit systems. In Stockholm 38 per cent of residents and 53 per cent of workers living in satellite new towns commute by rail.[11] Other

European cities have upgraded tram and light rail systems,[12] introduced electrified suburban railway lines (as in Copenhagen and Glasgow) or expanded underground metro systems (London and Paris). In the UK the Tyneside Metro was completed in 1981 at a cost of £300 million; the thirty-four-mile (55 km), forty-seven-station light rail system is integrated with bus services and park-and-ride facilities, and operates with a heavy subsidy. Cities such as Manchester and Sheffield have introduced electric trams running along existing roads and railway lines to limit costs, while

Croydon's 28 km tramlink system utilises converted sections of disused railway track. In London's Docklands a light rail system constructed to link the City with new office developments at Canary Wharf has since been extended east to Beckton to aid further regeneration of the area.

In the USA the Bay Area Rapid Transit (BART) system in San Francisco CA carries more than half of all CBD-bound work journeys. Other large cities such as Washington DC (METRO) and Atlanta GA (MARTA) have developed entirely new rapid rail transit (RRT) systems. More recently, light rail transit (LRT) has been seen as a low-cost alternative. Between 1981 and 1994 LRT systems were opened in San Diego CA, Buffalo NY, Portland OR, Sacramento CA, San José CA, Los Angeles CA, Baltimore IN and St Louis MO.[13] Significantly, these public-transit systems have not achieved the passenger volumes anticipated.[14] Nevertheless, such is the scale of concern over the impact of the automobile on the quality of city life that federal support for mass transit is ongoing, with the 1991 Intermodal Surface Transportation Efficiency Act (ISTEA) permitting the use of federal highway funds for other modes, including transit as an alternative to the single-occupancy vehicle form of urban travel. The Transportation Equity Act for the Twenty-first Century (TEA-21), enacted in 1998, is intended to further ISTEA's comprehensive flexible approach to urban transportation. Fundamentally, if public transport is to compete successfully with the private car it has to provide a comparable service. It has to be affordable, safe, comfortable, reliable, timely and flexible enough to cater for trips that are not necessarily centrally oriented. Integration between different public transport systems must be easy, leading to 'seamless travel' with smart cards to cover different modes and minimum waiting times at interchange stations. Such integration requires financial and political commitment.

TRANSPORT SYSTEM MANAGEMENT

Reaction against continued large-scale provision of highway infrastructure has resulted in literally hundreds of techniques for better use of existing transport facilities. These include channellisation of traffic, especially at intersections (e.g. segregation of left- and right-turning flows, and bus-only lanes), computer-controlled traffic signals, more traffic light-controlled junctions and mini-roundabouts to increase the flow that can be handled by a single intersection, one-way streets to reduce conflicting traffic movements at junc-

tions, limitations on street parking (e.g. clearways) to increase the **carrying capacity** of traffic routes, reserved lanes for 'high occupancy' vehicles, and reversible traffic lanes in which priority changes with time of day. All these management techniques can be easily implemented at relatively low cost and involve minor physical alterations. More sophisticated options based on the concept of intelligent vehicle highway systems (IVHS) involve the use of information technology, such as geographic information systems and global positioning systems, to improve traffic flow by means of on-board navigation. Advanced vehicle control systems (AVCS) under development include both collision-warning systems and devices that would automatically take over control in order to avoid a crash. The ultimate IVHS is the 'smart highway' that would enable all drivers to switch to autopilot, ensuring maximum throughflow without accidents.[15]

An alternative approach to traffic management in towns is to remove regulatory signals (e.g. traffic lights) and infrastructure (e.g. pedestrian barriers) in order to leave road users (drivers and pedestrians) uncertain about who has priority. Based on the Dutch concept of 'shared space', it is argued that providing road users with *less* information encourages greater care (e.g. by making drivers slow down to negotiate unmarked junctions). In the Dutch town of Drachten the removal of traffic lights at a major junction resulted in a significantly reduction in vehicle collisions, while on Kensington High Street, west London fewer pedestrians were injured after a 600 m stretch of railings was removed to allow people to cross where they liked.

Currently available strategies for transport system management also include the options of **road pricing**, auto-restraint policies and encouragement of ride-sharing:

Road pricing

Although in many countries road travellers pay to use inter-city roads, tolls for travel within urban areas have not been adopted widely. In Norway three cities, Oslo, Bergen and Trondheim, have introduced congestion charging,[16] as have Melbourne[17] and London[18] (Box 13.4). In Singapore the extent of traffic congestion stimulated the authorities to introduce road pricing as part of a package of measures. Constraints on traffic in the central area include the prohibition of all large vehicles with three or more axles during peak hours, increased parking fees, a revision of parking regulations to discourage commuter traffic but facilitate

BOX 13.4

The London congestion charge

London has pioneered road pricing in the UK. Road pricing in the capital takes the form of a cordon charge that drivers must pay when crossing entrance points to the eight-square-mile central city charging zone between 7.00 a.m. and 6.30 p.m. on weekdays. (Residents within the cordon receive a 10 per cent reduction, and there is total exemption for buses and taxis.) By 2005, after two years of operation, the charge had reduced traffic by 15 per cent and congestion by 30 per cent, and as a result disruption of bus services had fallen by 60 per cent. Since the introduction of the congestion charge 20,000 fewer vehicles enter the zone each day. Average speed of movement within the zone has risen from 8 mph to 11.mph; peak-period congestion delays fell by 30 per cent; bus delays fell by 50 per cent and taxi fares dropped by 20 per cent to 40 per cent owing to fewer delays. Bus ridership has increased by 14 per cent and Underground (subway) ridership by 12 per cent. There has been a 10 per cent increase in traffic on peripheral roads outside the congestion zone.

The system is expensive to operate but in 2003−4 raised a net £80 million for investment in the London transport system. Monitoring suggests that the charge has had no significant impact on the location of business or on property values, and an initial decline in footfall in the central retail area has been reversed.

Durham is currently the only other UK city with road pricing. The fact that Edinburgh chose not to proceed with its proposed system after it was rejected in a public referendum is evidence of motorists' opposition and the political difficulty of implementing road pricing arrangements.

short-term business and commercial trips, and the Area Licensing Scheme (ALS).[19] The latter, introduced in 1975, designates the centre of the city as a restricted travel zone. Between 7.30 a.m. and 10.15 a.m. only specially licensed vehicles are permitted into the zone. Licences are sold according to a sliding scale that ensures that company-owned vehicles pay the highest rate. Car-pool vehicles (defined as those carrying four or more persons) are exempt. Table 13.5 indicates the effect of the ALS on the modal split choice for work trips. In 1994 the ALS was extended to a full day, resulting in an immediate reduction of 9 per cent in traffic entering and leaving the restricted zone. In 1998 the ALS was replaced by a full system of electronic road pricing with 'congestion fees' deducted automatically as vehicles cross sensors installed along the route. The fee charged varies with the level of congestion, making it the world's first scheme to pass on real-time congestion charges to motorists.[20]

Other traffic management policies in Singapore include:

1. A 140 per cent tax (Additional Registration Fee) on all car imports.
2. An annual registration disc costing up to £1,500.
3. Restrictions on the number of new cars permitted (Vehicle Quota System).

From 1990 no new cars can be bought unless the purchaser has a government Certificate of Entitlement to Purchase, which is obtained via a public auction. The government decides each quarter on how many new cars it will allow, based on the number scrapped and road capacity, and potential owners submit a bid, with the number of successful bids equivalent to the number of new cars permitted. The price of the certificate is determined by the value of the lowest successful bid.[21]

TABLE 13.5 CHANGES IN MODAL SPLIT OF WORK TRIPS IN THE RESTRICTED ZONE, SINGAPORE (%)

Mode	Inbound home-to-work trips		Outbound work-to-home trips	
	Before ALS	After ALS	Before ALS	After ALS
Bus	33	46	36	48
Car	56	46	53	43
Car driver	32	20	35	23
Car pool	8	19	5	12
Car passenger	16	7	13	8
Motor cycle	7	7	7	6
Other	4	2	4	3

Advocates of road pricing favour it as an example of using market forces to achieve policy ends. Rather than mandating modification of travel directly, road pricing encourages it through monetary disincentives. Objections to road pricing have centred upon equity, political and practical issues. The last of these is the least serious constraint. The practical difficulties of implementing a system of road pricing can be overcome without excessive cost, and electronic toll-collection technology now makes it feasible to levy charges without interrupting traffic flow. In Orange County CA, one- and two-occupant cars are charged for using high-occupancy vehicle (HOV) lanes constructed with private funds down the median of an existing highway (Route 91). Future road pricing systems could be based on a global positioning system (GPS) satellites combined with in-vehicle computers and toll schemes for all roads. One of the strongest arguments against road pricing is that it could discriminate against lower-income car owners. Most significant, however, has been political opposition by the 'motoring lobby' and the general recognition by governments that the social benefits of road pricing could be offset by the political costs (Box 13.5). Nevertheless, the problem of urban traffic ensures that road pricing (or congestion pricing) remains on the political agenda.

Auto restraint

Road pricing is, in effect, one kind of auto restraint policy in that one of its goals is to promote the use of public transport by making car travel more difficult. Other restraint policies include prohibiting on-street car parking and the provision of additional downtown parking areas. The ultimate form of auto restraint is to ban cars from a section of the city, either wholly or partially, by prohibiting the entry of certain vehicles (e.g. according to licence plates) on certain days. Historic European cities (such as Vienna, Milan and Gothenburg) have limited traffic penetration in the interests of the liveability of the city centre.[22] In Gothenburg the central city is subdivided into five sectors, and while traffic is permitted entry within any sector, cross-sector flows are prohibited so as to exclude through traffic.

Ride-sharing

Three-quarters of the 84 per cent of the American labour force that travel to work by car are solo drivers. Nationally the average private vehicle in the US trans-

BOX 13.5

The political feasibility of high-occupancy vehicle lanes

High-occupancy vehicle (HOV) lanes may be provided either by the conversion of one or more existing general-purpose lanes or by the construction of a new lane. The add-a-lane strategy generates less opposition from motorists. The convert-a-lane approach meets ISTEA's goals by improving travel conditions for HOVs and decreasing the capacity for single-occupancy vehicles (SOVs). Inevitably, this leads to dissatisfaction among SOV users Opposition to this efficiency measure was illustrated in 1977 by the court decision in the case of the Santa Monica freeway HOV. Faced with chronic congestion on a 20 km section of the ten-lane freeway, California's Department of Transport (CalTrans) converted one of five lanes in each direction into an HOV lane. The initial increase in congestion in the four general traffic lanes led to a lawsuit against the project, and twenty-one weeks after its start a judge halted it and requested additional environmental impact studies. CalTrans responded by dropping the project, claiming that convert-a-lane schemes were 'not politically feasible'.

ports only 1.15 persons on its trip to the workplace. One way of increasing highway capacity is for transport planners to implement programmes to encourage ride-sharing through car- and van-pooling schemes. Research suggests that to maximise the possibility of adoption, such schemes should be targeted at particular populations. Many people are not amenable to giving up the freedom of solo driving or are constrained from participating by other daily activity patterns such as collecting children or shopping. It appears that the most effective programmes are those arranged through employers. Evidence from Phoenix suggests that employees favour van pools over car pools because their private cars need not be used, they need not drive in rush-hour conditions, the arrangements are made for them and they have an opportunity to socialise with colleagues.[23] Assisting employers to set up van pooling as an employee benefit is thus often more cost-effective than media campaigns aimed at the general public. The option of 'cashing out' free parking supplied by employers may be another inducement to ride-sharing. Under this scheme an employer could offer employees a choice of receiving a free parking space or a money payment equivalent to the cost of providing such a space. Employees who accept 'cash outs' can shift to ride sharing or public transit while those who continue to drive to work are paying an implicit charge for formerly free parking.[24]

Although all the strategies considered can contribute to ameliorating the urban transport problem, no single policy is sufficient to promote optimum use of the urban transport system. What is required is a package of self-reinforcing strategies that are integrated into the overall planning process for the metropolitan area. This means considering both transport and non-transport options.

NON-TRANSPORTATION INITIATIVES

Non-transport strategies seek to capitalise on the relationship between transportation, daily activity patterns and the aggregate land-use structure of the city. Two main approaches focus on patterns of work and on urban structure respectively.

Transport and work patterns

The problem of the commuting peak may be alleviated by promoting alternative work schedules (Table 13.6) which spread the journey-to-work period or reduce the number of workdays per week. Advances in telecommunications technology can also reduce the need for some trips during the business day and in some cases allow employees to work from home or from regional suburban 'work stations'. The principle of flexible working hours has been applied widely in West Germany, where more than half the labour force is permitted to work non-standard hours. The potential benefits for urban traffic movement may be considerable. For example, a study of the San Francisco–Oakland Bay bridge traffic corridor found that even small increases in the number of flexitime workers can have substantial impacts in alleviating peak-period congestion.[25] To date, however, urban transportation planning agencies have devoted only limited effort to promoting such strategies.[26]

The impact of telecommunications on traffic mitigation is slight as yet, with 1.4 per cent of Californian workers telecommuting on a typical workday, and 6.6 million American telecommuters in 1992,[27] although Nilles (1991 p. 425) estimated that by 2011 'there could be as many as 50 million telecommuters in the USA, saving as much as 380 billion passenger-miles of transportation use'.[28] Several US states, (such as Arizona, Florida, California and Oregon), now promote programmes in telecommuting as part of their transportation strategies.[29] In the UK, in 2005, 2.4 million (8 per cent of the work force) were teleworkers, an increase of 158 per cent since 1997.[30] The number of telecommuters in Tokyo is expected to grow to between 9 million and 14 million by 2010, reducing pressure on roads and public transport systems with associated savings of up to 25 per cent of annual spending on public transport.[31] On the other hand, a danger is that decentralised teleworking or even telecommuting centres could reinforce the trend towards sprawling edge cities, since living on the urban fringe or beyond would not be inconvenient from a commuting viewpoint.

TABLE 13.6 TYPES OF ALTERNATIVE WORK SCHEDULES

1. *Flexible work hours ('flexitime')*. Employee chooses his or her own schedule, with some constraints. Employee may be free to vary the schedule daily, vary the lunch breaks, or 'bank' hours from one day to the next or one pay period to the next. Typically all employees are required to be at work five days per week during designated 'core' periods (usually 9.30 a.m. to 3.30 p.m.)
2. *Staggered work hours*. Employee works a five-day week, but starting and ending times are spread over a wider time period than usual. Individual employee schedules are assigned by management; employees do not choose them
3. *Four-day (or compressed) work week*. Employee works the same total number of hours as in a typical five-day work week, but reports to work only four times per week. The four days may be the same each week, or the extra day off may rotate from week to week
4. *Job sharing/part-time work*. Employee works less than the standard forty-hour work week, either by working fewer than five eight-hour days, or by working less than eight hours per day. Job-sharing means that two or more people share the same office space and work responsibilities

Transport and urban structure

As we have seen, a symbiotic relationship exists between transport and urban structure. Thomson (1977) identified four general urban transport strategies based on the degree of car-ownership to be accommodated[32] (Figure 13.4). Although no city fits any single strategy, these archetypes offer general guidance on possible transport options for different forms of city:

1. *Full motorisation*. Since the motor car has many virtues as well as drawbacks, one approach is to structure the city to allow full use of the car. Small towns of up to 250,000 population can usually achieve this by upgrading radial routes and constructing inner ring roads around the town centre with adjoining car parks, but in large towns the provision of sufficient roads and parking space becomes impossible – unless one abandons

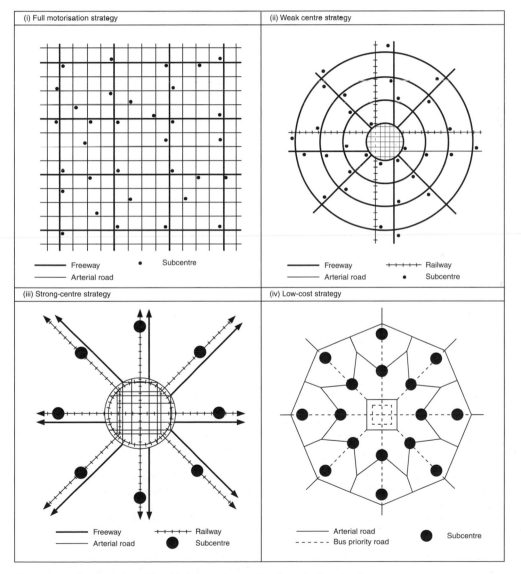

Figure 13.4   Models of the relationship between transport policy and urban structure
Source: M. Thomson (1977) *Great Cities and their Traffic* Harmondsworth: Penguin

the traditional monocentric city. If there is to be a city centre it will be smaller and could not fulfil a traditional multifaceted role, with employment, shopping and entertainment activities dispersed to outlying centres. The primary goal is to maintain high levels of car accessibility throughout the metropolitan area. Los Angeles comes closest to this model.

2. *Weak-centre strategy*. In cities where the decentralisation force has proved more dominant than those favouring the presence of a strong city centre, the traditional downtown has diminished in importance to become little more than another suburban centre. The archetype of this weak-centre strategy is a city with a radial road network and additional commuter rail service serving a small central area. The combination of ring and radial routes attracts industrial and commercial development to the interchange points and generates growth of strategic suburban centres. Boston provides a good example of the type.

3. *Strong-centre strategy*. The world's greatest cities owe much of their power to the concentration of activity in central areas which were established before the automobile began to exert an influence on urban structure. In such cities the transport system must be designed to maintain the strength of the centre. For this reason, the archetype comprises a radial network of road and rail routes, without high-speed orbital roads except close to the city centre itself. The dominant position of the central area will require a high-capacity public transport system, such as a short-stage, high-frequency underground railway, to distribute the large numbers of people entering the zone (Figure 13.4). This ideal is closely reflected in the urban structure of Tokyo.

4. *Low-cost strategy*. Many cities are unable or unwilling to pay the high cost of these strategies. In cities where a minority own a private car, substantial public expenditure on new roads or rail systems cannot be justified, socially or economically. Traffic problems require a low-cost approach that maximises the use of existing infrastructure through improved management. The archetype consists of a high-density city with a major centre served by numerous bus and tram corridors in which non-residential activities are concentrated. Hong Kong provides an example of this urban form.

TRANSPORT AND SUSTAINABLE URBAN DEVELOPMENT

As we have seen, the automobile has been a major agent underlying urban sprawl. Proponents of sustainable urban development advocate greater use of public transport and a more comprehensive approach to planning which acknowledges the fundamental relationship between transport and urban form. Common to most perspectives on sustainable urban development is the view that a solution to urban transport problems requires greater mixed-use and higher-density development, as well as integration of transport considerations directly into land-use planning with the aim of enabling individuals to sustain their mobility, but to do so with fewer vehicle trips.

In the Netherlands the urban transport problem is addressed within the framework of a national plan for sustainable development. The major aims are to ensure that:

1. Vehicles are as clean, quiet, safe and economical as possible.
2. The choice of mode of passenger transport must result in the lowest possible energy consumption and the least possible pollution.
3. The locations where people live, work, shop and spend their leisure time will be co-ordinated to minimise the need for travel.

These goals are to be approached via:

1. A series of measures to convert vehicles to clean running by, for example, establishing targets for the reduction of exhaust emissions.
2. A shift from single-occupancy vehicles to public transport for longer journeys or cycling and walking for shorter trips, by means of improved transit systems and restrictions on car use.
3. Concentration of residences, work areas and amenities to produce the shortest possible trip distances.

Public mass transit plays an important role in a number of cities that may offer lessons for elsewhere. Stockholm provides an example of the co-ordinated planning of rail transportation and urban development. Half the city's 750,000 inhabitants live in satellite communities linked to the urban core by a regional rail system. This transit-oriented multicentred built form is the outcome of post-Second World War regional planning that targeted population growth into rail-served new towns in a pattern prompted by Howard's garden-city principles. The contrast between the pattern of post-war urban growth in Sweden and that in the USA is significant. Like the USA, Sweden is one of the world's most affluent countries (with GDP *per capita* of US$17,900 in 1990) and has a high rate of car ownership (420 cars per 1,000 inhabitants), yet suburban growth follows a radically different path as a consequence of government intervention in urban planning at the regional level.

Tokyo is a powerful example of a 'strong centre' city, and without the contribution of rail mass transit, road congestion would undermine the functioning of the CBD. Over half the 1,250 miles (2,000 km) of suburban railway lines have been constructed by private-sector consortia that link mass transit with new-town development (Box 13.6). Tokyo's rail-oriented urban expansion is also promoted by various national and local government policies aimed at reducing car use. These include a range of taxes on car-ownership – purchase tax, an annual registration tax, a surcharge based on vehicle weight, high gasoline taxes, toll roads, and a requirement to provide off-street parking at one's residence – and incentives to promote transit riding, such as a tax-free commuting allowance of US$500 per month from employers which is fully deductible against corporate income taxes. The existence of a high-capacity public transportation infrastructure in Tokyo has served to ameliorate some of the problems of congestion and overcrowding within the dominant CBD. However, the development of a polycentric metropolitan structure, with both planned and unplanned subcentres and satellite towns, has resulted in dispersed population and employment growth and increased levels of car dependence in the outer suburban areas beyond the dense transit network.

Central co-ordination of transport and urban development is seen at its strongest in Singapore, where near-complete control over urban growth and design, and a pro-transit government, has produced one of the most efficient transit–land-use symbioses in the world.[33] The initial Concept Plan (1971) provided for decentralisation of population from overcrowded inner urban areas to a series of high-density new towns served by an efficient transit system comprising the Mass Rapid Transit (MRT) and feeder bus lines. The 1991 revised Concept Plan envisaged future urban growth based on further development of new towns in a hierarchy of centres linked by the MRT in a pattern of 'concentrated decentralisation' (Figure 13.5).

The concept of transit-oriented development has also emerged in the USA, where a 'new urbanism' is critical of the auto-oriented post-Second World War suburbs.[34] This advocates a type of suburban community designed in neo-traditional terms with a mix of land uses, moderate residential densities, pedestrian circulation, and small offices and shops clustered around a transit station (Figure 13.6). The first such community was developed at stations on the BART system (Box 13.7), but similar nodes have grown up

---

BOX 13.6

A transit-oriented community: Tama New Town, Japan

Tama new town is one of several publicly sponsored new towns in the Greater Tokyo region; others include Tsukuba Science City, Ryugasaki, Chiba and Kohoku. Tama new town is a joint venture of the Tokyo metropolitan government and the national Housing and Urban Development Corporation. The town currently has 170,000 inhabitants and 35,000 jobs, with a target population and labour force of 360,000 residents and 130,000 workers respectively. A 1991 survey found that 20 per cent of the employed residents had jobs within the town; of the remaining 80 per cent leaving the town for work two-thirds travelled to central Tokyo, with 70 per cent commuting by rail.

The new town is designed around twenty-one residential areas (neighbourhoods) each containing a centrally located junior high school and two primary schools. There is a variety of housing, with public housing consisting of mid-rise and high-rise apartments sited near rail stops and targeted at low- and middle-income households. These are surrounded by privately built single-family detached units sold at market rate. Four of Tama new town's railway stations form a focus for an urban centre comprising a retail plaza, offices, banks and institutional uses. These urban centres are notable for the conspicuous absence of park-and-ride lots. Even though most Tama residents own a car they bus-and-ride, walk-and-ride or bike-and-ride instead.

Source: **M. Bernick and R. Cervero (1997)** *Transit Villages in the Twenty-first Century* **New York: McGraw-Hill**

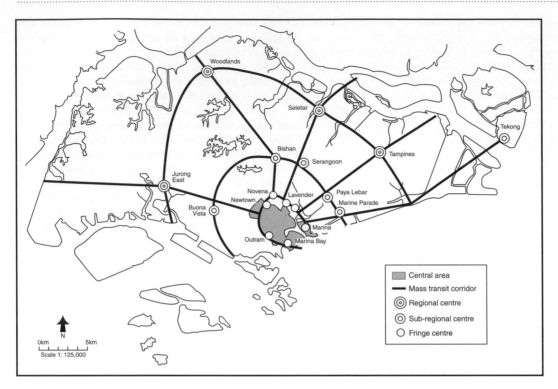

Figure 13.5   The mass rapid transit system and concentrated decentralisation in Singapore

BOX 13.7

Pleasant Hill CA: transit village

As the Bay Area Rapid Transit (BART) train travels through Contra Costa county it passes a series of post-Second World War suburbs: Orinda, Moraga, Lafayette, Walnut Creek, Concord. Here BART is enveloped in a sprawling environment of single-family homes and duplexes, one- to two-storey commercial buildings, and lots of hillside open space. The notable exception, however, is what until recently was the next to the last station on the line: Pleasant Hill.

At Pleasant Hill are the beginnings of a new type of suburban community: one with a mix of clustered housing, mid-rise offices, small shops, a hotel and a regional entertainment complex, all huddled around the transit station, and with a street configuration and landscaping that encourage people to walk to and from the station.

Housing, the primary land use in the area, lies slightly beyond the station core. In all, over 1,600 residential units built in the past five years lie within a three- to five-minute walk of Pleasant Hill station. Office and commercial developments are located even closer to the station. PacTel Corporate Plaza, housing the regional administrative offices of the telecommunications giant, is less than 100 yards (30 m) from the station entrance. At the station core much of the eleven-acre (4.5 ha) BART parking area has been converted into an entertainment and retail complex.

The Pleasant Hill station area illustrates an alternative to the low-density car-oriented suburban development that characterises virtually all of Contra Costa. It is a radical departure from business as usual in the suburbs. It is a transit-oriented development, or what local planners and elected officials refer to as a 'transit village'.

Source: **M. Bernick and R. Cervero (1997)** *Transit Villages in the Twenty-first Century* **New York: McGraw-Hill**

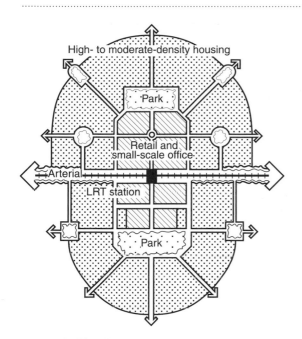

Figure 13.6   Plan of a transit-oriented community

along other transit routes,[35] including the line of the Washington Metrorail system in suburban Virginia and Maryland. Since being linked to the Metrorail network in 1979, the **transit village** of Ballston six miles (10 km) north-east of Tyson's Corner has grown into one of Arlington County's 'new downtowns' and has been characterised as a 'post-edge city'. Transit-oriented development may also aid regeneration of decayed inner-city areas, as at Fruitvale in Oakland, where a run-down commercial district has been transformed into a transit village by a combination of grassroots community action financed by federal and foundation grants which pump-primed sufficient development to attract private-sector investment.[36] However, although transit villages reflect many of the aims of sustainable urban development, they are likely to appeal to only a small proportion of the population, the majority of which will continue to favour the lifestyle associated with the **American dream** of a dispersed single-family residence on a large lot.

FURTHER READING

BOOKS

D. Banister (2005) *Unsustainable Transport: City Transport in the New Century* London: Routledge

M. Bernick and R. Cervero (1997) *Transit Villages in the Twenty-First Century* New York: McGraw-Hill

R. Cervero (1998) *The Transit Metropolis* Washington DC: Island Press

I. Docherty and J. Shaw (2003) *A New Deal for Transport* Oxford: Blackwell

A. Downs (2004) *Still Stuck in Traffic* Washington DC: Brookings Institution Press

S. Hanson and G. Guiliano (2004) *The Geography of Urban Transportation* New York: Guilford Press

C. Pooley, J. Turnbull and M. Adams (2005) *A Mobile Century? Changes in Everyday Mobility in Britain in the Twentieth Century* Aldershot: Ashgate

M. Thomson (1977) *Great Cities and their Traffic* Harmondsworth: Penguin

JOURNAL ARTICLES

R. Cervero and K.-L. Wu (1998) Sub-centring and commuting *Urban Studies* 35(7), 1059–76

J. Constantino (1992) The IVHS strategic plan for the United States *Transportation Quarterly* 46, 481–90

P. Gurstein (1996) Planning for teleworking and home-based employment *Journal of Planning Education and Research* 15, 212–24

J. Macedo (2004) Curitiba *Cities* 21(6), 537–549

A. Sorensen (2001) Subcentres and satellite cities: Tokyo's twentieth-century experience of planned polycentrism *International Planning Studies* 6(1), 9–32

C. Willoughby (2001) Singapore's motorization policies 1960–2000 *Transport Policy* 8, 125–39

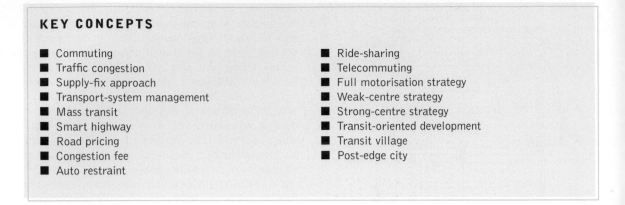

KEY CONCEPTS

- Commuting
- Traffic congestion
- Supply-fix approach
- Transport-system management
- Mass transit
- Smart highway
- Road pricing
- Congestion fee
- Auto restraint

- Ride-sharing
- Telecommuting
- Full motorisation strategy
- Weak-centre strategy
- Strong-centre strategy
- Transit-oriented development
- Transit village
- Post-edge city

STUDY QUESTIONS

1. Identify the principal dimensions of the urban transport problem. Relate them to your own experience of urban travel.
2. Examine the view that the best way to relieve traffic congestion in towns is to build more highways.
3. Make a list of the main advantages and disadvantages of private (car-based) transport and public (bus or rail) mass transit in cities.
4. With reference to particular examples, critically examine attempts by urban authorities to limit the use of private cars in cities. What is your view of the desirability and feasibility of such strategies?
5. Select one of the major models of the relationship between transport and urban structure discussed in the chapter. Explain its rationale and identify a city to which the model may be applied.
6. Think about the possible impacts of telecommuting on urban structure. How do you see these affecting patterns of urban transport and urban form by the years 2010, 2025, 2050?

PROJECT

Efforts to relieve urban traffic congestion by increasing the ridership of public transport are undermined by commuters' reluctance to leave their cars. Select one mode of public transport available in your city (e.g. commuter rail, bus or Underground) and undertake a survey of passengers to identify who uses the system and why, and the perceived advantages and disadvantages of this form of travel.

PART FOUR

Living in the City: Economy, Society and
Politics in the Western City

14

The Economy of Cities

Preview: the world/global economy; transnational corporations and the state; new industrial spaces; deindustrialisation and tertiarisation; the nature of work and the division of labour; feminisation; unemployment; informalisation; urban change within the global economy; world cities; transnational urban systems; economic restructuring in the Great Lakes region; the post-war restructuring of the UK space-economy

INTRODUCTION

The focus of this chapter is the urban economy. To understand fully the economy of cities we must locate our discussion within the framework of the wider phenomenon of post-Second World War global economic restructuring. This process is itself related to the more general concept of globalisation (see Chapter 1). In the context of urban economic change, particular importance is attached to the process of economic globalisation. The driving force behind economic globalisation is capitalism. This is evident in:

1. The growth of multinational and transnational corporations (TNCs).
2. The expansion of trade and foreign direct investment (FDI).
3. The emergence of a new international division of labour (NIDL).
4. The enhanced mobility of money capital across international boundaries.
5. The globalisation of markets for consumer goods.
6. The intensification of international economic competition with the rise of newly industrialising countries (NICs).

At the urban scale, economic globalisation is manifested on the one hand in the rise of global cities, and in the deindustrialisation and decline of older industrial cities on the other. Understanding the nature of the global economy is a prerequisite for understanding the changing economy of cities. We begin our discussion in this chapter by defining the concepts of a global/world economy and identify the changing structure of the global/world economy in the post-war era. We examine the role of TNCs and the state within the global economy and the rise of new production systems and industrial spaces as part of the transition to advanced capitalism. Deindustrialisation, tertiarisation, the changing nature of work and the division of labour, and problems of unemployment and informalisation are also discussed. The growth of world cities and of a transnational urban system is examined before attention is directed to the post-war restructuring of the UK space-economy, and parallel developments elsewhere.

THE WORLD/GLOBAL ECONOMY

We can differentiate between a global economy and a world economy. The latter is an economy in which capital accumulation in the form of trade between firms based in different nation-states proceeds throughout the world, and has existed in the West since the sixteenth century.[1] A global economy is one with the capacity to work as a unit in real time on a planetary scale.[2] Only with the advent of new information and communications technologies in the late twentieth century has the world economy become global, primarily in the context of the international financial system (Box 14.1).

BOX 14.1

The advent of megabyte money

Money was invented to replace barter in the exchange of goods and services. For centuries, money has served as a medium of exchange, a common unit of account and a store of value. In recent decades it has also become a commodity to be bought and sold. This new function of money was created by the US government's decision in 1971 to take the dollar off the gold standard and allow exchange rates to float in the international markets. With the collapse of the Bretton Woods agreement, which had pegged the value of the US dollar to gold (and other major currencies to the dollar at a fixed exchange rate) the value of national currencies was determined only by investor confidence. In 1995 US$1.2 trillion was exchanged every day in the currency markets, and 95 per cent of all currency transactions in the world are motivated by speculation. Time is critical to profit-making in the international money markets; it is the speed of the transaction, sometimes pre-programmed into a computer, that generates a profit or a loss, with the same capital often switching back and forth between economies in seconds. The creation of electronic 'megabyte money' was made possible by advances in telecommunications which link dealers in real time within the international financial system.[3]

The advent of money as a commodity has decoupled the 'real economy' from the 'money economy', the first and smaller of the two being where products are made, trade is conducted, research carried out and services are rendered. The other, far larger, money economy is concerned with speculation and the exchange of credits and debts. For the money markets, the long-term economic health of the real economy is incidental to the primary goal of electronic speculators, which is to realise a profit during the working day.

This short-termism can have an adverse effect on the operation of the real economy, with companies affected by rising interest rates, increasing commodity prices and declining exports due to a strong domestic currency. In extreme circumstances, companies and their jobs may be undermined regardless of performance, because of sudden changes in the financial environment in which they operate.

The main features of the capitalist world/global economy may be summarised as follows:

1. It comprises a single world market within which production is for exchange rather than use, with prices fixed by the self-regulating market. This means that for a period of time, more efficient producers can undercut others and increase their market share to the detriment of other producers *and places*.

2. Territorial divisions between states in the world economy lead to a competitive state system in which each seeks to insulate itself from the rigours of the world market (e.g. by trade barriers) while attempting to turn the world market to its advantage (e.g. by offering tax incentives to incoming industries).

3. The modern world economy comprises a core (characterised by relatively high incomes, advanced technology and diversified production) and a periphery. The core needs the periphery to provide the surplus to fuel its growth. **Core–periphery** relationships exist at all geographical scales, including, for example, differences in levels of prosperity between the 'sunbelt' towns of southern England and the older industrial cities of northern Britain.

4. The world economy has followed a temporally cyclical pattern of growth and recession characterised by Kondratieff cycles. These alternate phases of growth and stagnation show a high correlation with technological change and emphasise the regenerative effects of technological innovation. The first Kondratieff expansion phase (*c.* 1790–1815) was associated with the period of the original industrial revolution (with the introduction of mechanised textile production and improved iron production), while the fifth and latest highlights the role of microelectronics (in, for example, computing, industrial process control and telecommunications) in transforming the economy.

5. Finally, it is important to reiterate that every part of the world has its own particular relationship to the world economy based on locally specific modes of socio-economic organisation comprising a complex of management strategies, skill requirements, firm structures, labour organisation, infrastructural investments, consumption norms and government policies – all of the institutions and

institutional arrangements that impinge upon economic production, investment, consumption and employment.

THE WORLD ECONOMY IN THE POST-WAR ERA

The end of the post-war industrial boom in the early 1970s heralded a crisis for industrial capitalism. The system shock precipitated by the quadrupling of petroleum prices in 1973 as a result of the OPEC cartel has been indicted as a major cause of the economic downturn. However, as Hamilton (1984) explained, a number of other trends originating in the previous decade or before were also responsible for the recession.[4] These included the following:

1. A slowdown in economic growth and falling profits in the industrial core countries of the OECD was associated with falling levels of demand for capital goods (particularly in transport, steel and building).
2. Rising levels of inflation, which reduced profits and hampered capital accumulation, led to greater dependence on the banking sector for investment funds. High interest rates, however, restricted technological investment and so hindered competitiveness. Inflation also raised labour costs. The net outcome was the widespread depression of both capital-intensive (e.g. steel, ship-building, vehicles) and labour-intensive industries (e.g. textiles, clothing).
3. Increased international monetary instability, which took two forms:
   - Under- or overvaluation of exchange rates as a result of the transition in the early 1970s from fixed exchange rates to floating exchange rates (Box 14.1). In countries whose currencies were undervalued (e.g. Germany and Japan), industrial production was stimulated by increased demand for exports, whereas in those countries with overvalued currencies (e.g. oil and gas producers such as Norway and the UK) the loss of international competitiveness led to import penetration and a consequent decline in domestic industrial capacity.
   - Indebtedness among NICs and some underdeveloped countries as a result of massive borrowing from the 'petrodollar' surpluses of the OPEC states. This stimulated debtor countries

to increase exports of cheap manufactured goods to the core regions in order to obtain the necessary foreign exchange. This increased the competitive pressure on the labour-intensive sectors of the core economies.
4. The growth of new social values related to social welfare and environmental protection increased industrial costs and contributed to a higher tax burden for both producers and consumers.
5. The introduction of technological innovations in response to escalating energy and labour costs led to reduced demand in some traditional industrial sectors. For example, energy-saving designs in transport cut demand for steel, while innovations in micro-electronics reduced demand for electro-mechanical products.
6. A resurgence of political volatility (e.g. in the Middle East and South East Asia) reduced the area of stable business and constrained world trade.
7. The increasing intensity of international competition arose from the post-war liberalisation of trade, the spread of industrialisation to the periphery, the aggressive role of governments in NICs, and the post-1970 stagnation of world markets.

The net effect of these trends was to initiate a process of industrial reorganisation and structural change that signalled a transition from *industrial capitalism* to post-industrial or *advanced capitalism* (alternatively described as a move from the fourth to the fifth Kondratieff cycle).

TRANSNATIONAL CORPORATIONS AND THE STATE IN THE GLOBAL ECONOMY

Transnational corporations are major actors in the post-industrial global economy, directing investment and production on a worldwide basis with only limited consideration for national boundaries. By the early 1990s there were 37,000 TNCs with over 170,000 foreign affiliates. This complex of corporations accounted for over US$2 billion of foreign direct investment and totalled over US$5.5 billion in worldwide sales.[5] By 1999 there were 60,000 TNCs worldwide, with 500,000 foreign subsidiaries. The 100 largest non-financial TNCs control 15 per cent of global foreign assets and account for 22 per cent of total world sales for all TNCs. The economic power of some individual TNCs surpasses that of many nation-states. Of the eighty largest economic organisations in the world, thirty-four (43 per cent) are TNCs, all with head offices in the USA, Japan or Europe. TNCs are key players in the global economy.

The economic and political power of TNCs is manifested spatially in the creation of export processing zones in Third World countries. Within these enclaves TNCs benefit from low wage costs, government tax concessions and often freedom from national labour legislation. EPZs are labour-intensive manufacturing centres that involve the import of raw materials and export of factory products. Over ninety countries had established EPZs by the end of the twentieth century in response to problems of foreign indebtedness and as a cost-effective way to begin industrialisation. While the most common locations have been coastal, as in the case of China's special economic zones, one of the largest areas of EPZs is along the USA–Mexico border (see Chapter 5).

A feature of the capitalist global economy is that because the major economic actors operate within an area larger than that which any political entity can control, capitalists – financial institutions and TNCs – have a freedom to manoeuvre that is structurally based. The exercise of this freedom by 'footloose' capital can result in tension between them and territorially bounded national governments, and lead to adverse consequences (e.g. redundancies and economic decline) for less competitive places.

It is important to realise, however, that while the growth of a global economy has diminished the power of national governments to control their domestic economy, economic globalisation has not eliminated the economic significance of the nation-state. On the contrary, the deregulation and increasing global integration of financial markets (and corresponding loss of national autonomy) were promoted by nations that stood to gain most from a 'free market' in global finance.[6] The involvement of the state in the global economy is also evident in governments' attempts to aid 'their' firms in the global competition. Nowhere is this partnership between state and economy more strongly developed than in Japan, where the country's successful internationalisation of its production and financial capital is based on a close partnership between the Japanese state (the Ministry of International Trade and Industry, MITI) and industry. The state (MITI) played a crucial role in reorienting the economy away from heavy industry into electronics, cars and other consumer durables for world markets, and in directing the flow of Japanese capital into the USA, Britain and mainland South East Asia.[7]

Globalisation has altered rather than eliminated the economic role of the nation-state. Traditional economic policies managed within the boundaries of regulated national economies are increasingly ineffective, as key factors such as marketing policy, interest rates or technological innovation are highly dependent on global movements. However, while the power of global forces to influence urban growth or decline is profound, it is not omnipotent. Economic and political action undertaken at a national level or by regional groupings of nation-states can modify the effects of the global economic system. Regulatory and tax policies shape the environments that attract or repel investors; decisions about public investment determine whether infrastructure will be rebuilt or allowed to deteriorate; government procurement policy stimulates the private economy; and intergovernmental transfer payments can prevent the collapse of a local economy. Furthermore, the impacts of global processes are manifested in particular local contexts, the nature of which varies between places.

NEW PRODUCTION SYSTEMS AND NEW INDUSTRIAL SPACES

The flexible production systems of a post-Fordist economy and the technological innovation postulated as a basis for the fifth Kondratieff upswing come together in particular places or **new industrial spaces** variously referred to as technology parks, science cities or technopoles (Box 14.2).

The development of technopoles results from the clustering of 'specific varieties of the usual factors of production: capital, labour, and raw material, brought together by some kind of institutional entrepreneur, and constituted by a particular form of social organisation'[10] (Castells 1996 p. 391). The classic example is the development of 'Silicon Valley' in Santa Clara CA. From being an agricultural area with a population of 300,000 in 1950 the area was transformed into the world's most intensive complex of high-technology activity with a population of 1.25 million in 1980.[11] The raw material was knowledge produced by major centres of innovation such as Stanford University, Caltech or MIT. Labour came from the output of scientists and engineers from local universities. Capital was prepared to take the risk of investing in pioneering high technology either because of defence-related spending or for the potentially high rewards. The catalyst that energised these production factors was, in the case of Silicon Valley, the launching of the Stanford Industrial Park. Finally, over the years social networks of different kinds have contributed to the development of a local

Four main types of 'technologically innovative milieux'[8] can be identified:

1. There are industrial complexes of hightechnology firms in which R&D and manufacturing are linked. Some, such as Silicon Valley, are created entirely from the growth of new high-technology firms. Other complexes, such as Boston's Route 128, develop through the transformation of old industrial regions. These techno-industrial complexes arise without deliberate planning.

2. Science cities are research complexes with no direct link to manufacturing. The intention is to generate a high level of scientific excellence through the synergy created by agglomeration. Examples include the Siberian city of Akademgorodok and the Japanese science city of Tsukuba.

3. Technology parks aim to induce new industrial growth, in terms of jobs and production, by attracting high-technology manufacturing firms to a privileged space. These emanate from governments (e.g. Hsinchu in Taiwan) or universities (e.g. Cambridge) initiatives.

4. Technopoles may also be instruments of regional development and industrial decentralisation, the major example being the Technopolis programme initiated in 1984 in Japan.[9]

culture of innovation and dynamism that attracts capital and brainpower from around the world and encourages the circulation of ideas and a spirit of business **entrepreneurialism**. The combined effect has been to make Silicon Valley a self-sustaining innovative milieu of high-technology manufacturing and services.

Subsequent attempts have been made to duplicate the success of Silicon Valley by establishing technopoles in other parts of the world, including Munich (Germany), Grenoble (France), Cambridge and the M4 corridor to the west of London (England),[12] Bangalore (India),[13] Shenzen (China)[14] and Malaysia's multimedia supercorridor.[15] The most comprehensive approach is Japan's technopolis programme masterminded by the Ministry of International Trade and Industry (MITI). This aims to create twenty-six new science cities in the country's peripheral areas, in order to simultaneously promote new technologies and develop lagging regions.[16]

At the intra-metropolitan scale new industrial clusters centred on innovative, knowledge-based, technologically intensive activities such as computer graphics and imaging, software design and multimedia industries as well as technologically 'retooled' industries such as architecture and graphic design have been identified as key components of an emerging 'new economy' of the inner city. Such firms are attracted to the metropolitan core by the creative habitat, potential for 'knowledge spillovers' between companies, opportunities for social interactions across work and non-work life, and the cultural and environmental amenities of the locale. Examples of these new production spaces include Telok Ayer in Singapore, Multimedia Gulch in San Francisco, New York's Silicon Alley and the inner East End of London. While the nature of the new industrial clusters varies with local context general positive effects can include:

1. A contribution to the regeneration of local economies, with growth in employment and high-value output.
2. The preservation of heritage buildings through adaptive reuse.
3. The increased vitality of inner-city neighbourhoods.
4. Positive regional growth effects.

On the down side, critics have pointed to negative impacts similar to those emanating from the development of cultural-industry quarters (see Chapter 16). These include:

1. Community dislocation due to the physical encroachment of new industrial activities.
2. The displacement of existing businesses and residents unable to compete with the new industries' ability to pay enhanced rents.
3. Increased social polarisation due to gentrification and lifestyle differences between established residents and incoming workers.

DEINDUSTRIALISATION AND TERTIARISATION

The growth of a global economy is part of a transition to advanced capitalism. This change has been depicted as a shift from a post-war Fordist paradigm to a **post-Fordist** regime of flexible accumulation, as shown in Tables 14.1 and 14.2. A major element in this transition has been the expansion of the service sector of the

TABLE 14.1 THE POST-WAR 'FORDIST' EXPANSIONARY REGIME OF THE LATE 1940S TO EARLY 1970S

Characteristic	Key features
Industry	Monopolistic; increasing concentration of capital; steady growth of output and productivity, especially in new consumer durable-goods sectors; expansion of private and especially public services
Employment	Full employment: growth of manufacturing jobs up to mid-1960s; progressive expansion of service employment; growth of female work; marked skill divisions of labour
Consumption	Rise and spread of mass consumption norms for standardised household durables (especially electrical goods) and motor vehicles
Production	Economies of scale; volume, mechanised (Fordist-type) production processes; functional decentralisation and multinationalisation of production
Labour market	Collectivistic; segmented by skill; increasingly institutionalised and unionised; spread of collective wage bargaining; employment protection
Social structure	Organised mainly by occupation, but tendency towards homogenisation. Income distribution slowly convergent
Politics	Closely aligned with occupation and organised labour; working-class politics important; regionalist
State intervention	Keynesian-liberal collectivist; regulation of markets; maintenance of demand; expansion of welfare state; corporatist; nationalisation of capital for the state
Space-economy	Convergent; inherited regional sectoral specialisation (both old and new industries) overlaid by new spatial division of labour based on functional decentralisation and specialisation: regional unemployment disparities relatively stable

Source: R. Martin (1988) Industrial capitalism in transition: the contemporary reorganisation of the British space-economy, in D. Massey and J. Allen (eds) *Uneven Redevelopment* London: Hodder & Stoughton, 202–31

economy. Between 1960 and 1990 the service sector's share of global output, measured as a percentage of GDP, rose from 50.6 per cent to 62.4 per cent (Table 14.3). By 2000 70 per cent of output in 'high-income countries' derived from the service sector (Table 14.4). This appears to be a universal phenomenon, although the scale of tertiarisation is relatively slower in developing countries, where the proportionately smaller size of the service sector is explained in part by the existence of a larger informal sector that fulfils many of the functions undertaken by the formal service sector in advanced economies[17] (see Chapter 24). In terms of advanced or higher-order producer services the process of urban tertiarisation has led to the emergence of a number of 'urban service corridors' within Asian-Pacific city regions, including those centred on Seattle–Portland, Tokyo–Kyoto and Singapore–Kuala Lumpur.[18]

The related trend of deindustrialisation is a feature of developed economies (Tables 14.3 and 14.4). Whereas developing nations have witnessed manufac-turing growth (in terms of both absolute and relative output and employment), the 'industrial nations' have experienced a significant decline in manufacturing's share of total output and employment. In large part this trend reflects the replacement of the old international division of labour, based on sector differentiation (with, for example, primary-sector activity in the developing nations and higher value-added stages of manufacturing in developed nations) by a *new international division of labour* based on the separation of functions (in which the control and command functions are located in a network of global cities in the developed nations while physical production is increasingly dispersed to a host of developing countries where new technology can be allied to lower labour costs). Most recently the NIDL has affected the tertiary sector of the economy with the advent of 'off-shoring' whereby companies move white-collar jobs abroad (Box 14.3).

The expansion of the service sector is evident in cities at all levels of a national urban system from

TABLE 14.2 THE 'POST-FORDIST' REGIME OF FLEXIBLE ACCUMULATION FROM THE MID-1970S TO THE PRESENT

Characteristic	Key features
Industry	Rationalisation and modernisation of established sectors to restore profitability and improve competitiveness: growth of high-tech and producer service activities and small-firm sector
Employment	Persistent mass unemployment; general contraction of manufacturing employment, growth of private service-sector jobs; partial defeminisation (in manufacturing); flexibilisation of labour utilisation; large part-time and temporary segment
Consumption	Increasingly differentiated (customised) consumption patterns for new goods (especially electronics) and household services
Production	Growing importance of economies of scope; use of post-Fordist flexible automation; small-batch specialisation; organisational fragmentation combined with internationalisation of production
Labour market	Competitive; deunionisation and derigidification; increasing dualism between core and peripheral workers; less collective, more localised wage determination
Social structure	Trichotomous and increasingly hierarchical; income distribution divergent
Politics	Dealignment from socio-economic class; marked decline of working-class politics; rise of conservative individualism; localist
State intervention	Keynesianism replaced by free-market Conservatism; monetary and supply-side intervention rather than demand stabilisation; deregulation of markets; constraints on welfare; self-help ideology; privatising the state for capital
Space-economy	Divergent; decline of industrial areas (pre- and post-war); rise of new high-tech and product services complexes; increasingly polarised spatial division of labour; widening of regional and local unemployment disparities

Source: R. Martin (1988) Industrial capitalism in transition: the contemporary reorganisation of the British space-economy, in D. Massey and J. Allen (eds) *Uneven Redevelopment* London: Hodder & Stoughton, 202–31

TABLE 14.3 STRUCTURE OF WORLD OUTPUT, 1960–90 (% OF GDP)

	Agriculture				Manufacturing				Services			
	1960	1970	1980	1990	1960	1970	1980	1990	1960	1970	1980	1990
World	**10.4**	**6.9**	**5.6**	**4.4**	**28.4**	**26.1**	**22.4**	**21.4**	**50.6**	**56.4**	**57.0**	**62.4**
Industrial countries	**6.3**	**3.9**	**3.4**	**2.5**	**31.0**	**27.7**	**23.6**	**21.5**	**51.6**	**58.3**	**60.2**	**64.7**
USA	4.0	2.8	2.6	1.7	29.0	25.2	21.8	18.5	57.2	62.7	63.8	70.3
Japan	13.1	6.1	3.7	2.4	35.1	36.0	29.2	28.9	40.9	47.2	54.4	55.8
Europe	8.8	4.8	3.6	3.0	34.7	30.5	24.0	21.5	43.8	54.3	59.2	63.8
Developing countries	**31.6**	**22.4**	**14.7**	**14.6**	**15.6**	**17.8**	**17.6**	**20.7**	**42.8**	**48.1**	**44.2**	**49.8**
Latin America and Caribbean	16.5	11.9	9.5	8.5	21.1	22.8	23.5	22.3	50.6	55.3	54.5	57.5
Africa	45.8	33.1	25.1	30.2	5.8	8.6	7.7	10.9	37.5	44.0	39.9	41.6
Asia	38.0	23.9	17.8	17.4	14.4	15.1	15.4	20.9	37.1	42.5	37.9	45.9
South and South East Asia	44.0	35.6	24.7	18.9	13.7	15.6	19.8	22.9	36.2	41.9	43.5	46.6

Source: International Labour Organisation (1995) *World Employment 1995* Geneva: ILO

TABLE 14.4 WORLD OUTPUT BY SECTOR AND MAJOR REGION, 2000 (%)

Region	Agriculture	Industry	Services
World	3.9	29.8	66.3
High-income countries	1.9	28.8	69.5
EU	2.4	29.2	68.4
UK	1.1	28.7	70.3
USA	1.6	24.9	73.5
Japan	1.4	31.8	66.8
Middle-income countries	9.7	35.3	55.1
Low-income countries	25.4	30.94	43.7

Source: **World Bank World Development Indicators**

BOX 14.3

Offshoring: a new international division of labour

The impact of offshoring on manufacturing industry in the UK and USA is well documented. Now the trend is moving up the management ladder as globalisation and technology make it possible to move white-collar work to wherever the costs are lowest. In recent years many call centre jobs have gone overseas to cities (such as Bangalore, India) with an educated English-speaking low-cost work force.

Some see this process as an inevitable part of the creative destruction inherent in capitalism. It is argued that offshoring can be of benefit through:

1. Reduced costs and larger profits that may be re-invested in more innovative enterprises at home.

2. New and expanding subcontractors abroad create new markets.
3. Urban economies are forced to develop more dynamic sectors.

A report from Deloitte Consulting estimated that 2 million jobs in the financial services sector could migrate to lower-cost countries, such as India, in the next few years, including 730,000 from Europe. Looking further ahead, capital's continual search for lowest costs will inevitably see the tide of offshoring wash over countries like India as economies, such as that of China, acquire a competitive edge in the global market place.

Source: Sunday Times **5 October 2003**

---

those catering for local–regional markets to others serving a global market. In many cities a new economic core of banking and service activities has emerged to replace an older core based on manufacturing. At the heart of this new urban economy are the producer services provided to firms (e.g. legal, financial, advertising, consultancy and accounting services). Paradoxically, despite their use of the most advanced information technologies, producer services tend to be concentrated spatially in the downtowns of major cities. For example, over 90 per cent of jobs in finance, insurance and real estate (FIRE) in New York are located in Manhattan, as are 85 per cent of business service jobs.[19] This concentration in high-cost central cities is explained by the agglomeration economies provided by such locations, and the interdependence among highly specialised and innovative

service providers, and the desire to maintain close linkages with client firms. The twin processes of de-industrialisation and tertiarisation are evident in London's economy that has undergone a transformation from industrial to post-industrial over the past half century. Until the mid-1960s London was a major centre of light industrial production with one-third of the labour force employed in manufacturing. By contrast, although the city had long functioned as a national and international financial centre, the importance of finance and business services for overall employment was relatively small with only about 10 per cent of London's work force employed in this sector in 1961. Over the intervening period the relative importance of manufacturing industry and financial services has been reversed as London has emerged as one of the world's three key international financial centres.[20]

The changing structure of the London economy is shown in Table 14.5. Between 1981 and 2005 employment in banking, finance, insurance and business services grew from 15.9 per cent of total to 28.2 per cent; whereas employment in manufacturing fell from 19.2 per cent to 6.2 per cent. In addition to its direct contribution to the metropolitan economy the financial services sector helps sustain other industries such as transport and communications, restaurants and hotels while the high salaries earned add to aggregate spending power. Nevertheless, while manufacturing is no longer the engine of the London economy it remains an integral component in the machine.

This sectoral transformation in the economy is reflected in the changing nature of work in advanced capitalist society.

THE NATURE OF WORK AND THE DIVISION OF LABOUR

The processes of deindustrialisation and tertiarisation affect the mix of job opportunities available to urban residents. Fundamentally, whereas the mass-production techniques of Fordism did not require a well educated work force, in the post-Fordist era 'only a well-trained and highly adaptable labour force can provide the capacity to adjust to structural change and seize new employment opportunities' (OECD 1993 p. 9).[21] The advent of post-Fordist production strategies and the drive by firms to attract a flexible and skilled labour supply is creating a core–periphery division within the work force. The core comprises multi-skilled and well-trained employees in highly paid and reasonably secure jobs, whereas the peripheral labour force is most vulnerable to lay-off in times of economic downturn.[22] In labour-market terms this indicates a growing gap between what Drucker (1993) has referred to as *knowledge workers* and the rest.[23] A similar workforce division underlies Florida's (2000 p. 68) concept of the rise of a *creative class* of people, (knowledge workers and other creative professionals 'who add economic value through their creativity' This has given rise to the notion of 'creative cities' within the context of a post-Fordist economy characterised by flexible labour markets.[24]

Plate 14.1   Suburban commuters on their way to work flow past tourists on Westminster Bridge, London

TABLE 14.5 EMPLOYMENT IN LONDON

Sector	1981		1991		2005		Change 1981–91		Change 1991–2005	
	No. of jobs	% of total	No. of jobs	% of total	No. of jobs	% of total	No. of jobs	% change	No. of jobs	% change
Primary	57,275	1.6	41,364	1.3	4,800	0.2	−15,911	−27.8	−36,564	−88.4
Manufacturing	683,951	19.2	358,848	11.0	220,110	6.2	−352,103	−47.5	+92,633	+38.7
Construction	161,407	4.5	118,367	3.6	211,000	5.9	−43,040	−26.7	+92,633	+78.3
Services	2,655,288	74.6	2,736,165	84.1	3,129,000	87.8	+80,877	+3.0	+392,835	+14.4
Distribution, hotels and catering	686,598	19.3	645,955	19.8	562,300	15.8	−40,643	−5.9	−83,655	−13.0
Transport and communications	368,288	10.3	307,682	9.5	260,400	7.3	−60,606	−16.5	−47,282	−15.4
Banking, finance, insurance and business services	565,876	15.9	733,513	22.5	1,006,600	28.2	+167,637	29.6	+273,087	+37.2
Other, including public administration and health	1,034,526	29.1	1,049,015	32.2	1,299,700	36.5	+14,489	1.4	+250,685	+23.9
Total	3,557,921		3,254,744		3,564,900		−305,944	−8.6	+310,156	+9.5

The division of labour according to 'occupational status' (based on educational standards and occupational qualifications) is the principal form of labour market segmentation. Other important divisions of labour in post-industrial society include those based on the following:

1. *Full-time versus temporary work.* Although taking temporary work can be a voluntary response to the demands of competing activities (such as domestic work by females), it is also often involuntary, arising because of:
   - Employers' attempts to develop a 'flexible' labour force and reduce costs by paying lower wages to temporary workers and avoiding the health, pension and other benefits attached to permanent contracts.
   - The low wage levels in part-time employment, which require people to take on more than one job. In the USA in 1994 7 million people (6 per cent of the work force) held 15 million jobs.
2. *Gender.* The 'traditional' notion of waged work as a man's world and the home as the place for women has been overturned by the growth of female employment which has accompanied de-industrialisation and the rise of a service-based economy. In the UK women accounted for 38 per cent of the labour force in 1972 but 50 per cent in 1996. By the latter date 70 per cent of working-age women were engaged in the formal labour market. Despite the move towards equalisation of male and female economic activity rates, however, most women's experience of employment remains qualitatively different from that of most men. This is reflected in:
   - A higher involvement in part-time work. By 1995 part-time work accounted for 27 per cent of all officially recorded jobs in the UK, with women occupying 80 per cent of these.
   - The concentration of women in certain industrial sectors such as the service sector.
   - Income differentials, which may be explained in part by the female concentration in part-time low-status employment. Even in full-time employment, however, women receive, on average, 75 per cent of the wage of a male employee, although the difference in part reflects the role of overtime working in some sectors.
3. *Space.* The spatial segmentation of labour markets is evident in the inner-city 'jobs gap' whereby poor households with low skills tend to be anchored in declining neighbourhoods, distant from suburban areas of job growth.[25] Explanations for this 'spatial mismatch' are complex. Personal attributes, such as level of qualifications, coupled with shortages of suitable jobs in the local area are major factors. Other factors include the friction of distance combined with poor public transportation, and racial discrimination whereby 'race not space' may contribute to the spatial division of labour (see Chapter 15).

FEMINISATION

Over recent decades the feminisation of employment has been a feature of the UK economy. Between 1984 and 2004 while the employment rate of men remained stable at 78 per cent and 79 per cent respectively, that of women increased from 59 per cent to 70 per cent, to the extent that women constituted 49 per cent of the total UK work force. Notwithstanding this trend gender divisions remain in terms of the type of work undertaken.[26] Employment remains highly segregated by occupation and time. Women are over-represented in lower paid 'people-related' activities – the 5Cs of cleaning, catering, cashiering, clerical and caring – and are five times more likely than men to work part-time (40 per cent compared to 8 per cent). Consequently, women may experience a 'part-time pay penalty'. While full-time female workers earn on average 18 per cent less per hour than men, the corresponding figure for part-time women is 40 per cent.[27]

In addition to the general criteria of qualifications and skills women's participation and role in the labour force is influenced by a range of factors including multiple demands on their time that arise from balancing employment and family responsibilities ('social reproduction'). Decisions on whether to enter employment and the type of work sought are shaped by internal (e.g. the domestic division of labour) and external forces (e.g. cost and quality of childcare, public transport accessibility, and location of work opportunities).[28]

UNEMPLOYMENT

The greatest division in the labour market is that between the employed and the unemployed. Economic restructuring – deindustrialisation, tertiarisation, increased automation in both manufacturing and services, the demand for skilled labour, the 'feminisation' of labour markets, and requirements of 'flexibility' in post-industrial working practices – has contributed to a growth in unemployment and long-term unemployment.

Since the mid-1970s long-term unemployment has become a feature of most capitalist economies.[29] Typically, many of the long-term unemployed were previously in semi-skilled or unskilled jobs. During the 1980s youth unemployment emerged as a major problem, with the ranks of the long-term unemployed swollen by never-employed school leavers. In Glasgow, by 1987 54 per cent of unemployed males and 39 per cent of unemployed females had been out of work for more than a year, while 16 per cent of unemployed males and 6 per cent of unemployed females had not worked for over five years. At the same time, youth unemployment levels had reached 73 per cent for males and 52 per cent for females under the age of 25 years. These trends continued to characterise the urban economy during the last decade of the century.[30] The socio-spatial impact of unemployment is evident in deprived urban environments such as the Kirkby estate in Liverpool, where the local labour market 'is increasingly incapable of providing jobs in adequate numbers for those seeking them and is riven by a growing polarisation in the pay and conditions of the work that is available' (Meegan 1989 p. 198)[31] (Box 14.4).

INFORMALISATION

The growth in service-sector employment has not been sufficient to offset the loss of jobs in manufacturing. In addition, the type of labour required in many segments of the service sector (part-time, temporary and female) does not match the abilities of redundant male manufacturing workers. The organisation of remaining manufacturing industry has also changed, with a growth in small-batch production, high product differentiation and rapid changes in output. This has led to greater subcontracting and an increase in part-time and temporary work, piecework and industrial home-working (often under **sweatshop** conditions). In these circumstances employers may reduce their costs by circumventing employee benefits, labour unions and even the need to provide a workplace. The emergence of a 'downgraded manufacturing sector'[32] is an example of the informalisation common in Third World cities but which is increasingly evident in developed countries.[33] Los Angeles has become the largest centre for women's clothing manufacturing in the USA. A combination of global competition in the garment industry,

BOX 14.4

Unemployment on the Kirkby estate in Liverpool

Economic restructuring has resulted in a growing number of unemployed residents, with both long-term and youth unemployment significant. Registered unemployment is three times the national average, and according to a local resident, 'You never see anyone going to work in the morning.' Another Kirkby woman, talking about her 20-year-old son, stated, 'He's never had a job and wouldn't know how to go about getting one.'

In 1987 only 7 per cent of school leavers in the borough found jobs (compared with 28 per cent in 1979). The registered unemployment rate for young people has effectively been kept down by a combination of government training schemes and pupils 'staying on' at school. As the opportunities for full-time work has contracted, the local labour market has polarised between large private and public employers and a collection of small firms. The remaining large companies in the area enjoy increased dominance in the labour market and recruit selectively, often without advertising vacancies so as to avoid 'virtually unmanageable numbers of applications'. Pay and conditions in the small firms are poor, with wages often set below levels of remuneration on training schemes in the knowledge that the prospect of a 'real job' will attract workers.

The downgrading of the local labour market is reflected in the reduced standard of living for people on the estate. This is illustrated in increasing numbers of families in receipt of housing benefit (81 per cent in 1987) and free school meals (77 per cent). As an unemployed man stated, 'I'm always skint because I'm unemployed. I get depressed because I can't provide for my family. We have fights . . . A lot of people are in this situation. It's quite common.'

Source: R. Meegan (1989) Paradise postponed: the growth and decline of Merseyside's outer estates, in P. Cooke (ed.) *Localities* London: Unwin Hyman, 198–234

the city's association with fashion, its large population of immigrant workers, and its anti-union tradition and decline of unionisation, and the shift to flexible production have promoted growth of sweatshops. In Chicago the city's 'contingent economy' based on flexible, short-time and temporary employment expanded during the 1990s as secure decent-paying job opportunities declined. In some inner-city neighbourhoods the major local employers are temp agencies (both main-stream and 'backstreet') who specialise in the place-ment of day-labour through a network of hiring halls where prospective workers gather each day, as compo-nents in a 'regime of precarious employment'.[34]

Other forms of informal economic activity in industrial societies include:

1. The domestic or household economy, which pro-duces goods and services primarily for own con-sumption but which accounts for the great part of national child-care and dependent-adult-care services.
2. The neighbourhood or mutual aid economy in which goods and services may be exchanged on a non-commercial or non-monetary basis within a local community.
3. The alternative or counter-culture comprising a form of social economy (such as Amish or Old Order Mennonites) based on a rejection of the values of mainstream society.
4. The underground, hidden or black economy, which includes illegal activities, creative book-keeping for the purpose of tax avoidance, and unrecorded employment.

We shall return to consideration of informal economic activity later in our discussion of local responses to urban economic change (Chapter 16).

URBAN CHANGE WITHIN THE GLOBAL ECONOMY

For cities, the impact of the emergence of a global economic system is encapsulated in the distinction between the city as an autonomous self-governing polity (which existed in medieval Europe prior to the development of an economy based on the trade of mar-keted commodities) and present circumstances, under which city development is influenced to a significant degree by forces beyond its control. Today, investment decisions by managers in a TNC with headquarters in one of the 'command cities' of the global economic system can have a direct effect on the well-being of families living on different sides of the world. In order to confront such forces, cities must seek to position themselves and, increasingly, compete in global soci-ety. The fact that cities vary greatly in their capacity to meet the challenge posed by globalisation is reflected in the extent to which each can shape or simply react to global forces. This challenge is particularly acute

for older industrial cities that have been destabilised by the process of economic restructuring, which has accelerated since the early 1970s as part of the transi-tion to advanced capitalism.

The differential urban impact of globalisation may be illustrated by examining (1) the emergence of the **world city**, and (2) the problems of economic decline experienced by older industrial cities.

THE WORLD CITY

The structural changes that have occurred within the world economy have helped to reorder the relative importance of cities around the world. Some 'world cities' have become key command and control points for global capitalism. Such centres are distinguished not by their population size (as in the case of mega-cities), or their status as capital cities of large countries, but by the range and strength of their economic power.

An early attempt to define world cities by Hall (1966 p. 7)[35] was based on the premise that 'there are certain great cities, in which a disproportionate share of the world's most important business is conducted'. For Hall, such cities were characterised by their func-tion as major centres of political power and seats of national and international institutions whose major business is with government, including professional bodies, trade unions, employers' federations and cor-porate headquarters. They were also great ports and major airport hubs, leading banking and financial cen-tres, and cultural loci. On this basis, Hall recognised London, Paris, Randstad, Rhine–Ruhr, Moscow, New York and Tokyo as world cities.

Over the intervening period the meaning of the con-cept and the list of world cities have been revised. Braudel (1985)[36] used the term 'world city' to denote the centre of specific 'world economies'; Ross and Trachte (1983 p. 393) viewed world cities as 'the loca-tion of the institutional heights of worldwide resource allocation',[37] while for Feagin (1985 p. 1230) such cities are 'the cotter pins holding the capitalist eco-nomic system together'.[38] The economic bases of these definitions are clear. World cities are seen as places in and from which global business, finance, trade and government are organised. But the world city is also characterised by distinct social and cultural attributes (Box 14.5).

Friedmann (1986) identified a 'world city hier-archy' of thirty centres, each defined in terms of its role as a major financial, manufacturing and trans-port centre, a location for the headquarters of TNCs,

BOX 14.5

World city characteristics

1. The most inherent feature of the world city is its global control function in the world economy World cities occupy a position between the world economy and the territorial nation-state.

2. Since the institutions linking the world economy with the state are transnational corporations and international finance, world cities contain a disproportionate number of the headquarters of the world's largest corporations and international banks.

3. The presence of corporate headquarters and global financial institutions presupposes the rapid growth of a highly paid international elite, including a transnational producer service class (in law, banking, insurance, business services, accounting, advertising, etc.) engaged in the production and export of services across the globe.

4. The expansion of producer services leads to the refurbishment of existing office space and extensive new construction of both office space and high-class residential accommodation.

5. Because of the expanding economy, redevelopment and concentration of capital, world cities become major centres for international investment.

6. There has been a decentralisation of routine office jobs and recentralisation of control and management functions in the world city.

7. Increased global competition has led to de-industrialisation and decline in unionised blue-collar employment, with (to a much lesser degree) some selective re-industrialisation in high-technology, lower-wage, assembly-line jobs, non-unionised and often held by women or people belonging to minority ethnic groups.

8. The growth of the high-level service sector is accompanied by expansion of low-level, low-paid routine work in areas such as office cleaning, hotels, restaurants, domestic service, clerical duties, sales employment and tourism.

9. Much of the growth in industrial employment is associated with informally organised types of manufacturing, such as sweatshops, and industrial home work both in industrial (clothing/garment) and new (electronics) industries.

10. Depending on the degree of immigration controls, some world cities have experienced an influx of people seeking employment opportunities, leading to a sharp rise in income, occupational and social polarisation.

11. The polarisation of high-pay/high-skill and low-pay/low-skill employment is reflected in an increasing spatial separation of residential space according to occupation, race, ethnicity and income.

12. The growth of a high-income class provides the conditions for both new luxury residential building and the gentrification of older areas.

13. The growth in low-skill/low-pay employment combines with a surplus of labour in the city to inflate the informal or street economy, to which are added the numbers of unemployed, leading to the creation of an urban underclass.

14. Increasing economic and social polarisation combine to lead to increased lawlessness, expressed in racial conflict, violence and crime, despite increased expenditure on police and the growth of private security forces.

15. There is growing loss of local control over increasingly footloose capital and this is combined with increasing public expenditure to attract or maintain capital investment.

16. Heightened political conflict results from the contradictory demands of forces representing international capital and the needs of the local state.

17. Expansion in some growth sectors of the urban economy is accompanied by equally major decline in others, in keeping with the capitalist mode of production/creative destruction.

Source: adapted from A. King (1990) *Global Cities: Post-imperialism and the Internationalisation of London* London: Routledge

the number of international institutions present, rate of growth of business services, and population size.[39] Thrift (1989) recognised three main levels of world city:[40]

1. Truly global centres which contain many head offices, branch offices and regional headquarters offices of the large corporations and banks that account for most of the international trade – New York, London, Tokyo.

2. Zonal centres which have corporate offices of various types and serve as important links with the international business system, e.g. Paris and Los Angeles.

3. Regional centres which host many corporate offices and foreign financial outlets but are not essential links in the international business system, e.g. Sydney and Chicago.

Beaverstock *et al.* (2000) derived a roster of fifty-five world cities plus another sixty-seven cities showing evidence of world city formation[41] (Table 14.6). Despite differences in the detailed composition of lists of 'world cities', most analysts agree on the pre-eminence of London, New York and Tokyo as command centres in the global economy[42] (Box 14.6).

TRANSNATIONAL URBAN SYSTEMS

The concept of a world-city hierarchy focuses attention on linkages between individual cities within the global urban scene. Prime examples of links that bind cities across national borders are:

1. The multinational networks of affiliates and subsidiaries typical of major manufacturing firms and producer-services providers.
2. The global financial market created by deregulation and the advent of sophisticated electronic technology and telecommunications which allow traders to operate in 'real time' whatever their location.
3. The growing number of less directly economic linkages involving a variety of initiatives by urban

governments (such as designation of twin or sister cities) that amount to a type of 'foreign policy' by and for cities.

A key question for urban researchers is whether this rich network of linkages among world cities amounts to the formation of a transnational urban system. If one takes the view that global cities basically compete with each other for global business,[43] then they do not constitute a transnational urban system. If, on the other hand, one contends that world cities, beside competing, are also sites of transnational production processes with multiple locations,[44] then there is scope for the possibility of a 'systematic dynamic' binding these cities. This precept is best observed in the context of the global financial system, in which the three major world cities of New York, Tokyo and London play a different role in a series of processes which may be thought of as the 'chain of production' in finance. For example, in the mid-1980s Tokyo was the main exporter of the raw material known as money, while New York was the leading processing centre where through the invention of a range of new financial instruments (e.g. Eurocurrency bonds and interest-rate futures) money was transformed into 'products' that aimed to maximise the return on investment. London, on the other hand, was a major marketplace with an extensive international market able to centralise and concentrate small amounts of capital available in a large number of smaller financial markets around the world. For Sassen (1981) this suggested that these cities

TABLE 14.6 ROSTER OF WORLD CITIES

First-rank	Second-rank	Third-rank	
London	San Francisco CA	Amsterdam	Kuala Lumpur
New York	Sydney	Boston MA	Manila
Tokyo	Toronto	Caracas	Osaka
Paris	Zurich	Dallas TX	Prague
Chicago IL	Brussels	Dusseldorf	Santiago
Frankfurt	Madrid	Geneva	Taipei
Hong Kong	Mexico City	Houston TX	Washington DC
Los Angeles CA	São Paulo	Jakarta	Bangkok
Milan	Moscow	Johannesburg	Beijing
Singapore	Seoul	Melbourne	Rome
		Atlanta GA	Stockholm
		Barcelona	Warsaw
		Berlin	Miami FL
		Buenos Aires	Minneapolis MN
		Budapest	Montreal
		Copenhagen	Munich
		Hamburg	Shanghai
		Istanbul	

BOX 14.6

London as a world city

London is one of the great cities of the world. The strength of the Greater London economy is indicated by the fact that in 1985 the gross regional product amounted to US$78 billion, equal to 17 per cent of the UK's GDP and 4 per cent of the European Community's GDP, making London the twenty-fifth largest economy among the world's nations. London also contains the headquarters of most of the 190 UK-owned firms in the top 500 of European firms. The City of London is one of the three principal 'command centres' of the international financial system, along with New York and Tokyo.

London's position as a premier financial centre is based on a number of advantages, including the following:

1. A convenient time-zone location between Tokyo (nine hours ahead) and New York (five hours behind).
2. The use of English as the language of international commerce.
3. The large number, size and international character of British transnational corporations (TNCs), with forty-five of the 500 largest corporations in the world having headquarters in London.
4. The deregulation of the London Stock Exchange in the 'Big Bang' of 1986.
5. The existence of 'thick market externalities' whereby a firm's productivity benefits directly from the presence of competing firms engaged in the same or related activities.
6. The large size of the market for producer services.
7. The competitive advantage enjoyed by a long-established financial centre.

London's development as a world city and the accompanying restructuring of the urban economy have had a differ-ential social impact. This is seen most clearly in the nature of the labour market, which comprises the following types of employment:

1. *Elite professionals* engaged in well paid, relatively secure business service occupations in the fields of banking, finance, legal services, accountancy, technical consultancy, telecommunications and computing and R&D.
2. *Services to the elite:* less well paid occupations in real estate, entertainment, construction activities and domestic services, with some activities being seasonal.
3. *Services to international tourism:* jobs in hotels and catering services which are often poorly paid, part-time, casual and non-unionised.
4. *Manufacturing.* Following decentralisation, the remaining manufacturing industry is a mix of high-technology (office machinery), craft-based industry (e.g. leather goods) and information-processing and dissemination (printing and publishing).
5. *Informal economy* involving activities such as casual day labouring and street trading, which are often part of the 'black economy'.
6. *Marginalised populations*, including unemployed and homeless people, who are unable to compete successfully in the job market.

In social terms London and other world cities are marked by extremes of affluence and poverty, reflecting the differential distribution of the benefits of advanced capitalism.

Employment by industrial sector

Sector	1981		1991		Change 1981–91	
	No. of jobs	% of total	No. of jobs	% of total	No. of jobs	% change
Primary						
Agriculture, forestry and fishing	1,771	0.0	1,192	0.0	−579	−32.7
Energy and water supply	55,504	1.6	40,172	1.2	−15,332	−27.6
Total	57,275	1.6	41,364	1.3	−15,911	−27.8
Manufacturing						
Extraction of minerals other than fuel	72,838	2.0	35,439	1.1	−37,399	−51.3
Metal goods, engineering and vehicles	301,143	8.5	132,954	4.1	−168,189	−55.9
Other	309,970	8.7	190,455	5.9	−119,515	−38.6
Total	683,951	19.2	358,848	11.0	−325,103	−47.5
Construction	161,407	4.5	118,367	3.6	−43,040	−26.7
Services						
Distribution, hotels and catering; repairs	686,598	19.3	645,955	19.8	−40,643	−5.9
Transport and communications	368,288	10.3	307,682	9.5	−60,606	−16.5
Banking, finance, insurance, business services	565,876	15.9	733,513	22.5	167,637	29.6
Other	1,034,526	29.1	1,049,015	32.2	14,489	1.4
Total	2,655,288	74.6	2,736,165	84.1	80,877	3.0
Total employment	3,557,921		3,254,744		−305,944	−8.6

do not simply compete with each other for the same business, but represent an urban economic system based on three distinct types of locations.[45] Transnational urban systems in the form of 'co-operative clusters' have also emerged at a regional scale, as in the Hong Kong Pearl River delta region, and the Singapore–Johore–Riau (SIJORA) triangle.

The possibility of such a transnational urban system, based on cross-national ties between leading business and financial centres, raises the question of the nature of the links between world cities and their national urban systems. Although cities are embedded in the economies of their regions, those that are strategic sites in the global economy tend, in part, to disconnect from their local region. This may lead to growing inequality between cities that are integrated into the global urban hierarchy and those outside the system, which may become more peripheral.

The problems confronting cities outwith the mainstream of the global economy are considered next with particular reference to (1) economic restructuring in the Great Lakes region of the USA, and (2) post-war restructuring of the UK space-economy.

ECONOMIC RESTRUCTURING IN THE GREAT LAKES REGION

The processes of economic globalisation discussed earlier have shifted the focus of the US economy from the manufacture of goods to the conception, design, marketing and delivery of goods, services and ideas. Between 1970 and 2000 manufacturing employment in the USA fell by 3 per cent while services employment grew by 214 per cent. The share of American jobs in manufacturing fell from 22 per cent in 1970 to

---

BOX 14.7

Restructuring the Pennsylvania economy

Despite aggregate population growth in Pennsylvania's main metropolitan regions, their central cities, in tandem with older industrial cities throughout the USA, experienced severe population losses in the post-war period. Pittsburgh had lost nearly half of its 1950 population by 1990. Between 2000 and 2006 Pennsylvania's cities lost a further 3.3 per cent of their population while the state's outlying townships grew by 5.9 per cent, thereby contributing to the 'hollowing out' of metropolitan areas.

Suburbanisation has combined with industrial decline to weaken the urban economy. The de-industrialisation of Pennsylvania following the Second World War began with the decline of shipbuilding in the Philadelphia region in the 1950s, extending to textiles and apparel manufacturing in the Philadelphia–Allentown region, then coal mining around Scranton and Wilkes-Barre, and finally culminating in the collapse of the steel industry in Pittsburgh in the early 1980s. The employment effects were traumatic, with over half a million jobs lost from the four key industries between 1950 and 1990.

During the economically dynamic 1990s, Bureau of Labor Statistics data revealed slow growth in Pennsylvania jobs – of 0.96 per cent per annum

(compared with a national annual growth rate of 18.7 per cent). Between 2000 and 2006 the overall job growth rate in Pennsylvania slowed to 0.18 per cent per year to create +619,000 jobs (+1.1 per cent) over the period (i.e. one-third of the national rate) – a performance that ranked thirty-seventh among the states.

Pennsylvania, like the rest of the USA, has continued to shed manufacturing jobs and add service-sector jobs. Between 2000 and 2006 the state lost 22.2 per cent of its manufacturing jobs (−192,100, or −14.9 per cent), with greatest losses in metropolitan areas, such as Allentown. Economic growth areas include educational and health services (+14.9 per cent) and professional and business services (+10.6 per cent). The state needs to continue to diversify its traditional manufacturing-oriented economy by investing in promising high-value 'export' sectors such as life sciences, education, food processing, and business and financial services. Pennsylvania is no longer a leading national economic growth region. Lacking propulsive industries, its prospects of rapid expansion and release from slow growth are problematic, a scenario which is not unfamiliar in the northeast and Mid-west of the USA.

Sources: **S. Dietrick and R. Beauregard (1995) From front-runner to also-ran: the transformation of a once dominant industrial region: Pennsylvania USA, in P. Cooke (ed.) *The Rise of the Rustbelt* London: UCL Press, 52–71; Brookings Institution (2007) *Committing to Prosperity: Moving forward on the Agenda to renew Pennsylvania* Washington DC: Brookings Institution Press**

TABLE 14.7 EMPLOYMENT CHANGE IN GREAT LAKES METROPOLITAN AREAS, 1995–2005

Metropolitan area	Manufacturing employment change (No. of jobs)	Advanced service employment change (No. of jobs)	Sum of manufacturing and advanced service employment changes
Indianapolis IN	−11,500	46,000	34,500
Cincinnati OH	−24,400	55,000	30,600
Buffalo NY	−21,600	20,900	−700
Akron OH	−14,400	13,500	−900
Allentown PA	−17,000	15,800	−1,200
Ann Arbor MI	−9,500	5,900	−3,600
Scranton PA	−11,300	7,100	−4,200
Canton OH	−15,000	4,400	−10,600
Dayton OH	−21,300	9,400	−11,900
Milwaukee WI	−29,700	16,000	−13,700
Youngstown OH	−18,700	600	−18,100
Flint MI	−26,500	3,700	−22,800
Rochester NY	−37.400	11,900	−25,500
Cleveland OH	−52,700	20,500	−32,200
Chicago IL	−177,000	138,000	−38,600
Detroit MI	−87,700	31,500	56,200

11.5 per cent in 2000, while the national share of jobs in services increased from 19 per cent to 32 per cent. The processes of de-industrialisation and tertiarisation have impacted most on cities and regions with a traditional employment base in manufacturing, chief among these being the Great Lakes region.

The seven states of the Great Lakes manufacturing belt – Illinois, Indiana, Michigan, New York, Ohio, Pennsylvania, and Wisconsin – composed the heart of US manufacturing industry for most of the last century, still account for one-third of US manufacturing jobs, and comprise the only region on which all metropolitan areas of over 1 million population are 'manufacturing dependent' (defined as cities in which manufacturing's share of total metropolitan employment exceeded the national average share of 10.7 per cent in 2005).

The Great Lakes region and its cities have been especially susceptible to the currents of the global economy and, in particular, to competition from low-cost foreign-based manufacturers (Box 14.7). Between 2000 and 2005 more than one-third of the USA loss of manufacturing jobs occurred in the Great Lakes region. Most of the region's cities lost manufacturing employment, with Ann Arbor MI, Canton OH, Rochester NY and Youngstown OH recording declines of over 30 per cent. Flint MI lost 55 per cent of its manufacturing jobs over the period and correspondingly, along with

Dayton Oh and Youngstown OH, had fewer jobs in 2005 than ten years earlier. In contrast, most cities gained employment in high-wage advanced service industries (Table 14.7). However, this did not compensate for the loss of manufacturing jobs numerically or in terms of the different job skills required in the advanced service sector.[46] The case of the Pittsburgh steel industry is symptomatic. Between 1976 and 1996 manufacturing employment in the Pittsburgh region dominated by steel fell from 251,771 to 119,073 (−52,7 per cent), compared with a national decline in manufacturing employment of −0.7 per cent. This severe employment loss occurred mainly in the 1980s with the closure of large integrated steel plants due to globalisation of production and international competition.

Loss of jobs in the Great Lakes cities has been accompanied by loss of people. Between 1970 and 2000 the manufacturing work force in Flint MI fell by 65 per cent from 30,685 to 10,855 and the population declined by 35 per cent (from 197,000 to 125,000). Over the same period Detroit lost 65 per cent of its manufacturing work force (due largely to competition in the global automobile industry) that fell from 186,215 to 64,586; and 37 per cent of its population (from 1.51 million to 0.95 million). As a result, for the first time since 1920 the city's population was below 1 million.

THE POST-WAR RESTRUCTURING OF THE UK SPACE-ECONOMY

The process of economic globalisation has had a marked effect on the economies of most developed nations, and not least on the UK space-economy. The unparalleled prosperity of the long post-war boom (*c.* 1945–73) stimulated the Conservative Prime Minister Harold Macmillan to proclaim to the people of Britain that they had 'never had it so good'. Between the early 1950s and early 1970s labour productivity and real wages at least doubled. The fastest growth rates were in the new industries of electrical engineering, vehicle manufacture, and chemicals and petroleum products. The spread of mass-consumption norms, rising incomes and expansion of home and export demand for new, standardised consumer goods (such as cars, television sets and refrigerators) led to the adoption of mass assembly-line production methods and the growth of large, often multi-plant and multi-regional, firms capable of exploiting the economies of scale afforded by expanding markets. The expansion of manufacturing was accompanied by growth in the range of output of personal, business and public services. At the high point in the early 1950s Britain recorded a surplus on manufacturing trade equivalent to 10 per cent of GDP.[47] The buoyancy of the economy was such that rapid growth, rising prosperity and full employment combined to disguise the problems of the old industries of shipbuilding, coal, iron and steel, and heavy engineering. By 1960 Britain's share of total world exports had declined from 25.5 per cent in 1950 to 16.5 per cent – a trend that presaged further decline in the country's international competitiveness to 9.7 per cent in 1979. Import penetration had become a significant drain on the balance of payments even in sectors that during the 1950s had been dominated by domestic products (e.g. consumer durables). By the early 1970s it was evident that the post-war boom was coming to an end. This had fundamental consequences for the British economy and for Britain's cities.

Plate 14.2   In 1947 the shipyards of Glasgow on the river Clyde supplied 18 per cent of the world's tonnage. The decline of the industry over the following two decades was a stark manifestation of the deindustrialisation of the UK economy

Deindustrialisation

As we have noted, one of the most striking features of the transition to advanced capitalism has been the rapid and intense deindustrialisation of Britain's manufacturing base. Initially this took the form of relative decline (with manufacturing growth being less than that of the service sector), but with the advent of advanced capitalism there has been an *absolute* decline in manufacturing. Between 1966 and 1976 more than 1 million manufacturing jobs disappeared in net terms, a loss of 13 per cent. The decline affected most sectors of manufacturing: both traditional industries such as shipbuilding (–9.7 per cent), metal manufacturing (–21.3 per cent), mechanical engineering (–9.7 per cent) and textiles (–27.6 per cent) and the former growth sectors (and bases of the fourth Kondratieff cycle) of motor vehicles (–10.1 per cent) and electrical engineering (–10.5 per cent). In the West Midlands the net loss of 151,117 manufacturing jobs between 1978 and 1981 represented almost a quarter of all manufacturing employment in the region and helped redefine the once prosperous area as part of Britain's rustbelt.[48] In Coventry alone, between 1978 and 1982, 39,286 manufacturing jobs (39 per cent of the total) were lost.[49] The severity of the job losses in the West Midlands was in part due to the highly integrated nature of a local economy centred on metal and car industries which, during the 1960s and 1970s, were subjected to intense overseas competition. A further contributory factor was that national economic policies of the early 1980s, based on a free market and anti-interventionist philosophy, failed to mitigate (and arguably exacerbated) the effects of recession and uncompetitiveness in this and other sectors of British industry. By 1993 manufacturing accounted for only 18 per cent of jobs in the UK, compared with a figure of 38 per cent in 1971.

The problems of deindustrialisation in Britain have

BOX 14.8

Local economic restructuring in Glasgow

The Glasgow economy provides a stark illustration of the effects of the rise of post-Fordism, the advent of a global economic system, and the de-industrialisation of the UK economy. During the post-war decades the foundations of the local economy shifted from a traditional focus on heavy industry and manufacturing, in sectors such as coal, iron and steel, engineering and shipbuilding, to the present situation, where the leading component of the urban economy is the service sector (comprising producer, distributive, personal and 'non-marketed' services such as public administration).

Between 1981 and 1991 the manufacturing sector in Glasgow shed 38,869, or 44 per cent, of its jobs; by the latter date, service industries were employing 77 per cent of the city's work force. This process of local economic restructuring has been accompanied by fundamental changes in the nature of work and in the type of worker in demand. Whereas, in the past, unskilled assembly-line work offered employment for many Glasgow school leavers, (post)modern labour markets increasingly require employees with some degree of educational qualification. The shift in local labour demand is seen most strikingly in the growth of both long-term (more than a year) unemployment (which in 1993 accounted for 46 per cent of male and 31 per cent of female unemployment in Glasgow) and youth (under 25 years) unemployment, which accounted for 28 per cent of male and 44 per cent of female unemployment in the city. There is now over-whelming evidence of the link between the quality of educational opportunity and employment opportunity in post-Fordist society, and the salience of this relationship is all the greater for old industrial cities, such as Glasgow, which have been affected most severely by the transition to advanced capitalism.

Sector	1981		1991		2001		Change 1981–1991		Change 1991–2001	
	No.	%	No.	%	No.	%	No.	%	No.	%
Manufacturing	87,651	23.2	48,782	14.5	29,000	12.2	–38,869	–44.3	–19,782	–40.6
Construction	27,144	7.2	23,008	6.9	17,000	7.1	–4,136	–15.2	–6,008	–26.1
Services	256,819	68.0	259,614	77.1	184,000	77.3	2,795	1.1	–75,614	–29.1
Total	**377,831**	**100.0**	**336,544**	**100.0**	**238,000**	**100.0**	**–41,287**	**–10.9**	**–98,544**	**–29.3**

been felt most acutely in the peripheral regions of Clydeside (Box 14.8) and Tyneside, and in Lancashire, where the textile industry alone shed over half a million jobs.[50] The impact of deindustrialisation has had ripple effects beyond the manufacturing sector, with, for example, the coal industry adversely affected by the fall in demand from manufacturing (as well as from the power generators, due in part to the government decision to substitute cheaper oil and gas supplies in order to reduce energy costs for manufacturing industry). Similar effects were experienced in the German Ruhr[51] and, as we have seen, the old manufacturing belt of the north-eastern USA.

Tertiarisation

The growth of the service sector represented the other side of the coin, with over 3 million jobs created between 1971 and 1988 in areas such as R&D, marketing, finance and insurance (collectively referred to as producer services); transport and communications (distributive services); leisure and personal services (**consumer services**); and central and local government administration (public services). The growth of service-sector employment, however, has not been sufficient to cancel out the loss of employment opportunities in manufacturing. Evidence from individual cities provides a graphic illustration of the decline in non-service-sector employment (Table 14.8). In the Clydeside conurbation, manufacturing employment fell from 387,000 in 1961 to 187,000 in 1981. In the city of Glasgow, employment in manufacturing virtually collapsed from 227,000 in 1961 to 87,651 in 1981 (–61.4 per cent) and 48,782 by 1991 (–44.3 per cent). In Glasgow, between 1981 and 1991, while manufacturing shed 38,869 jobs, service sector employment increased by only 2,795. Nor are the new service-sector jobs necessarily suited to the skills of

Plate 14.3 The process of economic 'tertiarisation' is evident in the Isle of Dogs in London docklands, where an industrial economy has been replaced by financial services activities in **Canary Wharf**

TABLE 14.8 EMPLOYMENT CHANGE BY SECTOR IN MAJOR BRITISH CITIES, 1981–91 (%)

Sector	Glasgow	Newcastle	Birmingham	Leeds	Sheffield	Manchester	Liverpool	Cardiff
Primary	−16.3	−38.3	−18.6	−39.3	−52.0	−17.3	−64.8	−10.0
Chemicals, etc.	−67.4	−46.3	−48.3	−14.8	−73.0	−44.5	−41.9	−28.9
Engineering, etc.	−45.4	−49.0	−37.8	−26.2	−31.7	−32.8	−49.4	−12.2
Other manufacturing	−40.8	−37.2	−19.8	−26.9	−18.7	−48.6	−60.4	−3.7
Construction	−15.0	−41.1	−10.0	−14.0	−13.5	−26.2	−24.7	−24.4
Producer services	22.0	46.2	33.0	54.1	32.2	16.2	10.3	53.4
Distributive services	−25.3	−20.0	−6.4	9.8	11.1	−16.0	−36.9	−16.3
Personal services	4.6	16.7	22.2	19.3	32.5	−3.6	3.3	13.0
Non-market services	11.1	26.3	5.3	2.6	7.5	−0.7	−11.2	−1.0
Total	**−13.1**	**−4.5**	**−10.8**	**−0.6**	**−15.1**	**−12.9**	**−25.7**	**−0.7**

Source: M. Pacione (1995) Glasgow: The Socio-spatial Development of the City Chichester: Wiley

TABLE 14.9 EMPLOYMENT STRUCTURE OF MAJOR BRITISH CITIES, 2001 (%)

Sector	Glasgow	Newcastle	Birmingham	Leeds	Sheffield	Manchester	Liverpool	Cardiff	London
Manufacturing	12.3	10.0	21.8	15.2	17.7	12.7	10.5	11.7	8.5
Construction	7.3	5.5	7.7	7.6	6.5	5.5	6.4	5.9	5.7
All services	77.2	83.3	68.6	75.7	74.7	80.6	81.4	79.3	84.7
Distribution, hotels and catering	17.6	19.4	17.7	20.2	20.2	19.4	21.4	22.3	17.8
Transport and communications	9.4	7.0	6.6	6.9	5.5	9.0	7.0	6.4	8.3
Banking, finance and insurance	16.7	16.1	13.4	18.7	13.2	19.2	15.8	15.8	26.7
Public administration, education and health	28.0	33.7	26.8	25.4	31.8	27.0	32.1	26.0	23.5
Other services	55.0	7.1	4.1	4.4	4.0	5.9	5.2	8.7	8.4
Unemployment	5.1	4.4	5.3	3.0	3.9	5.2	6.6	2.9	7.0

redundant manufacturing workers. The result has been rising unemployment. In Glasgow in 1993, 40,261 males and 10,775 females were unemployed; 46 per cent of males and 31 per cent of females had been unemployed for more than a year, while 9 per cent of males and 5 per cent of females had been unemployed for over five years. Significantly, youth unemployment (i.e. among those aged 16–25 years) accounted for 28 per cent of male and 44 per cent of female unemployment.[52] These trends were repeated in all the large provincial cities of Britain following the onset of recession in the mid-1970s. By 2001, tertiarisation of urban economies was an established feature of the urban geography of the UK (Table 14.9).

One of the major factors underlying urban employment loss over recent decades has been the internationalisation of economic activity as private businesses have sought to respond to intense competition in stagnating markets by expanding the scope of their activities both functionally (through vertical and horizontal integration) and geographically (via the new international division of labour). This has produced a major paradox whereby the poor domestic record of British industrial capital in the post-war years is in contrast to the substantial growth that has taken place in the number of British **multinational corporations** with extensive investment and trading interests overseas. By 1970 all the largest 100 British manufacturing companies had become multinationals.[53] Although these investments abroad facilitated the penetration of foreign markets, the comparative lack of *domestic* investment by the leading sector of British industrial capital was an important factor underlying the failure of British industry to compete in the international economy. The internationalisation of Britain's economy has also occurred in part through inward investment. During the 1980s in both manufacturing and service industries there was an increased reliance on foreign-owned companies, and increased dependence of domestic employment upon decisions taken by non-UK companies.

State policy

Government policy has also played a major role in the restructuring of Britain's economy. The impact of government policy was clearly illustrated during the 1950s and 1960s, when a policy objective was to preserve the stability of sterling to enable it to continue to function as a top international reserve currency and as a major medium of international trade (and thereby maintain the key international role of the City of London).

This necessitated efforts to avoid deficits on the balance of payments in order to maintain confidence in sterling. Since this could not be achieved by imposing restrictions on the free flow of capital investment overseas, successive governments deflated the domestic economy in order to reduce demand for imports. At the same time, the need to maintain domestic levels of employment and to finance state expenditure limited the possibilities of such a policy, with the result that the UK economy experienced alternating periods of expansion and stagnation (stop–go). Not only did this fail to halt the long-term decline in the roles of sterling and of Britain in the international system but, as the periods of 'go' got shorter and those of 'stop' longer, it became increasingly difficult for firms to improve levels of investment, productivity and output.[54]

A further factor in the explanation of Britain's post-war economic problems was the defensive power of organised labour. For much of the period the better-organised workers established a significant degree of control over the labour process, which often led to restrictive practices concerning demarcation, manning levels, work rates and overtime as well as shopfloor resistance to the reorganisation of production. One consequence was that industrial capital was reluctant to re-equip and restructure, and often preferred to invest in lower-cost, non-unionised labour markets overseas. The commitment of post-war governments to the maintenance of full employment through Keynesian demand management and to the provision of a substantial welfare state further strengthened the power of organised labour (as well as being, in part, a reflection of that strength). During the 1950s and 1960s conditions of relative labour shortage aided labour to resist any reductions in real wages, thereby compounding the difficulties of industrial capital. In addition, financing of the state welfare system necessitated a rise in taxation levels. This impacted upon both industry (affecting profitability) and earned income and consumption (which fuelled trade union militancy and industrial disputes).

Urban impacts

The slow growth of the economy meant that by the late 1960s the British state was beset by both a crisis of industrial relations and an emergent fiscal crisis,[55] a combination that engendered descriptions of 'the British malaise'[56] and of the post-1973 era as a 'decade of discontent'.[57] A measure of the relative decline of the British economy in world markets was the reduction in the nation's share of world trade in manufacturing goods from 10 per cent by value in 1972 to under 7 per cent in 1988. Much of this fall was the result of competition from the rapidly growing Far Eastern economies, which challenged Britain in a variety of manufacturing industries, including the vehicle and textile sectors. There was also intense competition from Europe and Japan. At the end of 1986 unemployment had reached 3.25 million. By the end of the decade Britain was running a serious balance of payments deficit, with export prices above those of other West European economies. Recurrent fears were expressed, in both Labour and Conservative parties, about the uncompetitiveness of the British economy.

Much of the burden of recession was felt in the cities that had retained a disproportionate share of those industries most vulnerable to the demands of advanced capitalism. Urban areas with old factories employing outmoded production techniques and with uncompetitively low levels of labour productivity were affected most severely by capital reorganisation aimed at countering declining rates of profit. These measures included, in many urban areas, plant closures and the transfer of production to other locations with reduced labour inputs. Thus between 1966 and 1974 27 per cent of the job loss in manufacturing in Greater London was the result of firms relocating, 44 per cent was due to plant closures, and 23 per cent due to labour shedding by firms remaining *in situ*.[58] Furthermore, many of the firms that survived in inner-city areas imposed regimes of long working hours, low wages and poor conditions upon a work force drawn from exploitable social groups, including women and ethnic minorities.[59]

The loss of industry and jobs from inner urban areas was accompanied by decentralisation of population from overcrowded central areas to suburban locations, outer estates and New Towns. Between 1951 and 1981 the largest cities lost on average a third of their population.[60] The effects of these population shifts were heightened by the composition of migration flows, with the more able, affluent and self-sufficient departing by choice, leaving the elderly, young adults and those with below-average incomes to await the arrival of the redeveloper's bulldozer. In some of Britain's major cities the vacated spaces in the inner areas were occupied by immigrants. These economic, demographic and social trends contributed to a growing social polarisation within British cities and led to state recognition of an 'inner-city problem', a phenomenon that will be considered in detail in Chapter 15.

FURTHER READING

BOOKS

M. Castells (1996) *The Rise of the Network Society* Oxford: Blackwell

M. Castells and P. Hall (1994) *Technopoles of the World* London: Routledge

S. Deitrick and R. Beauregard (1995) From front-runner to also-ran: the transformation of a once-dominant industrial region: Pennsylvania USA, in P. Cooke (ed.) *The Rise of the Rustbelt* London: UCL Press

T. Hutton (2008) *The New Economy of the Inner City* London: Routledge

F. Lo and Y. Yeung (1998) *Globalisation and the World of Large Cities* New York: United Nations University Press

P. Newman and A. Thornley (2005) *Planning World Cities: Globalisation and Urban Policies* Basingstoke: Palgrave Macmillan

S. Sassen (1994) *Cities in the World Economy* London: Pine Forge Press

J. Short and Y.-H. Kim (1999) *Globalisation and the City* Harlow: Longman

I. Wallerstein (1974) *The Modern World System* New York: Academic Press

JOURNAL ARTICLES

J. Feagin (1985) The global context of metropolitan growth *American Journal of Sociology* 90(6), 1204–30

N. Flynn and A. Taylor (1986) Inside the rustbelt: an analysis of the decline of the West Midlands economy *Environment and Planning A* 18, 865–900

J. Friedmann (1986) The world city hypothesis *Development and Change* 17, 69–74

J. Peck and N. Theodore (2001) Contingent Chicago: restructuring the spaces of temporary labour *International Journal of Urban and Regional Research* 25(3), 471–96

A. Saxenian (1983) The genesis of Silicon Valley *Built Environment* 9, 7–17

R. Waldinger and M. Lapp (1993) Back to the sweatshop or ahead to the informal sector? *International Journal of Urban and Regional Research* 17(1), 6–29

KEY CONCEPTS

- Megabyte money
- World economy
- Global economy
- Kondratieff cycles
- Industrial capitalism
- Advanced capitalism
- Transnational corporations
- Globalisation
- Nation-state
- Fordism
- Post-Fordism
- New industrial spaces
- Technopoles
- Deindustrialisation
- Tertiarisation
- Division of labour
- Feminisation
- Unemployment
- Informalisation
- World city
- Transnational urban systems

STUDY QUESTIONS

1. What do you understand by the term 'globalisation'? Can you detect examples of the process in your own city?
2. Consider the relative influence of nation-states and transnational corporations within the global economy. Can you think of any examples of a transnational corporation impacting upon a local urban economy?
3. Explain the processes of deindustrialisation and tertiarisation and relate them to recent economic trends in a city with which you are familiar.
4. Explain the geography of the 'new economy' in the inner city.
5. Illustrate the changing nature of work in post-industrial societies and explain what is meant by the division of labour.
6. With reference to a city with which you are familiar explain the restructuring of its 'space-economy' in the post-war era.
7. Identify the principal characteristics of a world city and consider the validity of the concept of a transnational urban system.

PROJECT

Undertake a detailed case study of a technopole (e.g. Silicon Valley, Tsukuba, Cambridge). Examine its origins and development, the kinds of businesses involved, its position within the local, regional, national and global economy, and its likely future development trajectory.

15

Poverty and Deprivation in the Western City

Preview: theories of deprivation; the nature of poverty and deprivation; the underclass; dimensions of multiple deprivation; crime; ill health; ethnic status; gender; financial exclusion; the inner-city problem; problems of the outer city; the area-based approach to poverty and deprivation

INTRODUCTION

The problems of poverty and deprivation experienced by people and places marginal to the capitalist development process have intensified over recent decades. In the UK, during the 1980s, poverty increased faster than in any other member state of the European Community so that by the end of the decade one in four of all poor families in the Community lived in Britain.[1] Since 1979 the gap between rich and poor has widened, and by the mid-1990s over 90 per cent of the nation's wealth was owned by the richest half of the population. One in three of the poorest group were unemployed, 70 per cent of the income of poor households came from social-security payments and nearly one in five were single parents.[2] One in three of the child population and 75 per cent of all children in single-parent households were living in poverty.[3] By 2001 the wealthiest 1 per cent of the UK population owned 23 per cent of all wealth (compared with 21 per cent in 1976), while the poorest 50 per cent of the population owned 5 per cent of the nation's wealth (compared with 8 per cent in 1976).

The population living in poverty in the USA has also increased since the late 1970s, with 13 per cent below the official poverty line in 1980 and 15 per cent in 1993.[4] Although the majority of people in poverty were white, the poverty rate was higher among minority groups, with 33 per cent of African-Americans and 31 per cent of Hispanic Americans living below the poverty line compared with 12 per cent of whites.

While poverty rates fell to 24 per cent for African-Americans, to 23 per cent for Hispanic Americans and to 8 per cent for whites by 1999, the rates for minority groups remained significantly above those for the white population. The incidence of poverty was also reflected in family structure and gender, with 35 per cent of female-headed families living in poverty compared with 16 per cent of male-headed families, and 6 per cent of married couples.[5] In 1999 46 per cent of female-headed households with children under 5 years of age lived below the poverty line.[6]

In the UK most of the disadvantaged live in cities, large areas of which have been economically and socially devastated by the effects of global economic restructuring, the deindustrialisation of the UK economy and ineffective urban policies[7] (see Chapter 16). Similarly, in the USA, urban areas and central cities in particular contain disproportionate rates of social and economic problems associated with poverty and deprivation[8] (Table 15.1).

In this chapter we discuss the main theories proposed to explain the causes of deprivation. We examine the multi-dimensional nature of deprivation and the differential social incidence of multiple deprivation on different population groups. Using the concept of **territorial social indicators** we highlight spatial variations in poverty and deprivation within cities, focusing attention on the inner-city problem and problems of the outer city. Finally, we consider the value of an area-based approach for the analysis of urban poverty and deprivation.

TABLE 15.1 THE GEOGRAPHY OF POVERTY IN THE USA (% BELOW THE POVERTY LINE)

Type of family	USA	In metropolitan areas	In metropolitan areas		Not in metropolitan areas
			In central city	Not in central city	
All families	9.2	8.7	13.6	6.0	10.9
Families with children under 5 years of age	17.0	16.2	24.2	11.2	20.5
Female-headed families	26.5	25.1	31.3	19.3	33.0
Female-headed families with children under 5 years of age	46.4	44.6	50.3	37.6	54.8

THEORIES OF DEPRIVATION

Identifying the causal forces underlying deprivation is of more than academic importance, since the theory of deprivation espoused by policy-makers determines the nature of the response. Five main models have been proposed to explain the causes of deprivation, each pointing towards a different strategy. As Table 15.2 shows, theories of deprivation range from the concept of a 'culture of poverty' – which regards urban deprivation as the result of the internal deficiencies of the poor – to those that interpret deprivation as a product of class conflict within the prevailing social formation. The notion of a culture of poverty was first advanced in the context of the Third World and was seen as a

response by the poor to their marginal position in society.[9] According to this thesis, realisation of the improbability of their achieving advancement within a capitalist system resulted in a cycle of despair and lack of aspiration characteristic of the 'culture of poverty'. The related idea of transmitted deprivation focuses on the processes whereby social maladjustment is transmitted from one generation to the next (the **cycle of poverty**), undermining the ameliorative effects of welfare programmes. Particular emphasis is laid upon inadequacies in the home background and in the bringing up of children as causes of continued deprivation. The other three models take a wider perspective. The concept of institutional malfunctioning lays the blame for deprivation at the door of disjointed, and

TABLE 15.2 PRINCIPAL MODELS OF URBAN DEPRIVATION

Theoretical model	Explanation	Location of the problems
1. Culture of poverty	Problems arising from the internal pathology of deviant groups	Internal dynamics of deviant behaviour
2. Transmitted deprivation (cycle of deprivation)	Problems arising from individual psychological handicaps and inadequacies transmitted from one generation to the next	Relationships between individuals, families and groups
3. Institutional malfunction	Problems arising from failures of planning, management or administration	Relationship between the disadvantaged and the bureaucracy
4. Maldistribution of resources and opportunities	Problems arising from an inequitable distribution of resources	Relationship between the underprivileged and the formal political machine
5. Structural class conflict	Problems arising from the divisions necessary to maintain an economic system based on private profit	Relationship between the working class and the political and economic structure

Source: **Community Development Project (1975)** *Final Report Part I: Coventry and Hillfields CDP Information and Intelligence Unit* **London: HMSO**

therefore ineffective, administrative structures in which the uncoordinated individual approaches of separate departments are incapable of addressing the multi-faceted problem of deprivation. The theory of maldistribution of resources and opportunities regards deprivation as a consequence of the failure of certain groups to influence the political decision-making process. The final model, based on structural class conflict, stems from Marxist theory, in which problems of deprivation are viewed as an inevitable outcome of the prevailing capitalist economic order.

Although they are not mutually exclusive, each of the five theories does point to a particular policy response. Those subscribing to the culture-of-poverty thesis reject public expenditure on housing and other welfare items in favour of the concentration of resources on social education. Advocates of the concept of transmitted deprivation, while accepting the need for a range of anti-poverty programmes, emphasise the importance of the provision of facilities (such as nursery schools and health visitors) to assist child-rearing; while the solution to institutional malfunctioning has been seen in corporate management. Policies of **positive discrimination** are favoured mostly by those who view deprivation as a result of maldistribution of resources and opportunity. Such area-based policies are dismissed by those who subscribe to the structural class-conflict model of deprivation on the grounds that, being a product of the existing system responsible for deprivation, they are merely cosmetic, serving to 'gild the **ghetto**' without affecting the underlying causes of deprivation.

THE NATURE OF DEPRIVATION

A fundamental issue in the debate over the nature and extent of deprivation is the distinction made between absolute and relative poverty. The absolutist or subsistence definition of poverty, derived from that formulated by Rowntree (1901 p. 186),[10] contends that a family would be considered to be living in poverty if its 'total earnings are insufficient to obtain the minimum necessaries for the maintenance of merely physical efficiency'. This notion of a minimum level of subsistence and the related concept of a poverty line exerted a strong influence on the development of social-welfare legislation in post-Second World War Britain. Thus the system of National Assistance benefits introduced following the Beveridge report (1942)[11] was based on calculations of the amount required to satisfy the basic needs of food, clothing and housing

plus a small amount for other expenses. The same principle underlies the official definition of poverty in the USA, where the federal government identifies a range of poverty thresholds adjusted for the size of family, age of the householder and number of children under 18 years of age. Poverty thresholds are updated annually and adjusted for inflation in an attempt to provide an objective measure of poverty. (In 2007 the poverty threshold for a family of four was $20,650.)

If, on the other hand, we accept that needs are culturally determined rather than biologically fixed, then poverty is more accurately seen as a relative phenomenon. The broader definition of needs inherent in the concept of relative poverty includes job security, work satisfaction, fringe benefits (such as pension rights), plus various components of the 'social wage', including the use of public property and services as well as the satisfaction of higher-order needs such as status, power and self-esteem. An essential distinction between the two perspectives on poverty is that while the absolutist approach carries with it the implication that poverty can be eliminated in an economically advanced society, the relativist view accepts that the poor are always with us. As Figure 15.1 shows, poverty is a central element in the multi-dimensional problem of deprivation in which individual difficulties reinforce one another to produce a situation of compound disadvantage for those affected.

The root cause of deprivation is economic and stems from three sources.[12] The first arises from the

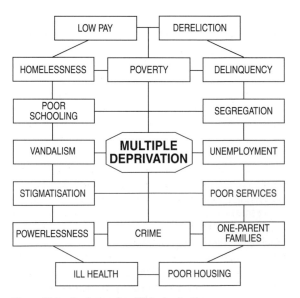

Figure 15.1   Anatomy of multiple deprivation

low wages earned by those employed in declining traditional industries or engaged, often on a part-time basis, in newer, service-based activities. The second cause is the unemployment experienced by those marginal to the job market such as single parents, the elderly, disabled and increasingly never-employed school-leavers. Significantly, since the 1960s, when poverty was largely age-related, the increasing number of unemployed and growing pool of economically inactive families (such as families with lone parents and or where the breadwinner suffers long-term sickness) have displaced pensioners as the poorest in society. The third contributory factor is related to reductions in welfare expenditure in most Western states as a result of growing demand and an ensuing fiscal crisis (see Chapter 17).

Significantly, the complex of poverty-related problems such as crime, delinquency, poor housing, unemployment and increased mortality and morbidity has been shown to exhibit spatial concentration in cities. This patterning serves to accentuate the effects of poverty and deprivation for the residents of particular localities.[13] Neighbourhood unemployment levels of three times the national average are common in economically deprived communities, with male unemployment rates frequently in excess of 40 per cent.[14] Lack of job opportunities leads to dependence on public support systems. The shift in regions such as Clydeside from heavy industrial employment to service-oriented activities, and the consequent demand for a different kind of labour force, has also served to undermine long-standing social structures built around full-time male employment and has contributed to social stress within families. Dependence upon social welfare and lack of disposable income lower self-esteem and can lead to clinical depression. Poverty also restricts diet and accentuates poor health. In both UK and US cities researchers have mapped the incidence of, and adaptive responses of low-income households to, 'food deserts' – areas of limited accessibility to foods, particularly those integral to a healthy diet.[15] Infant mortality rates are often higher in deprived areas, and children brought up in such environments are more likely to be exposed to criminal subcultures and to suffer educational disadvantage.[16] The physical environment in deprived areas is typically bleak, with little landscaping, extensive areas of dereliction, and shopping and leisure facilities that reflect the poverty of the area.[17] Residents are often the victims of stigmatisation that operates as an additional obstacle to obtaining employment or credit facilities. Many deprived areas are also socially and physically isolated, and those who are able to move away do so, leaving behind a residual population with limited control over their quality of life (Box 15.1). Below even this level of disadvantage are to be found the underclass.

THE UNDERCLASS

Myrdal (1962 p. 10) coined the term 'underclass' to describe those poor (Americans) being forced to the margins or out of the labour market by what is now known as post-industrial society.[18] Myrdal viewed the underclass as 'an unprivileged class of unemployed, unemployables and underemployed who are more and more hopelessly set apart from the nation at large and do not share in its life, its ambitions and its achievements'. For Williams (1986 p. 21), the underclass inhabit a Fourth World where 'all the disadvantages, inequalities and injustices of society [are] compounded among people, families and communities right at the very bottom of the social scale'.[19]

Recent debate over the emergence of a new urban underclass has focused academic and political attention on its causes. In essence, and reflecting the polarity in the explanations of deprivation discussed earlier, a key division exists between those who emphasise the causal importance of the behavioural characteristics of the people concerned[20] and those who stress the structural underpinnings of the phenomenon.[21] For Murray (1984) and others,[22] members of the underclass are identified most clearly by their behavioural traits of unemployment, criminality and, in particular, by the prevalence of unmarried mothers. Other researchers, while not disputing the composition of the underclass, focus causal attention on structural factors. Field (1989) includes in his definition of the underclass the frail elderly pensioner, the long-term unemployed person and the single parent with no chance of escaping welfare.[23] Clearly, while one viewpoint tends to blame individual failings, the other indicts the prevailing political economy in explaining the re-emergence of an urban underclass as an extreme form of the social polarisation that has accompanied the advent of post-industrial society.[24] Although the contribution of individual behaviour, albeit within a framework of constrained choice, cannot be ignored, particular attention has focused on the structural forces underlying the formation of an underclass. Figure 15.2 provides a cumulative causation model that incorporates both personal and structural factors in its explanation.

BOX 15.1

Life on a low income

I would like a day out to go shopping – just for food – to have the money and not get to the till and have to put things back.

I've lost all my friends. I don't go out much now, but a few years ago when I used to go and see my friends, as soon as they saw me the first thing that struck them was maybe I was coming to borrow something. Even if I had come to say 'hello'.

The kids live on things like fish fingers and, you know, the convenience foods that are really cheap to buy in bulk, and if I had money I would feed them differently. They seem healthy enough, but my conscience is pricked all the time because I feel I'm not doing the best for them.

This time last year, February, I was really down. I were going without heating. I had me coat on, I had a thick coat on, cardigans, socks. I daren't have the heating on.

When they turned the water tap off I felt very upset. I can't explain . . . I felt very ashamed at that time. I feel personally ashamed. I feel ashamed at myself. I couldn't manage to pay the water and the supply had been cut off.

I panic every time I get warning letters from everyone. I panic. Oh, they're going to break in, they're going to get keys to get into my place and when I go home I'm going to find everything gone.

I've got so much pride about asking people for money. I wouldnae ask them, like my dad or my grandparents. But I know I could have got it, but I wouldnae ask for it.

I don't go out any more. I only go to the paper shop and come back, about as far as I dare go . . . It frightens me, this area . . . I just walk in here and bolt and chain that door every time and lock it.

Children can be very mean to each other. I remember being teased at school because we couldn't afford things, and I don't want that to happen to my son.

Source: E. Kempson (1996) *Life on a Low Income* York: Joseph Rowntree Foundation

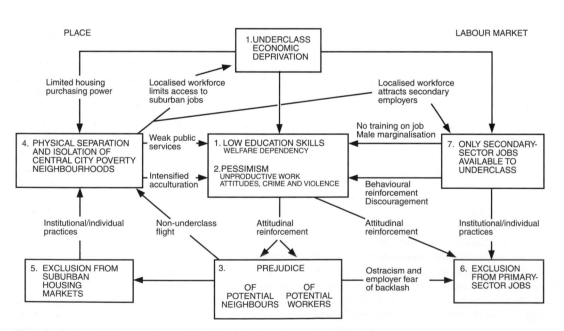

Figure 15.2   A cumulative causation model of the underclass phenomenon in US cities
Source: G. Galster and E. Hill (1992) *The Metropolis in Black and White: Place, Power and Polarization* New Brunswick NJ: Center for Urban Policy Research, Rutgers University

Plate 15.1 **The polarised city. Increasing polarisation within the postmodern city is demonstrated starkly by the sight of a homeless man sprawled on the riverside walkway in front of the gentrified warehouse apartments of Butler's Wharf in central London**

DIMENSIONS OF MULTIPLE DEPRIVATION

As Figure 15.1 indicates, there is a close relationship between poverty and deprivation and other dimensions of urban decline. We can illustrate this with reference to a number of particular issues.

CRIME

The emergence of mass unemployment in Britain since the mid-1970s coincided with an increase in recorded crime. The British Crime Survey confirmed that crime grew by 49 per cent, or 4.1 per cent per annum, between 1981 and 1991. This rise was dominated by property crime, which increased by 95 per cent over the period. The nature of the relationship between property crime and fluctuations in the business cycle is shown graphically in Figure 15.3, which reveals several clear trends for England and Wales. The first is a strong upward trend in property crime as Britain leaves full employment behind and enters an era of mass unem-

ployment. Second, and most striking, property crime, though rising strongly during the period, fell during each of the upturns in economic activity: during the 'Barber boom' (1972–3), the partial economic recovery of 1978–9 at the end of the Labour government, the most intense phase of the 'Lawson boom' (1988–9) and during 1993 in line with another economic upswing. Conversely, recorded property crime rose during each of the downturns in economic activity: during the recession in the mid-1970s following the OPEC price shock, during the early 1980s Thatcher–Howe recession (due to the twin economic shocks of monetary and fiscal deflation and the real sterling exchange-rate appreciation) and during the early 1990s Major recession. The association between property crime and unemployment is supported by records from police-force areas in England and Wales which show that 'the unemployment black spots in the country (Cleveland, Merseyside, Northumbria, Greater Manchester, South Yorkshire, West Midlands, Greater London and South Wales) are also crime black spots' (Wells 1994 p. 33)[25] and by findings that the unemployed are responsible for a disproportionate volume of

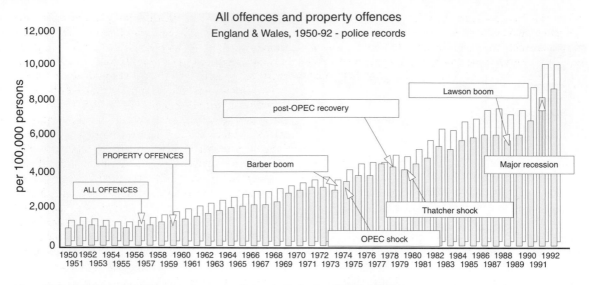

Figure 15.3   Incidence of all offences and property offences in England and Wales, 1950–92

crime.[26] Similarly in the USA during the economic boom of the 1990s serious crime declined significantly. The FBI crime rate data (per 100,000 population) recorded reductions in murder from 9.0 in 1994 to 5.5 in 2000; in rape from 43 in 1992 to 32 in 2000; and in robbery from 273 in 1991 to 145 in 2000. Most significantly, burglary decreased by over 50 per cent over twenty years from 1684 per 100,000 in 1980 to 728 per 100,000 in 2000.[27] In an attempt to identify factors conditioning criminal behaviour, Farrington and West (1988), in a longitudinal study of 400 predominantly white inner-city males born in 1953, found several factors at age 8–10 that significantly predicted chronic convicted offenders.[28] These were:

1. Economic/material deprivation, including low income, poor housing and unemployment periods experienced by parents.
2. Family criminality, including convicted parents and delinquent siblings.
3. Unsatisfactory parenting, either too authoritarian or too unbounded.
4. School failure.

A key predictor of persistent offending at age 14–18 was being unemployed at age 16, while for latecomers to crime in their 20s the only significant predictor was unemployment at age 18.

Although the association between unemployment and criminal activity is unequivocal, the question of the extent to which there is a *causal* relationship is a contentious issue. Clearly, one must avoid the suggestion that crime is predetermined by poverty or unemployment, not least since this slanders the majority of the unemployed who lead respectable law-abiding lives. Unemployment provides the motivation in the form of material deprivation, frustrated aspirations, boredom and anger (particularly in a society that extols material success), while the generally high level of material possessions enjoyed by the majority in employment provides the opportunity. The catalyst lies in the loosening of the individual's moral or ethical constraint on unlawful behaviour. This may be fostered by resentment on the part of the disadvantaged which questions the legitimacy of a social order that countenances mass unemployment and increasing income inequality. Kempson *et al.* (1994) have shown how the poor, lacking legal means to obtain a reasonable level of living, may as a last resort turn to criminal activities such as welfare fraud and burglary.[29] In urban areas where poverty and deprivation are spatially concentrated, the weakening of individual moral restraint on a more general scale can provide a fertile breeding ground for anti-social behaviour and may foster the development of a marginalised criminal subculture. This is illustrated most graphically by police-department designations of 'kill zone neighbourhoods' in some US cities, such as the infamous Robert Taylor low-income housing project in Chicago, where 300 shootings were reported in one two-week period during the summer of 1994. Fear of crime can have as great an effect on people's well-being as actual crime rates in a neighbourhood. For the most vulnerable subgroups living in high-risk areas the impact of fear of crime on daily living patterns and on the general quality of life can be profound[30] (see Chapter 19).

HEALTH

Although some of the factors that affect health, such as age, gender and genetic make-up, cannot be changed by public policy or individual choice, a number of 'external' factors are recognised to be of significance for health status. These include the physical environment (e.g. adequacy of housing, working conditions and air quality), social and economic factors (e.g. income and wealth, levels of unemployment), and access to appropriate and effective health and social services. There is a strong association between quality of housing and health.[31] In London the 're-emerging' infectious disease of tuberculosis is concentrated among the unemployed and those in rented accommodation,[32] while in the Bronx, New York, there is a clear relationship between childhood tuberculosis and residential crowding that is itself associated with household poverty, dependence upon public assistance, Hispanic ethnicity, larger household size and a high population of young children.[33] Lack of shelter or homelessness can have a direct impact on the individual's health. Bronchitis, tuberculosis, arthritis, skin diseases and infections, as well as alcohol- and drug-related problems and psychiatric difficulties, are all more prevalent among single people who are homeless. Even families occupying temporary bed-and-breakfast accommodation often find it difficult to maintain hygiene while washing, eating and sleeping in one overcrowded room.

One of the clearest indications that 'bad housing damages your health'[34] is the effects of inadequate heating and dampness. Insufficient warmth (due to poor housing design and excessive fuel costs), leading to hypothermia, is reflected in higher proportions of deaths among older people in winter than in summer. Dampness encourages the spread of dust mites and fungal spores that lead to respiratory illnesses,[35] and cockroaches thrive in the warm, damp conditions characteristic of many of the system-built tower blocks erected in British cities in the 1960s and 1970s. Other health problems related to inadequate housing include stress-related illness arising from poor sound insulation between neighbouring homes, lack of privacy and

overcrowding. As women generally spend more time in the home, the effects of bad housing often impact on them to a greater degree.[36] It is difficult, though, to separate out the impact of poor housing because of the interrelated nature of the effects of social, economic and environmental factors on individual well-being.

There is, however, unequivocal evidence of the fundamental importance of income and wealth status for health.[37] In the UK death rates at most ages are two or three times higher among the growing numbers of disadvantaged people than they are for their more affluent counterparts (Figure 15.4). Individuals experiencing poor socio-economic circumstances also have higher levels of illness and disability. Low income can impact upon family health in at least three ways: physiologically, with income providing the means of obtaining the fundamental prerequisites for health; psychologically, in that living with inadequate resources creates stress and reduces the individual's coping ability; and behaviourally, with poverty leading to health-damaging actions such as smoking and recourse to a low-nutrient diet. In the light of the body of evidence it is difficult to refute the conclusion that 'it is one of the greatest of contemporary social injustices that people who live in the most disadvantaged circumstances have more illness, more disability and shorter lives than those who are more affluent' (Benzeval *et al.* 1995 p. 1).[38]

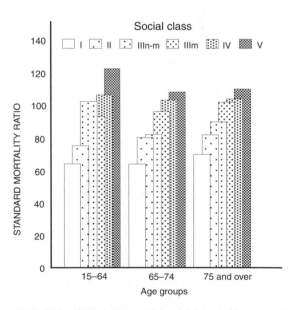

Figure 15.4 **Male mortality rates by social class and age group in England and Wales, 1976–81**

ETHNIC STATUS

An investigation of income and wealth in the UK found that the incomes of those belonging to ethnic minorities tend to be lower than those of the rest of the population.[39] Disaggregation by ethnic origin revealed that the Indian population was disproportionately concentrated in above-average income groups, whereas 40 per cent of the West Indian population and over half the Pakistani and Bangladeshi population were in the lowest income group. This is supported by Green's (1994) findings that over 60 per cent of the ethnic minority population were living in wards that were ranked in the worst 20 per cent nationally for unemployment. Over half the Bangladeshi population were living in wards that were in the most deprived 10 per cent nationally.[40]

The main reasons why people from ethnic minorities are more likely to be living in poverty are summarised in Figure 15.5. In the early post-Second World War era migrants were recruited to meet labour shortages in manufacturing industries (such as iron foundries and textile mills) and public services (such as transport and health), largely in manual occupations and often with low pay and minimum employment rights. The restructuring of the UK economy, and in particular the reduction in manufacturing employment and rationalisation of public-sector services, resulted in large-scale unemployment among ethnic minorities. Government policies to restrict employment rights and reductions in welfare support have also had a heavy impact on the ethnic-minority work force because of its concentration in low-paid sectors of the economy. Evidence that ethnic-minority groups have suffered disproportionately from the recession and effects of urban restructuring is provided by Amin (1992), who found that unemployment rates among ethnic- minority

groups increased more rapidly than those of the general population after 1979, reaching 22 per cent, or twice the national average, by 1984.[41] By 1990, while nationally 8 per cent of all men were unemployed, the figure for ethnic-minority groups was 14 per cent, and for Pakistani men aged 16–24 years it was 31 per cent. With 48 per cent of Pakistanis aged 16–24 and 54 per cent of Bangladeshis in the same age group having no formal qualifications, the culture of advanced capitalism represents a formidable challenge for many ethnic minorities.[42] The social and geographical segregation of the ethnic-minority population was confirmed by Owen's analysis of the 1991 census, which showed 'a clear pattern of the relative exclusion of minority ethnic groups from the successful parts of the British space-economy and a tendency for them to concentrate in areas of decline' (Owen 1995 p. 32).[43]

The socio-spatial segregation of poor members of ethnic-minority populations is particularly evident in the US city, where inner-city ghettos are inhabited predominantly by African-Americans and Hispanic Americans. The complex of disadvantage in the ghetto is illustrated by 'the dramatic increases in rates of social dislocation ... [and] the growing social problems – joblessness, family disruption, teenage pregnancy, failing schools, crime and drugs'.[44] This is confirmed, most graphically, by Duster's (1995) examination of 'incarceration versus matriculation', which revealed that more African-American males in the USA were in prison than were attending college on a full-time basis.[45] Although it would be wrong to assign such negative stereotypes to all ghetto residents, the concentration of high levels of poverty and black ethnic status in the ghetto represents an environment of profound social and economic disadvantage[46] (see Chapter 18).

GENDER

Although women constitute much the same proportion of the poor (60 per cent) in Britain today as they did in 1900[47] women's poverty has become more visible as a result of the growth in number of female-headed households. The two groups with the highest risk and longest duration of poverty are lone mothers and older women living alone.[48] In 1995 1 million lone mothers were receiving income support, including 650,000 who had been in receipt for at least two years. In the same year 1.2 million single women aged over 60, including 500,000 aged 80 or over, received income support. Most had been on income support for several years and few had any other source of income. Women

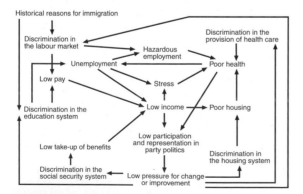

Figure 15.5   Major sources of poverty among ethnic minorities

in families with no wage-earner, low-paid women, non-employed women wholly dependent on others, and homeless women must all be added to the category of those vulnerable to poverty. Women's risk of poverty also reflects different life-cycle stages, with child-rearing and caring work when out of the labour market and low-paid and insecure employment when economically active, limiting their earning capacity and restricting their ability to provide for old age. Women may also experience the same deprivation to a different degree, with, for example, poor housing affecting women responsible for domestic labour and child care more than men. The intra-household distribution of resources can also affect a woman's experience of poverty.[49] In addition, women often shoulder the burden of 'managing' poverty, and in trying to stretch insufficient income to cover needs women may go without themselves (Box 15.2), or take on a burden of debt.[52]

FINANCIAL EXCLUSION

We have seen already, in Chapter 7, how financial institutions can affect the construction of urban space. The social implications of this process are illustrated graphically by the exclusion of poor urban residents from retail financial services provided by banks, building societies and insurance companies. In pursuit of the twin goals of profit maximisation and risk minimisation, and prompted by the recession of the early 1990s, the financial-services industry sought to 'restructure for profit'. This resulted in the retreat of financial services from poorer urban communities and concentration on those more affluent areas and consumers most likely to generate fees, investment accounts and account turnover, a process referred to as a 'flight to quality'.[53]

The withdrawal of banking services from low-income inner-city neighbourhoods has been mirrored in the USA[54] by the strategies of other financial institutions such as savings and loans institutions (building societies), which are in effect red-lining many inner-city neighbourhoods,[55] and insurance companies, which have abandoned similar areas through the simple device of charging premiums that are beyond the reach of most of the inhabitants. As a result, in such areas there are significant problems of uninsured houses, cars and businesses. Communities affected by **financial exclusion** are forced to depend on alternative informal sources of credit such as moneylenders and loan sharks. The poor must often operate within a local '**parasitic economy**' not only for credit but to obtain ordinary goods and services from grocery

---

BOX 15.2

Food poverty in the UK

In 1899 Seebohm Rowntree found that 9.9 per cent of the people surveyed in the city of York lived in a state of 'primary poverty' in that they could not afford to purchase a diet that contained the basic nutrients in the cheapest form. A century later Stilt and Grant (1997)[50] calculated that 21 per cent of the British population were living in such a state. This increase is explained only partly by an upward revision in ideas of what constitutes a healthy diet. The two principal factors are cutbacks in the 'food safety net' provided by the welfare state, which comprised direct food provision to children via school meals, and cash provision via the benefits system. Both have been reduced since 1979. The 1980 Education Act abolished nutritional standards for school meals and the 1986 Social Security Act reduced eligibility for free school meals. In addition, changes in retail structure, particularly

the advent of supermarkets, have altered the shopping and dietary options of the poor through the closure of local shops and an increase in the price of food as a result of additional travel costs.

There is ample evidence of the effect of food poverty on health. Food is regarded as a flexible item in the domestic budget of the poor, with meals being forgone owing to other demands on cash. The reduction of family food intake often affects women most, owing to a process of self-denial in which children and husbands are given priority. A survey by the Joseph Rowntree Foundation[51] described a picture of the poor juggling competing interests, shopping around for bargains and value for money, having to shop at the end of the day to get the pre-closing prices, and countless other tactics of surviving on a low income.

Source: T. Long (1997) Dividing up the cake: food as social exclusion, in A. Walker and C. Walker (eds) *Britain Divided* London: Child Poverty Action Group, 213–28

stores, supermarkets and gas stations where price mark-ups have become regularised and accepted as an inevitable part of daily life.[56] The absence of a formal financial infrastructure can also lead to deterioration of the local built environment, as owners find it difficult to obtain home-improvement loans, and to local economic decline due to the inability of small businesses to access loan capital.

The struggle between local communities and financial institutions is an unequal one but the 1977 Community Reinvestment Act set out the obligations of US banks and savings and loan associations to meet credit needs and make loans as well as take deposits in a community, including low-income areas. The Act gave local communities some political leverage in seeking to preserve access to financial services, but it does not force institutions to locate branches in low-income areas or engage in what some would regard as 'imprudent' lending. In addition, insurance companies are not covered by the Act. Nevertheless, the CRA has encouraged the growth of alternative financial institutions for the poor.[57] Two of the most important types are credit unions, which operate widely in both the USA and the UK and provide loans to members from a pool of savings (Box 15.3), and the US community development bank.[58]

One of the most successful community development banks is the Shorebank Corporation, based in the South Shore area of Chicago (Box 15.4). The unusual shareholder base, comprising charitable foundations, religious groups and 'concerned individuals', many of whom are based in the local area, enabled the bank to combine economic and social activities in a way not open to a 'normal' bank. Over twenty-five years the Shorebank Corporation has helped to stabilise the economy of the South Shore area, improved the quality of housing, fostered new businesses, and enhanced the internal and external image of the district – as well as making significant profits.

Another radical financial initiative with potential to alleviate financial exclusion stems from the experience of the Third World and is based on principles of the Grameen Bank, established in Bangladesh in 1974 to aid the reincorporation of marginalised people into the economy. Several schemes are in operation in the USA (Box 15.5). In addition ethnic banks owned and operated by minorities have emerged to serve market niches left unoccupied by mainstream banking services.[59] We shall return to the broad issue of local responses to urban decline in the next chapter. Here we focus attention on the socio-spatial manifestations of poverty and deprivation in cities.

BOX 15.3

The Lower East Side People's Federal Credit Union

Lower East Side People's Federal Credit Union (LESPFCU) was organised to replace the last commercial bank branch in a densely populated low-income neighbourhood of New York City. At its grand opening on 1 May 1986 the National Federation of Community Development Credit Unions (NFCDCU) presented the initial deposit of $5,000 to help launch the credit union. Six months later, Manufacturers Hanover Trust, which had closed its last bank branch in the area, made a deposit of $100,000. This deposit and other support allowed the credit union to hire its first staff members.

In its first five years LESPFCU built a deposit base of nearly $2 million. During this period it reinvested more than $1.8 million in loans to hundreds of borrowers. Most of these are female and minority (Hispanic and African-American) borrowers, and many are in receipt of welfare. The credit union now operates as a full-time institution with a staff of four and a growing range of services. LESPFCU serves nearly 3,000 members, the majority of whom have saving accounts of less than $500.

Source: R. Conaty and E. Mayo (1997) A Commitment to People and Place: The Case for Community Development Credit Unions London: New Economics Foundation

THE GEOGRAPHY OF DEPRIVATION

As we have seen, a strong relationship exists between poverty and deprivation, and other dimensions of urban decline. Analyses using territorial social indicators reveal that in some urban localities the intensity and socio-spatial concentration of problems are severe. Figure 15.6 shows the distribution of deprivation at the district level in the UK. This highlights the concentrations of disadvantage in inner London, the older urban industrial areas of South Wales, the West Midlands, Yorkshire, the north-east and north-west of England, and west central Scotland. At the metropolitan scale there is, in general, a gradient of deprivation that intensifies towards the 'black hole' of the inner city (Figure 15.7).

With reference to the USA, Rusk (1994) employed three indicators to identify thirty-five cities 'beyond the point of no return' that typically have lost 20 per cent or

BOX 15.4

The South Shore Community Development Bank

Between 1960 and 1970 many of the white residents of the South Shore district of Chicago moved to the suburbs. Most of the area's banks moved out with the whites, and by 1973 only three remained to serve a 70 per cent black population of 78,000. Two of these were closed by the government regulator and the third, the South Shore Bank, sought to open a branch in central Chicago with a view to running down its local branch. The bank had already switched its lending activities: of the $33 million taken from South Shore on deposit, only $120,000 had been returned on loan to customers living in the area.

In 1973 the South Shore Bank was bought by a company set up by a young team of idealistic bankers who sought to invest in the local area, first by offering mortgages for the purchase of good-quality single-family houses, then as capital accrued for the purchase and rehabilitation of multi-occupancy housing blocks. By the end of 1993 the bank had financed the renovation of more than 9,000 flats, over one-third of the district total.

The bank returned to profit in 1983 and has not made a loss since. In 1992 its net profit was $2.2 million on assets of $229.1 million, of which $161 million was out on loan. Losses were 0.4 per cent, a figure many commercial banks would envy. Overall, it pumped $41 million into the South Shore district during 1992, most of it deposited by savers outside its service area attracted by the work of the bank.

The success of the South Shore Bank may be attributed to its combination of economic and social goals, and in particular to the investment of a significant amount of pump-priming finance sufficient to attract other private investors, and a management committed to the economic revitalisation of a community and its residents.

Source: R. Douthwaite (1996) *Short Circuit* Totnes: Green Books

BOX 15.5

Grameen borrowing circles in Chicago

The remarkable success of the Grameen Bank, initiated in Bangladesh, has led to its replication in many countries of the Third World. The same principles underlie the Women's Self-employment Project (WSEP) in Chicago, which was set up in 1986 by a foundation and three professional women to help lone mothers on welfare with no credit records break out of poverty through self-employment.

Central to the Grameen method is the use of borrowing circles comprising small groups of women from the same area, each of whom wants to borrow capital to start or continue in a small business. Following a period of training in basic business skills, the group of five women is recognised as a circle. The circle then selects two of its members to receive the first loans, which may not exceed $1,500 each and are repayable over a year. Repayments on a circle's first two loans begin two weeks after the money is received, and if the three fortnightly payments are made on time, another two members qualify for their loans. The final member gets her loan after a further six weeks, provided all her colleagues' accounts are in order. The combination of peer pressure and peer support is the key to the borrowing circles' success.

In 1995 the WSEP made between 100 and 125 loans totalling $25,000 and distributed among seventeen circles. The repayment rate was 97 per cent and over the first decade of activity 600 businesses had been started, 85 per cent of which were still in operation.

Proponents of Grameen-style borrowing circles believe that self-employment helps women move away from low-wage service-oriented jobs that do not provide them with an opportunity for personal and economic advancement. In addition, it gives women a chance to provide their children with positive role models, to increase their own selfesteem, and to raise the quality of life of their families. It also keeps economic resources within the community.

Source: R. Douthwaite (1996) *Short Circuit* Totnes: Green Books

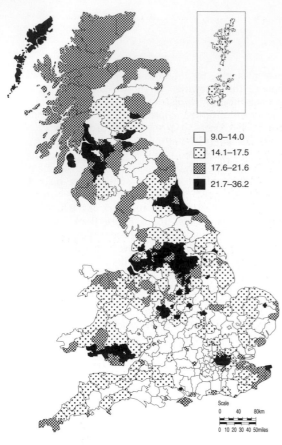

9.0–14.0
14.1–17.5
17.6–21.6
21.7–36.2

Figure 15.6 Deprived households as a percentage of all households in UK local authority districts
Source: P. Lee and A. Murie (1997) *Poverty, Housing Tenure and Social Exclusion* Bristol: Policy Press

THE INNER-CITY PROBLEM

As we saw in Chapter 3, the concept of the inner city as a locus for poor living conditions dates back to the era of nineteenth-century industrial urbanism, when the deprivation and depravity of Britain's inner cities were condemned by contemporary commentators. Variations in well-being continue to characterise the modern city. In Detroit in the mid-1970s a fourfold variation was recorded in infant mortality rates between the inner city and the suburbs, with inner-city rates equivalent to those of some Third World states.[62]

EXPLANATIONS OF INNER-CITY DECLINE

Table 15.5 summarises a number of possible explanations of inner-city decline. We have examined many of these processes and trends already, including ecological succession, white flight and the effects of suburbanisation, institutional exploitation, and economic restructuring. We have also noted that the relevance of each theoretical perspective for explaining inner-city decline is contingent upon the particular interaction of a variety of local and global processes; for example, the impact of market-driven suburbanisation is of greater importance in the US city than in either Canada or the UK. Nevertheless, a case could be made for the salience of each of these explanations operating to some degree in most Western cities.

CHARACTERISTICS OF INNER-CITY DECLINE

The 1977 White Paper on the inner cities[63] identified four basic components of the inner-city problem in the UK:

1. The first is the economic decline and unemployment associated with the contracting industrial base due to recession in the economy, deindustrialisation and the rundown of traditional inner-city services and industries (e.g. warehousing and dock-related activities); the closure of branch plants, often of multinational companies, and the associated ripple effects on the local economy resulting in the failure of small dependent firms; failure to attract new industry (owing to a series of disadvantages, including the high cost of industrial land, local taxes, shortage of suitable

more of their population since 1950, have an increasingly isolated non-white minority population in their central area, and have seen a dramatic decline in the purchasing power of central city residents[60] (Table 15.3). To these Waste (1998) adds the three cities of Los Angeles, San Francisco and New York City, which, along with Chicago and Philadelphia, form a group of 'urban reservation' cities characterised by concentrations of extreme-poverty neighbourhoods,[61] as exhibited in the demographic profile of south central Los Angeles (Table 15.4). Waste also identified fourteen cities with the highest rates of violent crime ('shooting-gallery cities'), thereby adding Atlanta GA, Dallas–Houston TX, New Orleans LA, Phoenix AZ, San Antonio CA and Washington DC, to produce a final list of forty-four 'adrenalin cities' containing 91.3 million people, or 37 per cent of the US population, and experiencing prolonged and chronic stress.

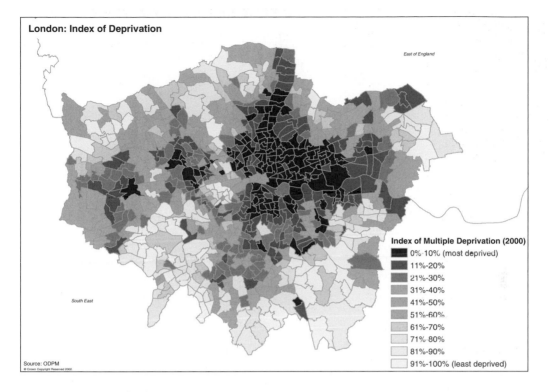

London: Index of Deprivation

Index of Multiple Deprivation (2000)
- 0%–10% (most deprived)
- 11%–20%
- 21%–30%
- 31%–40%
- 41%–50%
- 51%–60%
- 61%–70%
- 71%–80%
- 81%–90%
- 91%–100% (least deprived)

Source: ODPM
© Crown Copyright Reserved 2002

Figure 15.7   The geography of deprivation in London
Source: Office of the Deputy Prime Minister, Crown copyright reserved 2002

premises, problems of congestion and limited opportunities for expansion); and labour constraints, including a shortage of female workers.

2. A second highly visible feature of many inner-city areas is the physical dereliction and absence of amenities. A basic reason for this decay is age. Most of the inner areas of UK cities were built over a century ago and have not benefited from the continued investment and improvement that have been directed to the retail and commercial areas of the central business district (CBD). Environmental dereliction has also been exacerbated in many cities by public-sector activity. Land has been cleared, then left vacant for long periods as a result of inflated estimates of local-authority needs, and the stop–go nature of public-sector financing. This has affected the continuity of local-authority development plans and cast a pall of 'planning blight' over large parts of the inner city.

3. Social disadvantage is a third component of the inner-city problem. It characterises those who are poor as a result of the high levels of unemploy-ment and low-wage jobs available, as well as many of the infirm, elderly and ethnic minority groups. As well as individual effects this is also a collective phenomenon that affects all residents as a result of a pervasive sense of decay and neglect that diminishes community spirit, which may also be undermined by the incidence of antisocial activities such as crime and vandalism.

4. A fourth component of the inner-city problem is the concentration of ethnic minorities in parts of the inner city, which may lead to discrimination in job and housing markets and engender community and racial tensions, particularly in times of economic hardship. The demand for labour, primarily in manufacturing industries and public services, in some British cities during the 1950s and 1960s led to a marked increase in immigration. In the twenty years following the Second World War the non-white population of Britain increased eightfold to reach 595,000. Many of these immigrants were attracted to the low-cost housing areas of the inner city, particularly in English cities such as London,

TABLE 15.3 THE THIRTY-FOUR US CITIES PAST THE 'POINT OF NO RETURN'

City	Population loss, 1959–90 (%)	Non-white population in 1990 (%)	City-to-suburb income ratio (%)	MSA designation	1990 population
Holyoke MA	26	35	69	Springfield MA	529,519
Birmingham AL	22	64	69	Birmingham AL	907,810
Flint MI	29	52	69	Flint MI	951,270
Buffalo NY	43	37	69	Buffalo NY	1,189,288
St Louis MO	54	50	67	St Louis MO	2,444,099
Chicago IL	23	60	66	Chicago CMSA	8,065,633
Saginaw MI	29	28	66	Saginaw MI	399,320
Baltimore MD	23	60	64	Baltimore MD	2,382,172
Dayton OH	31	36	64	Dayton OH	951,270
Philadelphia PA	23	45	64	Philadelphia PMSA	4,856,881
Youngstown OH	44	35	64	Youngstown OH	492,619
Kansas City KS	11	38	63	Kansas City MO KS	1,566,280
Petersburg VA	7	74	63	Richmond VA	865,460
New Haven CT	21	47	62	New Haven CT	530,180
Milwaukee WI	15	39	62	Milwaukee WI PMSA	1,566,280
Atlantic City NJ	43	69	61	Atlantic City NJ	319,416
East Chicago IL	41	81	60	Chicago CMSA	
Gary IN	25	85	59	Gary IN PMSA	604,526
Bessemer AL	17	59	58	Birmingham AL	
Chicago Heights IL	19	50	57	Chicago CMSA	
Pontiac MI	17	52	55	Detroit CMSA	4,665,299
Elizabeth NJ	4	60	54	NY–NJ CMSA	18,087,251
Cleveland OH	45	50	54	Cleveland OH	2,759,823
Perth Amboy NJ	4	65	53	NY–NJ CMSA	
Hartford CT	21	66	53	Hartford PMSA	767,841
Detroit MI	44	77	53	Detroit CMSA	
Trenton NJ	31	59	50	Trenton PMSA	325,824
Paterson NJ	3	72	47	NY–NJ CMSA	
Benton Harbor MI	33	93	43	Benton Harbor MI	161,378
Newark NJ	38	83	42	NY–NJ CMSA	
Bridgeport CT	11	50	41	Bridgeport PMSA	443,722
North Chicago IL	26	47	39	Chicago CMSA	
Camden NJ	30	86	39	Philadelphia PMSA	
East St Louis IL	50	98	39	Chicago CMSA	
Totals: 34 cities				**23 MSAs**	**55,833,161**

Source: R. Waste (1998) *Independent Cities: Rethinking US Urban Policy* New York: Oxford University Press
Note: The total population in the MSAs described by Rusk as beyond the point of no return is 51,489,161, or 20.7 per cent of the census figure for the 1990 US population, which is 248,709,873

Birmingham, Wolverhampton, Bradford and Leicester. Although ethnic minorities in inner cities are affected by the same kind of disadvantage experienced by other residents, as a visually and culturally distinct group they are open to discrimination and are easy targets for those seeking scapegoats for the city's economic decline (see Chapter 18).

The structural underpinnings of the inner-city problem ensure its appearance in most of the older industrial cities of the West, but the interplay of global and local forces can place a different emphasis on the components of inner-city decline in different societies. While the general problems of Britain's inner cities are mirrored in the USA, the inner-city problem in the USA is also conditioned by particular social factors,

TABLE 15.4 DEMOGRAPHIC PROFILE OF SOUTH CENTRAL LOS ANGELES, 1990

Demographic characteristic	Measure
Total population	523,156
African-American population	55%
Latino population	45%
People per household	3.48
Population earning high school diploma	47.8%
Non-English-speakers	43.5%
Median household income	$20,357
Poverty	32.9%
Rent 35 per cent of income	47.9%
Median home value	$113,400

Source: R. Waste (1998) *Independent Cities: Rethinking US Urban Policy* New York: Oxford University Press

including, for example, a greater proportion of ethnic-minority residents.

The correlation between ethnic-minority status and residence in a US inner-city problem area is demonstrated clearly by the identification of 'underclass neighbourhoods'. These are defined as census tracts with above-average rates on four variables that indicate poor integration of potential workers into the mainstream economy. (These variables measure male detachment from the labour force, the percentage of households receiving public assistance, the percentage of households headed by women with children, and teenage high-school drop-out rates.[64]) In 1990 2.68 million people lived in such neighbourhoods (an increase of 8 per cent over 1980), with 57 per cent African-American residents, 20 per cent Hispanic and 20 per cent white. If we relax the definition of disadvantage and focus solely on 'extreme-poverty neighbourhoods' (comprising census tracts in which 40 per cent or more of residents have money incomes below the official poverty line), many of the areas are the same as those classified as having underclass status, but in 1990 there were nearly four times as many extremely poor census tracts as underclass ones. The number of people living in extreme-poverty neighbourhoods rose from 5.57 million in 1980 to 10.39 million in 1990. Three-quarters lived in the central cities, once again the majority being African-Americans. By 2000 the ethnic composition of extreme poverty showed little change in the proportion accounted for by white populations (from 20 per cent in 1990 to 24 per cent in 2000); a decline in black share in extreme poverty (from 47 per cent to 39 per cent); and an increase in that of Hispanic-Asian population from 24 per cent to 33 per cent. Evidence from black ghetto neighbourhoods in several US cities demonstrates the enduring nature of the problem of poverty (Table 15.6). Over the period 1990–2000 seven of the nine neighbourhoods shown in Table 15.6 experienced increases in families living below the poverty level, while all experienced population decline, increased housing vacancy, and increased incidence of 'high poverty' populations. In sum, all neighbourhoods fared worse in 2000 than in 1990.

The complex of problems experienced in the inner areas of many Western cities represents a major social challenge (Box 15.6). Although many of the difficulties can be attributed to industrial decline and

TABLE 15.5 A TYPOLOGY OF EXPLANATIONS OF INNER-CITY DECLINE

Explanation	Dominant process(es)
Natural evolution	Urban growth, ecological succession, down-filtering
Preference structure	Middle-class flight to the suburbs
Obsolescence	Ageing of built environment and social infrastructure
Unintended effects of public policy	Suburban subsidies, including construction of freeways and aids to new single-family home-ownership
Exploitation (1)	City manipulated by more powerful suburbs
Exploitation (2)	Institutional exploitation: red-lining by financial institutions; tax concessions; suburbanisation of factories
Structural change	Deindustrialisation and economic decline
Fiscal crisis	Inequitable tax burden; high welfare, social and infrastructure costs
Conflict	Racial and class polarisation

Source: L. Bourne (1982) The inner city, in C. Christian and R. Harper (eds) *Modern Metropolitan Systems* Columbus OH: Merrill

TABLE 15.6 POVERTY TRENDS IN BLACK GHETTO NEIGHBOURHOODS, 1990–2000

City/Neighbourhood	% population change	% below poverty 2000	% below poverty level change	% housing units vacant 2000	% change in housing units vacant	% change in ratio of high to low poverty [a]
Cleveland						
Fairfax	−13.1	35.7	−0.9	21.0	+3.6	+24.2
Hough	−19.2	41.3	−2.3	20.9	+0.6	+41.3
Philadelphia						
Fairhill	−22.8	57.1	+2.0	22.0	+5.8	+31.3
Hartranft	−7.0	33.9	+1.4	21.8	+7.8	+18.7
Chicago						
Englewood/Woodlawn	−16.8	43.8	+2.9	23.7	+8.7	+22.9
Baltimore						
Boyd Both	−7.6	38.3	+1.2	26.7	+9.0	+21.2
Broadway East	−11.2	39.0	+2.4	17.4	+11.7	+12.6
Washington						
Trinidad	−7.1	41.3	+1.8	18.8	+11.1	+19.7
Bellevue	−6.8	40.9	+0.3	17.6	+14.7	+23.9

Note: [a] High poverty is indicated by people with incomes two-thirds or greater below the poverty level; low poverty is indicated by people with incomes half or less below the poverty level

Source: Adapted from D. Wilson (2007) *Cities and Race* London: Routledge p. 14

BOX 15.6

In the depths of the 'inner city'

Peckham in south London lies within sight of the financial towers of Canary Wharf but has become a byword for the worst afflictions of inner-city Britain – guns, gangs, drugs, sink estates and social deprivation.

Valerie Theodule does not let her 11-year-old son play on the estate where they live. 'You can get killed for just looking at someone.' Some immigrant parents send their children home to Nigeria to be educated. Teenage mothers are often as hopelessly equipped for parenthood as their own parents were. There are children who have never been out of Peckham and teenagers who have never seen the river Thames, a ten-minute drive away.

An abandoned block of flats awaiting demolition had been taken over by junkies and squatters. Floors were littered with old mattresses, empty gin bottles, human excrement, and the detritus of crack dens. Some flats had been burned out. The block was fifty yards from a primary school.

Jason, an 18 year old Jamaican youth, quit school at 14, having never really learned to read or write. His two elder brothers are in prison on drugs and firearms charges. His girlfriend is four months pregnant. He sells drugs 'to survive', has ready access to guns (a submachine gun would cost £4,000), and is ambivalent about the prospect of ending up in prison. Teenagers like Jason come under pressure to join neighbourhood gangs partly for their own protection but also because, for some, these 'crews' represent surrogate families. As a local teacher put it, faced with the option of staying at school or a minimum wage job, it's difficult for kids to ignore the lifestyle of someone who leaves school at 16 without any qualifications and, within a couple of years, turns up at the school gates driving a £30,000 BMW.

Source: The Times 21 April 2007

unemployment, others relate to personal factors such as age, infirmity or ethnicity. Still others stem from the deteriorating physical environment and affect the standard of provision of housing, education, transport, health and other social services. It is important to emphasise, however, that although these problems are most apparent in inner-city areas, this does not mean that the underlying causes are geographical or that they are exclusive to the inner city.

THE OUTER-CITY PROBLEM

The 'inner city' is a generic term that may usefully be seen as a metaphor for wider social problems at the heart of which is the core issue of poverty. The dispersal of the 'inner-city problem' to other parts of the city is particularly evident in British cities, where the nature and incidence of 'urban disadvantage' have been affected by public urban-renewal programmes and associated population movements. In Glasgow the socio-spatial distribution of urban deprivation in 1991 revealed a major concentration in inner suburban areas such as Possilpark and Haghill, and in the four large peripheral council housing estates of Drumchapel, Castlemilk, Easterhouse and Pollok. This is an extension of the situation in 1981, which was in marked contrast to the position in 1971, when a much higher proportion of deprived areas were located in the inner city. In general over the period, and particularly during the 1970s, the traditional inner tenement housing areas that previously exhibited severe deprivation recorded a relative improvement in status. This was largely the result of a massive clearance and redevelopment programme undertaken by the local authority, combined in some areas with modernisation and new building aided by housing associations and private developers. This policy involved the large-scale relocation of residents in a general process of decentralisation. By 1981 the inner-city areas contained a much-reduced and ageing population living in improved accommodation. Conversely, the outer estates exhibited a younger demographic structure and, although the housing was generally well provided with the basic amenities, overcrowding was widespread. Serious social problems, such as unemployment and high proportions of single-parent households, were also prevalent. By 2001 demolition of areas of poorest housing, combined with further rehabilitation of stock by housing associations and some new private-housing construction, had resulted in a reduced incidence of multiple deprivation in the peripheral estates. Nevertheless the persistent geography of disadvantage in the city is reflected in the fact that the areas of greatest deprivation in 2001 were primarily those identified thirty years earlier (Figure 15.8).

The spatial changes in the incidence of deprivation over the period were accompanied by a redistribution in terms of housing tenure. Whereas in 1971 a high proportion of deprived areas included older, and frequently private-rental, properties (notably in the inner belt stretching from the East End to Govan), by 1981 deprivation had become increasingly concentrated in the public sector. Nationally, since 1970 council housing has become occupied predominantly by those on low incomes. Whereas in 1961 and 1971 fewer than half those in council housing were in the poorest 40 per cent of the population, by 1981 the proportion had increased to 57 per cent and by 1991 to 75 per cent[65] (see Chapter 11).

In US cities not all the households that participated in 'the secession of the successful' to the suburbs

TABLE 15.7 SUBURBANISATION OF INNER-CITY PROBLEMS IN THE US, 1980–90

Traditional inner-city problem (1980–90)	Suburban 'share' of problem
Decline in medium family household income	Income declined 35%
Population flight to outer-ring suburbs	Loss of 8% of population to outer-ring suburbs by inner-ring suburbs
Poverty	42% of all metro poor
White metropolitan poor	51.2% reside in suburbs
Hispanic-Latino metro poor	40% reside in suburbs
African-American metro poor	23.7% reside in suburbs
Suburban office vacancy rate	54.7% vacancy rate
Commercial growth absorption rate (years to absorb current vacant space at five-year average absorption rate)	4.3 years in suburbs versus 5.3 years for central business districts

Source: R. Waste (1998) *Independent Cities: Rethinking US Urban Policy* New York: Oxford University Press

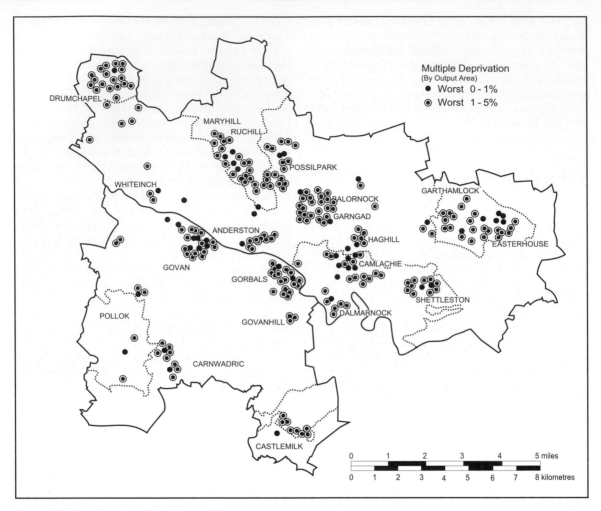

Figure 15.8   The geography of deprivation in Glasgow

escaped from the shadow of inner-city problems that have spread outwards from the urban core. As Table 15.7 shows, the suburban incidence of urban distress increased between 1980 and 1990, and by 1990 47 per cent of the urban poor in the USA lived in inner-ring suburbs. Over this period the inner-ring suburbs lost 8 per cent of their population to outer-ring suburbs as the suburbanising wave of white flight continued to ripple outwards (see Chapter 4). Significantly, during the1990s, a period of strong economic growth within the USA, the number of extreme-poverty neighbourhoods declined by 24 per cent nationally (representing a decrease of 2.5 million people) but poverty continued to rise in the older inner-ring suburbs of many metropolitan areas.[66]

DEPRIVATION AND THE AREA-BASED APPROACH

For urban geographers a key question is the value of an area-based approach to the study of deprivation. Critics have dismissed area-based research on the grounds that it does not offer an explanation of the causes of revealed patterns of deprivation. This is now generally accepted, and has informed more recent policy-oriented analyses of poverty and deprivation in which the identification of patterns is used as a basis for the critique of current policy aimed at alleviating multiple deprivation. Another argument against spatial analysis and spatial targeting of policies is that because of **ecological fallacy**, resources may be

Plate 15.2 A stolen car lies burnt out next to a fenced-in children's play area in front of a block of local authority flats in the Possilpark district of Glasgow

directed to areas in which a substantial percentage of residents do not require public assistance. It has also been argued that spatial targeting of resources ties people to declining areas instead of encouraging them to relocate to areas of greater opportunity.

Proponents maintain that an area-based approach is justified on several grounds. Some have argued for a degree of 'area effect' in accentuating if not actually causing deprivation. The Inner Area Studies of the 1970s concluded that

> there is a collective deprivation in some inner areas which affects all the residents, even though individually the majority of people may have satisfactory homes and worthwhile jobs. It arises from a pervasive sense of decay and neglect which affects the whole area. . . . This collective deprivation amounts to more than the sum of all the individual disadvantages with which people have to contend.
>
> (Department of the Environment 1977 p. 4)[67]

There are also 'additionality effects' from tackling problems in a defined area, including positive spillover for the area as a whole as well as adjacent areas, and efficiency gains associated with the administration of concerted action, as opposed to a 'pepper-pot' approach.

A real analysis can also be used to identify spatial concentrations of particular population groups with different policy requirements. These may include the elderly and the young (with implications for the provision and use of geriatric and paediatric services), unskilled workers (location of industry) and carless households (public-transport route planning). Analysis of the nature, intensity and geographical distribution of multiple deprivation permits comparisons within regions and cities as well as over time, and facilitates monitoring of the effectiveness of remedial strategies. Furthermore, while the long-term goal may remain a fundamental political-economic restructuring to address the roots of inequality in society, area-based policies of positive discrimination can provide more immediate benefits that enable some people to improve some aspects of their quality of life. As we shall see in the following chapter, despite doubts expressed over the efficacy of spatial targeting, area-based strategies have underlain government urban policy in the UK and the USA since the 1960s and continue to inform the construction of initiatives aimed at the social and economic regeneration of decayed urban environments.

FURTHER READING

BOOKS

E. Bright (2003) *Reviving America's Forgotten Neighborhoods* New York: Routledge

J. Dixon and D. Macorov (1998) *Poverty: A Persistent Global Reality* London: Routledge, 204–28

J. Gough, A. Eisenschitz and A. McCulloch (2006) *Spaces of Social Exclusion* London: Routledge

A. Madanipor, G. Cars and J. Allen (eds) (1998) *Social Exclusion in European Cities: Processes, Experiences and Responses* London: Kingsley

S. Mincy and S. Weiner (1993) *The Underclass in the 1990s: Changing Concept, Constant Reality* Washington DC: Urban Institute

S. Musterd and W. Ostendorf (1998) *Urban Segregation and the Welfare State* London: Routledge

M. Pacione (1997) Urban restructuring and the reproduction of inequality in Britain's cities: an overview, in M. Pacione *Britain's Cities: Geographies of Division in Urban Britain* London: Routledge, 7–60

JOURNAL ARTICLES

R. Atkinson and K. Kintrea (2001) Disentangling area effects *Urban Studies* 38(12), 2277–98

D. Fuller (1998) Credit union development: financial inclusion and exclusion *Geoforum* 29(2), 145–57

P. Lawless (2004) Locating and explaining area-based urban initiatives: new deal for communities in England *Environment and Planning C* 22, 383–99

O. Lewis (1966) The culture of poverty *Scientific American* 215, 19–25

A. Leyshon and N. Thrift (1995) Geographies of financial exclusion: financial abandonment in Britain and the United States *Transactions of the Institute of British Geographers* 20, 312–41

K. Newman and P. Ashton (2004) Neoliberal urban policy and new paths of neighbourhood change in the American inner city *Environment and Planning A* 36, 1151–72

Y. Shaw-Taylor (1998) Profile of social disadvantage in 100 largest cities of the US, 1980–1990/93 *Cities* 15(5), 317–26

KEY CONCEPTS

- Poverty
- Multiple deprivation
- Culture of poverty
- Positive discrimination
- Stigmatisation
- Underclass
- Socio-spatial segregation
- Financial exclusion
- Parasitic economy
- Flight to quality
- Credit unions
- Community development bank
- Territorial social indicators
- Inner city
- Outer city
- Area-based approach

STUDY QUESTIONS

1. Critically consider the five major models proposed to explain the causes of deprivation.
2. Examine the relationship between poverty and deprivation and other dimensions of urban decline.
3. To what extent would you agree that there is a direct relationship between wealth and health in contemporary urban societies.
4. Identify the causes and consequences of the inner-city problem.
5. Explain the concept of financial exclusion in disadvantaged communities and consider possible strategies to alleviate the problem.
6. Critically examine the value of an area-based approach to the study of urban poverty and deprivation.

PROJECT

The concept of financial exclusion suggests that poor urban residents are denied access to mainstream credit facilities owing to the withdrawal of financial agencies from certain parts of the city. Using information from local directories, plot the locations of different types of financial institutions (e.g. banks, building societies, insurance companies, credit unions). Are there underserved (and overserved) areas? Use census data for the city to map differing levels of social status. How does the distribution of financial services relate to the social geography of the city?

16

National and Local Responses
to Urban Economic Change

Preview: urban regeneration in the UK from the top down; urban policy in the USA; empowerment zones; privatisation; enterprise zones; urban development corporations; public–private partnerships; property-led regeneration; cultural industries and urban regeneration; urban tourism and downtown redevelopment; urban regeneration from the bottom up; local economic initiatives; progressive planning; the community/social economy; the black economy

INTRODUCTION

The differential use of space by capital in pursuit of profit creates a mosaic of inequality at all geographic scales from global to local. Consequently, at any one time certain countries, regions, cities and localities will be in the throes of decline as a result of the retreat of capital investment, while others will be experiencing the impact of capital inflows. At the metropolitan scale the outcome of this **uneven development** process is manifested in the poverty, powerlessness and polarisation of disadvantaged residents.

Although the global importance of capital in the process of urban restructuring is axiomatic, it is important to avoid the fallacy of economic determinism. Technical and economic processes, the material forces of production, are not the only determinants of urban growth and decline (see Chapter 1). As noted in Chapter 14, the policies of the national and **local state** can exert an important influence on urban change. Further, the dominance of market-oriented urban policy has not gone unchallenged, especially in those urban localities that remain meaningful places to their inhabitants but which are not considered profitable spaces by capital.[1] Grass-roots opposition to the adverse local effects of national urban policy has taken a number of forms aimed at strengthening local economies in ways that enable labour and local communities to obtain greater benefit from the process of urban restructuring.

In this chapter we focus attention on responses to urban decline, first from a 'top down' perspective of recent national urban policy initiatives, and then from the 'bottom up' with respect to local development and grass-roots strategies.

URBAN REGENERATION IN THE UK FROM THE TOP DOWN

As we discussed in Chapter 8, from the passage of the 1947 Town and Country Planning Act until the late 1960s urban problems and their solutions were viewed essentially in physical terms. The 'rediscovery of poverty'[2] in Britain's cities signalled an incipient urban crisis and led to a refocusing of urban policy. The spark that fired government action to tackle the emerging urban economic and social problems was Enoch Powell's 'rivers of blood' speech, in which he criticised the rate of immigration into Britain. This inflamed racial tension in those urban areas where black and Asian immigrants had settled in large numbers, and stimulated Harold Wilson's government to initiate the Urban Programme (Figure 16.1). This was intended to tackle 'needle-points' of deprivation by offering local authorities a 75 per cent grant towards the cost of projects in the fields of education, housing, health and welfare in 'areas of special need' (many of which contained concentrations of non-white residents). The kinds of projects fostered by the Urban

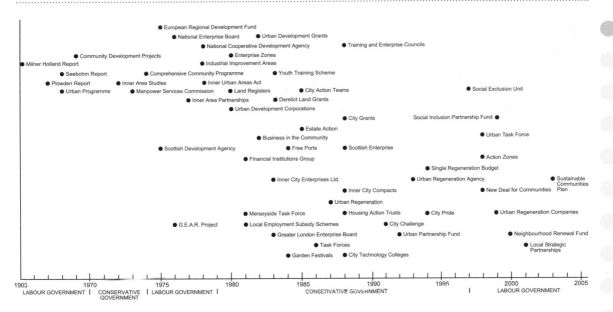

Figure 16.1   Major urban policy initiatives in the UK

Programme, however, did not address the fundamental issues of structural economic change that underlay the urban crisis. Official recognition that tackling the root causes of urban deprivation would require more than marginal adjustment to existing social policies was reflected in a major revision of urban policy in the 1977 White Paper on the inner cities.[3] One of the most important elements of the White Paper was the recognition that urban decline and poverty had structural causes located in economic, social and political relations that originated outwith the affected areas. This signalled more broadly based action on urban problems combining economic, social and environmental programmes and involving new organisational arrangements between central and local government.

The Labour government's response to the 1977 White Paper was passage of the 1978 Inner Urban Areas Act, the main element in which was the creation of seven partnerships between central and local government in an attempt to harness private capital for urban economic revival. The emphasis on improving the economic environment of cities was also promoted by a shift of policy from the New Towns programme to urban regeneration. The main vehicle for these measures was the expanded Urban Programme. In practice, a wide-ranging national and local attack on urban problems failed to materialise. Although the diagnosis implied long-term commitment of substantial financial resources to overcome the deep-rooted problems of the inner cities, doubts about the wisdom of 'excessive'

public expenditure had taken root in the polity, and the Treasury was wary of any significant long-term financial commitment. As a result, the rate support grant provided few additional resources to hard-pressed urban authorities at a time when the government's wider economic strategy meant that local authority spending was actually reduced. In this climate of retrenchment it proved difficult for urban local authorities to 'bend' their own main programmes towards the inner areas as they had been encouraged to do.[4] A second difficulty for urban policy was the growing strength of the idea that state action was an inefficient, ineffective and uneconomic method of dealing with the urban problem, and that the private sector offered a better solution.[5] Thus, in terms of the level of resources on the ground, the outcome of the 1977 White Paper was limited. Its main value lay in the identification of the structural underpinnings of the inner-city problem and its acknowledgement, albeit tentative, of the potential role of the private sector in urban renewal.

The election of a Conservative government under Margaret Thatcher in 1979 represented a watershed in British urban policy. Between 1979 and 1997 successive Conservative governments sought to reduce central government involvement in urban regeneration and to shift the policy emphasis from the public to the private sector.[6] The main role envisaged for the public sector was to attract and accommodate the requirements of private investors without unduly influencing their development decisions. This perspective underlay a number of new initiatives introduced by the 1980 Local Government Planning and Land Act, the most significant of which for urban policy were the concepts of enterprise zones and urban development corporations. The consistent belief of the government in the power of the private sector to initiate urban regeneration was also evident in a range of other schemes, which included derelict-land grants, urban-development grants, Inner City Enterprise, Business in the Community, city action teams, task forces, simplified planning zones, city technology colleges, British Urban Development, housing-action trusts, Estate Action, training and enterprise councils and local development companies (Figures 16.1 and 16.2).

Enterprise zones (EZs) formed a major plank in the Conservative strategy to remove state regulations on private capital investment and entrepreneurial activity. Firms within EZs were to enjoy a number of financial and administrative advantages, including freedom from local taxes for a ten-year period (the local authorities being compensated by the Treasury), tax allowances for capital expenditure on buildings and a simplified planning regime that granted automatic permission for approved types of development. The underlying assumption was that by encouraging companies to develop derelict urban sites 'a boost will be given to the entire local economy, leading to jobs and opportunities for the nearby residents'.[7] In practice, the efficiency of the multiplier effect has been less than anticipated, and despite advantages for capital, in particular the tax exemptions, EZs have not succeeded in engineering a fundamental revival of local economic activity.[8] Paradoxically, the most successful EZs (such as Corby and Clydebank) have proved to be not those where the public sector has withdrawn (as envisaged in the initial EZ model) but those where a single public agency controlled much of the land and could effect an integrated development strategy.[9]

The privatisation of urban policy was also reflected in the creation of urban development corporations (UDCs), invested with wide powers over land use and development and charged with the primary task of creating an environment attractive to private investment.[10] UDCs are managed by boards whose members are appointed by the Secretary of State. They normally include people with a working knowledge of the urban development area for which the UDC is responsible, as well as people with close links with the business community. The development corporations are funded direct from central government, but can also raise finance through the sale of assets (land and property) and by borrowing. They have a wide range of powers to acquire, reclaim and dispose of land, and to provide basic infrastructure. Their general remit is to secure the regeneration of their area by bringing land and buildings into effective use, encouraging the development of existing and new industry and commerce, creating an attractive environment, and ensuring that housing and social facilities are available to encourage people to live and work in the area. The economic bias of the approach has provoked criticism of the limited social content of development proposals and has highlighted the question of who are the intended beneficiaries of the development process (Box 16.1). The conflict

BOX 16.1

The London Docklands Development Corporation (LDDC): area, powers and objectives

The LDDC's area comprised 2,100 ha (5,200 acres), taking in portions of three east London boroughs (see table).

Borough	Docklands population, 1981	Proportion of LDDC area (%)
Tower Hamlets	21,000	33
Newham	10,000	50
Southwark	17,000	17

The LDDC's powers under the 1980 Local Government, Planning and Land Act are principally those of:

1. *Landowner,* with compulsory purchase powers and acquisition through vesting orders.
2. *Planning control authority,* but not the statutory plan-making authority.
3. *Initial developer,* making sites viable for subsequent development, mainly private.
4. *Manager of the Isle of Dogs Enterprise Zone,* established in 1982 for ten years.

The LDDC is not an education, health, housing or social services authority. Its objectives are to redevelop its area by investment in reclamation and infrastructure, together with business and community support for residential and commercial development. By co-operating with other authorities the area will become an increasingly pleasant and rewarding place in which to live, work and play.

Source: J. Hall (1990) Metropolis Now: London and its Region Cambridge: Cambridge University Press

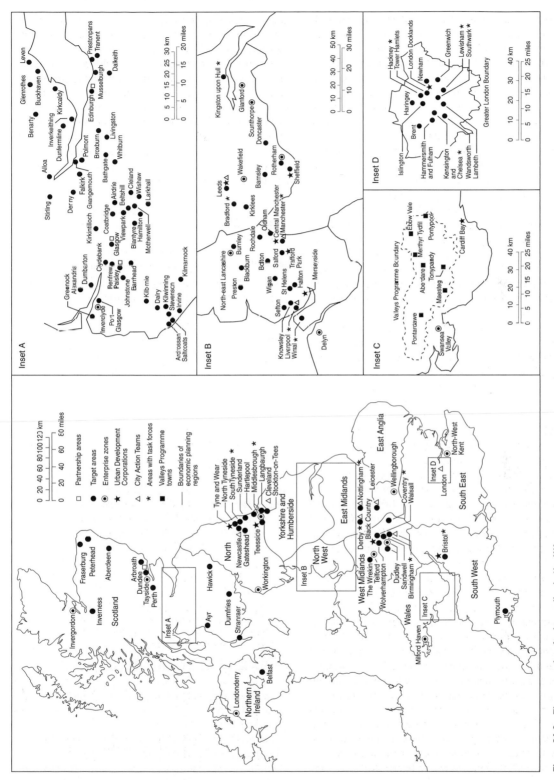

Figure 16.2　The geography of urban policy in the UK

between economic and social goals was particularly fierce in the London Docklands, where the strategic plan prepared by the local authorities employed a needs-based approach which emphasised the need to stem existing job losses, to attract new jobs that would match residents' skills, to use the vacant land to address the acute housing problems of east London, and to improve the general environment.[11] Central government favoured a 'demand-led' approach with the emphasis on creating a new economy attractive to firms and prospective residents from outside the area. In practice, the London Docklands Development Corporation (LDDC) switched the planning emphasis from attempting to provide manufacturing jobs towards office and warehousing schemes and retail complexes. In the field of housing, the UDC effort focused on construction for the private sector, while between 1981 and 1984 the waiting list for council housing in the borough of Newham rose from 2,650 to 9,112. In the Docklands area as a whole, during the 1980s the proportion of owner-occupied housing rose from 5 per cent to 36 per cent, while local-authority tenures fell from 83 per cent to 44 per cent. Although a greater attempt was made by the LDDC to accommodate local needs during the second half of the 1980s (with schemes for social housing and community development), the recession of the early 1990s and collapse of the property market saw a reversion to strict financial criteria in LDDC activity.[12]

Following its election in 1997 the New Labour government sought to replace the enterprise ideology of Conservative urban policy (see Chapter 8) with a 'Third Way' between free-market capitalism and classical social democracy; between individualism and *laissez-faire* on the one hand and corporate government intervention on the other. In terms of urban policies this represented a move away from government (via direct public-sector intervention) in favour of governance (via multi-agency working, negotiated agendas and consensus-seeking), where the state and local authorities 'steer, not row' the process of urban change.[13] Major policy instruments in this approach included a reshaped Single Regeneration Budget targeting 80 per cent of its resources on the sixty-five most deprived areas in Britain, with project funding conditional on the direct involvement of local communities. In similar vein, the New Deal for Communities launched in 1998 with a budget of £2 billion over ten years was designed to improve education and health, reduce crime and increase the number of people in work in thirty-nine neighbourhoods by means of local partnerships of residents, community organisations, local authorities and businesses. These strategies were supported by other initiatives such as Employment, Education and Health Action Areas, Local Strategic Partnerships, and the Neighbourhood Renewal Fund, designed to address the particular problems of inner-city areas of deprivation. The implementation and success of these Third Way programmes is predicated on communities having the capacity to engage as democratic agents in the decision-making process, and on the willingness of governments to devolve real powers to the local level (see Chapter 20). This 'capacity building' philosophy also underlay the Sustainable Communities Plan launched in 2003 with the aim of regenerating 'priority areas' (including the 20 per cent most deprived wards in England, former coalfield areas, and strategic areas of brownfield land) and responding to a growing demand for affordable housing in the growth areas of south-east England (Milton Keynes and the south Midlands, the London–Stansted–Cambridge–Peterborough corridor, Ashford, and the Thames Gateway). In addition, despite criticisms of the earlier 'Thatcherite' UDCs the UDC model was revived for use in selected areas, such as Thames Gateway.

In practice Labour's Third Way has produced a plethora of (sometimes overlapping) policies to effect urban regeneration. The policy package comprises a mixture of continuities from the recent past (evident in the use of the Private Finance Initiative to fund public projects[14] and similarity between new urban regeneration companies and earlier urban development corporations), and newer approaches, (with greater emphasis on partnership and active community involvement in decision-making). Three other significant features of UK urban policy are the continuing importance of area-based initiatives to alleviate poverty and deprivation[15] (see, for example, the Thames Gateway initiative[16]); transatlantic policy exchanges, with, for example, UK Action Zones reflecting US empowerment zone and enterprise community programmes,[17] and the growth of business improvement district schemes in British cities[18] (see Chapter 7); and growing emphasis on the principles of sustainable urban development.

URBAN POLICY IN THE USA

Federal involvement in the economic and social life of urban America, in terms of policy formulation, increased during the Depression of the 1930s, when Franklin D. Roosevelt's New Deal programme supported poor citizens by providing a minimum wage, unemployment insurance and social security, and middle-class Americans via programmes such as the Federal Housing Administration's scheme for below-market-rate mortgages. Following the Second World War the role of the federal government in urban life expanded even though policies were not specifically urban by designation. (We have already seen in Chapter 7 how FHA mortgage insurance and federal funding of the interstate highway system contributed to the suburbanisation of American cities.) During the 1960s the number of federal programmes aimed specifically at the city increased in response to the growth of urban problems ranging from poverty and civil-rights riots to the financial crises of urban governments. In 1965 the Department of Housing and Urban Development was established as part of Lyndon B. Johnson's Great Society. New federal programmes included the War on Poverty (which was intended to extend beyond the basic-income-security programmes of the New Deal era to job training for those unable to find work), the Community Action Programme (designed to facilitate the maximum feasible participation of residents in the local organisation of anti-poverty services) and the Model Cities Program (which aimed to produce the comprehensive renewal of urban slum neighbourhoods in 150 selected cities) (Figure 16.3).

The era of 'creative federalism' and its wide range of categorical urban programmes was ended by Richard Nixon's (1969–74) concept of **new federalism**. The stated policy objective was to allow localities to take spending decisions on the basis of local knowl-edge. Accordingly, the two main initiatives were a system of general revenue-sharing which provided cities with funds according to a needs-based formula, and the amalgamation of individual programmes into a **block grant**, the most significant of which was the Community Development Block Grant (CDBG), which consolidated seven categorical grants, including those for urban renewal and the Model Cities Program. In terms of social impact there is evidence that, owing to the high level of local discretion, up to 40 per cent of the CDBG expenditure occurred in wealthier urban areas.[19] For example, the town of Little Rock AR used $150,000 from the CDBG to build a tennis court in a wealthy area.

The introduction of the Urban Development Action Grant (UDAG) programme under the Carter administration aimed to encourage private investment in distressed communities by using federal grants to leverage private money into public–private-sector development schemes. The reorientation of urban policy towards private-sector-led urban economic development was extended by the subsequent New Right administrations of Ronald Reagan and George Bush, who, like Margaret Thatcher in the UK, believed in the power of unfettered market forces to create a prosperous economy within which benefits would 'trickle down' to most sections of society, leaving residual problems to be addressed by specific people-oriented as opposed to place-oriented policies. Programmes considered counter-productive were reduced or eliminated, including UDAGs and subsidised housing. The withdrawal of

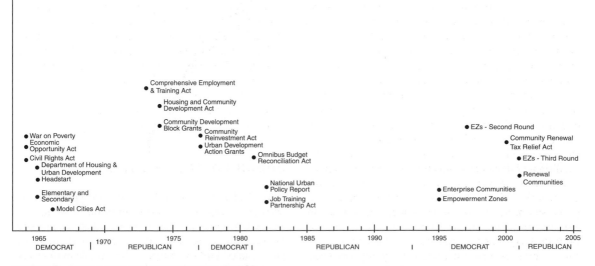

Figure 16.3   Major federal urban programmes in the USA

government intervention in urban policy and planning was reflected by a 59 per cent reduction in federal spending on US cities between 1980 and 1992.

Following the coming to power of the Clinton administration in 1993, urban policy was spearheaded by the concept of **empowerment zones** and enterprise communities.[20] Designated EZ/ECs receive federal investment plus tax incentives to help create employment. To be eligible for the programme, cities were expected to have secured a commitment of private-sector investment and to involve the community in the planning and implementation of the initiative (Box 16.2). By 2000 twenty-two EZs and forty-five ECs had been designated. A further eight EZs were ordered in 2002, together with twenty-eight Urban Renewal Communities in which tax incentives (but not direct federal funding) are available to stimulate local private-sector job growth and economic development. The Detroit URC area, for example, covers twenty-five square miles and contains 120,000 residents, 40 per cent of whom live in poverty.

The use of tax-based incentives to promote urban regeneration is a key element of US urban policy. Major initiatives involving taxation measures include EZs at federal level, ECs at state level, Tax Incremental Financing (TIF) at city level and business improvement districts at neighbourhood level. TIF is a mechanism that allocates future increases in property taxes from a designated area to pay for improvements within that area. Pioneered in California in 1952, TIF has been adopted widely by municipalities to finance infrastructure improvements and economic development in targeted areas.[21] The principle finds an echo in New Labour's thinking and may therefore be applicable also in the UK. The neo-liberal trajectory of US urban policy over recent decades was continued during the administration of George W. Bush with the introduction of legislation to involve faith-based and community organisations in provision of social services; re-authorisation of the 1996 Personal Responsibility and Work Opportunity (workfare) Act; and passage of the No Child Left Behind Act to address the problems of poor quality public schools.

PUBLIC-PRIVATE PARTNERSHIPS

The entrepreneurial thrust of urban policy during the 1980s was underwritten by co-operation between the government and the private sector (see Chapter 8). This approach owed much to experience in North America's cities, where the practice of public–private partnership in urban regeneration dates back more than half a century.

BOX 16.2

Detroit's Empowerment Zone plan

The Clinton administration's empowerment zone/ enterprise community (EZ/EC) initiative differed from previous proposals for enterprise zones, which relied almost exclusively on geographically targeted tax incentives to create jobs and business opportunities in distressed communities. The EZ/EC programme combines federal tax incentives with direct funding for physical improvements and social services, and requires unprecedented levels of private-sector investment as well as participation by community organisations and residents.

Within the 18.35 square miles (47.5 km²) of the Detroit Empowerment Zone, 47 per cent of the population are in poverty, the median family income is half that of the Detroit metropolitan area, and half the adults over 25 years of age do not have a high-school diploma. The area also exhibits high rates of morbidity, alcoholism and homelessness.

Empowerment zone status has brought a number of public and private financial benefits. These include a $100 million federal block grant for improvements to the physical and social infrastructure, tax incentives to employers who recruit labour from within the zone, and a pledge of $1.9 billion of private funding. Of the latter, $1 billion comprises a loan fund administered by thirteen local financial institutions, to make loans available to small businesses and home buyers who might otherwise experience difficulty in securing credit. The Detroit plan also includes eighty projects aimed at improving training provision in the zone, regeneration and social housing.

The combination of strong federal leadership, local partnership and consensus, and private-sector involvement offers a sound basis for urban revitalisation and, if successful, a potential model for practice elsewhere.

Sources: R. Hambleton (1995) The Clinton policy for cities: a transatlantic assessment *Planning Practice and Research* 10(3/4), 359–77; J. McCarthy (1997) Revitalization of the core city: the case of Detroit *Cities* 14(1), 1–11

Following the Second World War US city governments, faced with growing blight in downtown areas, and attracted by federal funding from the urban-renewal programme, joined forces with private developers in 'quasi-public' redevelopment corporations which were able to sidestep conventional procedures for municipal policy-making. By the 1960s these business–government partnerships had produced a range of downtown redevelopments, including Pittsburgh's Golden Triangle, Baltimore's Charles Centre and Minneapolis's Nicollett Mall. In the 1970s and 1980s, prompted by the problems of deindustrialisation and fiscal distress, city governments moved beyond single-project collaboration with developers. In the context of heightened inter-city competition for private investment, municipal governments, especially those with high-profile 'boosterish' mayors, became entrepreneurial, providing extensive subsidies and incentives to attract developers, and often becoming co-developers of more risky redevelopment projects.[22] Public–private partnerships became the cornerstone of economic development strategies of virtually all US cities–strategies that centred on the creation of a good business climate[23] (Box 16.3).

This trend is exemplified by the 'Rouse-ification' of downtowns across the USA. With its production of **festival market-places** in, for example, Boston (Faneuil Hall), New York (South Street Seaport), Baltimore (Harborplace), Milwaukee (Grand Avenue Mall) and St Louis (Union Station), the Rouse Company became the leading downtown developer in the country. As Levine (1989 p. 24) observed:

> Rouse projects, with their distinctive architecture and innovative linkage of entertainment and retailing, have been credited with changing the image of centre cities, stimulating spin-off downtown redevelopment, and rekindling investor confidence in downtown areas – all factors that have made mayors, anxious to promote growth and claim political credit for it, line up to coax Rouse to their cities.[24]

The distributional impact of these projects has been typically uneven, with, in most cities, redeveloped downtowns resembling 'islands of renewal in seas of decay'.[25] Baltimore's Inner Harbor is heralded as a national model of public–private waterfront reclamation, but during the course of its redevelopment the poverty rate increased in 90 per cent of the city's non-white neighbourhoods.[26] A principal reason for such contrasts is that downtown corporate centres based on advanced services and tourism often have only limited links with the local economy and rarely generate economic development in surrounding neighbourhoods. In addition, since the kinds of jobs created are unlikely to provide employment opportunities for urban poor and minority populations, many of the benefits of redevelopment are taken up by suburban commuters. The efficacy of relying on the 'trickle down' effect (see below),

BOX 16.3

Creating a better business climate in Houston TX

Private–public partnerships in which government played a supporting role to business were pioneered in Houston long before the term 'urban entrepreneurialism' was coined. During the 1970s and 1980s leading political figures in both the USA and Europe cited the unplanned free-enterprise economy and entrepreneurial philosophy in Houston as the reason for the city's remarkable material progress.

Since its founding in the 1830s the power structure in Houston has been dominated by business leaders who have been able to exercise control over governmental decisions, including key decisions such as planning and zoning. The policy approach favoured by the business elite involves control over political officials, the use of public expenditure for business goals, and limiting the scope of government regulation. Local elites have concentrated on creating and perpetuating a 'good business climate' free of outside intervention.

In 1984, faced with a declining energy-based economy, the leaders of the Houston Chamber of Commerce established the Houston Economic Development Council with a remit of economic diversification through corporate recruitment. Promotion of a good local business climate is central to the strategy, with emphasis given to low labour-force unionisation rates, a generous tax abatement policy and subsidies for new businesses.

This 'low tax/poor public service' approach has had an adverse effect on local social and environmental conditions. Until approval was given for a mass transit system in the late 1980s, Houston had one of the worst public transport systems in the country. Air pollution and water pollution from industries and motor vehicles are a major problem, while parts of the city experience subsidence due to the extraction of water from the subsoil.

Houston has a 'good business climate', but this reflects a particular interpretation of the concept. An alternative definition would focus greater attention on public service and 'quality of life' issues – the attainment of which would mean tax increases.

Source: R. Parker and J. Feagin (1990) A better business climate in Houston, in D. Judd and M. Parkinson (eds) *Leadership and Urban Renewal* Beverly Hills CA: Sage, 216–38

Plate 16.1   The 'Rouse-ification' of downtowns across the USA is exemplified by the redevelopment of Baltimore's Harborplace

rather than public targeting to encourage economic development in the most distressed urban neighbourhoods, is open to question (Box 16.4).

PROPERTY-LED REGENERATION

Property-based development has played a prominent role in urban regeneration projects undertaken by public– private partnerships. Property development refers to the assembly of finance, land, building materials and labour to produce or improve buildings for occupation and investment purposes.[27] Post-Second World War governments have sometimes used property as an instrument of macro-economic management, boosting or dampening aggregate demand by pumping public resources into, or withdrawing them from, house-building, roads, railways, schools and other physical infrastructure. Property has also been used extensively since the 1930s to promote regional development, mainly through public provision of advance factories, industrial estates and associated infrastructure. During the 1980s property-led regener-

ation assumed a central place in urban policy, with the key role of the private sector demonstrated most visibly in flagship projects such as the redevelopment of Canary Wharf in London's Docklands or the transformation of Baltimore's waterfront.[28]

However, property development alone is an insufficient basis for urban economic regeneration. Although property development and rehabilitation can improve a residential environment (see Chapter 10) and construction projects can provide scope for employment, a property-led approach fails to consider the crucial issues of:

1. The development of human resources, such as education and training, which have long-term effects on people's incomes and employment prospects.
2. The underlying competitiveness of local production.
3. Investment in infrastructure, such as transport and communications.

The value of property-led redevelopment is locally contingent. A strong role for property-based measures

BOX 16.4

A tale of two Baltimores

Baltimore MD has been identified as a prototype for urban renaissance. As a result of deindustrialisation and the decline of the port, by the late 1950s the city's economic distress was evident in the declining fortunes of its central business district. Demand for commercial and retail property was limited, with over 2 million ft² (186,000 m²) of loft and warehouse space vacant, and the waterfront had become an area of dereliction. The regeneration of downtown was placed in the hands of a private corporation. Based on the success of the Charles Center project, initialled in 1959, a second, more ambitious scheme for revitalisation of the Inner Harbor was approved in 1968. This plan called for large-scale public investment for infrastructure, a marina, World Trade Center, convention centre, new science centre, and an aquarium in order to pump-prime private-sector investment. Implementation of the central-city renewal programme has been entrusted to quasi-private corporations, such as the Center City Development Corporation and the Baltimore Economic Development Corporation.

By 1980 the Inner Harbor redevelopment was a commercial success and private investment flowed into the area. By 1985 a total investment of $2 billion had produced 4 million ft² (370,000 m²) of office space, 1 million ft² (93,000 m²) of retail/tourist space, 3,550 hotel rooms, 8,000 parking spaces and 1,200 housing units. The city, in defiance of conventional wisdom, has transformed itself from a declining manufacturing centre into a tourist destination.

The benefits of the redevelopment process have been distributed unevenly, however, producing a 'double doughnut' pattern of concentric rings. The central core contains the business, cultural and entertainment zone, which serves the whole metropolitan area, and attractive housing for the affluent. This is ringed by decaying, more populous neighbourhoods of the poor, dependent and largely black groups which are, in turn, surrounded by white middle and upper-class suburbs. This pattern illustrates graphically the failure of the 'trickle down' mechanism in the city.

Source: R. Hula (1990) The two Baltimores, in D. Judd and M. Parkinson (eds) *Leadership and Urban Renewal* Beverly Hills CA: Sage, 191–215

may be appropriate where there are problems associated with land or building conditions or where shortages of floor space restrict inward investment and indigenous growth. Nevertheless, uncontrolled property development may drive up local property prices, encourage land speculation and displace existing economic activities and populations unable to afford higher rents. More generally, the capital market's preference for short-term money-lending and speculation rather than longer-term investment can divert finance capital into property and away from more productive activities such as manufacturing.[29]

Nowhere has the impact of property-led urban redevelopment been more evident than in the global cities of London and New York, where, during the 1980s in particular, the growth of the financial services industry and the concomitant expansion of office space interacted to engineer a major restructuring of urban space.[30] By the end of the 1980s London was witnessing the largest office-building boom in its history as a result of the Conservative government's opening up of a once highly regulated property-development arena to speculative ventures. Much of the new development appeared in the traditional heart of the City, which sought to pro-

tect itself from a leakage of financial-sector activity to European rivals such as Frankfurt and Paris, as well as to the emerging Docklands, with its flagship development at Canary Wharf.[31] The regeneration of the London Docklands is part of a longer twenty-year strategy to redevelop the Thames Gateway subregion stretching thirty-two miles (52 km) from the old docklands to the Thames estuary[32] (Box 16.5).

CULTURAL INDUSTRIES AND URBAN RECONSTRUCTION

Many of the leading industries of the fifth Kondratieff wave (see Chapter 14) are part of the larger knowledge economy that includes the creative industries. Consequently, the cultural economy of post-industrial/postmodern cities has assumed increasing importance.[33] Cultural production embraces activities such as printing and publishing, film production, radio, television and theatre, libraries, museums and art galleries. High fashion, tourism and sports-related activities are also included in some definitions of the cultural industries (or creative industries) sector.[34] Such activities cut

BOX 16.5

Thames Gateway

The Thames Gateway is the largest brownfield urban development opportunity in Europe and one of four major designated growth areas in the south-east of England, (the others being the M11 corridor from London to Stansted–Cambridge; Milton Keynes; and Ashford in Kent). The Thames Gateway comprises parts of fifteen different local authority areas. It is home to 1.6 million people, and contains some of the most deprived communities in the country – as well as some of most affluent, related to the financial services sector developments at Canary Wharf. The Gateway boundary encompasses land on both sides of the river Thames, formerly occupied by land-extensive industries and port-related activities whose decline left a legacy of large-scale dereliction and contaminated land.

The UK government, in partnership with other agencies, including the Greater London Authority (GLA) and its functioning agencies (including the London Development Authority and Transport for London), the Thames Gateway Development Corporation, the London boroughs and national bodies such as the Housing Corporation and English Partnerships, is seeking to promote brownfield development, economic growth, environmental improvement and urban renewal in an integrated and sustainable manner. Specific objectives include construction of 120,000 new energy-efficient homes by 2016 with 80 per cent built on brownfield land and with 35 per cent affordable housing. Existing urban centres will be strengthened by new residential, commercial and transport developments. Over 180,000 jobs will be provided to offer local residents greater choice of work opportunity. Planned transport developments include extension of the Docklands Light Railway (DLR), new local stations on the Channel Tunnel Rail Link line, and improved public transport in east London. Environmental improvements will be based around a 50,000 ha green space network. Within the area six strategic development locations have been identified. One of these areas, the East London Gateway around Stratford and the lower Lea valley, received an additional boost with its selection as the site of the 2012 Olympic Games.

Plate 16.2   Heritage areas and features, such as the steam-powered clock in the Gastown district of Vancouver, play a key role in culture-based urban redevelopment

across the conventional production–consumption divide, blur the distinction between the functions of cities as centres of production and consumption, and illuminate the post-industrial/postmodern concept of flexible specialisation. Los Angeles represents the largest cultural-industries economy in North America, if not the world. In Britain the coincidence of production and consumption activity in 'consumption spaces' is seen in the regeneration of favoured areas such as Sheffield's cultural industries quarter, with the emphasis on media production,[35] and Nottingham's lace-market quarter, with its focus on fashion, design and media industries.[36] The development of cultural-industries quarters in contemporary cities has also involved the redevelopment of past urban landscapes as **heritage areas** as in the Gastown district of Vancouver, Albert Dock in Liverpool and the Merchant City in Glasgow. As urban places compete within the global economy for limited investment funds, their success often depends on the conscious and deliberate manipulation of culture in an effort to enhance their image and appeal.[37]

Critics of cultural-based strategies for urban renewal have pointed to a number of problem issues, including

TABLE 16.1 PROBLEMS AND OPPORTUNITIES OF CULTURE-LED REGENERATION

Problems	Opportunities
1. The need to project images conducive to inward investment has meant that most cultural projects tend to take the form of prestige art events or flagship developments that often cater to a select audience. If combined with a reduction of public funding for more community-oriented arts organisations, such policies can provoke antagonism	1. From an economic perspective the development of cultural industries can help create employment opportunities and diversify the structure of the local economy
2. In order to be economically viable, cultural initiatives require the creation of a critical mass of cultural activity. Often the appropriate conditions lie in the city centre and not in peripheral areas. The focus on the city centre can alienate communities in both outer and depressed inner-city areas unless equal consideration is given to developing local facilities	2. The reanimation of city centres through greater cultural provision and extended 'activity hours' can increase the areas used by city residents, especially during the evening, and in turn can help improve people's perceptions of safety in the city
3. Who benefits? Among obvious beneficiaries are visitors and tourists, traders, real-estate speculators and hoteliers. Distributional issues are important, as often the costs of cultural projects are borne by the citizenry as a whole	3. Public art policies can help improve the visual attractiveness of an area
4. Whose culture? In many instances the 'culture industry' creates a sanitised environment that has little to offer the 'lived practices' of local communities. The organisation of Glasgow's Year of Culture 1990 did not meet with universal approval in the city, with heated debate over the role and merits of indigenous working-class culture which some considered to have been marginalised in favour of an imported high culture of opera and ballet	4. Community-based programmes can be used to involve local people and provide a focal point for community pride and identity
5. The development of sites for cultural consumption may accelerate a process of gentrification to the detriment of existing residents, as in the development of the International Convention Centre in the Ladywood area of Birmingham and the creation of a loft market in SoHo in New York	

Sources: H. Lim (1993) Cultural strategies for revitalising the city *Regional Studies* 27(6), 589–95; P. Loftman (1991) *A Tale of Two Cities* Built Environment Development Centre Research Paper 6, Birmingham Polytechnic; S. Zukin (1988) *Loft Living: Culture and Capital in Urban Change* London: Radius

the uneven distribution of benefits and costs for the urban society as a whole (Table 16.1).

URBAN TOURISM AND DOWNTOWN REDEVELOPMENT

Tourism, along with finance and business services, is one of the fastest-growing components of the service sector. In many post-industrial cities the 'visitor economy' is of increasing importance and the promotion of tourism and leisure is a central element of downtown revitalisation strategies. Las Vegas, the archetypal post-modern twenty-four-hour consumption space, attracts over 46 million visitors annually. In Britain the city of Glasgow has used arts and special events (such as its designation as European City of Culture, 1990) to rebrand its image from that of a declining industrial city and increase visitor numbers.[38] In many cities disadvantaged multicultural districts on the fringe of city centres have been 'reconstructed' (i.e. made more accessible, safe, and visually appealing) and marketed as new destinations for leisure and tourism. Expressions of multiculturalism in the built environment, along with markets, festivals and other events in public spaces are presented as back-drops for consumption. Typical of such ethnoscapes are the Chinatowns and Latin Quarters of many US cities and the representation of Brick Lane in inner London, (the heart of the Bangladeshi community), as 'Banglatown' – an area that now caters for both the local ethnic minorities and tourists'[39]

Advocates of urban tourism development also identify wider benefits for cities. It is argued that advertising the city as a tourist destination and engaging in place promotion events, such as arts festivals, sports events and world fairs, benefits efforts to attract foot-loose economic activities. A revitalised inner city may also draw middle-class residents back into the centre, while new facilities constructed partly to attract tourists will also be available to local residents. Critics of tourist-based development contend that the industry provides only low-paid and seasonal work. However, many of the components of urban tourism, such as conferences, experience only minor seasonal variations. The industry also provides skilled employment at a management level, while the availability of low-skilled employment in centrally located tourist enterprises may benefit local populations. A further criticism concerns the possible diversion of public funds from services of direct benefit to resident groups, and the risk that cities may subsidise loss-making visitor attractions for the benefit of private businesses. In a number of US cities the force of these criticisms has been reduced by the use of a room tax or sales tax that transfers development costs to visitors. Calculation of the costs and benefits of urban tourism is complicated by the fact that the impact is spread across many sectors, from the visitor attractions to shops and transportation. Individual city authorities must decide whether the balance is right for them.

THE MARKET-LED APPROACH TO URBAN DEVELOPMENT

The hands-off market-led approach to urban development promoted by Conservative governments in the UK during the 1980s was criticised by the Audit Commission (1987 p. 1), which characterised inner-city policy as 'a patchwork quilt of complexity and idiosyncrasy' and described urban initiatives as lacking co-ordination and having no sense of strategic direction, with programmes frequently operating in isolation and often in competition with one another.[40] The government's response was to combine the resources of twenty different programmes into a Single Regeneration Budget (SRB) from 1994. This, however, marked a reduction of almost £300 million in the funds available, from £1.6 billion for SRB equivalent programmes in 1992–3 to £1.3 billion in 1996–7. According to Bradford and Robson (1995), 'the amount of money going into urban policy [was] minuscule compared with the size of the problems which are being tackled'.[41]

Justification for the market-led approach to urban regeneration centres on the concept of *trickle-down*, which argues that, in the longer term, an expanded city revenue base created by central-area revitalisation provides funds that can be used to address social needs. In practice, however, these funds are generally recycled into further development activity. Nowhere has a substantial **trickle-down effect** been demonstrated. The available evidence suggests that it is naive to expect a 'morally aware' private sector to effect the revitalisation of run-down urban areas. Private-sector investment decisions are founded largely upon self-interest and not philanthropy. The privatisation of urban development inevitably means accepting a policy of triage and concentrating on areas of greatest economic potential – with adverse consequences for other areas.

In order to address problems of deprivation and disadvantage, urban policy must possess both a social and an economic dimension. It must be concerned as much about the distribution of wealth as about wealth creation. Such social criteria did not figure prominently in top-down urban policy in the UK and the USA during the last quarter of the twentieth century.

URBAN REGENERATION FROM THE BOTTOM UP

The bottom-up perspective focuses on local economic development strategies initiated by urban authorities and popular grass-roots or neighbourhood-based efforts to capture the benefits of urban restructuring for local residents.

LOCAL ECONOMIC INITIATIVES

A wide range of local economic strategies have been employed in an attempt to combat the negative effects of the globalisation of capital. These range from conventional approaches, including supporting private enterprise (by the provision of sites, premises, infrastructure, grants and tax relief)[42] and **place marketing**[43] (Box 16.6) to more radical strategies that challenge the basis of the dominant political economy.[44]

Radical local economic initiatives seek to moderate the impact of uneven development on depressed urban regions by focusing attention on the social costs associated with the unfettered ability of corporate capital institutions to move investments between global locations in search of maximum profit. The specific objects of local employment initiatives are to:

1. to Provide the means for direct intervention in the local economy in order to prevent loss of jobs through plant closures or *in situ* redundancies.
2. to Plan for the long-term development of the local economy in order to secure future employment and production to meet social needs.

3. to Support the unemployed by breaking their isolation from the world of work and facilitating their return to employment by means of training and other support.
4. to Improve the quality of employment and increase the job opportunities for disadvantaged groups in the local labour market.
5. to Explore alternative forms of business organisation and to extend workers' control.[45]

Local enterprise boards were the main vehicle to advance this strategy of '**local socialism**',[46] which was followed by a number of Labour-controlled councils in the early 1980s. The most wide-ranging strategy of local economic intervention was formulated by the former Greater London Council.[47] However, the scale of the problems affecting London's deprived areas, and tension between the economic and social objectives of the strategy, meant that even this comprehensive initiative was incapable of combating the adverse structural trends affecting the urban economy. Following the abolition of the metropolitan counties in 1985, much of the momentum of local socialism and of local enterprise boards dissipated.

BOX 16.6

City marketing, festivals and urban spectacle

Marketing of places, with its emphasis on the projection of deliberately created images, has been a feature of the entrepreneurial mode of urban governance that has emerged since the 1970s as part of a competitive search for new sources of economic development, in response to de-industrialisation and the internationalisation of investment flows. Although 'civic boosterism' designed to attract inward investment has a long tradition, deepening socio-spatial inequalities within cities have meant that place marketing strategies may also be directed at an internal audience in order to legitimise regeneration strategies and foster social cohesion.

Glasgow was the UK pioneer in urban reimaging. To counter a negative image based on industrial decay, unemployment, violence, trade union militancy and poor environmental quality, Glasgow embarked on a concerted marketing campaign in the early 1980s. Commencing with a cartoon character (Mr Happy) and a slogan ('Glasgow's miles better'), the strategy advanced to focus on the arts and culture as a means of increasing tourism and conveying the image of Glasgow as a post-industrial city. In 1988 the city hosted a Garden Festival, in 1990 it was European City of Culture, and in 1999 City of Architecture.

The promotion of a postmodern urban image may involve physical reconstruction (e.g. through flagship buildings such as the Petronas Towers skyscrapers in Kuala Lumpur), re-emphasis of neighbourhood identity (e.g. Chinatown in San Francisco) or the creation of a twenty-four-hour city by replacing the rigid time rhythms of urban life under modernism by more flexible notions of the daily activity cycle.

Place marketing has an ideological context in that it presents a selective 'reading' of the city for a particular purpose. The projected image may be challenged by alternative readings. Place marketing can also have socially regressive consequences by shifting attention from socio-economic inequalities, and through the allocation of public expenditure to support flagship developments at the expense of local social investment. Finally, place marketing is a highly speculative venture: only a few cities can host the Olympic Games, and there is a limit to the number of convention centres required. Most cities, however, have no option but to compete in an effort to stay ahead of the game.

In relation to the overall scale of the urban problem, the impact of the local enterprise boards was marginal. Although in terms of cost per job to the public exchequer they rate as one of the most cost-effective agencies of urban regeneration, their greatest contribution was a symbolic one of demonstrating the existence of alternative economic strategies to those that dominated contemporary urban policy.

In contrast to local socialism some US cities have sought to capture the benefits of urban redevelopment for the public by operating as capitalist agents in their own right. Rather than simply taking an enabling role in downtown redevelopment the public sector has become the lead player, as focused on the return on investment from a project as the private sector. This approach, referred to as municipal capitalism, is characterised by the public sector undertaking high-risk projects utilising innovative financing mechanisms, such as speculative revenue bonds and tax increment financing. The public sector also routinely spearheads public–private development partnerships, and organises its redevelopment agencies as quasi-governmental entities able to operate outside the constraints of typical government agencies. For example, the quasi-public Battery Park City Authority in New York benefits from revenue generated by projects in its redevelopment area, while in San Diego the city's Ballpark Project is designed to underwrite redevelopment of the East Village district. The key difference between municipal capitalism and earlier public–private partnership arrangements is that the public sector is no longer simply an entrepreneurial agency facilitating downtown redevelopment but operates as an active capitalist agent seeking to obtain financial returns on its investments.[48]

PROGRESSIVE PLANNING POLICIES

One of the principal difficulties for local attempts to revitalise decaying urban environments is the understandable reluctance of capital to invest in areas perceived as lacking market potential. In some cases a degree of enlightened self-interest may convince firms to maintain or increase investment in an area (for example, a company with a sizeable investment in a plant may prefer to subsidise initiatives to promote local development as an alternative to relocation), but more generally some economic pressure is required to persuade capital to enter into a partnership with local authorities and communities. Three types of strategy have been employed to achieve this degree of leverage:

Equity planning

The basic concept underlying **equity planning** is that the unfettered operation of the capitalist system produces disadvantages for the poor. Concerned professional planners must therefore assume an interventionist role of politically committed activists to protect the interests of the poor in face of the profit-driven economic development and land-use decisions of the major economic and political actors (see Chapter 7). These 'advocacy planners' seek to alter the commercial and financial decisions that create urban social problems such as red-lining by banks, neighbourhood disinvestment, suburban relocation by industry and business, and socially costly redevelopment schemes.

Equity planning or 'pragmatic radicalism' was practised by the Cleveland OH city planning commission between 1969 and 1979 in order 'to provide a wider range of choices for the Cleveland residents who have few, if any, choices'.[49] For any proposed redevelopment project, the planning staff undertook a social-impact analysis to examine questions such as whether it would provide more jobs at liveable wages for city residents, the effect on the level and quality of public services, and whether it would support neighbourhood vitality or contribute to its decline. Projects that failed to meet these distributive criteria were opposed by the planning commission. Since 1979 Cleveland has reverted to a more conventional style of city planning and the power of private investors in local economic development has been reaffirmed[50] (Box 16.7).

Linked development

Linked development presents private-sector investors and developers with a constrained choice of investment opportunities in an attempt to induce a more socially and spatially balanced pattern of growth.[51] Policies may include, in the housing field, 'inclusionary zoning', which requires developers to set aside low-cost housing units in market-rate projects for allocation by a neighbourhood housing association (Box 16.8). Linked employment strategies could include a requirement on commercial developers to hire labour from specific geographical areas and social groups. This combination of regulation and partnership between public, private and community sectors can effect a more equitable distribution of the benefits of urban growth, but **linkage** policies are not a panacea for uneven urban growth. The extent to which the public sector can influence private investment decisions depends on the attractiveness of

BOX 16.7

Urban regeneration in Cleveland OH

Cleveland lies on the shores of Lake Erie in the north of the state of Ohio and forms part of the so-called 'rustbelt' of the north-eastern USA – an area that has witnessed massive industrial decline as a result of the collapse of basic manufacturing industries. Between 1970 and 1985 the Cleveland metropolitan area lost 86,000 manufacturing jobs, and a further loss of 50,000 was projected up to the year 2005.

Population loss, primarily as a result of migration to the suburbs, has accompanied economic decline. The city's population in 1950 was 914,000, compared with an estimated 500,500 in 1990. The forecast for 2005 was 321,800. The city constitutes one of the most racially segregated housing markets in the USA, with the vast majority of Cleveland's black American population living to the east of the Cuyahoga river and most of the white population to the west. Some 80,000 defective dwellings have been cleared in the city since 1960, but over one-third of the housing stock remains substandard. Neighbourhoods such as Hough to the east of the city centre are among the poorest in the country. The area has suffered from 'white flight' to the suburbs over the past forty years, the wholesale abandonment of properties, multi-occupation of older, poorer dwellings; poor-quality public-sector housing and high vacancy rates. In 1978 Cleveland was the first US city to default on federal loans, registering debts of $111 million. It became the butt of comedy shows, with references to the city as 'the mistake on the Lake'.

During the 1970s the main thrust of public policy in Cleveland was aimed at tackling directly the problems of the poorer neighbourhoods. Under planning director Norman Krumholz the idea of equity planning was put into practice. This meant explicitly attempting to influence public and private development and investment decisions across a range of initiatives in a way that would best benefit the residents of Cleveland's most deprived neighbourhoods. This approach was based on the belief that private investment decisions, if left to their own devices, would tend to favour the haves rather than the have-nots; and that traditional planning methods concentrating on land use could make only a marginal impression on the problems of people in the worst parts of the city.

Following changes in political control and administration in the 1980s the city council adopted the approach favoured by many cities in the US northeast. Emphasis has been placed on trying to attract private development and investment into the downtown area in the belief and hope that demand for services will be boosted, spending will increase, jobs will be created and the benefits will 'trickle down' to the neighbourhoods. A new convention centre has been built; older warehouses down by the river have been converted into bars and restaurants; there are plans for a new domed stadium for major sporting events; and the city's Rock and Roll Hall of Fame is now open. Nevertheless, there is considerable unease in some quarters that this focus on the city centre is at the expense of the poorer neighbourhoods, and that the tax concessions offered to induce private capital to invest in the city have reduced the amount of public funding available for essential services in deprived areas.

Source: C. Wood (1996) Renewal and regeneration, in D. Chapman (ed.) *Creating Neighbourhoods and Places*
London: Spon, 194–221

the city for capital; but rarely is there no potential for linkage. Most urban areas possess locations sufficiently attractive to capital for a premium to be exacted for the benefit of the community at large. While the type and value of any premium are matters for local political leaders to decide, this strategy offers a direct means of redistributing at least some of the profits of central- area development to less advantaged parts of the city.[52]

Popular planning

Popular planning means planning by local communities in their own neighbourhoods. It involves the formulation of planning proposals and their implementation by local community organisations, and rests on close collaboration between the community and the local planning authority that agrees to adopt the popular plan as official policy.

An example of popular planning in practice is provided by the redevelopment of the Coin Street area of Waterloo in Central London. In the late 1970s conflict arose between a commercial plan for a speculative development of offices and hotels and a community plan for social rented housing and local employment and amenities. Following several years of planning inquiries, protests and demonstrations, and legal action against the Secretary of State, the

BOX 16.8

Linkage policy in San Francisco CA

In 1981 San Francisco became the first US city to adopt linkage policies. The decision was based on several factors relating to a boom in downtown office construction, which had begun in the mid-1960s:

1. Growing community opposition to unfettered downtown development and in particular its effect on house prices and traffic consumption.
2. Pressure from community-based coalitions for the city to build and preserve affordable housing, and to improve the municipal transit system.
3. A need for the city to fund new revenue sources to offset property tax losses caused by the passage of California Proposition 13 in 1978.
4. The prospect of declining federal aid for housing and transit following the election of the Reagan administration.

Under the linkage policy, all developers of buildings exceeding 50,000 ft² (4,600 m²) in the central business district were required to provide new or rehabilitated housing or to pay an 'in lieu' fee of $5 per square foot to the city for housing. From its inception in 1981 to 1985 office developers subsidised 2,693 low- or moderate-income housing units, a figure that represented a contribution of 20 per cent to the city's affordable housing requirement.

commercial developers sold the land to the Labour-controlled Greater London Council, which sold it to a new community development group that financed the purchase with mortgages provided by the GLC and the Greater London Enterprise Board. This cleared the way for implementation of the popular plan for Coin Street.[53]

The Dudley Street initiative in Boston represents another example of locally controlled urban regeneration. A deprived community of 12,000 residents combined to create a politically effective organisation that enabled the community to form a partnership with City Hall in which the people took on the responsibility for redevelopment. The community gained control of the local land market through the Boston Development Corporation's granting it ownership in perpetuity of derelict land in the area. In addition, compulsory purchase of privately owned land was financed with a Ford Foundation loan, while land owned by the public sector was gifted to the community.[54]

The examples of Coin Street and Dudley Street demonstrate that while local communities can influence the development of their environment, success is often dependent upon the support of a sympathetic local authority prepared to intervene in the process of market-led urban redevelopment. This is evident in the failure of local communities to gain control of developments in the London docklands[55] and in Tolmer Square, north London.[56]

THE COMMUNITY/SOCIAL ECONOMY

All the 'bottom up' local economic development strategies examined so far have been embedded in the formal economy. An alternative approach to ameliorating local disadvantage is via the informal, complementary or community/social economy.[57] Essentially, the community/social economy (or 'Third Sector'), is made up of the voluntary, non-profit and co-operative sectors that are formally independent of the state. Particular attention has been focused on initiatives within the voluntary sector (Figure 16.4). These include the following.

COMMUNITY BUSINESS

The **community business** is another form of local economic initiative which stems from the fact that market forces are largely impotent in the most deprived urban areas.[58] It is essentially an organisation owned and controlled by the local community, with membership open to all residents, that aims to create ultimately self-supporting jobs for local people, and to use profits made from business activities either to create more employment or to provide local services or support local charitable work. Community businesses can be distinguished from other community enterprises, such as credit unions or housing co-operatives, by the fact that most are engaged in more than a single activity. In practice, although there are multiplier effects, including enhanced local spending power and some reduction in the pool of long-term unemployed, the economic impact of community businesses in urban Britain has been marginal to the scale of the problems facing depressed communities.[59]

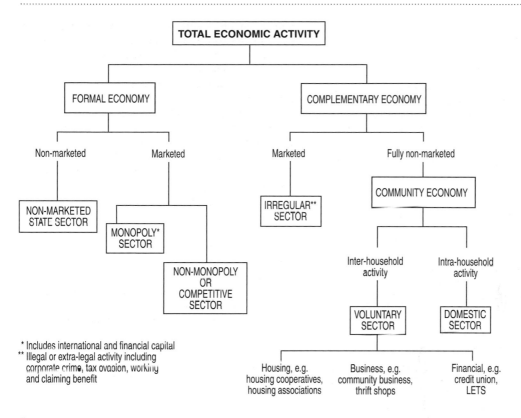

Figure 16.4 A general model of economic activity within a developed economy
Source: P. Ekins and M. Max-Neef (1992) *Real Life Economics* London: Routledge

COMMUNITY DEVELOPMENT CORPORATIONS

Community economic development has assumed major importance in the USA for groups traditionally concerned with advocacy and provision of community services[60] (see Chapter 20). An important vehicle has been the **community development corporation**, which first appeared in the late 1960s in response to civil disturbances in American cities. Community development corporations (CDCs) may be described as democratic firms that are accountable to residents of the local community by allowing them to become members for a small fee and to share equally in shaping the organisation's policy.[61]

A major contrast with the UK community business is that the CDCs generally engage in a wider range of activities, such as the promotion of credit unions, the supply of low-income housing, the provision of education and training initiatives, and support for new enterprise. The goal is for the various subsidiary activities to interact to ensure that the whole is greater than the sum of the parts. For example, if a grant is obtained to rehabilitate housing, by setting up a construction company to do the work the CDC can provide local employment and ensure that money circulates in the neighbourhood economy. This, in turn, may enhance the profitability of a new local shopping centre that the CDC is also trying to establish. Moreover, the subsidiaries of CDCs are often encouraged to operate in the wider marketplace. A construction or management company can obtain contracts from other organisations besides its parent body, thus contributing to the profitability of the main CDC, and in the process reversing the usual flow of money out of a deprived area (Box 16.9).

The growth of CDCs in US cities can be explained by a particular set of circumstances that includes the less interventionist role of the USA compared with the British government, the well-established tradition of private not-for-profit organisations providing for social welfare need, and access to a range of major non-governmental

BOX 16.9

The Watts Labor and Community Action Committee

Watts is a poor black neighbourhood situated in south central Los Angeles. The Watts Labor and Community Action Committee (WLCAC) was established in 1965, with funding from eleven international trade (labour) unions, to tackle the problems of job training for local youth and the lack of public social services in the area. By 1980 the organisation had grown into a multi-purpose community development corporation (CDC) with a budget of $17 million. Its services included several programmes for senior citizens, dial-a-bus, child care, food stamp sales and a variety of employment and training programmes. It had also built, rehabilitated or relocated over 500 units of housing, which it then managed. It had its own construction company and a small number of potentially profit-making subsidiaries, including a restaurant, supermarket, a petrol (gas) station and an appliance and tool centre.

By the late 1980s the organisation decided to participate in joint ventures with private developers to build housing and shopping centres in order to create a development and management company that could be sustained without subsidy. Crucially, however, the CDC maintained its social service and job training operations.

Like all CDCs, the WLCAC has to balance the need for a hard-headed business approach, consistent with commercial and economic development, with a socially sensitive approach to enable it to retain its consistency and fulfil its original aims.

Source: **A. Twelvetrees (1997)** *Organising for Neighbourhood Development* **Aldershot: Avebury**

funding sources such as foundations and corporations. This poses the question of the applicability of the CDC model elsewhere.[62]

LOCAL EXCHANGE TRADING SYSTEMS

Local exchange trading systems (LETS) have emerged as a community response to globalisation. In contrast to the 'local socialism' strategies employed by some UK authorities in the 1980s, LETS do not seek to challenge the hegemony of the capitalist economy head-on but instead attempt to develop a parallel, complementary form of social and economic organisation within the local

context.[63] A LETS effectively decouples from the global economic system in an effort to foster a local economic and social identity. This is achieved primarily through the use of a non-commodified 'local currency' that ensures that the circulation of LETS capital remains wholly within the local system (Box 16.10).

Since its introduction in the town of Courtenay BC in the early 1980s the LETS concept has spread throughout North America, Britain, Europe and Australia. Though economically insignificant in terms of national GDP, within a local context LETS have the potential to improve individual levels of well-being. A successful LETS can perform an enabling role both economically and socially by providing a substitute for the increasingly attenuated kinship networks that characterise contemporary society, by fostering community spirit in a locality, by enhancing individual self-esteem through the valuing of skills, and through the provision of interest-free credit by bringing individuals into the circle of exchange and thereby providing them with materials, services and an opportunity to improve their quality of life. LETS seek to stimulate both the local economy and the community.

THE BLACK ECONOMY

The black economy involves the paid production and sale of goods and services that are unregistered by or hidden from the state for tax, social security and/or labour-law purposes but which are legal in all other aspects. Contrary to the marginality thesis that views paid informal work as low-paid peripheral employment conducted for the purposes of economic survival by marginalised groups and areas, the black economy operates at both ends of the social spectrum. In addition to an economic motive, in lower-income neighbourhoods paid informal exchange is conducted also for and by close social relations to assist each other in a way that avoids any connotation of charity. Conversely, in higher-income areas such exchange is conducted more by self-employed people and firms for profit and is used primarily as a cheaper alternative to formal sector firms.[64] The black economy represents a difficult to measure, but pervasive, alternative urban economic space that exists between the legal and illegal economies (Box 16.11). There exists a 'grey' or variable relationship between the black economy and the underground criminal economy. In some US cities the underground economy organised by youth gangs employs hundreds in legitimate (e.g. laundromats) and illegitimate (e.g. drug dealing) activities. In Chicago, the Black Disciples gang runs

BOX 16.10

The anatomy of a LETS: the West Glasgow Local Exchange Trading System

All UK LETS are structured on broadly similar lines, with local variations in terms of currency used, volume of trading, membership profile and extent of community social activity. The structure of the West Glasgow LETS is typical.

The aim of the West Glasgow LETS is to enable 'those participating to give and receive all kinds of goods and services amongst themselves, without having to spend money.' The system is run on behalf of the membership by a steering group of at least four, elected annually at the AGM each May. This group appoints officers as necessary to carry out the tasks of running the LETS system. Members pay an annual fee (in local currency and/or sterling) to cover administrative expenses. Members also indicate which goods and services they can offer (and which they wish to acquire) and in return they receive a directory (updated every two months) that lists all the goods and services available in the system. They can then arrange to trade with each other, paying in the local currency (groats), although no member is obliged to accept any particular invitation to trade. No warranty or undertaking as to the value, condition or quality of services or items exchanged is expressed or implied by virtue of the introduction of members to each other through the directory.

Each member has an individual account with a cheque book and their transactions are recorded centrally by a bookkeeper. An open statement of each member's credit or debit balance is provided to all members on a regular basis and any member is entitled to know the balance and turnover of another member's account. Crucially, unlike a commercial bank, individuals' credit/debit balance does not affect their ability to trade. Any member can issue another member with credit from their account, subject to any limit that may be set by the steering group. On leaving the LETS, members with commitments outstanding are obliged morally to balance their accounts. *In extremis* the steering group has the power to seek an explanation from a member whose activity is considered to be contrary to the interests of the membership and to suspend or exclude a delinquent member.

No interest is charged or accrued to balances; the health of the system depends not on accumulated surpluses but on the circulation of 'capital' through trading. The price of each transaction is a matter of agreement between the parties. While general guidance suggests a normal rate of five groats per hour, some services, such as accountancy or childminding, may attract a higher rate. Part of the agreed price may be paid in sterling – for example, to cover the costs of materials purchased on the 'external' market – but only local currency units are recorded in the LETS accounting system. Members are individually responsible for their personal tax liabilities and returns within the formal economic system and, where appropriate, for their relationship with social security/benefits agencies.

In addition to trading activity, social events (such as musical evenings, markets, dinner parties and hill walks) are organised, with the aim of 'helping to overcome the problems, doubts and personal mistrust that might inhibit trading' and to foster the growth of community spirit in the locality.

Source: **M. Pacione (1998) Towards a community economy: an examination of local exchange trading systems in west Glasgow** Urban Geography **19(3), 211–31**

BOX 16.11

The black economy in urban Britain

In some of Britain's most deprived estates the black economy has almost entirely replaced over-the-counter trading. The Digmore estate in Skelmersdale, between Liverpool and Manchester, once had a supermarket and two banks but these have long since closed. Now a night club has become the general store for a community of 5,000. The estate's economy is run by a few locals. Bob trades in everything from detergents to clothes. State-of-the-art electronic goods are available on request, although these take time to get hold of. A request for spirits produced a bottle of whisky for £5.

Elsie, aged 60, has lived on the Digmore estate for half her life and thinks 'it's great living here. I haven't paid full price for anything in years'. In the local pub the barmaid says, 'The government doesn't have a clue. This is real life. When you live in a place like Skem, it's all about survival. If you can buy something cheap, you do it.'

Source: Sunday Times **30 April 2000**

a business earning $300,000 per day, with a CEO, board of directors, annual company picnic, and welfare benefit packages for members and their families in event of injury, incarceration or death.[65]

Having examined the principal economic dimensions of urban localities, in the next chapter we focus on urban service provision and examine the geographies of collective consumption in the city.

FURTHER READING

BOOKS

A. Cochrane (2007) *Understanding Urban Policy: A Critical Approach* Oxford: Blackwell

J. de Filippis (2004) *Unmaking Goliath: Community Control in the Face of Global Capital* London: Routledge

C. Johnstone and M. Whitehead (2004) *New Horizons in British Urban Policy* Aldershot: Ashgate

P. Jones and J. Evans (2008) *Urban Regeneration in the UK* London: Sage

W. Keating and N. Krumholz (1999) *Rebuilding Urban Neighbourhoods* London: Sage

A. Leyshon, R. Lee and C. Williams (2003) *Alternative Economic Spaces* London: Sage

J. Montgomery (2007) *The New Wealth of Cities: City Dynamics and the Fifth Wave* Aldershot: Ashgate

A. Scott (2000) *The Cultural Economy of Cities* London: Routledge

S. Ward (1998) *Selling Places: The Marketing and Promotion of Towns and Cities, 1850–2000* London: E & F. N. Spon

JOURNAL ARTICLES

A. Amin and N. Thrift (2007) Cultural economy and cities *Progress in Human Geography* 31, 143–61

M. Levine (1987) Downtown redevelopment as an urban growth strategy: a critical appraisal of the Baltimore renaissance *Journal of Urban Affairs* 9(2), 133–8

A. Merrifield (1993) The Canary Wharf debate *Environment and Planning A* 25, 1247–65

M. Pacione (1997) Local exchange trading systems as a response to the globalisation of capitalism *Urban Studies* 34(8), 1179–99

M. Pacione (2007) Sustainable urban development in the UK: rhetoric or reality? *Geography* 92(3), 246–63

S. Quilley (1999) Entrepreneurial Manchester: the genesis of elite consensus *Antipode* 31(2), 185–211

J. Rhodes, P. Tyler and A. Brennan (2005) Assessing the effect of area-based initiatives on local area outcomes *Urban Studies* 42, 1919–46

L. Sagalyn (1997) Negotiating for public benefits: the bargaining calculus of public–private development *Urban Studies* 34(12), 1955–70

I. Turok (1992) Property-led urban regeneration: panacea or placebo? *Environment and Planning A* 24, 361–79

KEY CONCEPTS

- Enterprise Zones
- Urban development corporations
- Privatisation
- Needs-based approach
- Demand-led approach
- The Third Way
- Business improvement districts
- Creative federalism
- New federalism
- Empowerment Zones
- Tax increment financing
- Public–private partnership
- Rouse-ification
- Trickle-down effect
- Cultural industries
- Market-led approach
- Local Enterprise Boards
- Local socialism
- Municipal capitalism
- Progressive planning
- Equity planning
- Pragmatic radicalism
- Linked development
- City marketing
- Inclusionary zoning
- Popular planning
- Community economy
- Community business
- Community development corporations
- Local exchange trading systems
- Local currency
- Black economy

STUDY QUESTIONS

1. With reference to a particular example, consider the impact of either an urban development corporation or an enterprise zone on local economic development.
2. Provide an assessment of public–private partnerships as a mechanism for urban regeneration.
3. Consider the value of the cultural industries in urban renewal.
4. With reference to a particular city examine strategies that have been employed to 'reinvent' the city's image in order to enhance its competitive position within the national-global economy.
5. Critically examine the effectiveness of local economic strategies to counter the adverse effects of global economic restructuring on disadvantaged urban areas.
6. Explain the rationale underlying either a US community-development corporation or a local exchange trading system and consider the potential of the approach for a local economy.
7. Based on your knowledge of urban economic change identify the strategies you would employ in a strategic plan designed to regenerate a declining city.

PROJECT

Make a list of the main policy initiatives introduced by government to address the challenge of urban economic regeneration. Alongside each policy, note examples of its impact on urban economies. Note both positive and negative impacts. Consider the political complexion of government in relation to type of legislation introduced. Can you identify any significant changes in policy orientation under different governments?

17

Collective Consumption and Social Justice in the City

Preview: collective consumption; welfare needs; welfare states; the theory of public goods; theories of public-service provision; the measurement of need; efficiency, equity and equality in public-service provision; patterns of inequity; the suburban exploitation thesis; deinstitutionalisation; educational services in the city; social justice and welfare

INTRODUCTION

The term 'collective consumption' was coined by Castells (1977) to indicate an increasing tendency, throughout most of the twentieth century, for governments in advanced capitalist societies to intervene in the provision of goods and services.[1] **Collective consumption** refers to all collectively organised and managed services consumed via non-market mechanisms and at least partly paid for from the public purse (see Figure 17.1 and Box 17.1). Individuals and households with limited personal resources (i.e. restricted access to private consumption goods and services) may well gauge their quality of life in terms of the **public**

goods and services available locally. Significantly, the availability and level of consumption of public services vary across space and between social groups in the city. The 'fairness' of the differential distribution of welfare services is a central issue in the concept of social justice.

In this chapter we examine geographies of collective consumption in the city. We consider different means of satisfying welfare needs, and the implications for collective consumption of recent changes in the nature of the welfare state. We then examine the theory of public goods and the effects of geography on the distribution of urban public services. We identify the main theoretical perspectives on public-service provision, ranging from market-based distribution to allocation on the basis of need. Attention then turns to the justness of patterns of collective consumption and the key issues of efficiency, equity and equality in public-service provision. We consider locational efficiency, personal and locational accessibility, and patterns of inequity in the consumption of urban services. The questions of territorial and social equity in public-service provision are then addressed at the metropolitan scale with particular reference to the 'suburban exploitation' thesis, and at the intra-urban level with reference to the situation of deinstitutionalised residents of the inner city, and the differential availability and quality of educational services in the city. Finally, we consider the value of the concept of social justice for the study of geographies of urban service provision.

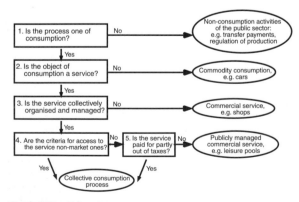

Figure 17.1 Determining whether a social process is an element of collective consumption

BOX 17.1

The importance of collective consumption

The importance of collective consumption in contemporary society is considerable. As Tietze (1968 p. 36) observed, 'modern urban man is born in a publicly financed hospital, receives his education in a publicly sponsored school and university, spends a good part of his life travelling on publicly built transportation facilities, communicates through the post office or the quasi-public telephone system, drinks his public water, disposes of his garbage through the public removal system, reads his public library books, picnics in his public parks, is protected by his public police, fire and health systems; eventually he dies, again in a hospital, and may even be buried in a public cemetery. Ideological conservatives notwithstanding, his everyday life is inextricably bound up with governmental decisions on these and numerous other local public services.'

Source: M. Tietze (1968) Towards a theory of urban public facility location *Proceedings of the Regional Science Association* 31, 35–44

WELFARE NEEDS AND PROVISION

According to the UN Universal Declaration of Human Rights:

> everyone has the right to a standard of living adequate for the health and well-being of himself and his family, including food, clothing, housing and medical care and the necessary social services and the right to security in the event of unemployment, sickness, disability, widowhood, old age or other lack of livelihood in circumstances beyond his control.
>
> (Pierson 1991 p. 127)[2]

There are, however, significant variations between and within countries and cities in the extent to which these goals are realised. There are also many different ways in which these 'welfare needs' may be satisfied. These include:

1. *Mutual support of family and friends*. This ranges from the 'extended family' networks found in many Third World settings to the informal recipro-

cal exchange of services, such as child-minding, among households in advanced capitalist society.
2. *Charitable and voluntary organisations*. These range from the mutual benefit societies formed in nineteenth-century cities to insure families against the risks of sickness, old age and death to the current diverse array of voluntary-sector agencies, including charitable trusts and pressure groups.[3]
3. *Private markets*. Private-sector employers may provide their workers not only with income but also with pensions and health insurance, while individuals and families can purchase education, health care and pensions from private-sector institutions.
4. *The state*. The belief that states have a responsibility to ensure that all citizens enjoy a decent standard of living translates into the direct public provision of services such as health care, housing and education on a non-market basis. This philosophy is encapsulated in the post-Second World War welfare state. This form of state welfare provision is also referred to as collective consumption.

THE CHANGING NATURE OF THE WELFARE STATE

The degree to which people rely on the state to meet their basic welfare needs varies with the political and economic context. Members of a middle-income urban family in Britain are likely to depend on the state-funded National Health Service (NHS) for their health-care needs, whereas in the USA a similar family is more likely to have a private health-insurance plan and to use privately owned health centres. Further, within British cities NHS health care is freely available to all, whereas in the US poor inner-city dwellers may be forced to rely on underfunded public hospitals that are in marked contrast to high-quality suburban facilities.[4]

The nature and level of public-service provision is changing throughout the world.[5] Recent decades have witnessed a degree of convergence towards the liberal welfare-state model. The key factor underlying this trend is the increasing fiscal strain placed on state welfare systems by rising demand for services and a declining resource base with which to meet the cost of providing them. The main forces leading to increased fiscal stress for welfare services stem from a complex set of demographic, cultural and economic factors

(Table 17.1). The ideology of the state also affects the provision of welfare services, with the introduction of restructuring policies (such as those depicted in Table 17.2), most advanced in countries dominated by New Right governments, as in the UK between 1979 and 1997 under Thatcher and Major, and in the USA from 1981 to 1993 under Reagan and Bush. The emphasis on greater individual responsibility and a reduced role for the welfare state continued in the UK and the USA under the Blair and Clinton administrations, in accordance with the ethos of 'no rights without responsibilities', as manifested in policies such as welfare-to-work (Box 17.2).

Despite the financial constraints placed on the welfare state, ensuring the availability of services to citizens (whether by direct provision or via forms of out-sourcing) remains the main purpose of urban local government. Public-service provision absorbs most of the finance provided to local government from taxes and central government disbursements and employs the vast majority of local government staff. The nature of public-service provision remains a major determinant of the well-being of different social groups in the city.

THE THEORY OF PUBLIC GOODS

The distinction between private and public consumption goods is central to the concept of collective consumption. According to Samuelson (1954), whereas private goods may be consumed by a single person or household and are amenable to distribution by private markets, public goods have characteristics that preclude this.[6] Examples of private goods are food, clothing and consumer durables, whereas typical public goods include national defence, street lighting and welfare services. Pure public goods have the three basic characteristics of:

1. *Joint supply*, which means that if goods can be supplied to one person they can be supplied to all other persons at no extra cost.
2. *Non-excludability*, which means that once the goods have been supplied to one person, it is impossible to withhold them from others wishing access to them. This, of course, means that people who do not pay for a service cannot be excluded from its benefits (the **free-rider** problem).

TABLE 17.1 PRESSURES ON THE WELFARE STATE

Demographic trends
   Increasing numbers of dependent groups (e.g. unemployed people, homeless people, old people, infirm elderly people, single-parent families, disabled people)
   Growth of the 'new poverty' through polarisation of incomes and large numbers on very low incomes
   Feminisation of the workforce (e.g. growth of low-paid jobs and increased need for day care)

Cultural factors
   Rising expectations and desire for higher standards
   Desire for diversity of provision
   Strength of pressure groups (e.g. on behalf of women, disabled people, old people, long-term sick people)
   Growth of consumer culture
   Increasing cultural pluralism and diversity
   Growth of 'New Age' movements (e.g. alternative and complementary medicines)
   Tax revolts

Intellectual/ideological developments
   Growth (rebirth) of New Right ideas
   Decline of faith in collective solutions to social problems

Economic trends
   Increasing proportion of the population dependent upon a smaller work force
   Inelastic sources of local revenue, 'stagflation' (high unemployment and high inflation) in the 1970s
   Increased competition between nations for mobile capital combined with enhanced sensitivity of national economies to capital flows
   Desire of transnational corporations for low social overheads

Source: S. Pinch (1997) *Worlds of Welfare* London: Routledge

TABLE 17.2 FORMS OF RESTRUCTURING IN PUBLIC-SECTOR SERVICE PROVISION

Strategy	Outcomes
1. Partial self-provisioning	Child care in the home; care of the elderly at home; personal forms of transport; crime prevention strategies such as neighbourhood watch schemes
2. Intensification: increases in labour productivity via managerial or organisational changes with little or no investment or major loss of capacity	The drive for efficiency in the health service; competitive tendering over direct labour organisations, house maintenance, garbage collection; increased numbers of graduates per academic in universities
3. Investment and technical change: capital investment into new forms of production, often with considerable job loss	Computerisation of health and welfare service records; distance-learning systems through telecommunications technology; larger plant
4. Rationalisation: closure of capacity with little or no new investment or new technology	Closure of schools, hospitals, nurseries, day-care centres; closure or reduction of public-transport systems
5. Subcontracting: of parts of the services sector to specialist companies	Privatisation or contracting-out of cleaning, laundry and catering within the health service; contracting-out of garbage collection, housing maintenance, public transport by local authorities
6. Replacement of existing labour input by part-time, female or non-white labour	Increased use of part-time teachers; predominance of women in the teaching profession
7. Enhancement of quality through increased labour input, better skills, increased training	Retraining of public-transport personnel; community policing
8. Materialisation of the service function so that the service takes the form of a material product that can be bought, sold and transported	Reliance on pharmaceuticals rather than on counselling and alternative therapies
9. Spatial relocation	Relocation from larger psychiatric hospitals into decentralised community-based hospitals
10. Domestication: the partial relocation of the provision of the functions within forms of household or family labour	Care of the very young and elderly in private houses following reductions in voluntary and public service provision
11. Centralisation: the spatial centralisation of services in larger units and the closure or reduction of the number of smaller units	Concentration of primary and secondary health care into larger units with the growth of large general hospitals and group general practices

Source: S. Pinch (1989) The restructuring thesis and the study of public services *Environment and Planning A* 21, 905–26

3. *Non-rejectability*, indicating that, once goods or a service have been supplied, they must be consumed equally by all, even those who may not wish to do so.[7]

Although some public goods exhibit these characteristics (e.g. laws or regulations relating to road safety), few goods and services satisfy completely the Samuelson–Musgrave properties, and the conditions are best viewed as representing an extreme theoretical case. Significantly, the major factors that undermine this theoretical purity, and thereby influence the distribution of public goods, are geographical in nature.

THE DISTORTING EFFECT OF GEOGRAPHY

The first of these geographical influences on the nature and availability of public goods stems from

BOX 17.2

Welfare-to-work

In the UK the 1997 New Labour government's description of the objective of welfare reform was 'Work for those who can and security for those who cannot'. In practice this involves sanctions and inducements to compel welfare claimants to accept paid work or training places either in the mainstream private sector or through a range of temporary subsidised 'workfare' options through Welfare-to-work. The same ethos underwrites the US Personal Responsibility and Work Opportunity Act of 1996 (re-authorised in 2002 with a new provision that Workfare workers no longer had to receive minimum wages). Workfare schemes have enabled many of the unemployed to enter the world of work. They have also transferred many from welfare rolls into usually low-paid and often temporary jobs, creating a new class of 'working poor'.

jurisdictional partitioning, or the political and administrative structure of an urban area. Every local political jurisdiction will have different financial resources, expenditure needs, and preferences for particular 'bundles' of public goods. Consequently, local governments vary greatly in the quantity and quality of public goods and services they provide. This means that the level of public-sector resources an individual receives is often dependent upon where they live. More specifically, inter-jurisdictional variations in collective consumption stem from the geographically variable tax base, which reflects variations in income and wealth of people in different jurisdictions. Inter-jurisdictional variations can also arise from the political decisions of residents that allow the formation of 'service clubs' in which the bundle of public goods provided reflects local collective preferences.[8] This can lead to 'fiscal migration' among jurisdictions as people move to obtain the service and taxation package most attractive to them.[9]

A second geographical factor that undermines the concept of a pure public good is *tapering*. As many public goods and services (e.g. libraries) are 'point-bound', individuals must bear the costs of travel to use the facility. In general, use declines with increasing cost, whether that cost is measured in time, effort or money. At some point costs may reach a level where the service is not utilised. In these circumstances the

frictional effect of distance undermines the criterion of non-exclusion in the supply of public goods. Similarly, under conditions where a public service is brought to the consumer (e.g. fire protection), distance may dictate that the level of service is less than that enjoyed by more proximate consumers. In this case geography undermines the criterion of joint supply.

Externalities also compromise the concept of a pure public good. Many land-use activities have externality effects that are not reflected in costs or prices, and that impact on the well-being of communities. Desirable facilities, such as good schools, promote positive externalities, whereas noxious facilities, such as a rubbish tip, produce a negative externality field.[10] As we see later, externality effects are not necessarily confined within local government boundaries. *Spillover* occurs when facilities provided in one local authority are consumed but not paid for by residents of other jurisdictions, obvious examples being the major shopping, cultural and leisure facilities provided by central cities and enjoyed on a regional basis by those living outside the city boundaries. Before examining the geographies of public-service provision, however, we must first consider the key questions of how the provision of public services should be financed and who should benefit from their provision.

THEORIES OF PUBLIC-SERVICE PROVISION

There are three main theoretical perspectives on public-service provision:

1. The *market-surrogate school* embraces market principles to determine which goods should be provided publicly. Public goods are defined as those desirable services that lack the necessary rate of profit to attract private provision. In distributional terms, public goods should be allocated according to the willingness of people to pay for them. According to this *public choice* approach (Box 17.3) local politicians who make decisions on the allocation of public services do so in response to the demands of their voter-consumers, expressed via public support for particular tax and expenditure options at election time. The most geographically explicit formulation of public-choice theory postulates that the electorate's demand for different bundles of public goods is registered by a process of residential selection, with individuals choosing to live in a local

BOX 17.3

Public choice theory

Public choice theory refers to a set of explanations for service allocations in cities. Local politics is envisaged as a 'market place' in which politicians respond to the demands of the public in a manner similar to that in which entrepreneurs respond to the preferences of consumers. While entrepreneurs seek to gain most profit, politicians seek to gain most votes by providing favourable tax and expenditure options to the public. According to Tiebout (1956),[11] the public responds to the various packages available through residential selection. The fragmented structure of US local government is a result of these competitive forces that sustain the existence of different types of community catering for different preferences.

A criticism of the public choice approach is that, like the neoclassical economics on which it is based,

the theory assumes that individuals are free to make rational decisions on the basis of their preferences. In practice, residential choice decisions may be based on a range of factors in addition to the quality of local services, including socio-economic status and stage in the life cycle (see Chapter 10). Nevertheless, there is evidence of people 'voting with their feet', particularly in the context of suburban migration from inner-city areas, and the level and quality of the tax–services package are a factor for many of those able to choose.

Public choice theory has found greater acceptance in the USA than in the UK, not least because of the greater fragmentation of local government administrative units, and consequent opportunities for 'fiscal mercantilism' or the attraction of desirable land uses and exclusion of others.

government jurisdiction whose mix of public services and taxes best meets their preferences. The fragmented structure of metropolitan government is viewed as a general outcome of households 'voting with their feet' in this way.[12] The 'fiscal migration' model contains a number of idealised assumptions (e.g. omniscience on the part of voters, no economies of scale in the provision of services, and a large number of municipalities). Also the role of factors (such as stage in the life cycle) other than the public-service–taxation package in household mobility is ignored. Most significantly, the model neglects (or perhaps highlights) the fact that many households have a relatively limited choice of residential location. For those unable to participate in the public choice market, the 'exit' strategy is not an option, resignation or 'learned acceptance' of their current situation being all that is available.

2. The *ideology-appeasement school* argues that public goods and services are provided by a dominant class in order to appease the interests of a repressed class and thereby maintain social order. This interpretation is central to the Marxist view of public finance that sees the public sector as essential for both political legitimation and capital accumulation.[13]

3. The *needs-assessment school* contends that public goods should be allocated according to need rather than ability to pay. While the egalitarian

principle is held in common with the ideology-appeasement view, the needs-assessment school does not interpret the organisation of public goods as a mechanism of class oppression but emphasises the goal of allocating public goods and services for the welfare of society as a whole. Inherent in this approach is the fundamental question of the measurement of need, and it is to this issue that we turn next.

THE MEASUREMENT OF NEED

Needs are defined in relation to the requirements of an individual (e.g. for food or medical treatment), but also in relation to the prevailing *mores* or norms of the society. While certain basic or physiological needs are common to all human beings, many 'needs' arise from specific social-geographical contexts. Thus a suburban American family may 'need' two or more cars to engage in the daily activity patterns of the household within the context of the spatially dispersed US city, but from the perspective of a typical British urban context it is more difficult to differentiate such 'needs' from 'wants'.

A useful classificatory device is Maslow's 'hierarchy of human needs' model (Box 17.4), in which needs are seen as ranging from basic physical requirements necessary for organic survival (e.g. food, clothing and shelter) to higher-order needs (such as those for love,

BOX 17.4

Human needs and wants

The satisfaction of needs and wants is the fundamental motivation of human behaviour. The degree to which this is achieved depends on co-operation and competition with others, and these processes may be of particular intensity within urban settings. It is important to draw a distinction between needs and wants. Needs, unlike wants, refer to an imperative, or something necessary to an individual, rather than merely the source of acquisitive desire. Need refers to some standard such as that to ensure physical survival. Thus basic needs involve a norm, falling below which results in harm. Clearly, while basic needs for food and shelter are common to all human beings, other needs are related to the social structure, such as the 'need' for a family to own at least one car. As we move away from minimum survival needs the definition of need becomes value-loaded and the boundary between needs and wants less distinct.

Maslow (1954)[14] identified a hierarchy of human needs:

1. *Physiological* needs, such as food and shelter.
2. *Safety* needs, which include protection from physical harm, psychic threats from others and personal privacy.
3. *Affiliation* needs, such as love, group membership, interpersonal interaction.
4. *Esteem* needs, personal integrity, esteem and self-neglect.
5. *Actualisation* needs, the fulfilment of one's capacities and perceived control of the environment.
6. *Aesthetic* needs, relating to personal concepts of beauty and our need to learn.

The particular need or combination of needs that underlies any specific behaviour depends on the individual and the context of the action.

respect and self-esteem). In general, lower-order needs must be satisfied before other needs are manifested. Clearly, the definition of other than basic needs is far from straightforward.[15] As we have seen, in the context of public-service provision the identification of need is influenced strongly by the political ideology of the ruling party, as witnessed by the conflicts over public expenditure needs between the Conservative central government and Labour-controlled local councils in

Britain during the 1980s. Need defined by political process has a direct influence on public-service provision, both through the criteria used to assess need and through the manner in which the assessment of needs is translated into arrangements for inter-governmental revenue-sharing. In statistical terms, need is commonly measured by means of 'social indicators' that gauge the social, economic and demographic health of different populations or areas. These may be single indicators (such as unemployment levels or pupil–teacher ratios) or composite measures based on a suite of relevant social indicators[16] (see Chapter 15). The socio-spatial assessment of need has underlain several major government programmes in the post-war era, including Medicaid in the USA and the Standard Spending Assessment formulas for the distribution of central-government grant aid to local authorities in the UK. The outcome of such evaluations of need directly affects the fairness of patterns of collective consumption in the city.

EFFICIENCY, EQUITY AND EQUALITY IN PUBLIC-SERVICE PROVISION

The justness of patterns of collective consumption may be measured by comparing the distribution of resources among political or administrative units with some normative criteria. The choice of criterion can range from locational efficiency to equity based on need but, in practice, usually involves some mix of both.

LOCATIONAL EFFICIENCY

The chief physical component of public-facility costs is that attached to overcoming the friction of distance. Public-facility planners seek locations that minimise production and distribution costs and maximise accessibility to citizens. For consumers, accessibility to public facilities confers opportunity and choice; enhances the use value of a residential property; minimises travel costs, thereby releasing more household income for expenditure on consumption; and can increase the exchange value of a property. Lack of physical accessibility can carry penalties above those arising from the cost and inconvenience of travel. Inaccessibility to health-care facilities has been shown to have an effect on therapeutic behaviour, with people living beyond a convenient distance (often as little as 0.75 km, half a mile) from their family doctor tending to make light of symptoms or endure discomfort and uncertainty.[17] Specific subgroups in the city – the elderly, handicapped, ethnic minorities, the young and women – have particular accessibility-related difficulties. Urban geographers have studied this location-accessibility problem in relation to a wide range of individual services including shopping facilities,[18] police protection,[19] primary medical care,[20] dental care,[21] social services,[22] recreational facilities,[23] educational resources[24] and public libraries.[25]

Two main approaches to resolving the location-accessibility problem have been employed. The first, focusing on *personal accessibility*, involves the identification and manipulation of space–time prisms based on the activity patterns of different population subgroups. The second, based on a modelling approach to *locational accessibility*, involves measures that weight units of separation (such as distance) against the number of destinations available.

PERSONAL ACCESSIBILITY

The time–space approach to personal accessibility studies the environment of resources and opportunities that surrounds each individual. The basic premise is that evaluation of accessibility levels should concentrate not on what people do or are likely to do but on what they are able to do. Each individual has their own 'action space' that limits the activities they can engage in. This action space is affected by three types of constraint:[26]

1. *Capability constraints*, which include the biological need to sleep and to eat (which limits the time available for travel), and the means of transport available (which influences the range of travel possible).
2. *Coupling constraints* that stem from temporal limits on activity as a result of business opening hours or scheduling of public transportation.
3. *Authority constraints* represented in laws, rules and customs that restrict access to certain places (e.g. laws governing alcohol consumption) and proscribe certain activities (e.g. highway speed limits).

The time–space budgets of some population subgroups permit more effective access to opportunities than do those of others. Consider the daily time–space prisms of two different women. Figure 17.2 depicts a hypothetical time–space prism for a suburban house-

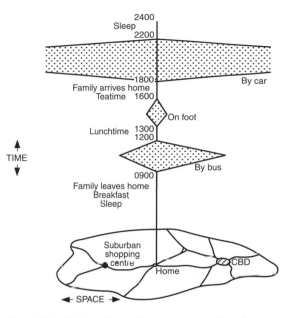

Figure 17.2 **The hypothetical time–space prism of a suburban housewife**

wife whose work is primarily home-centred. For certain periods of the day she may be confined to the home: until 9.00 by sleep and family commitments, from 12.00 to 13.00 by family lunchtime, from 16.00 to 18.00 when she welcomes children from school and prepares tea, and after 22.00 when she again prepares for bed. Contrast this with the daily time–space prism of a woman on **skid row**, Los Angeles (Figure 17.3). While her partner stays with their belongings at or near the current sleeping area, she is free to search out resources necessary for survival, such as legal aid, and to panhandle for money to buy food. The restricted daily activity space reveals the limited spatial mobility of homeless people and underlines the importance of geographical location of welfare services for those who are disadvantaged.[27] The challenge for accessibility planning is, first, to determine the dimensions of the action space of population subgroups, and second, to expand it or to place more opportunities within it.

LOCATIONAL ACCESSIBILITY

Although the final decision about the allocation of public resources is a political one, locational optimising models can play a useful role first by affording a means of evaluating different combinations of equity

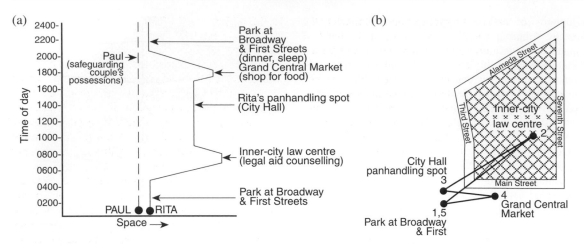

Figure 17.3 The daily time–space prism of a homeless woman on skid row, Los Angeles CA

Source: S. Rowe and J. Wolch (1990) Social networks in time and space: homeless women in skid row, Los Angeles, *Annals of the Association of American Geographers,* 80, 180–204

and efficiency, and second by providing evidence of 'better' solutions to support a case against inefficient or inequitable proposals. The outcome of an optimising model depends on its objectives and constraints (e.g. to minimise travel cost, maximise demand or maximise equity).[28]

EQUITY

Although the specification of locational efficiency is relatively straightforward, equity in public-service provision is more difficult to determine. As we have seen, some argue that equity is enhanced when services are distributed in proportion to taxes paid, while others contend that equity requires allocation of services in proportion to need. What is clear is that equity should not be confused with equality. Since citizens are not equal in either service needs or preferences, demonstrating that they receive equal services indicates little about how well they are served.

The concept of equity based on social need stems from the idea that:

one of the more important purposes of government in contemporary western society is to offset the burdens imposed on some groups by the operation of the market economy, so that a greater equality of life chances is achieved. An equitable arrangement is then one which promotes greater equality of condition. Services are equally distributed when

everyone gets the same services. They are equitably distributed when citizens are in a more equal life circumstance than before.

(Rich 1979 p. 152)[29]

Note that the goal is not to seek equality between individuals but to equalise the effects of society on different people. The inequality to be addressed is 'not inequality of personal gifts, but of social and economic environment' (Tawney 1952 p. 39).[30] Further, in order to gauge equity we must focus on outcome rather than output measures of services. Put simply, while the latter are the products of agency activities (e.g. the number of public-transit passengers carried), outcomes are the changes in relevant social conditions brought about by a service.

A major issue in public-service provision is the degree to which justness has to be traded off against efficiency. As the history of the public sector in Britain and the USA testifies, the importance attached to each of these concepts varies with the economic and political climate. In the USA the criterion of economy and efficiency of urban services prevailed in the immediate post-Second World War era, whereas equity and social justice at the local level were important goals of public programmes in the 1960s. With the election of New Right governments in Britain and the USA, 'value for money' re-emerged as a central measure of public-service provision. Agencies responsible for various aspects of collective consumption were encouraged to find ways of 'load shedding', most of which amounted to various forms of privatisation[31] (Table 17.2). Although this reduced the cost to the public purse, a major negative and socially regressive consequence stemmed from the fact that some citizens are better able than others to supplement reduced public services or to replace those transferred to the private market. Further, the services that the city continues to provide may be in a form that is more costly to consume or simply not accessible to some citizens. During times of resource scarcity and political conservatism in the public sector, strategies adopted by public officials to reduce production costs are likely to increase consumption costs (for example, users of services may have to complete innumerable forms or travel long distances), especially for politically and economically marginal groups. Particular problems are experienced by those living in the service-dependent ghettoes of North American inner cities and the welfare-dependent areas of some UK public housing estates (see Chapter 15).

INEQUITY: PATTERNED OR UNPATTERNED?

An issue that has attracted particular attention is whether there is a direct relationship between levels of access to urban services and the ecological character of neighbourhoods, and in particular whether there is a consistent bias in favour of certain social groups and areas in the city. Some degree of injustice in the provision of public services exists in all cities at all times, since the relationship between needs and resources is a dynamic one that is only rarely in balance. However, the 'underclass hypothesis' suggests that economically disadvantaged groups and areas are actively discriminated against in terms of service provision. According to Lineberry (1977), this view subsumes three overlapping hypotheses:[32]

1. The existence of a power elite capable of manipulating the distribution of public resources.

Plate 17.1 Protests over the effects of environmental pollution are evident in this residential neighbourhood of New Orleans LA

2. The 'race preference' hypothesis, which posits discrimination against ethnic minorities, such as non-white populations in the USA.
3. The 'class preference' hypothesis, which indicates discrimination against low-income communities in general.

Although qualified support for the underclass hypothesis has been found in the distribution of some services in some cities,[33] the weight of evidence points to a situation of 'unpatterned inequality', with most urban communities favoured by some service-delivery patterns and disadvantaged by others.[34] For Lineberry (1977), the decision rules adopted by local bureaucrats are the most important single factor in the distribution of local public services.[35] These decision rules almost inevitably favour some groups over others, but each rule may favour a different group so that there may be no consistent bias against any group or area across services. In similar vein, analysis of post-disaster federal programmes in the wake of the 1994 Northridge earthquake suggested that as a result of embedded structural constraints, federal assistance programmes favoured particular social groups – specifically homeowners versus renters and financially secure versus economically marginal households. Programme constraints underlying this 'patterned inequity' included complex bureaucratic procedures, a requirement for applicants to meet credit and income criteria, and designation of 'areas of greatest need' based on local political decisions over which marginal populations had little influence.[36] An alternative explanation of socio-spatial variations in public-service provision suggests that the 'unpatterned' distribution is the aggregate outcome of two different patterns: a socially regressive pattern of wealth-related services and amenities (such as education, fire and sanitation services) that favours middle-class areas, and a more socially progressive pattern of compensatory welfare-related services (such as day-care centres and drug rehabilitation clinics). As Box 17.5 illustrates, the 'patterned inequity' argument of the underclass hypothesis is also central to the concept of environmental racism. The fact that the concept of 'unpatterned inequality' does not relate well to the all-too-evident disparities in life quality within cities

BOX 17.5

Environmental racism

The concept of environmental racism refers primarily to the idea that people of colour and low-income groups (categories that often overlap) are disproportionately exposed to pollution, but also includes biases in natural resource policy and uneven enforcement of environmental regulations.

In terms of social/environmental justice, a fundamental question is: did residents come to the nuisance or was the nuisance imposed on them? In Cutter's (1995 p. 117) words, 'were the LULUs [locally undesirable land uses] or sources of environmental threats sited in communities because they were poor, contained people of colour, and/or politically weak? Or were the LULUs originally placed in communities with little reference to race or economic status and, over time, the racial composition of the area changed as a result of white flight, depressed housing prices, and a host of other social ills?'[37]

To answer this question requires examination of the historical social processes underlying the current spatial correlation between environmental degradation/pollution and particular economic/ethnic groups. In Los Angeles the area of East Los Angeles/Vernon contains hazardous and toxic metal-plating industries adjacent to the region's largest Latino *barrio*. Pulido *et al.* (1996) explain this juxtaposition in terms of the simultaneous evolution of industry, pollution and racism in the area.[38] In East Los Angeles/Vernon the concentration of Mexican population arose in conjunction with certain industries that depended on a pool of cheap labour. Over time the area became a stigmatised place. The negative image limited its redevelopment options and confirmed its racial and environmentally degraded character. This attracted further polluting industries. The patterns of 'environmental racism' evident in the present juxtaposition of hazardous industry and Latino population must therefore be understood not simply as the outcome of recent siting decisions but as part of a more general racialised process of urban and industrial development over several decades.

Source: adapted from L. Pulido, S. Sidawi and B. Vos (1996) An archaeology of environmental racism in Los Angeles *Urban Geography* 17(5), 419–39

reflects differences in personal consumption attributable to the *private wealth* of individuals. This is of particular significance in cities in liberal welfare states, where many urban services are purchased in the private sector (Box 17.6).

In the following sections we examine socio-spatial variations in the quality and quantity of collective consumption at several scales with particular reference to:

1. The fiscal relationship between central city and suburban jurisdictions.
2. The deinstitutionalised population of the inner city.
3. Levels of educational provision across the city.

THE SUBURBAN EXPLOITATION THESIS

Suburban residents who work in the central city benefit from a number of services, including police, fire, sanitation and road services provided by the city. Central cities also provide a range of cultural and recreational facilities that benefit the entire metropolis. Since user charges rarely cover the full costs of art galleries, museums and libraries this means that, in effect, central-city residents are subsidising 'free-riding' suburbanites. On the other hand, suburban residents contribute to the city's tax base, directly by their spending and indirectly because their workplaces add to the property-tax base of the city. Nevertheless, the net balance appears to favour the suburbs. The nature and extent of this cost–benefit (im)balance lies at the heart of the 'exploitation' thesis.

The suburban exploitation of central cities, a type of spillover effect resulting from the fragmented political structure of metropolitan areas, was first proposed by Hawley (1951), who found a positive correlation between *per capita* public expenditure of cities and the size of the suburban population in seventy-six US metropolitan areas.[39] As Brazer (1959) pointed out, however, there is no a priori reason why *per capita* expenditures should not be greater for central cities – among other reasons, because of their ageing infrastructure and older and poorer populations.[40] In terms of the suburban exploitation thesis it is the additional costs due to suburban residents that are of significance, and these are more difficult to determine. A few researchers, notably Neenan (1972), have concluded that the suburbs do make a net gain at the expense of the central cities, but that the magnitude of any benefit is dependent upon the set of public services included in the analysis and on the method of calculation employed.[41] In addition, the suburban exploitation thesis overlooks the tax and spending contributions of commuters to the public fisc of the central city. In the absence of detailed analysis of the costs and benefits of spillovers in individual cities it is not possible to

---

BOX 17.6

Differential social access to utility services

Privatisation of urban services previously provided by the public sector has had a differential impact on the well-being of social groups. This is manifested in socio-spatial variations in levels of connection with, and the quality of service received from, private agencies. These are of particular importance for utilities such as water, gas, electricity, sewerage and telephones. All households have to use these to some extent, irrespective of their income or the cost of the service.

Privatisation alters the relationship between utility companies and their domestic customers. A competitive market strategy advocates shedding of unprofitable and marginal customers and withdrawal from areas of limited commercial opportunity. Although utility prices in the UK are regulated, and universal maintenance of basic services is mandatory, marginal users, such as the poor, are often regarded by utility companies as an impediment to their primary corporate goal of profit maximisation in increasingly competitive international markets.

Since the privatisation of utilities in the UK increased charges have contributed to a rise in the incidence of 'water poverty' and 'fuel poverty' among the poor, with potentially serious implications for health. Commercial pressures have led to the introduction of pre-payment metering systems by utility companies, to reduce losses related to bad debt. A socially regressive consequence is that temporary 'voluntary' disconnections, often due to poverty, can go undetected by social service agencies.

Source: **S. Marvin, S. Graham and S. Guy (1999) Cities, regions and privatized utilities *Progress in Planning* 51(2), 88–165**

offer a definitive conclusion on the 'suburban exploitation' thesis.

The effect of political fragmentation on central city–suburban *fiscal disparities* is, however, more clear cut (see Chapter 20). Suburbanisation of population, commerce and industry has eroded the tax base of the central cities. While the centre has progressively lost many of its fiscally more profitable land uses, such as middle- and upper-income households, it has retained its traditional role of accommodating populations such as the elderly, low-income and migrant-transient groups with a greater need for public services across a range of fields. Prominent among such disadvantaged groups in many cities are the deinstitutionalised residents of the urban core.

DEINSTITUTIONALISATION

For most of the twentieth century and before, medical opinion advocated the isolation of the mentally ill in specialised institutions. Since the 1960s, however, a policy of deinstitutionalisation and the associated concept of **care in the community** have favoured the progressive discharge of patients from long-stay institutions. It was believed that the power of 'community' would help to reintegrate the mentally ill into mainstream society by providing a 'sense of belonging' as well as access to neighbourhood services, including mental-care facilities. In practice the policy met with only limited success in assimilating chronic mentally ill patients into urban society.[42] For many, their situation merely changed from one of segregation from society to one of segregation in society.

The city can be a difficult living environment for former mental patients confronted by marginalisation and social exclusion (see Chapter 19). Many end up in areas of the inner city where stress caused by poverty and isolation is likely to accentuate rather than alleviate their problems.[43] The difficulty of **assimilation** is compounded by the negative externality effects of mental-health facilities that generate the Nimby (not in my backyard) phenomenon.[44] While people are content to live close to a park, library or school, mental-health facilities rank alongside garbage dumps and prisons as undesirable neighbours that should be located 'elsewhere'. As Taylor (1989 p. 317) explained, community exclusion of the mentally ill emerges as 'a manifestation of individual and collective desires to protect territory with the aims of maintaining the use and exchange values of home and neighbourhood and, in a deeper but related level, as a

component within a process of reproduction that perpetuates the uneven distribution of life chances and advantages'.[45] Since resistance is strongest in suburban localities (with, for example, use of exclusionary zoning and criminalisation of homelessness to exclude 'undesirables'), mental-health facilities (such as day centres, cheap cafés and drop-in centres) tend to be concentrated in low-income inner-city neighbourhoods, reinforcing the spatial concentration of disadvantaged people in deprived environments.

EDUCATIONAL SERVICES IN THE CITY

The opportunity to obtain a sound education is a key element of human welfare. The absence of this opportunity constrains the life chances of those confronted by this form of disadvantage. We can examine the differential availability of urban educational opportunities from two perspectives. The first refers to the availability of educational facilities in terms of the physical accessibility of schools to local populations. The second concerns the quality of the educational environment as measured by students' performance in public examinations.

PHYSICAL ACCESSIBILITY

In addition to the direct educational benefits of accessibility to a good school, children living close to a school obtain extra user benefits through shorter journey times and greater opportunities to participate in after-school activities. Other general advantages include a number of user benefits primarily obtained by families with school-age children but which also include after-hours use of school facilities by the community for recreation or adult education. Schools may also provide indirect benefits to local residents by, for example, providing open spaces and quiet neighbours. The 'psychic income' derived from such advantages is embodied in property values that tend to be higher in proximity to a good school. Pacione (1989) employed a gravity model to identify levels of accessibility to secondary schools in Glasgow and discovered a correlation between levels of accessibility and social class. Areas of relative local overprovision of facilities were found in the middle-class 'west end', in contrast to the relative underprovision in the working-class areas east of the city centre and in the peripheral council-housing estates.[46]

Measures of accessibility can provide valuable information for urban managers, whether used to describe a current pattern of service provision or to evaluate a proposed restructuring of the service system. Although indices of accessibility offer a robust measure of the *spatial* aspects of the problem of educational opportunity, *social* barriers to a sound education can be as effective as physical distance, particularly for disadvantaged groups in society. The differential quality of educational opportunity in cities can have a direct impact on educational attainment and on subsequent life chances.

QUALITY OF EDUCATIONAL PROVISION

The changing nature of work (see Chapter 14) means that for school leavers with few skills or qualifications, entry into the labour force will be problematic. As Roberts (1995 p. 1) notes:

> qualifications are far and away the best assets with which young people can embark on their working lives [and] nowadays the way in which parents are best able to give their children good starts in life is by ensuring that they are well educated.[47]

Significantly, however, the opportunities for young people to gain a sound education are not uniformly available. In both the UK and the USA the quality of the urban educational environment, comprising school, home and neighbourhood, is generally below that enjoyed by more affluent suburban residents. Increasingly in many larger cities, particularly in the USA, 'urban education is primarily minority education' (Hill 1996 p. 131).[48] In the USA minority-group children constitute more than half the student population in all the ten largest urban school districts and, following a period of falling rolls due to middle-class flight from city schools, enrolments in many urban areas are growing as a result of immigration. Since the late 1980s New York, Los Angeles, Chicago and Miami have collectively added nearly 100,000 new school students annually who are either foreign-born or children of immigrants. As we saw in Chapter 15, many of the inner-city minority populations experience a suite of social and economic disadvantages that can compound and are compounded by the deficiencies of a poor education.

There is unequivocal evidence of the effect of socio-spatial context on educational attainment. Pacione (1997), in a study of the geography of educational disadvantage in Glasgow, revealed a direct relationship between levels of performance in public examinations and the socio-economic status of catchment populations.[49] More generally, in a conclusion redolent of the 'culture of poverty' thesis (see Chapter 15), Dorling (1995) identified a self-perpetuating process in which children's educational attainment is influenced strongly by the school they attend – which depends on where their parents live, which in turn depends on family income.[50] In a just society, accidents of birth and living environment should not exert undue influence on the quality of education available.

SOCIAL JUSTICE AND WELFARE

Social justice is a normative concept concerned with the question of who gets what where and how,[51] or, more accurately, who *should* get what where and how. It is concerned with the equitable allocation of society's benefits and burdens. Urban geographers examining collective consumption from a perspective of social justice seek to evaluate current socio-spatial distributions against an ethically defined norm, and to consider means of redressing situations of inequality. The urban geographer's concern with issues of social justice stretches back over several decades, from early studies of territorial social indicators and social well-being[52] to more recent analyses of urban injustice.[53]

As we noted in Chapter 2, the ethical or moral philosophy perspective of social justice runs counter to the postmodern critique of universal principles and, in particular, the rejection by postmodern theorists of a generally applicable moral basis for social behaviour, or political-economic action. This requires further comment. While we must be aware of the danger that the rich and powerful can seek to impose their view of what is right on the poor and weak, it is equally evident that all human beings have certain universal basic rights. Adherence to a crude form of postmodernism involving uncritical celebration of plurality and uniqueness not only obscures the need to address widespread and persistent problems of social injustice but undermines the possibility of even formulating an effective argument against particular injustices.[54]

As with the definition of need and poverty, the quantity and quality of public-service provision are socially determined (not least by the ability and willingness of the rich to meet the costs of welfare services for the poor). For three decades after the Second World War, irrespective of the party in government, the British political economy was characterised by a commitment to the provision of a substantial welfare state. The growing difficulties experienced by the UK economy in the process of global restructuring (see Chapter 14) meant that by the early 1970s the British state was confronted by a fiscal crisis that led to reductions in government expenditure. This also undermined the social democratic consensus on which the welfare state was founded. Increasing economic difficulties eroded public optimism and support for the welfare state concept and resulted in the replacement of the humanistic moral ideology of the social market with a utilitarian form of morality promoted by the New Right.

As we saw in Chapter 8, according to the moral ideology of the New Right, removal of state interference with the market mechanism allows full pursuit of such natural instincts as individual initiative, the acceptance of inequality and adoption of self-help.[55] Growing public sympathy for this form of morality during the 1980s and 1990s was manifested in lack of concern for the 'have nots' by the 'have lots', and a new mood of 'scroungerphobia' and antagonism towards the welfare state and its clients. Although there is some limited evidence of the emergence of a 'post-materialist' culture (seen, for example, in the rise of environmentalism over recent decades), it is unlikely that a 'politics of moderation'[56] will supplant the ethic of competitive individualism as the hallmark of advanced capitalist society. In terms of welfare and collective consumption this means that for those unable to compete successfully in market society the holes in the safety net have grown wider.

As we have seen, the outcomes of the processes of economic and social restructuring 'come to ground' in, or are *embedded* in, particular locales within the socio-spatial fabric of the city. In the next chapter we examine the nature of residential differentiation and community life in cities.

FURTHER READING

BOOKS

D. Harvey (1996) *Justice, Nature and the Geography of Difference* Oxford: Blackwell

A. Merrifield and E. Swyngedouw (1995) *The Urbanisation of Injustice* London: Lawrence & Wishart

C. Pierson (1991) *Beyond the Welfare State?* Cambridge: Polity Press

S. Pinch (1997) *Worlds of Welfare* London: Routledge

D. Smith (1994) *Geography and Social Justice* Oxford: Blackwell

JOURNAL ARTICLES

M.-P. Kwan (1999) Gender and individual access to urban opportunities: a study using space–time measures *Professional Geographer* 51(2), 210–27

M. Pacione (1989) Access to urban services: the case of secondary schools in Glasgow *Scottish Geographical Magazine* 105, 12–18

M. Pacione (1997) The geography of educational disadvantage in Glasgow *Applied Geography* 17(3), 169–92

M. Pastor, R. Morello-Frosch and J. Sadd (2005) The air is always greener on the other side: race, space and ambient air toxics in California *Journal of Urban Affairs* 27(2), 127–48

S. Rowe and J. Wolch (1990) Social networks in time and space: homeless women in skid row, Los Angeles *Annals of the Association of American Geographers* 80, 184–204

E. Talens and L. Anselin (1998) Assessing spatial equity: an evaluation of measures of accessibility to public playgrounds *Environment and Planning D* 30, 595–618

C. Tiebout (1956) A pure theory of local expenditures *Journal of Political Economy* 64, 416–24

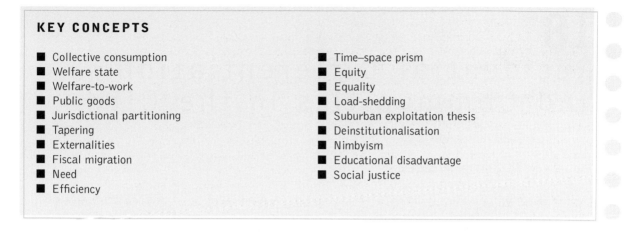

KEY CONCEPTS

- Collective consumption
- Welfare state
- Welfare-to-work
- Public goods
- Jurisdictional partitioning
- Tapering
- Externalities
- Fiscal migration
- Need
- Efficiency

- Time–space prism
- Equity
- Equality
- Load-shedding
- Suburban exploitation thesis
- Deinstitutionalisation
- Nimbyism
- Educational disadvantage
- Social justice

STUDY QUESTIONS

1. What do you understand by the term 'collective consumption'?
2. With the aid of relevant examples, illustrate how the distorting effects of geography on pure public goods influences the distribution of public services in the city.
3. Prepare daily time–space prisms for yourself and for other members of your family or friends. Compare and contrast the prisms and explain any differences in relation to the structure of the city in which you live.
4. Critically examine the validity of the suburban exploitation thesis.
5. Examine the value of the normative concept of social justice in studying the socio-spatial distribution of welfare services in the city.
6. Consider why for many communities in the contemporary city the grass is always greener on the other side.

PROJECT

Use local directories to map the distribution of urban services within a city of your choice. Are there any differences in the types of services to be found in different parts of the city? Can you identify any significant patterning of facilities – are there clusters in specific areas? If so, can these be related to the social structure of the city (for example, in terms of class or ethnicity)? Is there any evidence of bias in the distribution of facilities, or of the Nimby phenomenon?

With reference to a specific public service (such as primary health care, dental care, libraries or social services), select one or two service locations in the city (e.g. an inner-city and/or suburban location) and undertake a questionnaire survey of users to ascertain patterns of behaviour. Questions to ask would concern, among other things, how often they use the facility; how far they have travelled and by what means; what their personal characteristics are; the existence of any constraints on usage; and the perceived quality of the service). Are there any significant differences between users and use behaviour between the different service points?

Finally, consider whether the overall patterning of urban service provision in the city promotes or undermines the concept of social justice.

18

Residential Differentiation and Communities in the City

Preview: the identification of residential areas in cities; natural areas; social-area analysis; factorial ecology; sense of place; difference and identity in the city; the urban community; the bases of residential segregation; social-status segregation; lifestyle segregation; ethnic segregation; a typology of ethnic areas; ethnic areas in the US city; ethnic areas in the British city

INTRODUCTION

The development of cities, and in particular the onset of industrial urbanism, exerted a major influence on the nature of human association. For Tonnies (1887), urbanisation undermined a traditional rural way of life based on family, kinship and community (**Gemeinschaft**) that was replaced by an impersonal, contractual, self-centred lifestyle (**Gesellschaft**) characteristic of towns and cities. Individuality, not community, is the hallmark of urban life.[1] This perspective was reinforced by Durkheim (1893),[2] Spengler (1918)[3] and, in particular, Wirth (1938),[4] who believed that urbanites, regardless of social rank or ethnic status, inevitably react to their physical and social surroundings in a depersonalised 'urban' manner. This argument questions the possibility of community life in the modern city.

An alternative perspective is provided by those who support the idea of the socially cohesive community in cities. Jacobs (1961) envisaged the city as an inherently human place where sociability and friendliness are a natural consequence of social organisation at the neighbourhood level.[5] Her views are supported by studies of working-class communities and 'urban villages' in cities on both sides of the Atlantic.[6]

In this chapter we examine the main methods of identifying residential areas within the city, employing the statistical techniques of social-area analysis and factorial ecology, and the notion of sense of place. We explore the concepts of identity and community and the role of urban communities in advanced capitalist society, and examine the nature, bases and socio-spatial outcomes of residential **congregation** and segregation in terms of the key dimensions of social status, lifestyle and ethnicity.

THE IDENTIFICATION OF RESIDENTIAL AREAS

STATISTICAL DESCRIPTION

One of the earliest descriptions of residential areas in the city was undertaken by Booth (1902), who employed social surveys and mapping techniques to provide a detailed account of social conditions in London on an individual building scale.[7] Booth also pioneered the use of territorial social indicators in sub-areas of the city. By combining a variety of census measures into an 'index of social conditions' Booth believed that he had a generally applicable instrument for the measurement of social variation within cities. Booth's work was largely ignored by contemporary social researchers but provided an exemplar for subsequent studies of urban socio-spatial differentiation.

NATURAL AREAS

As we have seen in Chapter 7, the natural area was conceived by the Chicago ecologists as 'a geographical area characterised both by a physical individuality and by the cultural characteristics of the people who live in it'.[8] A classic depiction of natural areas is provided by Zorbaugh's (1929) study of two contrasting areas in Chicago's Near North Side, a district of 90,000 people close to the city centre and bounded by industrial land uses and the lake shore.[9] Running alongside Lake Michigan was the high-prestige area of the Gold Coast, which contained one-third of the households listed in the city's social register as 'of good family and not employed', behind which was the slum world of rooming houses, hobohemia, Little Italy and other ethnic quarters (Figure 18.1).

A defining quality of this approach was the detailed consideration of the rich texture of urban life. Despite subsequent criticism of natural areas as artificial units rather than meaningful community areas,[10] the mono-

graphs of the Chicago ecologists provided a means of understanding the city that combined empirical analysis with the experienced reality of urban life.[11] The ecologists' use of social indicators also paved the way for more sophisticated statistical methods of delimiting urban residential areas. A fundamental criticism of the natural areas approach, however, was that the identification of social areas in the city was not related to the broader social, economic and cultural changes taking place in society.

SOCIAL AREAS

The two main shortcomings of the ecological classification of social areas were the dependence on a few

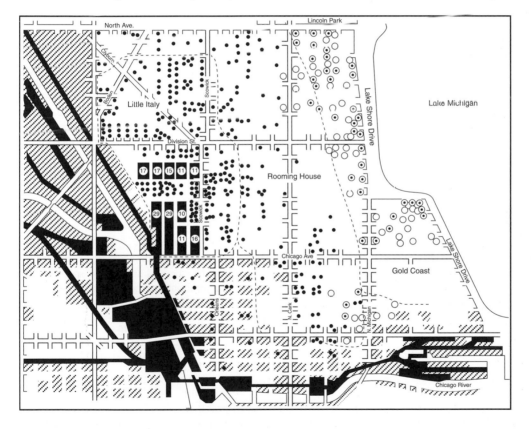

Railways
Industrial
Principal streets
Boundary of areas

● Family served
○ Contributors - $100 or less
◉ Contributors - Over $100

Figure 18.1 The Gold Coast and the slum in Chicago, 1929
Source: H. Zorbaugh (1929) *The Gold Coast and the Slum* Chicago: University of Chicago Press

key variables and the lack of a theoretical base to support the use of the selected variables. Shevky and Bell (1955) sought to overcome these deficiencies by using a multivariate classification procedure derived from an underlying theory of social change.[12]

Shevky and Bell (1955 p. 3) regarded the city as 'a product of the complex whole of modern society; thus the social forms of urban life are to be understood within the context of the changing character of the larger containing society'. They suggested that societal change, which they described as increasing societal scale or the change from traditional to modern lifestyles, had three main expressions or 'constructs'[13] (Figure 18.2):

1. *Social rank or economic status* referred to the tendency towards a more stratified society in terms of work specialisation and social prestige.
2. *Urbanisation or family status* described a weakening of the traditional organisation of the family as society became more urbanised.
3. *Segregation or ethnic status* suggested that over time the urban population would tend to separate into distinct clusters based primarily on ethnicity.

Having derived these constructs as key indicators of societal change, Shevky and Bell measured them by a number of diagnostic indices based on variables taken from the census. Social rank was measured by two indices related to occupation and education; urbanisation by three indices related to women in the labour force, female fertility and single family dwelling; and segregation by a simple percentage of specified ethnic groups in the population (Figure 18.2). Finally, the individual indices were standardised to run from 0 to 100, then averaged to form a 'construct score' for each census tract in the city.

The economic-status and family-status scores were each given a fourfold division to produce sixteen possible social-area types. In addition, each category could be further designated according to its composite segregation level above or below the city mean for the selected minority groups. In this way a typology of thirty-two potential social areas was identified. Importantly for the study of urban geography, since the emphasis of the study was on divisions in 'social space', census tracts were classified without reference to their spatial location in the city. Clearly, however, as Figure 18.3 shows, a map of urban social areas could be derived. In the case of San Francisco in 1950, high-economic-status tracts are found in hill and view locations away from industrial and port facilities; areas of nuclear-family status are displaced away from the urban cores; while areas of ethnic status emerged adjacent to the business zones and near industrial waterfront districts.

Social-area analysis has been applied in a large number of studies both as a tool for constructing an urban social typology and as a prelude to an ecological analysis of social traits such as crime or voting behaviour.[14] Despite questions about the arbitrariness of the computational procedure and the adequacy of the diagnostic indices to represent the theoretical bases, social-area analysis has been shown to represent the key elements of residential separation in the North American city, and has constituted an important contribution to the understanding of urban residential differentiation.

FACTORIAL ECOLOGY

The development of factorial ecology in the 1960s offered a means of constructing urban social areas based on a mathematically rigorous procedure and

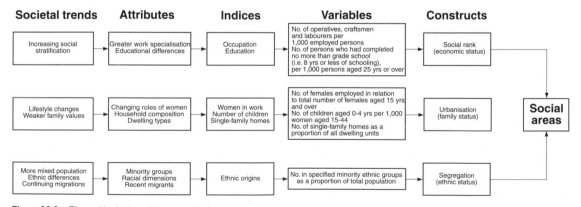

Figure 18.2   The method of social-area analysis

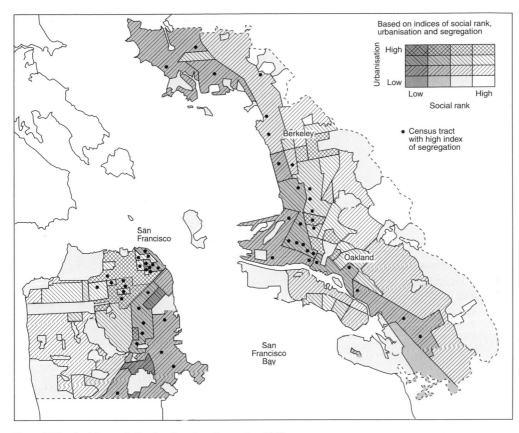

Figure 18.3   Social areas in the San Francisco Bay region, 1950
Source: E. Shevky and W. Bell (1955) *Social Area Analysis* Stanford, CA: Stanford University Press

using a larger set of diagnostic variables than the seven employed in social-area analysis. An additional difference is that whereas social-area analysis selected its dimensions (constructs) deductively (i.e. based on theory), factorial ecology does so inductively (i.e. by exploratory analysis of a data set).

Factorial ecology employs the multivariate statistical technique of factor analysis in order to derive a smaller set of diagnostic factors from an initial larger set of variables measuring the social, economic and demographic characteristics of census tracts in a city (Figure 18.4). Each factor may be seen as a 'supervariable' representing a highly correlated cluster of original census-based variables.[15] The meaning of each factor is determined by the character of the original variables with which it is associated most strongly. In a study of the Clydeside conurbation (Table 18.1), the first factor was associated with (loaded highly on) a number of variables indicative of low social status, and was labelled clearly as a measure of multiple depriva-

tion. Each census tract can be scored on each factor. Mapping the factor scores reveals the geographical distribution of the social conditions within the study space (for example, see Figure 15.9). The results of any factorial ecology will depend on the number and types of variables in the analysis, the units of observation employed, and the specific factorial procedure used. Nevertheless, factorial ecology is a powerful and versatile tool for the analysis of urban socio-spatial structure, and one that has been employed widely in both academic and commercial studies (Box 18.1).

The three general factors that have emerged commonly in factorial ecologies of North American cities are related to socio-economic status, family status and ethnic status. As Ley (1983 p. 78) concluded, 'thus the deductive Shevky–Bell formulation, which identified the same three characteristics as basic to American urbanism, is substantiated on the basis of census data by the inductive method of factorial ecology'.[16] The continuing relevance of these factors for understanding

Variables

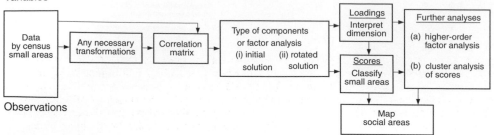

Observations

Figure 18.4   The method of factorial ecology

TABLE 18.1 FACTORIAL ECOLOGY OF CLYDESIDE: FACTOR STRUCTURES AND LOADINGS (%)

Variable	Loading
Factor I Multiple deprivation	
Male unemployment	0.7511
Council housing	0.4896
Households with more than 1.5 persons per room	0.5183
Household spaces vacant	0.4906
Households with a single person below pensionable age	0.5261
Households headed by lone parents	0.6617
Travel to work by bus 0.3189	
Household heads in low-earning socio-economic groups	0.3000
Young children living in multi-storey accommodation	0.6019
Households living above the occupancy norm	0.6935
Factor II Substandard housing	
Households lacking or sharing a bath or shower or inside WC	0.7341
Households with no central heating	0.7502
Factor III Elderly/infirm households	
Households living in one or two rooms	0.6541
Households comprising of a lone pensioner	0.8087
Household heads suffering long-term illness	0.7127

Source: **M. Pacione (1995) The geography of multiple deprivation in the Clydeside conurbation *Tijdschrift voor Economische en Social Geografie* 85(5), 407–25**

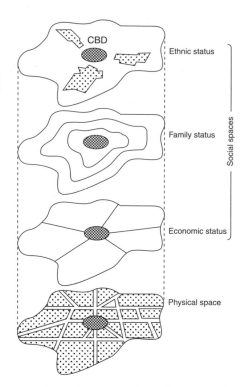

Figure 18.5   The components of urban ecological structure

Source: R. Murdie (1969) *Factorial Ecology of Metropolitan Toronto 1951–1961* Research Paper 116, Chicago: Department of Geography, University of Chicago

the internal structure of American cities is illustrated by White's (1987) study of twenty-one US metropolitan areas[17] and Erwin's (1984) comparative factorial ecology of thirty-eight American cities.[18] Erwin's

(1984) conclusion that each of these factors influences the structure of the city in a particular way echoes that of Berry and Rees (1969),[19] who reasoned that if *socio-economic status* were the sole factor, cities would tend to divide into sectors; if *family status* were dominant, the spatial order would be concentric zones; and if *ethnicity* were the major factor, the pattern would be one of multiple nuclei. In reality, as Figure 18.5 indicates, all three factors may operate simultaneously.

BOX 18.1

You are where you live

An address or, more accurately, a postcode (zip code) provides a wealth of information about residents and communities. By blending census data and consumer surveys, analysts can predict much about the social backgrounds and preferences of residents in any postcode area. In many countries, market researchers have produced detailed classifications of social space based on multivariate analysis of a large number of measures for each postcode area. Interestingly for the urban geographer, these classifications can also provide a typology of community or neighbourhood areas.

The Claritas Corporation in the USA has used a computer program called PRIZM (Potential Rating Index for Zip Markets) to cluster the nation's 36,000 zip codes into forty neighbourhood types or 'lifestyle clusters'. The neighbourhood types can also be ranked in terms of status, based on residents' household income, home value, education and occupation. At the top of the status ladder is ZQ1, labelled 'blue-blood estates', followed by ZQ2 ('money and brains') and ZQ3 ('furs and station wagons'). The lowest-status neighbourhood type, ZQ40, is labelled 'public assistance'. The characteristics of these ideal types are shown in the table.

Name/rank	Thumbnail demographics	Some high-usage products	Percentage of US population	Sample neighbourhoods	Median household income (MHI)	Percentage of college graduates in 1987
ZQ1 Blue blood estates	Super-rich, politically conservative suburbs	New York Times, Mercedes, skiing	1.1	Beverly Hills, Calif. (90212); Lake Forest, Ill. (60045)	$70,307	50.7
ZQ2 Money and brains	Ultra-sophisticated, politically moderate/ conservative posh in-town elegant apartments, swank townhouses of well-off academics, managers	New Yorker, Gourmet, classical records, Mercedes	0.9	Georgetown, Wash., DC (20007); Grosse Point, Mich. (48236)	$45,798	45.5
ZQ3 Furs and station wagons	New money, politically conservative executive bedroom communities	Forbes, Gourmet, country clubs	3.2	Plano, Tex. (75075); Reston, Va. (22091)	$50,086	38.1
ZQ4 Urban gold coast	Upscale, liberal/ moderate, high-rise renters	Tennis, New York Times, Atlantic Monthly, Mercedes	0.5	Upper West Side, Manhattan, New York City (10024); Rincon East, San Francisco, Calif. (94111)	$36,838	50.5
ZQ40 public assistance	The USA's poorest inner-city areas, mainly African-American singles and single-parent families	Malt liquor, burglar alarms, Essence, Jet, Chevrolet Novas	3.1	Watts, Los Angles, Calif. (90002); Hyde Park, Chicago, Ill. (60653)	$10,084	6.3

Source: **M. Weiss (1988)** *The Clustering of America* **New York: Harper & Row**

SENSE OF PLACE

The city has both an objective physical structure and a subjective or cognitive structure. Understanding the subjective or expressive interpretation of the city is essential, because meanings tell us not only about the places to which they refer but also about the people who articulate them and the social context in which they live. (This view underlies both the detailed urban ethnographic studies of the Chicago ecologists and the more recent postmodern perspective.)

Social area analysis and factorial ecology do not address explicitly the qualitative aspects of urban community. This may be examined through the concept of sense of place. This has two distinct but interlocking meanings:

1. The first refers to the intrinsic character of a place. Certain places are regarded as distinctive or memorable through their unique physical characteristics or **imageability**, or through association with significant events. San Francisco and St Peter's Square in Rome each possess a strong sense of place, a uniquely significant meaning even for those who have never been there.

2. The second refers to the attachment that people have to a place. In everyday life, individuals and groups develop a deep attachment to places through experience, memory and intention, a condition referred to as *topophilia*.[20] The strongest of such attachments are usually to the home area. The grief felt by urban working-class communities displaced by post-Second World War redevelopment schemes is well documented.[21] Even in a globalising world, people's daily lives are embedded and experienced in particular places (Box 18.2).

Urban geographers seeking to tap into these meanings of place have employed a range of approaches.

LITERARY SOURCES

Many novelists incorporate perceptive descriptions of places in their work, providing geographers with a qualitative and subjective insight into the meanings of local areas or neighbourhoods. However, using literature to shed light on a real-world concept such as urban community does pose certain problems, not the least of which is the selective representation of reality by the author. Descriptions can also vary according to the literary genre; realist literature perhaps portrays the world in some way closer to reality than romantic writing might. In addition to the filters of the novelist, readers may have different interpretations ('readings') of the text[26] and, as postmodern scholars point out, primacy cannot be attached to any one view.

Nevertheless, literature can contribute to understanding the meaning of places. Consider Glasser's (1986 p. 16)[27] description of life in the Gorbals district of Glasgow between the wars:

BOX 18.2

Territoriality

As Soja (1971 p. 9) points out, 'man is a territorial animal and territoriality affects human behaviour at all scales of social activity'.[22] Territoriality may be defined as the propensity of people to define certain areas and defend them. This implies that some space is of more value than others and that to give up that particular space is to incur a meaningful loss. The idea of a 'valued environment' underlies the concept of territory, and this clearly assigns a central role to the individual's perception and evaluation of their surroundings. This view is supported by Uexkull's (1957 p. 54) claim that territory 'is an entirely subjective product'[23] and Ardrey's (1972 p. 196) statement that a territory 'cannot exist in nature; it exists in the mind'.[24]

Human territories vary in terms of their organisation and scale. Rapaport (1977) presents a useful five-element ethological space model which describes a hierarchical arrangement of territories of increasing size.[25] These range from the egocentric 'personal space' through 'jurisdiction', 'territory' and 'core area' to 'home range' (defined as 'the usual limit of regular movements and activities'). In this typology the core areas (the space within the home range which is most commonly inhabited, used daily and best known) accords most closely with the notion of the urban neighbourhood.

We lived in a mid-Victorian tenement of blackened sandstone in Warwick Street, near the Clyde, in the heart of the Gorbals, a bustling district of small workshops and factories, a great many pawnshops and pubs and little shops, grocers, bakers, fishsellers and butchers and drysalters, tiny 'granny shops' – where at almost any hour of the day or night you could buy two ounces of tea, a needle, *Peg's Paper* and *Answers*, a cake of fireclay, a hank of mending wool – public baths and a wash-house, many churches and several synagogues. The streets were slippery with refuse and often with drunken vomit. It was a place of grime and poverty, or rather various levels of poverty and, in retrospect, an incongruous clinging to gentility, Dickensian social attitudes and prejudices.

In contrast to the working-class community described by Glasser, other novelists have identified the sense of 'placelessness' in which local community is undermined by modernity. Ford (1986) describes the transient nature of life for those on the margins of American society, occupying the motel strips of 'no place', a world where families and friendships fall apart, where nothing can be relied on, betrayal is commonplace and everyone is moving on in search of something better.[28]

Analysis of literary texts provides vivid descriptions of place-communities, of processes of social change and the meanings of neighbourhood, but, as critics point out, it does not provide a systematic body of replicable knowledge.

COGNITIVE MAPPING

The behavioural approach in urban geography (see Chapter 2) focused attention on the perception of places and led to the development of cognitive mapping as a means of identifying perceived neighbourhoods in the city.[29] One of the earliest applications of the technique, by Lee (1968), asked respondents to delimit their neighbourhood by drawing a line on a map.[30] Although individual delimitations were personal and idiosyncratic, there was sufficient congruence to allow a consensual boundary to be identified. Other studies have examined the neighbourhood cognitive maps of particular groups such as suburban housewives,[31] black youths,[32] children,[33] the elderly[34] and teenage gangs.[35] Although cognitive mapping can identify sub-areas in a city, it does not capture the sense of place associated with place-communities. To achieve this goal we need to move from cognitive mapping to 'the mapping of meaning'. The humanistic approach (see Chapter 2) attempts to reveal how shared values and common experiences transform a segment of physical space into a particular place.[36]

THE TAKEN-FOR-GRANTED WORLD

The humanistic approach advocates study not of cognitive maps produced by positivist social science but of the world of experience, and in particular the **taken-for-granted world** – that is, the sense of place and territorial bonding that people have and employ without much conscious thought. Researchers must seek to explore the lived experience of individuals, possibly using a phenomenological theory of social action[37] that entails asking the question 'What does this social world mean for the observed actor within

this world and what did he mean by his action within it?'[38]

The task of revealing the taken-for-granted assumptions that people have about their world is far from easy and is best approached by using multiple theories to achieve what Geertz (1980) has called 'thick description'[39] (in contrast to 'thin description', which merely reports the unexamined accounts of informants). This has led some urban geographers to renew interest in the ethnographic approach of the Chicago ecologists to provide detailed descriptions of city life.[40]

DIFFERENCE AND IDENTITY IN THE CITY

Difference, diversity and identity are key concepts in postmodern readings of the city. Differences in cities are constructed across many dimensions including class, ethnicity, lifestyle, gender, sexuality, age and able-bodiedness (see Chapter 19). During the 1970s and 1980s urban socio-spatial diversity was examined primarily from a political economy perspective that privileged socio-economic class as the dominant plane of division. Subsequently, postmodern discourse has focused on social dimensions of difference, on the diversity of cultures and cultural norms and on the notion of multiple identities. Identity (i.e. the view that people have of themselves) is shaped by a combination of social forces, individual personality and life experience. As a consequence, identities are not fixed but vary over time and across space. According to this 'cultural' perspective, the impermanence and conditional character of human identity allows the occupation of multiple identities within a constantly mutating social context. Thus, for example, during the course of a day a professor may interact with peers in a university, adopt a different identity as part of a crowd at a football game, and on returning home assume the identities of spouse and parent. Identity formation is affected by, and exerts influence upon, urban environments. Of particular significance for geographers, social differences often take on a spatial or territorial identity[41] that contributes to the construction of communities in the city.

THE URBAN COMMUNITY

Community is one of those terms the meaning of which everyone knows, but few can define. Hillery (1955) found ninety-four definitions in the literature up to 1953.[42] Most, however, agreed on three points:

1. Community involves groups of people who reside in a geographically distinct area.
2. Community refers to the quality of relationships within the group, with members bound together by common characteristics such as culture, values and attitudes.
3. Community refers to a group of people engaged in social interaction, such as neighbouring.

On this basis we can adopt an operational definition of community as a group of people who share a geographic area and are bound together by common culture, values, 'race' or social class. There is also general agreement that urban neighbourhood is smaller than community,[43] and that while community may have lost some of its spatial sense, neighbourhood remains a term applied solely to a local area.[44] Blowers (1973) links the terms 'neighbourhood' and 'community' using a continuum along which gradual change occurs in the relative importance of the spatial and social components; in this, 'the neighbourhood community' is seen as a close-knit, socially homogenous, territorially defined group engaging in primary contacts.[45]

Prolonged debate has ensued over the role of the local community in the modern city. One view contends that communal ties have become attenuated in contemporary industrial societies. This perspective underlay much of the theoretical writing of the Chicago school[46] and was summarised in Wirth's (1938) contention that increasing societal scale leads to increased dependence on formal institutions and decreasing dependence on local community structures[47] (Box 18.3). The idea of

BOX 18.3

Changes in the scale of society

The concept of societal scale used by Shevky and Bell (1955)[48] to indicate the relationship between urban form and societal change reflected the classic formulation of 'urbanism as a way of life' by Wirth (1938)[49] in which he defined cities in terms of their population size, density and heterogeneity and deduced the following socio-spatial consequences of each.

1. The effects of size:
   - The greater the number of interacting people, the greater the potential differentiation.
   - With dependence spread over more people, there is less dependence on particular individuals.
   - There is association with more people, but knowledge, especially intimate knowledge, of fewer.
   - Most contacts become impersonal, superficial, transitory and segmental.
   - There is more freedom from the physical and emotional controls of intimate groups.
   - There is association with a number of groups, but no allegiance to a single group.
2. The effects of density:
   - There is a tendency to differentiation and specialisation.
   - There is a separation of workplace and residence.
   - The functional segregation of areas take place.
   - The segregation of people occurs, creating a residential mosaic.
3. The effects of heterogeneity:
   - Without common background and knowledge, emphasis is placed on visual recognition and symbolism.
   - With no common value sets or ethical systems, money tends to become the sole measure of worth.
   - Formal controls of society replace the informal controls of the group.
   - Life has an economic basis, mass production-needing product and demand standardisation.
   - The growth of mass political movements occurs.

Although many of Wirth's detailed deductions have been criticised subsequently, the underlying premise that residential differentiation is related to societal scale has been demonstrated by social-area analyses that have shown that, in 'high-scale' (developed) societies, economic status and family status are independent dimensions of urban life, whereas in 'low-scale' (developing) societies the family remains the dominant institution and there are few alternative lifestyles beyond the extended family. These social trends are reflected in urban structure and patterns of behaviour.

'community without propinquity', in which interest-based rather than place-based communities are dominant, belongs to this school of thought.[50] The opposite view maintains that local communities have persisted in industrial bureaucratic social systems as an important source of support and sociability. Much of the evidence for this argument rests on the empirical demonstration of the continued vitality of urban primary ties in 'urban villages' such as Boston's North End, New York's Greenwich Village and the East End of London.[51] The terms 'community lost' and 'community saved' have been coined to describe these extreme viewpoints in the debate.[52] An intermediate perspective is provided by the concept of the 'community of limited liability', which recognises the local area or neighbourhood as one of a series of communities among which urbanites divide their membership.[53] This acknowledges that city dwellers do develop local attachments but are still prepared to withdraw physically or psychologically if and when local conditions fail to satisfy their needs and aspirations. An extension of this model (referred to as the 'community transformed') contends that although improved personal mobility and telecommunications have liberated communities from the confines of local neighbourhoods and dispersed network ties from all-embracing solidary communities to more narrowly based ones, nevertheless locality remains a significant context for social interaction. This is particularly so for relatively immobile social groups such as the young, elderly and poor.

COMMUNITY COHESION

The concept of community cohesion refers to the amount and quality of social relations and interactions in an urban neighbourhood and to the attraction to or identification with the area.[54] A number of attempts have been made to devise measures of community cohesion based on a mix of behavioural and perceptual variables.[55] Smith (1975) employed a multivariate index based on the four dimensions of:

1. The use of local facilities.
2. Personal identification with the neighbourhood.
3. Social interaction among neighbourhood residents.
4. Residents' consensus on certain values and forms of behaviour.[56]

A set of six measures was employed to construct a composite index of neighbourhood community cohesion in Glasgow:[57]

1. Personal attachment to the neighbourhood.
2. Neighbourhood-based friendships.
3. Participation in neighbourhood organizations.
4. Residential moves made within the neighbourhood.
5. Use of neighbourhood facilities.
6. Satisfaction with life in the neighbourhood.

Evidence of well defined, socially cohesive neighbourhood communities supports the view that individuals can be simultaneously members of both a (geographical) neighbourhood community and a more extensive (sociological) community without propinquity.

SUBCULTURAL THEORY

The decline of community in response to the increasing scale of urban life is a well-established premise. The subcultural theory of urbanism represents an alternative view that the diversity of the city creates an environment ripe for the emergence of subcultures. In a large city, people have a greater potential choice of social networks so that interest groups – environmentalists, gays or sport fans, say – can all find enough people with similar interests to form an association, or even a subculture, with its own values, beliefs and norms. Subcultures may be associated with certain parts of cities as in Chinatown or the Castro district of San Francisco. The existence of distinctive socio-spatial urban subcultures is the result of the processes of segregation and congregation of groups in the city.

THE BASES OF RESIDENTIAL SEGREGATION

The processes underlying neighbourhood change (see Chapter 10) lead to a sorting of the urban population into a mosaic of social worlds. The main dimensions of socio-spatial differentiation are based on:

1. *Socio-economic status*. The effects of social **class** on residential location, evident in the nineteenth-century industrial city, continue to influence the socio-spatial structure of the contemporary city. Whereas social constructs, such as the caste system in India, can determine social position in traditional societies, in Western society individual socio-economic status is determined largely by their economic power, which, for most people, is reflected in the nature of their employment.

2. *Family status and lifestyle*. The socio-spatial formation of the city is also a function of changing lifestyles, and of different stages in the life cycle. In industrial societies people may choose among several lifestyles, including:

   ■ *Familism*, in which child-rearing is the dominant feature. Although closest to the traditional pre-industrial lifestyle, the salience of the extended family relationship is much reduced in industrial society, and couples are more able to determine the size and timing of a family in order to enable them to participate in other lifestyles.

   ■ *Careerism*, in which people are mainly oriented towards the goal of vertical social mobility. Many may never marry; those who do, marry at an older age and, if having children, do so later in their married life.

   ■ Consumerism, in which people opt for patterns of consumption dominated by hedonistic self-gratification.

   ■ Differences in the choice of lifestyle are translated into choice of residential environment, with, for example, a traditional family looking for a suburban detached home with a garden, and a yuppie preferring a central-city loft apartment. Individuals are also influenced by their stage in the life cycle, which is also reflected in urban residential structure (Figure 18.6).

3. *Geographical mobility and membership of minority group*. The rise of urban industrial society was fuelled by the influx of migrants both from rural areas and from overseas. As we saw in Chapter 3, a large proportion of the residents of the growing nineteenth-century cities of Britain were not born in the city. In some inner quarters of London in 1881 over 60 per cent of the population were migrants,[58] while by 1900 Chicago had more Swedish residents than any other city except Stockholm.[59] The cultural difference of migrants often led to difficulties of communication with existing residents, resulting in varying degrees of residential segregation. Although for many migrant groups this has been reduced over time by processes of assimilation, for others (in particular, the more visible ethnic minorities) socio-spatial segregation remains a feature of urban life.

We shall examine each of these dimensions of segregation individually for analytical clarity while noting that, in reality, there may be significant overlap among the three.

SEGREGATION BY SOCIAL STATUS

Residential differentiation on the grounds of socio-economic status is a defining characteristic of cities. People of high social status have the income to select houses and neighbourhoods in accordance with their tastes, whereas the residential location decisions of lower-status households are constrained by their weaker market position.[60] The extreme manifestations of social-class segregation are represented by slums and status areas in the city.

Slums

We have already described the processes underlying the location and development of slums (see Chapter 10);

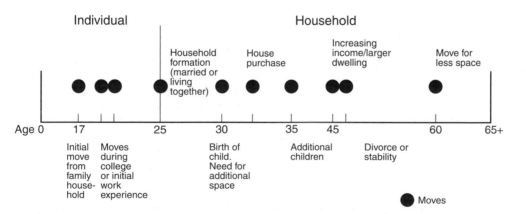

Figure 18.6 The relationship between life cycle and residential mobility

here we focus on the social order of the slum. The chief forms of social disorganisation in the slum are shown in Table 18.2. Since slum areas are the least desirable in the city, they attract the poor, who are unable to pay the higher rents demanded in better residential areas. The poor include welfare dependants, transients and members of ethnic minorities. The convenience of the area to downtown may also attract 'bohemians', as well as social deviants who welcome the anonymity provided by the slum's density of population and by its lack of organised resistance to their presence. The variety of people may also lead to the emergence of distinct sub-areas within the slum such as 'skid rows' (e.g. Harlem in New York)[61] and ethnic ghettos.[62] Despite the transient and socially dysfunctional nature of the slum population, many organisational features indicate a need for congregation. These include membership of teenage gangs, church communities and ethnic organisations, as well as patronage of particular bars that act as 'neighbourhood drop-in centres' where locals can get to know each other, exchange job information and borrow money before payday. For many residents the skid row environment becomes a way of life in which they feel comfortable and secure.

TABLE 18.2 FORMS OF SOCIAL DISORGANISATION IN THE SLUM

1. Poor educational facilities (run-down schools, few teachers, with insufficient qualifications, lack of facilities such as books and equipment)
2. High rates of unemployment (lack of skills, low levels of education)
3. Many dependent on welfare (for health care, food)
4. Lack of sense of community (transients, ethnic and racial diversity)
5. Family problems (high rates of divorce, separation, illegitimacy)
6. Personal degradation (drinking, drugs)
7. High crime rates (robbery, theft, violence)
8. Numerous delinquent gangs (anti-social and violent)
9. Opportunities for political graft (gang leaders, politicians)
10. Numerous religious sects (store-front churches, new religions)

Source: **L. Driedger** (1991) *The Urban Factor*
New York: Oxford University Press

Service-dependent ghettos

Another form of low-status urban area is the service-dependent ghetto.[63] Such a ghetto comprises vulnerable people such as low-income elderly people, mentally handicapped people, the physically disabled people and those who are chronically unemployed, who rely on cash income transfers from government and services in kind for their support. Studies of these non-labour-force groups in the USA have shown that up to half their income consists of the cash value of services in kind targeted for them by various public and private agencies.[64] The centrifugal population movements of the post-Second World War era have meant that in the US city these disadvantaged groups have increasingly been left behind in poverty-ridden central neighbourhoods. The same degree of central-city concentration is not apparent in Britain, where public urban renewal policy has served to disperse the service-dependent poor. However, many have been relocated on to peripheral housing estates, the most deprived areas of which have been characterised as 'cashless societies'. In the US community opposition to the deinstitutionalisation of mentally handicapped people, prisoners, juvenile delinquents or those with drug or alcohol abuse problems[65] and exclusionary local zoning aimed at service-dependent groups and facilities operate to ensure the continued spatial segregation of such groups. The concentration of service facilities attracts other populations in need of the services provided to these 'poverty zones', initiating a cycle of co-location that can ultimately result in the ghettoisation of those who depend on publicly provided services. This can have adverse effects both for those trapped within the area and, through negative spillover effects, for the surrounding non-dependent populations.

Status areas

The contrasts between Hampstead and Tower Hamlets in London, and between Beverly Hills and Watts in Los Angeles CA, are highly visible and generally acknowledged. Status areas reflect the residential preferences and economic power of upper and middle-income citizens. This is particularly evident in the US city, where, by the late 1980s, 40 million Americans, or one person in eight, lived in a **common-interest development** (CID) or **gated community**. These residential developments are planned and marketed as spatially segregated environments. The cost of the housing and amenities offered in individual gated communities virtually guarantees social exclusivity.

Plate 18.1   The imposing entrance to an exclusive guard-gated community in Nevada ensures the privacy of residents

The Urban Land Institute concluded that CIDs can achieve residential exclusion better than any other form of development.[66] Many gated communities are marketed to specialised groups, including retirees, singles, golfers and even nudists. Most are designed to sort home-buyers into different income groups.

The CID is a community in which residents own or control common areas or shared amenities and in which residents' rights and obligations are enforced by a private governing body or home-owner association to which all residents must belong. The use of compulsory-membership home-owner associations to enforce covenants, contracts and restrictions (CC&Rs) on home-buyers has spread since the US Supreme Court's *Shelley* v. *Kramer* decision of 1948, which outlawed racially restrictive real-estate covenants. The power of these organisations to control the social and physical character of a community is enshrined in by-laws that can be altered only by a unanimous vote of all the members. Covenants and restrictions may dictate minimum and maximum ages of residents, hours and frequency of visitors, colour of house paint, size of pets, number of children, parking rules and landscaping.[67] A

common theme is the collective emphasis on security via a mechanism of social exclusion (Box 18.4). In the Los Angeles metropolitan area 'fortress enclaves' have become a feature of suburban development[68] which, for some, presages an urban future of clusters of gated communities connected to each other and to enclosed shopping malls by a network of freeways.[69] Others see 'the malling of America' culminating in the incorporation of housing into shopping malls to create a 'city within a city'.[70] As we saw in Chapter 12, a number of cities have already developed integrated enclosed commercial–retail–leisure complexes connected by skyways and tunnels and segregated from the public space of the streets.

SEGREGATION BY LIFESTYLE

The 'them and us' or insider–outsider organisation of society is evident in the presence of groups segregated by their social behaviour or preferred lifestyle. In Britain, and elsewhere in Europe, gypsies are a social group that lacks both acceptability by the dominant

BOX 18.4

Gated communities

As the contrast between public and private space becomes more stark, the psychological lure of defended space becomes ever more enticing. Gated communities or common-interest developments (CIDs) are marketed by promising exclusiveness, high amenity and security. For example: 'Sailfish Point is an idyll celebrating your achievements. Its numbers add up to a lifestyle without compromise. . . . Jack Nicklaus designed our par 72 membership-only course to stimulate and challenge but not intimidate. . . . Yachtsmen alight in our private sea-walled marina. . . . The St Lucie Inlet puts boaters minutes from deep-sea fishing and blue-water sailing. . . . Natural seclusion and security are augmented by a guarded gate and twenty-four-hour security patrols' (*New York Times Magazine*, 18 November 1990, p. 13).

In some CIDs the concept of defensible space is promoted to an extreme degree. In Quayside FL laser beams sweep the perimeter, computers check the coded entry cards of the residents and store exits and entries against a permanent data file, and television cameras continuously monitor the living and recreation areas. Members of the security force can wake up or talk to any resident through a housing unit's television set. When the chief of the security police was asked if his employees would also be able to spy on people, he replied, 'Absolutely not. There's laws that prohibit that kind of thing.'

Source: D. Judd (1995) The rise of the new walled cities, in H. Liggett and D. Perry (eds) Spatial Practices London: Sage, 146–66

society and a 'home territory'. Those who retain their traditional itinerant lifestyle are classed as outsiders, treated with suspicion and denied places to live.[71] In the UK local authorities have found it difficult to conform with the law and provide designated sites for 'travelling people' in the face of opposition from local residents who associate gypsies with unacceptable lifestyles, lack of cleanliness and activities on the margins of legality.

Another group segregated by their lifestyle are those associated with the 'sex trade'. In many cities those employed in the sex industry are concentrated in particular areas, such as Soho in London and the red-light district of Amsterdam, for the practice of their trade, though not necessarily for their residences. Gays

and lesbians, distinguished by their sexuality from mainstream society, also occupy specific areas within cities. Places such as the Castro area of San Francisco or the West Hollywood district of Los Angeles have played a significant role in the evolution of a gay subculture. The commercial significance of the 'pink pound' in London's Soho[72] and the role of gay entrepreneurs in the development of Manchester's '**gay village**',[73] and in the gentrification of districts such as Martigny in New Orleans,[74] have created distinctive urban landscapes. While lesbian control of urban space is less visible, neighbourhoods such as Park Slope in Brooklyn have developed as lesbian residential spaces akin to the gay male districts of Manhattan's Greenwich Village or San Francisco's Castro.[75] The explicit social construction of place in such areas has been used by these groups as a means to resist domination and engineer a shift from a position of constraint and exclusion to one of choice and recognition[76] (Box 18.5).

ETHNIC SEGREGATION

The basis of ethnic segregation can be racial, religious or national, and its recognition may rest on distinguishing physical characteristics, cultural traits such as language or custom, or on a group identity due to common origin or traditions (Box 18.6). The categorisation may be employed both by the group itself and by the larger society of which it is a part.[77] Ethnic minority groups in cities are typically segregated (Box 18.7), but the extent of segregation depends on two factors. The first relates to the migrant status of the group, in particular the recency of migration. The second refers to the social distance that separates the ethnic minority from the **charter group**, or host society. For immigrants whose differences with the charter group are slight, separate identity may be of only temporary duration; whereas for those whose real or perceived differences are significant, separation in both social and physical space is likely to persist. The maintenance of an ethnic group as a distinctive entity depends on the degree to which assimilation occurs (Box 18.8).

Assimilation

A distinction can be made between behavioural and structural assimilation. The former describes a process of **acculturation** whereby members of a group acquire

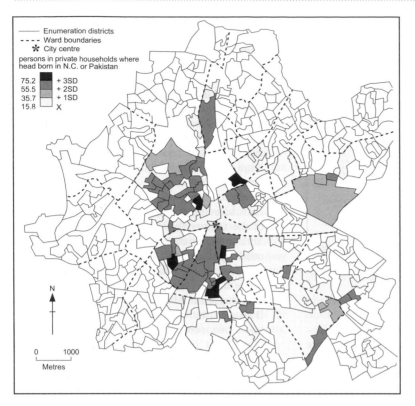

Figure 18.10   Distribution of Wolverhampton's population born in the New Commonwealth or Pakistan

South America, the Caribbean basin and Asia.[90] The migration stream was no longer directed to the large cities of the north-east but to cities across the nation. By 1990 63 per cent of the population of Miami was of Hispanic origin, mostly from Cuba and Nicaragua, while the 40 per cent Hispanic population of Los Angeles was primarily Mexican in origin (Box 18.11).

BLACK URBANISATION

One of the most significant internal migrations in American history has been the movement of African-Americans from the rural south to the industrial cities of the north. Whereas in 1900 76 per cent of the black population lived in rural areas and nearly 90 per cent in the south, by 1980 82 per cent of blacks were urban dwellers and 47 per cent lived in regions outside the south. Today the majority of moves by blacks are within and between metropolitan areas, and the growth of the black population within cities is due mainly to natural increase. Black population growth, however, has been contained in well defined areas, and, as a result of the forces of constraint

and choice, blacks remain the most segregated ('hyper-segregated') group in urban America.[91]

Detroit MI and Atlanta GA are among the most hyper-segregated cities in the USA. Detroit is the archetypal 'urban doughnut' – mainly black in the deindustrialised centre, and largely white on the job-rich periphery. In Atlanta the pattern of expansion of the black residential space over the period 1950–90 (Figure 18.11) illustrates the typical sectoral spread of the black area, first to the west from its confined location around the CBD and, more recently, also to the south. Northward expansion has been blocked by the constraints of high-priced housing and white resistance (Box 18.12).

BLACK SUBURBANISATION

In the early post-war decades much black suburbanisation was confined to areas just beyond the city limits and simply represented a spillover of the existing ghetto (Figure 18.12). Other suburban increases in black population have been due to the expansion of

BOX 18.11

The rise of Hispanic America

Hispanic peoples have overtaken the black population to become America's largest, youngest and fastest-growing minority. Through birth or immigration they now account for half of all new US citizens, and by 2050 more then 100 million Hispanics will make up a quarter of the population. Around the same time whites of European origin will become a minority, as they already are in Texas and California. In Dallas TX some districts, such as Oak Cliff, appear never to have seceded from Mexico in 1836. Everyone speaks Spanish, the signs are in Spanish, and the music is Mexican. Businesses offer money transfers, telephone calls and bus rides to Mexican cities, and food stores sell malanga, chayote macho, boniato root, and various forms of edible cactus. Dallas has two daily newspapers and a dozen weekly publications in Spanish. The city's leading radio station is Hispanic, as are many of the cable channels. Soccer pitches are replacing baseball diamonds in the parks.

Source: The Times **9 March 2007**

BOX 18.12

The tipping point

Despite strong evidence to the contrary, many residents believe that decline in house values is an inevitable result of the movement of a different ethnic group into a residential area. This perception is often reinforced by estate agents eager for housing transactions to occur. These trends underlie the phenomenon of blockbusting.

More generally, the tipping point refers to the proportion of minority households that white neighbourhoods will accept. When the critical level is reached, complete housing turnover to black occupancy is likely to occur. The precise level is highly variable, but rarely do white households tolerate more than 10–15 per cent of minority residents. The preference of whites to vacate areas when such levels are reached diminishes the prospects of a stable, racially mixed residential neighbourhood.

The process is reinforced by the self-fulfilling prophecy. When whites feel that their neighbourhood is about to change racially from white to black, they often decide to move. The move itself facilitates the change; that is, actual behaviour patterns produce the anticipated change once residents believe it is inevitable.

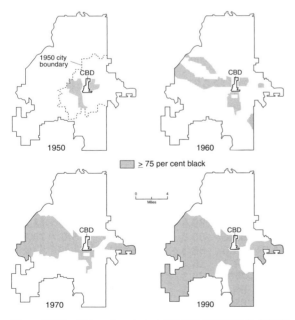

Figure 18.11 African-American residential expansion in Atlanta GA, 1950–90
Source: T. Hartshorn (1992) *Interpreting the City* 2nd edn, New York: Wiley

black residences in established outlying municipalities, many of which originated as unincorporated shack towns.[92] In effect, a dual housing market steered blacks into suburbs that were already predominantly black or which were no longer attractive to whites. These areas tended to be poor, fiscally stressed and crime-ridden compared to white suburbs.[93]

The Fair Housing legislation of 1968[94] removed the legal restriction of middle-class blacks to established urban ghettos and, although *de facto* housing discrimination remained,[95] there was an opportunity for those who could afford it to move into suburban neighbourhoods with better housing and educational opportunities for their children. Accordingly, though dwarfed by white suburbanisation, the number of blacks moving to suburbs increased during the 1970s. While the white suburban population increased by 13.1 per cent during the decade, the black suburban population increased by 42.7 per cent. The number of black suburbanites continues to increase. By 1990, of the 30 million blacks living in the USA, 17 million lived in the central cities and 8 million in suburbs. By 2000, of the black population of 34.7 million (12.3 per cent

In the late 1950s and 1960s London Transport, as well as hotels, restaurants and hospitals, employed West Indian labour. Other employment opportunities focused on the brick-making industry in Bedford, the foundries and engineering workshops of the West Midlands and the car plants of London, Oxford, Coventry and Birmingham. Later arrivals found similar employment niches in towns and cities such as Manchester, Oldham, Preston and Blackburn. Other major employment opportunities were provided in the textile and woollen industries of Lancashire and Yorkshire, and the clothing and footwear industries of the West Midlands. In these regions employers were seeking to reduce labour costs through capital investment and wage restrictions, and Asian migrants provided a willing work force to fill the vacancies created by the withdrawal of white male labour. As a result, significant concentrations of Pakistani and Indian populations developed in towns and cities such as Leeds, Bradford, Huddersfield, Manchester, Leicester, Nottingham, Oldham and Bolton.[103] The concentrated patterns of black and Asian settlement are in contrast with the dispersed nature of the Chinese, who generally opted for work in the restaurant–takeaway food sector, which required a more widespread pattern of settlement to prevent saturation of local markets. Principal destinations for the recent migrants from Eastern Europe have been London, the Midlands, Scotland and the east of England, (the later mainly related to employment in agriculture and food processing industries), with more generally significant numbers employed in the hospitality and catering sector.

Of greater importance than the fact of spatial concentration is the nature of the places. Minority ethnic groups are overrepresented in 'declining industrial centres' (such as Birmingham, Manchester, Glasgow, Newcastle, Liverpool and Sheffield), 'affluent commuter towns' (mainly located around London) and 'central London'. The greatest concentration of minority ethnic groups is to be found in inner London, where they accounted for 34 per cent of the population; but unlike in the USA there are no British towns or cities in which ethnic minorities constitute a majority population.

THE ETHNIC POPULATIONS OF LONDON

The distribution of foreign-born population resident in London in 2001 is shown in Figure 18.13. There are significant concentrations of Indians in Newham (12.1 per cent, compared with 5.3 per cent in 1991) and of Bangladeshis in Tower Hamlets (33.4 per cent

compared with 14.6 per cent in 1991). Caribbean- and African-born people are ubiquitous in inner London, but the areas of highest concentration of Caribbeans are Lewisham (12.3 per cent, compared with 4.5 per cent in 1991), Lambeth (12.1 per cent compared with 10.4 per cent in 1991) and Hackney (10.0 per cent compared with 5.6 per cent in 1991). High concentrations of African-born people are found in Southwark (16.1 per cent compared with 4.3 per cent in 1991), Newham (13.1 per cent compared with 4.4 per cent in 1991), Hackney (12.0 per cent compared with 4.0 per cent in 1991), and Lambeth (11.6 per cent compared with 4.5 per cent in 1991). These groups are less evident in outer London, apart from sectoral extensions of established inner-London hearths into Brent to the north-east, and Waltham Forest and Enfield to the north. The west London bias of Indians, particularly in the outer boroughs, is linked partly with their 'port of entry' at Heathrow, where an established nucleus attracts others, and to the availability of relatively cheap housing in the vicinity of the airport, where aircraft noise brought down prices and drove out indigenous residents.

The Bangladeshi population is the most segregated ethnic group in the city, and represents a partial restructuring of a rural community within an urban setting. Nearly all the Bangladeshis are from the province of Sylhet in Bengal, and in 1987 three-quarters of those under 15 years of age had no fluency in English. Their high level of segregation reflects recency of arrival, poverty and defensive clustering, as well as a positive desire to retain their cultural identity.[104]

Finally, we must also recognise that segregation occurs within groups classified as homogeneous by the census. A number of preferred locations exist for Caribbeans, depending on their home islands, with, for example, Jamaicans south of the river Thames, former Monserrat people in Hackney and Islington, Grenada folk in Ealing and Hammersmith, and St Lucians in Paddington and Notting Hill in Westminster.[105]

Several factors have contributed to the intra-urban pattern of ethnic segregation in London. During the 1960s and 1970s the take-up by immigrants of older property in the zone of transition was promoted by the planned decentralisation of London's population to New Towns (see Chapter 9). Later, in the 1970s and 1980s, the shift in emphasis of housing policy to rehabilitation (see Chapter 11) reduced the amount of accommodation available for immigrants, resulting in higher densities of non-white populations and overcrowding of property. A third factor was that as a result of council-house allocation rules (based on a

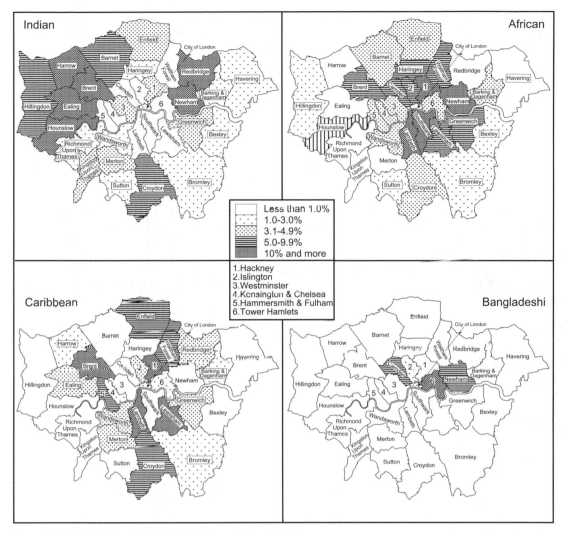

Figure 18.13 Distribution of immigrant ethnic populations in London

points system in which time spent on a waiting list was a crucial factor) immigrants were effectively disbarred from local-authority housing and became overwhelmingly renters of private property, or property-owners. There was also a degree of discrimination in operation in that local authority housing with design faults or on difficult-to-let estates unacceptable to white families was almost by default allocated to non-whites.[106] Clearly, council officials did not have to be 'racist' to effect such segregation, although there is some evidence that local-authority housing staff in Tower Hamlets displayed 'suspicion and resentment' and sometimes 'overt racism' towards Bengalis.[107] Those who opted to buy housing inevitably ended up in older,

cheaper and obsolescent properties deficient in basic amenities.

In addition to the foreign-born ethnic minorities, other groups display socio-spatial segregation. In the case of the Jewish community in London, relatively few of its members are not native-born, and their high level of assimilation indicates that the group is self-identified. This is expressed most distinctively by their concentration in Barnet, where 15 per cent of the population is Jewish, a majority of whom reside in only six of the twenty wards. The growth of Islamic fundamentalist beliefs within South Asian communities represents another example of an ethnic group seeking to proclaim its cultural identity.[108]

Plate 18.2   The Bangladeshi community of Tower Hamlets in inner London is the most segregated ethnic group in the city, many of its members having no fluency in the English language

Communities – whether defined by social status, lifestyle, ethnicity or a combination of these factors – enjoy different positions within urban society and urban space. This determines their degree of congru-ence and satisfaction with their living environ-ment. The issue of the liveability of the urban envi-ronment is the question to which we turn in the next chapter.

FURTHER READING

BOOKS

R. Bullard (ed.) (2007) *Black Metropolis in the Twenty-first Century* Lanham MD: Rowman & Littlefield

M. Davis (1992) *City of Quartz* New York: Vintage

R. Farley, S. Danziger and H. Holzer (2000) *Detroit Divided* New York: Russell Sage Foundation

E. McKenzie (1994) *Privatopia* New Haven CT: Yale University Press

D. Massey and N. Denton (1993) *Apartheid American Style* Cambridge MA: Harvard University Press

J. Palen (1995) *The Suburbs* New York: McGraw-Hill

D. Wilson (2007) *Cities and Race: America's New Black Ghetto* London: Routledge

JOURNAL ARTICLES

A. Collins (2004) Sexual dissidence, enterprise and assimilation *Urban Studies* 41(9), 1789–806

C. Dawkins (2004) Recent evidence on the continuing causes of black–white residential segregation *Journal of Urban Affairs* 26(3), 379–400

D. Ley and R. Cybriwsky (1974) Urban graffiti as territorial markers *Annals of the Association of American Geographers* 64, 491–505

W. Li (1998) Anatomy of a new ethnic suburb: the Chinese ethnoburb in Los Angeles *Urban Studies* 35(3), 479–501

D. Luymens (1997) The fortification of suburbia *Landscape and Urban Planning* 39, 187–203

D. Phillips (1998) Black minority ethnic concentration, segregation and dispersal in Britain *Urban Studies* 35(10), 1681–702

R. van Kempen and A. Ozvekren (1998) Ethnic segregation in cities *Urban Studies* 35(10), 1631–56

D. Varady (2008) Muslim residential clustering and political radicalism *Housing Studies* 23(1), 45–66

B. Wellman and B. Leighton (1979) Networks, neighbourhoods and communities *Urban Affairs Quarterly* 14(3), 363–90

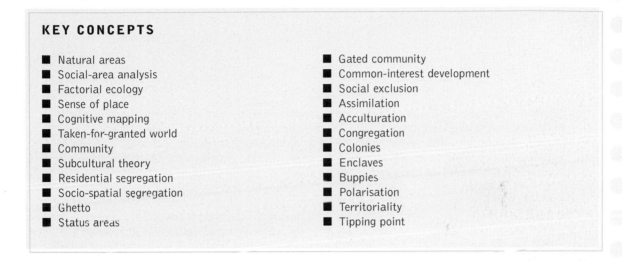

KEY CONCEPTS

- Natural areas
- Social-area analysis
- Factorial ecology
- Sense of place
- Cognitive mapping
- Taken-for-granted world
- Community
- Subcultural theory
- Residential segregation
- Socio-spatial segregation
- Ghetto
- Status areas
- Gated community
- Common-interest development
- Social exclusion
- Assimilation
- Acculturation
- Congregation
- Colonies
- Enclaves
- Buppies
- Polarisation
- Territoriality
- Tipping point

STUDY QUESTIONS

1. Explain the concept of sense of place and indicate how geographers can tap into this aspect of urban life.
2. Critically examine the use of social-area analysis and factorial ecology in understanding residential differentiation.
3. Consider the causes and socio-spatial outcomes of segregation within the city with reference to social status, or lifestyle or ethnicity. Reflect on how these dimensions might overlap.
4. With reference to particular examples, illustrate the functions of ethnic congregation in cities.
5. Examine the genesis and changing location of ethnic areas in either the USA or the UK city.
6. Gated communities are a common feature of metropolitan areas in the USA and, increasingly, elsewhere. Consider the advantages and disadvantages of this form of development for residents and for society in general.

PROJECT

In a city with which you are familiar, identify two contrasting neighbourhoods (e.g. in terms of social status, lifestyle or ethnicity). Carry out a sample survey of residents in each neighbourhood to determine the strength of the local community. This can be gauged by obtaining information on residents' activity patterns (e.g. social interaction and neighbouring behaviour, shopping, use of local facilities) and on how they feel about the neighbourhood as a place to live. You can gather this kind of information using a combination of closed questions (e.g. 'Where do you go for X?') and open questions ('What are the good/bad aspects of living here?'). The results of your sample survey can be aggregated to indicate the strength of community feeling in each area. How does this relate to the type of neighbourhood (e.g. working-class, ethnic area, status area)? Compare your results from the two selected study neighbourhoods and explain any differences identified.

19
Urban Liveability

Preview: urban quality of life; rating places; theories of urban impact; environmental hazards; landslides; earthquake; subsidence; urban climate; air pollution; flooding; noise pollution; socio-cultural impacts; crowding; urban legibility; fear in the city; site design and social behaviour; residential satisfaction; urban sub-areas; women in the city; the elderly in the city; the child in the city; the disabled in the city; towards the liveable city

INTRODUCTION

Growing concern for the future of cities and for the well-being of city dwellers, stimulated by trends in world urbanisation, the increasing number and size of cities, and the deterioration of many urban environments, has focused attention on the problems of living in the city. Central to this concern is the relationship between people and their everyday urban environments or life spaces. Understanding the nature of the person–environment relationship is a quintessential geographical problem. In the context of the built environment this can be interpreted as a concern with the degree of congruence or dissonance between city dwellers and their urban surroundings, or the degree to which a city satisfies the physical and psychological needs and wants of its citizens.[1]

Most attempts to assess the quality of different urban environments for people have employed objective measures derived either from primary field surveys[2] or from analyses of secondary data sets, often extracted from the census.[3] Collectively this line of research has contributed valuable insight into the rating of urban places in terms of 'quality of life'[4] and on questions such as the extent and distribution of substandard housing, and the incidence of deprivation in the modern city.[5] The objective perspective on the quality of urban life space has been paralleled by the development of an approach focused on the concept of **urban liveability**. In contrast to the objective definition of urban environmental quality, urban liveability is a relative term whose precise meaning depends on the place, time and purpose of the assessment and on the value system of the individual assessor. This view contends that quality is not an attribute inherent in the environment but a behaviour-related function of the interaction of environmental characteristics and person characteristics.

In order to obtain a proper understanding of the quality of urban life space it is necessary to employ both objective and subjective evaluations. In short, we must consider both the city on the ground and the city in the mind.[6] Accordingly, in this chapter we examine the major theories proposed to explain the influence of the urban environment on residents, then consider the impact of a number of environmental stressors, ranging from geophysical hazards to **daily hassles**, on quality of life in cities. The relationship between urban design and social behaviour is discussed, and the question of residential satisfaction explored. We then focus on the salience of urban sub-areas, and the relationship between urban structure and the social behaviour of different populations in the city. We begin at the inter-urban scale, before proceeding to an examination of liveability within cities.

RATING PLACES

The seminal work on inter-urban quality of life by Liu (1975) employed 132 variables related to economic, political, environmental, health, education and social conditions[7] to rank 243 US metropolitan areas. The majority of large Standard Metropolitan Statistical

Areas (SMSAs) with high quality of life were in the West, with substandard metropolitan areas concentrated in the south and north-east. For medium-sized cities the highest-rated places were in the Midwest and the lowest in the south. The best quality of life in small cities was found in the upper Great Plains and western New England, with the lowest quality of life in the south, east Texas and the north-east.

A similar procedure was employed by Boyer and Savageau (1981, 1985, 1989) to produce the popular *Places Rated Almanac*.[8] The difference between these and Liu's (1975) findings indicates the effect that choice and weighting of criteria can have on the rankings.[9] Notwithstanding the difficulty of comparability between indices, particular interest has been drawn to temporal changes in evaluations of the best places to live. Each year since 1987 *Money Magazine* has employed a set of indicators related to economy, health, income, housing, education, weather, transit, leisure and arts, to rank places in the USA in terms of liveability. The objective data collected on each of the component indicators was weighted according to the importance attached to it by a sample of magazine readers in terms of its contribution to quality of life. The top ten places to live in the USA are shown in Table 19.1.

The results of such national surveys are of more than popular interest.[10] Being rated as a good or bad place to live can have a significant impact on a city's ability to compete in the national and global market-place for inward investment, industry, tourism and new residents. A positive rating aids local place 'boosterism' strategies and enhances civic pride. Conversely, a poor rating can affect the local economy, offend residents and enrage civic leaders some of whom, as in the case of Yuba City CA have taken legal action to claim damages against the authors of unfavourable reports.

TABLE 19.1 THE TOP TEN PLACES TO LIVE IN THE USA, ACCORDING TO *MONEY MAGAZINE*, 2007

1. Fort Collins, CO
2. Naperville, IL
3. Sugar Land, TX
4. Columbia/Ellicott City, MD
5. Cary, NC
6. Overland Park, KS
7. Scottsdale, AZ
8. Boise, ID
9. Fairfield, CT
10. Eden Prairie, MN

In the rest of the chapter we move from consideration of the external image of urban places to focus attention on the differential quality of life space *within* the city.

THEORIES OF URBAN IMPACT

Five major theoretical perspectives have been advanced to explain the impact of the urban environment on residents. Each identifies a particular aspect of urban life and so contributes to an overall understanding of the person–environment relationship. All five are rooted in an appreciation of the key characteristic of cities – size – although each places a different interpretation on the effect of this dimension.

THE HUMAN ECOLOGICAL APPROACH

The essential premise of the ecological approach is encapsulated in Wirth's (1938) view that a person's behaviour is determined by the environment.[11] The individual either adapts or fails to survive. As we saw in Chapter 18, Wirth maintained that the size, density and population heterogeneity of the urban environment produced a particular socio-psychological adaptive response by the individual, characterised by the replacement of the primary social relations of *Gemeinschaft* by the formalised, contractual, impersonal and specialised relations of *Gesellschaft*. The behavioural end results are **anomie** (neglect of the normal rules that regulate social behaviour), **alienation** (the individual's sense of detachment from surrounding society) and deviance (crime and other social pathologies). According to this view, the city offers few advantages in exchange for stress, anonymity, alienation and personal and social disorganisation. This extreme interpretation has been criticised for the **environmental determinism** of the model, which reduces the urbanite to a position of virtual impotence, and for the overemphasis of the negative effects of urban living to the neglect of the virtues of size, density and diversity. Proshansky (1978), for example, has pointed out that the complexity of cities induces versatility in people.[12] Freedman (1975) introduced the 'density–intensity' hypothesis, which argues that the effect of urban density is to intensify an individual's *typical* response to a given situation – in other words, to make a lonely person lonelier or an active person more active.[13] Hawley (1981) has also developed a more positive view of the city from the basic ecological perspective.[14]

THE SUBCULTURAL APPROACH

The subcultural approach is a direct contradiction of the traditional ecological view that emphasises the similarities among groups of people rather than the differences that lead to the breakdown of close social ties. In Fischer's (1984) view, the urban environment enables people with different characteristics, interests, values and abilities to find other similar individuals.[15] Ethnic communities and neighbourhoods based on occupation, lifestyle and social class emerge as subcultures producing a city characterised by a mosaic of social works. Fischer's view is as polarised as that of Wirth. While Wirth gave insufficient attention to the close ethnic and occupational enclaves in Chicago between the world wars, so Fischer's emphasis overlooks the isolated people who fail to become part of any supportive community. Both models are correct in the sense of being supported by empirical evidence but neither is complete.

THE ENVIRONMENTAL LOAD APPROACH

The theory of **cognitive overload** argues that the urban setting tends to prevent people from responding 'normally' to new stimulations because their energy is dissipated in coping with an excess of environmental information.[16] To conserve 'psychic energy', people avoid all but superficial relationships. As a result, urbanites become close to some people and sensitive to some important parts of their environment, but at the expense of alienation from other aspects of the city. The theory does not imply that the city dweller is constantly bombarded with an unmanageable number of sensory inputs but rather that adaptation takes the form of a gradual evolution of behavioural norms that develop in response to frequent discrete experiences of overload. These norms persist and become generalised modes of responding. In contrast to the Wirthian view, which saw the individual being overwhelmed by the city, the concept of overload implies an ability to adapt and cope with the environment. The variety of coping strategies include (1) screening out inputs which are seen as less important (which may mean ignoring those in need of help); (2) erecting impersonal barriers (e.g. behaving in an unfriendly fashion or having an unlisted telephone number); and (3) allotting less time to each input (e.g. developing an aloof, 'business-like' attitude).

An alternative interpretation of the environmental load model postulates the concept of cognitive under-load. This suggests that many environment–behaviour problems result from too little stimulation.[17] As Krupat (1985 p. 125) states, 'monotony, whether in the form of dull urban design or rural isolation, can be just as stressful as any excess of stimulation'.[18] Some researchers have asserted that the lack of stimulation provided by modern urban settings leads to boredom and contributes to problems of juvenile delinquency and vandalism.[19] A fundamental question is whether there is an optimum level of stimulation. The concept of adaptation level sheds light on the problem.[20] This is the level of stimulation that is perceived as neutral (e.g. not too crowded and not too empty). Individuals at any given adaptation level evaluate the environment as a function of how much it deviates from neutral. Small deviations in either direction are experienced as pleasant but extreme deviations are considered to be stressful. This approach helps to explain why, as a function of different individuals' adaptation levels, contrasting evaluations are made of the same urban environment; why certain extreme conditions are generally experienced as unpleasant; and how perceptions and evaluations of a city may change over time.

THE BEHAVIOURAL CONSTRAINT APPROACH

Another potential consequence of undesirable environmental stimulation is loss of perceived control over a situation.[21] This first stage in the behavioural constraint model of environmental stimulation is followed by reactance and learned helplessness. Reactance is an attempt to reassert control over one's environmental stimulation (e.g. by moving from a crowded environment). If these efforts are unsuccessful, the ultimate consequence is learned helplessness.[22] That is, if repeated efforts to regain control result in failure, we stop trying and accept our situation (e.g. residence in an impoverished housing area). Psychologically, learned helplessness often leads to depression.

THE BEHAVIOUR-SETTING APPROACH

A behaviour setting is a discrete physical entity (e.g. a restaurant or schoolyard) in which a standing pattern of behaviour takes place (i.e. patterns that persist over time despite changes in the persons involved). In this model, environmental density is represented by the number of people available to populate or 'man' each of the behaviour settings. The number of settings increases with urban size but at a slower rate than population growth. At some point overmanning occurs, whereby the number of potential participants exceeds the capacity of the system. This argument clearly points to the virtues of small-scale environments. Bechtel (1971) characterises the city environment as overmanned and likely to foster passivity and apathy, and suggests the creation of more undermanned settings where people are more directly involved in the person–environment relationship.[23] As we shall see later, Newman's **defensible space** design incorporates some of the features of undermanned or optimally manned settings as an antidote to large, impersonal residential areas in which residents' indifference makes them more vulnerable to antisocial behaviour. The behaviour-setting approach has been employed to evaluate a number of built environments, as well as in studies of public participation in the urban social system,[24] and of time–space behaviour in the city.[25]

The five theoretical perspectives are not necessarily mutually exclusive and it is possible that each accounts for some of the negative effects that can result from an urban existence. All the viewpoints can be integrated into a general model built around the notion of stress, which is defined by Selye (1956) as increased wear and tear on the body as a result of attempts to cope with environmental influences[26] (Figure 19.1).

In this model the experience or perception of the city is represented as a joint function of the objective environmental conditions (e.g. population density, temperature or pollution levels) and the individual characteristics of the person (e.g. adaptation level, previous experience and time in the city). If the

perceived environment is outside the individual's optimal range (e.g. if it is overstimulating, contains too many stressors, constrains behaviour or offers insufficient resources), stress is experienced, which in turn elicits coping. If the attempted coping strategies are successful, adaptation and/or habituation occurs, though possibly followed by after-effects such as fatigue and reduced ability to cope with the next stressor. Positive cumulative after-effects would include a degree of learning about how to cope with the next occurrence of undesirable environmental stimulation. If the coping strategies are not successful, however, stress and/or arousal will continue, possibly heightened by the person's awareness that strategies are failing. Possible after-effects include exhaustion, learned helplessness, severe performance decrements, illness and mental disorders. Finally, as indicated in Figure 19.1, experiences feed back to influence perception of the environment for future events and also contribute to individual differences that affect future experiences.

ENVIRONMENTAL STRESSORS

Four general types of environmental stressor have been identified: (1) cataclysmic events, e.g. geographical hazards; (2) ambient stressors, e.g. air and water pollution; (3) stressful life events, e.g. death in the family; and (4) daily hassles, e.g. noisy neighbours.[27] The impacts of these physical and social stressors on city life are considered next, first with reference to a range of physical environmental stressors and then by examining a number of socio-cultural influences on urban behaviour.

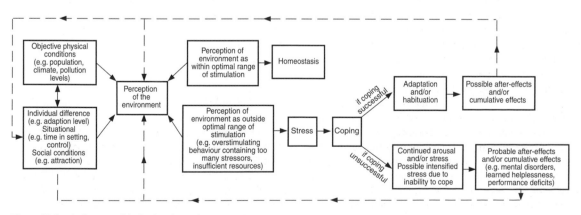

Figure 19.1   A stress model of urban impact

PHYSICAL ENVIRONMENTAL IMPACTS

The urban environment is a complex social, economic and biophysical system created by interaction between a human-made fabric and the physical characteristics of the landscape.[28] The hazards of urban living contingent upon the physical geography of the city vary with size and location but can include a host of events such as hurricanes,[29] flooding,[30] wind storms,[31] earthquake,[32] snow,[33] firestorms,[34] landslides[35] and ground subsidence.[36] Here we examine the incidence of and human response to several of these urban hazards.

Landslides

In those areas of northern Europe and North America affected by Quaternary glaciations, many cities are built on glacial drift material. Subsurface water can move at the interface between the drift and underlying rocks or at discontinuities between the products of different phases of glacial deposition. Should urban development take place on a steep hillside under these conditions, the consequent alteration in subsurface hydrology and in the pressure between the different materials in the glacial drift can result in land slipping. Other contributory factors that reduce slope stability include deforestation, construction and earthquakes. Although individual landslides are not usually as costly as major geophysical events such as earthquakes or floods, they are more widespread, and the total financial loss due to slope failures is greater than for most other geological hazards. Much of the damage associated with earthquakes and floods is due to landslides triggered by ground shake or water infiltration. In addition to the economic losses, significant loss of life can result from slope failure, as illustrated by the disasters in Saunders, West Virginia[37] and Aberfan, South Wales.[38]

In planning urban growth, therefore, it is essential to assess slope stability and to identify situations where human intervention is likely to trigger mass movement. Many planning authorities have general regulations prohibiting building on slopes exceeding a certain gradient. Some, such as those of the San Francisco region, have prepared slope stability maps to indicate the relative extent of landslide hazards. In practice, however, there is often a gap between identification and public notification of a hazardous locale, and the adoption of appropriate behaviour. As we shall see later, this is well illustrated by the experience of the mandated disclosure procedure introduced to reduce the earthquake hazard in California.

Earthquake

The earthquake hazard is most serious in Japan and the western USA, which form part of the Pacific seismic belt responsible for about 80 per cent of the world's earthquakes (Box 19.1). Nearly 90 per cent of the seismic activity of the continental USA occurs in Western Nevada and California, the latter state in particular being an area of pronounced immigration and population growth in recent decades. Several major earthquakes affected California over the past century, including the 1960 and 1989 earthquakes in San Francisco that killed 700 and sixty-two people respectively. In the 1994 Northridge LA earthquake fifty-seven people died, 60,000 housing units were destroyed and damage totalled $40,000 million.[40] Experts are agreed that the main question is when rather than whether another major earthquake will affect the area.

The Californian mandated-disclosure legislation was introduced in 1972 in an attempt to reduce earthquake risks by providing residents with information on the location of fault rupture zones. Specifically, the Alquist–Priolo Special Studies Zones Act required disclosures to prospective buyers if the location of a property (developed or undeveloped) was within one-eighth

BOX 19.1

The 1995 Kobe earthquake and its aftermath

In 1995 the epicentre of a 7.2 magnitude earthquake was directly under the city of Kobe, Japan. Over 6,000 people died in the event. Port facilities, freeways and railways were severely damaged. Almost 4,000 commercial, industrial and public buildings were heavily damaged or collapsed. In total 400,000 housing units in 190,000 buildings were made uninhabitable. Another 400,000 units were damaged. Total losses were estimated at US$89,000 million.[39] The displaced population lived in shelters for nearly a year, and around 100,000 were transferred to 48,000 temporary housing units assembled by the government and placed in parking lots and open sites outside central Kobe. Although the government issued a three-tier plan to build 125,000 housing units, government funds were used for only 3,000 units of public housing. The remaining recovery effort was left to private-sector initiatives. Because disaster insurance is virtually unobtainable in Japan individuals who lost their homes had to rely on savings and land value to finance any rebuilding.

of a mile (200 m) of a fault trace. In theory, on the assumption that most human behaviour is risk-aversive, such information should lead to an appropriate mitigation strategy by home-buyers – for example, avoidance of the area, or at least an attempt to reduce the risk through purchase of insurance. In an empirical study of two earthquake hazard zones in Berkeley and central Contra Costa County, Palm (1981) found that the disclosure legislation (based on maps prepared by the state geologist) had little measurable impact on buyer or market behaviour.[41] The explanation of such seemingly irrational behaviour must be sought in the value systems of buyers, the role of information agents (the property developers or real-estate agents) and the nature of the disclosure legislation. For most buyers the risk from an earthquake (as well as from other hazards) was considered to be a relatively unimportant factor in making a property purchase. Since most buyers intended to remain in a house for a relatively short time (three to five years), the decision to locate within a Special Studies (hazard) Zone and to forgo mitigation measures is seen as rational from an individual viewpoint. Generally, as Palm (1982 p. 273) observed, 'unless environmental hazards become translated into economic risk to individuals, hazard warnings not followed by severe disasters will probably not be heeded'.[42] A second contributory factor related to the purveyors of information is that real-estate agents may provide misinformation concerning the Special Studies Zones (possibly because of a genuine misunderstanding of the nature of the SSZ) or reinforce wishful thinking on the part of the buyer. ('All California is earthquake country!') Finally, failure of the disclosure legislation to specify the exact method and timing of the disclosure made it possible for agents to minimise the impact of disclosure on buyers. Real-estate agents were found to be disclosing at the least sensitive time (at the time of contract signing rather than at the initial viewing of the property) and to be using methods that conveyed the least amount of information about the zones.

The findings from behavioural studies of this kind carry a number of policy implications. Most important is recognition that 'mere provision of environmental information to homebuyers who are constrained by other aspects of the purchase process is insufficient as a hazard mitigation or consumer protection measure' (Palm 1981 p. 399).[43] An effective response requires direct action by local authorities in the shape of zoning or land-use regulation, more comprehensive definition and transmission of information on earthquake hazards and, in some cases, compulsory acquisition of dwellings in particularly hazardous environments. A key factor is that while the definition of hazard zones is essentially a technical matter, enaction of appropriate policy responses is an economic and political question.

Subsidence

Extraction of resources (such as ground water, oil, gas or coal) and/or the compaction of clays and gravels by the weight of building can lead to subsidence in urban areas. Problems due to subsidence include the following:

1. There is an increased risk of flooding of coastal areas by tides and storm surges necessitating expensive flood control works.
2. Regional tilting of land occurs which can affect the functioning of structures, such as canals and sewers, that rely on gravity for their operation.
3. Well-casing failure, which reduces and commonly destroys the productivity of water wells may occur.
4. Where subsidence occurs at shallow depth and buildings are founded on firm materials beneath compacting layers, utility connections may be broken, and building access and structural integrity affected.
5. Subsidence may lead to ground failure, seen in tension cracks and reactivated faults, as in the Houston area of Texas.[44]

Nelson and Clark (1976) describe how subsidence of up to 10 m (eleven yards) occurred in the Wilmington oilfield at Long Beach, Los Angeles.[45] Between 1937 and 1962, 913 million barrels of oil, 484 million barrels of water and 832 billion ft^3 (24 billion m^3) of gas were extracted from the underlying sediments to a depth of 6,000 ft (1,800 m). As the area was only a few metres above sea level, sea dykes, drainage systems and the elevation of harbour walls were required, until subsidence was eventually halted in 1966 by restoring the underground pressure by pumping salt water to replace the oil extracted. Around Phoenix AZ the extraction of ground water for irrigation and domestic use (including filling of swimming pools) has resulted in subsidence of up to 4 m (13 ft) in places. In Cheshire, England, salt extraction (initially by rock-salt mining and later by brine pumping) has created subsidence problems in adjacent towns as far as 8 km (five miles) from the production site.[46] Similar problems have been experienced in many large cities throughout the world.[47]

Urban climate

As a town develops, its presence exerts an influence upon the atmospheric environment which it eventually dominates. It then becomes possible to identify the urban climate as a distinct type of mesoclimate. On average temperatures in the city are usually about 1°C higher because of waste heat from houses, transport and industry, and the relatively low albedo (heat-reflecting properties) of buildings and paved surfaces (i.e. the urban heat island). Rainfall is also 5–10 per cent higher in the city because of greater air turbulence above urban areas and a higher concentration of dust particles (the urban dust dome) that can serve as hygroscopic nuclei. The incidence of cloud cover and of fogs is also higher over cities. On the other hand, because of friction around buildings, average wind speeds tend to be lower, although extreme gusts can be caused by groups of buildings, and sunshine duration is on average 5–15 per cent less.

The task in seeking to enhance urban liveability 'is partly that of preserving the advantages of urban climate, which are few, and of suppressing the disadvantages, which are many' (Parry 1979 p. 202).[48] Among such disadvantages are problems arising from urban-induced rainfall, including automobile accidents, bypassing of sewage-treatment plants by run-off, and higher costs of water management. Although the effect of urban areas on local climate seems destined to increase as the number of vehicles and fossil-fuelled power plants rises and urban growth proceeds, little has been done by public officials to manage urban weather anomalies or to adjust to their impacts. This is in part because social problems such as crime, poor housing and transportation present more pressing and more readily identifiable targets for action, and partly because, undeniably, it is more difficult to modify an existing built environment than to plan anew.

But several initiatives are possible in established urban areas. Amelioration of the urban-heat-island effect in summer can be aided by the provision of trees for shade, sprinkling the streets to increase evaporative cooling, and increasing the albedo by employing light-coloured paint on roofs and the replacement of open parking lots by garages. The utility of vegetation to improve urban climates is also widely acknowledged. As well as increasing cooling in summer by evapo-transpiration, trees and other plants reduce run-off and, if judiciously planted, will reduce particulate concentrations and noise pollution. Some of the undesirable wind eddies created by the tall buildings of the urban canyons can be avoided by sympathetic planning of street widths and orientations, and by employing appropriate building design. In cities where snow and ice are common winter problems, the channelling of waste heat under footpaths and streets, at least at critical points in the urban traffic system (such as bridges, slopes and pedestrian crossings), would be beneficial in combating the effects of snowfall and glaze storms (in which supercooled water falls on to a surface with temperature below 0°C and freezes on impact), which can paralyse a city. Few urban plans, however, explicitly take climatic considerations into account.

Air pollution

Atmospheric pollution *per se* is caused by gaseous (oxides of sulphur, nitrogen and carbon, hydrocarbons and ozone) and particulate (smoke, dust and grit) wastes emitted into the air. Particulates vary from unburned carbon to lead and radioactive compounds. Given adequate dilution and dispersion in the atmosphere, urban and industrial emissions of common pollutants may not be considered as giving rise to significant adverse effects. Pollution emerges as a problem when the waste assimilation capacity of the environment is exceeded by the volume of residuals output. Air pollution episodes during which concentrates reach socially unacceptable, economically costly and unhealthy levels may last for a few hours or several days. As Elsom (1983) explains, such episodes are typically of two types: smoky sulphurous smogs and photochemical smogs.[49] The former, which were the most common form of air pollution in the UK up the mid-1950s, result from the burning of fossil fuels. During cold anticyclonic conditions in winter, emissions of pollutants increase as the demand for space heating of buildings rises but the atmosphere, characterised by only light winds and a restricted mixing layer, has limited capacity for dispersing and diluting the emissions. The presence of large quantities of hygroscopic nuclei leads to the formation of fog as temperatures fall and relative humidity increases. During the worst conditions, in cities such as London, the droplets of dilute sulphuric acid were covered by a film of oily impurities that gave the urban fog its particular taste and colour (the 'pea-souper'). Although pollution concentrations within the smog largely emanate from local point sources, the urban wind circulation can spread the pollutants throughout the city. This occurs because of the urban heat island over the city centre which produces rising air (a low-pressure area) into which cooler air from the outskirts is drawn. The smog will remain for as long as the anticyclone persists.

Stationary sources of pollution are the main causes of smoky sulphurous smogs, but the automobile is chiefly responsible for photochemical smogs, which are a product of the action of sunlight on hydrocarbons and oxides of nitrogen present in vehicle exhausts. The photochemical smog is composed of ozone, aldehydes and peroxyacetyl nitrate (PAN). Collectively referred to as oxidants, these cause eye irritation and coughing, reduced visibility and damage to vegetation. Athens experiences some of the worst traffic pollution in Europe with the *nefos*, an ochre-coloured cloud containing dangerous levels of nitrogen oxide and ozone, while in California 97 per cent of populated areas have experienced reduced visibility and 70 per cent eye irritation.[50]

The motor vehicle is also the chief cause of lead pollution in the urban atmosphere. Lead is extremely toxic and accumulates in the body more rapidly than it is excreted, leading to damage to kidneys. It has also been implicated in slow-learning problems in children. Between 20 per cent and 50 per cent of the lead entering the body comes from the air we breathe, and people living near roads typically exhibit higher levels of blood lead than elsewhere in the city. Many governments have now introduced legislation to reduce the lead content in petrol. There are, however, a range of other metals (e.g. mercury, zinc, chromium and cadmium) whose potentially harmful effect on human beings has yet to be examined adequately.[51]

Flooding

The process of urban development carries with it a variety of hydrological consequences. In particular, urbanisation increases the magnitude of floods and reduces the average interval between serious flooding. The major mechanisms that cause this include:

1. The greater proportion of impervious surfaces, which tends to increase the total volume of storm run-off and reduce the amount of water which infiltrates into the ground.
2. The artificial 'improvement' (e.g. paving and straightening) of stream channels, which reduces the time lag between rainfall and channelled run-off.
3. Landscaping and subdivision of the land into building sites, which usually shortens the distance over which water flows before reaching a drainage way.
4. Human settlement of flood plains, which reduces the space available for storing flood waters in the

valley bottom so that water is forced to rise and flow more rapidly.

Urban areas differ greatly in the causes and combinations of flooding problems they experience but, in general, the problem of floodplain encroachment is more serious in North America than in Britain, where extensive floodplain development has been largely, but not wholly, prevented by the land-use regulations contained in the 1947 Town and Country Planning Act (Box 19.2).

The other three problem factors identified (collectively referred to as catchment urbanisation) represent the major cause of inland urban flooding in Britain. Parker and Penning-Rowsell (1982), for example, have described how urbanisation of the river Frome catchment upstream of the city of Bristol has contributed to a growing flood problem that affects the city centre, including a new shopping precinct, a large commercial district and a residential quarter.[52] The culverts used in the nineteenth century to channel the river Frome underground (to permit development) now suffer from silting and lack of maintenance just as flood flows due

BOX 19.2

Floodplain development in England

In the course of the year 2000–1 the value of homes and businesses planned on flood plains rose more than fivefold to £16 billion. Despite stricter guidelines from government to local planning authorities the Environment Agency, that is responsible for flood warnings, expected 342,000 new homes to be built in such areas in England and Wales between 2000 and 2004. The Royal Institute of Chartered Surveyors warns that people are buying homes unaware that they are in flood zones, which can make insurance difficult to arrange and could affect the value of the property.

On the outskirts of Burton upon Trent almost 100 houses are to be built by Westbury Homes. The Environment Agency objected to the development because parts of the site fall within a flood plain. Westbury Homes claimed that they had not known the site was in a flood plain when they planned the development. Local residents say that the area floods every year and, as the editor of the local newspaper observed, 'the area is not called Wetmore Hall farm for nothing!'

Source: The Times 27 October 2001

Plate 19.1 Flood impacts of Hurricane Katrina in New Orleans

to upstream urbanisation are increasing. In addition, as in other British cities (where the storm water and sewage systems are normally combined), the sewage disposal drains were built by the Victorians and are now of insufficient capacity and liable to collapse due to age and inadequate maintenance. In major cities like Liverpool and Manchester there is a twenty-year backlog of maintenance work. The rehabilitation and maintenance of water distribution systems is also a continuing concern in older US cities, such as Boston and Chicago.

The costs of an urban flood may include loss of life in addition to economic losses arising from damage to buildings and their contents, and costs accruing from disruption of economic activities and communications beyond the immediately affected area (Box 19.3). Human response to the flood hazard can be private, public or, more usually, some combination of the two. Individuals can either adopt a fatalistic attitude and accept the losses which are incumbent upon a flood-prone location or can take action to reduce the effects of the hazard. This can range from short-term adjustments,

BOX 19.3

Hurricane Katrina and New Orleans

Hurricane Katrina, the costliest ($81.2 billion) and one of the most deadly (1,836 deaths) natural disasters in the US history, made its second and third landfalls in the Gulf Coast region on 29 August 2005 as a powerful Category 3 hurricane (with a storm surge of a Category 5). By 31 August 80 per cent of New Orleans was flooded, with some parts of the city under 20 ft (6.1 m) of water. Four of the city's protective levees were breached, and over 150,000 properties were damaged or destroyed by wind, water and fire in the wake of Katrina. By 11.00 p.m. on 29 August there were reports of bodies floating on the water throughout the city. There was no clean water or electricity and some hospitals experienced diesel fuel shortages. In the immediate aftermath of the hurricane, looting, violence and other criminal activity was widespread, undertaken both by criminal gangs and by residents in search of food and water. Co-ordination of relief and rescue efforts was hampered by disruption of communications infrastructure, with roads damaged and telephones out of order. The repopulation and reconstruction of New Orleans represent a major ongoing challenge and underline the fragile balance between nature and human populations in flood-prone urban environments.

such as sandbagging or moving furniture upstairs, to longer-term strategies to spread the loss, such as insurance. Generally, however, studies have found that a large percentage of floodplain dwellers consistently underestimate the risk of flood. In a study of flood hazard awareness in Shrewsbury, Harding and Parker (1974) found that only 45 per cent of households interviewed acknowledged that there was a flood problem in their area.[53] Infrequent hazards such as floods are typically soon discounted by most people, and even in cases of repeated flood experience many floodplain dwellers ignore or play down the risk despite the opinions of experts. The general position in both Britain and the USA is of considerable indifference to flooding, which is perceived as an infrequent nuisance rather than as a major risk. In the USA the National Flood Insurance Programme has offered flood insurance to households and businesses since 1969 but only a quarter of those eligible have purchased policies. In the New Year's day flood in Reno NV in 1997 two people died and property damage to businesses and homes amounted to $200 million. Yet, four months after the event, only 625 flood insurance policies

were held by home-owners in Reno–Sparks, an area with a population of 25,000. The public costs of repairing flood damage and reducing the risk of inundation in flood-prone areas poses the question of whether flood (and other hazard) insurance should be mandatory.[54]

Noise pollution

Noise, defined as unwanted sound, is one of the most ubiquitous pollutants in contemporary urban areas. The adverse impacts of noise include (1) annoyance or stress which may contribute to psychological problems such as hypertension and neuroses, (2) physiological effects, e.g. hearing impairment; (3) effects on job performance; and (4) a reduction in property values. The effect of noise on urban residents is difficult to measure precisely, because tolerance levels vary between individuals. At the microscale, serious noise nuisance can be generated between apartments, offices and other multi-occupancy venues and, in the case of residential noise, may lead to social conflict between neighbours. Outdoor sources of urban noise include construction and street-repair work, industrial operations and transportation.

Significant alleviation of the road-traffic noise nuisance requires a range of measures, including technical improvements in vehicles and regulations to enforce their use; restrictions on traffic in certain areas (e.g. hospital zones) and at certain hours of the day; improvements in housing design (e.g. double or triple glazing); and greater attention to noise radiation in highway design (e.g. the use of vegetation buffer zones). The problem of airport noise is a function of the location and size of the facility but is an increasing hazard around many cities in the developed world. Residential areas under the flight path of major international airports such as London Heathrow or New York's JFK have to endure flights every few minutes at peak periods. Resolution of this problem requires a combination of technical improvements to jet engines to reduce noise at source, insulation of dwelling units near flight paths, land-use controls informed by accurate delimitation of noise exposure zones, and geographical research to determine the optimal aircraft flight paths over any urban area.

SOCIO-CULTURAL IMPACTS

A common type of socially induced stressor in the modern city is the result of 'daily hassles'. These are minor annoyances that, taken individually or as once-occurring events, would not tax the city dweller's coping mechanism. In the modern city, however, these events tend to be multiple, chronic and repetitive. Examples of common daily hassles include the urban commuting journey[55] and the bureaucratic hassle.[56] De Longis (1982) found the number and rate of hassles experienced to be a good predictor of an individual's overall health status and perceived energy level.[57]

Crowding

One of the most widely investigated stress agents is **crowding**,[58] yet the evidence of its impact is inconclusive. A major difficulty has been the failure to devise adequate measures of density and crowding.[59] Density is a physical, objective description of people in relation to space and is a necessary but not sufficient condition for crowding. Crowding is a psychological and subjective experience that stems from the recognition that one has less space than desired. The factors that determine the extent of crowding are related to the characteristics of:

1. *The physical environment*, e.g. density, site, design, temperature, colour, availability of resources, duration of exposure.
2. *The social environment*, e.g. the nature of the individual's relationship with proximate persons, and the individual's status position in the group.
3. *The task environment*, e.g. the degree of congruence between the behavioural goal of the individual and the environmental setting.
4. *The individual*, e.g. based on sex, age, ethnic or cultural norms as well as personality factors such as self-esteem, **personal space** preference and prevailing mood.

Other problems that confront those seeking to employ the concept of crowding in explaining certain urban ills include the fact that most of the evidence is based on aggregate data (e.g. persons per room or crimes per 100,000 population) and is open to the criticism of ecological fallacy. In addition, the relationships identified between crowding/density and certain pathologies are correlations only and cannot be taken to indicate that one event is the cause of the other. Furthermore, any relationship between a high-density environment and anti-social behaviour can be two-way, with, for example, criminals or delinquents seeking out a particular preferred environment. However, despite the difficulty of separating out the density/crowding effect as an independent factor in social malaise, there is sufficient evidence that crowding at least intensifies the influence of related environmental stressors.

For some urban residents, stress is caused by the effects of isolation, the polar opposite of crowding. Galle and Gore (1983), in a study of US cities with populations over 50,000, found a positive relationship between isolation and the death rate from suicide.[60] Given the increasing proportion of single-person households in Western society, such findings may carry significant social implications.

Urban legibility

A particular kind of stress is that associated with orientation and ease of movement in the city. As Ittelson *et al.* (1974 p. 246) observed, 'ease of movement is as much psychological as physical'.[61] People have to learn their way around the city, that is, they must invest their environment with order and meaning. In doing so they are transformed from 'people carrying maps in their hands to people carrying maps in their heads' (Krupat 1985 p. 68).[62]

Apart from Trowbridge's (1913) early ideas on imagery maps,[63] the seminal work on urban orientation was undertaken by Lynch (1960)[64] in his studies of the 'public image' of Boston MA, Los Angeles CA and Jersey City NJ. Lynch's basic proposition was that the quality of the city image was important to well-being and should be considered in designing or modifying any locality. According to this view, a successful landscape should possess the two desirable urban qualities of imageability (the ability of objects to evoke strong emotions in an observer) and **legibility** (the organisation of the elements of a city that allows them to be seen as a coherent whole). A city that is highly legible and imageable would contain individual structures and whole areas that are both distinct individually and clearly interconnected in a way that the citizens could appreciate.

Lynch identified five basic elements underlying the design of a legible urban environment (Figure 19.2). Although his work can be criticised on methodological grounds, e.g. the extremely small sample sizes (sixty respondents in total), the value of the underlying concept has stimulated a host of related investigations.[65] There is now a sufficient body of empirical knowledge on the urban image to suggest a number of principles of relevance for urban design (Box 19.4). The biggest problem is no longer identifying the components of urban imageability and legibility but translating them into practice. Lynch (1984), in a review of his own work, has admitted that findings have proved difficult to apply to actual public policy, the results of academic research being 'interesting but hard to put to use'.[72] A principal reason for this stems from the fact that the image of the city is idiosyncratic. Furthermore, despite evidence of the generality of certain design features, when one is dealing with a diverse area such as a city it is difficult to establish common problems and solutions. This scale factor and the need to consider the requirements of different population groups suggest that urban imagery techniques may be of greatest value in relation to sub-areas within the city.

Fear in the city

Fear is one of the most basic of human emotions that is linked to our inbred instinct for survival. Fear is the perception of a threat to some aspect of well-being concurrent with the feeling of inability to meet the challenge. Two principal sources of fear in contemporary urban environments are fear of crime and of terrorist attack.

Three main explanations for fear of crime have been advanced. These view fear of crime as the product of victimisation, as the consequence of a breakdown in social control, or as being mediated by the built environment. The first thesis relates fear of crime to experiences of victimisation or to a perceived risk of victimisation. The social control thesis posits that fear of crime is linked with individuals' ability to exercise control over their own life and the behaviour of others. Fear may be affected by community disintegration and poor neighbourhood amenities rather than high actual crime rates. In the environmental thesis fear of crime is embedded in the physical and social characteristics of place, and the familiarity of that place to the individual. People interpret the urban environment in terms of risk factors. Fear of crime is thus related to urban form and the ways in which urban spaces are utilised and given meaning. In practice, each of these perspectives contributes to understanding fear of crime.[73]

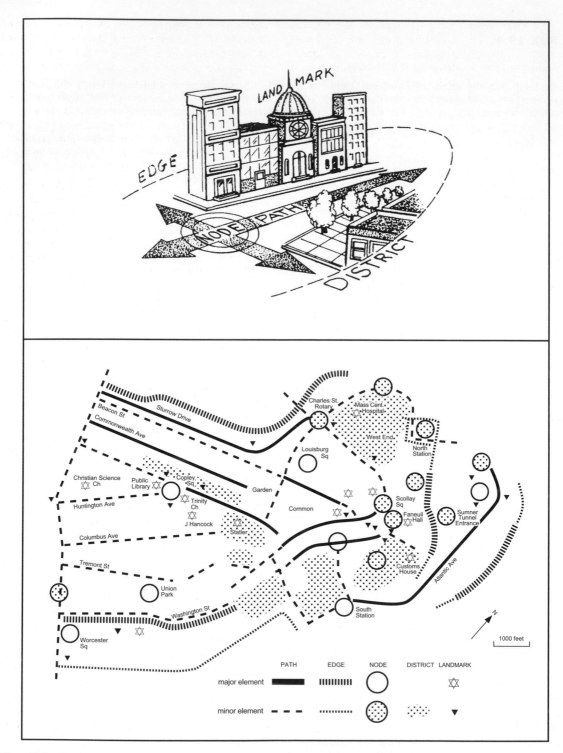

Figure 19.2   The elements of a legible city and their application to the urban form of Boston MA

Source: K. Lynch (1960) *The Image of the City* Cambridge MA: MIT Press

Terrorism may be defined as premeditated, politically motivated violence perpetrated against non-combatant targets by subnational groups or clandestine agents, usually intended to influence a wider audience. As a consequence of their heterogeneity, density of development and economic, social and symbolic importance cities are prime targets for terrorist attacks.[74] The indiscriminate nature of terrorism heightens the capacity for intimidation and public fear. Among a list of urban atrocities are the IRA bombings in London over the period 1973–97; the 1995 poison-gas attack in Tokyo; and the 2001 destruction of the World Trade Center and death of 2,752 people in New York City.

Fear of crime and violence influences the organisation of urban environments. Disaster contingency planning is now routine for public authorities as well as for many private firms. At a personal level an increased desire for security and control over public behaviour has led to the proliferation of gated communities; homes with concealed 'safe rooms'; 'stealth houses' built with an appearance that conceals the resident's wealth;[75] surveillance systems; and widespread privatisation of public space.

SITE DESIGN AND SOCIAL BEHAVIOUR

While rejecting the notion of architectural determinism, it is clear that the built environment can influence the behaviour and well-being of people by, for example, facilitating or discouraging interaction, fostering a sense of identity or alienating people from their surroundings. The general goal of the urban designer and planner is to organise physical space to facilitate certain forms of behaviour and to promote the satisfaction of human needs. According to Zeisel (1975), a behavioural approach to urban design can address six common human needs:[76]

1. *Security*, the need to feel safe in the residential environment.
2. *Clarity*, the need for ease of movement and a legible environment.
3. *Privacy*, the ability to regulate the amount of contact with others.
4. *Social interaction*, the need for sociopetal environments that facilitate desired interaction.
5. *Convenience*, the ease of accomplishing tasks at the domestic, neighbourhood and city scales.
6. *Identity*, the relationship between self and environment encapsulated in the notion of sense of place.

The success or otherwise of this approach depends on the level of communication between the planners and planned-for. Weak links between producers and consumers may contribute significantly to the socio-behavioural problems encountered in certain residential environments. As we saw earlier (Chapter 11), particular criticism has been levelled at the negative effects of urban-renewal programmes, particularly those carried out in US cities in the 1950s and 1960s. A classic example of the 'user-needs gap' in public housing is provided by the Pruitt-Igoe project. Built in 1954 in the inner city of St Louis, the fifty-seven-acre (23 ha) development housed 12,000 people in forty-three eleven-storey buildings. By 1970 twenty-seven of the buildings were vacant and two years later the entire project was demolished. Among the reasons that have been suggested for the failure of this project were the predominantly black racial composition, which made it as much a ghetto as the housing it replaced; the sheer scale of the project; isolation from surrounding neighbourhoods; poor administration; and inadequate funding. Design, however, was also a key factor. Although the physical design of the Pruitt-Igoe development was praised for having no 'wasted space', socially it proved to be a failure. Among the major design mistakes were: (1) sociopetal semi-public spaces and facilities around which neighbourly relations might develop were absent; (2) the high-rise building made it difficult for parents to supervise children; (3) the stairwells and elevators provided opportunities for criminal and other antisocial activities.[77]

Drawing on this and other experience,[78] Newman (1972) developed the concept of *defensible space* in an attempt to design more humane living environments.[79] Defensible space is seen as 'a model for residential environments that inhibits crime by creating the physical expression of a social fabric that defends itself'. The goal is to create an environment 'in which latent territoriality and a sense of community in the inhabitants can be translated into responsibility for insuring a safe, productive and well maintained living space' (Newman 1972 p. 3) (Box 19.5).

In order to test these ideas, Newman compared two adjacent but structurally different New York housing projects. The Brownsville project, built in 1947, had buildings of three to six storeys and housed 600 people at a density of twenty-nine per acre (seventy-two per hectare). The Van Dyke project, built in 1955, consisted of mainly fourteen-storey buildings that accommodated 76 per cent of its population. In terms of total population, density and social characteristics the

BOX 19.5

Newman's design guidelines for defensible space

To translate theory into practice, Newman suggested eight design guidelines for defensible space:

1. *Institutional appearance.* The structural materials and architectural design should be compatible with the surrounding neighbourhood to avoid stigmatisation of residents.
2. *Project size.* A density of fifty units per acre (125 per hectare) was suggested, especially where the development contains many children, to allow sufficient space for recreation.
3. *Building height.* Buildings should not exceed six storeys.
4. *Population density.* Density should be limited to a maximum of 200 persons per acre (500 per hectare) so that residents can distinguish neighbours from strangers.

5. *Public social spaces.* Areas within the project should be provided to promote social interaction and community feeling.
6. *Differentiation of transfer points.* Symbolic or real barriers should indicate the change in function and accessibility between public and private areas, thus promoting greater respect for privacy.
7. *Casual surveillance.* The high visibility of semi-public areas should deter intruders.
8. *Entryways.* Entryways should be clearly defined and provide differential access for residents and strangers (e.g. via security devices, such as entry phones).

Source: **O. Newman (1972) *Defensible Space*
New York: Macmillan**

projects were similar, but the incidence of crime was significantly different (Figure 19.3). This Newman attributed to the differences in design. Subsequent work by Newman and Franck (1982), in a survey of over 2,500 households in Newark NJ, San Francisco CA and St Louis MO,[80] and by Rubenstein (1980),[81] Taylor and Gottfredson (1986)[82] and Poyner (1983),[83] supported the efficacy of the defensible-space concept, while Holahan and Wandersman (1987)[84] have shown that preventive

design principles can also be applied to the most rapidly increasing type of housing in the USA: low-rise garden apartments and suburban town houses. In the UK, Coleman (1984) has applied defensible-space principles to a study of problem public housing estates in London and Oxford.[85] From an analysis of eighteen design variables she was able to recommend a series of design modifications aimed at reducing the disadvantages of life in such environments.

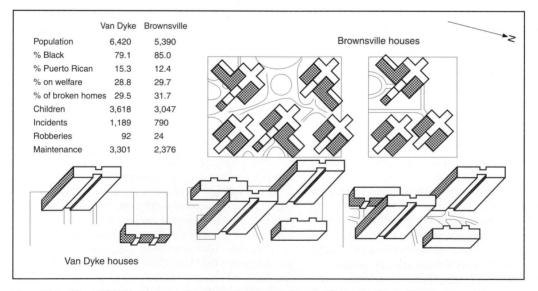

	Van Dyke	Brownsville
Population	6,420	5,390
% Black	79.1	85.0
% Puerto Rican	15.3	12.4
% on welfare	28.8	29.7
% of broken homes	29.5	31.7
Children	3,618	3,047
Incidents	1,189	790
Robberies	92	24
Maintenance	3,301	2,376

Figure 19.3   The relationship between crime and environmental characteristics, according to Newman
Source: O. Newman (1972) *Defensible Space* New York: Macmillan

Critics of Newman's argument have questioned whether the Brownsville and Van Dyke respondents were comparable on all relevant social characteristics, and have pointed out that since territorial behaviour and cognitions can vary according to social group, a single set of design solutions could not have the same effect for all groups. Much of the criticism stems from the seemingly deterministic nature of the model. Subsequent statements by Newman (1980) have sought to reduce this impression and emphasise the probabilistic basis of the theory.[86] It is now clear that physical design changes alone are unlikely to create a social fabric that eliminates crime and other unwanted activity but they can encourage and enable people to take an active role in their residential environments. Approaches that employ a combination of defensible-space design and social strategies (such as community policing and locally based management of problem estates) have the greatest chance of effecting improvement in poor residential environments. This socio-physical perspective underlies 'crime prevention through environmental design' strategies.[87]

RESIDENTIAL SATISFACTION

There is no general consensus definition of a decent home and a suitable living environment. English (1979) suggested that, other things being equal, houses are likely to be more popular than flats (apartments), central rather than peripheral areas, modern rather than old-fashioned dwellings, and 'respectable' rather than 'rough' neighbourhoods.[88] Francescato (1980) found that residents place great value on visual and auditory privacy, on the opportunities for personalisation of the dwelling unit both inside and outside, on a variety of shapes in buildings and landscapes, on easy access to the ground and to their parked cars, on having an enclosed piece of ground that they can call their own, and on the size of a development.[89] Other physical factors related to housing satisfaction are structural type, tenure type, the nature of the physical surroundings, and access to services and facilities.[90] As Duncan (1971) observed, some families have no need of a garden, while others enjoy tending a fair-sized area. Some wish to live close to a town centre for the convenience this brings; others do not mind a journey to work if they can live in more open surroundings.[91] Personal factors may also affect residential satisfaction, including previous housing experience, the degree of integration of the individual into society, the individual's reference group, the person's socio-psychological attitude to society in general, people's social customs and traditions, and the individual's aspiration level.[92] Finally, in addition to the characteristics of the house, neighbourhood and resident, the habitability of a residential setting can be affected by attributes of the management system – for example, the standard of garbage collection and other local services or, if rented housing, the effective response of the management authority.[93]

One way of advancing these various residential preferences would be to introduce greater public participation in urban design and planning. While the principle of public participation is increasingly

---

BOX 19.6

The concept of user-generated design

The process of user-generated design is intended to achieve greater involvement of residents in the planning of an urban environment. This advocates that at the first critical planning stage, referred to as programming, client and designer/planner should make explicit their assumptions and requirements for the project. The resulting detailed programme should specify the anticipated spatial and temporal patterns of living, giving information on detailed questions such as the arrangement of room walls and stairways and the materials to be used. Among the methods of obtaining the views of prospective users (e.g. those in a planned redevelopment area) are community meetings, questionnaire surveys and trade-off games. Zeisel *et al.* (1977) have employed this 'user-friendly' approach to design low-rise housing for the elderly.[97]

In general, designers/planners should offer several alternative project plans for resident review and be prepared to modify designs in the light of user feedback. Clearly, no single design solution can incorporate all that is good and avoid all that is negative, nor can it satisfy completely all members of a user group. The key is to make informed trade-offs to arrive at the solutions whose net outcomes are most positive. With this sort of collaboration in advance of construction, the questions, needs, uncertainties and assumptions of all relevant parties can be tested so that the physical product is as responsive to residents' needs as possible. Moreover, because the procedure explicitly states how each element of the design is supposed to work, it allows post-occupancy evaluation, which can lead to modifications and which can be fed into future design decisions.

accepted by policy-makers in Western Europe[94] and the US,[95] in practice many planners and decision-makers pay only lip service to the ideal, often seeing it as 'an inefficient wedge in the wheels of government' (Alterman 1982 p. 296).[96] As a result, the user-needs gap is not reduced significantly (Box 19.6). We shall return to the general question of public participation in the political process in Chapter 20.

URBAN SUB-AREAS

We have seen how the reciprocal relationship between environment and social behaviour leads to the delimitation of sub-areas in the city. In the process, these areas are imbued with symbolic meaning and become as real a part of the urban scene as the sub-areas defined by land values or other objective criteria. Such areas are most readily defined when physical boundaries correspond with symbolic meanings, but such a correspondence is not essential. Perceived areas exist in every city because their presence is collectively acknowledged by people within and beyond their boundaries and is reinforced by behaviour patterns. Examples of such areas include ethnic enclaves such as Chinatown in San Francisco, racially defined areas such as Watts in Los Angeles, areas defined by lifestyle such as sections of Greenwich Village in New York, gang turfs, urban slums, exclusive suburbs, or high-status areas such as Beacon Hill in Boston. Each of these is primarily a social construction that arises because of shared values and beliefs. An example of how this urban imagery can impose negative impacts on people's behaviour and well-being is the way in which some areas become stigmatised, with the inhabitants *en masse* being labelled as troublesome, work-shy or unreliable, thus making it more difficult for them to compete in local housing and job markets, or obtain goods under hire-purchase agreements.

Subjective areas in the city are important elements in urban liveability because they can restrict people's movement and guide behaviour. Milgram *et al.* (1972) produced a 'fear map' of New York City identifying areas of danger for non-residents,[98] while Pacione (1993)[99] identified perceived danger areas for male and female residents of a peripheral estate in Glasgow (Box 19.7). Ley (1974) mapped gang areas in the Munroe district of north Philadelphia and found them to be direct determinants of people's spatial behaviour, actual travel patterns of residents reflecting the 'stress topography', with many preferred routes increasing journey length by over 50 per cent.[100] Another graphic

illustration of the effect of a perceptual barrier on people's behaviour is the religious divide between Catholic and Protestant communities in west Belfast, where trip distance is of secondary importance to the wish to avoid the territory of the other group.[101] These studies indicate how the differential perception and partitioning of urban space influences the quality of life for particular social groups. The question of differential liveability for different urban populations is considered next.

WOMEN IN THE CITY

The advent of industrial urbanism introduced a separation of home and workplace, a division between men's and women's jobs, and a distinction between public and private space. Specifically, from the mid-nineteenth century onwards a redefinition of women's role in society emerged which prioritised the care of home, husband and children. The institutionalised separation of social roles on the grounds of gender is reflected in and affected by urban structure. In particular, home location can exert a strong influence on female employment opportunities. The large-scale suburbanisation of the early post-Second World War era in North America restricted middle-class women to predominantly residential environments. Post-war suburbs were seen as worlds of women. City and suburb were interpreted as dichotomous urban spaces characterised by related distinctions between public–private, production–reproduction and masculine–feminine. The subsequent diffusion of office activities to suburban locations reinforced this gendered division of urban space by providing specific 'female' work opportunities in areas such as back-office data-processing. Other employment activities, including child-minding and small-scale retailing, may also confine women's activity patterns to within the social sphere.[102] In addition, working wives often choose jobs that are portable in order to fit more readily into their spouse's job-related residential moves.[103]

The validity of such a gender-based differentiation of urban space is diminished by the declining importance of the nuclear family and increased female participation in the labour force. By the early 1990s only one-quarter of all American families consisted of a married couple with dependent children, and almost as many were headed by women, while the labour-force participation rate of women in the core 'home-making' years (between the ages of 25 and 44) had risen to almost 75 per cent. Nevertheless, gender continues to

BOX 19.7

Danger areas on a Glasgow housing estate

Identification of dangerous spaces and analysis of their characteristics can contribute to the design of policies aimed at fear reduction. The fear-engendering properties of perceived danger areas for males and females in the peripheral Easterhouse estate in Glasgow are summarised in the table.

Locations	% of respondents	
	Male	Female
Parks	7.6	19.7
Stigmatised neighbourhoods	34.0	14.9
Bridges/overpasses	0.0	11.9
Playing fields	6.8	9.5
Schools	8.2	7.1
Peripheral roads	0.0	20.8
Town centre	43.4	16.1

Although the cognitive maps of danger areas for males and females exhibit a degree of congruence, significant gender-related differences are evident. Generally, females fear the extensive open spaces of parks and playing fields after dark. These areas are regarded as the haunts of muggers, sex offenders and drug-taking youths. For similar reasons, females also avoid certain peripheral roads adjacent to open countryside. Unlit bridges and poorly lit streets also invoke fear and avoidance behaviour after dark. Football pitches known to be the location of gang fighting are also avoided, as are schools, where the physical layout of buildings offers dark doorways and blind corners as potential loci for assault. The fear maps of male respondents are less spatially discriminating. Male territorial behaviour tends to be based on larger-scale classifications of space, with certain neighbourhoods within the estate stigmatised and regarded as 'no-go areas'. (For example, Blairtumnock and Provanhall are widely regarded as dangerous areas.) The temporal dimension of dangerous space is marked by the changing image of the town centre, which is regarded by both males and females as safe during daylight hours when it is busy and well policed but as a major danger zone after dark, when it is perceived as a locus for groups of drug-taking youths. This type of detailed analysis is a prerequisite for the enhancement of urban liveability through the production of locally sensitive fear reduction policies.

shape people's use of urban space in several ways, not least in relation to employment and residential location.

There are also gender differences in patterns of mobility in urban areas. Men generally travel farther to work. The need for many women to deal with the responsibilities of home and work means that it is more common for women to combine a journey to work with other tasks such as shopping or taking children to and from school. Significant numbers of women restrict their employment options to those that facilitate such multipurpose trip patterns. The gender difference in commuting distances is also related to the fact that, in no-car or single-car households, women tend to make more journeys by public transport as a result of limited access to a family car. The predominantly radial layout of transit lines further restricts choice of jobs for those who are dependent upon public transport.[104]

Differences in the ways in which men and women use the city also include spaces for leisure and recreation. The settings of many leisure pursuits are male preserves, and female sociability tends to be more restricted by access and opportunity constraints, family commitments and custom. (In some communities it would not be 'right' for an unaccompanied female to enter a public bar.)[105] For many working-class women in British cities the bingo hall is a common venue at which to socialise on a regular basis with other women.[106] Shopping is another activity that reflects the institutionalised gendering of urban space. Whether viewed as a domestic necessity or form of recreation, shopping malls are used disproportionately by women.

Clearly, women and men do not follow wholly separate and spatially discrete lives. However, despite increasing flexibility in the social roles of the sexes, their choices, constraints and activity patterns differ significantly. For some feminist writers urban structure carries a masculine bias based on a patriarchal assumption that men will be the primary wage earners, women will be responsible for unpaid family work, and the environment will be organised primarily to facilitate these social roles.[107] This has significant implications for the liveability of the city for half the urban population.

Plate 19.2   Protected public space is provided in a park on Manhattan Island, New York City, with the play area fenced and designated 'children and their guardians only'

ELDERLY PEOPLE IN THE CITY

Elderly people represent one of the most rapidly increasing groups in society, and the majority live in urban areas.[108] Factors found to be of importance for residential satisfaction among elderly populations include (1) accessibility to a pharmacy and doctor; (2) participation in group activities; (3) proximity and access to shopping centres; and (4) the security, safety and friendliness of the neighbourhood.[109] Activity patterns vary in frequency, purpose and locational context, and reflect a range of factors including age, sex, ethnicity, health, personality, lifestyle and economic status. A key issue for a liveable environment is the extent to which activity patterns are a result of individual preference or are imposed by constraints such as lack of access to a car, absence of public transport or fear of crime.

Although the majority of people retire *in situ*, an increasing number of elderly who can afford to do so are relocating to areas of high amenity, such as the US sunbelt or the south coast of England. For some, housing in a special-purpose retirement community may be the preferred option, and in the USA, retirement housing is an important niche component of the housing market.[110] In retirement communities such as Sun City AZ restricted covenants on house sales ensure that at least one member of a household is over 55 years of age and none is under the age of 18 years. The local political influence of elderly residents is evident in their decision to secede from the local school system on the grounds that they need not pay for a service they did not use. The geographical concentration of elderly residents and more general expression of 'grey power' places particular demands on urban service provision and underlies the need both for an age-sensitive reading of the urban landscape and for responsible authorities to create user-friendly, liveable environments for all groups in society.

YOUNG PEOPLE IN THE CITY

A fundamental prerequisite for urban living is the ability to navigate a complex environment. This skill must be learned by new residents, visitors to a city, and children. According to Piaget and Inhelder (1967), spatial learning proceeds through four stages from infancy to adolescence and beyond:[111]

1. The *sensory-motor* stage (up to 2 years of age) is that during which a child defines its place in the world through tactile senses.
2. In the *pre-operational* stage (between 2 and 7 years of age) children acquire knowledge of some topological properties of space such as proximity, separation, enclosure; recognise home as a special place with strong emotional attachment; and develop elementary notions of territoriality.
3. The *concrete operational* stage (from 7 to 11 years of age) is marked by appreciation of concepts such as reversibility and by an increasing ability to recognise connections between topological properties in an integrated system.
4. In the *formal operational thinking stage* (from 11 years on), development advances from local solutions to specific spatial problems to logical solutions to all classes of problem.

According to this constructivist approach, spatial learning is based on field experience. Newcomers to a city may follow a similar progression in acquiring a working knowledge of the environment, albeit condensed into a shorter time scale.

In terms of the physical components of the urban landscape, Siegel and White (1975) suggest that landmark recognition is followed by knowledge of paths or routes between the landmarks, with route knowledge progressing from topological representation to metric representation.[112] Subsequently, sets of landmarks and paths are organised into clusters based on metric relationships within each cluster and the maintenance of general topological relationships between clusters. In the final stages an overall co-ordinated frame of reference develops, characterised by metric properties throughout, leading to survey knowledge. For children, primary landmarks or 'anchor points' include home, school, play area or shopping centre. Children's development of spatial knowledge may also be influenced by constraints imposed on their movement by adults in the interests of safety.[113]

Significantly, teenagers often interact with their urban environment in ways contrary to adult expectations, giving rise to conflict between age groups (for example, over the 'proper' use of a public space such as a community playground). The teenagers' use of space within the urban environment (e.g. 'hanging out') is an important element in the development of youth cultures characterised by shared patterns of consumption and lifestyle as defined by choice of clothing, hairstyles, musical tastes and other recreational interests. Such cultures are manifested most visibly in the formation of teenage gangs that develop an identity through shared characteristics (e.g. based on race, class or gender) and common interests (e.g. sport or music), dress (e.g. a 'uniform' of baseball cap, baggy trousers and trainers) and meeting places (e.g. a local park or shopping mall). Gang areas such as street corners, shopping centres and vacant lots are imbued with cultural values and meanings not shared with adults. In this way young people can create their own cultural locations and micro-geographies within 'adult-produced' urban environments.

DISABLED PEOPLE IN THE CITY

Conventional urban settings pose particular difficulties for the disabled citizen. Problems of access to buildings and facilities and in use of public transport are symptomatic of a hostile urban environment for many disabled people.[114] Although the primary causes of disability lie in the physical or mental limitations of the affected individual, disability is also socially constructed, being a product of a society that does not make adequate provision for the needs of its impaired members. Design changes that would assist wheelchair users include ensuring that lift controls can be reached, providing lowered drinking fountains and wash basins, dropped kerbs, wider parking spaces and doors, and ramps as alternatives to stairs. For visually impaired people, improvements include the installation of sensory or tactile tiles at the edge of a pavement (sidewalk), Braille maps of a local area, and the removal of obstacles to pedestrian access. However, public-sector resource constraints and voluntaristic rather than mandatory legislation have limited the widespread introduction of such enabling urban design.

Failure to accommodate the particular needs of the disabled urban dweller may be interpreted as institutional discrimination[115] or as a reflection of the position of a minority group within the social structure. Most fundamentally, it represents differing interpretations of the causes of disability and the extent to which the emphasis is placed on disabled people or disabling environments.

TOWARDS THE LIVEABLE CITY

Population-based suggestions for the liveable city have included Plato's 5,040 citizens (a total of around

30,000 people, including women, children and slaves); Howard's (1902)[116] 32,000; over 200,000 in the later UK and USSR New Towns; Le Corbusier's (1947) 3 million;[117] and the Goodmans' (1960) suggestion of between 6 million and 8 million.[118] This range indicates the futility of trying to define a liveable city in terms of size. There is no single optimum size. As we have seen, the experience and quality of urban life vary between individuals.

Whereas the Greeks thought that the good city was one in which all the free men could participate in face-to-face government, in modern times criteria of liveability have more usually emphasised economic factors such as job opportunities, good housing, schools and shopping facilities, efficient transport systems and sound urban finance.[119] Several writers have sought to leaven this economic viewpoint by consideration of social or 'human' concerns.[120] These have emphasised the need to ease orientation and movement in the city, to reduce the stresses caused by pollution, crowding, poor housing and stimulus overload, and to design a built environment that is responsive to the varying needs of residents. Clearly, in order to attain the goal of a liveable city, a wide range of social, economic and physical factors must be satisfied. Not all of these fall within the power of urban geographers, planners and designers to regulate. As we have seen, the city is not a closed system but is linked to regional, national and international systems that impinge upon the quality of the urban life space.

However, those urban components that can be manipulated positively should not be overlooked. An important step towards enhancing urban liveability is for urban geographers and others to acknowledge the subjectivity of the objective environment. Much of the debate over urban liveability and quality of life takes place within the framework of the urban political system. The question of the differential socio-spatial distribution of power and the conduct and outcomes of urban politics and governance will be considered next.

FURTHER READING

BOOKS

I. Douglas (1983) *The Urban Environment* London: Arnold

R. Imrie (1996) *Disability and the City* London: Chapman

E. Krupat (1985) *People in Cities* Cambridge: Cambridge University Press

J. Little, L. Peake and P. Richardson (eds) (1988) *Women in Cities* New York: New York University Press

K. Lynch (1960) *The Image of the City* Cambridge MA: MIT Press

J. Mitchell (1999) *Crucibles of Hazard: Megacities and Disasters in Transition* New York: United Nations University Press

O. Newman (1972) *Defensible Space* New York: Macmillan

T. Skelton and G. Valentine (1998) *Cool Places: Geographies of Youth Cultures* London: Routledge

D. Walmsley (1988) *Urban Living* Harlow: Longman

JOURNAL ARTICLES

J. Dunstan (1979) The effect of crowding on behaviour *Urban Studies* 16, 299–307

C. Kirby (1995) Urban air pollution *Geography* 80, 375–92

G. Laws (1993) The land of old age: society's changing attitudes toward urban built environments for elderly people *Annals of the Association of American Geographers* 83(4), 672–93

H. Matthews, M. Limb and B. Percy-Smith (1998) Changing worlds: the microgeographies of young teenagers *Tijdschrift voor Economische en Sociale Geografie* 89(2), 193–202

M. Pacione (1982) Evaluating the quality of the residential environment in a deprived council estate *Geoforum* 13(1), 45–55

M. Pacione (1982) The use of objective and subjective indicators of quality of life in human geography *Progress in Human Geography* 6, 495–514

M. Pacione (1990) Urban liveability: a review *Urban Geography* 11(1), 1–30

R. Palm (1981) Public response to earthquake hazard information *Annals of the Association of American Geographers* 71, 389–99

A. Warnes (1994) Cities and elderly people *Urban Studies* 31(4/5), 799–816

KEY CONCEPTS

- Liveability
- Quality of life
- Environmental determinism
- Environmental load
- Behavioural constraint
- Behaviour setting
- Stressors
- Environmental hazards
- Urban heat island

- Daily hassles
- Fear
- Crowding
- Urban legibility
- Architectural determinism
- User-needs gap
- Defensible space
- Stress topography
- 'Grey power'

STUDY QUESTIONS

1. Examine the major theoretical perspectives advanced to explain the impact of the urban environment on residents.
2. Select one environmental hazard and study its effect on cities and urban life.
3. With the aid of appropriate examples examine the impacts of fear on people's behaviour in cities.
4. Consider the relationship between site design and social behaviour, and examine the concept of defensible space.
5. With reference to particular examples, illustrate the relevance of subjective areas for urban living.
6. Examine the differences in liveability experienced by different population groups in the city.

PROJECT

Residential satisfaction varies between social groups and across space within the city. Select either a particular social group (e.g. elderly people, women or a minority ethnic or lifestyle population or a particular location (e.g. public housing or private housing) and use a questionnaire–interview technique to examine residents' satisfaction with their home, neighbourhood and city environments. Different factors will be of importance at each level of analysis. By studying more than a single social group or area you can engage in comparative analysis. Identify the key components of residential satisfaction at each level of analysis. How well are these satisfied by the current residential environment? What social and spatial variations are revealed regarding residential satisfaction in the city?

20

Power, Politics and Urban Governance

Preview: the role of local government; constraints on local government; the spatial structure of local government; city growth, annexation and incorporation; the consequences of fragmented government; the politics of secession; metropolitan government; power in the city; informal influence; voting; public participation; urban social movements

INTRODUCTION

'Power' refers to the capacity of an individual or group to command or influence the behaviour of others. Power vested in people who are selected or appointed by a socially approved procedure (such as democratic election) is regarded as legitimate, and is often referred to as authority. Power may also be exercised through social pressure or persuasion or by use of economic or even physical force. A key factor in determining the outcome of power struggles – of 'who gets what, when, and how'[1] – is the system of urban **governance**.

In this chapter we shall focus attention on forms of political power and the dialectical relationship between the operation of power and the socio-spatial structures of the city. We shall examine the role of local government and the constraints on local government autonomy, including the influence of different interest groups and the relationship between central and local government. We shall consider the internal structure of and spatial organisation of urban government in the UK and the USA. We shall then change the scale of analysis to focus on the nature and distribution of power within the city.

THE ROLE OF LOCAL GOVERNMENT

In principle, local government promotes the basic values of:

1. Liberty from central authority and abuse of central power.
2. Popular participation in government which is encouraged by the proximity of decision-makers and citizens.
3. Efficiency in government, which is advanced by a scale of organisation that permits locally sensitive provision of public services and functions.

The diverse activities of local government include:

1. Providing public services.
2. Acting as an agent of central government, as when it enforces state legislation.
3. Formulating policies and plans such as those relating to local development.
4. Representing the locality in dealings with other governments, as when seeking financial aid.
5. Resolving conflicts between competing local interests, for example over the location of facilities.
6. Regulating private-sector activities, as in land-use zoning and building control.

In practice the importance of local government activities varies, depending on the prevailing philosophies of central and local government, attitudinal and ideological differences among political parties and elected officials at each level, and on the nature of the relationship between local political and economic interests.

CONSTRAINTS ON LOCAL GOVERNMENT

Gurr and King (1987) identified two main forms of constraint on local-government automony.[2] These are limits imposed by local economic and social conditions (Type I constraints) and limits imposed by higher levels of government (Type II constraints).

TYPE I CONSTRAINTS

Economic and social constraints derive from a variety of sources. Most fundamentally, the health of the local economy may limit the tax base and therefore the potential for local revenue-raising. In contrast to most European nations, North American cities rely on local property tax and increasingly, in the USA, sales tax for revenues. In the US cities set their own rates and assessments. The absence of a uniform rate encourages communities to compete to attract businesses and increased tax revenue. The local imperative to maximise the tax-revenue base is traditionally a distinction between North American and European cities, although it may be less so in future with the decline of central grant equalisation strategies in Europe. In the absence of central-government grants, poorer areas may find it more difficult to finance a given level of public services.

A second form of Type I constraint refers to the blocking power of dominant interests in the community, which may impose their views on the political decision-making process. The *elitist theory* of urban power identified by Hunter (1953)[3] in Atlanta GA found that most important city decisions were made by informal consensus between economic elites drawn mainly from business and industrial circles. Formal governmental decision-makers, including the mayor, were only peripheral actors until the stage of implementation was reached. In contrast, the *plural model* of community power advanced by Dahl (1961)[4] holds that power is dispersed, with different elites dominant at different times over different issues. Decision-makers in one issue area, such as education, may be less influential in another, such as urban redevelopment. In Dahl's model, business elites of the kind Hunter identified are only one among a number of types of influential power clusters. As we saw in Chapter 7, from the mid-1970s the advent of political economy models of urban power focused attention on the actions of growth-oriented alliances. The urban growth machine model envisaged a broad coalition of groups with a common interest in urban growth including business leaders, public officials and politicians, and organised labour.[5]

Regime theory, a variant of this pro-growth perspective, develops Dahl's notion of fluid overlapping alliances among local business and political leaders in order to achieve desired solutions to particular problems.[6] A regime is defined as the informal arrangements by which public and private interests function together in order to be able to make and carry out governing decisions. The regime forms through a meshing of the interests of a number of groups co-operating behind an agenda to achieve a set of policies. While not all members will necessarily want the same outcomes from the regime, all perceive it as in their interest to remain within the coalition. Globalisation has led to formation of new 'global growth coalitions' comprising externally oriented economic interests (such as tourist industries), elements of city government that benefit from the attraction of 'world city' functions (e.g. the CBD) and higher-level political actors (e.g. central-government agencies). The physical product of such 'regimes' is evident in developments such as Sydney's opera house and London's Canary Wharf. The intensification of social and economic change in the post-industrial city and construction of new socio-political groupings, in addition to traditional class- and ethnic-based cleavages, has served to make urban regimes more complex and volatile. Local political cultures defining the 'proper' role of local government may also limit the range of possible local policy action.[7] The presence of active local community groups[8] or grassroots urban social movements[9] may constrain local government policy and action, while, more generally, local governments must be responsive to public opinion, group demands and electoral imperative.

The structure of local government itself may impede effective action. As we shall see later, the fragmentation of local government in the USA represents a barrier to effective policy action for metropolitan areas. In particular, the fragmentation of urban government isolates central-city communities with the highest concentration of poor residents from the wealth of the suburbs and limits the scope for redistribution. Spatial fragmentation also enables wealthier suburbs to engage in exclusionary zoning. Although the structure may be seen as socially regressive and inefficient in terms of metropolitan government, it may be defended on the grounds of local democracy. Further, public choice theorists view metropolitan fragmentation as a desirable structural device to promote efficiency since it resembles a market in which a large number of sellers (local governments) compete for buyers (citizen-consumers) by offering various packages of goods and services to appeal to different consumer tastes at various prices (taxes). The major types of metropolitan government along with the advantages and disadvantages of each are shown in Table 20.1.

TYPE II CONSTRAINTS

Local governments are also constrained by constitutional and statutory limitations imposed by higher levels of government.

Internal structure

In Britain the internal structure of local authorities is set largely by central-government statute. Each local government must consist of a council composed of members elected from wards within the local authority. The mayor is appointed from the body of councillors (Figure 20.1). The geography of local government in Britain is shown in Figure 20.2. In the USA, local-government structure is governed by state law,

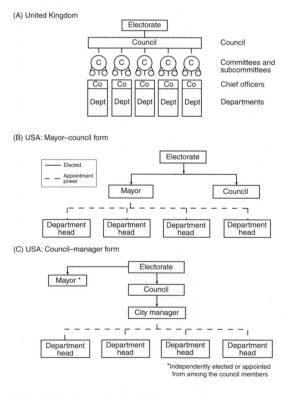

(A) United Kingdom

(B) USA: Mayor–council form

(C) USA: Council–manager form

*Independently elected or appointed from among the council members

Figure 20.1   Forms of urban local government

although in most states electorates are given some discretion under **home-rule** provisions to choose their own form of government. The two most common forms are (Figure 20.1):

1. The *mayor–council* form, featuring a separately elected mayor as the executive head of government and an elected council as the legislative branch. This is the most widespread form of city government in the USA and the predominant form in large cities. Under this arrangement, mayors have significant executive powers, including powers of appointment and powers to intervene directly in the conduct of city government, to veto the city's budget and to initiate legislation.
2. The *council–manager* form, comprising an elected council and professional city manager hired by the council to administer the government. This structure is common in many medium-size communities. Under this system the mayor has less power and the key actor is the city manager, who has appointment power, is responsible for preparing the city budget, developing policy recommendations and overseeing city government.

It is suggested that council–manager systems, which emphasise the professional manager or the effective executive, afford less attention to citizen demands and concerns than do mayor–council systems, where such demands can be expressed more directly through the political process. According to Bollens and Schmandt (1970 p. 116), council–manager governments seem best suited to 'relatively homogeneous white-collar communities where the political representation of diverse interests is not an important factor'.[10]

Service provision

Central government in Britain and state government in the USA specify certain services that local governments must provide, and in some cases also specify the level and quality of service. Equally important are the constraints on services that local councils may wish to provide but cannot. Under the *ultra vires* rule, British local authorities are prohibited from providing services that Parliament has not authorised specifically. In the USA city governments in states with strict adherence to Dillon's Rule (which states that cities can do only what state legislatures expressly permit them to do, or what is 'fairly implied' or 'indispensable') are in the same position as British urban authorities. Cities in states

TABLE 20.1 TYPES OF METROPOLITAN GOVERNMENT

Type of government	Roles	Advantages	Disadvantages
Small independent councils	Local representation, planning and regulation. Purchase of community services	'Public choice' for local residents. Payment limited to selected services	Affluent communities and industrial areas opt out of metropolitan service funding. Exclusionary zoning
Voluntary co-ordination of local councils	Exchange of information. Regional advocacy. Preparation of region-wide studies and advisory plans	Voluntarism, co-operation and consensus-building. Development of regional consciousness	Lack of powers of implementation, control or co-ordination. Record of irreffectiveness
National/state/provincial administration and service provision	Preparation of metropolitan planning strategies. Supply of power, water and transport services	Potentially strong implementation powers and finances. Capacity for wide contextual view	Indirect accountability and poor political control. Poor recognition of metropolitan-scale problems and impacts. Practical difficulties in achieving interdepartmental co-ordination
Special boards and districts performing specific functions and services	Schools, fire, highways, health, water, sewerage and drainage, power, light, gas and transit	Easy and popular to establish as new needs become evident. Elected boards can maximise local participation	Lack of co-ordination, planning capacity or impact control. Lack of accountability of appointed boards
Metropolitan governments nominated from local councils (indirectly elected)	Strategic environmental transport and functional planning. Water supply, waste disposal, transit. Regional open space, social housing	Regional-scale co-ordination of planning, services and powers. Good articulation with state/provincial and local governments. Flexibility to expand boundaries in response to growth	Lack of direct electoral mandate and relatively low public profile. Reliance on powers from above and political support from below. Dependence on consensus-building
Metropolitan governments (directly elected)	Vary from metropolitan and transport planning and waste disposal to very wide range of powers and services	High public profile and role. Powers to exact compliance from junior jurisdictions. Capacity to take on new functions and extend effectiveness	Conflicts of interest and jurisdiction with local and central governments. Difficulties in adjusting boundaries to match expanding metropolitan extent because of fears of aggrandisement
Regional governments	Vary from environmental management and resource policy to wide-ranging powers and services	Reflect changed scale of modern societies' social interaction, communities of interest and centre–periphery relations. Useful intermediaries between local authorities and national governments	Involve radical changes in existing administrative systems, for which effective local consensus would be difficult to achieve. Hard to introduce in federal systems because of likely superseding of existing states/provinces

Source: **P. Heywood (1997) The emerging social metropolis: successful planning initiatives in five New World metropolitan regions** *Progress in Planning* 47(3), 159–250

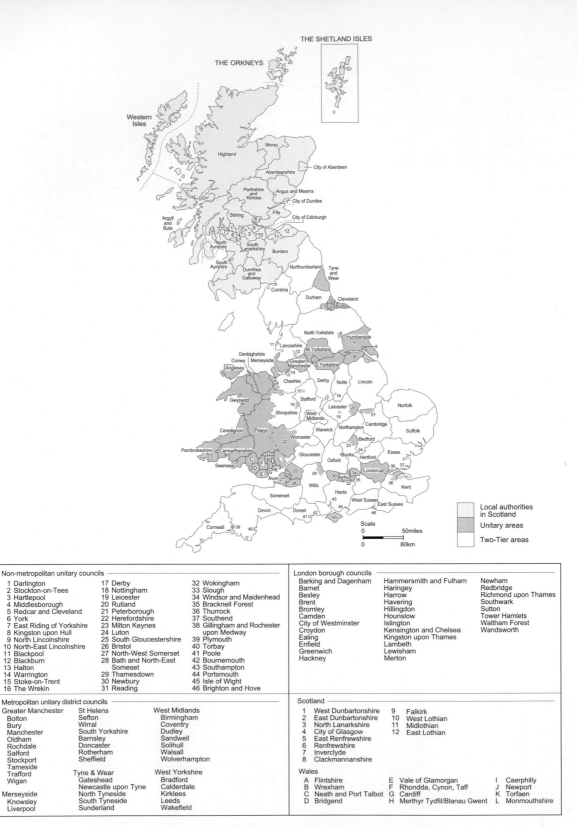

Figure 20.2 Spatial structure of local government in the UK

that permit broad home-rule powers, on the other hand, can provide any service not explicitly prohibited by the state government. The result is a complicated patch-work of service provision in US metropolitan areas.

Tax and expenditure limitations

Whereas in both the UK and the US local authorities are limited in the amount of taxation they can impose, the motivation behind such constraints differs. In the USA, where the electorate can vote on local tax limits in an annual referendum, spending limits are seen as a device to check the 'tax and spend' proclivities of elected politicians and affirm the legitimacy of direct democracy (Box 20.1). In the UK the main aim of cen-tral control over local expenditure is not local account-ability but to ensure that local authorities do not pursue financial policies that run counter to national economic policy. In 1985 Labour councillors in Liverpool who set their budgets late in an attempt to resist central-government spending restraints were sur-charged individually. Since they could not pay, each was declared bankrupt and hence disqualified from holding public office.

Grant aid

The provision of grant aid enables central government to exercise control over local government. Grants can be either for a specific purpose (e.g. to fund a public-transport system) or for general purposes. In the UK most grants are general grants and the main objective is to control the aggregate amount spent by local govern-ment. By contrast, in the USA virtually all federal grants to state and local governments and 90 per cent of state grants to local governments are for special pur-poses. This enables the higher tier to influence the activ-ity of local government. Thus federal law prohibits a recipient from using federal funds in a way that dis-criminates against minority groups, while federal grants for education require all school districts receiving funds to offer bilingual education to any group of twenty or more students speaking the same foreign language.[11]

Policy guidance

Whereas administration circulars are employed in the USA only as a legally binding means of implementing federal laws and grant programmes, in the UK a

BOX 20.1

The California tax revolt

The tax revolt was a distinctly American phenome-non that relied on instruments of citizen initiative and public referendum that are virtually unique to the USA. The main factors behind the revolt in California in 1978 were suburban growth accom-panied by political conservatism, and escalating property values and hence taxation. In Los Angeles, the heart of the tax revolt, 500,000 homes were reassessed in 1978 for property tax purposes as part of the region's triennial reassessment cycle. The assessment, based on market value, increased by an average of 125 per cent. In the same year the state generated a budget surplus of $5 billion.

Opponents of what they saw as a 'tax and spend' regime gathered 1.25 million signatures to place Proposition 13 on the California ballot. The initia-tive was approved by 65 per cent of the voters. The effect was to roll back property taxes to 1 per cent of market value, limit year-to-year increases in property assessments to 2 per cent per year as long as home owners occupied the same house, stipu-lated that any future increase in state taxes must receive approval by two-thirds of the state legisla-ture, and allowed cities, counties and local special districts to introduce new taxes or raise existing tax rates only if they could secure two-thirds voter approval at a local referendum.

Similar taxpayers' rebellions were successful in Tennessee, Texas, Michigan, Massachusetts, Arizona, Hawaii, New Jersey and large numbers of local jurisdictions. The groundswell of opposition to increased taxation contributed to the election in 1980 of the Reagan administration, committed to cutting government spending. The longer-term legacy of the tax limitation movement has been to restrict local government revenues and curtail social expenditure in areas such as public housing and anticyclical employment initiatives.

Source: **G. Peterson (1996) Intergovernmental financial relations, in G. Galster (ed.)** *Reality and Research: Social Science and US Urban Policy since 1990* **Washington DC: Urban Institute, 205–26**

steady one-way flow of circulars links central and local government. These documents may be advisory, explanatory or mandatory. In addition to these regular policy-related circulars, as we saw in Chapter 8, local-government autonomy is constrained significantly by national urban policy and planning.

THE SPATIAL STRUCTURE OF LOCAL GOVERNMENT

THE UK

Local governments in Britain are the creation of Parliament, which by simple majority vote can restructure local government or even abolish units of local government as in 1986 when the Conservative government of Mrs Thatcher abolished the Labour-controlled Greater London Council, along with the metropolitan counties in six other conurbations. In the USA it would be unconstitutional for the federal government to impose its will in this way. States, on the other hand, do have supreme authority over local governments, as provided by the Dillon ruling of 1868, but would not consider using their constitutional powers to reorganise local government in such a radical manner. In England the current system of local government is based on a single tier of unitary district councils which encompass the major urban areas, and a set of two-tier county and district authorities elsewhere. Unitary authorities are responsible for local government in Scotland and Wales (Figure 20.2). The removal of the upper tier of local government in 1996 (regions in Scotland and counties in England and Wales) has posed difficulties for the planning of functional urban regions. This may require the creation of *ad hoc* joint authority organisations with executive powers to co-ordinate functions such as transport in larger urban regions.[12]

THE USA

Local government in the USA is organised in a series of overlapping territories. At the highest level are the fifty states, which are responsible for a wide range of functions, many of which are delegated to local jurisdictions. Below the state level there are four main types of local government:

1. The *county* is the basic unit of local government. Counties were established early in the settlement of the original colonies and were used as a framework for settling the interior, providing a regular grid pattern of administrative areas across much of the country. In rural areas where there are no incorporated cities, county government acts as the general-purpose local government. Typically it regulates land use, licenses businesses, and provides police, fire and other services. In urban areas cities usually assume the basic general-purpose local-government function for their residents, and counties serve this purpose only for unincorporated territory.

2. In twenty-one states, mostly in the north-east, counties contain a network of *townships* which perform a limited number of functions such as the provision of street lighting and road maintenance.

3. The *municipalities* are of greater importance within the counties. These are densely populated areas that have been legally incorporated to provide local-government services independent of those of the encompassing county. Incorporation is sanctioned by the state government following a public petition and poll of affected voters. Many municipalities are cities or towns. The functions of the municipal government are either laid down in an individual **charter** that describes the form, composition, powers and limitations of city officials, or are established generally in state law for all cities. An important variation is the home-rule charter under which the precise definition of city powers is left to the electorate within limits set by the state constitution. About three-quarters of large US cities operate under home-rule provisions. A large municipality may completely replace the county, to become an urban county.

4. *Special districts* are the most widespread type of local government (Table 20.2) and are established to perform a specific function such as provision of

TABLE 20.2 NUMBER OF LOCAL GOVERNMENTS IN THE USA, 1942–2002

Tier	1942	1957[a]	1967	1977	1987	1997	2002
Local governments	155,067	102,341	81,248	79,862	83,186	84,955	87,900
County	3,050	3,050	3,049	3,042	3,042	3,043	3,034
Municipal	16,220	17,215	18,048	18,862	19,200	19,279	19,431
Township and town	18,919	17,198	17,105	16,822	16,691	16,656	16,506
School district	108,579	50,454	21,782	15,174	14,721	14,422	13,522
Special district	8,299	14,424	21,264	25,962	29,532	31,555	35,356

Note: [a]Adjusted to include units in Alaska and Hawaii, which adopted statehood in 1959

fire protection, hospital services or water supply. Directors of special districts are not accountable to city or county officials because special districts are separate legal entities. Their boundaries do not necessarily conform to those of any other local government unit and often overlap the boundaries of the city and each other. The most common type of special district is the school district. The separate existence of these single-function authorities can be a major attraction for suburbanites wishing to control the social and demographic character of school catchments (Box 20.2). More generally, special districts are valued because they can operate separately from the financial constraints of other local governments. For example, residents who want better street lighting in a city that has already reached its debt limit might form an independent street-lighting district that could sell bonds to finance the new system unrestricted by the city debt limit. Special districts vary considerably in size. Some cover a few city blocks (e.g. a historic preservation district). One of the largest special districts is the New York Port Authority, which has jurisdiction in both New York and New Jersey, and in 1987 had more than 9,000 employees and an annual turnover of $1.3 billion. The number of special districts has risen consistently, with a 59 per cent increase between 1957 and 2002.

CITY GROWTH, ANNEXATION AND INCORPORATION IN THE USA

Whereas the layout of a system of counties was a requirement in all US states, municipal government was established only when the affected populations considered it necessary. The original municipalities obtained independent government for the whole of the built-up area, and as towns and cities grew, municipal boundaries

were extended accordingly. In some cases boundary extension was necessary prior to urban development to enable utilities to be provided, and in the nineteenth century cities often annexed large areas of undeveloped land. Until the late nineteenth century there was little opposition to the territorial claims of the expanding municipalities, but increasingly the residents of urban fringe areas concluded that their interests were better served by incorporating as a separate municipality and resisting **annexation** by the central city.[13] The key factor in this shift of opinion was the system of local taxation. Since the main contribution to local revenue comes from property taxes, it was in the interest of commercial and industrial property owners and higher-income households to locate in municipalities that had lower expenditure needs and therefore lower rates of tax. Since the late nineteenth century the process of **incorporation** has been of major importance in determining the socio-spatial composition of the US metropolis.

Defensive incorporation, designed to preserve the status quo, is commonplace. This is evident in the number of municipalities that mushroomed in Nassau County NY in the inter-war period (Table 20.3). One of these, Lake Success, was incorporated as a municipality in 1926 to keep out the growing number of weekend trippers by giving residents control of lakeside property and access roads. In St Louis County MO the number of municipalities doubled between 1940 and 1950 as communities opted for municipal status to protect themselves from annexation by others. Some of these new municipalities were tiny; in 1946 residents of a tract of only eleven acres (4.5 ha) chose to incorporate as McKenzie Village, and by 1951 twenty-six of the county's municipalities had areas of less than 100 acres (40 ha). In some instances suburbanites wanted to gain the tax revenues available from the post-war suburbanisation of commercial land uses. The only way for home-owners in unincorporated areas to retain the tax receipts from a new

TABLE 20.3 NUMBER OF MUNICIPALITIES IN SELECTED US COUNTIES, 1920–70

County	1920	1930	1940	1950	1960	1970
Suffolk NY	12	26	27	27	27	29
Nassau NY	20	47	65	65	65	66
Oakland CA	14	24	24	25	38	38
St Louis MO	15	20	41	84	98	95
Orange CA	9	13	13	13	22	25

Source: **J.C. Teaford (1997)** *Post-Suburbia: Government and Politics in the Edge Cities* **Baltimore, MD: Johns Hopkins University Press**

BOX 20.2

Segregated education

The belief that mixed or multi-racial schooling may have a regressive effect on white children led to several means of preserving the white public school system. These included:

1. The provision of 'separate but equal' school systems.
2. Use of a 'neighbourhood schooling' policy and delimitation of catchments which, given the residential segregation of blacks, ensured that most black or white children attended a school dominated by their own race.
3. The creation of separate suburban school districts whose racial and social character could be controlled via the zoning mechanism.

The first of these strategies was employed widely in the southern states until it was successfully challenged in the case of *Brown* v. *Board of Education of Topeka* (1954). Although this decision required school boards to remedy the situation as soon as practicable, there was considerable 'playing for time', and it was not until 1971, in the case of *Swann* v. *Charlotte-Mecklenburg Board of Education*, that bussing was ruled necessary to achieve desegregation in areas with little spatial mixing of the races.

The two other strategies have been employed, particularly in northern metropolitan areas. The diagram illustrates ways in which the combination of

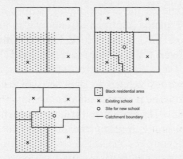

Black residential area
× Existing school
○ Site for new school
— Catchment boundary

the neighbourhood schooling principle and judicious arrangement of catchments can affect the racial composition of school areas. Actions to eradicate such practices have generally been brought under the Fourteenth Amendment clause, which meant that plaintiffs must prove not only that such practices were in existence but that they had been enacted with intent to discriminate on racial grounds. The

first important decision was in 1973, when segregation in part of Denver was found to be a deliberate result of school board policies Subsequent decisions of the Supreme Court have ruled unconstitutional many of the local managerial practices that had been operated to segregate black and white pupils.

The third strategy, the creation of special school districts, does not infringe the US constitution. Thus, if a school district is ordered to desegregate, parents who do not wish their children to attend integrated schools can either transfer them to the private sector or migrate to another school district. The resulting central-city-to-suburbs movement ('white flight') means that racial integration in metropolitan schools can be effected only by amalgamating separate school districts. This was the basis of the plaintiffs' argument in the case of *Millicken* v. *Bradley* (1974), which referred to the Detroit metropolitan area. The court ruled against such a solution, however, deciding that local autonomy was more important than its segregation consequences. The Supreme Court's stance against inter-district solutions has protected the position of those involved in white flight to the suburbs in order to avoid integrated schools. The court has also declined to intervene in regard to inter-district disparities in school financing. This can result in major discrepancies in the quality of education received, since there is ample evidence that white middle-class pupils receive higher-quality teachers and more capital equipment than schools with lower-class and/or black enrolments.

Debate over racially integrated schooling was reignited in 1999 by evidence from the US Department of Education that the level of integration had fallen below that achieved a quarter of a century earlier. A number of major school districts are in the process of ending their desegregation plans, including Boston, Buffalo, Fort Lauderdale, Jacksonville, Las Vegas, Minneapolis, Cleveland, San José, Oklahoma City and Seattle. It is argued that since the *Brown* v. *Board of Education* ruling of 1954 social, demographic and ethnic changes in US cities (with, for example, a concentration of ethnic minorities in inner cities) make it increasingly difficult to retain a significant proportion of white children in some schools. For some commentators, whereas the initial desegregation policy was appropriate to the time, the limits of what can be achieved by judicially mandated 'social engineering' of school catchments may well have been reached.

Source: **part from R. Johnson (1984)** *Residential Segregation, the State and Constitutional Conflict in American Urban Areas* **London: Academic Press**

shopping centre or factory was to incorporate in
advance of annexation by a neighbouring municipality.
This led to some remarkable incorporation struggles in
the immediate post-war era (Box 20.3).

BENEFITS OF MUNICIPAL INCORPORATION

Once incorporated, a municipality acquires statutory
powers which convey a number of advantages to residents. These include the following:

1. Residents must pay taxes only to cover the
   costs of services within their own municipality.
   Affluent suburban residents can therefore avoid
   sharing the costs of social consumption for the
   inner-city poor.
2. Residents of independent municipalities can insulate themselves from the political activities of larger
   places in which their interests may be outvoted.

3. The land-use planning powers of municipalities
   enable residents to determine the socio-economic
   composition of the area and to exclude undesirable
   land uses. Poor households can be kept out through
   exclusionary zoning (Box 20.4). Figure 20.3
   depicts the political geography of the Denver
   Standard Metropolitan Statistical Area (SMSA)
   and illustrates the impact of differential zoning.
   Small exclusive residential suburbs inhabited by
   high-income families and zoned for a single dominant land use are represented by Bow Mar,
   Columbine Valley and Cherry Hills. Most of the
   latter is zoned for a minimum building lot of two
   and a half acres (1 ha) and very little of the municipality is zoned for a density of less than an acre
   (0.4 ha) per house. However, not all exclusively
   residential municipalities are zoned for such low
   densities and therefore high-income residents.
   Glendale is zoned almost entirely for apartments.
   Federal Heights has large areas zoned for mobile
   homes, and Mountain View is occupied by
   medium-income groups (with an average dwelling
   value one-fifth of that in Cherry Hills). The larger
   middle-income suburbs to the west of Denver seek
   to attract, rather than repel, non-residential land
   uses in order to use their contributions to the property-tax base to hold down residents' tax bills and
   subsidise the provision of municipal services. In
   this they are competing for property-tax revenue
   both with each other and with the central city.

NEGATIVE CONSEQUENCES OF FRAGMENTED GOVERNMENT

The problems posed by a fragmented local government
structure are most evident in the USA, where the diversity of local jurisdictions and plethora of intergovernmental relationships forms a complex web. One consequence
of this political balkanisation is the development of major
fiscal disparities between different local governments,
and in particular between central city and suburban
authorities. The fiscal crisis of the core cities stems from:

1. Increasing demand for urban services, including
   the provision of social welfare.
2. A rise in overall urban expenditure due to an ageing infrastructure and need to service the central
   business district (CBD).
3. Mounting opposition to urban taxation.
4. The inability of most cities to tap the economic
   resources of the whole metropolitan area.

BOX 20.4

Legal challenges to exclusionary zoning

Most legal challenges to *exclusionary zoning* in the USA have been brought under the 'equal protection clause' or Fourteenth Amendment of the constitution. In general, however, the Supreme Court has refused to hear such cases, arguing that zoning is a right granted to local governments by *state* constitutions and not by the federal constitution. The court has indicated that it is prepared to intervene where legislation contradicts 'fundamental rights' or where it has drawn 'suspect classifications' between sets of individuals. Fundamental rights are those explicitly or implicitly protected by the US constitution. Explicit rights include freedom of assembly, speech and religion. Examples of implicit rights are the right to travel and to privacy. The major disputes have centred on the range of implicit rights. The case of *Lindsey* v. *Normet* (1971) found that decent housing was not a constitutional right, and generally the Supreme Court has adopted a relatively narrow definition of fundamental rights. The Supreme Court's decisions on 'suspect classifications' have also worked against those seeking access to suburban housing. In cases such as *James* v. *Valtierra* (1971) the court has refused to accept wealth as a suspect classification and has dismissed claims that discrimination by wealth is *de facto* racial discrimination.

The difficulty of seeking redress by means of the Fourteenth Amendment was emphasised by the finding in *Arlington Heights* v. *Metropolitan Housing Authority* (1977) that 'proof of a racially discriminatory motive, purpose, or intent is required to find a violation of the Fourteenth Amendment' – in other words, differential *impact* is insufficient evidence of *intent*. One further hurdle, demonstrated in the *Warth* v. *Seldin* (1975) judgement, restricts resort to the Court to those individuals and corporate bodies able to prove personal injury as a result of the claimed discriminatory acts. The purpose of the 'standing' doctrine is to ensure that the courts are not bogged down with frivolous suits, but the effect has been to exclude third parties. In the Warth case a number of plaintiffs alleged that the zoning ordinance of Penfield NY excluded persons of low and moderate income from the town. However, they personally had not been rejected by the town in applying for a building permit, and the developers whose rights might have been violated had not applied for building permission because on past evidence they 'knew' the application would be denied. In general, therefore, the 'voice' strategy via the Supreme Court has failed to alter the social geography of the suburban USA.

An alternative source of redress for non-resident black plaintiffs denied low-income housing in suburban municipalities is the 1968 Fair Housing Act. The important difference between the standards of proof of discrimination under the Fourteenth Amendment and the Fair Housing Act is documented by Clark (1981).[14] Most significantly, whereas the former requires evidence of *intent*, the latter requires only a demonstration of discriminatory *effect*. However, although the 1968 Fair Housing Act outlawed discriminatory practices, it has achieved little 'opening up of the suburbs'. Eliminating discrimination does nothing to overcome the basic economic inability of African-Americans to afford suburban housing. If the aim is integration, then positive action to provide low-income housing is required. Some progress in this direction was made by the 1974 Housing and Community Development Act. This made the granting of community development block grants contingent upon a community submitting an acceptable Housing Assistance Plan (HAP) which assessed the needs of low-income persons residing or expecting to reside in the community as a result of existing or projected employment opportunities.[15] Clearly, though, small affluent municipalities can decide to forgo federal aid rather than meet these conditions.

More successful challenges to exclusionary zoning have been made in the state courts, which may interpret their own constitutions differently from Supreme Court interpretations of the federal constitution. Pennsylvania rejected exclusionary zoning in a series of cases between 1965 and 1975 and, in the *Mount Laurel* decision (1975), New Jersey required expanding communities to accept their 'fair share' of regional low and moderate-income housing needs. Legislation to counter the effects of exclusionary zoning has also been passed in New York and Massachusetts, although, as elsewhere, dogged resistance by suburbanites has reduced its impact. Between 1969 and 1979 only 3,600 low and medium-income units were actually built under the 'anti-snob zoning' law in Massachusetts, with, significantly, most designed for the elderly as opposed to family housing. A negative 'stick' approach such as the threat of state intervention clearly is not sufficient to overcome resistance. Positive 'carrots' such as state or federal guarantees to cover the local costs of servicing such developments, thereby relieving pressure on the municipal property tax, may be necessary. Such strategies, however, will not overcome racial prejudice or fears over possible reduction in the exchange value of middle-class housing in the affected suburbs.

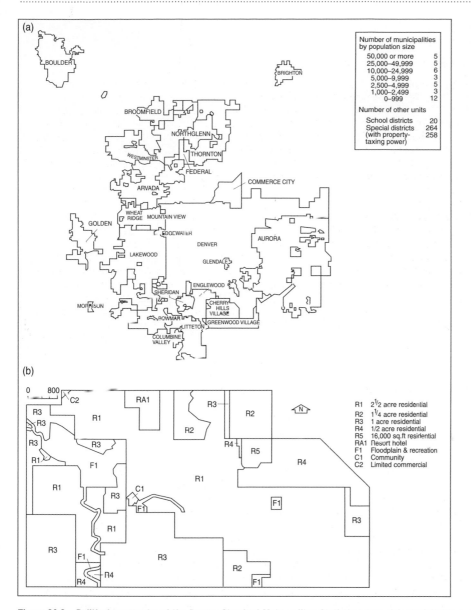

Figure 20.3   Political geography of the Denver Standard Metropolitan Statistical Area: (a) municipalities of the Denver urban area and (b) land-use zoning in Cherry Hills

Source: R. Johnston (1984) *Residential Segregation: The State and Constitutional Conflict in American Urban Areas* London: Academic Press

As we saw in Chapter 17, at the heart of the central-city–suburban fiscal disparity is the debate over the extent to which suburban residents benefit from city services while making little contribution to their costs. That such debate is not confined to the metropolitan areas of the USA is evident from Box 20.5.

Other difficulties that arise from the fragmented structure of US local government include the following:

1. Voters have little control over the special district, the most widespread unit of local government.
2. Many ostensibly local issues such as land-use decisions can have area-wide effects, as in the spillover of pollution from one jurisdiction.
3. No single entity has the power to exercise effective leadership, making it difficult to plan for the metropolitan region as a whole.

Plate 20.1 The affluent residential suburban municipality of Cherry Hills to the south of Denver CO is occupied by high-income households and zoned exclusively for low-density residential use

BOX 20.5

The central-city–suburban fiscal debate in the Glasgow region

Glaswegians are facing a Band D council tax of £1,057 in two years' time while residents in neighbouring East Renfrewshire and East Dunbartonshire at the same banding will have to fork out less than £500, according to a new study. The staggering differential of up to 131 per cent being projected is bound to fuel fresh demands for the city's constricted boundaries and tax base to be re-examined before such inequities are embedded in the system.

Already there are demands from Glasgow and other councils representing areas of deprivation that the Scottish Office formulas for grant distribution should be reviewed to take greater account of social need. Senior Glasgow councillor Des McNulty, who has recently been urging the government to make additional financial arrangements to support the city's metropolitan services, said the projected variance was 'frightening'. He declared, 'The situation that Glasgow finds itself in cannot be left any longer. We are looking for urgent responses from the present government, and we'll be looking for an urgent response from any new government.' However, he emphasised the situation should not mean a fight among local authorities. In his view it was a fight between local government and central government.

Also anxious that the issue does not develop into a feud is Councillor Charles Kennedy, leader of East Dunbartonshire, but he is not prepared to envisage a permanent subsidy for his larger neighbour. Pointing out that many authorities, including his own, had agreed, with varying degrees of reluctance, to help ease the transition for Glasgow and others, he said it was on the understanding that this would end after three years. 'Glasgow has a problem,' he said. 'But they don't seem to recognise the fact that we are not here to be net contributors to Glasgow's problems. We have difficulties of our own.'

Source: The Herald 5 November 1996

THE POLITICS OF SECESSION

Secession tendencies infiltrate the fragmented structure of local government in the USA. For some, the politics of secession privileges private values over collective social values. Citizens with the necessary resources can choose to move from one jurisdiction to another that suits their ideals. This is manifested most visibly in 'privatopian' housing developments separated from the rest of the world by walls and by privatised services and methods of governance. The home-owner associations that govern these CIDs (see Chapter 18), essentially are private governments wielding substantial powers but without the constitutional rights and responsibilities normally associated with municipal government. Increasingly, urban politics in the USA involves a struggle between separate places rather than a struggle between groups and individuals within places. In some instances this involves moves by whole communities to secede from the structure of local government.

In Los Angeles, since the early 1990s, districts such as the San Fernando valley in the north and the San Pedro/Wilmington area in the south have sought to secede from the city. These movements have generated opposing interpretations. Secessionists argue in the tradition of US rights-based liberalism that citizens (i.e. property-owners and tax-payers) are entitled to rule their own affairs on as local a scale as possible. Opponents characterise secession as a class-based (and strongly racialised) movement of social separation expressed in political terms that seeks to avoid responsibility for poorer neighbouring communities.[16]

For urban geographers, the secession debate highlights the spatial dimension of urban politics. A local-scale example of a 'secession of the successful' from the larger urban polity concerned an attempt by retirees in the Sun City West CID in north Phoenix to de-annex themselves from the local school district to avoid paying taxes to support a service they do not use.[17] This conflict between Anglo middle-class senior citizens and neighbouring working-class Latino communities underlined the social consequences of the political fragmentation of urban space and, for some, supports the case for metropolitan government.

METROPOLITAN GOVERNMENT

The main strategies proposed to tackle political fragmentation are:

1. Consolidation of local governments.
2. Transfer of functions to higher-level governments.
3. Creation of special-purpose authorities.
4. Provision of services through inter-municipal agreements.
5. Establishing a two-tier system of metropolitan government (Box 20.6).

The functional advantages of consolidating metropolitan government into a single authority were highlighted in the USA by the Advisory Commission on Intergovernmental Relations (1974).[18] However, the traditional power of municipal authorities and antipathy to 'big government' largely prevented implementation of this strategy. Paradoxically, a two-tier system of metropolitan government was introduced in Britain in 1975, only to be dismantled in 1996 with the introduction of unitary authorities for most urban areas.

In the US federal government requirements or incentives have persuaded most metropolitan statistical areas (MSAs) to establish some form of metropolitan-wide planning organisation. These 'councils of governments' (e.g. the Association of Bay Area Governments) are strictly voluntary and advisory, and local governments are not legally required to follow their recommendations. A more ambitious attempt at metropolitan area government is represented by the two-tier structure established in Toronto in 1953, in which the metropolitan-wide first tier, Metro, is governed by representatives from the pre-existing local-government units. The long-term existence of Metro is testimony to its functional utility. A growing problem, however, is that the built-up area of the metropolis has spilled over the Metro boundary, which has remained unchanged owing to opposition from the City of Toronto (for which such an enlargement would be unacceptably centralist) and from the provincial government (for which enlargement would be unacceptably decentralist). In 1998 the seven existing local-government units within Metro Toronto were amalgamated into a single megacity of 2.4 million people. While the structure of local government in the remainder of the Greater Toronto Area (GTA) was left unchanged, the provincial government recommended the formation of a Greater Toronto Services Board to develop an infrastructural co-ordination strategy to guide urban growth and management across the GTA. In practice this body, with no revenue-raising powers and an executive split between city and suburban authorities, each with their own goals, is unlikely to be able to effect resource redistribution within the GTA. Furthermore, the continuing suburbanisation of population and the emergence of a multi-nodal metropolis are likely to

BOX 20.6

Strategies to resolve political fragmentation

There are several ways of dealing with a fragmented local government structure:

1. *Consolidated government.* The creation of a single unit of government for a metropolitan area eliminates political fragmentation. The strategy is difficult to implement, owing to opposition from local politicians and residents There is a danger that the needs and interests of localities within the metropolis may not be well served.

2. *Transfer of functions.* Public service pressure on local finances is reduced, allowing municipalities to provide fewer but better services. Such transfers weaken local government by making it less important in public service provision, and there is no guarantee that functions transferred will be organised effectively for the metropolitan area.

3. *Special-purpose authorities.* These can reduce problems related to the efficiency and effectiveness of public services, since the size and configuration of authorities can be tailored to functional requirements. Providing special-purpose authorities does add to the number of government bodies in a metropolitan area and may distance functions from community control, as well as making co-ordination between functions difficult.

4. *Inter-municipal co-operation.* Co-operation can reduce negative spillover effects and preserve municipal autonomy and identity but can do little to address the problem of fiscal disparities. A major flaw is the voluntary basis of co-operation, which tends to be between municipalities of similar social, economic, cultural and fiscal character.

5. *Two-tier government.* Spillovers and boundary problems are reduced and scale advantages gained by an area-wide upper tier of government, while small-scale democracy and local control can be retained and local needs met by the lower-tier units of government. The main difficulty lies in implementation and, in particular, in ensuring a good working relationship between the two levels of government.

Source: **I. Barlow (1991)** *Metropolitan Government*
London: Routledge

enhance the electoral power of the outer municipalities in provincial policy-making and increase the social and financial disparities between inner and outer zones of the GTA.[19]

A more successful example of resource redistribution within a metropolitan region is the 189-municipality Minneapolis–St Paul MN Twin Cities area, which has created an inner-city–suburban tax-sharing mechanism, the Fiscal Disparities Plan (FDP), under which 40 per cent of all increases in commercial and industrial property taxes are placed in an area-wide FDP fund. The annual fund is distributed to each of the municipalities on the basis of their population and the ratio between the *per capita* valuation of property within each jurisdiction and the *per capita* valuation of property in the Twin Cities region as a whole. In 1991 the FDP fund represented $290.5 million, over 30 per cent of the region's assessed property valuation, and resulted in tax transfers to 157 jurisdictions and net contributions from thirty-one, thereby reducing the gap between rich and poor municipalities within the region.

A two-tier form of metropolitan government was introduced in Dade County, encompassing Miami, in 1957. The comprehensive urban county plan gives the county government a powerful and integrated role over an area of 2,054 square miles (5,300 km³) and twenty-seven municipalities. A similar structure is the Unigov administration, introduced in 1969 for the city of Indianapolis and the outlying areas of Marion County IN.

The differing objectives, institutional forms, tactics and outcomes of inter-jurisdictional co-operation in the USA are illustrated, conceptually and in relation to particular examples, in Figure 20.4. Thus the Metro Denver Network, a coalition of public and private organisations in the six-county Denver region, has economic development as a desired outcome. It seeks to attain this by reducing competition among member jurisdictions (socio-political change) through strategies characterised by low formality and local autonomy, and which engender low political resistance. On the other hand, the Louisville–Jefferson County Compact established in 1985 sought to address issues of mutual gain and redistribution by pursuing fiscal equalisation within the region. These goals clearly affect the autonomy of local jurisdictions. The political feasibility or 'achievability' of each objective is also indicated in Figure 20.4. Unsurprisingly, resource redistribution is the most difficult to attain. In general, opposition to greater transfer of power to metropolitan government emanates from those in higher-tier government, who rarely wish to have their authority challenged by a major urban region,

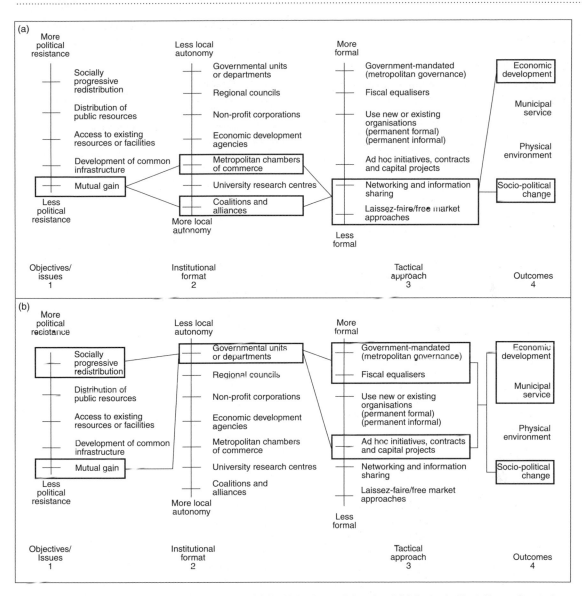

Figure 20.4   Forms of inter-jurisdictional co-operation: (a) the Metro Denver Network and (b) the Louisville–Jefferson County Compact

and from lower-level municipal governments that guard their powers and privileges jealously[20] (Box 20.7).

From a broader perspective metropolitan governance projects in US city regions may be viewed not only as efforts to reconnect central cities with their suburbs but as a response to the myriad of economic, social, political and spatial forces that have influenced urban development over recent decades. Table 20.4 indicates the main consequences for cities and regions of three principal restructuring processes and identifies

resultant problems and putative solutions employed in various metropolitan contexts. This framework also illustrates the heterogeneity of metropolitan regionalist projects. It also reveals the wide range of actors engaged in attempts to reconfigure metropolitan governance, that includes political elites, national and local state agencies, major local and regional firms, local chambers of commerce, organised labour, environmental movements and diverse locally based citizen groups.[21]

BOX 20.7

Metropolitanism versus municipalism

The battle between metropolitanism in the shape of county government and municipalism raged fiercely in St Louis County MO.

In 1980 40 per cent of St Louis County was unincorporated territory and therefore taxable by the county, which, following a Missouri Supreme Court ruling of 1963, could block any annexation. In 1983 this 'Graeler decision' was overturned, and municipalities were not slow to see the opportunity to enhance their tax base at the county's expense. Threats of annexation also stimulated defensive incorporation, which again eroded the county's tax base. Thus the municipality of Bridgeton sought to annex the Riverport and Earth City commercial areas in the unincorporated territory of Maryland Heights, while Overton and Creve Coeur both wanted to annex the West Port Plaza shopping centre from Maryland Heights. Consequently, in 1984 Maryland Heights residents voted to become a municipality of twenty square miles (52 km^2) and 26,000 inhabitants, thereby depriving the county of $8 million of annual tax receipts.

The renewal of annexation and incorporation struggles prompted a metropolitan citizens' group (Confluence St Louis) to propose the elimination of all unincorporated territory in the county, and allowing the county government to serve as a regional government. The proposal was taken up by the County Planning Department, which recommended the rationalisation of existing local government units from ninety municipalities to twenty-one cities. This was opposed vehemently by the St Louis County Municipal League, which represented seventy-seven cities and villages. The final plan to emerge in 1988 proposed a reduction from ninety municipalities to thirty-seven, with enhanced powers for the county to produce a regional land-use plan enforceable on the municipalities. This ignited furore among municipal representatives, but before the proposal could be voted on it was rejected by the US Supreme Court on the grounds that the process of developing the plan had been a violation of the Fourteenth Amendment, since it had restricted membership of the planning board to property owners.

Municipalism had triumphed over metropolitanism. Once again the residents of 'post-suburbia' had demonstrated their devotion to a suburban ideal and aversion to the centralisation associated with the big city.

Source: J. Teaford (1997) Post-suburbia Government and Politics in the Edge Cities Baltimore MD: Johns Hopkins University Press

POWER IN THE CITY

So far our discussion has focused on the structure of urban government and conflicts between different jurisdictions. Power is also distributed unevenly within cities, where a power struggle is waged between a host of formal and informal interest groups, each seeking to influence the nature and actions of the local state. A useful distinction is between agents or interest groups that operate within the prevailing government structure, including both informal influences and formal 'top-down' public participation strategies, and more radical 'bottom up' pressure groups or urban social movements.

INFORMAL INFLUENCE

Informal networks of influence operate outside the formal mechanisms of representative democracy to channel the interests of particular sections of the local population to policy-makers. Dunleavy (1980)[22] identified three areas in which informal influence networks are significant:

1. In the UK local party organisations play an important role in linking council groups with party-affiliated interests. This occurs via formal party structures (e.g. trade-union involvement in the Labour Party), through overlapping memberships (such as between the Conservative Party and the local chamber of commerce) or via the social organisations (e.g. political clubs) associated with local parties. All these serve to structure access to local political elites. Given a situation of one-party control in urban local government, this can result in a real definition of 'insider' and 'outsider' interest groups in particular localities. In most US municipalities the electoral system is non-partisan, and political parties are relatively unimportant at the local level. Interest-group political activity is of greater significance. Citizen

TABLE 20.4 METROPOLITAN RESTRUCTURING PROCESSES IN THE USA

Restructuring and rescaling process	Consequences for cities and regions	Resultant governance problems	Putative metropolitan solutions	Examples
The spatial reconstruction of urban form: the deconcentration and reconcentration of metropolitan settlement spaces and production complexes	The rise of the edge city and the exopolis Intensified metropolitan jurisdictional fragmentation Continued population dispersal and industrial deconcentration The spreading of urban problems into suburban zones Urban sprawl	Spatial mismatch between public resources and social needs Inefficient delivery of public services Increased spatial concentration of poverty and minority populations in city cores Severe traffic congestion Environmental destruction	*Regional growth and environmental management:* Comprehensive regional land-use planning Smart growth projects Construction of regional or metropolitan growth boundaries New forms of environmental legislation	Various state laws passed in New Jersey, Oregon, Florida, Maryland and California State-level approval of new planning powers for metropolitan agencies in Portland, Atlanta, and Minneapolis-St Paul
Global economic restructuring: the globalisation, (re)territorialisation, and localisation of various factions of capital	Processes of de- and re-industrialisation and the shift towards lean production Intensified inter-urban competition for mobile capital investment at regional, national, continental and global scales	Capital flight, unemployment and derelict industrial sites Deskilling of local labour supplies Decay of local industrial infrastructure Increased local fiscal constraints and declining tax revenues from locally collected taxes	*Regional economic development policies:* Co-ordinated regional industrial and labour policy Regional investments in human capital and infrastructure Regional land-use planning Regional place-marketing	Bay Area Council (San Francisco) Metropolis Project (Chicago) Greater Philadelphia First Greater Houston Partnership
Neo-liberal state restructuring: the 'destructuring' and reconstitution of state policies coupled with the upscaling and downscaling of state functions	Federal devolution. Lean government, entrepreneurial states and revanchist cities Intensified city-suburban fiscal disparities The shift from welfare to work Increased class- and race-based socio-spatial polarisation Ghettoisation of poverty	Local fiscal crises Lack of funding for key social services: affordable housing, schools, public transport, infrastructure improvements, etc. Expansion of repressive functions of the local state, (policing, prisons) Social unrest (e.g. riots)	*Regional revenue-sharing and redistributive arrangements:* Tax-base and revenue-sharing measures Region-wide provision for low-income housing, public transit and infrastructural development Regional enforcement of laws against racial discrimination in the housing market	Tax-based sharing schemes in Minneapolis-St Paul Regional asset districts in Denver Efforts to introduce regional 'fair share' low-income housing in Minneapolis-St Paul

Source: adapted from N. Brenner (2002) *Decoding the newest 'metropolitan regionalism'* in the *USA Cities* 19(1), 3–21

groups are often formed to defend neighbourhood space against unwanted intrusion. Home-owner or rate-payer associations campaign for lower taxes, while, as we have seen in our discussions of growth coalitions and urban regimes, business groups are among the most active and most influential agents at the local level.

2. Patronage represents the second main line of informal influence. In most localities it is possible to identify a 'burgher community' composed of major party council members, some party activists and members of partly affiliated interests, cemented together by a middle-class bias and overlapping membership of local government and other state agencies (e.g. regional and area health authorities). The influence of this community extends widely through the social structure because of the prevalence of political nominations for positions such as justice of the peace (magistrate) and membership of school management boards. In addition to bodies on which politically affiliated elites are formally specified, the influence of the burgher community can extend to other groups such as voluntary welfare agencies, which are often dependent on council grants and hence keen to please on policy and appointments.

3. Some degree of corruption (e.g. favours for friends in council-house allocation or planning permission) is inevitable, and at an individual level does not have a great impact on public policy. A more fundamental threat to the legitimacy of local government is posed by the direct economic exploitation of the system by large corporations (e.g. in the determination of land and property values and the construction of public infrastructure) with corruption networks often being organised by local politicians.[23] In the city of Birmingham, for example, a regional building firm maintained a 'Christmas list' of 2,000 local-authority employees, including a corrupt city architect, and in the period 1966–73 received £110 million worth of public housing work.

Political patronage and corruption were fused in the system of political machines and bosses that ran many US cities in the nineteenth and early twentieth centuries. The machine was an informal structure of influence and power that mobilised votes and distributed the benefits of office to supporters. The city boss was the head of a hierarchical organisation that reached down to ward and precinct levels, where captains were responsible for organising the vote. In the context of the nineteenth-century urban USA this also functioned as an informal welfare system, providing immigrants with jobs and shelter in return for loyalty and disciplined voting. Although the heyday of machine politics in the USA was in the nineteenth century, a powerful political machine functioned in Chicago until the mid-1970s (Box 20.8).

VOTING

The election of public officials through voting ostensibly represents a direct link between public preferences and public policy. In democratic societies all citizens have equal voting rights. This does not mean that all citizens have an equal voice in deciding public policy. Many citizens fail to exercise their right to vote in local elections or referendums. Low voter turnout is a feature of urban politics in the UK and the USA. On average, 30 per cent of those eligible vote in US municipal elections, and fewer in referendums,[24] while the corresponding figure for local elections in Britain is 40 per cent.[25] The sensitivity of the political system to the interests of some sections of society is reduced further because non-voters are not distributed randomly throughout the population. In general, women vote less than men, young people and the retired are less likely to go to the polls than those in the intervening age group, people with lower incomes and lower educational qualifications tend to vote less than the affluent and well educated, and recent immigrants to an area are less inclined to vote in local elections than long-term residents.[26]

Bias may also be introduced into urban politics by structural manipulation of the geographical boundaries of constituencies. The two main methods of engineering electoral injustices are by **malapportionment** and **gerrymandering**. Electoral malapportionment is the deliberate creation of constituencies of different sizes for the benefit of one party. The English 'rotten boroughs' of the early nineteenth century, where the MPs occasionally outnumbered the voters, were the most notorious examples. As the hypothetical situation in Figure 20.5 illustrates, deliberate malapportionment involves devaluing an opponent's votes by creating constituencies of above average size in areas where your opponent has an electoral majority and below-average constituencies where you are powerful, thus maximising your opponent's 'excess' votes.[27]

Gerrymandering has been a relatively common strategy to bias the democratic process in favour of a ruling party. Busteed (1975) has noted how it has been used to

BOX 20.8

The Daley machine in Chicago, 1955–76

Mayor Richard J. Daley came from a lower-class Irish immigrant family. During the 1950s, when machines in other cities were in decline, Daley revitalised the Chicago machine. As central boss, Daley spent part of each working day in activities designed to aid his re-election and ensure machine control. A city official unique to Chicago was the 'director of patronage', who maintained a list of all city employees, including their political sponsor. Mayor Daley had personal control over 25,000 jobs. The Daley machine was largely responsible for reviving Chicago's central business district (CBD) in an era of suburban decentralisation by enticing large-scale businesses into the city, including the headquarters of the Sears Corporation. The machine also effectively blocked dispersal of low-income racially segregated housing, believing that ethnic concentrations provided a controllable voting bloc.

Daley's political machine began to decline before his death in 1976, owing to the following factors:

1. The increased scope of government meant that fewer people depended on city machines for favours and rewards as government began supplying welfare services. Also, party machines lost control of many benefits as merit-based bureaucracy undermined patronage jobs.
2. Competing institutions grew in strength, including labour unions and single-issue pressure groups.
3. Business interests no longer found the machine useful, since the growth of federal government meant that they could deal direct with central agencies.

In 1983 the anti-machine (black) candidate Harold Washington was elected mayor. However, reports of the death of the machine may have been premature, since, following Mayor Washington's death, Daley's son, Richard M. Daley, was elected to office.

Source: B. Phillips (1996) City Lights New York: Oxford University Press

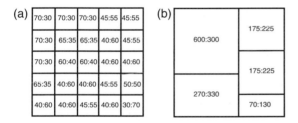

Figure 20.5 Hypothetical example of malapportionment showing (a) the distribution of votes for parties X and Y in twenty-five areas and (b) a set of five constituencies designed to maximise seat winning for party Y

Source: R. Johnston (1979) Political, Electoral and Spatial Systems Oxford: Oxford University Press

dilute Catholic political strength in Londonderry,[28] and Figure 20.6 provides a graphic illustration from a 1971 proposal by the majority Democrats for a reapportionment of electoral district boundaries in Philadelphia. Although each district conforms well to the equal population criterion, the irregular geography was designed to produce a partisan electoral result. For example, district 196 in the north of the city runs for four and a half miles (7.2 km) in the clear intent to submerge the Republican vote of the inner suburbs beneath the Democratic strength in the inner city. Perhaps the most blatant example of partisan cartography was the attempt by the town of Tuskegee AL to disenfranchise black voters by rearranging the town boundaries from a square shape to a twenty-eight-sided figure. Some US cities, such as Detroit MI, Miami FL and Seattle WA, employ an at-large voting system to elect council members, but a majority, including New York, Los Angeles CA and Chicago IL, use single-member districts that open up at least the possibility of gerrymandering of constituency boundaries.[29]

The main problem with gerrymanders is that whereas the result is usually apparent, the intent is difficult to prove. The Tuskegee case apart, legal challenges to electoral district boundaries on grounds of racial vote dilution have frequently failed because plaintiffs were unable to show evidence of discriminatory intent on the part of the redistricting authorities, and because no accepted measure of gerrymandering has been developed. One possible solution is to remove the districting process from the political arena and entrust the task to an independent non-political body, as in the UK. However, partisan solutions can still arise in two ways. The first entails a process of benign neglect or 'silent gerrymandering' under which no action is taken on the committee's recommendations

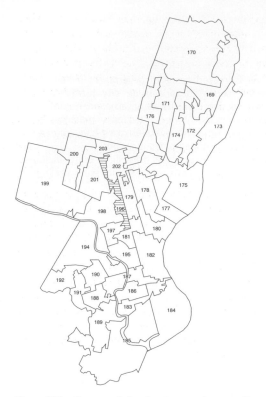

Figure 20.6   Gerrymandering in a proposed reapportionment of electoral district boundaries in Philadelphia
Source: D. Ley (1983) *A Social Geography of the City* New York: Harper & Row

if redistricting would be disadvantageous to the party in power. The second stems from the fact that although the neutrality of the bodies such as the English Boundary Commission is beyond doubt, the rules under which they operate are politically defined and may introduce biases into the procedure. Even the use of 'scientific' algorithms does not exclude subjective influences.[30] Computer-based redistricting systems are powerful tools for designing a set of feasible redistricting plans, but they do not resolve the question of whether they are acceptable politically.[31]

PUBLIC PARTICIPATION

In most Western nations the twin processes of globalisation and professionalisation of decision-making have reduced the power of local electorates, which has in turn led to calls for public participation and local democracy. The basic dilemma stems from the perpetual tension between the need for effective

administration and the need for maximum accountability. While the former invokes centralist tendencies in urban government, the latter demands greater decentralisation. Proponents of citizen involvement in government decision-making envisage the concept being operationalised in the democratisation of resource-allocation strategies, decentralisation of service-systems management, deprofessionalisation of bureaucracies, and demystification of planning and investment decisions. These goals are at the same time anathema to many elected officials who question the value of participation.

A key question in the participation debate is the amount of power that is devolved to citizens. Several classifications of the power relationship have been advanced, including Arnstein's (1969) ladder of citizen participation (Figure 20.7).[32] The eight rungs of the ladder can be grouped into three broad categories relating to:

1. Non-participation forms that merely give an illusion of power.
2. Degrees of tokenism based on one-way 'top down' strategies of informing, consulting and placating.
3. Degrees of real citizen power.

PARTICIPATION STRATEGIES

Many of the conventional methods utilised have advanced the cause of citizen participation only slightly, being more concerned with disseminating information to the public. Such techniques include

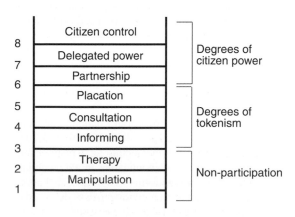

Figure 20.7   Arnstein's ladder of citizen participation
Source: S. Arnstein (1969) A ladder of citizen participation *Journal of the American Institute of Planners* 35, 216–24

exhibitions, public meetings, publication of surveys and reports, media publicity, ideas competitions, referendums and public inquiries.[33] Approaches with more potential for citizen involvement include area management and advocacy planning.

THE NEIGHBOURHOOD APPROACH TO URBAN MANAGEMENT

Underlying the neighbourhood approach is the notion of public learning, whereby local governments develop an enhanced ability to perceive and understand the interrelated and changing nature of problems in a community (problem recognition), as well as an adaptive capacity to respond to these changing difficulties in a timely and effective manner (problem response). In relation to this concept, the neighbourhood area approach has the twin aims of (1) bringing local government closer to the people, and (2) tuning actions to the needs of particular areas within a city. In short:

> area approaches involve gearing the planning and/or management of policies to the needs of particular geographical areas within the local authority and may involve delegating administrative and/or political responsibility for at least part of this work to the local level.[34]

A number of local authorities in Britain, Western Europe and the USA have introduced area-management schemes.[35] The 'little city halls' programme in Boston was built on the concept that Boston was a city of ethnically and culturally diverse neighbourhoods – with, for example, the North End and East Boston being loci of the Italian community, Roxbury of the black population and Charleston mainly Irish. Seventeen little city halls were established throughout the city. The main goals of the programme were to:

1. Provide a range of direct services and information to the public.
2. Respond to claimants.
3. Act as a catalyst for increased citizen participation.
4. Improve delivery of city services at the local level.

On balance, assessments of the programme were favourable,[36] a chief success being improved public access to city hall by providing a concerned local information and complaint service combined with some direct service provision. Drawbacks of the scheme included a tendency for political patronage to influence staff appointments and operational backlogs in dealing with complaints. A more fundamental criticism of an area-management approach is that 'the model is largely a liberal response to radical demands for greater power which tries to mellow dissidents with formal, procedural, and voluntaristic change while neglecting to make any significant changes in the allocation of power or authority'[37] (Nordlinger and Hardy 1972 p. 372). A graphic indication of continuing dissatisfaction on the part of some minority groups in Boston was a 1986 referendum on a proposal that the black Roxbury neighbourhood should secede from the city. Similar criticisms may be directed at the system of community councils introduced in British cities as a means of ascertaining and conveying the views of local communities to relevant public bodies. Interpreted in pluralist terms, community councils represent a widening of political accessibility, but this is taking place within constraints set from above. A structuralist perspective would emphasise the lack of any real delegation of power to the local level and view the scheme as a useful means of social control by which grass-roots pressure for social change may be diverted into a concern with social justice.[38]

ADVOCACY PLANNING

The roots of advocacy planning have been traced to the growing rate of delinquency and vandalism in the USA city of the late 1950s, which several analysts attributed to citizen alienation and anomie. For Cloward and Ohlin (1960),[39] anomie resulted from the disparity perceived by low-income youth between their legitimate aspirations and the social, economic, political and educational opportunities made available to them by society. The solution lay in reforming society with the involvement and participation of low-income citizens, and the major agent for this change was seen as the planning professional. When urban riots erupted in the 1960s these arguments were adopted by government and translated into the 1964 Economic Opportunity Act, which embraced the principle of 'maximum feasible participation'.[40]

In essence, advocacy planning means the provision of professional planning expertise and services to minority groups lacking the financial ability to purchase such services. The concept was popularised by Davidoff (1965), who envisaged a plural form of planning (as opposed to, for example, a unitary master plan) in which all the special interests of society would be represented by planners who meet 'in the political arena' with their clients' proposals so that the best plan would emerge.[41] This perspective recognises that 'the public interest' is not a matter of science but rather a matter of politics. Advocacy planning, in its various forms, is an attempt to move citizens' participation forward from merely reacting to agency plans to proposing their own concepts of appropriate goals and future action.

Advocacy planning has taken many forms, but three examples of the original concept are the Architect's Renewal Committee of Harlem (ARCH), Urban Planning Aids of Boston (UPA) and the Community Design Centre (CDC) of San Francisco. The ARCH organisation, for example, was formed in 1964 by a group of white professionals who felt that their expertise could be of service to the Harlem community.[42] The limited power of the advocacy strategy has been criticised by one of the founders of ARCH,[43] and in general, fundamental system change has proved to be beyond the powers of even the most successful advocacy planners.[44] Despite its limitations, however, advocacy planning continues, particularly in relation to community and economic development. Some of its supporters have suggested that it should be undertaken by planners operating within the established city planning system without the necessity of a client group. This approach has been labelled pragmatic radicalism (see Chapter 8).

POPULIST POLITICS

In Britain, where local politics is closely tied to the national party structure, the formation of new political groupings to contest local elections is not common. By contrast, in North America the dominance of non-partisanship in municipal government allows independent interests to enter the political arena more easily, as demonstrated by the election of a populist/liberal majority on the Santa Monica CA city council in 1981.[45] Non-partisanship in US cities most often exists where the structure of 'reformed' government has produced a commission or city-manager form of administration. In many of the large North American cities, however, the apparent independence of politicians disguises a conservative, business-oriented outlook.

Under particular circumstances, social and economic trends can combine to generate a momentum that is sufficient to give rise to a popular alternative political movement as a challenge to the status quo. One example of such an event was the emergence of The Electors' Action Movement (TEAM) in Vancouver. This liberal urban-reform party was founded in 1968 and assumed

political control of the city between 1972 and 1978. Based on an ideology of producing a liveable city, TEAM challenged and temporarily replaced the prevailing growth- boosterism basis of Vancouver politics.[46] The TEAM ideology achieved some significant successes, most notably the redevelopment of the False Creek inner-city area, but it also produced several outcomes that were socially regressive for the working class and opened the administration to charges of elitism. This was particularly the case in the private land market, where the administration could wield only indirect influence. Thus, for example, the beautification and pedestrianisation of the main downtown shopping avenue was carried through despite protests from a number of shopkeepers. The subsequent rent increases by landlords forced some of the retailers to relocate. The False Creek redevelopment itself also resulted in some negative externalities for low-income residents in the neighbouring area of Fairview Slopes, whose homes were demolished to make way for expensive townhouse developments. This elitism, albeit unintended, together with internal divisions, the constraints of the national and provincial political structure, and a downturn in the local economy, combined to defeat TEAM in the 1978 elections. The demise of TEAM as a party, however, did not herald the extinction of the TEAM ideology. The political vacuum was filled by the Committee of Progressive Electors (COPE), a democratic socialist party which by 1990 held half the seats on the city council. For two decades from 1970, TEAM and COPE offered electors a credible alternative to the pro-growth NPA (Non-Partisan Association) city government.

As the experience of TEAM and other formal participation initiatives demonstrates, the struggle between citizen interests and those inherent in the capitalist urban development process is an unequal one. The main problem that has confounded attempts at municipal democratisation is that in most instances a level of influence rather than power has been redistributed (Figure 20.7). The limited ability to exert a meaningful influence on urban decision-making through formal channels of participation (that is, representative democracy and government-mandated citizen participation) has resulted in the rise of radical 'bottom up' protest groups and urban social movements.

URBAN SOCIAL MOVEMENTS

Consensual community participation – gaining 'inside' access to policy-makers – offers a direct formal route to government officials but also raises the possibility of co-optation. For urban protest groups there are clear advantages in operating outside the formal system via a strategy of *conflictual participation*. These include the strengthening of internal group solidarity, and the ability to gain concessions from government through fear, sympathy or successful mobilisation of public groups to which elites are normally attentive. Viewed in this context, participation and resistance are dialectically linked and represent a combination that may offer the best hope of success to alienated neighbourhoods. This is illustrated in a study of the political impact of Community Action Agencies (CAAs) in the USA. These stemmed from a federal initiative, launched in 1964, to promote neighbourhood-based services and community organising with at least one-third representation of the poor on CAA governing boards. Austin (1972), in a survey of twenty cities, found that only in the one-third of agencies where neighbourhood participation was adversarial

Plate 20.2   A poster offers a concise social commentary outside the US Mint in Denver CO

was there any relationship between resident action and urban change.[47] Direct action to obtain decisions favourable to disadvantaged groups and neighbourhoods is often necessary, because, as Reidel (1972 p. 219) notes, 'no one gives up power to others unless he no longer needs it, can no longer sustain it for personal reasons, or is forced to do so'.[48]

Urban social movements are present in every Western society, including, for example, Spain,[49] Italy,[50] the Netherlands,[51] (West) Germany[52] and Australia,[53] as well as Britain[54] and the USA.[55] Such groups are distinguished by their grass-roots orientation and non-hierarchical mode of organisation, their distance from and non-involvement in formal politics, and their emphasis on direct action and protest.[56] The main goals of urban social movements are of three types, relating to:

1. *Collective consumption issues.* These include movements for the provision of housing and urban services where absent, and movements over access to housing and urban services (e.g. public housing rents, eligibility rules, bus fares).
2. *The defence of cultural identity.* This includes movements opposed to physical threats to a neighbourhood (such as urban renewal), and against social threats such as the introduction of undesirable families or facilities. (This can include socially regressive movements such as those designed to maintain racial segregation.)
3. *Political mobilisation.* The aim of this type of movement is to achieve control and management of local institutions. Demands typically include participation in planning decisions, self-management of public housing schemes, and control over council spending within a neighbourhood.

In the USA, in the late 1960s, the pursuit of black civil rights, growing unemployment and inflation, the cost of the Vietnam War, which drained funds from federal social programmes, student protests against the capitalist ideology, and the negative social effects of the federal-backed urban renewal and highway-construction programmes combined to give rise to vociferous urban protest movements. The challenge posed to the dominant social order by inner-city residents had its most striking manifestation in the ghetto riots – a form of protest that also occurred in several minority ethnic areas in British cities during the 1980s.[57] The US protest movement had a variety of forms, generally represented in the work of militant community-based organisations. Aided by the Community Action Program funded by the federal Office for Economic

Plate 20.3   Street demonstrations by Hispanic-Americans in Chicago protesting against proposed new immigration laws

Opportunity, thousands of popular organisations arose in the central cities of large metropolitan areas. The most widespread inner-city urban movements have been mobilisations against urban renewal, either to protect a neighbourhood from demolition or to obtain adequate relocation and compensation. Several detailed case studies of this process in individual cities are available.[58] Many of these are based on the concept of the 'right to the city'.[59] As formulated originally by Lefebvre[60] this contends that, in contrast to **representative democracy** in which citizens have indirect influence over processes of change, people should participate directly in decisions that produce urban space – **participatory democracy**. Fundamentally, Lefebvre argues that the right to the city requires that a city is produced to meet the needs of the users of urban space, not its owners. This privileges the role of urban residents in decision-making. Clearly this radical perspective on urban change runs counter to

the dominant **liberal democratic** practice of urban governance in most Western societies (Box 20.9). The use of urban space by residents is seen as a key element of the right to the city. In the USA the role of public spaces (e.g. streets and squares) as spaces for representation for disadvantaged and marginalised groups was demonstrated by the nationwide urban protest in 2006 by Hispanic-Americans against the introduction of stricter immigration laws. In the UK the introduction of the community charge (poll tax) by the Conservative government of Mrs Thatcher prompted widespread popular opposition and civil disobedience in the form of non-payment campaigns and street demonstrations, which ultimately led to the replacement of the *per capita*-based poll tax by a new household-based council tax. More recent foci for local public protests have included campaigns over local hospital closures, and domination of local retailing by major chain stores such as Tesco.

BOX 20.9

The Right to the City

The South Lake Union (SLU) district of Seattle is undergoing significant redevelopment. The aim is to transform the area from a low-density mix of warehouses, car dealerships, light manufacturing and some moderate-income residents into a biotechnology based technopole. This vision is being progressed by a partnership between a private real estate company and municipal and state government. The growth coalition views SLU as an empty space in which redevelopment will cause minimum disruption and produce maximum economic benefit. It is argued that for the project to succeed in a competitive global market place it is necessary to set aside the traditional 'Seattle way' of extensive public debate over development.

The Right to the City discourse rejects this viewpoint, focusing instead on the needs of the 1,500 residents and users of a number of local social service agencies. The fundamental issue concerns who has the greater right to decide on the future of this part of the city: current residents who may be displaced; the inhabitants of the wider region, who may benefit from any economic growth; people in other neighbourhoods who object to the use of public funds in one area; or a coalition of citizen groups who may oppose the redevelopment on the grounds that any benefits will boost the exchange value of the private investors rather than the use value of local residents? In essence: who has the greater right to the city?

Source: M. Purcell (2006) Urban democracy and the local trap *Urban Studies* 43(1), 1921–41

It may be argued that, over recent decades, the radical political motivations of 1960s urban social movements have been diluted by the process of **neoliberalism** that has led to replacement of grassroots organisations pursuing an agenda of social welfare through conflictual political participation by large corporatist organisations (e.g. CDCs) focused on service delivery and physical revitalisation of places.[61] This changed emphasis is apparent, for example, in The Woodlawn Organisation in Chicago in which a shift from a grassroots community organisation to a neighbourhood not-for-profit development corporation was accompanied by a shift of strategy from confrontation to co-existence with other agencies of urban governance.[62] Neo-liberalism, however, has also exposed new spaces of conflict in US cities with, for example, debate over the role and funding of mass transit between those campaigning on social equity grounds for greater support for transit-dependent urban poor, and those who, from a perspective of economic efficiency, view the primary role of mass transit as relief of traffic congestion. One example of a grassroots response to the neo-liberal agenda is the Bus Riders Union that emerged in Los Angeles during the 1990s to fight for better bus services in the urban core.[63]

The issue of tenants' rights in the USA affords another example of a modern urban social movement in operation. By 1980 building-level tenants' groups existed in every US city, stable state-wide tenant organisations existed in New York, New Jersey, Massachusetts and California, and a National Tenants' Union had been established. Significantly, landlords have also developed greater cohesiveness in order to oppose the threat of rent control and other housing regulations contrary to their interests. In 1978 the National Rental Housing Council (renamed the National Multi-Housing Council in 1980) was formed to provide local landlord groups with advice on media campaigns, legal tactics and arguments against rent control and pro-tenant demands, as well as to lobby Washington. As a consequence, 'the burgeoning self-consciousness and activism among both tenants and landlords at the local, state and national levels has made tenant–landlord conflict a significant feature of America's political landscape' (Dreier 1984 p. 255).[64]

Notwithstanding the achievements of tenants' groups (Box 20.10) they face several built-in limitations. These include their relative poverty compared with landlord groups, the considerable influence of real-estate interests in local and national politics, and traditional low levels of voter registration and political participation. The tenants' movement will most likely continue to focus on local and state-level issues such as rent-control laws, eviction regulations, improved enforcement of housing codes and changes

BOX 20.10

The New Jersey Tenants' Organisation

One of the most successful tenants' groups is the New Jersey Tenants' Organization, which is based on an alliance of low- and middle-income tenants. The organisation was started in 1969 in the middle-income suburbs of New York City by residents of large apartment complexes, and by the occupants of public and private slum housing in Newark and Passaic. By the end of 1970 the NJTO had organised forty-three rent strikes involving 2,000 tenants in the middle-income apartment complexes in protest against proposed rent increases of 20–40 per cent. In response to five of the strikes, landlords dropped the planned increases, in thirty cases the increases were spread over several years, and in eight cases tenants successfully bargained for better conditions. The rent strike tactic was particularly effective since New Jersey law does not allow tenants to be evicted if they agree to pay the rent to the court. Direct action in the form of rent strikes, demonstrations and pickets was only one part of a three-pronged strategy. The NJTO also recognised the importance of developing the political power of tenants as a voting bloc and in endorsing pro-tenant candidates for local and state elections. In addition, the NJTO pursued litigation to establish protection for tenants engaged in direct action and to strengthen landlord–tenant law. The NJTO action proved so successful in furthering tenant rights that similar activity has taken place in other cities.

in tenant–landlord law. It cannot, however, ignore the fact that to a significant extent federal decisions in the fields of tax laws, subsidies, revitalisation programmes, interest rates and public housing budgets determine the nature of local housing problems and the resources available to solve them.

These conclusions on the future role of tenants' organisations can equally be applied to other 'bottom up' social movements. Generally, in order to develop a political voice at the national level where major policy decisions are made, it may be necessary for individual groups to broaden their perspective and seek to form coalitions with other single-issue interest groups, such as state-level trade unions, senior citizens' organisations and environmental and women's organisations. The example of the Australian Green Ban Movement (an alliance between residents' action groups opposed to certain types of development and the Builders' Labourers' Federation) demonstrates the potential power of broad-based social movements.[65] The Los Angeles Alliance for a New Economy (LAANE) is a multiracial and cross-class coalition of labour unions, religious groups and community organisations. Pursuing an agenda based on 'growth with justice' LAANE successfully pressured Universal Studios into agreeing to pay improved wages and health benefits to 8,000 new service workers involved in the expansion of the City Walk theme park. Forming alliances without surrendering identity and penetrating the political system while reserving the right of direct action are among the major requirements (Table 20.5) for a shift from grass-roots pressure to grass-roots power in the shaping of urban policies.

TABLE 20.5 CHARACTERISTICS OF SUCCESSFUL GRASS-ROOTS GROUPS

1. Full-time, paid, professional staff
2. Well-developed fund-raising capacity
3. Sophisticated mode of operation, including:
   - Neighbourhood street organising
   - Advanced issue-research capacity
   - Information dissemination and exposé techniques
   - Negotiation and confrontation skills
   - Management capability in the service delivery and economic development areas
   - Policy and planning skills
   - Lobbying skills
   - Experience in monitoring and evaluating government programmes
4. Issue growth from the neighbourhood to the nation
5. A support network of umbrella groups, technical assistants, action research projects, organiser training schools
6. Expanding coalition building with one another, with public-interest groups and with labour

Source: J. Perlman (1976) Grassrooting the system *Social Policy* 7(2), 4–20

FURTHER READING

BOOKS

P. Hamel, H. Lustiger-Thaler and M. Mayer (eds) (2000) *Urban Movements in a Globalising World* London: Routledge

T. Herschel and P. Newman (2002) *Governance of Europe's City Regions* London: Routledge

D. Hill (2000) *Urban Policy and Politics in Britain* London: Macmillan

D. Judd and T. Swanstrom (1998) *City Politics* London: Longman

H. Kriesi, R. Koopmans, J. W. Duyvendak and M. G. Giugni (eds) (1995) *New Social Movements in Western Europe* London: UCL Press

L. Sharpe (1995) *The Government of World Cities: The Future of the Metro Model* Chichester: Wiley

JOURNAL ARTICLES

S. Arnstein (1969) A ladder of citizen participation *Journal of the American Institute of Planners* 35, 216–24

K. Dowding (2001) Explaining urban regimes *International Journal of Urban and Regional Research* 25(1), 7–19

P. Dreier (1984) The tenants' movement in the United States *International Journal of Urban and Regional Research* 8, 255–79

C. Drennan (2006) Social relations spatially fixed: construction and maintenance of school districts in San Antonio, Texas *Geographical Review* 96(4), 567–93

S. Nunn and M. Rosentraub (1997) Dimensions of interjurisdictional cooperation *Journal of the American Planning Association* 63(2), 205–19

G. Rodriguez (1999) Rapid transit and community power: West Oakland residents confront BART *Antipode* 31(2), 212–28

G. Webster (1997) The potential impact of recent Supreme Court decisions on the use of race and ethnicity in the restructuring process *Cities* 14(1), 13–19

G. Williams (1999) Institutional capacity and metropolitan governance: the Greater Toronto area *Cities* 16(3), 171–80

KEY CONCEPTS

- Power
- Urban growth machines
- Regime theory
- Mayor–council government
- Council–manager government
- Dillon's rule
- Incorporated cities
- Unincorporated territory
- Townships
- Municipalities
- Special districts
- Annexation
- Defensive incorporation
- Secession
- Metropolitan government
- Patronage
- Gerrymandering
- Malapportionment
- Citizen participation
- Advocacy planning
- Populist politics
- Urban social movement
- Representative democracy
- Participatory democracy
- The right to the city
- Neo-liberalism

STUDY QUESTIONS

1. Examine the relative powers and role of central and local governments in the UK and/or USA.
2. Compare and contrast the spatial organisation of local government in the UK and USA.
3. With reference to particular examples, explain what is meant by the term 'gerrymandering'.
4. Explain the significance of the process of incorporation in the formation of urban areas in the USA.
5. Prepare a list of the arguments for and against metropolitan government.
6. Critically examine the role and potential of urban social movements as a means of transferring political power to the grass-roots.
7. With the aid of relevant examples critically examine the concept of the 'right to the city'.

PROJECT

Undertake a study of the structure of government in your city. How is the city government elected? How are power and responsibility distributed within the government? What is the committee/decision-making structure? How does the electorate make its voice heard between elections? How is the city's annual budget determined? What are its spending priorities? Examine the geography of government expenditure – have some areas of the city benefited more than others? What is the nature of the relationship between the city government and surrounding local authorities? Is there any form of formal or informal urban region/metropolitan co-operation? What do the people (politicians, businesses and citizens) think about the organisation, policies and quality of the city government?

PART FIVE

Urban Geography in the Third World

Plate 9 The squatter city: Rio de Janeiro

Plate 10 The postcolonial city: Singapore

Plate 11 The contested city: Belfast

Plate 12 The multicultural city: London

Plate 13 The emblematic city: Sydney

Plate 14 The unregulated city: Rome

Plate 15 The defensive city: New York

Plate 16 The emerging global city: Shanghai

21

Third World Urbanisation within a Global Urban System

Preview: world urbanisation; urbanisation and development; modernisation theory; dependency theory; world systems theory; third world urbanisation in historical context; stages of colonial urbanisation; peripheral urbanisation; exo-urbanisation; urbanisation by implosion

INTRODUCTION

Since 1950 urbanisation has become a worldwide phe-nomenon (see Chapters 4 and 5). Although the pace of change has varied considerably between countries and regions, virtually every country of the Third World (Box 21.1) has been urbanising rapidly. Evidence of a slowdown in the rate of growth of some of the largest cities[1] and of polarisation reversal or spatial deconcen-tration into polycentric metropolitan forms does not contradict the conclusion that the Third World is becoming increasingly urbanised. In this chapter we identify a number of significant differences between urbanisation in the First and Third Worlds, and review the main theories of the link between urbanisation and economic development. The process of Third World urbanisation and its effects on urban structure are then analysed in historical context from the early period of mercantile **colonialism** to the present-day situation characterised by high levels of urbanisation and phe-nomena such as exo-urbanisation.

URBANISATION IN THE FIRST AND THIRD WORLDS

Urbanisation in the Third World exhibits a number of important contrasts with the earlier process in the First World:

1. It is taking place in countries with the lowest levels of economic development, rather than the highest, as was the case when accelerated urbanisation began in Western Europe and North America.
2. It involves countries in which people have the lowest levels of life expectancy at birth, the poor-est nutritional levels, the lowest energy consump-tion levels and the lowest levels of education.
3. It involves greater numbers of people than it did in the developed world.
4. Migration is greater in volume and more rapid.
5. Industrialisation lags far behind the rate of urban-isation, so that most of the migrants find at best marginal employment in cities.
6. The environment in cities of the Third World is usually more healthy than in their rural hinter-lands, unlike in the industrial cities of the West. Urban fertility is greater in Third World cities and net reproduction rates are higher than they ever were in most of the industrial countries.
7. Massive slum areas of spontaneous settlements characterise most large cities of the Third World.
8. Rising expectations mean that pressures for rapid social change are greater than they were in the West.
9. Political circumstances conducive to revolution-ary take-overs of government are often present as a result of the recent colonial or neo-colonial status of most of the Third World nations.
10. Most of the Third World countries have inherited an intentionally centralised administration, with the result that government involvement in urban development is more likely in these countries today than it was in the nineteenth-century West.

BOX 21.1

Defining the Third World

The term 'Third World' has both political and economic connotations. In a political sense it was
employed, especially during the Cold War, to identify
a threefold division of the world into a First World of
capitalist industrial market economies, a Second
World of centrally planned socialist-communist
states, and a Third World of initially non-aligned
states (which eventually became client states of one
camp or the other). In a socio-economic sense the
Third World refers to countries that generally,
although not uniformly, failed fully to develop economically after independence.

For some, the collapse of the Soviet Union and the
demise of communism in Eastern Europe undermine
the rationale and cohesive power of the term 'Third
World'. For others, the disappearance of the Second
World merely serves to focus stronger attention on
the relationship between the First and Third Worlds

within a world system Although the blanket term
'Third World' obscures often considerable cultural,
economic, social and political differences between
and within individual states, it has advantages over
common alternatives such as 'the South' or 'developing countries'. The South is essentially a geographical term that ignores the fact that some developed
states – Japan, Australia and New Zealand – are in
the region, while the concept of developing countries
implies a process that is far from automatic.

While the World Bank categorises the countries of
the world into five groups based on GNP *per capita*,
as the UN Human Development Index indicates,
income is only one measure of level of development.
For these reasons, and notwithstanding the difficulties of precise spatial definition and internal variations within the region, the terms 'less developed
realm/countries' and 'Third World' are preferred.

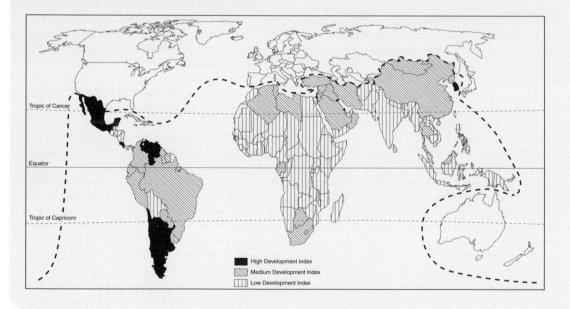

Notwithstanding these general differences between the
urbanisation process in the two world realms, it is
essential to recognise that urbanisation is not a uniform process that all countries go through in the
course of 'development'. Although there are similarities at a general level, the process of urbanisation in
different parts of the world is the result of a complex
interplay of global and local social, economic,

political, technological, geographical and cultural
factors. We are dealing with several fundamentally
different processes that have arisen out of differences
in culture and time. These processes are producing
different results in different world regions (see
Chapter 5). A common underlying factor, however,
concerns the relationship between urbanisation and
economic development.

THEORIES OF URBANISATION AND DEVELOPMENT

Three principal paradigms have been proposed to explain Third World urbanisation and underdevelopment.

MODERNISATION THEORY

In the developed world, economic development and urbanisation were linked in a process of industrialisation and **modernisation**. Accordingly, in the immediate Second World War decades, Western theorists emphasised the necessity of urban industrialisation for the economic development of Third World countries. **Modernisation theory** viewed Third World development as a convergent and evolutionary process in which diffusion of economic and cultural innovation from the West would move less developed societies towards the kind of advanced economic, social and political structures that prevailed in Western Europe and North America. This philosophy was encapsulated in Rostow's (1959) stages of economic growth model[2] (Box 21.2). Cities, especially primate cities, played a central role in this developmental model, functioning as the key portals through which innovation would be transmitted to the rest of the society.

Modernisation theory has been criticised on the grounds that:

1. It ignored the political and economic diversity among Third World countries, which makes comparative analysis of development along a single trajectory highly problematic.
2. It failed to consider the role of cultural factors in the development process.
3. It gave insufficient attention to the nature of the relationship between industrially developed countries and the Third World.

DEPENDENCY THEORY

By the 1970s the continued economic stagnation of much of the Third World made it apparent that

---

BOX 21.2

Rostow's stages of economic growth

Rostow's theory helped to popularise the idea of a 'take-off' into self-sustained economic growth as the key stage of the Third World development process. His theory of economic growth is based on the concept of a five-stage process through which all developing societies must pass to attain the status of a liberal democratic system akin to the USA or the UK.

1. *The traditional stage*. In this stage the expansion of production beyond a limited ceiling is difficult because of primitive technology, a fundamentally agricultural economy and a rigid hierarchical social structure.
2. *Preconditions for take-off*. This stage is triggered by an external stimulus (such as innovation in agriculture or industrial production). The economy becomes less self-sufficient and local as trade and improved communications facilitate the growth of both national and international economies. These processes are related socially and politically to the emergence of an elite group able and willing to invest wealth rather than use it solely for immediate personal consumption.

3. *Take-off*. The first of two crucial characteristics is that investment as a proportion of national income rises to at least 10 per cent, thus ensuring both that increases in *per capita* output outstrip population growth and that industrial output increases appreciably. Second, as a result, political and social institutions are reshaped in order to permit the pursuit of growth to take root. This stage lasts, typically, about twenty years. According to Rostow, Britain reached this stage between 1783 and 1803, the USA in 1843–60, Japan in 1878–1900 and India from 1950.
4. *The drive to maturity*. A period of consolidation follows in which modern science and technology extend to most branches of the economy, the investment rate remains high at 10–20 per cent of national income, political reform continues and the economy is able to compete internationally.
5. *The age of mass consumption*. Such is the productive power of the society that three strategic choices are available. Wealth can be concentrated in individual consumption (the USA), channelled into a welfare state (the UK) or used to build up global power (the former USSR).

the region was not conforming to Western models of urbanisation based on the 'generative' role of cities[3] and the notion of inter-societal convergence. **Dependency theory** explained this by emphasising the hegemony of Western nations over the world economic order and the consequent ability of these states and multinational corporations to exploit peripheral areas.[4] As we see below, this process began with the European colonisation of the Third World, which established an 'interdependent but unequal relationship' between coloniser and colonised (Spybey 1992 p. 225).[5] Dependence formulations gained wide circulation largely as a result of research in Latin America by Frank (1967), who argued that the Iberian conquest absorbed pre-Columbian America into the world system of capitalism.[6] Latin America was therefore linked into a chain of surplus extraction (initially in the form of silver, gold, cotton and sugar) which both 'developed' Spain and Portugal, and later Britain, and 'underdeveloped' Latin America. The fundamental premise of dependency theory was that development and underdevelopment were different outcomes of the same process.

Cities played a central role in this process. The chain of surplus extraction established by colonial powers created centres and peripheries both on a world scale and within Latin America. At a world scale Latin America supplied the metropolitan countries with raw materials and imported manufactured products through major port-cities. At a regional scale the large cities acted as centres exploiting provincial towns and rural areas (Figure 21.1).

Dependency theory enjoyed a period of intellectual prominence from the early 1970s to mid-1980s. Since then, critics have drawn attention to the following inadequacies:

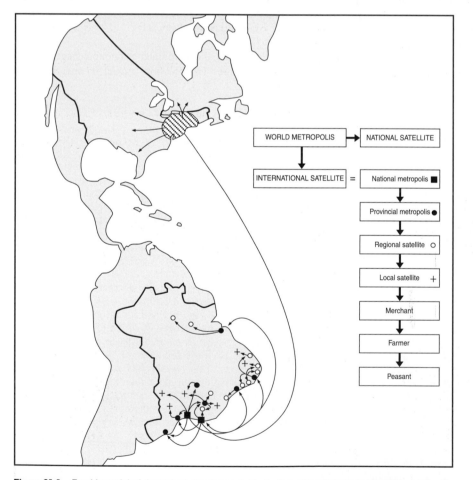

Figure 21.1   Frank's model of dependent development, illustrated with reference to the USA and Brazil
Source: J. Dickenson, B. Gould, C. Clarke *et al.* (1996) *A Geography of the Third World* London: Routledge

1. It focuses excessively on the operation of the international capitalist system and pays insufficient attention to internal cultural and political forces operating in 'dependent' countries.
2. It was implicit in the Frankian argument that the process of surplus extraction had impoverished Latin America, but this was not uniformly the case. For example, the investment of British capital in Argentina and Uruguay created railways, urban infrastructure and the basis of cattle, wool, mutton and cereal production.
3. Dependency theory was unable to predict or even explain the economic rise of the newly industrialising countries of East Asia in the 1970s.

WORLD-SYSTEMS THEORY

The political-economy perspective of dependency theory was extended by Wallerstein (1974), who, in recognising that the world system was more complex than was suggested by the polar contrast of dependent and metropolitan countries, identified three types of national position within the world system:[7]

1. A dominant core of North America and Western Europe.
2. A semi-periphery of richer Third World states with mineral exports and/or limited industrialisation for export.

3. A periphery of poor countries that had been in the past or continue to be exploited as a result of their involvement in the global economy (Figure 21.2).

Significantly, and unlike in the earlier dependency theory, **world-systems theory** holds that countries may change their relative position and are not doomed to perpetual underdevelopment. In addition, the size, role and characteristics of individual cities reflect the world position of their society.[8] Arguably the most important conclusions to emerge from these theoretical debates over Third World development and urbanisation are, first, acknowledgement of the interdependence of economic and urban development across the globe and second, recognition of the need to employ an 'historical-structural' perspective to interpret the contemporary urban geography of the Third World. We shall employ these perspectives in the remainder of our discussion of the changing nature of Third World urbanisation.

THIRD WORLD URBANISATION IN HISTORICAL CONTEXT

From the sixteenth century onwards, Britain and other European powers directly influenced the urbanisation of developing countries through imperialism and through investments in transport and infrastructure

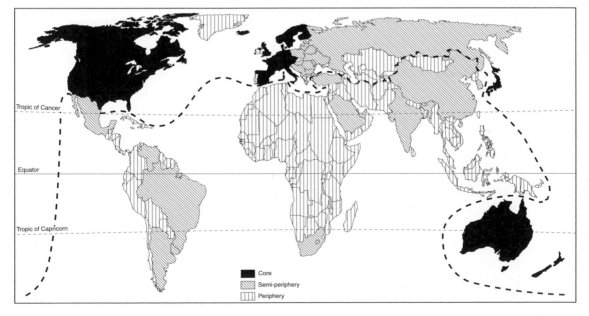

Figure 21.2   Wallerstein's world systems model

designed to facilitate the extraction of primary products.[9] The most obvious effects of European and, later, US expansion in the Third World were the creation of cities and, in some cases, the destruction of existing urban cultures. In Latin America the cities of the Incas and Aztecs (see Chapter 3) were demolished and replaced by Spanish settlements. Many cities were constructed in new locations, with most of today's major metropolises, including Lima, Buenos Aires, Bogotá, Caracas and La Paz, founded by the Spanish. Similarly, most of Brazil's main cities, such as Rio de Janeiro, São Paulo and Salvador de Bahia, were established by the Portuguese.[10] In non-Yoruba areas of West Africa some of the fort settlements established by the British, French and Portuguese (including Dakar, Accra and Lagos) became the primate cities of the twentieth century.[11] In the rest of sub-Saharan Africa the main urban centres of Johannesburg, Cape Town, Nairobi and Salisbury were all founded by Europeans.[12] Similarly, in Asia, European coastal cities such as Calcutta and Bombay became major metropolitan centres.[13] Only in China and parts of the Arab world were existing economic and cultural systems strong enough to resist the European impact on urban form.[14]

In general, European cities prospered throughout the Third World. However, the nature of colonial impact varied:

1. It varied according to the nature of the colonised areas. While some were backward subsistence economies, others contained nations that were larger and more sophisticated than the colonial powers.
2. It varied according to the methods of the colonising power. All sought economic profit, but methods of exploitation varied, with the British preferring indirect rule through advisers and the French direct incorporation of territories into a central metropolitan administrative system.
3. Above all, the mix of colonial and pre-colonial influence varied over time (Box 21.3).

BOX 21.3

Determinants of colonial urban form

Although the concept of the colonial city embraces a variety of urban types and forms, it may be generalised as a distinct settlement form resulting from the domination of an indigenous civilisation by colonial settlers.[15] Ten major factors have been identified as determinants of colonial/postcolonial urban form:

1. *The motives for colonisation*, e.g. trade (mercantilism), agricultural settlement or strategic acquisition.
2. *The nature of pre-colonial settlement*, e.g. rural villages or permanent urban areas.
3. *The nature of imperial or colonial settlement*, e.g. imperial control requiring military security with little if any permanent settlement, or explicit colonialism involving significant levels of permanent settlement.
4. *Relations between the colonisers and the indigenous population*, e.g. extermination (as in Australia or the USA), assimilation (as in Hispanic America after the conquest) or an intermediate form of accommodation (as in much of Africa).
5. *The structure of any indigenous settlements* that were either destroyed, ignored, added to by accretion or incorporated within a new planned city. Where no centres already existed, new colonial cities were established, sometimes for colonists alone, sometimes for colonists and indigenes in separate quarters, and sometimes for all groups without formal segregation.
6. *The nature of the anti-colonial struggle* and the degree to which the new leadership identified with existing politico-administrative centres, especially the capital city.
7. *The role of the ex-colonial elite* and the degree to which it retains economic dominance.
8. *The policies pursued by the independent state* with respect to national integration, ethnic and class conflict, and the nature of the country's insertion into the world economy.
9. *The nature of the modes and forms of production*, including state policies towards the private, state and informal sectors, and the nature of access to and control over the means of social reproduction.
10. *The extent of urban planning and legislative change* under either capitalist expansion policies or some variant of socialist centralisation and transformation.

Source: **D. Simon (1992)** *Cities, Capital and Development* **London: Belhaven Press**

STAGES OF COLONIAL URBANISATION

Table 21.1 illustrates the structural evolution of colonialism. Although the illustrative time scale refers specifically to Asia, the successive stages identified can be applied more generally to the Third World.

MERCANTILE COLONIALISM

Initial exploration from Europe was undertaken by individuals seeking objects with inherent value, such as gold and silver. Subsequently the search shifted to commodities that were valued within the European trading system, such as silks, spices or sugar. Most of the commodities were the natural products of the country, and their local collection was left in the hands of existing traders, whose networks remained intact and were simply incorporated into the new European markets. Overall, therefore, there was little need for an extensive European presence in the foreign country. The limited extent of European settlement was due partly to the efficiency of local trading networks, but also to the fact that mercantile colonialism was based on the private company rather than state enterprises. Permanent company representatives were few in number and tended to be confined to spatially restricted concession areas within existing indigenous cities.[16] As trade expanded and profits grew, Europeans sought a greater presence in order to oversee the collection of

the trade goods. The need to protect warehouses led to the presence of troops armed with superior weaponry. Later the demand for commodities of reliable quality led to the gradual incursion of Europeans into the production process, a step that extended their influence well beyond the trading concession area.

In broad terms the mercantile colonial period had little impact on Third World urban systems. Europeans were usually confined to small areas of indigenous cities that were already divided along ethnic/occupational lines so that the introduction of a European enclave made little difference to urban structure and function. No new urban hierarchies were created during this period and only in Latin America[17] did settlements of purely colonial origin, such as Buenos Aires and Lima, appear.

THE TRANSITIONAL PHASE

By the end of the eighteenth century European interest in overseas ventures began to moderate for several reasons. First, extensive European wars in the Napoleonic period soaked up finance for overseas investments and occupied many of the adventurers on whom mercantile colonialism depended. Second, the shift into production rather than just trade began to increase the cost of colonial activities beyond the capacity of the companies involved. Several of the largest went into liquidation during this period and their operations were

TABLE 21.1 STAGES OF COLONIAL URBANISATION IN ASIA

Chronological phases	Major features of urbanisation
Pre-contact	Small, organically patterned towns predominate
1500 Mercantile colonialism	Limited colonial presence in existing ports. Trade usually in natural products of local region
1800 Transitional phase	Reduced European interest in investment overseas. Greater profits to be made in the industrial revolution
1850 Industrial colonialism	European need for cheap raw materials and food. Colonialism takes territorial form, new settlement patterns and morphology created
1920 Late colonialism	Intensification of European morphological influence. Extension to smaller towns in hierarchy. Increased ethnic segregation
1950 Early independence	Rapid growth of indigenous populations through migration in search of jobs. Expansion of slum and squatter settlements
1970 New international division of labour	Appearance of multinational corporation factories. Further migrational growth of cities. Increasing social polarisation

Source: **D. Drakakis-Smith (1987)** *The Third World City* **London: Methuen**

taken over by national governments. Finally, the key characteristic of the transitional period was the interest of European investors in the greater profits to be made from the industrial revolution at home rather than mercantile colonialism. It was in this context that Raffles, Light and Elliot found the British government less than enthusiastic about the development prospects of their acquisitions of Singapore, Penang and Hong Kong respectively.

INDUSTRIAL COLONIALISM

Investor reluctance did not last for long. The rapid growth of the European industrial revolution led to increasing demands for raw materials and food for the burgeoning urban work force. By the third quarter of the nineteenth century European capital was again flowing overseas, but on this occasion the principal agent of organisation was the state. The accumulation

Plate 21.1 Sir Stamford Raffles, the founder of Singapore, is overlooked by a potent symbol of post-industrial society

of raw materials and food required more than a trading toehold. It depended on the acquisition of territory and the organisation of production in order to keep costs as low as possible.

This had a major impact on urbanisation in the colonies. King (1976)[18] suggests that the economic motivation of colonialism was translated into urban form through three main intermediate forces:

1. *Culture*. The social, legal or religious values of the colonial power produce a set of institutions that determined the physical expression of the culture. The various cultural elements of Victorian British society were reflected in the built form of its colonial settlements, seen in the presence of churches, theatres, men's clubs and in the application of building standards and planning regulations.
2. *Technology*. The main technological difference between the colonial power and the colony was the extensive use of inanimate sources of energy instead of animal and human portage. New transport infrastructure, including railways and wide boulevards, destroyed the old medieval street patterns of many existing cities. As a more complex urban system developed, new social organisations were needed, including police forces, transport companies and urban management agencies. In some cases, such as that of Delhi, existing urban morphologies were abandoned in favour of new districts based on European technology (Box 21.4).
3. *Political structure*. The factor that enabled the wholesale introduction of these new technologies and cultural values was the political control exercised by the Europeans. Nineteenth-century colonial society comprised a two-tier relationship between the dominant colonisers and the subordinate colonised. As the colonial elite controlled the economy and the municipal government, the city could be shaped according to the wishes of a small proportion of its total population.

The industrial colonial period witnessed a major European impact on urban development and morphology. This era saw the origins of contemporary urban primacy as economic and political power were concentrated on certain cities at the expense of others (Box 21.5). Within **colonial cities**, similar patterns of social, economic and spatial segregation were being reinforced. Functional specialisation of economic activity along ethnic and class lines meant that the foreign component of trading was dominated by

BOX 21.4

Delhi: the evolution of an imperial city

The pre-colonial city

When the British arrived in Delhi the city was still the Mughal capital of north India, although it was in the hands of rebels and its population had fallen to about 150,000. Almost all lived within the city wall, which was 5.6 miles (9 km) long and enclosed an area of about two and a half square miles (6.5 km²). At the heart of the city were the political, religious and commercial foci of the royal palace, the Jama mosque and Chadni Chowk respectively. The remainder comprised an organically patterned pre-industrial city of narrow, twisting lanes and mixed land uses – not dissimilar to those of medieval Europe.

The period of coexistence, 1803–57

In its early colonial years Delhi was not a major administrative or commercial centre, merely a district military post for the Punjab. There were no more than a few hundred Europeans and, as the situation was very stable, the military cantonment was located to the north-west of the city, with most of the British living in the city itself in an area adjacent to the royal palace (where a puppet emperor still reigned), previously occupied by the Mughal aristocracy. Western technology was barely in evidence, apart from some institutional architecture, and most of the British lived in the same manner as the indigenous elites. There was little social contact between the races but little conflict either. So, although there was political dualism, the retention of the court meant that it was not profound and was barely reflected in lifestyle or technology. The principal contrast, therefore, was cultural.

Colonial consolidation, 1857–1911

The pressures of the nineteenth-century colonial expansion erupted in India in the Mutiny of 1857 in which the Mughal emperor was implicated and dethroned. The consequent sharpening of military control resulted in the more forceful imposition of political power and cultural values. The royal palace became Delhi Fort, and a free-fire zone 550 yards (500 m) wide was created around it and the city walls. The military cantonment occupied the northern third of the city and civilians were moved out to the civil lines to the north of the walls, where several physically imposing buildings were constructed as institutional symbols of power. The indigenous population was thus confined to the remaining area of the old city, which, although crowded, was still functionally and socially sound. Culturally, however,

Old Delhi and its population became increasingly isolated as the British withdrew to a distinct area to the north, separated by the military zone, police lines, newly constructed gardens and, above all, by the railways. Poor communications had been blamed for many of the problems of the Mutiny, and by 1911 eight separate railways had been linked to the city, further eroding the area available to the indigenous population, which had by then reached some 230,000.

Thus the full utilisation of political power had, within the space of some fifty years, transformed the appearance of Delhi, with cultural contrasts spatially accentuated through the medium of superior military and transport technology. By 1911 almost a quarter of a million Indians were crammed into one and a half square miles (4 km²) of the old city, while a few thousand British enjoyed the relatively open spaces of the northern districts.

Imperial Delhi, 1911–47

Railway technology enabled Delhi to be chosen as the new centralised capital of a consolidated India – the jewel in George V's imperial crown. It was not until 1921, however, that the physical site was moved southwards to a new location in the crook of the Aravalli hills. New Delhi was planned on a vast scale; as early as 1931 it covered thirty square miles (78 km²). This was due to the incorporation of new technological developments, the motor car and the telephone, which in theory enabled such large distances to be bridged. In fact such technology spread only slowly and erratically, and for many years communications depended on 'people power'.

There was no provision for manufacturing growth in the new capital, although the city had acquired several important processing industries such as flour milling. But the most extraordinary example of colonial influence in urban planning was the minute residential stratification of New Delhi into rigid spatial categories based on the infinitely complex combination of Indian, British, military, civil and colonial ranking systems. This resulted in several designated zones, within which occurred further stratification of size of house, garden and facing direction in relation to the social mecca of Government House.

The old city received some improvements in its drainage and water supply systems but, as in most colonial cities, urban planning was primarily reserved for the expatriate zones. Old Delhi became even more physically and socially isolated by more

BOX 21.4 – continued

gardens and a spacious new business district. As a result, in-migration led to massive overcrowding, deterioration of the urban fabric and overspill into rapidly growing squatter areas to the west and east of the old city. Not surprisingly, by the Second World War death rates in Old Delhi were five times those in the new capital.

Independence did not change this contrast very much. Ten years later Old Delhi still contained 60 per cent of the city's population at an average density of 16,000 per square mile (41,300 per square kilometre). But even this was but a foretaste of what was to come later as the capital's population quadrupled to its present size of 13 million.

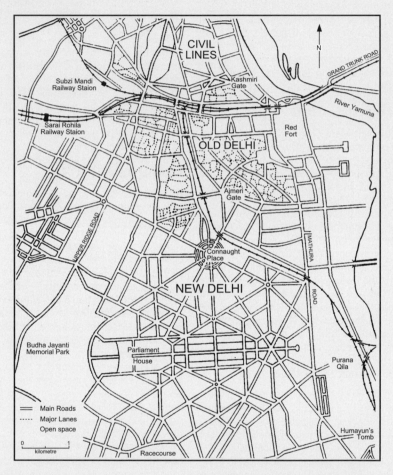

Source: **D. Drakakis-Smith (1993)** *Pacific Asia* **London: Routledge**

Europeans and their institutions, local assembly and distribution of the goods were in the hands of expatriate non-Europeans, and local production was the function of indigenous populations under expatriate supervision. This functional distinction was reflected by ethnic residential segregation, with European districts effectively separated from non-European zones by a *cordon sanitaire* of open spaces and railway lines to provide *Lebensraum* for the minority population.

BOX 21.5

Urban primacy in the Third World

In many Third World countries most large-scale modern activities are located in a single major city which dominates all others. Thus Montevideo has seventeen times the population of Salto, Uruguay's second city, and Bangkok fifty times that of Thailand's second city, Chiang Mai.

The relationship between urban primacy and economic development is inconclusive. Some researchers believe that primate cities in the Third World are 'parasitic' and retard development. Their reasons are:

1. The cost of supporting a large city, especially in an agriculturally based economy, drains resources from development activities.
2. The power of the primate city to attract migrants from rural areas, many of whom remain unemployed or underemployed and constitute a non-productive population and a cost to the national economy.
3. The concentration of government investment in the primate city reduces the growth prospects of other parts of the country.
4. The focus of national political and economic power in a primate city creates major social divisions between urban and rural areas.
5. As cities grow beyond a certain size, diseconomies of scale may arise, including increased land and service costs, extended transport lines and environmental problems such as congestion and pollution.

An alternative view is that primate cities provide an environment necessary for development. To compete successfully in the global market, Third World countries must develop the financing, marketing and management patterns that the world economic community demands. Only in the primate cities is the necessary infrastructure to be found, along with the skilled labour. Primate cities also offer agglomeration economies that attract enterprises and stimulate a cycle of growth.

LATE COLONIALISM

The inter-war years were marked by a major economic recession that resulted in erratic demand for the primary products of the colonies, and profits became unreliable. In some areas this led to a search for economies of scale, and small-scale producers began to be ousted by mechanisation and land reforms. This stimulated an accelerated drift of rural labour to the towns, where it was only partially absorbed by a still-limited factory-based production system. Squatter settlements also began to appear, although the principal morphological change in cities occurred in European quarters, where colonial architecture and land-use planning reached their most grandiose form.

In certain colonies an important demographic transformation took place, marked by the accelerated migration from the European recession of large numbers of blue- and **white-collar** workers. One consequence of this influx was that the slowly expanding group of educated indigenous residents found it difficult to break into middle-income occupations in either commerce or administration. Paradoxically, many drifted to Europe, where the experience of political turmoil transformed some into activists who eventually played a role in the independence struggles in their own countries.

THE EARLY INDEPENDENCE PERIOD

As Drakakis-Smith (1987) explained, the post-Second World War decades saw independence spread rapidly through most of Asia and Africa, although in those areas where inter-war European settlement had been substantial (such as Algeria or Java) decolonisation was often slow, and contested by colonists who had little to gain from returning to a war-ravaged Europe.[19] Once the colonial powers had departed, there was a surge of indigenous people into the cities, attracted by the prospect of jobs in administrative and commercial positions from which they had been excluded. However, in the early years of independence such jobs were relatively few, given continued European control of commercial enterprises and sluggish demand for primary materials from a shattered European industrial economy.

Ironically, one of the major problems in Europe was a shortage of labour, a commodity which Third World cities had in abundance. The result was encouragement for this 'surplus' labour to move to employment in north-west Europe. Initially the migration was from former colonies to former powers with, for example, Algerians moving to France and Indians to Britain, but the trend soon spread to other poor countries, especially those around the Mediterranean (e.g. Turkey). For European industrialists and governments this migrant labour was an economic gift. It was abundant, non-unionised, cheap, docile and willing to take on menial tasks, and for two decades the 'economic miracles' of several West European states were built on this exploited labour. For their part, Third World governments were happy to encourage such moves since they helped slow urban population growth, increased foreign-exchange revenues through remittances sent home and, they hoped, trained some of their work force. In fact, the benefits for the Third World labour-exporting countries were less than anticipated. They lost large numbers of their younger and most able workers; only a minority received useful skills training; and savings and remittances were mostly used to set up small consumer businesses in the big cities on the return of the migrant, thus accentuating the problem of rapid urbanisation.

Overall, the early independence period saw little change in the economy of most Third World cities, which continued to be dominated by commercial and trading activities controlled by expatriate firms. The biggest change was in the emergence of a huge sub-sector of the urban population unable to find waged employment and whose poverty excluded them from obtaining adequate housing, education and health care. These households looked 'inwards' for survival and created their own informal economic system in which a meagre income could be earned in a variety of legal, semi-legal and illegal activities collectively referred to as 'petty commodity' production (see Chapter 24).

NEO-COLONIALISM AND THE NEW INTERNATIONAL DIVISION OF LABOUR

The incorporation of Third World labour into the world economic system began to change in the late 1960s. In Europe the migrants had become more organised, less cheap and therefore less welcome to either employers or other workers as a global recession took effect. In response, many European and North American companies began to switch their points of production into Third World cities, where cheap labour still existed and could be guaranteed by authoritarian governments reliant on the West for political support.

While the international division of labour is not a new phenomenon (throughout the colonial era the metropolitan power undertook manufacturing of raw materials produced by the colony), the major factor underlying the new international division of labour (NIDL) was the rise of finance capital comprising investment funds accumulated by organisations (such as banks and insurance companies) concerned with the management of money. The reasons why the bulk of these investment funds have gone to organisations operating in the Third World include the following:

1. Costs of production have risen in Europe and North America.
2. Cheap labour is available in Third World cities as a result of rural–urban migration (see Chapter 23).
3. A large informal sector (or reserve army of labour) is present in the Third World urban economy which helps keep demand for wage increases down.
4. Technological advances have permitted the spatial segregation of the production and management processes. The use of computers, satellite links and containerisation has made it possible for the labour-intensive parts of the production process to be located in the Third World while specialist management, research and development are retained in the home country.
5. The NIDL has also been encouraged by international agencies and national governments anxious to bring employment to burgeoning Third World cities in order to forestall possible political instability.

The impact of the NIDL on Third World cities has been mixed:

1. Most important, the changes have been selective, with only a small number of Third World countries, such as Taiwan and South Korea, rapidly expanding their industrial economy. However, as labour costs rise in these locations so the multinational corporations look elsewhere for new supplies of cheap urban labour.
2. Generally, it is the already large cities that have received the bulk of investment (see Box 21.5).
3. The social impact has been seen in class formation. A waged working class has appeared in many cities. Being more privileged than many other citizens, they form a conservative labour group. By contrast, the informal sector of the urban economy continues to grow, leading in some instance to fears of political instability. Another significant social phenomenon has been the greater incorporation of women into the urban work force.

THE ERA OF GLOBALISATION

We can add a further phase to the 'stages of colonial urbanisation' model. Covering the period from 1980 onwards, this is characterised by the onset of economic liberalisation (i.e. the progressive opening up of markets and promotion of global free trade) and the emergence of a global economic system in which cities are key elements of capital accumulation. As discussed in Chapter 14, the position of cities in the global economic hierarchy is determined by the interplay of global (e.g. FDI) and local (e.g. government policy) forces. Cities that perform 'gateway' functions, (typically large port-cities), are exposed to the global political economy.[20] In such cities globalisation has sharpened the distinction between modern internationally oriented and traditional locally oriented sectors of the economy (see Chapter 24). Global economic competition among cities has induced strategies (such as EPZs) and reduced local–national influence over urban development. In addition, freed from the regulatory framework of colonial spatial planning, urban form and land use are now influenced, to a large degree, by market forces.[21] The reorganisation of the urban space economy has also produced new geographies of inclusion and exclusion.

These are revealed most visibly in the distinction between central urban areas consisting of modern office blocks, luxury apartments and associated retail and leisure facilities and informal settlements lacking basic infrastructure and services that accommodate a majority of the urban population (see Chapter 25). While each city is embedded within a particular local context, the appearance of phenomena such as gated communities,[22] urban sprawl,[23] gentrification[24] and the privatisation of public services,[25] that mirror recent developments in many Western cities, underlines the influence of globalisation on urban development in the Third World. Furthermore, in several cities tertiarisation is overtaking industrialisation as the dominant economic process,[26] heralding the potential emergence of post-industrialism in parts of the Third World.

Fundamentally, the urban transformation of the Third World in the post-war period has led to unprecedented demands for basic services and infrastructure that most governments have been unable or unwilling to meet. These and related social, economic and political difficulties are characteristic of the phenomenon of peripheral urbanisation

PERIPHERAL URBANISATION

The model of peripheral urbanisation is an extension of dependency/world-systems theory which employs a political economy perspective to provide a generalised description of the impact of global capitalism on national urban systems in the Third World. The expansion of capitalism into peripheral areas is seen to generate a strong process of urbanisation. This may be depicted in a number of stages:[27]

1. First, cityward migration increases, owing to the disruption of pre-capitalist forms of agriculture by commercial agriculture, and the competitive impact of cheap imports on traditional craft industry.
2. Second, surplus generated in rural/peripheral areas is extracted by national bourgeois groups and representatives of foreign capital interests based in the main urban centres. The process of extraction leads to the expansion of the main transportation and market centres, and to the rapid growth of the national capital and main ports.
3. Third, the growth of manufacturing within the national economy concentrates production still further in the largest cities, stimulates the growth of a national state bureaucracy to encourage the process of industrialisation and leads to the concentration

Plate 21.2   The Burj al Arab (Arab Tower) hotel in Dubai is a symbol of diversification from an oil-based to a service-based economy

of higher-income groups in the major centres where the surplus is accumulated.

4. Fourth, labour is attracted to the largest cities in search of work and produces a surplus, through both wage labour and petty commodity production, which supports the expansion of the capitalist sector.

5. Fifth, the state acts to support industrial expansion by providing infrastructure in the main urban centres, and by legitimising the continued functioning of the capitalist system through the provision of social services to selected groups.

6. Sixth, as metropolitan development accelerates, private capital begins to deconcentrate to areas within the metropolitan region but outside the main city in order to avoid rising land prices, labour costs and negative externalities such as traffic congestion. The state may encourage the process of deconcentration or introduce measures to promote decentralisation (Box 21.6).

The model of peripheral urbanisation is a useful general description of urban growth processes in the contemporary Third World. Like all political-economy perspectives, however, there is an inherent risk that a deterministic acceptance of the power of global capitalism may obscure the diversity and potential of local responses to external forces. The importance of acknowledging a dialectic relationship between global and local forces for understanding the process of urban development in the Third World is evident in the concept of exo-urbanisation.

EXO-URBANISATION

As we have seen, transnational corporations (TNCs) use foreign direct investment (FDI) as a means of organising production processes across national boundaries by creating an international intra-firm division of labour (NIDL). According to dependency-theory formulations based on Latin American experience, the negative effects of FDI are manifested in external economic dependence and weak links between foreign and local firms, and overurbanisation as a result of the concentration of investment in the primate city.

An alternative, more positive perspective on FDI has emerged from the experience of the newly industrialised countries (NICs) of South East Asia. In some regions, most notably the Zhujiang (Pearl river) delta of south China, this has resulted in a pattern of exo-urbanisation characterised by labour-intensive and assembly-

BOX 21.6

Jabotabek, Indonesia: an extended metropolitan region

The metropolitan Jakarta region encompasses the Indonesian capital and primate city of Jakarta and its three neighbouring regencies of Bogor, Tangerang and Bekasi. This extended metropolitan region is known as Jabotabek (an acronym of Jakarta, Bogor, Tangerang and Bekasi), and the three-regency area surrounding Jakarta is known as Botabek (Bogor, Tangerang, Bekasi). The population of Jabotabek is projected to grow from 17 million in 1990 to 30 million by 2010, with most of this growth occurring in the Botabek area. As a result of urban pressure, land uses in the areas surrounding Jabotabek reveal a mix of both rural and urban characteristics. This is manifested in the growing diversification and commercialisation of agricultural activities, an increasing influx of foreign investment and development of the infrastructure, particularly improved transport and communications. These changes have resulted in a growing intensity of rural–urban linkages and the blurring of traditional urban–rural distinctions. In Asia such regions have been termed *desakota* (a term combining the Indonesian words for village, *desa*, and town, *kota*).[28]

Attempts to control the explosive growth of Jakarta have included a 'closed city' policy to stem the inflow of migrants, and a transmigration programme to encourage resettlement to less populous islands, both of which have had only limited effect. At the metropolitan scale, policies have aimed to encourage deconcentration of population, commerce and industry into secondary growth centres in Botabek. Planned measures included restrictions on certain types of manufacturing investment in Jakarta, the decentralisation of central government agencies, the promotion of competing centres outside Jakarta and a strategy of land consolidation to regulate and direct urban growth.

A basic problem encountered in many Third World cities, however, has been the lack of a comprehensive, legally enforceable metropolitan area development plan, and consequent inability of government to control the development activities of the private sector.

manufacturing types of export-oriented industrialisation based on the low-cost input of large quantities of labour and land.[29] This form of 'urbanisation from above' based on the growth of industrial cities through central planning and investment is complemented by a phenomenon of 'urbanisation from below'.[30] This has been characterised by population in-migration, urban growth in predominantly smaller cities and towns, and a major transformation of the countryside landscape with the development of an intensive mixture of agricultural and non-agricultural activities (Box 21.7).

Foreign direct investment has been a major influence on the post-reform urbanisation of China. Three phases of FDI may be identified. In the initial phase, 1979–85, foreign investment, mainly from Hong Kong and Macau, began to flow into limited areas such as four special economic zones at Shenzen, Zhuhai, Shanton and Xianen, and fourteen open cities. Most FDI was in labour-intensive manufacturing sectors. In

the second stage, 1986–91, FDI was encouraged by new legislation allowing the establishment of wholly foreign-owned enterprises and co-operative ventures. While still concentrated in the manufacturing sector, geographically FDI extended to other areas, including Shanghai Pudong. In the third phase, from 1992 onwards, FDI increased rapidly and began also to flow into the tertiary sector, such as real estate, with highly visible impacts in Chinese cities.[31]

Spatially, this foreign investment-induced form of urbanisation has given rise to the kind of extended metropolitan region that McGee (1991)[32] has characterised as a *desakota* (see Box 21.6), examples of which have been identified around cities such as Jakarta,[33] Manila[34] and Bangkok.[35] Although the 'rural urbanisation' of the

BOX 21.7

Exo urbanisation in the Pearl River delta region

Until the late 1970s, external forces did not exert a significant influence on urbanisation in China. Following the 'Opening and Reform' initiated in 1978, however, the pace of urbanisation and urban growth has quickened, and in some parts of the country, notably in the Pearl River delta, foreign direct investment has emerged as a driving force behind urbanisation. This has given rise to the phenomenon of exo-urbanisation, with the following characteristics:

1. Intensive inward flows of foreign investment, especially from Hong Kong, have become the major dynamic for the structural transformation of an agrarian economy to an export-oriented industrial economy, closely integrated with the world economy. Under this 'front shop–back factory' model the Pearl River delta acts as the back factory for traditional Hong Kong manufactures, while Hong Kong concentrates on the upper-end 'front shop' activities of the industrial economy (marketing, research and design, finance, management).

2. Labour-intensive and low-skill manufacturing has predominated in the new economy, leading to large-scale rural–urban population migration of largely unskilled females, and the expansion of the built-up area for industrial and urban uses at the expense of agricultural uses.

3. These small-scale, processing-type industrial investments are biased towards small cities and

towns and have thus brought about a rapid growth of smaller urban places and the relative decline of the primate city – a pattern referred to as 'concentrated dispersal' (see Chapter 4).

4. The trade-creative (activities based on imported raw materials and exported products), export-oriented foreign investment with a border orientation has resulted in a large number of trans-border flows of people, goods and information which has been characterised as 'triangularised urbanisation' centred on the Pearl River delta, Hong Kong and Macau.

5. The dependence on foreign direct investment means that the pattern of exo-urbanisation in the Pearl River delta is volatile. Future development patterns will reflect the region's continued ability to attract inward investment and to resist increasing competition from other regions (e.g. the Yangtze delta) and countries (such as Vietnam, Indonesia and Thailand).

6. In terms of physical structure the processes of economic integration between Hong Kong and the south China mainland have given rise to the 'Pearl City', a functionally integrated economic system and mega-urban region stretching from Hong Kong to Guangzhou and linking a constellation of smaller cities such as Dongguan, Shunde, Zhuhai and Macau and the Shenzhen Special Economic Zone with a combined population of over 40 million.

Sources: V. Sit and C. Yung (1997) Foreign investment-induced exo-urbanisation in the Pearl River delta, *China Urban Studies* 34(4), 647–77; R. Chan (1996) Urban development strategy in an era of global competition: the case of south China *Habitat International* 20(4), 509–23; Y. Yeung (1997) Planning for Pearl City *Cities* 14(5), 249–56

Pearl River delta region displays many of these features, Sit and Yang (1997) highlight a number of differences.[36] The most significant is that, in contrast to the Bangkok and Jabotabek extended metropolitan regions, in the Pearl River delta the core city of Guangzhou suffers from relative decline, and growth in the smaller centres and surrounding rural areas is unrelated to the dispersal or spread effect of the core city.[37] This can be explained by the nature of the FDI involved in the Pearl River delta. In contrast to the large-scale capital-intensive investment typical of TNC involvement elsewhere in the Third World, in the Pearl River delta FDI is mainly of small and medium size, and is in the form of labour-intensive manufacturing enterprises. For these, the availability of cheap land and labour in smaller cities and towns is a major attraction. The investment climate in smaller cities is also more attractive, since local officials are more flexible and willing to accommodate foreign investors' needs with less bureaucratic intervention than may be the case in the larger cities, where remnants of the centrally planned economy can still persist.

The phenomenon of exo-urbanisation underlines the varied forms of urbanisation and urban growth, illustrates how international forces and, in particular, flows of FDI can affect local economic development and urbanisation, and emphasises the necessity of understanding global forces within the local context (the global–local nexus[38]) in order to understand the complexity of contemporary processes of urbanisation.

URBANISATION BY IMPLOSION

As we shall see in Chapter 23, rural–urban migration has fuelled urbanisation and urban growth through-out the Third World. Yet in many countries of the developing world rural populations remain high (see Chapter 5). While these populations represent a reservoir of potential migrants to cities, in some rural regions *in situ* population growth is producing densities that equal or surpass the widely accepted urban threshold of 400 persons per square kilometre (1,000 persons per square mile). These are comparable to the population concentrations found in the exurbs of North American cities. The process of in-situ urbanisation ongoing in Fujian Province, China since the 1980s has been achieved mainly through the transformation of rural settlements into urban or quasi-urban places through the growth of manufacturing and other non-agricultural activities in the form of township or village enterprises. While this pattern bears some similarities to the model of extended metropolitan areas (desakota regions) it is mainly initiated by the rural communities and is a result of bottom-up rural development rather than the outwards expansion of large cities.[39] For some observers, these rural regions of high population density (such as the belt stretching from west Bengal to the outskirts of Delhi, in India), created by a process of 'urbanisation by implosion', represent a hybrid settlement system or **ruralopolis** that is spatially urban but economically, socially and institutionally rural/agrarian.[40] The problems facing ruralopolises include familiar 'urban' difficulties such as pressure on land, marketisation of land, increasing homelessness, and a pressing need for infrastructure facilities and services including land-use planning. In some regions, including much of Bangladesh, urban implosion in the countryside is advancing to envelop the expanding cities, creating a particular challenge for governments.

FURTHER READING

BOOKS

D. Butterworth and J. Chance (1981) *Latin American Urbanisation* Cambridge: Cambridge University Press

D. Drakakis-Smith (1987) *The Third World City* London: Methuen

A.K. Dutt, F.J. Costa, S. Aggarwal and A.G. Noble (1994) *The Asian City* London: GeoJournal Library

J. Gugler (1997) *Cities in the Developing World* Oxford: Oxford University Press

J. Gugler (2004) *World Cities beyond the West* Cambridge: Cambridge University Press

J. Gugler and W. Flanagan (1978) *Urbanisation and Social Change in West Africa* Cambridge: Cambridge University Press

M. Montgomery, R. Stren, B. Cohen and H. Reed (2004) *Cities Transformed: Demographic Change and its Implications in the Developing World* London: Earthscan

A. O'Connor (1983) *The African City* London: Hutchinson

R. Potter and S. Lloyd-Evans (1998) *The City in the Developing World* Harlow: Longman

JOURNAL ARTICLES

H. Dick and P. Rimmer (1998) Beyond the Third World city: the new urban geography of South East Asia *Urban Studies* 35(12), 2302–21

A. Gilbert (1993) Third World cities: the changing national settlement system *Urban Studies* 30(4/5), 721–40

L. Lu and Y. Wei (2007) Domesticating globalisation, new economic spaces and regional polarisation in Guangdong Province, China *Tijdschrift voor Economische en Sociale Geografie* 98(2), 225–44

V. Sit and C. Yung (1997) Foreign-investment-induced exo-urbanisation in the Pearl River delta, China *Urban Studies* 34(4), 647–77

M. Qadeer (2004) Urbanisation by implosion *Habitat International* 28, 1–12

KEY CONCEPTS

- Urbanisation
- Development
- Modernisation theory
- Dependency theory
- World-systems theory
- Colonialism
- Mercantile colonialism
- Industrial colonialism

- Independence
- New international division of labour
- Peripheral urbanisation
- Exo-urbanisation
- Rural urbanisation
- Urban primacy
- Extended metropolitan region
- Urbanisation by implosion

STUDY QUESTIONS

1. Compare and contrast the main characteristics of the urbanisation process in the First and Third Worlds.
2. Explain why the urbanisation of the Third World can be understood only by considering the region's incorporation into the world/global economy.
3. Identify the major historical stages in Third World urbanisation.
4. What do you understand by the term 'peripheral urbanisation'?
5. With references to specific examples, consider the phenomenon of exo-urbanisation.

PROJECT

Using data on the urban populations of countries in the annual report of the World Bank, examine levels of urbanisation in different countries and regions of the Third World. Relate the degree of urbanisation to other indicators of 'development' (such as GNP, housing conditions, employment, infant mortality, energy use or pollution). You may plot your results on a base map of the world if you wish. You can also make use of statistical techniques (such as correlation and regression analyses) to test the nature and strength of the relationship between factors. How would you explain the differential levels of urbanisation across the Third World?

22

Internal Structure of Third World Cities

Preview: cities of Latin America; Africa's cities; cities of the Middle East and North Africa; the city in South Asia; the South East Asian city; the Chinese city

INTRODUCTION

Third World urbanisation is not a single, uniform phenomenon. Rather it comprises several distinctive processes occurring in different parts of the less developed world that reflect the relationship between global economic forces and local cultural context. This diversity is evident at the intra-urban scale in the differing forms of city that characterise different realms within the Third World. Just as there is no single process of Third World urbanisation, so there is no such single entity as the 'Third World city'. In this chapter we examine the varied internal structure of cities in the major regions of the Third World.

CITIES OF LATIN AMERICA

Little remains of the indigenous urban civilisation, which was largely destroyed by the Spanish conquest. The influence of the pre-Columbian culture on the urban geography of Latin America is limited to the tendency to build a European settlement on the same geographically favoured sites, as is evident in the locations of inland cities such as Bogotá and Mexico City. In general, however, the advent of mercantile colonialism reoriented the inward-looking indigenous urban system onto port-cities such as Lima and Rio de Janeiro. The colonial legacy influenced much more than the physical location of cities, however. As Griffin and Ford (1983) commented, more than in any other culture region, Latin American cities share a common

urban structure that derives from their colonial roots and that persists to the present day.[1]

During the colonial period the Spanish American city was controlled by the Laws of the Indies, which dictated a gridiron street pattern developed around a central plaza, around which all major government offices, the majority of commercial activities and most social amenities were clustered. Consequently, employment opportunities were also concentrated in the city centre. Residential proximity to the centre was a symbol of social status. This pattern, which reflects the layout of Sjoberg's pre-industrial city (see Chapter 3), remains a characteristic of many smaller Latin American cities. In larger cities, urbanisation and urban growth, especially since the Second World War, have been accompanied by several factors that have 'modernised' Latin American cities and transformed traditional urban forms. Griffin and Ford (1980) proposed a model of the Latin American city, subsequently updated by Ford (1996), that seeks to combine traditional elements of urban structure with the effects of modernising processes.[2] As Figure 22.1 shows, the model is characterised by a downtown area, a commercial spine and associated elite residential sector, and a series of concentric zones in which residential quality decreases with distance from the city centre.

Downtown and inner city: the downtown area is divided into a 'modern' central business district (CBD) and 'traditional' market district. The distinction is seen most visibly in the physical contrasts between small street-oriented businesses and self-contained

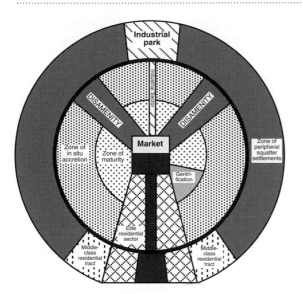

Figure 22.1   A model of the Latin American city

Source: L. Ford (1996) A new and improved model of Latin American city structure *Geographical Review* 83(3), 437–40

megastructures. Unlike in most large North American cities, the CDD has remained the principal employment, commercial and entertainment node of the Latin American city. The relative dominance of the CBD is explained in part by reliance on public transport lines that continue to focus on the CBD, and in part by the existence of a relatively affluent middle-class population in the inner rings of the city. The CBD is surrounded by an 'inner city', but this area is not typified to the same extent by the population and economic decline evident in North American cities. The Latin American inner city exhibits economic viability, with commercial districts providing cheap foodstuffs, clothing, wholesale and retail markets, and a wide variety of goods manufactured in small-scale workshops and sweatshops, many of which form part of the informal economy (see Chapter 24). The area accommodates stable working-class communities with urban cultural traditions akin to that of areas such as the Cockney East End of London or the Jewish working-class district of the Bronx in New York. It also acts as a reception area for new migrants, which helps to maintain the resident population.[3] While the inner city represents the principal concentration of urban problems in North America or the UK, this is not the situation in Latin America.[4]

Commercial spine and elite residential sector. The dominant element of city structure is the commercial/industrial spine surrounded by an elite residential sector

that extends outwards from the CBD of all Latin American cities. This 'tree-lined boulevard' contains the city's most important amenities, including private hospitals, hotels, museums, theatres and nearly all the professionally built upper-class and upper middle-class housing stock. Land-use and zoning controls are enforced strictly, to maintain the quality of the area. Within the Hoytian spine, the housing market operates as it does in North American cities, with dwellings in once fashionable neighbourhoods closer to the city centre, vacated by the wealthy elite, filtering down to the upper middle class. Increasingly, upper middle-class apartments and houses are built on the fringes of the elite residential sector, partly owing to increasing demand in this market and partly as a buffer against the squatter settlements near by. The elite residential sector combines Western-style amenities with a traditional Latin American desire for centrality. It also represents a morphological response to the limited ability of city authorities to extend urban services, the relatively recent acquisition of 'suburban values' by the elites, and the limited availability of mortgage finance coupled with conservative lending practices and high interest rates. In many Latin American cities a suburban mall or competing 'edge city' development runs along the outer end of the spine. Often this incorporates walled residential areas or gated communities that provide elite residents with a secure, comfortable living environment segregated from the mass of the population.[5]

Industry. An industrial sector, often following the line of a railway or highway, culminates in a suburban industrial park that accommodates space-extensive factories and warehouses.

Periférico. The industrial park and mall are connected by a ring road. Although most large Latin American cities now have some form of peripheral highway system, adjacent development is still limited by the difficulties involved in expanding infrastructure and upgrading outlying squatter settlements. While the ring road may not completely encircle the city, within the elite sector the *periférico* may form a boundary between older-established communities and newer, planned unit developments.

Other residential zones. Away from the commercial–elite spine, and in contrast to the model proposed by Burgess for the North American city, the urban structure comprises a series of zones in which socioeconomic status and housing quality decrease with increasing distance from the central core. These are:

1. *Zone of maturity.* This is an area of better residences comprising filtered-down former elite

dwellings and self-built housing that has been gradually upgraded over time by residents unable to participate in the housing market of the elite residential sector. The area is likely to be fully serviced, with paved streets, lighting, sewerage, schools and public transportation. Dwelling-unit density is similar to that in more peripheral zones, but because the population is older, there are fewer children and population density may be lower. In many cities, old colonial residences around the *zócalo* (plaza) have suffered from lack of investment as the wealthy have moved along the spine, leaving such dwellings to become rented apartments.[6] As in Western cities, part of this zone may be undergoing gentrification, often aided by government recognition and legislation designed to protect historic townscapes.

2. *Zone of* in situ *accretion*. In contrast to the relative stability of the previous zone, this zone is in a constant state of flux as residents move in and depart according to lifestyle and status. The zone is characterised by a variety of housing types, sizes and quality, and many houses have unfinished rooms or second storeys. Some districts are 'completed' and similar to the zone of maturity, while others are 'under development', either by self-builders or by government-sponsored housing projects. The level of service provision is variable, with truck delivery of water and butane gas for heating and cooking common. Although Griffin and Ford (1980)[7] envisaged that such zones would gradually improve over time, the rate of change will depend on the health of the national economy, which determines the city's ability to provide infrastructure, and the economic prospects of its residents.

3. *Middle-class residential tracts*. These areas are typically located as close as possible to the elite sector and the *periférico* in order to ensure access, status and protection. Although not depicted in the model, middle-class housing tracts and/or government housing projects may also be sited in the zone of *in situ* accretion near the suburban industrial park.

4. *Zone of peripheral squatter settlements*. This accommodates the impoverished migrants to the city and is the worst section of the city in terms of housing quality and public service provision (see Chapter 25). In addition to the main squatter areas, Latin American cities also contain sectors of disamenity (Figure 22.1). These areas have not consolidated over time through *in situ* accretion and remain areas of slums and rented tenements

(as in the *favelas* of Rio de Janeiro). Major industrial and environmentally polluting activities are also likely to be located here.

Crowley's (1995) urban land-use model also contributes to our understanding of Latin American city structure by emphasising the relative absence of government control over land use and subsequent organic mix of residential, commercial and industrial functions that is characteristic of Mexican and Central American cities in particular[8] (Figure 22.2).

These models serve as useful pedagogic devices, elements of which are evident in many Latin American cities (Box 22.1).

AFRICA'S CITIES

Africa is the least urbanised of the continents (see Chapter 5) yet exhibits the greatest variety of urban forms. This diversity stems from the distinctive indigenous urban traditions, particularly in North and West Africa, and from the urban imprints of the colonial powers.

The United Nations (1973)[9] proposed a general model of the African city based on the existence of an indigenous core, and the distribution of different ethnic groups according to density gradients which assigned low-density land use to the administrative and residential requirements of the colonial elites and high density to indigenous populations (Figure 22.3). Criticism of the model focused on its failure to recognise the postcolonial transformations of African cities, characterised by a greater mixing of economic and residential land uses. In a more comprehensive analysis O'Connor (1983)[10] identified seven types of African city:

1. *The indigenous city*. Indigenous cities were constructed in the period prior to European colonisation in accordance with local values and traditions. In south-west Nigeria the Yoruba city of Ife dates back to the tenth century, while at least ten others with populations exceeding 50,000 (such as Ibadan) existed before colonial rule.[11] Elsewhere in tropical Africa, Addis Ababa is the largest extant example of an indigenous city.

2. *The Islamic city*. Though influenced by an urban tradition brought across the Sahara, most **Islamic cities** were built by Africans, with local initiatives dominant in their early growth. Found across much of the Sahara, this type includes Tombouctou, Katsina and Sokoto.

BOX 22.1

The structure of the Latin American city: the case of Mexico City

The Spanish colonial city, built on the remains of the Aztec city of Tenochtitlán, was based on a rectangular grid plan focused on a central *zócalo* which contained a large plaza, national palace and cathedrals, all of which remained prominent after independence in 1821. In the early twentieth century the city's growth largely accrued around the *zócalo*, with elite residences being built westwards near Chapultepec Park. In the 1930s, when the population had reached 1 million (from 345,000 in 1900), many members of the elite built residences in Lomas de Chapultepec, along the tree-lined Avenida de la Reforma. Since then, as a result of rapid population growth, the city has constantly expanded its boundaries to encompass many surrounding municipalities. After the Second World War the city expanded in all directions, but particularly to the east on to the dry bed of Lake Texaco. This unstable eastern district, which was mainly occupied by lower-income settlements, was the main area of destruction in the 1985 earthquake.

The central business district has expanded westwards for over a mile along the Paseo de la Reforma, known locally as the *zona rosa* (pink zone), and accommodates the city's main commercial offices, exclusive retail outlets, theatres and hotels, serving as a dynamic centre for the wealthy. The historic *zócalo* still retains an important religious and administrative function but has filtered down the socio-economic ladder, and many colonial buildings have been subdivided into rented apartments (*vecinidades*).

The elite westerly commercial/residential spine extends to the suburban development of Bosques de las Lomas with its American-style shopping malls, golf and health clubs and private security patrols. Within the sector, nearer the city centre, areas such as Polanco are filtering down to upper middle-class residents.

Established areas of squatter settlements known as *colonias proletarias* (people's neighbourhoods), constructed in the 1950s, have now become part of the formal city structure and are reasonably well serviced, forming zones of maturity around the city centre. Elsewhere, new *colonias proletarias* are growing on the urban periphery close to the airport and major industries, and lacking basic services. Areas of slums, known as *ciudades perdidas* (lost cities), also constitute part of these developments. Some of the peripheral *colonias proletarias* are themselves the size of cities; for example, in the 1980s Naucalpán in the north-west had over a million residents.

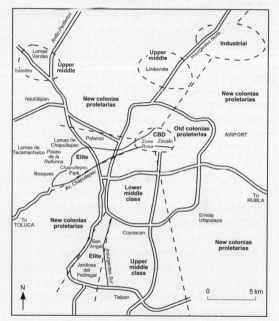

Sources: P. Ward (1990) *Mexico City* London: Belhaven Press; R. Potter and S. Lloyd-Evans (1998) *The City in the Developing World* London: Longman

3. *The colonial city*. Established by Europeans, mainly in the late nineteenth and early twentieth centuries, colonial cities comprise the majority of urban centres in tropical Africa and include most of today's capital cities. From the earliest years, immigration has ensured an African majority in the population. Although many decisions affecting city structure are made locally, they are still constrained by the inherited colonial framework and by contin-uing ties with the international economic system. Income is replacing ethnicity as a basis of residential segregation, with 'Westernised' Africans dominating the formerly European residential zones.

4. *The European city*. Founded primarily in southern and eastern Africa, for example Nairobi, Lusaka and Johannesburg, these settlements were established by and principally for Europeans. African in-migration and permanent residence were

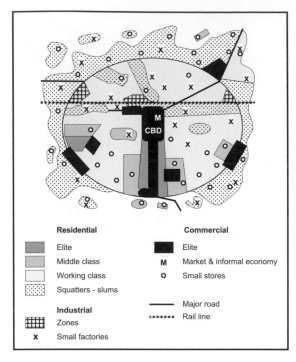

Residential

Elite

Middle class

Working class

Squatters - slums

Industrial

Zones

x Small factories

Commercial

Elite

M Market & informal economy

o Small stores

——— Major road

••••••• Rail line

Figure 22.2 A model of Latin American urban land use

Source: Modified from W. Crowley (1995) Order and diversity: a model of Latin American urban land use *Yearbook of the Association of Pacific Coast Geographers* 57, 28

constrained, subject to the labour requirements of the Europeans. Normally the African population lived in segregated areas on the urban fringe (see the following discussion of the apartheid city).

5. *The dual city*. In a dual city, two or more of the above types are combined, as in Kano, where a walled Islamic city is surrounded by a modern colonial-type city, or Khartoum–Omdurman, where the Islamic and colonial city elements are separated by the river Nile.

6. *The hybrid city*. A hybrid city is one that comprises indigenous and alien elements in roughly equal proportions (as in the dual city) but in which the parts are integrated. This urban type has increased since decolonisation as cities expand and become more integrated. Examples include Accra, Kumasi and Lagos.

7. *The apartheid city*. South Africa's apartheid city represented a unique form of urban social segregation that dominated the national urban system for most of the second half of the twentieth century. The roots of the apartheid city lay in the concept of 'separate development' and in early British colonial policy, which favoured 'native' reserves

and urban segregation of Africans. Davies's (1981) model of the colonial 'segregation city' is a useful representation of the pre-apartheid city in South Africa.[12] This depicts a highly but not wholly segregated urban space (Figure 22.4). The Group Areas Acts of 1950 and 1966 and the 1953 Reservation of Separate Amenities Act extended the principle of racial segregation to produce the urban structure of the apartheid city, designed to minimise interracial contact. In this the living environments of the different races mirrored their socio-political positions (Figure 22.5). Despite the collapse of the apartheid regime, the legacy of almost half a century of apartheid urban policy remains a strong influence on urban form in South Africa[13] (Box 22.2).

CITIES OF THE MIDDLE EAST AND NORTH AFRICA

Since the first urban settlements appeared in the Middle East (see Chapter 3), towns have remained important nodes of population and culture. The religion of Islam has dominated the region since the seventh century. The pervasive influence of Islam on the lives of people in the Middle East led Western orientalists to propose a model of the Islamic city with an urban structure that reflects religious principles (Figure 22.6):

1. The lack of any corporate bodies in a society composed of state and subjects obviated the need for public buildings.

2. The city has a major mosque, called the Friday mosque, which as well as being a place of worship also provides a range of welfare and education functions. As cities grew, so did the number of local mosques, built in some districts out of earshot of existing ones.

3. The bazaar or *suq* is a key element of the Muslim city. *Suqs* comprise a series of small, contiguous market stalls located in a maze of passageways that are often covered over by vaults or domes. The *suqs* display a degree of functional specialisation, with complementary trades such as leather-sellers next to shoemakers. (Compare the medieval city, discussed in Chapter 3.)

4. The irregular street pattern reflects both a lack of a civil planning authority to prevent the encroachment of houses onto public thoroughfares and a response to local climate by maximising shade.

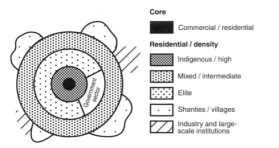

Core

■ Commercial / residential

Residential / density

▨ Indigenous / high

▩ Mixed / intermediate

⬚ Elite

⬚ Shanties / villages

▨ Industry and large-scale institutions

Figure 22.3 A general model of the African city

Source: United Nations (1973) *Urban Land Policies and Land Use Control Measures* volume 1 New York: United Nations

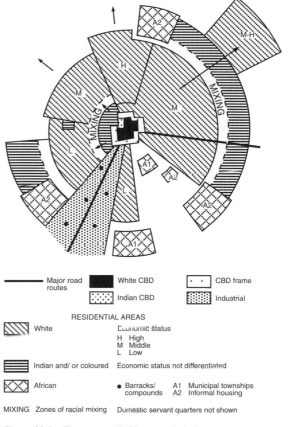

▬▬ Major road routes

■ White CBD

⬚ Indian CBD

⬚ CBD frame

▨ Industrial

RESIDENTIAL AREAS

▨ White Economic status
H High
M Middle
L Low

▤ Indian and/ or coloured Economic status not differentiated

▨ African ● Barracks/ compounds A1 Municipal townships A2 Informal housing

MIXING Zones of racial mixing Domestic servant quarters not shown

Figure 22.4 The pre-apartheid segregated city

Source: A. Lemon (1991) *Homes Apart* London: Chapman

Plate 22.1 Most of the housing in the indigenous part of the Nigerian city of Kano is built in the form of walled compounds with dwellings constructed of a mix of local (mud walls) and imported (corrugated iron roofs) materials

5. The residential fabric is composed of a compact structure of open courtyard houses in which all rooms face onto the interior courtyard. As well as providing thermal relief, this dwelling form reflects the importance attached to privacy in Islamic society, especially for women. This is also promoted by the use of culs-de-sac that restrict the number of people needing to approach the house, and L-shaped entrances that block views into a house.[14]

Although these morphological elements remain distinctive features of cities in North Africa and the Middle East, the concept of the Islamic city has been questioned on the grounds that many of the structural characteristics are climate-related rather than religion-based, and may be found in many pre-industrial cities. Within the modern Middle Eastern city the old Islamic city or *medina*, built originally for a pedestrian society, can pose major problems for city planners, particularly if seeking to improve vehicular accessibility.[15] The *medinas* also tend to house a population of poor and relatively recent migrants, as wealthy residents have abandoned the area in favour of more modern living environments (Box 22.3).

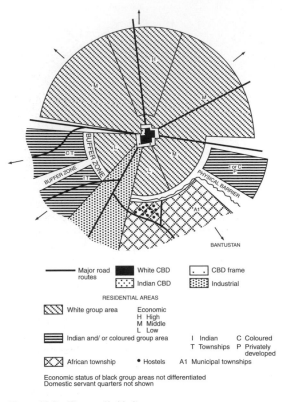

Figure 22.5   The apartheid city

Source: A. Lemon (1991) *Homes Apart* London: Chapman

THE CITY IN SOUTH ASIA

As in other regions of the Third World, contemporary urban forms in South Asia reveal the imprint of both indigenous and colonial forces. Two basic models depict the effect of these forces on the form of the South Asian city: the colonial-based city and the bazaar-based city. The *colonial-based* city model reveals features characteristic of colonial foundations elsewhere but also reflects the particular colonial methods of the British in the Indian subcontinent (Figure 22.7):

1. Colonial cities in the subcontinent generally have a waterfront location accessible to ocean-going ships. This facilitated trade and any necessary military intervention, and provided the initial growth point of the city.
2. A walled fort adjacent to the port afforded protection for the colonists. It also often accommodated factories processing agricultural raw materials for

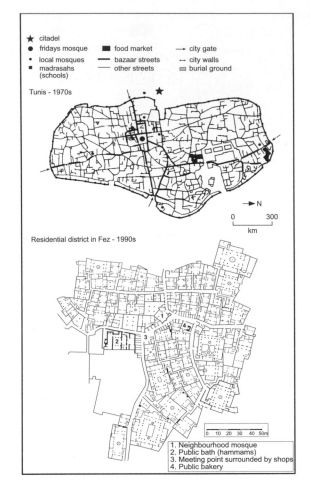

Figure 22.6   Structure of the Islamic city

Sources: adapted from S. Lowder (1983) *Inside Third World Cities* London: Croom Helm; S. Bianca (2000) *Urban Form in the Arab World* London: Thames and Hudson

export to the mother country, thereby providing a nucleus of the colonial exchange system.

3. An open space (*maidan*) around the fort for security.
4. Part of the open space between the fort and European town was reserved for military parades and recreational facilities, such as a racecourse or cricket ground.
5. Beyond the open area was a native town developed to service the fort and the colonial administration. This was an unplanned area of overcrowded, insanitary living conditions.
6. A Western-style CBD contained the major commercial and administrative functions, institutions and public buildings, and low-density residential

BOX 22.2

Towards the post-apartheid city

The form of the future post-apartheid city in South Africa is constrained by the social and physical structures created under apartheid. Some signs of change are evident in the emergence of inner-city 'grey areas': white areas of the apartheid city in which non-whites have taken up residence. The phenomenon of 'greying' first became apparent in the 1980s, and reflected the chronic housing shortage experienced by those non-whites who, though wealthy enough to buy, build or rent houses in middle-class white areas, were precluded from the market on grounds of race. Demand for housing close to the central business district (CBD) was fuelled by a growing trend for black people to gain entry to commercial-sector employment previously reserved for whites. Access to cheap accommodation and informal job opportunities in and around the CBD also attracted poorer residents. The residential quality of grey areas ranges from houses occupied by affluent black professionals to subdivided apartment blocks managed by slum landlords.

In the longer term it is likely that the capital gains to be made from the redevelopment of low-quality apartment stock will hasten the displacement of poor inner-city residents in favour of (black and white) 'yuppies' within a non-racial residential structure for the high-rise inner-city area.

To a lesser degree, a trend towards racial integration is seen in the incursion of small numbers of non-whites into all-white suburbs. However, in politically and socially conservative areas this process has encountered resistance by white residents. In these areas, repeal of the Separate Amenities legislation was met with local by-laws demanding exorbitant deposits from non-residents wishing to use facilities such as the local library, and an insistence that public swimming pools could be used only by holders of (high-cost) season tickets. Although such 'exclusionary zoning'

practices may be made illegal, in practice discrimination on economic grounds can have the same effect as overt racial discrimination. Conversely, on the other side of the coin, residential segregation between whites and non-whites may also be maintained by the construction of up-market enclaves for black, 'coloured' or Asian households, similar to the 'voluntary apartheid' in some US suburbs.

The spatial impact of continued economic and racial segregation within the fabric of the post-apartheid city is most evident in the presence of the black townships (such as Soweto). Repeal of all apartheid legislation and introduction of more progressive urban policies will not eliminate such areas in the short term. Indeed, continued expansion of racially homogeneous areas is an inevitable consequence of the poverty of the mass of the black population. Change is more likely to take the form of *in situ* improvements by occupants, and perhaps more of the upgrading and resale of shelter that already occurs where the private sector sees an opportunity to profit from this form of 'gentrification'.

The only certainty in the post-apartheid city is that the numbers of homeless and jobless people will increase, to ensure that a 'squatterscape' will continue to be part of the urban geography The post-apartheid city of the early twenty-first century is a place of continuing segregation. A black elite, along with a skilled and well organised labour aristocracy, will gain access to better residential areas, but for the mass of Africans in the townships and 'shack-lands', and for many 'coloureds' and Asians, little if any improvement can be anticipated. The prospect is one of clear class divisions augmenting the racial segregation inherited from the past to produce a post-apartheid city characterised by what some have referred to as 'deracialised apartheid'.

areas. A bazaar style of commercial area developed in the native town.

7. The planned European town comprising spacious bungalows along tree-lined avenues, developed away from the native settlement.

8. At an intermediate location between the 'black town' and 'white town' developed colonies of Anglo-Indians. The fact that they were Christian and the offspring of mixed European–Indian marriages ensured that they were not accepted by either the

native or the 'pure-bred' European communities.

9. As the colonial city expanded, new living space for the elites was provided by peripheral developments, mostly undertaken by private co-operative housing associations and specially designated improvement trusts supported by city revenues.

Calcutta, Mumbai and Madras represent classic examples of these colonial-base cities in India. As colonial power over the subcontinent tightened, the capital of

Plate 22.2   The King Abdullah mosque occupies a prominent position in the Islamic city of Amman, Jordan

BOX 22.3

Cairo, Egypt

With a population in 1995 of 9,656,000, Cairo is the largest city in the Middle East. It is also one of the world's most densely populated cities. Population expansion due to rural–urban migration and natural increase of the resident population have overloaded urban services and led to acute housing shortages. Much of the old city or *medina* is an urban slum accommodating those without the economic means to move, and new migrants to the city. In the *medina* and older working-class districts, population density reaches over 260,000 per square mile (100,000 persons per square kilometre). The extent of overcrowding and shortage of housing is most evident in the 'cities of the dead' on the eastern edge of the old city, where tomb houses unserved by any municipal utilities and designed for temporary occupation by visitors to the cemetery have been occupied by rural migrants. An entire urban culture has also developed among the 500,000 rooftop dwellers who occupy

shacks built on the roofs of the city's apartment buildings. Provided the structures are not constructed of permanent material, they do not require a building permit and are not illegal. Overcrowded and substandard housing is also found in the old industrial zones in the port area on the east bank of the Nile and to the north of the city, where urban expansion is threatening rich agricultural land. The Cairo elite occupy the previous western section of the city, as well as the Nile islands of Rawda and Gezira, and the east bank district containing foreign embassies and exclusive apartment complexes. In contrast to the socio-spatial admixture of pre-modern Cairo, the morphology of the modern city has been determined largely by the Western principle of economic status leading to greater residential segregation. The vast majority of the population occupy poor-quality environments in an impoverished yet expanding metropolis.

British India moved to Delhi in 1911, where a new European city was appended to the historic city (see Box 21.4).

The traditional *bazaar-based* city is widespread in South Asia and retains features that pre-date the colonial era (Figure 22.8). This model comprises a number of concentric zones:

1. The city grows normally in response to a trade function originating from agricultural exchange, location as a transport node, temple site, or due to an administrative role. The urban core is a crossroads or *chowk* around which develop the houses of the richer merchants, who often live above or behind their shops and warehouses.
2. In the bazaar, which provides the central-place functions of the city, retail activities are dominated by the sale of basic necessities such as foodstuffs and clothing, and functional specialisation is common. Overnight accommodation is also available in the bazaar through traditional

public non-profit inns (*dharmasalas*) or Western hotels in medium-sized and large cities.
3. Beyond the inner core the rich live alongside, but not in the same structures as, their servants.
4. The houses of the poor occupy a third zone in which the demand for and price of land are low.
5. As the city grows, ethnic, religious and caste neighbourhoods are formed in specific areas, the location of which depends on the time of settlement and availability of developable land. The 'untouchables' always occupy the periphery of the city, although other housing may develop beyond their neighbourhoods as the city expands. In Hindu-dominated areas of India the Muslims always form separate residential neighbourhoods.[16]

The colonial and bazaar-based city models provide useful representations of the forces underlying the different urban traditions in the subcontinent, but in practice many Indian cities reveal a combined impact in their contemporary urban form (Box 22.4).

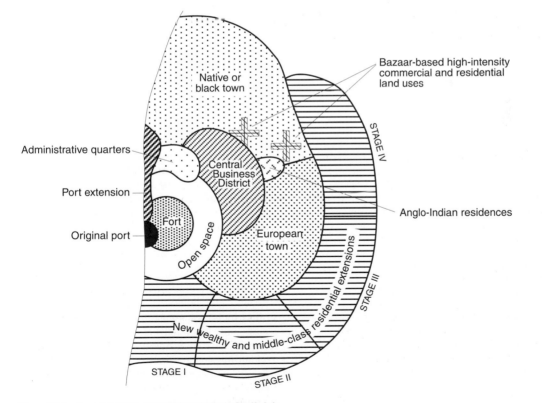

Figure 22.7   A model of the colonial-based city in South Asia

Source: S. Brunn and J. Williams (1983) *Cities of the World* New York: Harper & Row

THE SOUTH EAST ASIAN CITY

External cultural influences have had a major impact on the urban geography of South East Asia. Over the centuries, Indians, Chinese, Arabs, Europeans, Americans and Japanese have all shaped the region's cities.[18] The period of European influence dates from 1511, when the Portuguese captured the port of Malacca, but the major colonial impact on the urban system was experienced during the nineteenth century, when European investment created or boosted coastal primate cities such as Manila, Jakarta, Singapore, Saigon (now Ho Chi Minh City), Hanoi, Bangkok and Rangoon.

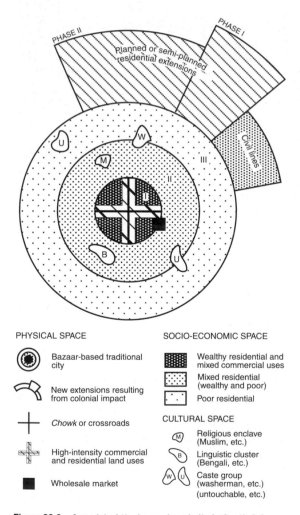

PHYSICAL SPACE

Bazaar-based traditional city

New extensions resulting from colonial impact

Chowk or crossroads

High-intensity commercial and residential land uses

Wholesale market

SOCIO-ECONOMIC SPACE

Wealthy residential and mixed commercial uses

Mixed residential (wealthy and poor)

Poor residential

CULTURAL SPACE

(M) Religious enclave (Muslim, etc.)

(B) Linguistic cluster (Bengali, etc.)

(W)(U) Caste group (washerman, etc.) (untouchable, etc.)

Figure 22.8   A model of the bazaar-based city in South Asia

Source: S. Brunn and J. Williams (1983) *Cities of the World* New York: Harper & Row

The ethnic diversity caused by the influx of foreign migrants is a prominent element in McGee's (1967) model of the South East Asian city[19] (Figure 22.9). Commercial zones are differentiated by the ethnicity of the entrepreneurs, whether alien (Chinese or Indian) or Western. A high-class residential sector extends outwards from the government zone. Squatter settlements are located on the urban periphery, along with more recent suburbs. The growth of the city is spreading urban influences into the surrounding countryside, producing a *desakota*, or extended metropolitan region. Another significant feature is the spontaneously evolving traditional villages (**kampungs**) which occur throughout the city, having been absorbed by urban growth.[20] These include both planned legal *kampungs*, designed for those displaced by urban development, and illegal squatter settlements.

These characteristics have been incorporated into Ford's (1993) model of the Indonesian city,[21] which identifies nine major zones (Figure 22.10):

1. *Port-colonial city zone.* The port-colonial city zone remains a major morphological component in most Indonesian coastal cities. Even when new port facilities have been constructed elsewhere the old port normally retains some functions, while the adjacent Dutch colonial district is a visible, if functionally marginal, element of the cityscape.

2. *Chinese commercial zone.* Chinese people typically constitute 10–40 per cent of city residents, and ethnic Chinese capitalists now control much of the Indonesian economy. The Chinese commercial zone is a distinct high-density district of traditional shop-houses and new shopping plazas that extends from the colonial city out along the inner part of the commercial spine.

3. *Mixed commercial zone.* The mixed commercial zone is a zone of ethnically and functionally mixed activities and architectural diversity which is the economic heart of the city.

4. *International commercial zone.* The international commercial zone typically runs along the main monumental boulevard and contains the major office buildings, convention centres, luxury shopping outlets, hotels and entertainment foci. Immediately alongside the zone are *kampungs* in which low-income housing is under threat from commercial expansion of the zone.

5. *Government zone.* Relatively distant from the colonial city, government offices are set in public or semi-public open spaces that act as the lungs

Plate 22.3   A squatter settlement in the shadow of luxury hotels in Mumbai

of the city. Elite residences are often located near by.

6. *Elite residential zone*. The elite residential zone is provided with modern urban services and land-use controls. Planned gated luxury residential communities have been developed around rural *kampungs*.

7. *Middle-income suburbs*. Middle-income suburbs are a relatively new phenomenon, since this income group was limited in numbers until recently. Most have arisen on either side of the central spine away from both the elite residential areas and the *kampungs*.

8. *Industrial zone*. While heavy industry has played a minor role in most Indonesian cities, since the import substitution programmes of the 1970s port facilities, suburban industrial parks and satellite cities have emerged to exert an impact on future urban structure.

9. *Kampungs*. These unplanned low-income residential areas constitute a significant element in the urban structure of the Indonesian city (Box 22.5).

An important distinction between the models of McGhee (1967) and Ford (1993) is that the latter incorporates the effects of globalisation on the structure of the South East Asian city. This replaces the notion of a unique regional urban form with that of a regionally distinctive variation on a more general process of urbanisation that is producing similar urban outcomes in major cities in both the First and Third Worlds.[23] This is evident in Jakarta, where gated residential communities, condominium developments, shopping malls, freeways and suburbanisation of industry and housing are indicative of a convergence of urban land use patterns between First and Third World cities[24] (Box 22.6).

Mumbai, India

The combination of colonial, indigenous and post-Independence urban elements is evident in cities such as Mumbai (Bombay). In this modern Indian 'megacity' the original colonial fortified town developed first into a multi-functional commercial and administrative centre, then metamorphosed into a 'Western' form of CBD characterised by tertiary and quaternary-sector land uses. The principal docks and associated functions are concentrated on the eastern waterfront of Bombay Island. In residential terms, following Independence in 1947 the former exclusive European areas were occupied by a Westernised Indian elite, the colonial-era residential segregation on religious and ethnic lines being replaced by segregation on religious and socio-economic lines. The open spaces or maidens became esplanades lined by upper and middle-class apartments. In addition to an admixture of small-scale retail, wholesaling, service and backyard industries, the bazaar area is characterised by internal segregation of land uses. Certain streets specialise in providing particular commodities and

services, and different ethnic groups occupy specific niches in the urban economy (with, for example, Hindus and Jains in textile cloth and yarn markets). Shopkeepers typically live above their premises, with employees living either in the shops themselves or on the pavements.

The subsequent post-Independence development of Mumbai has introduced further 'Western' characteristics into the urban landscape. Population growth and rapid suburbanisation have led to attempts to incorporate contiguous territory into the metropolitan area, although the city remains underbounded in comparison with the functional metropolitan region. This leads to problems of growth regulation and urban planning. In an effort to ameliorate growing congestion several 'counter-magnets' have been initiated, as at New Mumbai and Bandra-Kurla.[17] As in most large cities of the Third World the most visible sign of rapid urbanisation and urban growth is the ubiquitous interstitial spread of slums (*zopadpattis*) and squatter settlements.

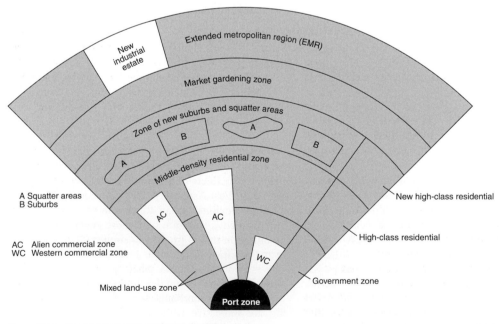

Figure 22.9   A model of the South East Asian city
Source: T. McGee (1967) *The Southeast Asian City* New York: Praeger

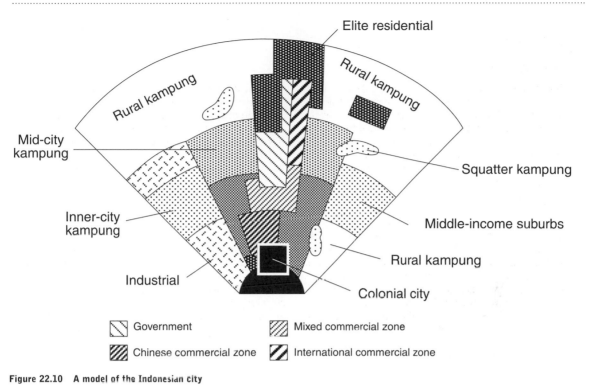

Figure 22.10   A model of the Indonesian city
Source: L. Ford (1993) A model of the Indonesian city structure *Geographical Review* 83(2) 374–96

THE CHINESE CITY

Chinese cities have changed more in the past fifty years than in the course of the preceding 500 years. We can identify five major epochs in Chinese urban development:

1. *Early traditional urban form* (AD 202–618). Many early cities were planned rectangular walled settlements established primarily for administrative and military purposes. Trade was controlled and limited to designated spaces. Internal walls delimited functionally differentiated spaces, with narrow straight streets linking walled compounds.
2. *Late traditional urban form* (AD 618–907). It became common for cities to be established for economic rather than administrative purposes. Existing cities also gained commercial functions both on existing streets and in new districts outside early walls. New walls were often constructed to encompass the extended urban area.
3. *The treaty port era* (1842–1949). The Treaty of Nanking that ended the opium wars between

China and Britain marked the beginning of a new era in Chinese urban form by permitting foreigners to live and carry out business in treaty ports (such as Shanghai) that were extraterritorial enclaves adjacent to Chinese cities. Sections of cities were remade along Western lines, with new factory districts, commercial areas and residential neighbourhoods of single-family homes. During the treaty port era and the republic era (1911–49) cities expanded in terms of population, physical extent, trade and industrial production. Western architecture became increasingly prevalent. City walls were modified or removed to accommodate growth pressures and to relieve congestion. An influx of migrant labour led to the formation of districts of poorly serviced informal housing on the outskirts of cities.

4. *The Maoist city* (1949–78). In the aftermath of the 1949 revolution the national government set out to rebuild the cities in accordance with socialist ideology (see Chapter 8). Cities were to be transformed from consumer cities to producer cities. Industry came to dominate urban life, with newly

BOX 22.5

The *kampungs* of Jakarta, Indonesia

Approximately two-thirds of all urban Indonesians live in *kampungs*, which, in general, may be regarded as largely unplanned and primarily low-income residential areas. Ford (1993)[22] distinguishes four types of *kampungs:*

1. *Inner-city kampungs.* Located mostly between the original colonial city and the new inland cores, these are typically old, high-density areas with severe environmental problems. Densities reach as high as 260,000 per square mile (100,000 per square kilometre) in Jakarta, with a high percentage of the population sharing space in *pondoks*, or traditional rooming houses. Health problems are rampant owing to uncovered sewers, flooding is a major problem in the rainy season, and traffic noise and congestion add to the unpleasant conditions. Population is declining in many inner-city *kampungs* as a result of commercial intrusions but overcrowding remains high, with access to employment a main attraction.
2. *Mid-city kampungs.* These are less densely populated, with typically 20,000–40,000 people per square kilometre (52,000–104,000 per square mile). Environmental conditions are better, with many benefiting from government *kampung* improvement schemes (KISs). Some mid-city *kampungs* are close enough to fashionable residential districts and the commercial spine of the city to benefit from the provision of urban services and the employment opportunities.
3. *Rural kampungs.* Some of these remain essentially agricultural, others are being engulfed by the city. Though population densities are lower than in other *kampung* areas, rural densities are sufficiently high to blur the boundary between urban and rural, forming a *desakota* region (see Box 21.6). Services are scarce, streets unpaved and traditional building materials intermixed with more substantial modern dwellings. The population is less transient than that of the inner-city *pondoks*.
4. *Temporary squatter kampungs.* These are scattered throughout the metropolitan area and are associated with disamenity sites such as marshland or with areas in transition to other uses. Although many areas have existed for a long time, they are officially temporary, since they are not eligible to participate in KISs. Since people with no legal tenure have little inclination to improve the area, conditions are appalling.

BOX 22.6

Private cities in Greater Jakarta

Lippo Karawasi (Lippo Village) and Lippo Cikarang (Lippo City) are two of nearly twenty new town developments throughout Greater Jakarta. Both were planned in the early 1990s, by the Lippo consortium of private companies, and are now fully functioning cities of 30,000 residents each. The plan is for both to accommodate 1 million people by 2020. Not simply suburban real-estate developments but private cities, they contain a full array of urban facilities and systems that compare favourably with those found in corporate city developments in the USA, such as Irvine CA. Much of the labour force in these upper middle-class private cities comes from the surrounding *kampungs*, with movement between the two realms controlled by each city's private security force.

developed urban areas shaped around new centres of industrial employment. The basic building block was the *danwei* (work unit), a walled compound containing a workplace, housing, social services and recreation facilities. New developments consisted of multiple *danwei* laid out on a grid pattern. Older neighbourhoods were reorganised around the concept of neighbourhood units that functioned similarly to the *danwei*. As residents were assumed to have no great need to travel outwith their neighbourhood, 'unproductive' infrastructure, such as transport developments, was limited in proportion to the size of the cities. In many Chinese cities, a large paved space was created, (as in Beijing's Tiananmen Square), to cater for mass rallies and ceremonies.

5. *The contemporary 'great international city' (post-1979).* The contemporary city is a direct product of the economic-reform process that introduced foreign investment, private and semi-private enterprises, economic competition and

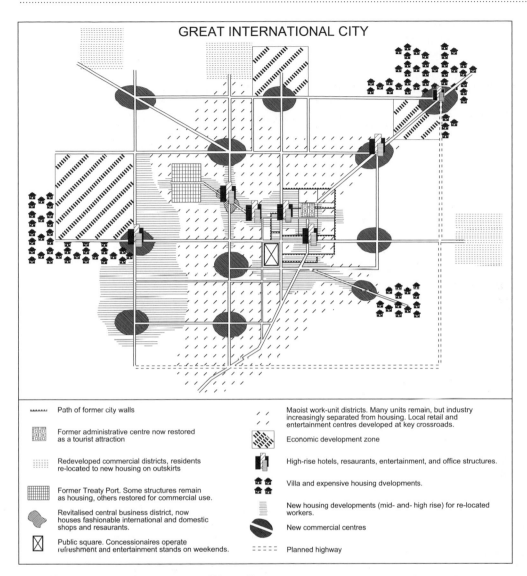

Figure 22.11   A model of the contemporary Chinese city
Source: P. Gaubatz (1998) Understanding Chinese urban form: contexts for interpreting continuity and change *Built Environment* 24(4), 251–70

speculation, and real-estate markets. The most fundamental structural change has been a trend towards the separation of housing and workplace (Figure 22.11). The second is the re-establishment of specialised commercial and business districts in cities, creating an urban commercial network and urban hierarchy. A third major structural change is the establishment of industrial development zones designed to attract FDI. These new residential, commercial and industrial clusters are increasingly linked by modern transport infrastructure that includes expressways, subways and light rail systems akin to those found in Western cities. Other physical symbols of the convergence of urban forms in Chinese and Western cities are evident in the introduction of landmark office towers, international hotels, foreign chain stores, shopping malls, and in the common processes of economic tertiarisation, suburbanisation and increasing socio-spatial polarisation.[25]

As we shall see in the next chapter, many urban dwellers, especially those in low-income residential areas, are migrants to the cities of the Third World.

FURTHER READING

BOOKS

S. Brunn, J. Williams and D. Zeigler (2003) *Cities of the World* Lanham MD: Rowman & Littlefield

A. Dutt, A. Noble, G. Venugopal and S. Subbiah (2003) *Challenges to Asian Urbanisation in the Twenty-first Century* Dordrecht: Kluwer

J. Kasarda and A. Parnell (1993) *Third World Cities* London: Sage

L. Ma and F. Wu (2005) *Restructuring the Chinese City* London: Routledge

T. McGee (1967) *The Southeast Asian City* New York: Praeger

D. Smith (1992) *The Apartheid City and Beyond* London: Routledge

JOURNAL ARTICLES

A. Borsdorf and R. Hidalgo (2008) New dimensions of social exclusion in Latin America: from gated communities to gated cities: the case of Santiago de Chile *Land Use Policy* 25, 153–60

A. Christopher (2005) The slow pace of desegregation in South African cities *Urban Studies* 43(12), 2305–20

M. Coy and M. Pohler (2002) Gated communities in Latin American megacities: case studies in Brazil and Argentina *Environment and Planning A* 29, 355–70

V. Dupont (2004) Socio-spatial differentiation and residential segregation in Delhi *Geoforum* 35, 157–75

P. Gaubatz (1998) Understanding Chinese urban form *Built Environment* 24(4), 251–70

L. Ford (1993) A model of Indonesian city structure *Geographical Review* 83(2), 374–96

L. Ford (1996) A new and improved model of Latin American city structure *Geographical Review* 86(3), 437–40

G. Krausse (1978) Intra-urban variation in *kampung* settlements of Jakarta *Journal of Tropical Geography* 46, 11–46

M. Lopes de Souza (2001) Metropolitan deconcentration, socio-political fragmentation and extended suburbanization: Brazilian urbanization in the 1980s and 1990s *Geoforum* 32(4), 437–47

KEY CONCEPTS

- *Laws of the Indies*
- *Zócalo*
- Colonial city
- Indigenous city
- Apartheid city
- Segregation city
- Post-apartheid city

- Islamic city
- Medina
- Bazaar-based city
- *Desakota*
- *Kampungs*
- Maoist city

STUDY QUESTIONS

1. Examine the model of the Latin American city discussed in the chapter and consider how well this relates to the urban geography of any particular city in the region.
2. Critically consider the concept of the Islamic city.
3. Describe the socio-spatial structure of the apartheid city in South Africa and consider the future development of this urban form following the introduction of majority rule.
4. With reference to a particular city in the region, consider the validity of the model of the Indonesian city presented in the chapter.
5. Examine the changing form of the Chinese city.
6. To what extent would you agree with the view that in an era of globalisation there is an increasing convergence in the character of cities in the First and Third Worlds?

PROJECT

Prepare a poster display to illustrate the major models of Third World urban structure. A poster presentation should be visually stimulating and, in contrast to an essay or project report, the poster should contain only enough text to explain the diagrams. The skill lies in providing the reader/viewer with a concise diagram-based explanation of the theme under review. The poster will be based on a set of annotated models of urban structure in the different regions of the Third World. These will be set within a framework of text consisting of a short introduction explaining the goals of the poster, brief explanations highlighting key points in each model, and a succinct conclusion summarising the advantages and disadvantages of the models and perhaps drawing comparisons between models of Third World urban structure and/or with models of urban land use in cities of the developed world, considered earlier in the book.

23

Rural–Urban Migration in the Third World

Preview: why people move; migration models; structural determinants of migration; who migrates; migration strategies; circular migration; long-term migration; family and return migration; reverse migration; migration controls; accommodating migrants; dispersed urbanisation; rural transformation

INTRODUCTION

The three main sources of urban population growth are net migration, natural increase and the administrative reclassification of urban areas, with the first two being of greatest importance. In general, the relative contribution of migration and natural increase to urban growth varies with level of urbanisation (see Chapter 4). At an early stage of development, when levels of urbanisation are low and rates of both urban and rural natural increase are moderately high, net migration generally contributes more to urban population growth than natural increase. At an intermediate stage of urbanisation, natural increase predominates. At a later stage, with a high level of urbanisation and low rate of natural increase, the balance reverts in favour of net migration. Although some Third World countries are now in the intermediate stage and many large cities are growing more from natural increase, migration remains a major factor in urban growth and, in view of the size of the rural reservoir of potential migrants, will continue to be so for the foreseeable future. In this chapter we identify the major motives underlying rural–urban migration in the Third World and examine the key structural determinants of migration flows. We consider the characteristics of migrant populations and the different types of migration strategies adopted. We then assess the differing policy responses to large-scale rural–urban migration, ranging from efforts at preventing flows via the imposition of migration controls to various means of accommodating the influx of population.

WHY PEOPLE MOVE

Rural–urban migration is a human response to the geography of uneven development. Excluding involuntary relocation by refugees, most people move for economic reasons[1] (Table 23.1). Several models have been proposed to account for migration. The classical economic models of the 1960s saw migration as a response by individuals to place differences in job opportunities and wage rates. Migration was viewed as a mechanism for eliminating such differentials by matching labour supply and demand, and thereby contributing to national economic development.[2] A second, neo-classical economic perspective also represented migration as a process in which individuals make a rational economic choice based on the current or likely future benefits of a move.[3]

More recent interpretations, based on a political economy approach, focused attention on the structural forces underlying spatial variations in economic opportunity. This historical-structural perspective emphasises the importance of *level of development* in determining the scale and form of migration, and stresses that the significance of the various development-related factors underlying migration differs between countries and over time.[4] This thesis is linked directly to the incorporation of Third World countries into the global economy. The urban bias inherent in the colonial capitalist development of the Third World introduced considerable differentiation into what had been frequently quite egalitarian societies, with most people having access to some land. The traditional

TABLE 23.1 PRINCIPAL REASONS FOR MIGRATION FROM RURAL NORTH-EAST THAILAND

Principal reason	No. of respondents citing reason	% of respondents citing reason
To earn more money for the household	138	52.9
To earn more money for self	57	21.8
To earn more money for parents	31	11.9
To further education	12	4.6
To earn money to build a house	10	3.8
To earn money to invest in farming	4	1.5
For fun	3	1.1
To earn money to purchase land/land title	2	0.8
To earn money to repay a debt	1	0.4
To earn money to pay for hired labour	1	0.4
To see Bangkok	1	0.4
To earn money to get married	1	0.4
Total	261	100.0

Source: M. Parnwell (1993) *Population Movements and the Third World* London: Routledge

socio-economic structure was undermined by the labour needs of the capitalist mode of production (e.g. in mines or on plantations). These were often met by restricting peasants' access to land resources, and by coercing people (either directly through forced-labour systems or indirectly through taxation) into migration to work as waged labourers in the capitalist sector. As the influence of capitalism spread, populations that previously had supplied most of their own needs were confronted by a new need to earn money to operate in the capitalist economy. Further, as incorporation proceeded, rural populations came to experience and recognise their relative deprivation. While some sought to improve their position by staying in rural areas (adaptation), many saw better prospects in the cities (migration). Following independence, national attempts to achieve rapid economic growth, principally through the intensified exploitation of natural resources, the commercialisation of agriculture and industrialisation, had a powerful impact on the level and pattern of migration, with rural people attracted to urban areas in increasing numbers.

Zelinsky's (1971) hypothesis of the *mobility transition* was an early attempt to relate the changing nature of migration to levels of development.[5] As Table 23.2 shows, each stage of the mobility transition may be related to a corresponding stage in the demographic transition (see Box 4.1). Although the universality of the model has been questioned, the basic assumption of systematic changes in the nature of mobility over time retains analytical value.[6] Using Zelinsky's model as a base, Brown and Sanders (1981)[7] provided a

model to illustrate the changing importance of the major development- related factors underlying migration (Figure 23.1):

1. Early migration streams are seen as the result of a chain effect, origin-pushed and oriented towards opportunities in the informal small-scale enterprise labour market. At this level of development, rural–rural migration is as likely as rural–urban movement.
2. As development proceeds, migration flows by better-off social classes are pulled by education and modern-sector employment opportunities, but retain a significant chain dimension owing to rudimentary transportation and communication systems. At the same time, migration by less well-off social classes maintains its origin-push motivation, orientation towards the informal labour market, and chain characteristics. Rural–urban flows increase.
3. Finally, as development reaches a relatively advanced level, migration of all social classes is oriented towards formal modern-sector employment. Formal communication channels are the primary sources of information, thus reducing and, in many instances, eliminating the chain dimension. The dominant pattern of migration is urban–urban, rather than rural–urban.

The historical-structural basis of the 'development paradigm of migration' highlights the fact that migration is not a single process but a human response to

TABLE 23.2 THE DEMOGRAPHIC AND MOBILITY TRANSITIONS AMONG MODERNISING POPULATIONS

The demographic transition	*The mobility transition*
Phase A The pre-modern traditional society 1. A moderately high to quite high fertility pattern that tends to fluctuate only slightly 2. Mortality at nearly the same level as fertility on the average, but fluctuating much more from year to year 3. Little, if any, long-range natural increase or decrease	**Phase I The pre-modern traditional society** 1. Little genuine residential migration and only such limited circulation as is sanctioned by customary practice in land utilisation, social visits, commerce, warfare or religious observations
Phase B The early transitional society 1. Slight, but significant, rise in fertility, which then remains fairly constant at a high level 2. Rapid decline in mortality 3. A relatively rapid rate of natural increase, and thus a major growth in size of population	**Phase II The early transitional society** 1. Massive movement from countryside to cities, old and new 2. Significant movement of rural folk to colonisation frontiers, if land suitable for pioneering is available within the country 3. Major outflows of emigrants to available and attractive foreign destinations 4. Under certain circumstances, a small, but significant, immigration of skilled workers, technicians and professionals from more advanced parts of the world 5. Significant growth in various kinds of circulation
Phase C The late transitional society 1. A major decline in fertility, initially rather slight and slow, later quite rapid, until another slowdown occurs as fertility approaches mortality level 2. A continuing, but slackening, decline in mortality 3. A significant, but decelerating, natural increase at rates well below those observed during Phase B	**Phase III The late transitional society** 1. Slackening, but still major, movement from countryside to city 2. Lessening flow of migrants to colonisation frontiers 3. Emigration on the decline or may have ceased altogether 4. Further increases in circulation, with growth in structural complexity
Phase D The advanced society 1. The decline in fertility has terminated, and a socially controlled fertility oscillates rather unpredictably at low to moderate levels 2. Mortality is stabilised at levels near or slightly below fertility with little year to year variability 3. There is either a slight to moderate rate of natural increase or none at all	**Phase IV The advanced society** 1. Residential mobility has levelled off and oscillates at a high level 2. Movement from countryside to city continues but is further reduced in absolute and relative terms 3. Vigorous movement of migrants from city to city and within individual urban agglomerations 4. If a settlement frontier has persisted, it is now stagnant or actually retreating 5. Significant net immigration of unskilled and semi-skilled workers from relatively underdeveloped lands

TABLE 23.2 CONTINUED

The demographic transition	The mobility transition
	6. There may be a significant international migration or circulation of skilled and professional persons, but direction and volume of flow depend on specific conditions
	7. Vigorous accelerating circulation, particularly the economic and pleasure oriented, but other varieties as well
Phase E A future super-advanced society	**Phase V A future super-advanced society**
1. No plausible predictions of fertility behaviour are available, but it is likely that births will be more carefully controlled by individuals – and perhaps by new socio-political means	1. There may be a decline in level of residential migration and a deceleration in some forms of circulation as better communication and delivery systems are instituted
2. A stable mortality pattern slightly below present levels seems likely, unless organic diseases are controlled and lifespan is greatly extended	2. Nearly all residential migration may be of the inter-urban and intra-urban variety
	3. Some further immigration of relatively unskilled labour from less developed areas is possible
	4. Further acceleration in some current forms of circulation and perhaps the inception of new forms
	5. Strict political control of internal as well as international movements may be imposed

Source: W. Zelinsky (1971) The hypothesis of the mobility transition *Geographical Review* 61, 230–1

Development phases		Factors of migration						
		Wages and/or job opportunity differentials		Education and other amenity pulls	Origin push	Migration chain effects	Formal communication channel effects	Pattern of migration
Brown and Sanders (1981)	Zelinsky (1971)	Modern sector	Informal and/or small-scale Enterprise Sector					
Early move towards modernisation	Early transitional society	Affects only a small segment of the population	Significant for all social classes	Affects only a small segment of the population	Significant for all social classes	Significant for all social classes	Affects only a small segment of the population	Rural to rural and rural to urban
Later move towards modernisation	Late transitional society	Significant for the more well-off social classes	Significant for the less well-off social classes	Significant for the more well-off social classes	Significant for the less well-off social classes	Significant for all social classes but more so for the less well-off	Somewhat significant for the more well-off social classes	Increase in rural to urban
Modernisation	Advanced society	Significant for all social classes	Affects only a small segment of the population	Significant for all social classes	Significant for the less well-off social classes but trend towards affecting only a small segment of the population	Significant for the less well-off social classes but trend towards affecting only a small segment of the population	Significant for all social classes	Urban to urban
Trend of each factor's role in migration over the development process		Time	Time	Time	Time	Time	Time	
Preponderence of each factor in society at large		These graphs would parallel the above						

Figure 23.1   Changing roles of development-related factors in migration

Source: L. Brown (1991) *Place, Migration and Development in the Third World* London: Routledge

changing local circumstances within a global economic system. This emphasises that an understanding of Third World migration requires appreciation of both structural forces (such as world economic trends, government policy and technological innovation) and conditions within individual households (e.g. socio-economic status, age and sex). In the following section we examine a number of key development-related factors that influence the scale and pattern of migration. We then consider the question of who migrates by reviewing a range of factors operating at the level of the individual family that underlie migration.

STRUCTURAL DETERMINANTS OF MIGRATION

A large number of factors associated with the development process influence patterns of migration in the Third World. They may be divided into rural *push* factors and urban *pull* factors.

RURAL PUSH FACTORS

1. *Population-growth rates*. One of the most common explanations for out-migration is the high rate of population growth in rural areas. This informed economic models of migration in the 1950s and 1960s, in which migration was seen as a movement of surplus labour. However, population growth alone is not the main cause of emigration. The effects of demographic pressure must be seen in conjunction with the failure of other processes to cater adequately for the needs of a growing rural populace.[8] One of the most pressing needs is access to land.

2. *Pressure on land*. Migration is often a direct response to a situation in which the amount of land available is no longer sufficient to support a family. In rural Mexico population growth and subdivision of plots over several generations has resulted in families farming insufficient land for their needs. As Shaw (1976) demonstrated in

Plate 23.1   Kejeta bus station in Kumasi, Ghana, is a port of entry for many rural–urban migrants in search of an improved lifestyle

TABLE 23.3 AVERAGE ANNUAL RATES OF RURAL OUT-MIGRATION IN LATIN AMERICA (%)

Distribution of farms	Distribution of land	
	Less than half held by latifundios	More than half held by latifundios
Less than half are minifundios	0.56	1.60
More than half are minifundios	1.48	2.33

Source: **R. Shaw (1976)** *Land Tenure and the Rural Exodus in Chile, Colombia, Costa Rica and Peru* **Latin American Monographs, Second Series 19, Gainsville, FL: University Presses of Florida**

Latin America, the problem is compounded by the concentrated ownership of land[9] (Table 23.3). The average rate of rural out-migration was highest in countries such as Mexico and Peru, where more than half of all agricultural land was held as *latifundia* (estates of more than 500 ha, 1,200 acres) and where more than half of all farms were *minifundia* (less than 5 ha, twelve acres).

3. *Land quality*. The quality or suitability of land for agriculture also affects migration. Colonisation of the abundant land resources of Amazonia has failed, in many places, to assuage the land shortage because of the unsuitability of the land for many of the crops being planted. Similarly, in many parts of Asia, population growth and shortages of cultivable land are forcing people onto marginal ecological zones or into increasing the intensity of land use, with a resulting decline in soil fertility.

4. *Agricultural inefficiency*. The effects of rural population growth are compounded by the slow pace of economic and technological change in the rural sector generally, and agriculture in particular. The persistence of inefficient farming practices and the scarcity of investment capital among peasant farmers limit the capacity of the agricultural sector to provide the cash required for participation in the market economy. The limited development of the non-agricultural sector of the rural economy also restricts local employment opportunities. In such circumstances, labour migration provides an invaluable source of cash income for poor rural households struggling to satisfy even their basic subsistence needs from the land.

5. *Agricultural intensification*. The intensification of agriculture and the introduction of modern farming practices have helped to absorb rural population growth (e.g. in Java and the Indian Punjab) but have also often had the opposite effect by replacing agricultural labourers with mechanical and technologically intensive farming systems. In Malaysia thousands of paddy farmers have been displaced as the government sought to improve productivity in the rice-growing sector by investing in major irrigation schemes and consolidating fragmented paddy farms. In the absence of alternative rural employment, landless labourers have moved to the cities in search of work. In Latin America programmes of agrarian reform have released peasant farmers from 'feudal' systems of tied labour that had hitherto restricted migration.

URBAN PULL FACTORS

1. *Wage and employment differentials*. The principal cause of rural–urban migration is the higher wages and more varied employment opportunities available in the city. There is ample evidence that patterns of migration switch in response to changes in income differentials between destinations.[10] Comparison of urban and rural standards of living is complicated by:
   - The need to consider differences in the cost of living, which is often much lower in the country owing to cheaper food, energy, transportation and housing.
   - the fact that levels of collective consumption are higher in the city, where the quality of education and health care is better and the availability of services such as clean water and electricity is greater.
   - the need for disaggregation, since average urban wage rates have little relevance for the unskilled rural migrant.

Nevertheless, it is generally the case that migrants' prospects of economic advancement are better in the city. Although many urban dwellers live in desperate conditions, most consider themselves better off economically than before their move to the city. Even some of the poorest street-traders in Jakarta who could barely feed their families reported that they were better off in the city, their income levels having increased by two-thirds.[11] Similarly, some of the poorest rural migrants to Delhi found twice as many days' work in the city, resulting in income levels two and a half times those enjoyed in the village.[12]

2. *Future prospects*. The coexistence of large-scale rural–urban migration and rising levels of urban poverty and unemployment led some analysts to question the link between urban job opportunities and migration. Todaro (1969) suggested that this apparent paradox could be explained by migrants taking a longer-term view of a prospective improvement in their standard of living.[13] People were seen to be willing to endure short-term difficulties in the hope of better prospects of economic gain and improved welfare in the longer term, even if only for their children.

3. *Bright lights*. The social attractions of the city have been suggested as a non-economic factor to explain rural–urban migration. However, most migrants do not have the financial means to avail themselves of the city's attractions. Although the 'bright lights' may influence a migrant's choice among several potential destinations, the concept does not explain the initial decision to migrate. In addition, for many migrants the bustle of the city presents an intimidating environment that holds fewer attractions than their familiar home community (Box 23.1).

WHO MIGRATES?

The process of rural–urban migration is selective in terms of migrants' age, sex and socio-economic background. These characteristics determine their prospects in the city and, therefore, affect the decision to move or stay. At one end of the socio-economic spectrum, migrants come from impoverished backgrounds, such as village India, and are ill prepared for any but the most menial tasks in the city. At the other extreme are members of privileged rural minorities with a level of education sufficient to gain entry to a career in public administration, commerce or the professions.

Young adults predominate among economic migrants. They are usually unmarried, but even when married they have less of a stake in the rural area than their elders. They frequently lack control over

---

BOX 23.1

Not so bright lights: a migrant's view of Bombay

I asked [a textile worker] if he didn't prefer to live in Bombay. Wouldn't he miss the excitement if he went back to live in Sugao (a village in the Deccan)? What kind of question is that, he said. There is no question about it. Of course I would live at home if I could make enough money there. Waving his arm to encompass the dirty pavement and the roaring pedestrian and vehicular traffic outside the door where we were sitting, he said, There's no excitement here, the air smells of the mills, the food is bad, and there is nowhere to go for a walk that isn't as crowded as this. He was right. It did smell of the mills. We were in the heart of the textile area of the city, and the nearest beaches, which are the lungs for congested Bombay, would be packed with the city's population strolling shoulder to shoulder. In Sugao, he continued, at least the fields are open and the breeze from the hills is fresh, the bhakari made in one's own home tastes so much sweeter than what the khanawal bai [the woman who prepares meals for pay] throws on your plate here. Bombay, the queen of India's cities, with its thriving commerce, cosmopolitan and heterogeneous population, more open society and huge entertainment industry, has little to offer one in his economic class. Working long days in the textile mill, he earns barely enough to maintain himself in the city and his family in the village in a very modest lifestyle. He has little surplus income with which to splurge on the city's luxuries and allurements.

Source: H. Dandekar (1986) Men to Bombay, Women at Home: Urban Influence on Sugao Village, Deccan Maharashtra, India 1942–1982 Ann Arbor MI: Center for South and Southeast Asian Studies, University of Michigan

---

resources, particularly land, and wield little power in local affairs. Not yet firmly committed to an adult role in the local setting, they enjoy an advantage in the urban environment and are more adaptable to the urban milieu. Even if migration entails accepting marginal earnings initially, the potential rewards are highest for the young embarking on an urban career path. The relatively young age of rural migrants contributes to high levels of natural population increase in cities through their higher fertility and lower mortality rates.

In addition to altering the age structure of the urban population, migration often affects the sex ratio in large cities, a trend with important demographic and employment consequences. In cities of South Asia and Africa, where women move usually as dependants and unmarried females remain in the village, men predominate in migration flows. In contrast, in Latin America and the Philippines the majority of rural–urban migrants are women. This reflects a cultural ethos that, in contrast to an emphasis on early marriage and childbirth in most Third World countries, exalts the status of the single woman. Faced with limited rural job opportunities, many look to the cities for work in activities such as domestic service.[14]

Stage in the family life cycle is also a determinant of migration. This is evident in the pattern of 'relay migration' practised in the Mexican village community of Toxi, where at different stages in a family's life cycle different members take on the responsibility of migration. When children are very young it is usually the father who engages in seasonal forms of migration that keep him away for only a few months at a time. As children grow up, they take over responsibility for migration, starting with the eldest, with responsibility passing to younger siblings as the older ones marry and raise their own families. Throughout the course of a generation relay migration is responsible for maintaining the contribution of externally generated income to the household economy while, at the same time, minimising disruption to the family. Relay migration illustrates that the decision to migrate is rarely an individual one. Much rural–urban migration is part of a family strategy to ensure the viability of the rural household[15] (Box 23.2).

Even when individuals migrate alone, others are typically involved in implementing the move, in assisting the migrant to adapt to the urban environment and secure a foothold in the urban economy. The extended family is of key importance in this process. Frequently a wide range of relatives can be drawn on to help pay for an education for the future migrant, provide a home for children who are sent to town to attend school, offer the newly arrived migrant shelter and food, and take care of parents, wife and children who stay behind. In a sample of migrants to Bombay, over three-quarters had relatives living in the city and more than half cited this as an important reason for choosing Bombay over another city. Nine out of ten reported that they had been assisted by relatives or friends upon arrival; two-thirds received free accommodation and food, and help in finding a job.[16] In some situations the rural–urban link is so well developed that a potential migrant can wait in the home village until their urban contact signals a job opportunity. The extended family acts as a major agent of Third World urbanisation.

MIGRATION STRATEGIES

Once a migration decision has been made (Figure 23.2), several possible strategies are available to the migrant. We can identify four principal forms of rural–urban migration in the Third World.

CIRCULAR MIGRATION

In much of Asia, Africa and Oceania the preponderance of men over women in rural–urban migration reflects a widespread tendency for male migrants to leave wife and children in the rural area. This practice is aided by extended family support, with the assistance of male kin in farm tasks and the protection they afford often being essential if the wife is to manage the farm and hold her own in a male-dominated society. Circular migration of individuals, whether single or separated from their families, has several economic advantages. Employers save on wages and fringe benefits, and public authorities face less demand for housing and infrastructure. Migrants also accrue benefits that make them accept family separation. Living costs in the city are high, while urban earning opportunities for women are often limited. A wife and children remaining on the farm can grow their own food and perhaps even raise a cash crop. Where land is under communal control, as in much of tropical Africa, there

---

BOX 23.2

The culture of migration in Mexico

All my life I have farmed [a 1.5 ha plot of irrigated land], and that land provides well for us. We harvested nearly eighteen months' worth of maize last year. But that wasn't enough. My family needed more help. I have three young children in school, and my two eldest sons [18 and 21 years of age] – they helped me in the field, but it wasn't enough. I sent them to the United States; that was a dangerous trip. But they went. I borrowed the money from friends to send them. They are there now: they live in Santa Monica, California. They are helping us out from there, paying for school and for the house.

Source: adapted from J. Cohen (2004) The Culture of Migration in Southern Mexico Austin TX: University of Texas Press, 30

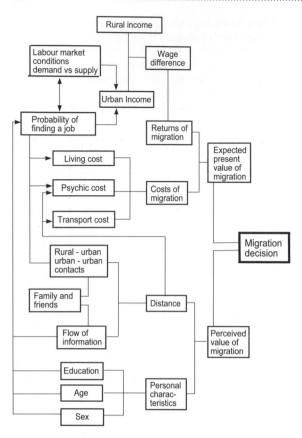

Figure 23.2 The migration decision process

Source: D. Shefer and L. Steinvortz (1993) Rural-to-urban and urban-to-rural migration patterns in Colombia *Habitat International* 17(1), 133–50

LONG-TERM MIGRANTS

Under a strategy of long-term migration by men, short visits to the family left behind in the village replace the extended stays that characterised circular migration. As a result, what had been an economic cost for employers, in the form of a labour force characterised by high turnover and absenteeism, became an increased social cost for workers. This includes a greater strain on their relationship with their wife, children, extended family and village community. The frequency of home visits varies with employment conditions and distance. Fast, cheap transport may allow monthly or even weekly commuting, but in a large country, such as India, many long-distance migrants can visit their families only during their annual leave. One means of offsetting such lengthy separations is to reverse the pattern, with the wife visiting the spouse in town, sometimes for an extended period. Children may also stay with the father in order to take advantage of better educational facilities in the city.

FAMILY MIGRATION AND RETURN MIGRATION

As the period of urban employment has lengthened, the family separation entailed by long-term migration of men has become less acceptable. For some migrants, increases in urban wages are sufficient to offset the significance of rural income forgone if the family moves to the city. Urban earning opportunities for women are also less constrained. In some areas of tropical Africa the breakdown of communal control over land has transformed it into an asset that can be realised. All these factors have made family migration more common.

Although family migration is usually for a long term, typically the working life of the head of household, it is not necessarily seen as a permanent move and is often followed by return migration. This may be made attractive by strong ties with members of an extended family and with the village community. Also, most migrants, even when they manage to support a family in the city, enjoy little economic security. Few qualify for welfare benefits, such as a pension, and for many urban migrants the solidarity of the village provides an alternative meagre, but reliable, form of social security. In addition, where there is still plentiful land under communal control, the migrant can maintain his position in the rural community, even during an extended urban career, and remain assured of access to land on his return.[20]

is no compensation for those who give up farming. A wife who comes to town gives up an assured source of food and cash to join a husband on low wages.

Circular migration was common in tropical Africa in colonial times and was actively promoted by employers seeking cheap labour. It is still practised in some regions,[17] and in exceptional circumstances groups of migrants may control a particular niche in the urban economy to such an extent that its members can maintain a pattern of circular migration as they replace one another on the job. This has been reported in the case of ice-cream vendors in Jakarta[18] and construction workers in Manila.[19] More generally, the viability of the circular migration strategy has been undermined by the growth of substantial urban unemployment, which means that the search for a job in the city may take several months with no guarantee of success. In these circumstances a migrant who has secured regular employment has good reason to hold on to it.

PERMANENT URBAN SETTLEMENT

In parts of Africa, much of Asia and virtually all of Latin America migrants have little prospect of maintaining access to agricultural land, owing to population pressure and institutional constraints. Wholly dependent on their urban earnings, they have no plans to return to rural life. Their priorities are urban-based. They press for the provision of social security for workers; search for sources of earnings outside the vagaries of waged employment (e.g. by setting up their own business); and strive for ownership of a home, which provides both shelter and an opportunity for rental income. As we shall see later (Chapter 29), action in pursuit of such goals has contributed to the strength of urban social movements particularly in Latin America.

REVERSE MIGRATION

We can add a fifth model to the four principal forms of rural-urban migration. In marked contrast to the general trends of urbanisation and rural-urban movement some writers[21] have identified a process of counter-urbanisation in some African countries over recent decades. In Zambia the percentage of urban population declined between 1990 and 2001 from 42 per cent to 40 per cent. Whereas in the West counter-urbanisation is driven by choice, associated with affluence and a desire for suburban living, in Zambia it has been linked to economic stagnation, urban unemployment, and a need to access rural areas for survival. In some cases circular migration has replaced permanent urban relocation as people strive to retain an economic foothold in both urban and rural areas. When such adaptive strategies prove insufficient many former urbanites are forced to relocate to rural areas.

THE POLICY RESPONSE

Four main policy approaches have been employed in the attempt to control the growth of Third World cities and reduce levels of regional inequality (Table 23.4). These:

1. Aim to limit the growth of large cities through migration controls.
2. Seek to absorb the flow of migrants in existing urban areas.
3. Attempt to redirect migration flows to alternative destinations.
4. Aim to transform the rural economy and thereby slow the rate of rural–urban migration.

MIGRATION CONTROLS

Many city administrations in the Third World have sought to limit rural–urban migration and even to reverse the flow by employing a series of measures including administrative and legal controls on population movement, police registration schemes and direct rustication programmes to relocate urban dwellers in the countryside.

These strategies have had most success in socialist states. In China, in order to slow the rate of urbanisation and reduce emerging income and welfare differentials between urban and rural areas, the government in 1958 introduced a population registration system (*hukou*) that classified people as either urban or rural residents. Limiting the number of citizens who could be registered as urban dwellers restricted the numbers moving to cities opportunistically in the hope of finding work. The registration scheme was supported by rationing of food, which was available in cities only to those in possession of an urban household registration document. In the absence of a significant black market these measures proved highly effective in slowing the rate of rural emigration, albeit at a cost to personal liberty. Migration controls were complemented by rustication programmes that resulted in the movement of 30 million urban residents (including 17 million young people) into the countryside during the period 1966–76.[22] These measures continued until the late 1970s, when the policy was relaxed to support China's industrialisation strategy, by permitting the 'temporary' migration of rural dwellers to work in urban areas.[23] The distinction between temporary migrants (who do not possess an urban *hukou* and are therefore excluded from the better jobs) and permanent migrants (state-sponsored and selected migrants who are granted urban citizenship) is manifested in the 'floating population' of urban China that, in 1997, amounted to 100 million persons or 25 per cent of China's urban population.[24] The marginalisation of this group, particularly in the housing and labour markets, has created what amounts to a new urban underclass (Box 23.3). Whether the relaxation of controls on rural–urban migration since the 1980s will lead to the kind of urban growth rates and problems confronting other Third World cities remains to be seen.

Outside the socialist Third World, the only country in which controls on migration have had a significant effect is South Africa, where for several decades

TABLE 23.4 PRINCIPAL POLICY RESPONSES TO RURAL–URBAN MIGRATION IN THE THIRD WORLD

Policy approach	Rationale	Types of policies and programmes
Negative	Emphasises the undesirability of migration and seeks to erect barriers to population movement and to forcibly 'deport' migrants	Closed city; pass laws; deportation of beggars, the homeless and those in marginal occupations; bulldozing of squatter settlements; enforced resettlement from urban to rural areas; sedentarisation of nomads; registration systems; employment controls; restricted access to housing; food rationing systems; benign neglect
Accommodative	Accepts migration as inevitable, and seeks to minimise the negative effects in both origin and destination places	Slum upgrading; sites and services; urban job creation; labour-intensive industrialisation; minimum wage legislation; urban skills training; urban infrastructural investment; improved social welfare; improvements in transportation; relieving congestion
Manipulative	Accepts migration as inevitable and even desirable in some cases but seeks to redirect migration flows towards alternative destinations	Colonisation; land settlement; land development; polarisation reversal; growth poles; urban, industrial and administrative decentralisation; information systems; management of contact networks
Preventive	Rather than dealing with the symptoms of migration, attempts to confront the root causes by tackling poverty, inequality and unemployment at source, and reducing the attractiveness of urban areas to potential migrants	Land reform; agricultural intensification; agricultural extension; rural infrastructural investment; rural industrialisation; rural minimum wages legislation; rural job creation; improving rural–urban terms of trade; reducing urban bias; increased emphasis on 'bottom-up' planning; propaganda in favour of the rural sector

Source: **M. Parnwell (1993)** *Population Movements and the Third World* **London: Routledge**

BOX 23.3

China's urban floating population

An increasingly visible urban underclass exists in sharp contrast to the dynamism of 'market socialism'. Some of the migrants are easily identifiable in the cities: hanging around outside rail and bus stations; working on construction sites; and operating stalls or providing services on the streets. Permanent urban residents view the migrants as 'country cousins' who look, speak and behave differently. They are unsophisticated and, most important, they are poor and to be avoided wherever possible. To some extent these perceptions are nourished by the lifestyles of the floating population. Those who work are forced to accept jobs that are too undesirable and low-paying for regular city residents, and as a result they often live in areas that are only marginally habitable. Others in search of work are forced to wander the streets and gather in public places, creating images of vagrancy and homelessness that are unsettling to the city's permanent residents.

Source: **C. Smith (2000) The floating population in China's cities, in T. Cannon (ed.)**
***China's Economic Growth* London: Macmillan, 91–114**

apartheid policy limited the movement of black Africans into the cities.[25] Elsewhere, controls on population movement have been less effective, because legal restrictions are difficult to enforce, licences are easily forged, monitoring entire cities is costly, and people who are evicted can return.[26] In Manila the government hoped to deter migrants by restricting access to education and housing, but the measures were administratively difficult to enforce and invited corruption. In Nairobi repeated demolitions of squatter settlements may have influenced the pattern of urban growth but have not substantially discouraged in-migration. In Lima and Jakarta street vendors have been forcibly removed from the city, while in the Republic of the Congo, Niger, Tanzania and Zaire (now the Democratic Republic of Congo) there have been periodic expulsions of unemployed migrants from cities (Box 23.4). These measures have caused personal grief and resentment but have had little impact on urban growth. As long as a significant gap persists between rural and urban living standards, people will continue to evade migration controls in the hope of enhancing their quality of life.

ACCOMMODATING MIGRANTS

In some cities there is acknowledgement of people's need and right to move to the city, and attempts are made to ameliorate the hardship of migrants. It has been suggested that the welfare of the migrant population could be improved by introducing regulations on minimum wages and working conditions, but since a majority work outside the formal sector of the urban economy, such legislative steps are unlikely to reach the mass of the poor. Of more direct relevance are schemes to provide training, start-up capital and marketing assistance for hawkers, and policies that legitimise squatter settlements. In general, however, the scale of urban poverty and the limited financial resources available mean that only a minority of cities pursue such programmes. The accommodation approach has also been criticised for dealing with the symptoms rather than addressing the causes of rural–urban migration. It may even attract more migrants into the cities, thereby exacerbating the problem in the longer term.

DISPERSED URBANISATION

An alternative strategy is to accept the inevitability of rural emigration but seek to redirect flows away from the large cities. A common approach, based on growth-pole

BOX 23.4

Migration controls in Jakarta, Indonesia

The metropolitan region of Jakarta is the major focus of rural–urban migration in Indonesia. In 1970 Jakarta was declared a 'closed city'. Under this policy, which is nominally still in effect, migrants first have to obtain permission to leave their home area They must report the specific dwelling in which they will be staying in the city and are required to deposit a sum equivalent to twice the return fare to their home village (a sum that represented an almost insurmountable barrier for the poorest of the poor). Within a period of six months of arrival, migrants had to provide a certificate from an employer to indicate they had a job. Those of school age had to produce a certificate to show they were properly enrolled in a school. As part of the city closure package, students wishing to enroll at an educational institution had to have a certificate stating that no similar course was available closer to their place of origin. Moreover, from 1975 all primary and secondary schools have been closed to students whose parents are not registered citizens of Jakarta. Migrants who satisfy these requirements have their deposit refunded and are given an identity card acknowledging them as citizens. Those who cannot comply with the regulations are returned to their village. In its attempt to stem the flow of migrants, the government also declared sections of the city off limits to *betjaks* (bicycle rickshaws), and prohibited pavement (sidewalk) vendors and hawkers from working in particular areas – both jobs that typically attracted large numbers of new migrants. In addition, periodic checks of identity papers led to illegal residents being deported from the city.

Source: **United Nations (1989)** *Population Growth and Policies in Mega-cities: Jakarta,* **Population Policy Paper 18, New York: Department of International Economic and Social Affairs, United Nations**

theory,[27] employs a combination of carrot (e.g. tax incentives and infrastructure provision) and stick (e.g. relocation of public administrative functions) to stimulate the economy of designated medium-size secondary cities,[28] thereby providing alternative employment opportunities for migrants. Industrial decentralisation strategies have been employed with some success in a number of Third World cities, including Mumbai,[29] Mexico City[30] and Seoul[31] (Box 23.5). More generally, however, the

Plate 23.2   Emergent social polarisation in Beijing is manifested in the housing of (*upper*) migrant workers and (*lower*) middle-class suburban dwellers

strategy has proved less effective, owing to the reluctance of companies and personnel to relinquish the amenities of the large city, the insufficient level of incentives, changing priorities and government economic policy, and, in particular, policy conflict between the goals of urban decentralisation and national economic development. In addition, the limited employment created by decentralisation of capital-intensive industries means that although lagging regions may benefit from dispersal strategies, the poor within the region may not benefit sufficiently to dissuade them from migration to the large city. Critics have also suggested that secondary cities merely serve as stepping-stones for migrants to the primary city, and that greater attention should be paid to the national settlement pattern as a whole. In accordance with the precepts of central place theory (see Chapter 6), it is argued that the existence of a primate city settlement system and the absence of an integrated system of central places obstruct the emergence of a more geographically balanced pattern of development.[32]

Most Third World governments acknowledge the difficulties in developing intermediate cities in peripheral regions as counter-magnets to the primate city, and have instead sought to promote decentralisation of population and industry to satellite towns located around the main urban centres. In 1997 twenty new towns had been approved or were under construction around Jakarta alone.[33] Many Third World new towns, however, as a result of spatial proximity, tend to grow towards the major metropolitan centre, thereby defeating the aim of separate development. In addition, the promotion of a polycentric metropolitan region with satellite subcentres perpetuates the concentration of migration streams on the economically more dynamic capital region. Most fundamentally, decentralisation policies only indirectly address the causes of rural–urban migration.

TRANSFORMING THE RURAL ECONOMY

Any solution to the urban problems associated with rural–urban migration must take account of the condition of the rural population. As we have seen, most migrants move for economic reasons, thus any policy that transforms the rural economy will affect the scale and pace of urban development. Policies to redistribute land to the poor may slow urban growth by raising agricultural incomes. Few Third World governments, however, possess the political will or ability to implement proposed land-reform policies. Other rural programmes may have the opposite effect: the green revolution and other attempts to raise agricultural productivity through incentives to commercial farming accentuated landlessness and stimulated rather than reduced the flow of citybound migrants.[34] The effects of rural industrialisation strategies have also been mixed. In Thailand the government has promoted the 'industrialisation of the countryside' in the peripheral north and north-east regions, whence the majority of cityward migrants originate. This has involved the modernisation of cottage industries and the establishment of 'putting out' systems where rural households perform part of the manufacturing process for urban-based industries. The resulting reduction in out-migration contrasts with Indian experience, where attempts to develop small-scale cottage industries in rural areas have tended to increase out-migration because they improved villagers' skills and their prospects in urban labour markets.

As we have seen, as well as income differentials, perceived welfare differences between town and country also influence migration decisions. The Sri Lankan government has reduced the rate of migration to

BOX 23.5

Promoting secondary city growth in South Korea

In 1960 Korea's urban structure resembled that of many contemporary Third World countries. Other than Seoul, only eight cities had a population larger than 100,000 and only Pusan and Taegu had more than half a million. The national capital dominated both the economy and the urban system. Manufacturing employment was highly concentrated in Seoul and other large cities, and only a few secondary cities had more than one-fifth of their labour force working in industry. During the period 1960–80, as part of an attempt to generate more equitable growth, the government made deliberate efforts to deconcentrate manufacturing and commercial employment from the Seoul metropolitan area, which continued to grow rapidly despite policies to restrict its growth. By 1980 both the urban structure of Korea and the occupational composition of secondary city economies had changed markedly. The number of cities with a population of 100,000 rose to thirty and manufacturing employment increased in the larger secondary cities, which also began to exhibit a high degree of functional specialisation.

The promotion of secondary city growth through economic deconcentration was accompanied by policies to slow the expansion of the capital city. The two are necessarily related, since reducing rural–urban migration to the primate city is difficult to achieve in the absence of secondary cities able to provide high population threshold activities that would normally locate in the primate city. Growth restraint strategies in Korea included restricting the expansion of higher education institutions and the construction of new high schools in Seoul, and improving educational services in regional centres. The government also employed zoning regulations, construction permits and financial incentives for industrial relocation to make it more difficult for large industries to locate in Seoul. Since 1994 these measures have been complemented by a congestion charge levied on large-scale office and commercial developments in Seoul (with the revenue used to provide incentives for development elsewhere in the capital region), and a quota system for allocating factory and university building in a new 'Growth Management Zone' beyond the 'Overconcentration Restriction Zone' around Seoul.

Although the capital city has continued to grow, it has done so at a slower rate. In addition, the promotion of secondary cities has shifted the national urban system away from a primate city structure towards a more balanced hierarchy or rank–size distribution. Nevertheless, continuing growth pressure on the Seoul metropolitan area means that decentralisation strategies aimed at promoting balanced regional development remain at the top of the political agenda.

Sources: **D. Rondinelli (1983) *Secondary Cities in Developing Countries* Beverly Hills CA: Sage; K. S. Kim and N. Galient (1998) Regulating industrial growth in the South Korean capital region *Cities* 15(1), 1–11**

Colombo and other urban centres by providing a package of rural social-welfare measures that include free medical services and education, income support in the form of guaranteed prices for poor farmers, and improved housing for lower-income groups.

As Parnwell (1993)[35] concludes, the countries that have been most successful in influencing the pattern and level of migration have been those with either the financial resources to fund adequate development programmes (e.g. South Korea and Malaysia) or the political authority (e.g. China and Cuba) to implement their migration policies. Given that the majority of Third World governments do not possess the resources, authority or political will to influence the pattern and process of migration, it seems inevitable that rural–urban migration will remain a feature of Third World societies for the foreseeable future. This raises the key issue of what happens to the migrants once they reach the city. In the next chapter we examine the economic dimensions of this question.

FURTHER READING

BOOKS

M. Montgomery, R. Stren, B. Cohen and H. Reed (2004) *Cities Transformed: Demographic Change and its Implications in the Developing World* London: Earthscan

M. Parnwell (1993) *Population Movements in the Third World* London: Routledge

D. Rondinelli (1983) *Secondary Cities in Developing Countries* Beverly Hills CA: Sage

N. Shrestha (1996) *A Structural Perspective on Labour Migration in Underdeveloped Countries* Cheltenham: Elgar

R. Skeldon (1990) *Population Mobility in Developing Countries* London: Belhaven Press

JOURNAL ARTICLES

C. Beauchemin and P. Bocquier (2004) Migration and urbanisation in Francophone West Africa *Urban Studies* 41(11), 2245–72

S. Chant (1998) Households, gender and rural–urban migration *Environment and Urbanization* 10(1), 5–21

J. Cuervo and D. Hin (1998) *Todaro* migration and primacy models *Cities* 15(4), 245–56

C. Gu, R. Chan. J. Liu and C. Kesteloot (2006) Beijing's socio-spatial restructuring *Progress in Planning* 66, 249–310

J. Harris and M. Todaro (1970) Migration, unemployment and development *American Economic Review* 60, 126–42

F. Kruger (1998) Taking advantage of rural assets as a coping strategy for the urban poor *Environment and Urbanization* 10(1), 119–34

E. Mobrand (2006) Politics of cityward migration: an overview of China in comparative perspective *Habitat International* 30, 261–74

M. Woube and O. Sjoberg (1999) Socialism and urbanisation in Ethiopia 1975–90 *International Journal of Urban and Regional Research* 23(1), 26–44

KEY CONCEPTS

- Mobility transition
- Secondary cities
- Migration controls
- Demographic transition
- Development paradigm of migration
- Relay migration

- Circular migration
- Long-term migration
- Return migration
- Family migration
- Reverse migration
- Dispersed urbanisation

STUDY QUESTIONS

1. Explain why people move from the rural areas to the cities of the Third World.
2. Identify the main migration strategies underlying the rural–urban movements of population.
3. Critically examine Third World governments' responses to the phenomenon of rural–urban migration.
4. With reference to a particular Third World city, examine the nature and consequences of rural–urban migration.
5. Evaluate the success of migration controls as a strategy to reduce rural-urban migration in Third World countries.

PROJECT

Construct a flow diagram depicting the rural–urban migration process in a Third World country. This can be done in several stages. First, think of yourself as an impoverished villager considering a move to the city. Second, make a list of the structural (e.g. investment in the country by transnational corporations), local (e.g. land availability) and personal (e.g. age, sex, education) factors that will influence the migration process. It will help you organise this information if you think in terms of rural 'push' factors and urban 'pull' factors. Third, consider the different migration options available (e.g. circular or permanent migration). Fourth, begin constructing the diagram by locating the migration factors in boxes at appropriate points in the diagram. Insert lines to indicate any linkages between factors (e.g. on the 'push' side between land shortage and agricultural mechanisation, on the 'pull' side between urban job prospects and presence of family or friends in the city). Don't forget to insert any 'feedback' linkages (e.g. between savings accumulated in the city and relief of family poverty in the village). You may have to adjust the positioning of some boxes as the diagram takes shape. Finally, once you are satisfied that you have identified all relevant factors and linkages, prepare a 'fair copy' of your diagram.

24

Urban Economy and Employment in the Third World

Preview: the evolution of the Third World urban economy; the structure of the urban economy; the upper circuit; lower-circuit activity; employment relations in the urban economy; labour-market structure; job access; unemployment, underemployment and misemployment; women in the labour force; child labour; the household economy

INTRODUCTION

The economy of cities in the Third World is based on peripheral capitalism. This mode of production consists of two interrelated parts: a capitalist sector integrated into the world economy, and a range of petty capitalist forms of production oriented more towards the domestic economy. These have been described as a 'firm-centred economy' and a 'bazaar economy',[1] or the formal and informal sectors.[2] Santos (1979) refers to the upper circuit and lower circuit, in order to highlight the dependence of the traditional informal sector upon the modern formal sector.[3] The well-being of individuals and households is dependent on their position within this dual-sector or bipolar urban economy.

In this chapter we explain the evolution of the two sectors of the Third World urban economy, and we examine the structure of each circuit, with particular reference to the nature of lower-circuit activity. We consider employment relations and labour market structure and problems of job access, and the position of women and children in the urban labour force. We then shift the scale of analysis to examine the household economy and the coping strategies of the poor.

THE EVOLUTION OF THE THIRD WORLD URBAN ECONOMY

The contemporary character of the two circuits of the urban economy reflects the incorporation of Third World national economies into the world economic system. This process has been conceptualised as a series of modernisations, defined as the diffusion of an innovation from a core region (developed countries) to a peripheral subordinate region (Third World countries). This perspective does not embrace the now discredited notion of modernisation as a unilinear process of social and economic change through which all societies pass in the course of development. Santos (1979) identifies three major modernisations and their effect on the Third World urban economy:[4]

1. A *commercial modernisation* (from the late sixteenth century to the industrial revolution), a mercantile period during which an international division of labour, strengthened by the metropolitan administrative structure, began to develop. As colonial powers systematically appropriated the wealth of Third World colonies, an impoverished countryside coexisted with urban areas characterised by a limited market for consumer goods, and a demand for labour commensurate with their almost exclusively commercial and administrative functions.

2. An *industrial modernisation* (from the mid-eighteenth century to the mid-twentieth) during which urban areas capitalised on their already privileged position through the development of modern land and maritime transport, thereby facilitating large-scale metropolitan capital accumulation.

3. A *technological modernisation* (from the mid-twentieth century to the present). The contemporary mode of modernisation, often referred to as globalisation (see Chapter 1), is marked in economic terms by the increasing power of transnational corporations (TNCs) to control patterns of trade and investment and therefore influence the nature of the urban economic structure in many Third World states. Technological modernisation in the Third World creates only a limited number of jobs, given the capital-intensive nature of the industries. Even much of the resulting indirect employment is generated in the core countries or for expatriates working locally. Industry is increasingly incapable of meeting the growing local demand for work, meaning that in Third World cities a high proportion of the people have neither stable employment nor income.[5] The lack of job opportunities in the modern sector for most urban dwellers has resulted in the growth of a large number of small-scale activities in the Third World city. Another impact of globalisation, the spread of information through the mass media, has had a major impact on consumer preferences through an 'international demonstration effect' or 'Coca-colonisation'. This tends to raise the general propensity to consume, thereby acting as an obstacle to capital formation and development, and can also reduce demand for local products in favour of imported goods. The presence of a mass of people with very low wages or depending upon casual work for a living, alongside a minority with high incomes, creates different patterns of consumption in urban society.

Taken together, these differences in modes of production and consumption create the two circuits of the urban economy, each of which largely produces, distributes and consumes a particular set of goods and services. One of the circuits is the direct result of modernisation and includes activities created by technological progress, while the other is the indirect result of modernisation involving individuals who benefit only partially, if at all, from technological progress.

THE STRUCTURE OF THE URBAN ECONOMY

Most simply, the upper circuit consists of capital-intensive activities such as banking, export trade and industry, modern urban industry, trade and services,

and wholesaling and transport. The lower circuit comprises non-capital-intensive forms of manufacturing, non-modern services generally provided at the retail level, and non-modern and small-scale trade (Figure 24.1).

THE UPPER CIRCUIT

Within the upper circuit we may differentiate three types of activities according to their degree of integration with the city.

1. *Integrated.* Modern urban industry, trade and modern services are integrated elements since they are activities peculiar to both the city and the upper circuit.
2. *Non-integrated.* Export-oriented industry and trade are non-integrated activities, for, although they are situated in the city to benefit from locational advantages, outputs are consumed outside the city in another part of the world economy, and most activities are controlled by external interests. Banking is included in this category, since it acts as a link

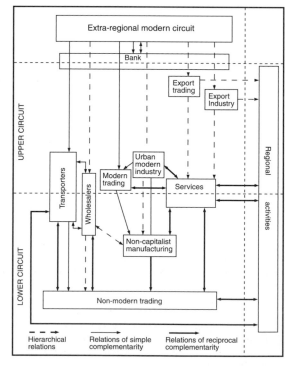

Figure 24.1   The structure of the Third World urban economy
Source: M. Santos (1979) The Shared Space London: Methuen

between the modern activities of the city and larger cities both within the country and abroad.

3. *Mixed activities.* Wholesaling and transportation are mixed activities, being linked to both upper and lower circuits of the urban economy. The wholesaler, for example, is both an integrating element within the upper circuit and the apex of a pyramid of intermediaries reaching down to the itinerant hawker and street vendor.

The fundamental organisational differences between the activities of the upper and lower circuits are summarised in Table 24.1. Although this schema is useful generally, it should not be taken to indicate a separation or dualism between the two sectors. Members of one social stratum may consume outside the corresponding circuit, although this consumption will be partial and occasional. For example, the self-employed repairers of audio equipment in the lower circuit are often dependent on imported supplies, and individuals more directly attached to the lower circuit may occasionally sell their labour in the upper circuit. A number of studies have explored the linkages between the two sectors of the urban economy[6] (Box 24.1).

THE NATURE OF LOWER-CIRCUIT ACTIVITY

Despite difficulties in measuring the exact scale of the informal sector,[8] it is estimated that 50 per cent of the labour force in the Third World work within it, compared to 3 per cent in developed nations.[9] In sub-Saharan Africa the informal sector accounts for almost 80 per cent of all non-agricultural employment. In Asia, Latin America and North Africa the corresponding figures are 65 per cent, 51 per cent and 48 per cent respectively. In addition to its contribution to employment the informal sector contributes significantly to national economies in terms of income, with the sector's share of gross domestic product (GDP) in these regions being 41 per cent, 31 per cent, 29 per cent and 27 per cent.[10] The informal sector is heterogeneous with respect to both its activities and its work force (Table 24.2), but a common characteristic is the small scale of activities. This is evident in a number of ways, including the following:

1. *Fractionalisation of activities.* All lower-circuit activities, and particularly commerce, are carried out in a very large number of small-size enterprises. The micro-scale proliferation of commercial activities is explained, on the one hand, by the

TABLE 24.1 THE NATURE OF THE TWO CIRCUITS IN THE THIRD WORLD URBAN ECONOMY

	Upper circuit	Lower circuit
Technology	Capital-intensive	Labour-intensive
Organisation	Bureaucratic	Primitive
Capital	Abundant	Limited
Labour	Limited	Abundant
Regular wages	Prevalent	Exceptionally
Inventories	Large quantities and/or high quality	Small quantities and poor quality
Prices	Generally fixed	Negotiable between buyer and seller (haggling)
Credit	From banks, institutional	Personal, non-institutional
Profit margin	Small per unit; but with large turnover	Large per unit; but small turnover
Relations with customers	Impersonal and/or on paper	Direct, personalised
Fixed costs	Substantial	Negligible
Advertisement	Necessary	None
Reuse of goods	None (waste)	Frequent
Overhead capital	Essential	Not essential
Government aid	Extensive	None or almost none
Direct dependence on foreign countries	Great; externally orientated	Small or none

Source: M. Santos (1979) *The Shared Space* London: Methuen

BOX 24.1

Linkages between the two circuits of the Third World urban economy

Economic linkages between the two circuits may be *forward* (involving the use of informal-sector products as input for a formal-sector production process, or the sale of finished products and services of the informal sector in the formal sector) or *backward* (the supply of materials, equipment or goods from the formal sector to the informal sector, as when wholesalers supply street vendors).

Beneria (1989) describes linkages between workers in the informal sector of the urban economy and a multinational corporation producing electrical appliances in Mexico City.[7] The company employs 3,000 workers but sends out 70 per cent of its production to 300 regular and 1,500 occasional subcontractors. One of the subcontractors employs 350 workers and sends out 5 per cent of its production. Typical for this subcontractor's subcontractor is a sweatshop operating illegally in the owner's home. The sweatshop employs six workers between the ages of 15 and 17 on a temporary basis, and sends out work for a fluctuating number of home workers. Average monthly wages ranged from 12,000 pesos for manual workers at the multinational corporation to 8,500 pesos at the main subcontracting firm, 6,000 pesos at the sweat shop and 1,800 pesos for the home workers.

Clearly, the informal-sector participants in such relationships are in a subordinate position. Formal-sector enterprises benefit from subcontracting in several ways:

1. They may avoid having to pay a legal minimum wage, and avoid wage increases demanded by a unionised work force.
2. They avoid responsibility for whether or not social security contributions and other fringe benefits are paid to the workers by employers operating in the informal sector.
3. They are able to adjust output to fluctuations in demand by varying the amount of subcontracting rather than the size of their own labour force, thereby avoiding the costs associated with redundancy payments during a recession and training when demand revives.

adaptation to a limited and irregular consumption pattern. Sale on the micro-retail level permits the poor customer with a small and uncertain income to obtain small quantities of goods. The poorer the population, the smaller the scale of commercial activities.

2. *Work in the home.* Often both artisans and merchants work at home, even if some also have a stall in the market or elsewhere in the city. Use of the home as a workplace saves time and money that are often indispensable for the entrepreneur's survival. Moreover, artisans may avoid taxes, while tradeswomen are able at the same time to engage in other activities, including motherhood. Use of the home fosters neighbourhood linkages, with local customers sure of being served at any time. Work in the home also makes possible the long hours characterising the lower circuit.[11]

3. *Family employment.* This is common in the small enterprises of the lower circuit. Use of family labour permits increased output without the need for more capital. If small enterprises were to pay wages (and possibly employees' social benefits and taxes), they would be less competitive. In some cases, especially when demand is uncertain, the transformation of a family enterprise into a capitalist enterprise would lead to bankruptcy.

4. *Limited stocks.* The small quantity of stock held by lower-circuit enterprises and the day-to-day renewal reflects the limited capital and access to credit of the small business, and also the consumption patterns of the lower circuit, where, as we have seen, daily micro-purchasing is typical in poor districts of Third World cities.

5. *Hawkers.* These constitute the lowest level in the fractionalisation of commercial activities, being the final link in the chain of middle-men between importers, manufacturers or wholesalers and consumers (Figure 24.1). Traders supply hawkers with merchandise on credit. Although the activity is clearly a response to poverty, it also reflects the needs of the commercial and manufacturing activities of the upper circuit. Merchants and small manufacturers use hawkers in order to avoid taxes, to employ minors or old persons, to reach customers not having the time or inclination to frequent large stores, or to dispose of unsold or unsaleable merchandise such as clothing no longer in fashion.[12]

6. *Self-employment.* In the cities of the Third

residents of poor urban districts buying locally to avoid the cost of transportation to the modern trade sector, often located in the city centre. On the other hand, small-scale commerce is an

TABLE 24.2 TYPES OF INFORMALLY ORGANISED MARKET-ORIENTED ACTIVITIES IN THE THIRD WORLD URBAN ECONOMY

Activity	Examples	Customary location	Persons per operation
Home industry	Manufacture: food for vending, handicrafts, clothing Services: washing and ironing Trading: retail Trading: retail	Own household	Predominantly female, including unpaid family labour (one to three)
Street economy	Trading: food sales, vending Services: shoeshining, portering, transport, entertainment	Street: ambulatory, but also fixed location	Both men and women (one to three), including some unpaid family labour
Domestic service	Maids, cooks, gardeners, nannies, chauffeurs	Employer household, including live-in arrangements for some staff	Several per high-income household, both men and women
Micro-enterprise	Manufacture: shoes, tailoring, metalworking Services: electrical and radio repair; plumbing; car repair	Rented space, but may also operate out of own home	Owner-manager, plus several employees (fewer than ten; average three to five)
Construction work	Day labourers, bricklayers, electricians, carpenters	On-site	Individually recruited for specific projects

Source: **J. Friedmann (1992)** *Empowerment* **Oxford: Blackwell**

World, particularly those experiencing strong rural–urban migration, income is derived largely from non-wage-earning activities. Some of these non-wage earners are landowners and others live on rental income, but the majority are the self-employed of the informal sector.

EMPLOYMENT RELATIONS IN THE THIRD WORLD URBAN ECONOMY

Recognising the links between the two sectors of the Third World urban economy, and the fact that 'informal' forms of employment practice (e.g. part-time, contractual and low-paid work) may occur in the capitalist upper circuit, has meant that recent analyses have shifted attention from the organisational structure of the two sectors to the different forms of employment relationships within the economy as a whole. Bromley (1988) has suggested a continuum of employment relationships[13] (Figure 24.2). The two extremes, described as 'career wage work' and 'career self-employment', have relatively high levels of stability and security. The intermediate relationships have relatively low levels of stability and security and are referred to collectively as casual work. Thus short-term wage work covers those paid for a specific period of time, disguised wage work includes out-workers paid for piecework, and dependent work refers to those ostensibly self-employed but who must pay a supplier in order to remain in business (e.g. a street hawker or taxi driver). This formulation does not invalidate the concept of a bipolar urban economic structure but rather provides a richer basis for analysis of urban labour markets.

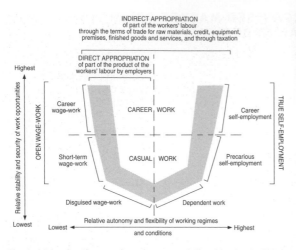

Figure 24.2 Continuum of employment relationships in the Third World urban economy

LABOUR MARKET STRUCTURE

Labour markets are structured into segments made up of jobs of different kinds with different levels of reward, and separated by barriers that limit mobility between segments. A number of labour-market disaggregations have been proposed, but a classification of direct relevance to the Third World urban economy identifies the following categories:

1. *Protected wage work* in which contracts and legal constraints operate and jobs are protected from market forces by restrictions on entry.
2. *Competitive regular wage work* in which entry is relatively open and market forces operate but employment is nevertheless continuous and perhaps subject to contract.
3. *Unprotected wage labour*, a heterogeneous category that includes much casual labour, domestic service and wage workers in petty trade, and is characterised by insecurity and/or irregularity. Various forms of disguised wage labour are also included (Figure 24.2).
4. *Self-employment and family labour* engaged in small-scale production.
5. *Marginal activities*, which range from peripheral low-productivity work, such as shoe-shining and hawking, to semi-legal and illegal activities.

Each mode of production distributes the labour force across these categories and in doing so contributes to socio-spatial variations in poverty and affluence.[14]

Plate 24.1 An elderly herb seller in an informal market in Hong Kong

JOB ACCESS

Segmentation of the labour market is dependent upon the existence of forces that control access to jobs of different types. These include both institutional barriers and the characteristics, credentials and resources of individuals. Some barriers are particular in that access to certain jobs is obtained through interpersonal networks that reflect social structures based on kinship, caste, community or ethnic origin. Other restrictions on access to work depend on more general criteria. They include educational qualifications, skills and experience, personal characteristics likely to appeal to employers (e.g. initiative, docility, age, sex), access to capital (which is especially important for

Plate 24.2   Recycling is an important form of informal employment and income in Shenzhen, China

self-employment but also relevant where migration costs must be met) and access to a market for output from self-employment.

Significantly, for those who fail to gain access to the better segments of the urban labour market it is by no means certain that they will find easy-entry low-income jobs available. Much of the informal sector is also highly protected by entry rules. As a result, many urban dwellers face a lifetime of casual work eking out a living by taking on whatever menial jobs they can find. The more open entry points into the urban labour market include portering, casual construction work and domestic service (Box 24.2). At the bottom of the scale are marginal activities such as begging, prostitution, theft and other illegal operations.

Discrimination can be a potent force in determining access to urban employment. There is clear evidence of sex discrimination in the form of large wage differ-entials even for virtually identical jobs.[15] Ethnic discrimination is also a major factor. In India caste determines entry to particular labour market segments (Box 24.3).

UNEMPLOYMENT, UNDEREMPLOYMENT AND MISEMPLOYMENT

For several decades, urban labour markets throughout the Third World have experienced an excess of labour with limited skills. This has led to a growth in open unemployment. Some writers maintain that in countries where few qualify for unemployment benefits, unemployment is not usually representative of those in the most desperate living conditions. It is argued that only the not-so-poor family can support an unemployed member (e.g. an educated offspring) during an extended search for employment within the protected labour force.[16] For the poor, unemployment is a luxury they cannot afford. Although one should not dismiss the problem of unemployment among middle-income groups, the bulk of evidence underlines the concentration of unemployment among the poor residents of Third World cities.[17]

A second element of the urban labour surplus in the Third World is underemployment. This is seen in three distinct forms:

1. Fluctuations in economic activity may occur during the day (e.g. at markets), over the week or month (e.g. in recreational services) or seasonally (as in tourism). As activity ebbs, casual labour is laid off and many self-employed people are left without work.
2. Workers may be so numerous that at all times a substantial proportion are less than fully employed (e.g. street vendors).
3. Hidden unemployment may occur where solidarity groups continue to employ all their members rather than discharging them when there is insufficient work to keep them fully occupied. Such generated employment is typical of family enterprise but social ties (e.g. based on common origin or shared religion) can also promote a commitment to maintain every member of a community.

Misemployment refers to activity that contributes little to social welfare and includes begging as well as the hangers-on of the powerful and affluent, a role that is institutionalised in the inflated bureaucratic structures of many contemporary Third World societies.[18]

WOMEN IN THE LABOUR FORCE

The urban labour market is segmented along gender lines. In general, women occupy a less favoured position, being found disproportionately in the least remunerative and/or lowest-status occupations.[19] Neoclassical economic theory attributes male–female differentials in earnings to differences in productivity

---

BOX 24.2

Daily activity pattern of a part-time domestic worker in Dhaka, Bangladesh

Morning	Afternoon
5.15–6.00 Wash face, make and bake roti for children and husband. Prepare to go to first workplace	*2.00–4.00* Walk back home, cook rice or roti for children, with lentils or cooked vegetables brought from second workplace. May take rest or talk with children. Prepare to go to third workplace.
6.00–6.15 Leave house and walk for twenty minutes to first workplace.	*4.00–4.15* Walk for fifteen minutes to third workplace
6.30–11.00 Wash utensils, plates and cutlery, dust rooms and wipe floors, eat breakfast, go to daily market for fresh items, return, wash clothes and put out to dry. Prepare to go to second workplace. Monthly pay: 200 taka.	*4.15–6.30* Wash utensils and clothes, wipe floors, prepare tea, take children to playground, make beds, may have tea. Prepare to go to fourth workplace. Monthly pay: 150 taka.
11.10–2.00 Walk for five minutes to second workplace. Wash utensils, help prepare food, pick children up from school, wipe floors, wash clothes, have a bath, have some left-over food, may take some home for own children. Monthly pay: 200 taka.	*6.30–8.00* Walk few minutes to fourth workplace. Help prepare and serve dinner, clean rooms, make beds, wash utensils, may be given some food to take home. Monthly pay: 100 taka. Walk home, cook rice and something for the family, chat with family, discuss family problems with husband, rest.

Source: **S. Huq-Hussain (1995) Fighting poverty: the economic adjustment of female migrants in Dhaka** *Environment and Urbanization* **7(2), 51–65**

BOX 24.3

Waste-picking as a survival strategy in Bangalore, India

Most of the estimated 25,000 waste pickers in the southern Indian city of Bangalore are members of the lower castes. A large proportion of the labour force are women, since, despite the low returns and the health risks, waste-picking offers one of the few ways in which women from lower castes can earn an income and also meet their household and child-rearing responsibilities.

Waste pickers collect materials that form the main input for both small and large-scale recycling operations but their share of the total profit is tiny. There is no formal contract between the pickers and the factory, materials passing through a network of dealers, each of whom seeks to maximise his profit by reducing the amount paid to the pickers, cheating them through undercounting or tying them in to a labour agreement through high-interest loans. Prices are also affected by market prices for raw materials, including waste imported from countries of the developed world. In times of inflation, falling prices, poor weather and competition the precarious position of waste pickers is attenuated.

Waste-picking by mothers means that most children do not attend school, their labour being required for the survival of the household. Girls usually accompany their mother and continue waste-picking even after marriage. Boys work alone or with friends; many leave home to survive on the streets, either through continuing waste-picking or through other casual sources of income. Waste-picking is tiring, heavy work and pickers are exposed to illnesses and infection, which, combined with inadequate washing facilities in the slums, and limited access to medical care, form a potent threat to their health and that of their children. Yet waste-picking is often the only economic opportunity open to those groups whose work and life reflect their low status in Indian society.

Sources: M. Huysman (1994) Waste picking as a survival strategy for women in Indian cities *Environment and Urbanization* 6(2), 155–74; C. Hunt (1996) Child waste pickers in India *Environment and Urbanization* 8(2), 111–18

due to gender-based differences in human capital (with women being less valuable because of a lack of physical strength, limited education and training, and family responsibilities that may give rise to greater absenteeism and labour turnover). These factors, however, explain only part of the wage gap. Feminist theories have emphasised the importance of socio-cultural factors in restricting women's access to and progression in the labour market. These include the favouring of male children over female offspring in human-resource development (Box 24.4). The classification of certain types of employment as 'women's work' and the lower pay and security associated with such activities accentuate the gender segregation of the urban labour market.[20] Also, in traditional societies a woman who moves out of her accepted family role in order to take a job may be seen as 'loose' and subjected to sexual harassment. For those most in need of work, leaving the job is not an option. Particular attention has been focused on the traditional allocation of housework and child care (reproduction) to women, and the limiting impact of this gender division of labour on women's ability to participate in non-domestic (production) work. The extent of the additional burden on women is illustrated by the finding that married women in Malaysia who do housework and are in paid employment outside the home spend, on average, 112 hours per week working, while the equivalent figure for the USA is fifty-nine hours.[21] Furthermore, contrary to the common notion of the male breadwinner, in many Third World cities, particularly in Latin America and parts of Africa, up to half of households are female-headed.[22] Even where men and women both contribute to household income, women typically contribute a larger part of their earnings to household expenses.

Women have increasingly been absorbed into wage labour within the formal sector as part of the NIDL. In countries such as Mexico, India and the Philippines, export processing zones (EPZs) and free-trade zones specialising in light manufacturing and data-processing have led to a 'feminisation of the labour market as a result of employers' demands for a low-cost, flexible and passive work force'.[23] Despite their insistence on non-union agreements, deregulation of any existing national labour laws and long hours of work, most TNCs tend to better local wages and working conditions, and for many women such jobs are preferable to traditional alternatives such as domestic service.

BOX 24.4

Gender, work and human development in Manila, Philippines

Maria is a widow aged 37. She has four children: two daughters aged 20 and 15, and two sons aged 16 and 12. She came to Manila when she was 6 years old with her parents, who were both fish-paste vendors. Maria attended school until she was 11 but then had to leave because her parents could not afford to keep all their children at school and decided that the boys made the better investment.

For a while Maria looked after her younger brothers and sisters in their squatter house in the huge Tondo settlement; gradually she began to help her parents, and it was while selling the fish paste that she met her husband, Eduardo, who was already a vendor of sweets, newspapers and candies from a more or less permanent pitch. In the years that followed their common-law marriage she not only helped her husband in his business but also had nine pregnancies, five of which ended in miscarriage. Her husband died two years ago and she took over the business herself. As a result, her working day is now extremely demanding.

It begins at 4.00 a.m., when she wakes and cooks breakfast for her family, usually any left-overs from the night before, and then goes to the public market, where she buys the variety of goods that she sells. Her stall is near the entrance to the main Port Authority harbour area in Tondo, where her patrons are the dock workers themselves, many of whom also live in Tondo and know Maria and her family. Maria sells from 7.00 a.m. to 11.00 a.m., from 2.00 p.m. to 5.00 p.m. and most evenings to 11.00 p.m. In the intervening periods she prepares her family's meals, and while she is at home her younger daughter looks after the stall. Maria's net earnings, despite all the long hours, amount to only £3 or £4 per day. She cannot afford to pay the full licence fee for vending, which costs £60, so she pays the police about 80p per month instead.

To ensure that her sons' education continued, Maria took her younger daughter out of school soon after her husband died. Both boys are doing well and Maria is confident that her eldest son, Ramon, will stand a good chance of obtaining a job at a foreign factory in the export processing zone when he leaves school later this year. If he does, the family's financial situation should improve dramatically, but until then it is sustained by the work of the two girls.

Rosa, the older daughter, has been working in domestic service now for six or seven years. She works for a middle-class family, where the head of the household is a bank clerk. She began as a laundress and general cleaner but is now the family cook, and gets on very well with her employers. Her salary is £70 per month, near the top end of the scale, and although she lives at home she eats at her employer's and is sometimes able to bring home some food for her own family. Rosa knows that she is good at her job – that is why her employers pay her reasonably well – and she has ambitions to capitalise on this by going abroad to work. She knows that government regulations will make sure that 75 per cent of her salary is remitted back home but her real goal is to marry abroad. To this end she has advertised in a newspaper that circulates among potential employers and husbands in Australia. Maria knows of her plans and fully supports them, for it will mean further financial security for the family if her daughter is successful. However, Rosa is competing against many professionally qualified women in the overseas job market and her chances of success depend on her looks. She has invested a lot of money in a good photograph for her advertisement and is hoping for the best.

Maria, the younger daughter, works mainly at home. She does much of the housework, while her mother is at her stall, and in addition she works from time to time for a nearby government factory. Work is available only at certain times of the year and Maria junior is dependent on the agent to put the sewing her way. However, a few of her mother's sweets for the agent's children have proved to be a good investment in this respect. The factory is locally owned and the pay is poor, but Maria junior hopes that the experience she gains will help her land a full-time job with a bigger company at some point in the future.

In short, times are hard at present but hopes are high that in the future the investment in the boys' education will provide financial security. All members of the family are struggling against a system that exploits poor people, but the women are doubly disadvantaged, since they have all made substantial sacrifices in order to fulfil the expectations and demands of a strongly patriarchal society.

Source: **D. Drakakis-Smith (1992)** *Pacific Asia* **London: Routledge**

In most Third World cities, however, the industrial labour force is small and the majority of female workers still find employment in the informal sector. Here, as in the formal sector, women are commonly found in low-waged occupations such as food preparation, petty commodity production, street trading or working in subcontracting enterprises[24] (Box 24.5).

CHILD LABOUR

Child labour is common in the Third World, where most children are required to make a contribution to the household economy.[25] Child labour takes many forms from paid work in a factory and other types of wage labour such as street-vending to bonded labour in which a child must work to pay off a family debt. Child labourers are popular with employers, since they work long hours for low pay and are unable or unwilling to complain about poor conditions. Many work in hazardous activities such as brick-making, construction and mining, do not receive adequate nutrition or health care, and have little or no formal education[26] (Box 24.6).

Child labour is a graphic representation of urban poverty and deprivation. The meagre earnings of children make a significant difference for impoverished families, and for many female-headed households the contribution of the firstborn is essential to family survival.[27] In extreme cases, children may be sold into conditions little better than slavery. Child prostitution is an increasing problem in cities of Asia; in Bombay alone an estimated 100,000 are exploited in this way. In most Third World cities there is a growing

Plate 24.3   Daba-wallahs deliver home-prepared lunches to office workers in central Mumbai

BOX 24.5

Sweatshop labour in a Bangladeshi clothing factory

The garment industry is the biggest employer of women in Bangladesh and accounts for more than half the country's export market. The conditions the women in the factories work under are horrendous. Dangerous and under-maintained machines. Working hours that stretch past dawn and dusk. Few breaks. No respect for health and safety. Appalling wages. To Western eyes it is exploitation. To them it is freedom.

It is true that conditions in the factories are atrocious. There are no paid holidays: even on the holiest Islamic days, women – who tend to adhere more ardently to the tenets of Islam than Bangladeshi men – must work. They work beneath harsh strip lighting, damaging their eyes. Fires are a perennial problem. They work for at least twelve hours each day. They have only fifteen or twenty minutes for lunch breaks; the lucky ones will get perhaps ten minutes more in the evening, when they huddle together in a too-small room to talk about their families and their problems.

In the fasting weeks of Ramadan, when no food or water is allowed to pass their lips between the hours of sunrise and sunset, the women must work in sweltering heat, and they must work well past sunset, starving themselves for periods way beyond what is required by their religion. Most of the women are employed on a casual basis, picked up and dropped as orders come and go. Many are forced to find work at different factories each day; they stand at the doors of the factories each morning, waiting in hope of work.

These women are given their wage packets each month or each fortnight; rarely are they paid a daily wage for their day's work, and they are ripe for exploitation. Only around 22 per cent of women in Bangladesh are literate. They do not have the skills necessary to calculate how much money they should be paid for the work they have done. They are seldom paid correctly by factory owners; overtime is merely work they do for nothing.

Despite these problems the waiting list for jobs at the garment factories is long. Nurnun waited six months before she received her first job, making shirts for a British high-street chain. 'The work is very hard and the supervisor shouts at us all the time, especially when we have an order due and they ask us to make more shirts than it is possible to do. I must be very careful not to make any mistakes, or I will lose my job and no one else will employ me. But I like to work in the garment factory and the 1,200 taka (£20) I earn each month means that my family can live well.'

It is the dream of many Bangladeshi women to work in the garment factories. They see the work as a way out of absolute poverty and some are beginning to realise that, through the factories, they can begin to obtain a degree of freedom from their often tyrannical husbands.

Source: The Herald 18 November 1997

population of street children who have left home or have been abandoned.[28] Legislative attempts to regulate child labour are largely ineffective, being almost impossible to enforce in cities where the mass of the population are engaged in a daily struggle to satisfy their basic needs.

THE HOUSEHOLD ECONOMY AND COPING STRATEGIES

To understand the coping mechanisms by which poor urban households survive, it is necessary to complement macro-scale neoclassical economic analysis of the Third World urban economy with detailed consideration at the micro-scale of the household economy. In Friedmann's (1992) model of the 'whole economy' the household is seen as the basic economic unit – a miniature political economy that makes decisions on a continuing daily basis concerning the best use of available resources[29] (Figure 24.3).

Within this framework, households have three sources of monetary income, deriving respectively from formal work, informal work and net intra-family transfers. The first two are part of the market economy and the third is part of the moral economy. The total amount of disposable income (indicated by $ in Figure 24.3) may be reduced by taxes (some of which may be returned to the household via public transfers such as health care) and by capital expenditure to support informal production. The household also has a moral obligation to make income transfers to non-household family (e.g. to support a new migrant to the city), although these may be recipro-

BOX 24.6

Child labour in Pakistan

Thousands of Pakistani children struggle with rough tools and nimble fingers to stitch footballs in hundreds of factories in the central Punjab city of Sialkot, the main centre for making footballs for export.

They begin on low-grade balls before graduating to international-class footballs. The basic materials are imported from Europe. Factories cut them up, print on the logos and laminate them before sending them to villages, where children often work in mud-brick workshops. They assemble the balls by leaning on tweezer-shaped pieces of wood. Holding a large needle in each hand, they punch holes in the material and pull through the white twine. Their ascent from child to family breadwinner is rapid, and much before its time. Most join the work force before reaching 10 years of age. The small wages they earn by working ten to twelve hours a day provide their family with only basic necessities.

More than a million children are employed in the carpet industry in Pakistan and thousands work in football factories. Labour laws, including social security, do not apply to them. A large number are bonded labour, because their parents cannot pay their debts to factory owners. Not only do the children provide cheap labour, they are also considered

ideal workers because of their small and delicate fingers. According to some employers, children can work in stooping and squatting positions more effectively than adults.

A few years ago Pakistan introduced a law banning child labour, but it could not be implemented because of economic and social conditions. Families that often have five or six children depend on their wages to keep the family alive. They could not afford to send the children to school, even if education were free. A survey that examined the reason behind child labour revealed that the parents placed a low value on education. They did not see education as necessarily a means to a job. A large number of parents also complained that the quality of education was poor. Most preferred to send their children to work rather than to school or to wander the streets.

Most Pakistani economic analysts agree that child labour cannot be eradicated until the social and economic factors behind it are addressed. Some economists believe that a better education system, placing emphasis on vocational training, might be a positive step. But they also give a warning that, if there are still no jobs available, such a move would not change the situation.

Source: The Times 29 October 1997

cated at a future time. In addition to money income, the household's livelihood also depends on in-kind contributions from domestic and communal work. Although not represented explicitly in the model, in many low-income communities informal financial systems, such as rotating credit associations, can also play an important role in the household economy.[30]

The model identifies three kinds of household expenditure, on:

1. Consumption proper, such as food.
2. Investment in household durables, e.g. housing (H&DG).
3. Investment in human resources, e.g. the education of children (HR).

Expenditures also include time inputs to domestic work (D) such as shopping, cooking, cleaning, nursing,

child-rearing, construction and repair, water and firewood collection, and vegetable gardening, and to productive activities within the neighbourhood or community economy (CE), including the construction of schools, making improvements to roads and the distribution of hot meals to the destitute.

The **whole-economy model** enables us to trace changes in household coping strategies in response to changes in internal (e.g. illness, birth, death) or external (e.g. job loss) circumstances. In times of prolonged economic crisis:

1. Women may enter the labour force (at lower rates of pay than men) or start to engage in informally organised activities (e.g. taking in washing or dress-making), thus reducing the time available for domestic work in the household.
2. Expenditure on household durables, housing construction and maintenance as well as

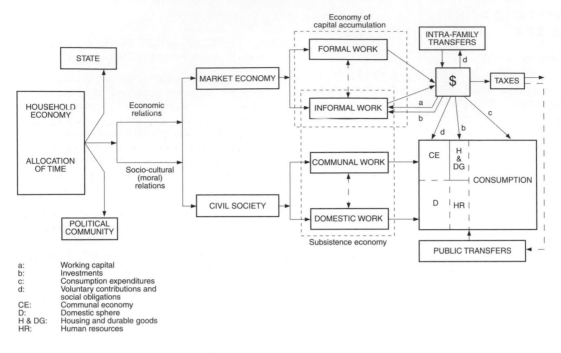

a: Working capital
b: Investments
c: Consumption expenditures
d: Voluntary contributions and
            social obligations
CE: Communal economy
D: Domestic sphere
H & DG: Housing and durable goods
HR: Human resources

Figure 24.3   The household economy in the Third World city
Source: J. Friedmann (1992) *Empowerment* Oxford: Blackwell

human-resource development (e.g. schooling), may be reduced. Survival becomes the goal, not future development.

3. *Per capita* consumption of food may be reduced, although not necessarily equally among household members, with women, especially young girls and children, generally last to eat.

4. Contributions to the community economy, especially in work effort, may increase, or households may seek to reduce their consumption costs by participating in joint enterprises such as community gardens or communal kitchens.

Reverse tendencies may be observed when the market economy picks up. As the emphasis in household decisions moves from subsistence to accumulation, household investment in durables and human resources increases. Informal activities that are marginal to the household economy are abandoned, while those that are linked into the formal economy through subcontracting are likely to prosper. Enthusiasm for community enterprises also declines, allowing more time to be spent on domestic work, but some community obligations continue to be honoured, both for reasons of genuine neighbourhood solidarity, and to maintain trust and reciprocity within the moral economy in case another crisis occurs in the future.

By articulating the dialectical links between rational (economic) and socio-cultural (moral) relations, formal and informal activities, life space and economic space, and the economy of capital accumulation and the economy of subsistence, the 'whole-economy' model offers a useful corrective to macro-scale models of the urban economy and contributes to our understanding of how poor households cope with their poverty.

FURTHER READING

BOOKS

J. Allen (1995) Crossing borders: footloose multinationals? in
J. Allen and C. Hamnett (eds) *A Shrinking World?* Oxford:
Oxford University Press, 55–102

S. Chant (1999) Informal sector activity in the Third World city,
in M. Pacione (ed.) *Applied Geography: Principles and
Practice* London: Routledge, 509–27

A. Portes, M. Castells and L. Benton (eds) (1989) *The Informal
Economy* Baltimore MD: Johns Hopkins University Press

J. Pryor (2003) *Poverty and Vulnerability in Dhaka Slums*
Aldershot: Ashgate

C. Rakodi and T. Lloyd-Jones (2002) *Urban Livelihoods:
A People-centred Approach to Reducing Poverty* London:
Earthscan

M. Santos (1979) *The Shared Space* London: Methuen

J. Thomas (1992) *Informal Economic Activity* Hemel
Hempstead: Harvester Wheatsheaf

JOURNAL ARTICLES

H. Beazley (2002) Vagrants wearing make-up: negotiating
spaces on the streets of Yogyakarta, Indonesia *Urban
Studies* 39(9), 1665–83

M. Bose (1999) Women's work and the built environment:
lessons from the slums of Calcutta, India *Habitat
International* 23(1), 5–18

A. Conticim (2005) Urban livelihoods from children's
perspectives *Environment and Urbanization* 17(2),
69–81

P. Daniels (2004) Urban challenges: the formal and informal
economies in mega-cities *Cities* 21(6), 501–11

M. Hays-Mitchell (1994) Street vending in Peruvian cities
Professional Geographer 46(4), 425–38

G. Iyenda (2005) Street enterprises, urban livelihoods and
poverty in Kinshasa *Environment and Urbanization* 17(2),
55–67

S. Mahmud (2003) Women and the transformation of domestic
spaces for income generation in Dhaka bustees *Cities*
20(5), 321–9

D. Mead and C. Morrison (1996) The informal sector elephant
World Development 24(10), 1611–19

W. Mitullah (1991) Hawking as a survival strategy for the
urban poor in Nairobi *Environment and Urbanization*
3(2), 13–22

KEY CONCEPTS

- Formal sector
- Informal sector
- Modernisation
- Hawkers
- Protected wage work
- Unemployment

- Underemployment
- Misemployment
- Hidden unemployment
- Petty commodity production
- Household economy
- Whole-economy model

STUDY QUESTIONS

1. Explain the structure of the Third World urban economy.
2. Identify the major characteristics of the lower circuit in the Third World urban economy.
3. Consider the view that the informal sector of the urban economy represents a panacea for the economic problems of Third World cities.
4. Examine the role of women and/or children within the urban labour force of Third World countries.
5. Explain the nature of the household economy and illustrate the coping strategies of the urban poor.

PROJECT

Undertake a review of recent issues of major journals covering Third World urban themes (e.g. *Environment and Urbanization, Habitat International, International Development Planning Review,* formerly *Third World Planning Review*). Prepare a series of case studies that illustrate the diversity of economic activity within Third World cities and the role of different population groups within the urban economy. These might include studies of the activities of a transnational corporation, a modern manufacturing industry, child labourers, street hawkers or garbage pickers.

25

Housing the Third World Urban Poor

Preview: sources of housing for the urban poor; housing sub-markets; squatter settlements; formative processes; development typologies; the consolidation of squatter settlements; security of tenure; housing finance; urban land markets and the poor; government housing policies in the Third World city

INTRODUCTION

Access to decent, affordable housing is a basic requirement for human well-being, yet in most large cities of the Third World much of the population occupies the most rudimentary forms of shelter. An indication of the difference in housing quality between high- and low-income countries is provided in Table 25.1. In terms of floor area per person the city average for Dhaka was 3.7 m² (40 ft²) in 1990, compared with over 60 m² (650 ft²) in Washington DC.[1] City averages, however, obscure high levels of overcrowding within cities and, in particular, within low-income areas. In Karachi the living space in many of the informal settlements (*katchi abadis*), where 40 per cent of the population lives, is between 2 m³ and 3 m² per person while those living in larger town houses or apartments enjoy between 22 m² and 33 m² per person.[2] An equally wide gap exists between rich and poor countries in terms of levels of urban service provision (Table 25.1).

TABLE 25.1 INDICATORS OF HOUSING QUALITY IN HIGH- AND LOW-INCOME COUNTRIES

Cities in:	Floor area per person (m²)	Persons per room	% of permanent structures	% of dwelling units with water connection to their plot	Government expenditure per person on water supply, sanitation, drainage, garbage collection, roads and electricity (US$)
Low-income countries	6.1	2.47	67	56	15.0
Low- to mid-income countries	8.8	2.24	86	74	31.4
Middle-income countries	15.1	1.69	94	94	40.1
Mid- to high-income countries	22.0	1.03	99	99	304.6
High-income countries	35.0	0.66	100	100	813.5

Source: United Nations Centre for Human Settlements (1993) *The Housing Indicators Programme* volume 3 *Preliminary Findings* Washington, DC: United Nations

The failure of government housing programmes to provide affordable housing has forced the mass of the Third World urban population into cheaper alternative forms of shelter, which range from inner-city slum tenements and peripheral squatter settlements to the pavements (sidewalks) of major cities (Box 25.1). The shortage of adequate housing is exacerbated by high rates of population growth. During the 1980s in low-income Third World countries nine new households were formed for every new permanent dwelling built. Over the same period, governments typically spent only 2 per cent of their budget on housing and community services.[3] These trends leave no doubt that the lack of decent affordable housing for the urban poor represents an ongoing challenge. In this chapter we identify the main sources of housing for the urban poor and examine the nature of different housing sub-markets. We focus on the processes underlying the formation of squatter settlements and consider the development trajectories of such areas and their relationship with the city as a whole. We then turn to examine housing policy and critically evaluate a range of governmental responses to the problem of low-income housing provision in Third World cities.

THE MAIN SOURCES OF HOUSING FOR THE URBAN POOR

Figure 25.1 provides a general framework for understanding the main sources of housing in the Third World. Housing is classified as conventional if it is constructed through the medium of recognised formal institutions (e.g. banks and planning authorities) and in accordance with established legal practices and standards. In general this corresponds with an industrial mode of production that utilises wage labour, is capital-intensive and employs relatively sophisticated technology. Non-conventional housing is that which does not comply with established procedures, is usually constructed outside

BOX 25.1

Pavement dwellers in Mumbai, India

Pavement dwellers live in small shacks made of temporary materials on pavements (sidewalks) and utilise the walls or fences that separate building compounds from the pavement and street outside. They can be seen in most large cities in India but are mostly concentrated in Mumbai (formerly Bombay) and Calcutta.

Most households living in pavement dwellings have at least one household member employed – and hardly any make their livelihood through begging. Far from being a burden to the city's economy, they are supplying it with a vast pool of cheap labour for the unpleasant jobs that organised labour does not like to do. Most are employed in the informal economy as petty traders, hawkers, cobblers, tailors, handcart pullers, domestic servants and waste pickers. They can work for such low wages only because they are living on pavements, which minimises their housing costs, and incur no overheads on either shelter or transport (most walk to and from work).

The inhabitants of pavement dwellings come to live there initially as a temporary measure until they can locate and afford better housing. Unfortunately, most are never able to acquire better housing and live out their lives on the footpath. In one ward almost all the families have been living on the pavement ever since their arrival in Bombay – which could be as much as thirty years previously.

In a sample of 375 pavement dwellers, half lived in huts of less than 54 ft² (5 m²). Given the size of the hut, the pavement in front of the shelter becomes an important part of the domestic space – and is where most children eat, sleep, study and wash. Most pavement dwellers had to obtain water from housing near by (probably where the women work as domestic helps). For toilets, most used public toilets, for which a nominal charge had to be paid, while 30 per cent used the railway tracks. When mothers of children under 6 were asked about the main problems they faced in staying in their current location, problems of water and sanitation were the ones most frequently mentioned.

Pavement dwellers also had many characteristics in common with most other low-income households in terms of their incorporation within the city. Around two-thirds had their names on an electoral roll and just over half the children attended school. Three-quarters of the children had been born in municipal hospitals. Just over a third of all households said they had savings and 28 per cent had loans with banks.

Source: S. Patel (1985) Street children, hotel boys and children of pavement dwellers and construction workers in Bombay: how they meet their daily needs *Environment and Urbanization* 2(2), 9–26

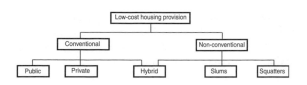

Figure 25.1 Major sources of housing for the Third World urban poor

Source: D. Drakakis-Smith (1981) *Urbanisation, Housing and the Development Process* London: Croom Helm

the institutions of the formal building industry, is frequently in contravention of existing legislation, and is almost always unacceptable to prevailing middle-class standards. Production is often through the labour of the individual or household intending to occupy the house, although petty capitalist construction firms also operate within this sector.[4]

Conventional and non-conventional housing represent the poles of a continuum linked by a category of hybrid dwellings that incorporate features from both of the main sectors. Hybrid housing contravenes fewer legal standards than non-conventional housing and, as a result, the units may become socially and politically more acceptable. In cities where building standards are stringently defined but loosely enforced, hybrid housing can be extensive.[5] In addition four other types of housing are present in most Third World cities:

1. *Public housing.* During the 1960s and 1970s, many Third World governments initiated large public housing programmes or extended existing ones. During the 1980s, however, support for public housing diminished as a consequence of recession and cutbacks in public expenditure linked with debt crises and structural adjustment. Also public housing programmes had often failed to meet building targets despite the expenditure of large sums, and the small size, poor design and maintenance problems made units unpopular with many tenants. A fundamental problem of public housing programmes in many countries was the high unit costs caused by the adoption of Western standards and building techniques, which meant that, even with subsidy, rents were beyond the reach of a large proportion of the low-income population. Few Third World governments now have large public housing programmes building rental accommodation, and in most countries where there had been a large volume of public housing for rent, as in Singapore and Hong Kong, much of the stock has been sold to former tenants.

Plate 25.1 Large-scale public housing projects in Singapore have provided improved accommodation for a majority of the population

2. *Private housing.* Although the private sector builds most of the conventional housing in Third World cities, relatively few of the new units fall within the financial reach of the poor. Since the profits to be made from low-cost housing construction are relatively small, the contribution of the private sector to housing the poor is likely to remain a minor one. Furthermore, construction of housing for lower middle-income families is unlikely to lead to downward filtering of vacated dwellings since the general shortage of accommodation in Third World cities ensures that any increase in production is absorbed by the middle class with little release of 'surplus' housing. The majority of the poor live in slums or squatter settlements.

3. *Squatter settlements.* Squatting has strict legal connotations in most countries and refers to either the occupation of land without the permission of the owner or the erection or occupation of a building in contravention of existing legislation. This

juridical interpretation helps to overcome the variety of forms of squatter settlements and defines the nature of the power relationship between squatters and urban authorities. The illegal status of squatter settlements provides government with complete legal justification for any 'remedial' action against squatters. It also generates insecurity among squatters, which inhibits their participation in city life, including access to services and amenities to which they are legally entitled, and may encourage clandestine occupations that further complicate the squatters' relationship with the urban authority.

4. *Slums*. These are defined as legal permanent dwellings that have become substandard through age, neglect and/or subdivision into smaller occupational units such as rooms or cubicles. Slums can develop either as a result of inadequate maintenance by a landlord, often prompted by rent-control legislation, or by the internal subdivision of buildings by residents in order to accommodate new arrivals or newly created households. Slum dwellings shelter large numbers of urban poor in the Third World.

These categories offer a general guide to the types of housing available. More detailed insight into the housing status of the Third World urban poor is gained by examination of the major housing sub-markets.

HOUSING SUB-MARKETS AND THE URBAN POOR

Within any city a number of different sub-markets exist to satisfy the needs and financial abilities of different populations. As we saw in Chapter 10, housing decisions are based on trade-offs between a host of factors, including cost, size, quality and location. For low-income households in Third World cities, accommodation within easy reach of jobs is frequently the most important consideration. Some will want temporary accommodation only, others a more permanent arrangement. The housing priorities of a young single person intending to save money by spending as little as possible on accommodation differ from those of a family with several children and a stable, if low, income. Renters and owner-occupiers also exhibit different preferences and priorities (Box 25.2). Moreover, needs change over time.[6] These considerations are reflected in the existence of different housing sub-markets.

Plate 25.2   Homeless people in Calcutta find temporary shelter in drainpipes

LOW-INCOME RENTAL HOUSING

Table 25.2 illustrates the different kinds of rental housing used by low-income groups. These include the following categories:

1. At the bottom of the range, a tenant may rent a bed in a room rather than the room, and may even rent the bed by the hour. The 'hotbed' system in Calcutta permits three persons to use the same bed over a twenty-four-hour period.
2. Central tenement areas offer cheap rental accommodation for those whose main source of income is in the city centre and who are unable to afford the cost, in time or fares, of public transport to peripheral squatter areas.
3. In many cities the bulk of low-cost rental housing is provided by home-owners in the informal sector, since the returns from commercial renting are not sufficiently high to encourage large-scale

BOX 25.2

Factors underlying choice of residential location among low-income groups in Abidjan

For both owner-occupiers and tenants, housing opportunities, access to areas of economic activity and potential employment, and the presence of relatives were the most important attractions for people moving to the precarious settlements. The table shows the main factors that attracted owners and tenants, based on interviews with 620 people.

Attraction	Owners	Tenants	All
Land available	196 (39.2)	—	196 (31.6)
Accommodation available	23 (4.6)	8 (6.7)	31 (5.0)
Presence of relatives	89 (17.8)	37 (30.8)	126 (20.3)
Easy access	78 (15.6)	—	78 (12.6)
Close to employment	107 (21.4)	29 (24.2)	136 (22.0)
Low cost of living	3 (0.6)	44 (36.7)	47 (7.6)
Community lifestyle	4 (0.8)	2 (1.7)	6 (1.0)
All	500 (100)	120 (100)	620 (100)

Note: **Numbers in parentheses are percentages**

Proximity to one's job, the most important factor for 22 per cent of those interviewed, has about the same weight as the presence of relatives (20.3 per cent). However, it is logical that owners should be attracted by the availability of land (39.2 per cent), whereas tenants are more likely to look for neighbourhoods with a low cost of living (36.7 per cent) or where their relatives live, or close to areas of economic activity. Living close to their place of work may reduce but does not remove the high cost of transport. According to various studies, transport represents between 6 per cent and 12 per cent of the expenses of heads of household in informal housing. Whether transport and proximity to places of work, or competitive rents and property opportunities, appear among criteria for choosing neighbourhoods, these elements show that low-income groups' residential strategy cannot be reduced to a simple equation. Choice is generally based on a combination of several interacting factors.

Source: **United Nations Centre for Human Settlements (1996)** *An Urbanising World* **Oxford: Oxford University Press**

---

capital investment in the formal sector. Higher levels of renting are found typically in more established squatter settlements with a greater degree of security from dispossession.[7]

4. In addition to rental of rooms and dwellings, sub-markets also operate for renting of house plots, roof space or back-yard space where a temporary shelter may be erected. In some cities there is even a rental market for people sleeping on pavements or in public spaces, with small payments made to officials or others demanding protection money.

SHARERS

In addition to tenancy arrangements that involve cash payments of rent, there are also sharers or 'free-rent tenants' who are permitted to occupy a dwelling with the owner or tenant. Sharers are typically the children or employees of the owner or principal tenant, and may often be seen as 'disguised renters' by making an in-kind contribution to household expenditure through domestic work or child care. In Mexico Gilbert (1990) found that newly arriving migrants to cities rarely moved directly into home ownership, most tending to follow the transition from sharing to renting to ownership; while in the Nigerian capital city of Abuja, sharing of dwelling units was a common strategy adopted by both middle- and low-income households to counter a limited housing supply and to reduce housing costs.[8]

LOW-INCOME OWNER-OCCUPIED HOUSING

A precise definition of owner-occupation is difficult in many Third World cities, where a considerable proportion of the population live in shelters of which they are *de facto* but not *de jure* owner-occupiers. In such cases the term 'possessor-occupier' is more appropriate. Many are illegally occupying the land on which the house is built. Others are closer to being *de jure* owner-occupiers, having purchased the land from a landowner, and it is the shelter or subdivision that is

TABLE 25.2 TYPES OF RENTAL HOUSING USED BY LOWER-INCOME GROUPS IN MANY THIRD WORLD CITIES

Types of rental accommodation	Common characteristics	Problems
Rented room in subdivided inner-city housing (tenements)	Often the most common form of low-income housing in early stages of a city's growth. Buildings usually legally built as residences for middle- or upper-income groups but subdivided and turned into tenements when these move to suburbs or elsewhere. Advantage of being centrally located, so usually close to jobs or income-earning opportunities. Rent levels sometimes controlled by legislation. Infrastructure (e.g. paved streets, pavements, (sidewalks), piped water, sewers) available. Access to schools and hospitals. Certain Third World cities never had sufficient quantity of middle/upper housing suited to conversion to tenements to make this type of accommodation common	Usually very overcrowded and in poor state of repair. Whole families often in one room, sometimes with no window. Facilities for water supply, cooking, storage, laundry and excreta/garbage disposal very poor and have rarely been increased or improved to cope with much higher density of occupation brought by subdivision. If subject to rent control, landlord often demanding extra payment 'unofficially'. Certain inner-city areas with tenements may be subject to strong commercial pressures to redevelop them (or their site) for more profitable uses
Rented room in custom-built tenements	Government-built or government-approved buildings specially built as tenements for low-income groups; sometimes publicly owned. Common in many Latin American cities and some Asian cities, and usually built some decades ago. Some quite recently constructed public-housing estates fall into this category, although it is now rare for governments to sanction private-sector tenement construction	Similar problems to above, in that original building never had adequate provision for water supply, cooking, ventilation, food storage, laundry, excreta and garbage disposal. Inadequate maintenance common
Rented room or bed in boarding, rooming house, cheap hotel or pension	Often most in evidence near railway station or bus station, though may also be common in other areas, including illegal settlements. Perhaps common for newly arrived migrant family or single person working in city to use there. Single persons may hire bed for a set number of hours each day so more than one person shares the cost of each bed. Usually relatively cheap and centrally located	Similar problems to above in terms of overcrowding, poor maintenance and lack of facilities. A rapidly changing population in most establishments prevents united action on part of users to get improvements
Renting room or bed in illegal settlement	In many cities, rented rooms in illegal settlements represent a larger stock of rental accommodation than in tenements that are legally built. May take form of room or bed within room rented in house or shack with de facto owner-occupier; may be rented from small- or large-scale landlord even though it is within an illegal settlement	Problems in terms of quality of building and lack of infrastructure (paved roads, pavements (sidewalks), storm drainage) plus site often ill suited to housing as in squatter settlements and in illegal subdivisions. Also, insecurity of tenure which is even greater than for de facto house/shack owners

Renting a plot on which shack is built	The renting of plots in illegal subdivision or renting space to build a shack in some other person's lot, courtyard or garden is known to be common in certain cities; in some cities, space is even rented to people to build a shack on the flat roofs of houses or apartments. Its extent in these and other Third World cities is not known	Similar problems to those listed above in terms of insecure tenure and lack of basic services and infrastructure. Additional burden on household to build, despite no tenure and no incentive to improve shack
Renting room in houses in lower to middle-income or formal-sector worker districts	Declines in purchasing power for many lower-middle income or formal sector worker households has encouraged them to rent out rooms to supplement their incomes and to help pay off loans or mortgages on their homes	Probably relatively good quality compared to above options. Tenant–landlord relationship not subject to contract. Such rooms frequently in areas at a considerable distance from concentrations of employment
Employer housing for cheap labour	Some large enterprises provide rented accommodation for their work force. This is common on plantations but also evident in some cities	The quality of this housing is usually very poor with several people crowded into each room and very inadequate provision of basic services. Rules often prevent families living there, so workers' families have to live elsewhere
Renting space to sleep outside	Where there are large numbers of people who sleep outside or in public places (e.g. temples; railway stations or graveyards), local officials or protection gangs may demand payment informally, especially in the best locations	The problems are obvious: not only the insecurity and lack of shelter and basic services, but also the need to pay for this space, and pay people who have no right to demand such payments

Source: **J. Hardoy and D. Satterthwaite (1989)** *Squatter Citizen* London: Earthscan

illegal, not the land occupation. The variety of types of 'owner-occupied' housing is shown in Table 25.3. Building a house or shack in an illegal subdivision or squatter settlement is the main source of new housing in most large Third World cities.[9]

SQUATTER SETTLEMENTS

Since they cannot afford the cheapest legal house or apartment, most citizens have no choice but to build, buy or rent an illegal dwelling. In most Third World cities 70–95 per cent of all new housing is built illegally and it is common for two-thirds of the population to occupy houses and neighbourhoods developed in this way. In Nairobi informal settlements occupy 5.8 per cent of the land area but house 55 per cent of the city's population,[10] including Kibera, Africa's largest slum, while in Rio de Janeiro 394 *favelas* house over 1 million people, or 10.3 per cent of the population.[11]

The form of squatter dwellings and their location in the city vary. Most cities have squatter communities in their central areas adjacent to the major employment sources. In general these are occupied by more established families, with migrants unable to secure a place in the inner-city squatter and slum areas forced on to the periphery. Although the outer squatter areas are often further from main employment foci, there is greater availability of building land and relative freedom from restrictive municipal legislation. These locations are also attractive to larger households with less dependence on job proximity, and there is ample evidence of centrifugal movement from older, crowded central areas to peripheral squatter settlements over time (Figure 25.2).

FORMATIVE PROCESSES

There are many ways in which individual families become residents of squatter communities, either as renters or as 'owner-occupiers', but two main processes may be identified. The first is the large-scale squatter invasion, found most frequently in Latin America.[12] The second, slower process of infiltration by individual households or small groups occurs widely throughout the Third World.[13]

Land invasion

Land invasions by squatters tend to be well organised and planned operations. Most of those involved have

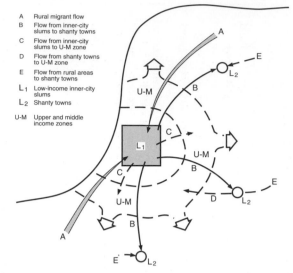

Figure 25.2   Migration patterns among low-income neighbourhoods in the Third World city

Source: R. Potter and S. Lloyd-Evans (1998) *The City in the Developing World* Harlow: Longman

experience of urban life and are aware of the precautions that must be taken to acquire and subsequently retain possession of the land. Participating families have also usually progressed to a stable if low income and are less reliant on proximity to employment. Prior association through employment or residence is common. Normally some form of tacit approval has been secured from political groups interested in the future support of the residents. This is especially important in the immediate post-invasion period, when eviction and reprisals are most likely. Other precautions involve selection of the invasion date, with the night prior to a public holiday particularly favoured, since it allows several days for consolidation and usually means that the police are deployed elsewhere in the city.

The actual invasion and consolidation process varies according to the physical conditions of the site, and the political climate. Demarcated building plots are allocated by the organising committee with the intention of settling the land as quickly as possible. The earliest constructions on individual plots comprise simple huts, but subsequently, with increased security, permanent building materials are used. Building is usually undertaken with the assistance of families and friends, although for specialised work experienced labour may be hired from the petty capitalist sector. As we shall see later, the subsequent development of the squatter settlement into a settled

TABLE 25.3 EXAMPLES OF 'OWNER OCCUPATION' HOUSING USED BY LOW-INCOME GROUPS IN MANY THIRD WORLD CITIES

Types of owner occupation	Common characteristics	Problems
Building house or shack in squatter settlement	As the city grows and number of people unable to afford a legal house or house site increases, illegal occupation of land sites on which occupants organise construction of their house or shack usually becomes common. Advantage of what is usually a cheap (or free) site on which to build – although as the settlement develops, a monetised market for sites often appears and land sites can be expensive in better-quality, better-located settlements. The extent to which households actually build most of their house varies considerably; many lack the time to contribute much and hire workers or small firms to undertake much or all of the construction	Lack of secure tenure; settlement often subject to constant threat of destruction by government. Lack of legal tenure inhibits or prevents use of site as collateral in getting a loan to help in construction. No public provision of water, sanitation, roads, storm drainage, electricity, schools, health-care services, public transport – or, even where government does make provision, it is long after settlement has been built and is usually inadequate. Poor-quality sites are often chosen (e.g. subject to flooding or landslides) since these have lowest commercial value and thus give the best chance of avoiding forceful eviction
Building house or shack in illegal subdivision	Together with housing built in squatter settlements, this represents the main source of new housing in most large Third World cities. Site is bought or rented from land-owner or 'middleman' who acts as developer for land-owner. Or where customary law is still common, access to a site through the permission of the appropriate chief, who acts for the 'community'. Governments often prepared to tolerate these, while strongly suppressing squatter occupation. Often relatively well-off households also organise their house construction on such illegal developments. As in squatter settlements, the extent to which people build their own houses varies considerably	Comparable problems to those above except that land tenure is more secure and land-owner or developer sometimes provides some basic services and infrastructure. The site is also usually planned (although so too are some squatter settlements). The better-located and better-quality illegal subdivisions are also likely to be expensive. If the city's physical growth is largely defined by where squatter settlements or illegal subdivisions spring up, it produces a haphazard and chaotic pattern and density of development to which it will be very expensive to provide infrastructure and services

TABLE 25.3 CONTINUED

Types of owner occupation	Common characteristics	Problems
Building house or shack in government site and services or core housing scheme	An increasing number of governments have moved from a concentration on public-housing schemes (which were rarely on a scale to make any impact) to serviced sites or core housing schemes. Very rarely are these on a scale to have much impact on reducing the housing problems faced by lower-income groups	Public agency responsible for scheme often finds it impossible to acquire cheap, well-located sites. Sites far from low-income groups' sources of employment chosen, since they are cheaper and easier to acquire. Extra cost in time and bus fares for primary and secondary income earners can make household worse off than in squatter settlement. Eligibility criteria often bar women-headed households. Regulations on repayment, building schedule and use of house for work or renting rooms often make many ineligible and bring considerable hardship to those who do take part
Invading empty houses or apartments	Known to be common in a few cities; overall importance in the Third World is not known	Obviously insecure tenure since occupation is illegal. May be impossible to get electricity and water even if dwelling was originally connected
Building or developing a house or shack in a 'temporary camp'	Many examples known of governments who develop 'temporary' camps for victims of disasters or for those evicted by redevelopment – usually on the periphery of the city. Many become permanent settlements	Land and house tenure is often ambiguous: the provision of basic infrastructure and services at best inadequate, at worst almost non-existent; the location is often far from the inhabitants' main centres of employment

Source: **J. Hardoy and D. Satterthwaite (1989)** *Squatter Citizen* **London: Earthscan**

community is a complex process that is related to the socio-economic status and motivation of the inhabitants and dependent upon the attitude of the government towards illegal settlement (Box 25.3).

Infiltration

The slower process of squatter infiltration is more typical of Asian or African urbanisation.[14] This may include a family's cultivating a piece of marginal land for several seasons before erecting a dwelling. Early settlers may subsequently invite others to join them. Although this process does not involve the degree of organisation evident in land invasions, entry of families to a plot of land or an existing squatter settlement may still be subject to regulation through having to obtain formal permission from a tribal elder or leader of the settlement.[15] Infiltration is aided by the **pirate subdivision** of land, as for example in Kano, where owners of peri-urban farmland partitioned it into building plots for sale in direct contravention of planning regulations.[16]

DEVELOPMENT TYPOLOGIES

In view of the differences in the formative processes and nature of squatter communities, a number of typologies have been derived in an attempt to understand this complexity. An early classification by Stokes (1962) distinguished between *slums of hope* and *slums of despair*, and implied that the success or otherwise of the community was largely in the hands of the squatters.[17] A different approach identifies the changing needs of squatters and how their priorities change with different stages in their development. Turner (1969) proposed a threefold classification of individuals into bridgeheader, consolidator and status-seeker, each corresponding to successive stages in income growth.[18] This taxonomy assumes that as economic security increases, there is a revaluation of needs and priorities which is translated into housing preferences (Figure 25.3).

More recent neo-Marxist interpretations have focused more explicit attention on the constraints that influence the squatters' choice of shelter, pointing out

Plate 25.3　Self-built housing exemplifies the poverty and innovativeness of many squatters in São Paulo

BOX 25.3

The development of San Martín, Buenos Aires, Argentina

In late 1981 some 20,000 people invaded 211 ha (521 acres) of abandoned private land in two outer districts of Buenos Aires. These invasions began in September when an initial relatively small and well organised squatter invasion prompted many other people to join in. As one invader explained, basically they want a roof over their heads, and it should be noted that a large proportion of these people do not come from the shanty towns but are tenants who cannot afford current rents, people without jobs and also new couples who have been living in overcrowded, uncomfortable conditions with their relatives and who want a place of their own.

Towards the end of November 1981, as word spread about the invasion, some 3,000 people entered the settlements, with their belongings, in the course of just five days. The government tried to bulldoze them but was successfully resisted – largely by women and children who stood in front of the bulldozers. Support for the squatters soon grew – especially from priests and Church groups and certain lawyers. The police then tried to isolate the new settlements by forming a cordon with 300 vehicles and 3,000 men. One of the squatters described how they got round this. 'We worked at night. On the houses too. We knew that the police did not have the authorisation to tear down houses; their function was to keep us from building them. The people who were going to enter the settlement waited outside the police circle; when it loosened up, because they were asleep and there were less of them, we lined up behind the first opening, we brought in the material, the wood. All the neighbours helped. In two hours we built one house; later the neighbour who just arrived helped to raise others. By morning, when the police entered, five or six houses had been built.' In the autumn of 1982 the security forces gave up their attempt to maintain the cordon.

San Martín was one of six new settlements (*barrios*) formed by this invasion. Despite the illegality, the new settlers were determined not to live in a *villa miseria* (the name given to squatter settlements in Buenos Aires; literally translated, it means 'misery settlement'). To avoid this, they organised the layout of the site so that space was left for access roads and for community facilities. A democratic system of planning and organising the settlement soon developed. At the lowest level – the level of the block – 'block delegates' were elected (one representative and two deputy representatives) who served for six months on the 'board of delegates'. Committees were also elected to supervise and co-ordinate activities and petitions. Each neighbourhood had a co-ordinating committee which was represented on what came to be called the Co-ordinating Committee of Neighbours of the West Quilmes Settlement.

The co-ordinating committee published a newsletter, *Nuestra esperanza* ('Our hope'), and quickly took charge of negotiations with public agencies for electricity, health care and public works. Initially they had little success. Illegal connections were made to the electricity grid and to the mains water supply; the electricity company had agreed to supply the settlement but the police kept the company out. The need for piped water was particularly pressing, since the nearby stream, a possible alternative source, was contaminated with wastes from local paper, weaving and meat-packing plants.

Conditions in San Martín and the other *barrios* remained very poor, since the local government refused to pave the streets, install sewers or drains, or provide health care. Diarrhoea epidemics, which were especially serious in the summer months, led to the community organising a campaign to demand health care. When this failed, the inhabitants organised for themselves a basic level of health care and built a health centre with help from the local Catholic parish. They also built their first school.

With the election of a democratic government in 1984 the municipal authorities became more sympathetic to the squatters' needs and there are no longer fears of eviction. Some medicines have been provided and vaccinations made available; on occasion, garbage has been collected – although the longer-term problems of having no drains and sewers, or paved roads and pathways, remain.

Source: B. Cuenya, D. Armus, M. Di Loreto and S. Penalva (1990) Land invasions and grassroots organisation: the Quilmes settlements in Greater Buenos Aires, Argentina *Environment and Urbanization* 2(1), 61–73

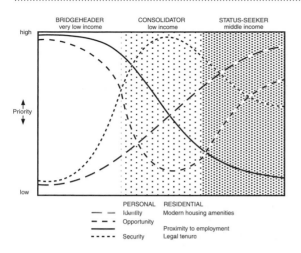

Figure 25.3 Turner's typology of squatters
Source: J. Turner (1969) Architecture that works *Ekistics* 27(158), 40–4

that, given their inferior position in society, it is at best a highly 'constrained choice' and for the very poor no choice at all. In Mexico City Ward (1976) drew a distinction between static and stagnating squatter settlements (*ciudados perdidas*) and consolidating settlements (*colonias paracaidistas*).[19] The difference was explained in terms of factors that inhibited the auto-improvement of the former, including relatively high rents and insecurity of tenure, which acted against housing investment and discouraged development of group solidarity and the emergence of political leaders. This poses the general question of what forces aid or obstruct the consolidation of informal settlements.

THE CONSOLIDATION OF SQUATTER SETTLEMENTS

Many settlements that originated as unserviced agglomerations of huts gradually develop to become recognised suburbs in the Third World city. This process of consolidation is dependent on a number of factors. These include:

1. *Security of tenure*. Without a high level of confidence that they will be permitted to retain the land, no family will willingly invest time and money in consolidating their dwelling. Where the poor rent a building plot, security depends on the length of lease and ability to maintain rental payments. Where land has been occupied through invasion, security depends primarily on the attitude of the government.[20]

2. *Access to credit*. The extent to which and terms under which finance is available exert a major influence on housing development. Low-income families, without a stable source of income or collateral on which to secure a loan, do not satisfy the requirements of formal-sector financial institutions. With limited government provision of low-interest loans, most low-income families resort to personal savings, assistance from friends or kin or loans from employers or informal credit unions.[21] The unavailability of housing finance for a large proportion of low- and middle-income groups with a capacity to save, invest in housing and repay loans is a major constraint on improving their housing quality, and extends the time scale of settlement consolidation.

3. *Price of construction materials*. As urbanisation proceeds, there are fewer opportunities to cut trees or make bricks from local clay, and more materials have to be obtained from the commercial market. The rising costs of building materials, due to either monopoly practices or production shortfalls, will impede the process of consolidation.

4. *Infrastructure provision*. The rate of settlement consolidation also depends on the extent to which public authorities are able to provide infrastructure and services to informal settlements. This may reflect the city's tax base, physical difficulties or political patronage, but in general, intra-urban variations in levels of service provision are linked to income levels.

5. *Land availability*. The availability of land is a critical factor determining the ability of the poor to construct and consolidate their own settlements. This depends on the nature of urban land markets and on government policy on land.

These fundamental issues are examined in detail below.

SECURITY OF TENURE

A person or household has security of tenure when they are protected from involuntary removal from their land or residence, except in exceptional circumstances, and then only by means of a known and agreed legal procedure, which must itself be objective, equally applicable, contestable and independent. Security of tenure is a fundamental requirement for the progressive integration of the urban poor in the city. It is an important catalyst in stabilising communities, improving shelter conditions and encouraging investment in home-based economic activities which, in turn, play a major role in poverty alleviation, reducing social exclusion and improving access to urban services.

Just as there is a variety of existing tenure systems[22] so there exists a range of policy options to provide security of tenure, either on an individual or on a collective basis, the use of which depends on local context. These include:

1. *De facto* recognition, but without legal status (this guarantees against displacement);
2. Recognition of security of tenure, but without any form of tenure regularisation (the authorities certify that a settlement will not be removed);
3. Provision of temporary (renewable) occupancy permits;
4. Temporary non-transferable leases;
5. Long-term leases (that may or may not be transferable);
6. Provision of legal tenure (leasehold or freehold).

Conventional responses regarding access to land and housing for the urban poor are based mainly on the regularisation of irregular settlements, emphasising tenure legalisation and the provision of individual freehold. This can be an expensive and time-consuming process, especially in contexts where administrative capabilities are limited, land ownership information is incomplete, centralised land registration procedures are complex, and corruption is commonplace. Nevertheless, it remains the goal of most tenure regularisation programmes. An alternative approach to providing tenure security involves an 'incremental regularisation' process of tenure upgrading that stresses user rights to land and property rather than outright ownership. It is argued that this approach is simpler to enact, gives communities time to consolidate their settlements and resolve land ownership disputes (e.g. between a community and 'legal' landowners), maintains community social ties during and, more important, after the tenure upgrading process and lessens market pressures on a settlement (e.g. to sell freeholds to 'downward-raiding' middle-income groups). Under this system occupancy rights may be upgraded, even to long-term leases or freehold status, as appropriate.

HOUSING FINANCE

The majority of low-income households in the Third World city do not satisfy conventional criteria for mortgage finance. They are unable to service the debt in terms of the amount and requirement for regular repayments, often do not hold legal title to the property (which cannot be used as collateral), while the small loans sought are less profitable (owing to transaction costs) and therefore unattractive to commercial financial institutions. Accordingly, the bulk of home finance in Africa, Asia and Latin America comes from outside the commercial financial institutions. Households use their own savings, sweat equity, barter arrangements and other 'informal' sources to build homes over an extended period, typically five to fifteen years. Schemes to provide micro-finance for housing are intended to operate within this 'precarious' investment environment by providing small loans (typically $300–$3,000) at market rates of interest over short terms secured by forms of collateral that recognise the paralegal ownership of land or property.

Micro-finance of housing has similarities to the micro-finance of enterprise practised by organisations such as the Grameen Bank in Bangladesh. Both involve small amounts of capital, have the potential to rotate (i.e. once an initial loan is repaid the borrower is eligible for another loan), and often operate through specialised non-governmental organisations (NGOs). There are also major differences, however, in that housing loans involve a longer-term risk for lenders. In addition, the kind of collective loan guarantee through community peer pressure is less effective in a housing lending programme spread over a larger urban area and longer time span than the typical six to twelve months for micro-enterprise finance. Consequently, whereas micro-enterprise is well established, there are as yet fewer programmes or organisations dedicated principally to micro-financing housing.

Significant top-down government-sponsored attempts to provide subsidised housing finance for low-income communities include the national fund for low-income housing (FONHAPO) established in Mexico in 1981 and the Community Mortgage Programme created in the Philippines in 1988. This latter programme provides loans to allow community associations to acquire land on behalf of their members, improve the site, develop individual titling of the land, and provide individual housing loans for home improvement or house construction. The community association is responsible for collecting repayments and for ensuring that the loan continues to be serviced until the community loan has been individualised. By 1994 40,000 households had received a loan, averaging 21,000 pesos. Repayment rates were around 65 per cent. An essential part of the programme is the involvement of a government agency or NGO to act as an intermediary helping residents to form an association, apply for a loan, and manage the programme.

An example of a successful bottom-up micro-finance initiative was established in 1988 by an NGO (PROA) in the Bolivian city of El Alto.[23] Using funds provided by a large mutual savings association (Mutual La Paz), PROA provides US$175,000 per month in loans. Foreclosure and late repayment rates are below those of Mutual La Paz's overall mortgage portfolio. The average loan is US$3,750 and repayment terms are adjusted to suit the borrower's ability to pay. The scheme benefits all involved. Mutual La Paz raises funds in the capital markets at 10 per cent and lends under the programme at 15 per cent; PROA receives a 5 per cent fee for acting as local management agent; and the low-income households gain access to housing credit tailored to their particular needs. A potential difficulty is the dependence of PROA on the continued provision of funds from a commercial financial institution. An alternative might be for the NGO to create a micro-credit financial institution specifically for housing, with the capacity to secure deposits and issue loans within low-income communities.

URBAN LAND MARKETS AND THE POOR

The response of the poor to their exclusion from the formal land market and to the inefficiencies of public action on land allocation has been the development of illegal or informal land markets. These have provided the land for most new housing in Third World cities over recent decades (Table 25.4). Illegal land markets provide space for housing at a cost that is affordable by many low-income households and with the advantages of immediate possession and no bureaucracy. There is no doubt that the system is exploitative, since a high proportion of housing plots have no service provision and, in many instances, the occupier pays a substantial sum yet does not receive secure tenure. Nevertheless, the housing conditions of the poor would be much worse without the illegal land market.[24]

Illegal land markets normally operate within limits that do not threaten the interests of governments, major landowners or real-estate companies. In many cities the illegal land market involves people who are also key actors in legal land markets, including developers, politicians and staff of government agencies. There is also a wide variety in the nature of illegality. All squatter settlements are illegal, but not all illegal settlers are squatters, since they may occupy the land with the permission of the landowner. Illegality may occur in the occupation of the land, in the registration of ownership, in the way the site is subdivided, in

TABLE 25.4 PROPORTION OF CITY POPULATION IN EXTRA-LEGAL HOUSING IN VARIOUS THIRD WORLD CITIES

City	Population (million)	% of population in extra-legal habitat
Jakarta	8	62
Manila	5.6	40
Delhi	7.5	40
Karachi	8	50
Ankara	2	51
Addis Ababa	1.6	85
Dar es Salaam	1	60
Cairo	5.7	54
Lusaka	0.8	50
Mexico City	16	50
São Paulo	13	32
Bogotá	5.5	59
Lima	4.6	33
Caracas	3	34

Source: R. Baker and J. Van der Linden (1992) *Land Delivery for Low-income Groups in Third World Cities* Aldershot: Avebury

the use to which the land is put or in the nature of the building on it.

Table 25.5 illustrates the various ways in which the poor obtain land for housing in the Third World city. Semi- or non-commercial mechanisms operate where customary patterns of land tenure still apply, as in parts of West Africa. Under this arrangement, land for a house may be obtained as a gift from a tribal elder. In a survey of households living in informal settlements in Abidjan, 61 per cent obtained their land in this manner.[25] This form of land acquisition is becoming increasingly commercialised, however. The reciprocal gift given to traditional leaders or farmers when a land plot is received (e.g. 'the price of a drink') has become more substantial and occasionally reaches a size comparable to the market value of the land. Furthermore, in places where land held under customary tenure by indigenous land-owning communities or families is being depleted, (as, for example, around Enugu, Nigeria[26]). There is an increasing reluctance to sell suitable building land to strangers due to a need to accommodate family or community members. The range of informal means of obtaining land is evident in Guayaquil, Ecuador, where settlers acquired building plots via eight different routes, ranging from physically opening up the mangrove (14 per cent) to buying a plot with a house already built (11 per cent). Significantly, two-thirds made some form of payment to obtain their plot.[27]

TABLE 25.5 PRINCIPAL MEANS BY WHICH PEOPLE OBTAIN LAND FOR HOUSING IN THIRD WORLD CITIES

Formal	Informal and/or illegal
Commercial	
Public or private residential development of serviced sites available through purchase or renting	Purchase of plot on illegally subdivided public land Purchase of plot in illegal subdivision
Land purchase with approval obtained for its use for housing	Purchase of house site formed by subdivision of existing plot Renting of land site on which a shelter (often only a temporary shelter) can be built Purchase or renting of permission to develop a house on a plot without tenure rights to the plot
Semi- or non-commercial	
Government-subsidised site and services schemes	Settlement on customary land with permission of traditional authority or farmer – although the size of 'gifts' given for this may reach a level where this is better considered as commercial
'Regularisation' of tenure in what were illegal settlements	Squatting on government land
Inheritance or gift	Squatting on marginal or dangerous land which has no clear ownership Squatting on private land 'Nomadic' squatters who use land site temporarily

Source: United Nations Centre for Human Settlements (1996) *An Urbanising World* Oxford: Oxford University Press

In most Third World cities the person who sells the plot of land for illegal development is the landowner or a developer acting on the owner's behalf, or a developer who subdivides public land with the tacit approval of government. Studies of illegal land markets in Africa, Asia and Latin America have identified the central role of land agents in the development of illegal subdivisions. These brokers include politicians, administrative personnel and customary landowners, as well as professional subdivision companies and 'pirate subdividers' who, without official sanction, subdivide (usually government) land that they do not own.[28] Over time, illegal developers can become important figures in the community by exercising political influence to obtain civic amenities for the subdivision. As Box 25.4 illustrates, the development of an illegal subdivision can be a complex process involving a wide range of agents, actors and motives.

INFORMAL SETTLEMENTS: GENERATIVE OR PARASITIC?

The attitude of governments to illegal settlements varies from hostility to suppression through degrees of tolerance to occasional support. Much depends on how the inhabitants acquired the land, with most governments being more tolerant of illegal subdivisions than of squatter settlements.

More generally, the attitudes of urban analysts towards the residents of informal settlements have shifted over recent decades, from essentially negative interpretations[29] to a more positive view that sees squatter communities as making a major contribution to the city by adding to its labour resources, contributing individual enterprise, including that engaged in **recuperative production**, and consuming some of the city's production while housing themselves at little direct cost to government.[30] Considered in this positive light, the efficiency of squatter communities in satisfying many of their own needs may be seen to benefit the capitalist sector of the Third World city. This rebuts earlier notions, based on the 'culture of poverty' thesis,[31] that characterised squatters as marginal to the life of the city and locked into a 'cycle of poverty' from which they cannot escape. This view underlay Stokes's (1962) concept of 'slums of despair'.[32] The myth of squatter marginality was exposed by Perlman (1974 p. 3), who demonstrated the high level of integration between squatter communities and urban society in Brazil in her conclusion that squatters 'are not economically marginal but exploited, not socially marginal but rejected, not culturally marginal but stigmatised, and not politically marginal but manipulated and repressed'[33] Thirty years later, the same

BOX 25.4

The development of an illegal subdivision in Karachi, Pakistan

Yakoobabad is a settlement on the north-western fringes of Orangi township in Karachi. By 1976–8 most of the plots in settlements adjacent to Yakoobabad had been occupied and the value of the remaining vacant plots had increased beyond what poorer groups could afford. The west Karachi sub-dividers felt that the time had come to colonise new land. The plan to colonise the area was first conceived by a well established developer whom we shall call Mr X. He was also involved with the development of adjacent sites and, some years previously, had begun work on a plan for the development of Yakoobabad.

People who had been settled in the neighbouring areas kept informing Mr X of their friends and relations who needed plots of land. Many social welfare organisations and public-spirited people also approached him, asking for help in settling refugees, widows and the destitute. At some point in 1977 Mr X drew up a list of 100 families and made informal representations on their behalf to government officials. They agreed to the development of Yakoobabad. The developers say that Karachi Municipal Corporation officials were to be paid Rs 2,000 (about US$10) for each plot sold, and the police also collected this amount for each construction undertaken in the settlement. These negotiations took place in the evenings in the tea shops of Orangi.

The subdivider then laid out the settlement with about 2,000 plots on a gridiron plan. Roads were levelled by informally hiring (at a reduced rate) tractors and bulldozers from the Karachi Municipal Corporation. Space was set aside for a mosque and school, and plots on the main road were left for communal purposes. It is reported that informal representations were made by government officials, and local informants state that about 30 per cent of all plots were set aside for speculative purposes, with the subdivider agreeing to sell these at an appropriate time on behalf of government officials. Additional speculative plots were held by the subdivider.

In 1979 there were local elections. Residents have told us of how the subdivider gave small sums of money to elected councillors and also tried, successfully in some cases, to make them office-bearers of his welfare organisation. (It is common for illegal subdividers to formally register a social welfare society formed by residents. This strengthens their ability to push for service provision to the area, thereby increasing the value of the land.)

The subdivider engaged people with donkey carts to first supply water to the residents. The water was acquired illegally from the water mains in Orangi. The subdivider paid for the first supply but after that the residents bought direct from the water vendors. Electricity is provided commercially to a few households from a privately operated generator. This is illegal but can function because of police protection. The subdivider is pressing people to apply for regular electricity connections. When a sufficient number of applications have been collected, pressure for electricity connections will be put on the Karachi Electricity Supply Corporation through the subdivider's welfare organisation. The subsequent increase in the value of the property will benefit the subdivider and government officials holding speculative plots in the area.

Source: United Nations Centre for Human Settlements (1996) *An Urbanising World* Oxford: Oxford University Press

author concluded that for the *favela* dwellers in Rio de Janiero at least parts of the notion of marginality had been transformed from myth to reality as their hopes for a better life for their children were undermined by lack of jobs and economic opportunity as a result of reduced demand for labour in the construction and domestic service sectors; replacement of labour intensive jobs with fewer high-skilled ones; and the entrenched stigma associated with living in a *favela*.[34]

GOVERNMENT HOUSING POLICY

Housing policies in any society are the outcome of interplay between different interest groups. Some

indication of government attitudes to low-income housing supply in Third World cities, and different policy responses to the problem, is provided in Table 25.6. It should be noted that although this portrays a progression of stages from non-action to more enlightened responses, this should not be taken to represent a uni-directional model of development. Government policies may move in either direction depending upon local circumstances that include the pattern of urban land ownership (which can influence attitudes towards squatting), the rate of economic growth (which can induce pressure for central-area redevelopment schemes) and political ideology (which reflects the balance of power in society). To understand this complexity it is useful to group

TABLE 25.6 DIFFERENT ATTITUDES BY GOVERNMENTS TO HOUSING PROBLEMS IN CITIES AND DIFFERENT POLICY RESPONSES

Stage 1	Stage 2	Stage 3	Stage 4	Stage 5
Government attitude to housing				
Investment in housing provision a waste of scarce resources. Problems will be solved as the economy grows	Government worried at rapid growth of city populations and of more rapid growth of tenements and illegal settlements. Seen as 'social problem'; squatter settlements often referred to as a 'cancer'	Recognition that the approach in stage 2 is having very limited impact. Still seen as 'social problem' but with political or social dangers if not addressed. Recognition that squatter settlements or other forms of illegal development are 'here to stay'	Recognition that the approach tried in stage 3 is having very limited impact. Recognition that people in slums contribute much to cities' economies – providing cheap labour and cheap goods and services, with the so-called 'informal sector' being a key part of city economy and employment base	Recognition that major institutional changes are needed to make the approach first tried in stage 4 effecti ve. Recognition that improving housing conditions demands multi-sectoral approach including health-care and perhaps food programmes. Recognition that low-income groups are the real builders and designers of cities and that government action should be oriented to support ng their efforts
Government action on housing				
No action	Special institutions set up to build (or fund) special public-housing programmes supposedly for lower-income groups. Slum and squatter eradication programmes initiated, often destroying more units than public agencies build	Public-housing programmes with increasingly ambitious targets. First sign of sites and services (or core housing) projects. Reduced emphasis on slum- and squatter-eradication programmes	Reduced emphasis on public-housing programmes. Far more emphasis on slum and squatter upgrading and serviced site schemes. Ending of squatter-eradication programmes	Governme rt action to ensure that all the resources needed for house construction or improvement (cheap, well-located sites, building materia s, technical assistance, credit, etc.) are available as cheaply as possible
Government action on basic services				
Very little action; not seen as a priority. Richer neighbourhoods in cities only ones supplied with basic services	Initial projects to extend water supply to more city areas	Water supply (and sometimes sewers or other sanitation types) included in site and services schemes and upgrading programmes	Major commitment to provision of water supply and sanitation	Strengthening of local/city governments to ensure widespread provision of water supply, sanitation, storm drainage, garbage removal, roads and public transport to existing and new housing developments. Health care also provided; link between poor housing conditions and poor health understood. Perhaps supplemented by cheap food shops or school meal project to improve nutrition

Government action on finance Discourage housing investment; considered waste of resources	Set up first publicly supported or guaranteed mortgage/housing finance agency	Attempt to set up system to stimulate saving and provide long-term loans for low-income groups	Improve efficiency of formal housing-finance institutions to allow cheaper loans; flexible attitude to collateral and to small loans for land purchase and house upgrading. Encourage and support informal and community finance institutions to serve those not reached by formal institutions	
Building and planning codes No action	No action. Unrealistic standards in public housing one reason why unit costs are so high	Recognition that these are unrealistic in that low-income households cannot meet them. Not used in site and services and upgrading schemes	Most public programmes to provide services, land, etc., not following existing laws/codes on standards and norms because they are unrealistic	Building/planning standards reformulated – advice and technical assistance as to how health and safety standards can be met
Government action on land No action	No action	Cheap land sites made available in a few site and services projects	Provision of tenure to illegal settlements. Recognition that unregulated land market a major block to improving housing and conditions	Release of unutilised/ underutilised public land and action to ensure sufficient supply of cheap, well-located sites plus provision for public facilities and open space

TABLE 25.6 CONTINUED

Stage 1	Stage 2	Stage 3	Stage 4	Stage 5
Government action on building materials				
No action	No action	Acceptance of use of cheap materials in low-income housing which are illegal according to existing building codes	Government support for widespread production of cheap building materials and common components, fixtures and fittings – perhaps supporting co-operatives within each neighbourhood for production of some of these	Recognition that government support to community groups formed by lower-income residents is a most effective and cost-effective way of supporting new construction and upgrading
Government attitude to community groups				
Ignore them	Ignore or repress them	Some 'public participation' programmed into certain projects	More acceptance of low-income people's rights to define what public programmes should provide and to take a major role in their implementation	
Impact on problems.				
None	None or negative	Usually minimal although certain projects may be successful	Substantially larger impact than in previous stages but still not on scale to match growing needs	Impact becoming commensurate with need

Source: J. Hardoy and D. Satterthwaite (1989) Squatter Citizen London: Earthscan

government responses to the provision of low-income housing into four broad categories: indifference, reactionary responses, Westernised responses and innovative responses.[35]

INDIFFERENCE

Various explanations have been advanced for the limited investment of Third World governments in low-income housing. For Dwyer (1975), some urban authorities adopt a policy of inaction in the hope that migrants, whom they consider to be the cause of the housing problem, will return to their rural origins.[36] The continuing expansion of Third World cities and growth of illegal settlements indicate the futility of this approach. Other explanations for government indifference include a fear of the latent political power of squatters, which forestalls any public reaction to their acquisition of land. This is equally unlikely. Even in Latin America, where squatters have mobilised in defence of their settlements, only rarely do they become enmeshed in more general political action. More pragmatic explanations for government inaction refer to administrative inefficiencies and a lack of information on land ownership. In Cameroon fewer than one-fifth of urban plots are demarcated on cadastral maps, and registration of land titles can take up to seven years.[37] In Tanzania, between 1978/9 and 1991/2 the Dar es Salaam city council received 261,668 applications for land plots but was able to allocate only 17,751, satisfying a mere 7 per cent of applicants.[38] Recourse to 'master planning' often accentuates inaction. The master plan for Dhaka covers only 40 per cent of the built-up area and was produced in 1959 for an anticipated population of 2 million (compared with a present figure of over 6 million that is growing by 250,000 per year). In most Third World countries the conception of urban development as 'planning–servicing–building–occupation' is *reversed* for most of the population. Even where planning information is available, its effective use can be hindered by a corrupt political process in which the urban poor have little power (see Chapter 29). Nominal improvement of squatter settlements and partial recognition of tenure in the run-up to elections are both common strategies of most political groups in the Third World. Further, much of the land in the large cities is held by individuals or institutions for speculative gain.[39] The inability of urban local authorities to appropriate a sufficient share of the benefits of economic growth through taxation undermines their capacity to improve the living environments of low-income citizens, even if the political will were there.

REACTIONARY RESPONSES

Reactionary responses regard urban growth and the spread of informal settlements as problems to be resolved by government action. Measures employed are either preventive or remedial:

1. *Preventive measures*. Most involve migration controls aimed at stemming or deflecting the movement of people into cities. In Turkey the economic problems arising from the drift of population into cities in the 1950s induced the government to encourage direct out-migration to Western Europe as a palliative. Internal controls on rural–urban migration are more direct but, as we have seen, are difficult to enforce (see Chapter 23).
2. *Remedial measures*. In cities where migration controls have not been enacted, or have failed to reduce migration flows, the most common reactionary response has been the forcible relocation of 'surplus' urban population to rural areas.[40]

Governments usually justify evictions in one of three ways:

1. The first refers to urban improvement or 'city beautification'. Authoritarian governments with little concern for citizen participation have commonly employed mass evictions in advance of major international events such as the 1974 Miss Universe contest in Manila, the 1988 Olympic Games in Seoul or the 2008 African Nations Football Championships in Kumasi.
2. A second way in which governments justify evictions is to claim that slums and squatter areas are repositories of crime. Health concerns are also often cited as reasons for their eradication, even though the health of the former residents may suffer from having to move into overcrowded accommodation elsewhere or onto hazardous marginal land. Environmental protection has also been given as a reason for removing squatters from land required for other uses.[41]
3. A third justification for eviction is redevelopment to use the land for more profitable activities.[42] In this case, underlying reasons and official justifications are likely to coincide. Centrally located areas and other strategically placed sites increase in value as the city's economy develops. In many cities this has led to demolition of cheap inner-city tenements, a trend that helps to explain the population decline of central areas in cities such

as Delhi, Bombay, Karachi, Bangkok, Santiago and Lima. Similarly, squatter settlements that developed on what was the urban periphery may occupy land whose commercial value has risen with city expansion. Developers' profits can be even higher if they can avoid the costs of rehousing those evicted. The illegal status of many informal settlements permits the bulldozing option to be exercised (Box 25.5).

For displaced residents the uncompensated loss of a home and possessions and the need to reconstruct shelter, often on an unserviced plot, are compounded by the disruption of community networks that provided a safety net in case of ill health or job loss. Those evicted often lose their main source of income when forced to move away from employment locations and are faced with increased transport costs to find work. In many cases injury or death may be caused by the violent nature of evictions. Most significantly, few eviction schemes have a long-term effect on the squatter population of the city. Most families eventually return and resume residence in the area from where they had been cleared.

BOX 25.5

Forced relocation in Phnom Penh, Cambodia

In many instances families are simply loaded into trucks at short notice and dumped in the countryside with no support. In Phnom Penh in 1990, 310 families who had been living on temple land for ten years were deposited on a stretch of road along the top of a flood-prevention dyke to the north of the city. The people were provided with no building materials, other than what they could carry with them, no financial compensation, and no welfare aid. They were expected to live at the side of the road, with no shelter, at the height of the monsoon rains. The surrounding area had no readily available local employment, no land suitable for cultivation and no public or community facilities. The site was nine miles (15 km) from Phnom Penh, four miles (7 km) from local markets and bus services, and three miles (5 km) from the nearest school. The land was subject to seasonal inundation from the Mekong river and would require substantial capital investment to make it suitable for housing.

Source: J. Audefroy (1994) Eviction trends worldwide – and the role of local authorities in implementing the right to housing *Environment and Urbanization* 6(1), 8–24

In Bangkok **land sharing** has been developed as an alternative to forcible eviction of slum dwellers.[43] This involves a negotiated agreement to partition the land into two parts, one for use by the landowner and the other for use by the present occupants of the site. From the squatters' perspective, land sharing is a preferred alternative once it becomes clear that continued occupation of the whole site is impossible. By removing the threat of eviction it allows them to remain within reach of their main employment sources and to maintain the community intact. It also ensures a formal land tenure agreement and hence an incentive to invest in their houses. From the landowner's viewpoint a land-sharing agreement may be a realistic alternative to a protracted struggle to regain the land. It clears some land for immediate development, reduces uncertainty in development schedules and avoids legal costs involved in the prosecution of squatters. From the government's point of view land sharing offers a source of land for the urban poor without public finance and allows the government to remain neutral between landowners and slum dwellers, while maintaining order and the rule of law. The principles involved in land sharing are shown in Box 25.6, and an example from one of five projects in Bangkok is illustrated in Figure 25.4. Land sharing schemes often involve relocation of slum dwellers but in contrast to forced removal through eviction relocation involves negotiation with residents and payment of compensation. In addition to provision of new housing and tenure security successful relocation schemes also include follow-up social welfare support to assist the new community to consolidate.[44] In an effort to extend the principles of in-situ upgrading in partnership with the poor the Thai government initiated, in 2003, a national slum and squatter upgrading and secure tenure programme based on provision of infrastructure subsidies and housing loans to low-income communities to enable them to play a major role in redevelopment, and to continue to live in their preferred neighbourhoods.[45]

WESTERNISED RESPONSES

Government involvement in low-income housing has produced a wide range of public housing programmes in Third World cities. Many of these were genuine attempts to alleviate housing stress among the poor; others were driven by less generous motives. Until the 1980s most schemes were based on Western concepts of planning and design which, in general, made the resulting housing expensive to construct and rent.[46] Major initiatives may be categorised as follows.

BOX 25.6

Principles of the land-sharing strategy in Bangkok, Thailand

1. *Community organisation*. Negotiations for land sharing require that slum dwellers mobilise to counter the threat of eviction, to enlist the support of outside organisations, and to create the indigenous leadership necessary to represent the community in negotiations. Community participation is also required in planning for the site reconstruction, in the allocation of plots, in the demolition of existing structures and sometimes in rebuilding the houses.
2. *A land-sharing agreement*. This requires a binding agreement to partition the land. Usually the land parcel with the best development potential is allocated to the landlord. Other parts are allocated to the existing residents for re-housing themselves. Such an agreement must guarantee secure land tenure on the parcels allocated to the residents, and may specify the necessary payments and time schedules for implementation.
3. *Densification*. Rehousing the existing community on a smaller site requires increased residential densities. If the original density in the slum was already high, the new density will be even higher, unless some of the residents are excluded by the new scheme.
4. *Reconstruction*. The increase in residential density and the need to clear part of the site usually necessitates the reconstruction of houses, unless original densities are low enough to permit infilling of vacant plots on the site. Rebuilding may require new forms of construction, using more permanent or more solid materials, to achieve the required densities and to upgrade the quality of houses.
5. *Capital investment*. Reconstruction requires capital from the domestic savings of the residents or loans from outside sources. To be economically feasible, land-sharing schemes cannot rely on massive subsidies and must arrange for housing within the people's ability to pay. This may occasionally require cross-subsidies within the land-sharing scheme, utilising some of the development gains partially to offset housing reconstruction costs.

Source: **S. Angel and S. Boonyabancha (1988) Land sharing as an alternative to eviction *Third World Planning Review* 10(2), 107–27**

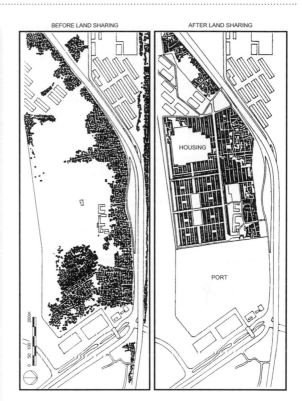

Figure 25.4   The Klong Toey land-sharing project in Bangkok
Source: S. Angel and S. Boonyabancha (1988) Land sharing as an alternative to eviction *Third World Planning Review* 10(2), 107–27

Tokenism

The goals of token housing programmes are political rather than social. They are intended to be visible symbols of government concern for the urban poor rather than meeting the real needs of low-income groups. For this reason, buildings occupy prominent locations in the urban landscape, usually taking the form of high-rise blocks irrespective of their fiscal or cultural suitability.

Misplaced philanthropism

Many socially motivated housing schemes fail to assist the urban poor because of the high costs, which result from inappropriate adherence to Western housing standards, often established in the colonial era, and the use of high-technology construction methods. Large subsidies are required to bring the housing within the financial reach of the poor, and few Third World governments are able or willing to provide these. As a consequence, most schemes are eventually occupied by those families able to pay a 'fair rent' on a regular basis, which excludes those on an irregular and low income. In Malaysia the Rifle Range project, begun in 1968 on the outskirts of Georgetown, employed an expensive prefabricated construction technique which meant that even with government subsidy the lowest monthly rent was beyond the reach of the intended residents. By the mid-1970s the government appeared to have accepted the scheme's failure as a low-income housing project and raised the income threshold for entry from M$200 to M$500. This led to an influx of middle-income families for whom the flats provided cheap, adequate but temporary accommodation while they saved enough to enter the private market. By 1980 only 25 per cent of households could be considered low-income, and many were in rent arrears and in danger of eviction. In Dar es Salaam a regulatory framework of bureaucratic administrative procedures, unrealistic planning standards (e.g. relating to plot size), and overly stringent planning regulations (e.g. on land-use control) stemming from the colonial era has fostered a climate of corruption in the land development process and inflated the cost of access to formal (legal) forms of shelter, with adverse consequences for the urban poor and rapid growth of unplanned 'spontaneous settlements'.[47] In Bangkok the government has provided financial and legislative support to encourage private-sector development of low-cost housing. This strategy has extended the availability of affordable housing to larger numbers. However, it is important to recognise the distinction between low-cost and low-income housing. The buyers of low-cost housing in Bangkok uniformly have monthly incomes above the average for the city. Many low-cost units were purchased for speculative purposes and remained empty. At the same time the population of slums and squatter areas has increased and, as a consequence of the property boom, many central informal settlements have been redeveloped and squatters relocated to the outskirts of the city.[48]

Large-scale public housing

As we have noted, while most Third World governments embarked on large public housing programmes in the decades after the Second World War (akin to stage 3 in Table 25.6), by the 1980s few were still involved in this sector. The major problems for public housing schemes in the Third World derive from the following factors:

1. Unit costs were so high that few apartments were built relative to needs. The dilemma was that if little or no subsidy was given (to enable more units to be constructed), the new housing could be afforded only by relatively well-off households. Alternatively, if a sufficient subsidy was provided to facilitate entry by low-income groups, relatively few schemes could be built. In 1972 the Karachi Development Authority planned to construct between 30,000 and 40,000 flats to rehouse residents from an urban redevelopment scheme. After eight years, when the project was discontinued owing to lack of funds, only 800 flats had been completed, with none going to the intended beneficiaries.

2. Middle- or upper-income groups benefited more than low-income groups. In addition to the problem of relatively high rents that excluded low-income residents, some public housing schemes were restricted to certain people such as military personnel or mid-level civil servants, able to qualify for mortgages from commercial sources. In other instances, application procedures automatically ruled out many poor households. Female-headed households are often not eligible, even though half the households in low-income areas are headed by women. In others, proof of formal, stable employment is required.

3. Housing designs and locations were ill matched to the needs of poorer groups. Many of the units produced that did reach poorer households were unpopular with residents. Many schemes were located on the urban periphery far from sources of employment. In some cases apartments designed on the Western concept of a small nuclear family were too small for extended families. Also families required work space at home. Western designs also proved ill suited to tropical and subtropical climates in which high humidity and rainfall led to repair and maintenance requirements beyond the capabilities of public authorities.

New towns

The primary function of Third World new towns has been as a catalyst for regional development, most evident in the construction of new capital cities, such as Brasilia. The other traditional function of British new towns, decongestion of crowded urban areas, has received less emphasis. A major problem has been that most Third World new towns have been located too close to the major city to act as counter-magnets for either migrants or existing urban populations. The result has been to reduce the new towns to dormitory suburbs for middle-class residents. In Malaysia the new town of Petaling Jaya, six miles (10 km) south-west of Kuala Lumpur, was designed to be a satellite city of 70,000 linked with the redevelopment of squatter areas in the central city. Following independence in 1957, growing demand for middle- and upper-income housing fuelled speculative developments in the new town. Petaling Jaya is now an overwhelmingly middle-class suburb from which white-collar residents commute to office jobs in the city, passing *en route* blue-collar workers from Kuala Lumpur travelling to take up industrial jobs in the 'new town'.

Urban renewal

Few cities in the Third World are in a position to undertake urban renewal to improve housing conditions, since the general shortage of housing militates against any deliberate reduction in stock through slum clearance. Furthermore, in practice most urban renewal schemes are undertaken for commercial reasons rather than for social-welfare objectives.

INNOVATIVE RESPONSES

The underlying premise is that the residents of informal settlements will be able to improve their own housing if some public assistance is provided. In brief, the self-help that exists within certain squatter settlements could be transformed into **aided self-help**.[49] Closely related to the idea of aided self-help is the notion of *progressive development*, whereby families can build at a pace and to a standard consistent with their needs and financial resources. In the early 1970s the World Bank began to finance shelter projects based on this concept. Two main approaches have been employed. The first refers to the upgrading of existing squatter settlements through the provision of services and rationalisation of house and street layout. The second entails provision of serviced sites, sometimes with a partially completed unit, where the beneficiary is responsible for building or completing the unit.

Upgrading

This can involve the improvement of dwellings but more usually refers to the insertion of basic infrastructure into a neighbourhood. The principal objectives of upgrading *in situ* are to reduce both the costs of housing improvement for the squatter and the disruption caused by clearance to a peripheral resettlement scheme. In practice the benefits vary according to the way the government promotes the scheme and the extent to which goals other than the welfare of squatters enter the equation.

The Kampung Improvement Programme (KIP), begun in Jakarta in 1969, is widely acknowledged to have been successful in its own terms of making simple if marginal improvements in the living standards of as many residents as possible. By the early 1980s the KIP had improved more than 500 *kampungs* and provided basic services for 3.8 million residents. The programme triggered increased private investment in home improvement within upgraded areas, leading to increased property values. To achieve its goals in as short a time as possible, in 1974 control of the KIP was placed in the hands of a separate project-management unit. This provoked criticism of a lack of popular participation in the upgrading programme. Some observers linked this with subsequent problems of maintaining the improved infrastructure, but in general the KIP has enjoyed wide community support.[50] Since 1989 greater efforts have been made to involve community participation and self-help in *kampung* development, not least in order to increase the effectiveness of public investments.[51] Working along similar lines, *favela* Bairro, the largest-scale squatter settlement upgrading programme in Latin America, was initiated in 1994 with the aim of improving housing conditions and reducing poverty in squatter settlements in Rio de Janiero.[52] Evidence from this, and other integrated projects,[53] illustrates the potential benefits of housing rehabilitation schemes that also support home-based enterprises that help local people into work.

Site and services schemes

Under this type of initiative the land is prepared and a service infrastructure provided. Lots are then sold or leased, and the new residents either build a house themselves or contract the work out. Government subsidies may extend to building materials and/or cash loans. From a resident's point of view the main attraction of site and services housing is the security of tenure it affords, together with adequate infrastructure and freedom to build at one's own pace, once the requirement to construct a basic core house has been met. On the other hand, the peripheral location of most site and services schemes often results in their being

TABLE 25.7 POSITIVE AND NEGATIVE FEATURES OF ILLEGAL LAND SUBDIVISION FOR LOW-INCOME HOUSING

Positive
1. They are generally created on the periphery of the city
2. They are planned on a gridiron pattern and according to official planning regulations; thus, they do not pose problems with regard to the official public plans
3. Water is made available either on payment or through civic agencies
4. Transport needs are taken care of
5. The cost is affordable even to the poorer section of the population. In some cases no profit is charged, because some of the commercial plots are kept in reserve for disposal at high commercial rates to subsidise the development costs of smaller residential plots
6. Technical advice for construction and sanitary problems is informally available
7. Immediate possession is handed over to the purchasers, with little or no paperwork
8. If needed, credit is also made available in cash or kind
9. The availability of artisanal skills is assured
10. Protection from eviction is guaranteed in 90 per cent of cases, as the developers are in very close liaison with the police and government agencies

Negative
1. They cannot manage to obtain a bulk sewage-disposal system, as it requires major financial outlay and complicated techniques
2. Although there is no imminent threat of eviction, the title to land remains a major issue, as the developers grab either government or private land without paying the owners. Without title papers, getting official loans or legally selling the plots poses a problem; besides, the illegal position of the inhabitants is a cause of much manipulation, oppression and abuse for financial and/or political ends
3. External roads are not developed, as major investments are required for the purpose and government agencies refuse to take up this work because they consider the colonies to be illegal and unauthorised areas

Source: **T. Siddiqui and M. Khan (1994) The incremental development scheme** *Third World Planning Review* **16(3), 277–91**

some distance from major sources of employment, which can strain the ability of the household to make regular repayments. In some instances middle-income families may pressure low-income households to sell their tenure rights or their lease, even though these are not intended to be circulated commercially. In the Chaani site and services development in Mombasa the standards of housing demanded and conditions that had to be met to qualify for a government loan were beyond the majority of low-income households. As a result, the greatest benefits of the project accrued to middle-income groups, including staff from the government agency responsible for administering the project, while displaced low-income households have relocated as squatters elsewhere.[54]

Site and services schemes have been implemented throughout the Third World with varying degrees of success.[55] Since 1990 World Bank policy has shifted away from direct provision of housing by government towards an enabling role for public authorities that envisaged greater involvement of the private sector in housing provision. This change was influenced by the failure of many site and services schemes to recover costs. This problem was compounded by the inability of Third World governments to ensure a sufficient supply of land in the face of opposition from private landowners who demanded high prices and, if necessary, could delay the whole project by challenging land expropriations in the courts.

Critics of aided self-help schemes have pointed to operational difficulties associated with poor administration, lack of enthusiasm by local authorities, an unwillingness to lower building standards, and an inability to prevent a middle-class take-over of improved dwellings. More fundamental criticism has been directed at the concept itself, which the political left views as an excuse for non-action by governments on fundamental social issues of progressive taxation and land reform.[56] From this perspective, self-help projects serve to maintain an unequal status quo. Further, encouraging low-income communities to address their problems through self-help while high-income areas expect government to provide services and facilities at a high standard is regarded as discriminatory and exploitative. A more general criticism is that site and services schemes alone are insufficient to help the mass of the urban poor. In the absence of large-scale job-creation programmes, too few people

BOX 25.7

Site and services schemes in Pakistan: lessons in failure and success

Metroville, Karachi
Initiated in 1973 as one of the early World Bank-sponsored site and services schemes, the target was to settle 200,000 people. Under the first scheme 4,133 plots with installation walls were prepared, but after ten years only 700 were occupied. Most had come into the possession of the middle class, with the low-income households originally targeted settled in private subdivisions surrounding the Metroville site. As a result of excessive costs for the poor, Metroville is today a middle-class suburb.

The Incremental Development Project, Hyderabad
This set out to provide housing plots that were affordable by the low-income groups which made up 70 per cent of the city's population. This was achieved by the project's buying public land from the state, subdividing it and distributing plots under conditions that imitated the private system. Thus, initially, service provision was minimal, with only a communal water supply and public transport to the city provided. Later, as people settled in sufficient numbers, the service system would be developed gradually in consultation with the settlers. The total price of a fully serviced eighty-square-yard (67 m²) plot is Rs 9,600 (low-income households having an income of around Rs 2,000 per month), with only a down payment of Rs 1,100 required to secure a plot. The project is fully self-financing. Bureaucracy is kept to a minimum and there is an adequate supply of land plots at the same price as was fixed when the scheme opened in 1986. Residents are responsible for infrastructure maintenance. A loan scheme for housing construction has been set up on the principle of the Grameen Bank, with applicants required to organise themselves into groups of ten which are collectively responsible for the debts of members. Public participation fosters a commitment to ensure the success of the scheme. By 1990 the population of the Khuda ki Basti (God's Colony) settlement had risen to over 2,000 families.

Source: **T. Siddiqui and M. Khan (1995) The incremental development scheme** *Third World Planning* **Review 16(3), 277–91**

will be able to participate in even the cheapest of self-help schemes. To reach the bulk of the low-income population, aided self-help schemes would need to be accompanied by structural reform of the urban land market, taxation system and planning policies, which currently operate in favour of a minority of the population in Third World cities (see stage 5 in Table 25.6). The most successful aided self-help projects have been those in which governments have exercised control over the urban land market, as in Tunisia,[57] Botswana[58] and Pakistan.[59] Public–private–people partnerships are an offshoot of the wider 'enabling' approach to housing provision. Their, as yet, limited emergence is related to the difficulty of creating the necessary institutional framework for housing markets that can balance the profit-driven motivation of the market and the need for low-income housing. In addition, in many Third world cities tortuous bureaucratic procedures, low levels of literacy and awareness, and the abject poverty of a majority of urban residents inhibits them from participation in PPPs. Thus, in Kolkata, India although housing produced under the PPP model has been impressive in terms of cost and quality, it has been miniscule in terms of numbers.[60]

Public authorities in the Third World may also learn lessons from the development strategy of the illegal sub-dividers, which is successful because it is compatible with the actual socio-economic conditions of the poor (Table 25.7). In addition to ensuring an adequate supply of land for low-income housing and supporting the extension of credit to borrowers with less conventional forms of collateral, governments might usefully consider the benefits of limited bureaucracy and the application of more 'flexible' planning arrangements, including less restrictive development-control procedures, that facilitate shelter construction by small-scale developers at lower standards (while ensuring structural safety and public health) over an extended period.[61] The incremental development approach adopted in Hyderabad (Box 25.7) and upgradable sites projects in Indian cities[62] are examples of public initiatives on low-income housing that address directly the reality of urban poverty in the Third World. Such projects acknowledge that 'in the Third World city incremental building is the functional substitute for the incremental paying that takes place in countries with mortgage systems that reach the working class' (Brennan 1993 p. 89).[63] Rather than depending on an 'enabling' approach to provision of low-income housing based on market-led development and a trickle-down mechanism,[64] government housing policies might usefully be informed by the experience of the informal sector in terms of land supply, housing standards, administrative simplicity, public participation and time scale for development. Plurality of provision is a prerequisite for addressing the housing problems in Third World cities.

FURTHER READING

BOOKS

K. Datta and G. Jones (1999) *Housing and Finance in Developing Countries* London: Routledge

M. Davis (2006) *Planet of Slums* London: Verso

E. Fernandes and A. Varley (1998) *Illegal Cities* London: Zed Books

P. Jenkins, H. Smith and Y. Wang (2007) *Planning and Housing in the Rapidly Urbanising World* London: Routledge

K. Mathey (1992) *Beyond Self-help Housing* London: Mansell

P. McAuslan (1990) *Urban Land and Shelter for the Poor* London: Earthscan

G. Payne (2002) *Land Rights and Innovation: Improving Tenure Security for the Urban Poor* London: ITDG Publishers

JOURNAL ARTICLES

J. Audefroy (1994) Eviction trends worldwide – and the role of local authorities in implementing the right to housing *Environment and Urbanization* 6(1), 8–24

F. Beijaard (1995) Rental and rent-free housing as coping mechanisms in La Paz, Bolivia *Environment and Urbanization* 7(2), 167–81

A. Gilbert (2000) Housing in Third World cities *Geography* 85(2), 145–55

M. Greene and E. Rojas (2008) Incremental construction: a strategy to facilitate access to housing *Environment and Urbanization* 20(1), 89–108

R. Keivani and E. Werna (2001) Modes of housing provision in developing countries *Progress in Planning* 55, 65–118

R. Neuwirth (2007) Squatters and the cities of tomorrow *City* 11(1), 71–80

G. L. Peattie (1990) Participation: a case study of how invaders organise, negotiate and interact with government in Lima *Environment and Urbanization* 2(1), 19–30

C. Rakodi (1995) Rental tenure in the cities of developing countries *Urban Studies* 32(4/5), 791–811

C. Rakodi and P. Withers (1995) Sites and services: home ownership for the poor? *Habitat International* 19(3), 371–89

P. Smets (2006) Small is beautiful, but big is often the practice: housing microfinance in discussion *Habitat International* 30, 595–613

C. Stokes (1962) A theory of slums *Land Economics* 38(3), 187–97

KEY CONCEPTS

- Conventional housing
- Non-conventional housing
- Squatter settlements
- Free-rent tenants
- Possessor-occupier
- Land invasion
- Infiltration
- Security of tenure
- Micro-finance
- Pirate subdivision
- Slums of hope
- Slums of despair
- Eviction
- Land sharing
- Pavement dwellers
- Aided self help
- Site and services scheme
- Incremental development

STUDY QUESTIONS

1. Examine the major sources of shelter for the urban poor in Third World cities.
2. With reference to a particular example, identify the formative processes and development of a Third World urban squatter settlement.
3. Consider the question of whether informal settlements are generative or parasitic elements in the Third World city.
4. What do you understand by the 'informal land market' in Third World cities?
5. Critically examine the response of Third World governments to the housing problems of the urban poor.
6. Explain what is meant by an incremental development approach to the provision of low-cost housing, and evaluate the potential of the approach in Third World cities.
7. Examine the view that plurality of provision is a prerequisite for addressing housing problems in Third World cities.

PROJECT

Develop a statistical profile of housing conditions in a selection of cities across the world. This may be achieved by extracting relevant data from the publications of the United Nations and World Bank (e.g. the findings of *The Housing Indicators Programme*, published jointly by the United Nations Centre for Human Settlements (Habitat) and the World Bank in 1993). Indicators that could be usefully included in your profile include national *per capita* income (as a general guide to high- and low-income cities); the proportion of dwelling units with a piped water supply; the proportion constructed of permanent building materials; floor area per person; persons per room; government expenditure per person on sanitation, garbage collection and electricity supplies; percentage of unauthorised housing; housing affordability (i.e. price–income ratio); rent levels per square metre; land and construction costs; the availability of housing credit. An appropriate suite of indicators should be collected for a representative selection of cities. You can then present a statistical profile of housing conditions in a particular city and/or compare housing conditions across cities in different world realms.

26

Environmental Problems
in Third World Cities

Preview: urban environmental problems; the domestic environment; water supply; sanitation; indoor pollution; overcrowding; the workplace environment; the neighbourhood environment; the city environment; air and water pollution; toxic waste; natural and human-induced hazards; the city's ecological footprint

INTRODUCTION

Urban agglomeration brings many potential advantages and disadvantages. The concentration of population and production in cities offers economies of scale that reduce the unit costs of providing infrastructure and services such as piped water, sanitation, refuse collection, electricity, drains and paved roads. It also reduces the unit cost for health, police and fire services, and the provision of facilities such as schools and libraries. Industrial concentration in cities lessens the cost of enforcing regulations on environmental and occupational health and pollution control. The urban concentration of households and enterprises also makes it easier for public authorities to collect taxes and charge for public services.

The extent to which a city capitalises on the advantages of agglomeration depends, to a large extent, on the power and quality of urban governance. This is dependent upon the nature of local political-economic structures as well as on factors beyond the direct influence of municipal governments, such as a stagnant national economy and international debt burden that can undermine the economic base necessary for the development of sound governance.[1] The absence of effective governance exacerbates the multitude of environmental problems of urban agglomeration that confront the majority of Third World cities[2] (Table 26.1). In this chapter we examine the nature and incidence of a range of urban environmental problems at a number of geographical scales, from the internal environments of home and workplace to the neighbourhood and city levels.[3]

THE DOMESTIC ENVIRONMENT

The various forms of poor-quality housing in the Third World city discussed in Chapter 25 share a number of characteristics that contribute to poor environmental health. The first of these is the lack of safe and sufficient water supply. The second is the presence of pathogens or pollutants in the human environment because of a lack of basic infrastructure such as sewers, drains or services to collect and safely dispose of solid and liquid wastes. The third is the overcrowded living conditions, which increase the transmission of airborne infections, as well as the risk of domestic accidents. We can examine these in detail.

WATER SUPPLY

A lack of readily available drinking water and the absence of sewage connections can result in many debilitating and easily prevented diseases being endemic among poorer households. These include diarrhoea, dysenteries, typhoid, intestinal parasites and food poisoning. When combined with undernutrition they can weaken the body's defences to the extent that measles, pneumonia and other common childhood diseases become major causes of death.[4]

TABLE 26.1 MAJOR ENVIRONMENTAL PROBLEMS IN THIRD WORLD CITIES

Context	Nature of hazard or problem	Specific examples
Within house and its plot	Biological pathogens	Water-borne, water-washed (or water-scarce), airborne, food-borne, vector-borne, including some water-related vectors (such as Aedes mosquitoes breeding in water containers where households lack reliable piped supplies). Note that insufficient quantity of water may be as serious in terms of health impact as poor water quality. Quality of provision for sanitation is also very important. Overcrowding/poor ventilation aids transmission of infectious diseases
	Chemical pollutants	Indoor air pollution from fires, stoves or heaters
		Accidental poisoning from household chemicals
		Occupational exposure for home workers
	Physical hazards	Household accidents – burns and scalds, cuts, falls
		Physical hazards from home-based economic activities
		Inadequate protection from rain, extreme temperatures etc
Neighbourhood	Biological pathogens	Pathogens in waste water, solid waste (if not removed from the site), local water bodies. Disease vectors such as malaria spreading *Anopheles* mosquitoes breeding in standing water or filariasis spreading *Culex* mosquitoes breeding in blocked drains, latrines or septic tanks. If sanitation is inadequate, people may defecate on open sites, leading to faecal contamination, including sites where children play. If drainage is inadequate, flooding will also spread faecal contamination. If a settlement is served by communal standpipes, latrines and/or solid-waste collection points, these need intensive maintenance to keep them clean and functioning well
	Chemical pollutants	Ambient air pollution from fires, stoves, from burning garbage if there is no regular collection service. Air pollution and wastes from 'cottage' industries and road vehicles
	Physical hazards	Site-related hazards, such as housing on slopes with risks of landslides, sites regularly flooded, sites at risk from earthquakes. Traffic hazards. Noise. Health hazards to children if open sites have wastes dumped there because of no regular service to collect household wastes
Workplace	Biological pathogens	Overcrowding/poor ventilation aiding transmission of infectious diseases
	Chemical pollutants	Toxic chemicals, dust
	Physical hazards	Dangerous machinery, noise

TABLE 26.1 CONTINUED

Context	Nature of hazard or problem	Specific examples
City (or municipality within a larger city)	Biological pathogens	Quality and extent of provision of piped water, sanitation, drainage, solid-waste collection, disease control and health care at city or municipal level are a critical influence on extent of the problems
	Chemical pollutants	Ambient air pollution (mostly from industry and motor vehicles, with motor vehicles' role generally growing), water pollution, hazardous wastes
	Physical hazards	Traffic hazards, violence, 'natural' disasters and their 'unnaturally' large impact because of inadequate attention to prevention and mitigation
	Citizens' access to land for housing	Important influence on housing quality directly and indirectly (such as through insecure tenure discouraging households investing in improved housing and discouraging water, electricity and other utilities from serving them)
	Heat island effect and thermal inversions	Raised temperatures a health risk, especially for vulnerable groups (such as the elderly and very young). Pollutants may become trapped, increasing their concentration and length of exposure
City region (or city periphery)	Resource degradation	Soil erosion from poor watershed management or land development/clearance, deforestation, water pollution, ecological damage from acid precipitation and ozone plumes
	Land or water pollution from waste dumping	Pollution of land from dumping of conventional household, industrial and commercial solid wastes and toxic/hazardous wastes. Leaching of toxic chemicals from waste dumps into water Contaminated industrial sites. Pollution of surface-water and possibly ground-water from sewage and storm/surface run-off
	Pre-emption of loss of resources	Fresh water for city, pre-empting its use for agriculture; expansion of paved area over good-quality agricultural land
Global	Non-renewable resource use	Fossil fuel use, use of other mineral resources, loss of biodiversity
	Non-renewable sink use	Persistent chemicals, greenhouse gas emissions, stratospheric ozone-depleting chemicals
	Overuse of finite renewable resources	Scale of consumption that is incompatible with global limits for soil, forests, fresh water

Source: adapted from D. Satterthwaite (1999) *The Links Between Poverty and the Environment in Urban Areas of Africa, Asia and Latin America* New York: United Nations University Press

In 1994 300 million urban dwellers in the Third World lacked access to safe drinking water[5] and had no alternative but to obtain supplies from streams or other surface sources, which in urban areas are often little more than open sewers, or from private vendors where quality is not guaranteed (Box 26.1).

As well as water quality, the quantity available to a household and the price that must be paid are also important influences on family health. Where there is a public water supply, via a well or standpipe, the quantity used per person will depend on the time and energy needed to collect and carry it back to the home. Often there are 500 or more persons for each commercial tap, and water may be available only for a few hours a day.[6] Time spent queuing for use of the tap or transporting buckets of water takes up time that could be devoted to earning income. To meet all its needs, a family of six requires the equivalent of thirty to forty buckets of water each day. Rarely is this available. Where public agencies do not provide water, as in most illegal settlements, the poor may obtain a supply from private water sellers, often at prices 100 times those paid by richer households for publicly provided piped water. In Guayaquil, Ecuador, 600,000 people (35 per cent of the population) are dependent on private vendors, who charge a price 400 times higher than that paid by consumers connected to the public water system.[7] In general, water vendors serve between 20 per cent and 30 per cent of the Third World urban population. The UN Environment Programme (UNEP) emphasised better urban governance as the key to more sustainable use of water and proposed a six-point integrated strategy for managing urban water resources. This requires:

1. City-wide water audits by local authorities.
2. Policies to halt pollution of water sources and to protect watersheds.

Plate 26.1 Water collection for domestic use is a labour-intensive and time-consuming daily task for many poor residents in Nairobi

BOX 26.1

The inadequacy and inequity of water supply in Lagos, Nigeria

Deficiencies in water and sanitation provision continue to provide some of the most striking manifestations of Lagos's worsening infrastructure crisis. Less than 5 per cent of households have piped water connections (a fall from around 10 per cent in the 1960s) and less than 1 per cent are linked to a closed sewer system (principally hotels and high-income compounds). Even those with piped connections must contend with interruptions due to power supply failures affecting the city's water works. The rest of the city depends on wells, boreholes, water tankers, various illegal connections, street vendors and, in desperation, the 'scooping' of water from open drains by the side of the road. Inhabitants of slum settlements often face a stark choice between either polluted wells or expensive tanker water distributed by various intermediaries at high and fluctuating prices, making the management of household budgets even more precarious. When municipal authorities do attempt to extend water supply to poorer neighbourhoods, they are often met with violence and intimidation from water tanker lobbies, 'area boys' and other groups who benefit from the unequal distribution of water and the 'micro circuits' of exploitation which characterise slum life: the city's water corporation must consistently confront the 'water lords' who intentionally vandalise the network in order to continue charging exorbitant rates to the poor. People's daily survival is based on careful distinctions between different kinds of water suitable for drinking, cooking and washing, with much time and expense devoted to securing household water needs. Regulatory authorities also struggle to cope with the proliferation of 'pure water' manufacturers producing small plastic sachets of drinking water sold throughout the city which have been associated with the spread of water-borne disease. The sellers of these 'pure water' sachets – thousands of mostly young Lagosians – weave their way between lines of slowly moving or stationary traffic as the need for potable water has become part of the city's burgeoning informal economy.

Source: M. Gandy (2006) Planning, anti-planning and the infrastructure crisis facing metropolitan Lagos *Urban Studies* 43(2), 2371–96

3. The use of new technologies to minimise loss of water through leakages and illegal connections.
4. Socially sensitive pricing policies.
5. The involvement of individuals and community groups to devise innovative ways of recycling waste water.
6. An integrated strategy for demand management in each city, including educating users of the need for water conservation.

SANITATION

In addition to an adequate supply of clean water, the removal and safe disposal of excreta and waste-water are a critical environmental health requirement. Absence of drains or sewers to take away waste water and rainwater can lead to waterlogged soil and stagnant pools that can convey enteric diseases such as typhoid and provide breeding grounds for mosquitoes, which spread malaria and other diseases. In 1994 almost 600 million urban dwellers in the Third World lacked adequate sanitation, and this figure was expected to rise to 850 million by 2000.[8] Moreover, official estimates tend to overstate the extent of provision for sanitation. For instance, 40 per cent of Kumasi's population is dependent on public latrines and is, therefore, officially classified as served with sanitation, even if many defecate in the open because they cannot afford the wait or the fee, or avoid the use of poorly maintained facilities that can represent a health hazard.[9]

Many cities in Asia and Africa have no sewers, and most human excrement and waste-water ends up untreated in watercourses, gullies and ditches. In cities with sewers, rarely does the system serve more than a small proportion of the population (Table 26.2). In Dhaka water-borne sewerage serves only 20 per cent of the population. Around 30 per cent use a total of 50,000 septic tanks which overflow into the streets and drains during the rainy season, 10 per cent use pit latrines, 5 per cent use bucket latrines, and the remaining 35 per cent either use surface latrines or do not have any facilities.[10] In Indian cities defecation in the open air is common practice, since one-third of the population (80 million people) have no latrine of any kind, while another one-third rely on bucket latrines (Box 26.2). In some cases, these conditions have stimulated community-based toilet-building projects,[11] as in Mumbai, where the local community of Sabzi Mandi, aided by the National Slum Dwellers' Federation, the Mahila Milan (Women Together) Alliance and the UK DFID, designed, built and maintains a toilet facility in collaboration with the city government, which provides the sewer and water connections.

INDOOR POLLUTION

Where open fires or inefficient stoves are used for cooking or heating, smoke and fumes from coal, wood and other biomass fuels can cause serious respiratory problems. Carbon-monoxide poisoning is a hazard in poorly ventilated dwellings, as are the dangers arising from exposure to carcinogens in emissions from biomass fuel combustion. Those people in the household who are responsible for tending the fire and cooking (generally women and girls) are at greatest risk. Infants and children may also suffer greater exposure because they remain with their mother. Combined with malnutrition, this can retard growth, leading to smaller lungs and higher prevalence of chronic bronchitis.

OVERCROWDING

Many health problems affecting poorer groups in Third World cities are associated with overcrowded living

TABLE 26.2 HOUSEHOLD ENVIRONMENTAL CONDITIONS IN SELECTED CITIES

Environmental health indicator (%)	São Paulo	Jakarta	Accra	Within Accra		
				Poor	Middle-class	Affluent
Water: no water source at residence	5	13	46	55	14	4
Sanitation: share toilet facilities with over ten households	3	13	48	60	17	2
Solid waste: no home garbage collection	5	27	89	94	77	55
Indoor air: main cooking fuel wood or charcoal	0	2	76	85	44	30
Pests: flies in kitchen	20	38	82	91	56	18

BOX 26.2

Environmental problems in a low-income community in Karachi, Pakistan

Living conditions in the Ghera community in the Chanisar Goth district of south Karachi are typical of many Third World slum areas, with overcrowded housing and inadequate provision of water and sanitation. Narrow lanes run between the homes. Some households with a little extra space have built latrines. They allow other householders to use them for one rupee a time. Children squat down in the lanes to relieve themselves, adding to the stench from open drains. In addition to this problem is that of inadequate water supply. Chanisar Goth has piped water, but the municipality supplies the area for only three hours a day. Different communities receive water at different times and for different periods. Ghera is supplied for thirty minutes during the evening at variable times. The Ghera community has six taps, three of them installed in communal areas while the remaining three were reported to be located in the local leaders' houses. There is an acute shortage of water, and most households buy water from Pathans who live in lower middle-income areas near the settlement, and whose houses are supplied for the full three hours. The water table in the Goth is so high that the piped water is contaminated. Taps are left running for fifteen minutes before the water runs clear. Overcrowding is another problem, with at present 200 houses with one or two small rooms each and an average household size of ten persons.

Source: **J. Beall (1995) Social security and social networks among the urban poor in Pakistan** *Habitat International* **19(4), 427–45**

conditions. In many low-income areas there is often less than 1 m^2 of floor space per resident. Infectious diseases such as tuberculosis, influenza and meningitis are readily transmitted from one person to another, and spread is often aided by low resistance among the inhabitants as a result of malnutrition. Vaccine-preventable 'childhood' diseases such as mumps, measles, diphtheria and whooping cough also spread rapidly in overcrowded urban conditions. Measles is one of the most common causes of infant and child mortality in the Third World city. Cramped conditions, inadequate water supplies and unhygienic facilities for preparing and storing food also exacerbate the risk of food contamination, leading to a high incidence of acute diarrhoea and food-borne diseases, including cholera, botulism and typhoid fever. The combination of overcrowding and poor-quality housing also increases the risk of accidents in the home.

THE WORKPLACE ENVIRONMENT

Environmental hazards arising in the workplace are a major problem in Third World cities. They include dangerous concentrations of toxic chemicals and dust, inadequate lighting, ventilation and space, and lack of protection for workers from machinery and noise. The effects of workplace hazards are heightened by the general lack of social security, with little or no provision by most employers of sick pay or compensation if workers are injured or laid off.

Certain industrial activities have long been associated with high levels of risk for their work force, as in factories extracting, processing and milling asbestos, chemical plants, the cement, glass and ceramics industries, the iron and steel industries, factories manufacturing rubber and plastics products, and the textile and leather industries. Some of the most common environment-related occupational diseases are silicosis, byssinosis, lead and mercury poisoning, pesticide poisoning, noise-induced hearing loss, and skin diseases. In Bombay many of those working in the asbestos industry suffer from asbestosis, many employed in the cotton mills experience byssinosis (brown lung) and one-quarter of those in the gas company suffer from chronic bronchitis, tuberculosis and emphysema. In the informal sector of the Third World urban economy many small enterprises use chemicals that should be used only with special safety equipment. The dangers to health are illustrated by the increase in leukaemia cases among Turkish leather workers following the introduction of a cheaper glue containing benzene.[12]

Certain groups are particularly at risk from occupational hazards. Many light industries prefer to employ young women to assemble products and often use hazardous chemicals, such as PCBs, without adequate safeguards. This can pose a threat to pregnant women and their unborn children. The widespread use of child labour exposes minors to the risk of accident, injury and industrial disease (Box 26.3).

THE NEIGHBOURHOOD ENVIRONMENT

To the risks associated with home and workplace must be added environmental dangers arising from the sites on which many poorer households live.

BOX 26.3

Environmental hazards for children at work in India

Brass moulding

A report on the workers who make brass instruments in Roorkee (Uttar Pradesh) mentions that 400 children are engaged in this industry, mainly in packing and moulding. In the moulding, gas escapes and is inhaled by the young workers. They are also burned by sparks and accidental spillage from molten metal.

Carpet weaving

In Kashmir's carpet weaving industry some 6,500 children between the ages of 8 and 10 work in congested sheds in long rows behind giant looms. The air is thick with particles of fluff and wool, and 60 per cent suffer from asthma and tuberculosis.

Agate processing

The agate processing industry in Khambat, Gujarat, employs some 30,000 workers, including many children. A study of 342 workers included thirty-five children and found that half the children had lung diseases, while five had pneumonoconiosis.

Match industries

Match factories in or around Sivakasi employ as many as 45,000 children of between four and fifteen years of age. A twelve-hour day is common, working in cramped conditions with hazardous chemicals and inadequate ventilation. A supervisor commented, 'We prefer child labour. Children work faster, work longer hours and are more dependable; they also do not form unions or take time off for tea and cigarettes.'

Glass industries

In and around Firozabad there are some 200,000 people working in glass industries; roughly a quarter are children. Among the environmental hazards are exposure to silica and soda ash dust, excessive heat, accidental burns, accidents caused by defective machines or unprotected machinery, and cuts and lacerations caused by broken glass.

Source: **J. Hardoy, D. Mitlin and D. Satterthwaite (1992)** ***Environmental Problems in Third World Cities*** **London: Earthscan**

DANGEROUS BUILDING SITES

Tens of millions of urban inhabitants in Africa, Asia and Latin America occupy land sites susceptible to natural hazards. Clusters of illegal housing often develop on marginal land such as steep hillsides, flood plains or desert land, or on the most unhealthy or polluted land sites around solid-waste dumps, beside open drains and sewers, or in and around industrial areas with high levels of air pollution. Many low-income settlements also develop on sites subject to high noise pollution, close to airports or major highways.

In many instances there is a complex interaction between natural hazards and human actions. In Caracas 600,000 people live on slopes with a high risk of landslide. Until the 1960s most slope failures were associated with earthquakes. Since then, increased deforestation, which has accompanied the construction of *barrios* on the slopes, has increased the likelihood that rainfall can trigger landslides.[13] In 1999 hundreds of people were killed and thousands made homeless by mudslides that swept through hillside shanties (Box 26.4). In Petropolis City, Brazil, the incidence of landslides, mudslides, rock falls and floods has increased with rapid urbanisation of steep hillsides, leading to 300 fatalities in the past two decades.[14]

Rarely do the poor occupy such sites in ignorance of the dangers. These sites are cheap or can be occupied without payment because they are dangerous and therefore unattractive to alternative uses.

THE GARBAGE HAZARD

In most Third World cities between one-third and one-half of the solid wastes generated remain uncollected and left to accumulate on wasteland and in the streets to pose a serious human-induced environmental hazard. In Jakarta 40 per cent of the solid waste is uncollected, much of it ending up in watercourses and along the roadside, where it clogs drainage channels and causes extensive flooding in the rainy season. Uncollected garbage attracts disease vectors (rats, cockroaches, mosquitoes, flies), while leachate from decomposing and putrefying garbage can contaminate water sources. These environmental hazards are greatest in the poorest neighbourhoods of the Third World city. In Dhaka 90 per cent of the slum areas have no regular garbage-collection service. Since the poorest areas are also those worst served by sanitation facilities, the uncollected solid wastes usually contain a significant amount of

BOX 26.4

The 1999 mudslides in Caracas, Venezuela

In the week before Christmas the Blandin shanty town in Caracas, Venezuela, was virtually obliterated by a river of mud, rocks and water that cascaded down the mountainside. The Díaz family lost eleven members. One of the survivors, Alexis Díaz, told how incessant rainfall had raised a normally shallow stream into a raging torrent that engulfed his neighbourhood. The half-exposed body of Alexis's brother lay face down in the mud on the floor of what had been his living room. His brother's pregnant wife had gone back into the house to get the children and was buried with them. Ten people trapped on the second floor of a nearby house were killed as the flood of debris swept the structure away. Daylight revealed the full extent of the disaster as survivors struggled knee-deep in mud to search for the bodies of the dead and to recover household belongings from the ruins. Mangled cars and bodies were buried deep in the mud. Huge boulders blocked roads and tree trunks jutted from the walls of houses. Personal effects – clothes, toys and family photographs – lay scattered on the ground.

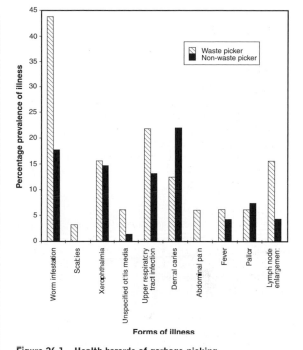

Figure 26.1   Health hazards of garbage-picking

Source: C. Hunt (1996) Child waste pickers in India *Environment and Urbanization* 8(2), 113–18

faecal material, creating a major health risk for both children playing on open sites and those engaged in garbage picking (Figure 26.1).

THE CITY ENVIRONMENT

The most serious citywide environmental problems in the Third World are the following:

AIR POLLUTION

In developed countries, until the clean-air legislation of the 1960s and 1970s most air pollution was caused by the combustion of coal or heavy oil by industry, power stations and households, which produced a mixture of sulphur dioxide, suspended particulates and inorganic components. These were the source of the infamous London pea-soup 'smogs' of the 1950s. They remain the main source of air pollution in many Third World cities. China is home to sixteen of the world's twenty most polluted cities and is the world's second-largest producer of greenhouse gases (after the USA). Over the period 1996–2006 the concentration of pollutants in China's air increased by 50 per cent. Chemical analysis of particulate matter from Beijing, where a high proportion of all energy is produced by coal-burning, identified high levels of organic compounds, including the carcinogen benzopyrene.[15] In many Third World cities the concentrations and mixes of air pollutants are high enough to cause illness in more susceptible individuals and premature death among the elderly, especially those with respiratory problems.

More recently, largely as a result of growing automobile use in Third World cities, photochemical ('oxidising') pollutants have become a major problem (the 'Los Angeles smog complex'). Among these, oxides of nitrogen from petrol-fuelled motor vehicles and hydrocarbons (e.g. from petrol evaporation and leakage from gas pipelines) are particularly important. Secondary reactions in the air between nitrogen dioxide, hydrocarbons and sunlight cause the formation of ground-level ozone, which is present in photochemical smog and can impair health when in high concentrations. Carbon monoxide, formed by the incomplete combustion of fossil fuels, is also a common air pollutant, as is lead, especially in Third World cities, where leaded fuel is still used. Relatively low concentrations

Plate 26.2   Apartment blocks in central Beijing are shrouded in a smog created by the widespread use of coal as a domestic and industrial fuel

of lead in the blood may contribute to higher risks of heart attacks and strokes in adults, and can impair a child's mental development. In Bangkok newly born babies have been found to exhibit signs of lead poisoning due to their mother's exposure to airborne lead during pregnancy.[16] Airborne lead can also contaminate soil near busy roads, affecting crops and finding its way into the food chain.

Although most cities have some problems of air pollution, the scale of the problem, the relative importance of different pollutants and the contribution of different sources vary greatly from city to city and often from season to season. A cautionary exemplar of the potential environmental and health impacts of uncontrolled air pollution is provided by the Brazilian city of Cubatão (Box 26.5).

WATER POLLUTION

The main sources of water pollution are sewage, industrial effluents, storm and urban run-off, and agricultural run-off that can penetrate supplies of drinking water.

Virtually all Third World cities cause serious water pollution, and in many cases urban rivers are literally open sewers (Box 26.6). In Bangkok most of the canals are so severely polluted that they are anaerobic (i.e. all dissolved oxygen has been depleted by organic wastes).[17] In Jakarta all the rivers are heavily polluted by discharges from drains and ditches carrying untreated waste-water from households, commercial buildings and institutions, industrial effluent, solid wastes and faecal wastes from overflowing or leaking septic tanks. Water-related diseases such as diarrhoea, typhoid and cholera increase in frequency downstream as the rivers pass through the metropolitan area. Sea-water and sediments in Jakarta Bay are affected by the pollution load of discharging rivers. High concentrations of heavy metals, such as cadmium, mercury and lead, have been recorded and can enter the food chain through fish and shellfish.

In many Third World cities, such as Mexico City[18] and Dakar,[19] a shortage of fresh water compounds the problem of dealing with liquid wastes, especially sewage and industrial effluents. Hundreds of urban centres that developed in relatively arid areas, such as Lima, have grown beyond the point where adequate supplies can be

BOX 26.5

Air pollution in the Valley of Death

The city of Cubatão in Brazil, close to São Paulo and to the major port of Santos, was long known as the 'Valley of Death'. The city contains a high concentration of heavy industry, which developed rapidly under Brazil's military government from 1964 – including twenty-three large industrial plants and dozens of smaller ones. These included a steel plant, a pulp and paper plant, a rubber plant and several metallurgical, chemical and petrochemical industries. Most were Brazilian-owned (including five owned by the federal government), although they included some multinational plants, for instance a Union Carbide fertiliser factory and the French Rhodia company.

There was little or no attempt on the part of government or companies to control pollution. High levels of tuberculosis, pneumonia, bronchitis, emphysema and asthma, and a high infant mortality rate, were all associated with the very high levels of air pollution. At certain times it became common for young children to go, almost daily, to hospital to breathe medicated air. The Cubatão river was once an important source of fish but industrial pollution severely damaged this. Toxic industrial wastes were also dumped in the surrounding forests, contaminating surface water and groundwater which was used for drinking and cooking. Vegetation in and around

Cubatão suffered substantially from air pollution and from acid precipitation, arising from locally generated air pollution, to the point where landslides occurred on certain slopes as vegetation died and no longer helped retain the soil. Hundreds of the inhabitants of a squatter settlement were killed in late February 1984 after a pipeline carrying gasoline (petrol) leaked into the swamp under the settlement and then caught fire. For many years there was little or no protection for the workers in many of the industries and many came to suffer from serious occupational diseases or disabilities arising from their exposure to chemicals or waste products while at work.

Conditions in Cubatão have improved since 1983, when the state of São Paulo's environmental body began to impose environmental controls and fined many industries while closing others down because they contravened the environmental regulations. The local authorities have sponsored a project to help poorer households move to more healthy sites and to build good-quality housing for themselves. But for many people this is too late, since they are already permanently disabled. The longer-term impact on health of pollutants already in the local rivers or leaking into groundwater is also uncertain.

Source: **J. Hardoy, D. Mitlin and D. Satterthwaite (1992)** *Environmental Problems in Third World Cities* London: Earthscan

tapped from local or even regional sources. Many other cities face major financial problems in expanding supply to meet demand. Nearly 80 per cent of Jakarta's residents use underground water, supplies of which have become steadily depleted. In low-lying northern parts of the city the extraction of water has led to land subsidence, which has increased the city's susceptibility to flooding and allowed saline water to penetrate the coastal aquifers and pollute central city wells, nine miles (15 km) inland.[20] Similar problems have been encountered in Bangkok, which sank by 5.2 ft (1.6 m) between 1960 and 1988 owing to groundwater extraction.

TOXIC WASTE

The range of toxic or hazardous wastes includes materials that are highly inflammable (e.g. many solvents used in the chemical industry), highly reactive (that can

explode or generate toxic gases when coming into contact with water or other chemicals), contain disease-causing agents (e.g. sewage sludge or hospital wastes), are lethal poisons (such as many heavy metals) or are carcinogenic (e.g. asbestos). These require special care in handling to ensure that they are isolated from contact with humans or the natural environment. Despite this, in most of the Third World, toxic wastes are disposed of as liquid wastes which run untreated into sewers or drains or directly into rivers, or are placed on land sites with few safeguards to protect those living near by or nearby water sources from contamination.

As well as the toxic output from Third World cities themselves, either from indigenous activities or from the transfer of 'dirty industries' by multinational corporations,[21] local environmental and health problems may be compounded by the export of hazardous wastes from business or municipal authorities in the developed world to Third World states, where storage

BOX 26.6

River pollution in Dar es Salaam, Tanzania

The Msimbazi river, which runs through Dar es Salaam, is an open sewer. It receives effluent from textile companies, including chemicals such as heavy metals, dyes, caustic soda, bleaches and acids. Food-processing industries produce waste water containing fats, blood and organic materials that have a high bio-chemical oxygen demand (BOD). This enters the river untreated, as does raw sewage from residential areas. In addition to a high BOD, the abundance of nitrogen compounds in the industrial effluent promotes algal blooms that cover the surface of the river, depriving aquatic micro-organisms of dissolved oxygen and creating anaerobic conditions. The smell of gases such as hydrogen sulphide, ammonia and methane is evident all along the river and especially at the coast.

The polluted river directly and indirectly affects human health. Children swim in the river and some people use the water for agriculture during the dry season, as well as for washing and drinking. The coliform count in the river on entry to the city is relatively low at 75–100 per 100 ml of water. When it leaves Dar es Salaam the count is between 250,000 and 400,000. This is over a thousand times the coliform count considered safe just for swimming, let alone human consumption.

Source: **M. Yhdego (1991) Urban environmental degradation in Tanzania** *Environment and Urbanization* **3(1), 147–52**

Plate 26.3   Low-lying coastal cities such as Jakarta are prone to the hazards of flooding

costs represent only a small fraction of the cost of meeting domestic regulations for disposal of toxic waste.

NATURAL AND HUMAN-INDUCED HAZARDS

In addition to the site-specific risks from natural and human-induced hazards experienced by low-income urban dwellers, many cities occupy locations that pose threats to a large proportion of the population. In the event of a disaster, the concentration of population in major cities can mean a heavy death toll and substantial property loss.

The earthquakes that struck Mexico City in 1985 and their aftershocks killed 10,000 people, injured 50,000 and made 250,000 homeless. Damage amounted to US$4 billion.[22] The 1986 earthquake in San Salvador caused 2,000 deaths, injured 10,000 and resulted in total damage of US$2 billion.[23] Urban floods also represent a major hazard in the Third World. Flooding often arises as a result of the extension of urban areas unaccompanied by the development of storm drainage systems, as in Benin City[24] and Nairobi.[25] In Lusaka the combination of a period of heavy rain and lack of drainage infrastructure left 50,000 people homeless after the 1989 floods.[26] Urban floods can contaminate water supplies and are associated with epidemics of dysentery and other waterborne (e.g. ascariasis) and water-washed (e.g. leishmaniasis) diseases. Outbreaks of leptospirosis (usually caused by drinking water infected by rat urine) have been associated with floods in Rio de Janeiro and São Paulo, with those living in the poorest areas particularly at risk. Threat of inundation is especially severe in rapidly growing coastal cities, including Bangkok, Jakarta, Shanghai and Alexandria, which are particularly susceptible to the combined effects of sea-level rise and land subsidence.[27] The Bangladeshi city of Dhaka is affected regularly by flooding triggered by heavy monsoon rains. The magnitude and duration of the 1998 inundation were unprecedented, with half the city under water. The human consequences of such catastrophic events range from loss of life and property to outbreaks of disease, unemployment and increased levels of crime (Box 26.7). Large-scale industrial accidents can also have a devastating effect on densely populated cities. The release of methyl isocyanate gas from a chemical factory in Bhopal, India, in 1984 caused the death of more than 3,000 people, with another 100,000 seriously injured and 200,000 people evacuated.[28]

BOX 26.7

The anatomy of a disaster: the 1998 flood in Dhaka, Bangladesh

In 1998 the Bangladeshi capital city of Dhaka was flooded for over two months. The inundation wreaked havoc on the urban infrastructure and brought misery to the mass of the population. The full impact of the event is revealed in the chronology below.

10 September. Breaches have developed at thirty-six points in the twenty-mile (32 km) Dhaka Narayangang-Derma (DND) embankment, threatening the entire capital city with inundation as the water rises daily. At Chaschara, the most vulnerable point, the height of the protection wall built with over 20,000 sandbags has been raised by another 6 in. (15 cm). Local residents have started to leave the area but flooding has severed road links between Dhaka and the rest of the country.

12 September. The DND embankment remains vulnerable. A ring dam has been erected around a major crack in the wall. A moderate shower created panic among people living in the DND area. Sandbags are being kept ready and bamboo fences erected for emergency protection of the embankment. Wives and children of British diplomats in Dhaka have been sent to London and Bangkok as flood-water swamped their residences.

Two local people were killed when they came into contact with live electricity wires while shifting belongings from their flood-affected homes. Uncollected garbage is rotting in Dhaka's submerged streets, creating a health hazard.

13 September. Some food traders are taking advantage of the flood situation to raise the price of essential food items Flood victims living on the city's periphery have been hit hard. People living on makeshift platforms or the roofs of their *kutcha* (thatched or mud-built) houses are under constant threat from robbers who arrive in hired boats to steal from marooned people at gunpoint.

BOX 26.7 – continued

14 September. As the floodwaters start to recede, reported cases of diarrhoea increase (as a result of polluted drinking water), as well as the incidence of skin diseases, hepatitis, dysentery, typhoid, respiratory tract infections and malaria. Most educational institutions remain closed. Demand for wage labour in the city has declined, leading to high unemployment. Construction, a major source of work, has virtually ceased, factory production is much reduced, and rickshaw pullers are unable to negotiate flooded streets. The loss of wage-earning opportunities has led to increased levels of malnutrition, especially among children.

15 September. Most road links between Dhaka and the rest of the country remain cut off. In the capital over 3,000 patients a day are seeking treatment for diarrhoea, most of them children. After having been submerged for over two months, leaks have been found at more than 300 places in the cooking-gas supply line, affecting over 100,000 households. The electricity supply is damaged and the city's deep tube wells are still under water, resulting in a shortage of safe drinking water.

16 September. With the main dump flooded the city is littered with 40,000 tons of piled-up garbage. Blocked drains are slowing the escape of floodwater. Demand for oral saline and water purifying tablets exceeds supply, leading to profiteering.

19 September. Half the underground sewer lines in the city are leaking and mixing with the floodwater. Over 60 per cent of the sewage system and more than 40 per cent of the city's water supply are in need of repair. Incidences of robbery, mugging and looting have increased. Large numbers of slum dwellers who have taken refuge in the city's schools are reluctant to return to their homes until their shanties are rebuilt. The diarrhoea outbreak is nearing epidemic proportions because of the scarcity of drinking water and lack of awareness of basic hygiene.

21 September. With the gradual retreat of the floodwater the number of malaria-carrying mosquitoes in Dhaka has greatly increased. Flood-affected low-lying areas are ideal breeding grounds for mosquitoes, and the piles of rotting garbage hasten the breeding process. Discarded polythene bags continue to clog the city's drainage system.

22 September. Floodwaters have receded from most parts of the city but the lack of safe drinking water is a major problem. Cracks in the supply pipes have allowed floodwater to pollute the system. Forty-four of the city's 236 deep tube wells have also been affected and residents are advised to boil all drinking water. Layers of sludge left behind by the floods still await cleaning from most parts of the city.

28 September. Dhaka City Corporation has initiated a 'food for work' programme to rid the city of the accumulated garbage and sludge that have blocked all the surface and underground drains, creating serious waterlogging and a threat to public health. Destitute people from flood-affected rural areas are moving into the city in search of food and work. Refugees, who are refusing to leave their temporary shelter in the city schools until their shacks are rebuilt, are delaying the full resumption of educational services.

More than 2 million people in Dhaka were driven from their homes by polluted floodwaters that created a host of environmental, economic and social problems for the city and its residents After more than two months of disruption the life of the city returned to normal – until the next flood.

THE CITY'S ECOLOGICAL FOOTPRINT

Cities impose an environmental impact on their hinterlands and, in some cases, on ecosystems far beyond the immediate region, owing to their demand for renewable resources, such as water, fossil fuels, land and building materials, which cannot be met from within the city's boundaries. Cities are also major producers of wastes, much of which impact upon the surrounding region (see Chapter 30).

The more populous the city the greater its 'ecological footprint', which may be defined as the land area and natural resource capital on which the city draws to sustain its population and production structure.[29] In the past, the size and economic base of any city was constrained by the size and quality of the resource endowments of its surrounding region, and a city's ecological footprint remained relatively local. Today city-based consumers and industries based in wealthy nations have the capacity to draw resources from far beyond their immediate regions and have increasingly appropriated the carrying capacity of rural regions in other nations, with little apparent regard for the environmental impact of their actions.

In the Third World the urban imprint is generally less far-reaching, but nevertheless exerts a fundamental influence on ecosystems within the city region (Box 26.8). Among the major challenges for most Third World cities is the need to address the range of environmental problems that has been termed the **brown agenda**.[30] This focuses the attention of urban analysts and decision-makers on the primary requirement of providing safe, sufficient supplies of water to households and enterprises, and making provision for the collection and disposal of liquid and solid wastes in order to combat the effects of pathogens that underlie the high levels of morbidity and mortality characteristic of most Third World cities. In Chapter 27 we shall focus attention on health conditions in Third World cities.

BOX 26.8

The ecological footprint of Jakarta, Indonesia

The city of Jakarta casts an ecological shadow over an extensive area. The major problems include severe water pollution from both urban and rural agricultural uses, unnecessary loss and degradation of prime agricultural land through urban expansion, potentially serious erosion problems developing in the uplands, extensive loss of natural habitat, and severe threats to the remaining areas of natural forest, coastland and marine eco-systems. Other environmental problems in Jakarta include mercury poisoning in Jakarta Bay and the absence of firm government measures to deal with the mounting levels of toxic waste.

These problems are not simply the result of either a unidirectional spread of urbanisation into agricultural land or the movement of rural households into ecologically critical uplands. Rather, they are the outcome of negatively reinforcing impacts of both rapid urbanisation and the rapid expansion of rural land use in coastal, upland and forest areas in the region reaching beyond the Jakarta agglomeration, and along the Jakarta–Bandung corridor.

In a region such as Jakarta, environmental problems go beyond the categories of simple negative externalities and threaten the very sustainability of development. One example is the extraction of groundwater at an accelerating rate by a multiplicity of users.

The result has been that seawater has now intruded nine miles (15 km) inland, creating a zone of salinised groundwater reaching to the centre of the city.

The major dilemma facing efforts to improve the quality of life in Jakarta and other metropolitan core regions in Asia is that the key parameters that need to be controlled are neither contained within Jakarta nor are they subject to substantial manipulation by the spatial allocation of infrastructure in the capital city region.

The main factors propelling the accelerated growth of the Jakarta metropolitan region are the mechanisation of agriculture and a decline in public spending on rural construction (both working to accelerate rural–metropolitan migration); the fall in the prices of outer-island exports (which has reduced the economic pull to direct migration away from Java); import substitution and, more recently, export-oriented manufacturing policies that have worked to polarise manufacturing employment within the Jakarta agglomeration.

Land-use management within the region must be dramatically improved if the negative impacts of land-use changes and conflicts are to be reduced, to allow an environmentally sustainable development process.

Sources: J. Hardoy, D. Mitlin and D. Satterthwaite (1992) *Environmental Problems in Third World Cities* London: Earthscan; F. Steinberg (2007) Jakarta: environmental problems and sustainability *Habitat International* 31, 354–65

FURTHER READING

BOOKS

J. Hardoy, D. Mitlin and D. Satterthwaite (2001) *Environmental Problems in an Urbanizing World* London: Earthscan

G. McGranahan, P. Jacobi, J. Songsore, C. Sujadi and M. Kjellen (2001) *The Citizens at Risk: From Urban Sanitation to Sustainable Cities* London: Earthscan

H. Mann and S. Williams (1994) *Environment and Housing in Third World Cities* Chichester: Wiley

C. Pearson (1987) *Multinational Corporations, the Environment and the Third World* Durham NC: Duke University Press

M. Pelling (2003) *The Vulnerability of Cities* London: Earthscan

World Health Organisation (1994) *Water Supply and Sanitation Sector Monitoring Report 1994* Geneva: WHO

JOURNAL ARTICLES

H. Akbar, J. Minnery, E. van Horen and P. Smith (2007) Community water supply for the urban poor in developing countries *Habitat International* 31, 24–35

V. Brajer and R. Mead (2004) Valuing air pollution mortality in China's cities *Urban Studies* 41(8), 1567–85

A. Daniere and L. Takahashi (1999) Poverty and access: differences and commonalities across slum communities in Bangkok *Habitat International* 23(2), 271–88

A. de Sherbinin, A. Schiller and A. Pulsipher (2007) The vulnerability of global cities to climate hazards *Environment and urbanization* 19(1), 39–64

S. Hasan and M. Adil Khan (1999) Community-based environmental management in a megacity: considering Calcutta *Cities* 16(2), 103–10

J. Leitmann, C. Bartone and J. Bernstein (1992) Environmental management and urban development *Environment and Urbanization* 4(2), 131–40

W. Rees (1992) Ecological footprints and appropriate carrying capacity *Environment and Urbanization* 4(2), 121–30

KEY CONCEPTS

- Environmental health
- Land subsidence
- Dirty industries
- Carrying capacity

- Brown agenda
- Environmental hazards
- Pollution
- Ecological footprint

STUDY QUESTIONS

1. Identify the major environmental problems at the domestic, neighbourhood and city levels in Third World metropolitan areas.
2. Examine the nature, incidence and impact of the environmental hazards present in a Third World city of your choice.
3. With reference to specific examples, consider the mechanisms and levels of provision of urban services (e.g. water, sanitation, garbage collection) in Third World cities.
4. Examine the problems of water supply in Third world cities.
5. With reference to particular examples discuss the vulnerability of Third World cities to the flood hazard.
6. Explain what is meant by the concept of a city's ecological footprint.

PROJECT

The 'brown agenda', focusing on environmental hazards, pollution problems and poverty, is a major concern for the growing urban population of the Third World. Consult back issues of a national daily newspaper for the past three months (available in your local library or via the World Wide Web) and identify reports on issues related to the brown agenda. Classify the information by city location and type of problem. You will have to devise a suitable classification scheme (which may involve categories such as landslides, lack of shelter, urban crime, inadequate infrastructure, infant mortality, disease, poverty). Construct a bar graph to show the relative incidence of different types of environmental problem, as well as individual graphs to show the frequency of problems in individual cities. What are the major problems reported? How do they compare with urban problems in cities of the First World?

27

Health in the Third World City

Preview: the epidemiological transition; factors determining health status; an urban penalty; intra-urban variations in health; diseases of the urban poor; primary health care; malnutrition; integrated approaches to health care

INTRODUCTION

Many of the environmental hazards discussed in Chapter 26 exert a direct or indirect effect on the health and well-being of urban dwellers. Here we focus on the causes, differential incidence and socio-spatial consequences of ill health within Third World cities. We examine the diseases of the urban poor and the nature and effectiveness of the principal health-care responses.

THE EPIDEMIOLOGICAL TRANSITION

As societies develop, health generally improves and life expectancy increases. The major causes of illness and death also change. Consequently, differing disease profiles may be identified for developed and underdeveloped countries respectively. This is encapsulated in the concept of the **epidemiological transition**, which proposes that, with development, pandemics of infection are replaced as the chief cause of death by degenerative diseases brought about by people's lifestyles. According to Omran (1971), during the early stages of the transition (typified by the situation of underdeveloped countries), infections, parasitic diseases and nutritional diseases are the main causes of morbidity and mortality.[1] By contrast, in more developed societies such conditions are responsible for only a small amount of ill health and death, although conditions such as measles and chickenpox, which mainly affect children, often persist. At a later stage

of the epidemiological transition only pneumonia and influenza are major infections, and then mainly among older age groups. In advanced societies, chronic degenerative conditions associated mainly with older adulthood, such as heart diseases and cancers, are responsible for most morbidity and mortality.

The epidemiological transition provides a useful general model of the relationship between disease and development. However, individual countries vary in terms of their experience of the transition, with many newly industrialising countries (NICs) accelerating through the stages towards a 'Western' disease profile and other Third World states exhibiting a profile characteristic of underdevelopment. Many Third World countries are currently experiencing a mixed stage of the transition (between 'the age of receding pandemics' and 'the age of degenerative diseases') in which, on average, mortality and morbidity from degenerative diseases are increasing while those from infectious diseases remain high. The global disease of HIV/AIDS is a particular problem in many Third World cities (Box 27.1).

Furthermore, major differences in disease incidence and aetiology occur within Third World countries, with the epidemiological transition proceeding at different rates in rural and urban areas, and between different population groups within cities. While affluent urbanites may be developing a Western pattern of mortality, the lack of environmental-health provision means that the poorest urban dwellers remain subject to a host of infections and dietary diseases.

BOX 27.1

HIV/AIDS: the modern plague

The AIDS epidemic began in Uganda in the late 1970s. By 1997 over 30 million people worldwide were infected by HIV, which causes AIDS – that is, one in every hundred adults in the 'sexually active' age group of 15 to 50. The figure of 30 million also included 1.1 million children under the age of 15 years.

HIV/AIDS is mainly a sexually transmitted disease but can also be contracted through infected blood transfusions, injection syringes and between mother and child at birth. HIV destroys the human immune system, leaving the victim open to infections that can lead to death. In terms of income and wealth, AIDS is associated with poverty both cross-nationally and within nations. More than 90 per cent of HIV-infected people live in the Third World; many of them do not know they are infected. In Zambia and Malawi groups of antenatal clinic attendees had infection rates of 25 per cent, while in Francistown, Botswana, rates of 40 per cent have been reported. In the city of Mumbai, India, infection rates as high as 51 per cent have been recorded among commercial sex workers. In many African countries a young person has a lifetime risk of 30 per cent of contracting the disease.

In Asia and Africa the survival period with AIDS is typically shorter than in Western Europe and North America, and reflects the lower health status of the population, the greater number of infectious agents in the environment and the unavailability of expensive palliative treatments.

A price cut in HIV/AIDS drugs by five major pharmaceutical companies in 2000 was a step in the right direction but is expected to have limited impact in sub-Saharan Africa, where 70 per cent of the world's 33.6 million HIV-infected people live. Aside from the costs there are other barriers to the health improvement of this population. The effectiveness of the drug depends on a balanced and adequate diet, which is often impossible because food is unavailable or too expensive. In addition, AIDS treatment typically requires a strict regimen of medication that must be taken to a set schedule that may not be possible due to unavailability of supplies. Furthermore, many people who are HIV-positive die from other causes, such as pneumonia, because they lack the medicines to fight these curable diseases that attack their weakened immune systems.

Sources: T. Barrett (1999) HIV/AIDS, poverty, exclusion and the Third World, in M. Pacione (ed.) *Applied Geography: Principles and Practice* London: Routledge, 528–36; UN Centre for Human Settlements (2001) *Cities in a Globalising World* London: Earthscan, 112.

FACTORS DETERMINING HEALTH STATUS

An individual's level of health is affected by a complex set of personal, environmental and structural factors. These include the following:

1. *Intermediary factors*. These refer to the knowledge and behaviour of a person, household or community in addressing their own health problems. The level of education of a mother and her ability to read and comprehend health educational literature have been shown to exhibit an inverse relationship with levels of child mortality. This also illustrates the interrelationship among the factors underlying morbidity and mortality, since the extent to which women have access to education depends on both underlying factors (such as provision of schools) and structural factors (such as the strength of the national economy, and public attitudes to the education of women).

2. *Underlying factors*. These include aspects of the physical environment (such as house site and location), socio-economic factors (such as income and housing tenure), infrastructure provision (e.g. the type and availability of health care) and biological factors related to age and gender.

3. *Structural factors*. These include the laws, codes, norms and practices of national and city governments, and the health of the national and city economy. For many poor Third World countries these political-economic forces represent the main barriers to improved health.

AN URBAN PENALTY?

The notion of an urban health penalty emerged in nineteenth-century England when analyses of mortality data revealed that urban death rates were significantly higher than rural rates. Public health measures, such as a supply of clean water and sanitation, contributed to a decline in urban mortality after 1890.[2] Evidence from the Third World, today, indicates a slower decline in infant mortality rates in large cities compared with smaller towns and villages. In many cities of sub-Saharan Africa infant mortality rates have increased over recent decades.[3] In addition, sustained economic recession, urban growth, deteriorating physical environments and the HIV/AIDS epidemic contribute to the sub-Saharan 'urban penalty'.

More generally, the overcrowded living conditions in low-income urban communities foster the spread of communicable diseases. The interaction of multiple risk factors in urban environments is illustrated in the case of acute respiratory infections that are endemic rather than epidemic, affect younger groups and are more prevalent in urban than rural areas. Frequency of inter-personal contact, density of population, and the circulation of infective and susceptible people promote transmission of disease organisms. Poor urban groups are at greatest risk. Further, a constant influx of migrants who may be susceptible to infection and possible carriers of virulent new strains of infective agents promotes the transfer of nasopharygeol micro-organisms. The urban poor suffer a 'double penalty' of risk from both old and new epidemiological profiles.[4]

INTRA-URBAN VARIATIONS IN HEALTH

At an aggregate level, city health statistics generally seem better than those for rural areas of the Third World. This is misleading, however, since squatter populations are often not included in official statistics. Even when these settlements are included, health conditions in low-income areas are disguised by the figures for healthier and better-served middle- and high-income areas of the city.

Disaggregated data for different urban sub-areas reveal marked intra-urban socio-spatial variations in health status. In São Paulo levels of infant mortality ranged from twenty-nine per thousand in the more affluent areas to sixty-one per thousand in the poorest areas.[5] In Manila the incidence of severe malnutrition and the infant-mortality rate were found to be three times higher in the slums than in the rest of the city, while the rate for tuberculosis was nine times higher.[6] In Accra the incidence of diarrhoea in children under 6 years of age was twenty-two per thousand among the poorest quintile of the population and nine per thousand among the richest quintile.[7] In one Bombay slum the rate for leprosy was twenty-two per thousand, compared with a city average

of seven per thousand.[8] The effect of the slum environment on health is illustrated by the causes of infant mortality in Porto Alegre, Brazil. In the non-shanty areas, most infant deaths were caused by neo-natal problems of gestation, delivery or immediate afterbirth and only 25 per cent by pneumonia, influenza, infectious intestinal diseases or septicaemia. In contrast, environmental agents caused 51 per cent of infant deaths in the shanty towns, reflecting a much greater exposure to hazardous living conditions.

DISEASES OF THE URBAN POOR

Before the HIV/AIDS epidemic, tuberculosis was the main cause of death among adults in developing countries. The interaction of HIV and TB and the spread of multi-drug-resistant strains of TB raise concerns over a resurgence of tuberculosis, which kills one in four of the adult population of the Third World.[9] High-density low-income urban populations are potentially at risk. The socio-environmental conditions in such areas have also led to the emergence or re-emergence of vector-borne diseases, including malaria, filariasis, dengue, typhus and Chaga's disease (Box 27.2).

Urban malaria is a major health problem in slums and squatter settlements where the carrier (the anopheles mosquito) breeds in stagnant pools of water caused by rain or a lack of sanitation or drainage. Epidemics of other vector-borne diseases such as dengue haemorrhagic fever are associated with the need of households to store water in iron drums or earthenware containers which provide ideal breeding conditions for the mosquito *Aedes aegypti*, the vector of dengue and yellow fever. Tuberculosis is prevalent in the slums and shanty towns, and malnutrition is widespread. The scarcity of clean water and lack of sanitation make diarrhoeal diseases a major health problem, while a variety of intestinal parasites, such as ascaris (roundworm) and trichuris (whipworm), are usually present. The crowded living conditions also increase the risk of meningococcal meningitis, and lead to a high incidence of preventable infections in children such as measles, whooping cough and polio (Box 27.3).

As well as physical health problems, social and psychological difficulties may arise as a result of the disadvantaged position of the urban poor. Chronic stress may arise in those who feel depressed, cheated, bitter, desperate, isolated and vulnerable; who are worried about debts or jobs and housing insecurity; and who feel devalued and alienated from wider society.[10] In the city the protection afforded by rural local

BOX 27.2

Aetiology of Chaga's disease

Chaga's disease (American trypanosimiasis) is a parasitic infection that affects 20 million people in Mexico, Central and South America. It causes 50,000 fatalities annually and debilitates many more, making it a leading cause of death, behind acute respiratory infections, diarrhoeal diseases and HIV/AIDS.

Although the disease may be passed on through blood transfusions and breastfeeding, its main vectors are nocturnal beetles (*Triatoma redurii* and *Rhodvius prolixus*) whose bites transmit a parasite (*Trypanosoma cruzi*). These beetles breed in cracks in the walls of homes, in thatched roofs and in spaces between wooden boards. They thrive in dark, poorly ventilated, humid environments. There is no cure, and attempts to control the disease involve preventive measures. Fumigation is temporarily effective but long-term intervention must be related to improvements in living conditions. In addition, as the disease is often asymptomatic for fifteen to twenty years, education is important. As the disease mainly affects poor people who have more immediate survival concerns, awareness-raising efforts must accompany schemes to upgrade the housing environment.

Source: **UN Centre for Human Settlements (2001)** ***Cities in a Globalising World*** **London: Earthscan.**

BOX 27.3

Health problems in low-income settlements in Third World cities

In Chheetpur, a squatter settlement in the city of Allahabad with some 500 people in 1984, 55 per cent of the children and 45 per cent of the adults had intestinal worm infections, while at the time of the survey 60 per cent had scabies. Most inhabitants had a food intake of less than 1,500 calories a day; among infants and children up to the age of 4, 90 per cent had an intake significantly below the minimum needed. Over a period of fourteen years 143 children's deaths had been recorded; malaria was the most commonly identified cause of death, followed by tetanus, injuries from accidents or burns, and diarrhoea, dysentery or cholera. Malnutrition was an important contributing factor in many of these deaths. The settlement's site is subject to flooding in the rainy season, and lack of drainage means that stagnant pools exist for much of the year. Two standpipes serve the entire population's water needs and there is no public provision for sanitation or the removal of household wastes.

In San Martín, a squatter community in one of the municipalities that ring the central city of Buenos Aires (see Box 25.3), clinical tests on a small sample of inhabitants showed that more than half had intestinal worms. Many children were underweight and malnutrition was widespread, especially in terms of protein, vitamin and mineral intake. Physical examination of a larger sample found 15 per cent of children with inflammations in the upper breathing passage; among adults who underwent physical examination, one in four men and one in ten women had chronic bronchial afflictions. Diarrhoea was a major problem, especially in the summer. Although 60 per cent of the households obtained water from piped public supplies – since the inhabitants simply tapped into a nearby mains – the quantity of water available was frequently inadequate and many households could not afford the cost of piping water into their houses. Most of the 40 per cent who did not have access to piped water used water pumped from wells; tests on water quality found that first-level groundwater had high levels of bacteriological contamination. There was no site drainage apart from some open drains dug by inhabitants.

Source: **J. Hardoy and D. Satterthwaite (1989)** ***Squatter Citizens*** **London: Earthscan**

communities and the extended family is less readily available. Single-parent households, often female-headed, are common, and with the need for women to work, children are often neglected. In addition, as we have seen, children may have to contribute to family income by working under sweatshop conditions, exposed to accidents and exploitation. The failure of some migrants to adapt to city life can lead to problems of alcoholism and depression.[11]

PRIMARY HEALTH CARE AND THE URBAN POOR

In spite of the concentration of health resources in the cities compared with the rural areas, and the relative proximity of hospitals and other medical facilities, for those who live in the slums and shanty towns of the Third World, standards of health care fall below a

Plate 27.1   Toilet facilities are often non-existent in the slums of Mumbai, India

reasonable minimum. Hart's (1971)[12] **inverse care law**, according to which those in greatest need of care have the worst access to it, operates to marked effect in the Third World city.[13] The description of health services in the slums around Nairobi is typical (Box 27.4). Furthermore, the health-care services that are available tend to emphasise curative rather than preventive medicine, meaning that the underlying causes of ill health are unlikely to be addressed adequately.

Primary health care (PHC) has emerged as the favoured response to the health problems in Third World cities. At the heart of the PHC approach are the principles of equity in distribution, community involvement, a focus on prevention, use of appropriate technology, and a multi-sectoral approach that acknowledges the multiple aetiology of health problems. The main constituents of a PHC strategy are education about diseases and their control, the provision of safe water and basic sanitation and attempts to ensure maternal and

child health, including family planning, immunisation against major infectious diseases, treatment of common diseases and injuries and the provision of essential drugs.[14] Some have argued that this form of 'comprehensive PHC' is idealistic and unattainable in the Third World city and that, faced with the vast number of health problems of varying severity, priorities must be identified for reasons of practicality, cost and effective use of available resources.[15] This form of 'selective PHC' (SPHC) typically focuses on paediatric conditions, such as measles, whooping cough, neo-natal tetanus and diarrhoeal diseases. A lower priority is usually accorded to conditions such as polio, typhoid, respiratory infections, meningitis and malnutrition, since of this group only polio can effectively be controlled by medical intervention. The lowest priorities, in SPHC terms, are accorded to diseases such as dengue, filariasis and amoebiasis, since their control is difficult, largely socio-environmental and hence costly and

BOX 27.4

Health-care provision in the slums of
Nairobi, Kenya

Looking at the health facilities around the slum areas, we see that they are not enough for the people living in the slums. That means that people have to queue, often half a day or more. Let us take for example a dispensary, where people queue half a day. Then they face the problem that there are no medicines. So now they are waiting for nothing, getting told to go to the chemist to buy the medicine. But that is something they cannot afford. To underline this fact we quote the *Sunday Standard* from 21 April 1985: 'Medical services at the Kenyatta National Hospital Nairobi are coming to a standstill following an acute shortage of a wide range of drugs, particularly antibiotics. An increasing number of both in and out-patients at the hospital, most of whom are poor people, are being forced to buy prescriptions from private chemists, while others are referred to private hospitals for such simple services as blood tests.'

If the mother goes to the clinic for family planning, infant welfare and child welfare, then she needs to go three mornings, because the clinic staff do family planning one day, infant welfare another day and child welfare a third day. There is of course queuing again, and so this mother spends a further three half-days. When is she actually to earn her living?

Another problem in the health facilities is that the slum dwellers are often treated in a very rough way. The reason for that is that the facilities are often understaffed, so that the nurses are overworked and tired, and they do not explain to their patients what their problem is. Second, they are frustrated when they want to give treatment but there is no medicine. On top of this, unfortunately some nurses or clinical staff members feel superior towards the slum dwellers. All these factors do nothing to encourage the people in the slums to go to the health facilities for their services.

Source: J. Harpham, T. Lusty and P. Vaughan (1988) In the Shadow of the City Oxford: Oxford University Press

safe sanitation.[16] Without these wider initiatives, effective treatment of many conditions, particularly diarrhoeal and respiratory infections, may result only in cured children becoming rapidly reinfected.

Critics of SPHC contend that the selective nature of the approach, and the tendency to focus on problems of paediatric and maternal illness, exclude much of the adult population, especially males. Others have argued that SPHC programmes negate the principle of community participation, are too technological, too cost-oriented and reflect a narrow definition of health (the absence of disease) compared with the WHO definition (a state of complete physical, mental and social well-being, not merely the absence of infirmity in an individual).[17] In practice, each country designs PHC schemes to suit its own needs. Some schemes have significant community involvement,[18] while others are organised in a top-down manner and adopt an SPHC approach.[19]

An example of a selective PHC policy is the UNICEF Child Survival and Development Revolution in which the emphasis is on:

1. Growth monitoring through the use of growth charts on which serial weight-for-age readings are marked. It has been shown that these charts can help to identify and draw attention to children at high risk from the synergistic effects of infection and malnutrition.
2. Oral rehydration therapy (ORT), which is a safe salt and sugar solution effective in preventing death from dehydration caused by acute watery episodes of diarrhoea.
3. Breast-feeding, which, besides providing an ideal food, protects the infant from infection and acts as a natural form of contraception.
4. Immunisation against the six preventable childhood infections of measles, whooping cough, tetanus, polio, diphtheria and tuberculosis.
5. Food supplements for malnourished children.
6. Family spacing through birth control to improve the health of mother and child.
7. Female literacy, which is known to be related to infant mortality and fertility rates, independent of socio-economic status.

MALNUTRITION

Shortage of food is one of the most important reasons for the high rates of disease and mortality among the urban poor. Malnutrition is fundamentally a reflection of poverty that is accentuated by the greater monetisation

requiring continuing efforts. This focus on interventions of proven efficacy acknowledges that the complex poverty syndrome of malnutrition, gastroenteric diseases and respiratory infection will not yield to specific programmes but requires social and environmental improvements such as better housing, potable water and

of urban economies. This ensures that urban rates of malnutrition are typically higher than those in rural areas, where the poor can cultivate some of their own food.[20] In addition, although wages are higher in the city, so too are costs, with the result that the poor have a smaller proportion of their income available for food. Further, in the highly competitive situations of the city, women are often forced out to work and may have less time for food preparation. They may also resort to early weaning or often not breast-feed at all, leaving infants in the custody of child-minders or young children who are unable to prepare food properly.

In the slum and squatter areas of Third World cities, severe malnutrition can be well established in the first six months of life. In addition to the decline in breast-feeding, low birth weight also often culminates in severe malnutrition during infancy, which results in greater susceptibility to infection and increased risk of neo-natal death. Child and maternal malnutrition is particularly common in large, closely spaced families. Infectious diseases also have an impact on nutritional status. In the slum area of Cité Solcil, Haiti, 40 per cent of children had contracted measles by the age of twelve months. These children were more likely to be malnourished and had a mortality rate three times as high as did infants without the disease.

COMBATING URBAN MALNUTRITION

The most direct means of alleviating malnutrition is to reduce the gulf in income levels between rich and poor in the Third World city. Given that large-scale redistribution in favour of the poor is unlikely, it is clear that even with some growth in the income of the poor and increased food production, other, complementary measures will be required to improve the nutritional status of the poor. These include:

1. *Improving the quantity and quality of food.* Possible measures include distribution of supplementary food rations, the provision of cheap food shops and consumer co-operatives, the provision of special nutrient foods, advice on food preparation, targeting food hand-outs to children and pregnant and lactating mothers, and the provision of crèches for working mothers. Each of these initiatives offers some advantages, but the success of any scheme is dependent upon a proper understanding of the needs and lifestyle practices of the recipients. Even when the distribution of supplementary food rations is well organised the results can be disappointing, owing to failure to appreciate the family's feeding practices. Take-home rations intended for a malnourished child are likely to be shared among the family or, if they are provided directly at a feeding station, the child's share of the normal family ration will be reduced. In such circumstances it may be preferable to regard the family unit as vulnerable and provide a larger supplement of cheap staple food rather than a small quantity of more expensive supplement.

2. *Legislation.* National or municipal legislation can influence the quality of diet in a number of ways, ranging from controlling advertising of junk foods to instituting a legal entitlement to maternity leave or the provision of workplace crèche facilities.

3. *Education.* The influence of nutritional education is often small because the educators do not understand the community they seek to serve. Often, recommended foods or cooking methods are too expensive or impractical. Mothers may be exhorted to boil all water used with no thought of the cost in fuel and time this entails.

4. *Breast-feeding.* The decline in breast feeding is due to social and economic pressures and the availability of alternative artificial feeds. A combination of education on the health benefits of breast-feeding and enabling strategies, such as the provision of workplace facilities, is required to reverse this trend.

5. *Reduction of infection in children and mothers.* This can be achieved by general and specific measures. The former includes improved housing, sanitation and water supply in addition to health education. The latter involves a number of SPHC measures to ensure early treatment of infection during pregnancy, treatment of anaemia in pregnant women, immunisation (especially against measles) and early treatment of diarrhoea with oral rehydration solutions.

6. *Provision of family-planning advice.* In the past, aggressive government family-planning policies motivated by demographic rather than health considerations have not been well received by low-income urban communities. A decentralised service involving local people and local facilities is more likely to be successful.

7. *Informal food supplies.* In Third World cities a significant proportion of food is provided by the informal sector, notably by street vendors. This has the advantages of income generation for the producers and low-cost food for consumers. A community-organised scheme to encourage local

Plate 27.2   Street vendors provide a significant proportion of food in Third World cities such as Hong Kong

food-processing and marketing could help keep scarce money circulating in the local economies of poor areas. This might be promoted by government provision of training, credit and hygiene education.

8. *Urban agriculture.* Given that one of the biggest problems for the urban poor is the necessity of buying food, urban agriculture can make a direct contribution to combating malnutrition. The diverse range of food and fuel production within many Third World cities includes aquaculture in tanks, ponds and rivers; livestock raised in back yards, along roadsides, or within utility rights of way; orchards, including vineyards and street trees; and vegetables and other crops grown on rooftops, in back yards, on vacant lots, alongside roads and canals, and in many suburban small farms.[21] The particular form of urban agriculture depends on, among other things, the natural resource base in and around the city, formal and informal rules governing who can use open land and on what terms, and time and economic constraints on

cultivators. In addition to a basic need for food, the principal reasons for engaging in urban agriculture are to grow vegetables as a dietary supplement, to augment cash income through sale of part of the crop, and to cut food expenditure, thereby releasing cash for other uses.[22] Urban agriculture is not confined to low-income households – in Harare 80 per cent of households in a middle-class suburb grew food crops in their gardens[23] – and in some cases the entry costs, in the form of equipment, seeds and facilities, may exclude the poorest households or newest arrivals in the city without access to cultivable land.[24] In the city of Rosario, Argentina an Urban Agriculture Programme that provides training to poor urban families and identifies vacant public and private spaces, includes an 'urban kitchen gardens' project that has helped improve **food security** in the city, generated income for poor families, and transformed uncultivated land into productive spaces. Since the programme began in 2001 almost 800 urban kitchen gardens have been set up providing employment to 5,000 families, with another 10,000 families directly linked to the production of vegetables that feed 40,000 people in the city of 1.3 million.[25]

In Latin America urban agriculture has been estimated to save 10–30 per cent of household food costs, representing a saving of 5–20 per cent of the total household income among the poorest families,[26] while the poorest urban households in Tanzania met one-third of their food needs by subsistence production.[27] In Chinese cities there is a long tradition of urban agriculture (Box 27.5). Within the, albeit extensive, administrative boundaries of the Shanghai city region, a belt of farms supplies 80 per cent of all vegetables shipped to the central urban core. Public authorities in China have encouraged synergy between urban core and rural hinterland, in contrast to the situation in many countries, where this potential is lost as agricultural land on the urban periphery is held unused for speculative purposes or developed illegally for housing. Governments can support urban agriculture through grants and establishing legal title to land, and by facilitating temporary use of wasteland in the city. Since 1978 owners of unused land in Manila have been obliged to cultivate it or allow another person to grow food on it. One of these community gardens supplied 800 squatter families with 80 per cent of their vegetables from an area of 1,500m² (0.37 acre). Despite the economic and health benefits of urban agriculture in many cities the practice is under increasing threat from problems of tenure insecurity and urban encroachment.[28]

INTEGRATED APPROACHES TO PRIMARY HEALTH CARE

Integrated approaches seek to bring together a range of health and other initiatives to produce an outcome that is greater than the sum of the parts. Health improvement is seen as only part of an integrated approach, the goal of which is the total development of the community. This resembles the concept of comprehensive PHC. Despite its potential for building systematic linkages between physical improvements, social services and resident participation, the integrated community development approach is an exception in Third World cities.

Community participation is a key principle in an integrated approach. Environmental improvements, such as the reconstruction of a slum community, may require some of the residents to give up part of their plot or building to allow improved streets and drainage lines to

---

BOX 27.5

Urban agriculture in Chinese cities

Fish and livestock, vegetables and fruit can be grown or raised together in small urban courtyards. One example comes from the courtyard of Yang Puzhong, located on the outskirts of Bozhou, in Hebei Province. This courtyard is only 240 square yards (200 m²) in size At the centre is a fish pool twenty-four square yards (20 m²) in size and 6.5 ft (2 m) deep where carp, black carp, grass carp, loach and turtle are raised. Around the pool is a grape trellis that produces more than 1,000 kg of grapes a year. On one side of the pool are a pigeonhouse and a pigsty with about forty pigeons and eight pigs. Above the pigsty is a chicken house with twenty chickens, and, on top of the chicken house, a solar water heater. Below the pigsty is a methane-generating pit. On the other side of the pool is a small vegetable garden that supplies the family with vegetables all year round. Chicken droppings are used to feed pigs, night soil is used to generate methane for cooking, and liquid from the methane pit is used to feed the fish – with the remainder used as manure for farmland or as a culture medium for mushrooms.

Intensive cultivation is also evident in many other cities. For instance, in Shanghai, China's most populous city, increasing numbers of back yards, roofs, balconies, walls and vacant spaces near houses are being used to develop such agro-forestry systems as orange tree–vegetable–leguminous plant, grapevine–gourd and melon–leguminous plant, and Chinese tallow tree–vegetable–leguminous plant.

Source: J. Hardoy, D. Mitlin and D. Satterthwaite (1992) Environmental Problems in Third World Cities London: Earthscan

---

be installed. Failure to consult local communities in advance often leads to a subsequent lack of co-operation and problems with ongoing maintenance of the new infrastructure. On the other hand, when community workers ascertain when and where the people want improvements and act as liaison between residents and project personnel, the changes are more likely to be welcomed, understood and longer-lasting. Fully integrated programmes can also generate community spirit and promote local self-help.[29] The potential of a grass-roots-based integrated community development programme is illustrated by an urban upgrading project undertaken in one of the poorest neighbourhoods (*kebele*) of Addis

Ababa. The residents of *kebele* 41 must seek to survive in the cash economy in a social setting lacking many of the traditional supports of the extended family. The project mobilised local labour and resources and channelled action through the existing *kebele* structure to develop an integrated programme based on health initiatives, physical upgrading and community development (Figure 27.1). Another example of a comprehensive community-based integrated development programme is the participatory planning approach to urban upgrading employed in Kitale, Kenya.[30] Focused on three slum communities the programme was designed to address a complex of problems relating to the lack of safe drinking water and sanitation, and high levels of youth unemployment. In the slum area of Kipsongo five new sanitation blocks were built using stabilised soil blocks produced by unemployed youth who were given a block press and

trained in the technique. A protected spring is managed by a women-led Water and Sanitation Committee who are also establishing a seedling nursery and commercial vegetable garden to provide income and help protect the stream's catchment area. These case studies demonstrate the significant benefits that can be derived from adopting pro-poor participatory slum upgrading strategies that use employment-intensive approaches to foster local economic development.

Effective introduction and implementation of integrated community development initiatives requires co-ordinated action at a number of scales (Table 27.1) and is dependent upon close links between service providers and intended recipients. The complex of poverty-related problems in the most deprived urban communities, including low levels of education, limited resources and unfamiliarity with urban power

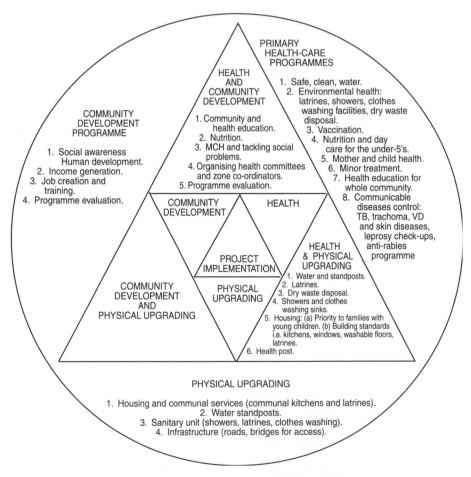

Figure 27.1 An integrated approach to community health in Addis Ababa, Ethiopia
Source: T. Harpham, T. Lusty and P. Vaughan (1988) *In the Shadow of the City* Oxford: Oxford University Press

Health risk	Action at individual and household level	Public action at neighbourhood or community level	Action at district or city level	Action at level
Contaminated water – typhoid, hepatitis, dysenteries, diarrhoea, cholera, etc	Protected water supply to house; promote knowledge of hygienic water storage	Provision of water supply infrastructure; knowledge and motivation in the community	Plans and resources to undertake or support action at lower levels	Ensure that local and city governments have the power, funding base and trained personnel to implement actions at the household, neighbourhood, district and city levels. Review and, where appropriate, change legislative framework and norms and codes to allow and encourage actions at lower levels and ensure that infrastructure standards are appropriate to the needs and the resources available. Support for training courses and seminars for architects, planners, engineers, etc., on the health aspects of their work
Inadequate disposal of human wastes – pathogens from excreta contaminating food, water or fingers, leading to faecal–oral diseases or intestinal worms (e.g. hookworm, roundworm, tapeworm, schistosomiasis)	Support for construction of easily maintained w.c. or latrine that matches physical conditions, social preferences and economic resources. Washing facilities to promote hand-washing	Mix of technical advice, equipment installation and its servicing and maintenance (the mix is dependent on the technology used)	Plans and resources to undertake and support action at lower levels. Trained personnel and finances to service and maintain them	
Waste-water and garbage – waterlogged soil ideal to transmit diseases such as hookworm; pools of contaminated standing water, conveying enteric diseases and providing breeding grounds for mosquitoes, spreading filariasis, malaria and other diseases. Garbage attracting disease vectors	Provision of storm and surface-water drains on house plot and spaces for storing garbage that are rat-, cat-, dog- and childproof	Design and provision of storm and surface-water drains. Advice to households on materials and construction techniques to make houses less damp. Consider feasibility of community-level garbage recycling/ reclamation	Regular removal or provision for safe disposal of household wastes (including support for community schemes) and plan framework and resources for improving drainage	

TABLE 27.1 CONTINUED

Health risk	Action at individual and household level	Public action at neighbourhood or community level	Action at district or city level	Action at level
Insufficient water for domestic hygiene – diarrhoea diseases, eye infections (including trachoma), skin diseases, scabies, lice, fleas	Adequate water supply for washing and bathing. Provision for doing laundry at household or community level	Health and personal hygiene education for children and adults. Facilities for laundry at this level, if not within individual houses	Support for health education and public facilities for laundry	Technical and financial support for educational campaigns. Co-ordination of housing, health and education ministries
Disease vectors or parasites in house structure with access to occupants, food or water, e.g. rats, cockroaches, mosquitoes or other insects (including Chaga's disease vector)	Support for improved house structure, e.g. tiled floors, protected food storage areas, roofs, walls and floors protected from disease vectors	Technical advice and information – part of adult and child education programme	Loans for households to upgrade shelters. Guarantee supply of cheap and easily available building materials, fixtures and fittings	Ensure building codes and official procedures to approve house construction or improvement are not inhibiting individual, household and local government actions.
Inadequate-sized house and poor ventilation – helps transmission of diseases such as tuberculosis, influenza and meningitis (aerosol drops), especially when many households share premises. Risks of household accidents increased with overcrowding; impossible to safeguard children from poisons, open fires and stoves	Technical and financial support for house improvement or extension and provision of cheap sites with basic services in different parts of city to offer poorer groups alternatives to their current shelters	Technical advice on improving ventilation and lessening indoor fumes and smoke. Education on overcrowding-related diseases and accidents	Loans for upgrading (including small ones with flexible repayment terms); support for building advice centres in each neighbourhood	Support for nationwide availability of building loans, cheap materials (where possible based on local resources) and building-advice centres. Produce technical and educational support material
Children playing in and around house site constantly exposed to hazards from traffic, unsafe sites (e.g. on slopes or with open drains) or sites contaminated with pollutants or faeces	Child-care services to allow care and supervision for children in households where all adults work	Provision within each neighbourhood of well-drained site, separated from traffic, kept clean and free from garbage and easily supervised for children's play. Ensure first-aid services are to hand	Support given to neighbourhood-level play, sport and recreation facilities	Support for city/local governments with information and advice on recreation and play provision for child development

Indoor air pollution because of open fires or poorly designed stoves – exacerbates respiratory illness, especially in women and children	Posters/booklets on improved stove design and improving ventilation	Ensure availability of designs and materials to build improved designs	Consider extent to which promotion of alternative fuels would lessen problem	
House sites subject to landslides or floods as a result of no other land being affordable to lower-income groups	Regularise each household's tenure if danger can be lessened; relocation through offer of alternative sites as last resort	Action to reduce risks of floods/landslides or to reduce potential impact; community-based contingency plan for emergency. Encourage upgrading or offer alternative sites	Ensure availability of safe housing sites that lower-income groups can afford in locations accessible to work	National legislation, and financial and technical support for interventions by local and city governments in land markets to support action at lower level. Training institutions to provide needed personnel at each level
Illegal occupation of house site or illegal subdivision with disincentive to upgrade, lack of services and mental stress from fear of eviction	Regularisation of each household's tenure and provision for piped water, sanitation and storm and surface drains	Local government working with community to provide basic infrastructure and services and incorporation into 'official city'	Support for incorporating illegal subdivisions and for providing tenure to squatter households	
Nutritional deficiencies and low income	Reduce intestinal worm burden and worm transmission. Support for income-generating work within the house	Food supplements/meals or community kitchens. If land is available, support food production	Support for local enterprises and appropriate nutrition programmes	Structural reforms, funds for nutrition programmes and other measures to improve poorer groups' real incomes
No or inadequate access to curative/preventive health care and advice	Widespread availability of simple primer on first aid and health in the home plus home visits to promote its use	Primary health-care centre; emphasis on child and maternal health, preventive health and support for community action and for community volunteers	Small hospital (first referral level) and resources and training to support lower-level services and volunteers	Technical and financial support for nationwide system of hospitals and health-care centres. Preventive health campaigns (e.g. immunisation) and nationwide availability

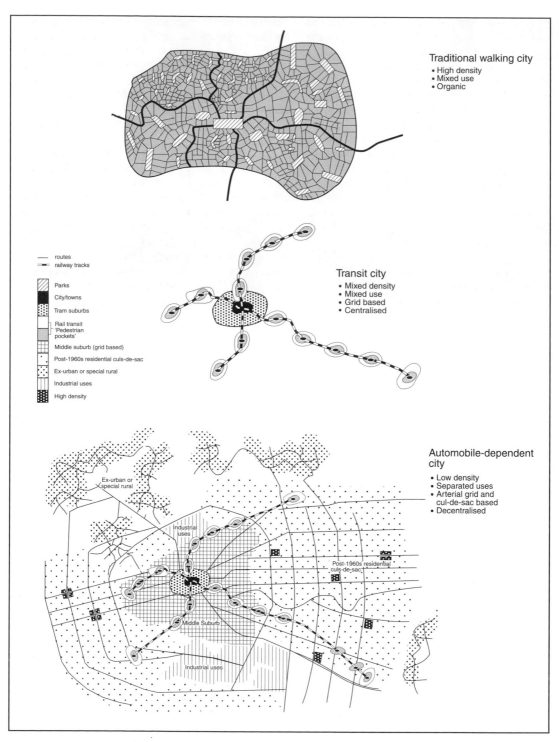

Figure 28.1   The relationship between urban form and dominant mode of transport

Source: United Nations (1996) *An Urbanising World* Oxford: Oxford University Press

rapid urban expansion driven by in-migration leads to many new arrivals being forced to live at increasing distances from the job opportunities of the central city.[4] The significance of informal economic activities also produces a pattern of travel demand with spatial and temporal characteristics different from those generated by the more formally organised economic activity of the Western city. The polarised distribution of income within the Third World city also affects levels of mobility and patterns of travel. The diverse demand for urban transport cannot be satisfied by the kind of high-capacity radial transport corridors that serve Western cities.

The relationship between land use and transport is encapsulated in the distinction between the 'walking', 'tracked' and 'rubber' cities[5] (Figure 28.1):

1. The traditional walking city generally has 100–200 persons per hectare (250–500 per acre) and is a compact 'pre-industrial' urban form in which activities are in close proximity and easily accessible on foot or with simple hand- or animal-drawn carts. Streets are often narrow and congested, and, apart from a small administrative core, functions are not segregated spatially. Walking is the primary form of transport, and may continue to be so in the core even after the city has evolved beyond that form.
2. The tracked or transit city, with 70–100 persons per hectare (170–250 per acre), is exemplified by the nineteenth-century growth pattern in European and North American cities, where railways and trams extended the range and capacity of urban transport, removing the need for residence and workplace to be in proximity. As towns expanded, suburbanisation and commuting to work became the norm. Many cities, such as London and Chicago, developed along radial routes with residential property development around stations. Functional zonation of land uses also became more distinct.
3. The rubber-tyred motor vehicle heralded the advent of the automobile-dependent city, with densities of ten to twenty persons per hectare (twenty-five to fifty per acre). The automobile freed people from the tracks, widened the choice of residential location and led to a decline in the attraction of traditional urban cores. As Figure 28.1 shows, many cities comprise a small walking city centre and a distinctive transit city with higher density and very different transport patterns, surrounded by an automobile-dependent city with uniformly low-density suburbs and high car use.

The transport problems of Third World cities derive in large measure from their rapid transition from the 'walking' to the 'rubber' city form often without the development of the tracked transport associated with the expansion of cities in advanced economies. In most Third World cities, expansion has been based on road transport, but in a situation where vehicular traffic competes for road space with walkers, hand carts and animal-drawn vehicles, which take up considerable space in relation to their carrying capacity, and consequently determine the overall speed of movement over large areas of the city. The speed of the transition from 'walking cities' to the automobile era has meant that, in contrast to cities of the industrialised world, few Third World cities had an opportunity to develop an urban public transport system. The compact urban form of many Third World cities would appear to be well suited to public transport. However, increasing urban sprawl reduces the advantage of public transport and limits access, especially for those who occupy peri-urban settlements beyond the range of existing systems.

PUBLIC TRANSPORT

Availability of personal transport is a function of income. In the poorest areas of Kuala Lumpur almost 40 per cent of journeys are made on foot.[6] For those without access to personal transport, a variety of forms of public transport and informal-sector provision offer alternatives to walking.

BUSES

Given that few cities in the Third World have developed high-capacity urban rail systems, the bus is often an essential element in the public transport system. In many cities, however, the provision of public transport by the formal sector has not kept pace with population growth, with the deficit between supply and demand widening as the city expands.

A number of factors determine the operational effectiveness of urban bus fleets in the Third World. Labour costs are low, but fuel costs usually represent a high proportion of total costs, with oil often having to be imported. The intensity of operation, high loadings, congested road space and environmental factors, such as temperatures, rainfall, dust and road damage, all increase operational and maintenance costs in situations where foreign exchange may be in short supply and lead times for obtaining essential spares and equipment are often measured in months or years rather than days or weeks. Many vehicles in the fleet may be out of operation for long periods of time.

Revenue is limited by the low-income market served and often by government-imposed ceilings on fares introduced for socio-political reasons.[7] There may also be significant leakage in revenue collection systems through staff fraud, fare evasion and physical problems of fare collection, as illustrated by the twelve persons/square metre (ten per square yard) recorded on Bangkok buses.[8] Since the revenues of most publicly owned bus, rail or metro systems do not cover costs, there are no reserves available to improve maintenance or the quality of service, or to invest in route expansion.

The spatial structure of the built-up area also presents problems for the provision of a comprehensive public transport system. In most Third World cities there are districts into which buses cannot go. These include historic city centres where thoroughfares are too narrow, and informal settlements developed on hazardous sites at some distance from a motorable road. The unplanned manner of urban expansion also makes it difficult to organise a cost-effective public transport service. In contrast to the deficiencies of many traditional bus services, as discussed below, a modern bus rapid transit system can play an important role in a mass transportation strategy for many Third World cities, as well as providing a possible model for some congested cities in the West (see Chapter 13).

INTERMEDIATE PUBLIC TRANSPORT

In many cities the deficiencies of formal public transport services are overcome by the provision of informal private-sector services, variously referred to as **paratransit** or intermediate public transport (IPT). By the end of the 1980s the informal sector had a 40–80 per cent share of public transport in most capital cities of sub-Saharan Africa. In many cities of South and South East Asia, where a formal municipal bus system is largely absent, IPT provides the bulk of public transport[9] (Table 28.2).

There are a great variety of vehicle types in IPT, including both non-motorised and motorised forms. The simplest motorised from of paratransit is the small-engined tricycle such as the *samlor* of Thailand, the auto-rickshaw common in many Indian cities, and the *tuk-tuk* of Indonesia. Vehicles such as the *dolmus* of Turkey are essentially saloon cars or vans with little modification, while the *silors* of Thailand are pick-up

trucks converted for passenger use with bench seats, a metal or canvas roof and a tailgate replaced with a passenger boarding step.[10] At the top end of the range are imported mini-buses that include the *tro-tro* of Ghana and *gbakas* of Côte d'Ivoire. These have also become the main forms of IPT (referred to as the Public Light Bus) in Hong Kong.

The small size of IPT vehicles gives them a number of advantages over commercial public transport. Vehicles load and unload more speedily, with reduced stopping time and passenger waiting, and can better negotiate narrow congested streets. Paratransit provides a frequent and flexible service. Although some services operate fixed routes, deviation is often normal, with precise routes determined by passenger demand. The small size of vehicle and the possibility of starting with a second-hand one reduce the initial capital investment and place the enterprise within the reach of local private initiative. Typically, the majority of vehicles are owner-driven or 'hired' from family members, although there are instances where multiple vehicle ownership and leasing have developed, or where individual owners group into co-operatives.

The owner-operator and small firm characterise the IPT sector, in which competition can be fierce, administrative overheads are minimised, and there is a flexible and rapid response to shifts in demand. Services operate only if profitable and are reduced at times of low demand. Drivers may have to work long hours to make a living, but the small-vehicle intimacy of driver and passenger reduces fare avoidance and fraud. On the other hand, the IPT sector has been criticised for severe overloading of vehicles, dangerous driving habits (such as poor lane discipline and double parking) and lack of proper maintenance, which contribute to a poor safety record. In addition, although units are individually small, the number of IPT vehicles in a city is often a major contributory factor to urban road congestion.

In many Third World cities there has been a demand for greater government regulation of IPT operations on the grounds that they represent a cause of rather than a solution to urban transport problems. In Hong Kong the IPT services were a technically illegal though essential element in urban transport provision until 1969 when regulatory measures were introduced to protect passengers, ease congestion and provide better co-ordination between different forms of public transport.[11] In Harare 'pirate' taxis were legalised in 1980; now vehicles have to be registered, routes are marked on vehicles, insurance, roadworthiness and vehicle capacities are checked, and, at least in theory,

TABLE 28.2 MODAL SHARES OF URBAN TRANSPORT TRIPS IN ASIAN MEGA-URBAN REGIONS (PERCENT)

City Region	Transport Mode					
	Walking	Non-motorized Vehicles	Paratransit	Public Transit	Motorcycles	Private Cars
Tokyo	8	0	0	53	17	22
Bangkok	1	5	5	40	17	32
Jakarta	23	2	3	25	13	34
Manila	12	3	39	13	3	30
Beijing	12	48	6	20	2	12
Kolkata	15	9	40	6	10	20
Delhi	20	12	53	8	0	7
Dhaka	40	20	8	20	4	8
Mumbai	15	3	28	9	20	25

Note: Non-motorized vehicles include bicycles, rickshaws, and pedal-powered *betjaks.* Paratransit includes motorized vehices such as 'baby taxis', tempos, jeepneys, auto-rickshaws, motorized tricycles, *helijaks, tuktuks, samlors,* and various kinds of two- three- or four-wheeled vehicles. Public transit includes buses, trams, heavy rail transit, light rail transit, subways, commuter rail, and bus rapid transit systems. Motorcycles include motorized two-wheeler private vehicles. Private cars include taxis, rental cars and limousines.

Source: Adapted from A. Laquian (2005) *Beyond Metropolis: The Planning and Governance of Asian's Mega-Urban Regions* Baltimore: Johns Hopkins Press, p243.

Plate 28.2   The *tuk-tuk*, essentially a motor scooter fitted with a rear seat and covered by a plastic-coated canopy, carries up to three passengers through the traffic chaos of Bangkok

the number per route is controlled. The difficulty of regulating the number of IPT vehicles is evident in Manila, where an official total of 28,000 jeepneys is thought to represent less than half the actual number operating in the city. Even when there is a measure of government regulation, much of the IPT operates on the margins or outside the regulatory codes, which are often only weakly enforced.

Some argue that the informality of the IPT system and its ability to operate outside established regulatory frameworks represent a major advantage in providing transport services to the mobility-disadvantaged residents of Third World cities. Supporters of paratransit vehicles believe that, in the context of the Third World city, IPT has characteristics that should be encouraged rather than controlled. It employs an appropriate technology, utilises an abundant labour supply and minimises the use of scarce resources related to capital and managerial skills. Fundamentally, IPT appears to provide users with services appropriate to their demands. Rimmer (1984) has pointed to the apparent paradox that while IPT-type services were being encouraged as a partial solution to urban transport problems in some advanced economies (in the form of taxis, community buses, dial-a-ride schemes and car pools) they were being actively discouraged in several Third World countries.[12]

A third perspective sees conventional and IPT services as complementary parts of an integrated public transport system in which IPT may provide specialist line-haul services (e.g. contract services for workers, additional peak-hour capacity, and services to supplement the decline in off-peak provision by conventional transit), as well as extension of conventional services (e.g. short-distance services in low-density residential areas, and feeders to high-density routes with recognised interchange facilities to link the associated routes). Clearly, this level of co-ordination between formal and informal urban transport services requires effective government intervention to maximise the benefits and minimise the costs associated with each.

NON-MOTORISED TRANSPORT

Walking is the most basic form of non-motorised transport used by the urban poor for a large proportion of journeys, but there is also a wide range of human-powered vehicles for passenger and freight that operate as public transport. This includes bicycles, rickshaws, hand carts, cycle-based rickshaws and cycles with sidecars. In addition there is also widespread use of animals for direct carriage of goods and people, and for drawing vehicles. Non-motorised forms of transport are a conspicuous feature of Third World urban areas, with, for example, an estimated 30,000 rickshaw pullers in Calcutta and 10,000 rickshaws in Phnom Penh.[13]

The bicycle is a common mode of personal transport in many Third World cities (Box 28.3), but the pedal cycle can also be the basis of public transport of people or goods, with many local variants of the bicycle with sidecar and the tricycle.[14] The *cyclos* of

BOX 28.3

The bicycle in Third World cities

Bicycles are the chief form of personal transport in China, with over 5 million in Beijing (with a population of 12.4 million in 1995), where 30 per cent of work journeys are made by bicycle. This mode of travel was influenced by a pre-1980 prohibition on private car ownership, a workers' travel allowance that could be accumulated to purchase a bicycle, the relatively flat urban terrain and the inadequacy of public transport in the face of rising demand and an urban road system that cannot easily accommodate large vehicles.

There are also high levels of bicycle ownership in many Indian cities. There are more than 1 million cycles in Delhi, with the number growing at 4 per cent per year. In cities such as Pune, Kanpur, Lucknow and Delhi bicycles account for 35–50 per cent of the traffic on main corridors, with one junction in Delhi having a flow of over 7,500 bicycles per hour.

By contrast, although income levels in many African urban areas are comparable with those of many Indian cities, bicycle ownership and use are generally less significant, with only 6–15 per cent of households having access to one. This may be explained in part by the small population and small market in most African countries, which preclude a domestic bicycle manufacturing industry, but cultural factors may also be at work suggesting that, in contrast to India, where bicycle riding appears to have status value, in Africa it is considered to be demeaning, and possibly more dangerous than other modes.

Source: **D. Hilling (1996)** *Transport and Developing Countries* **London: Routledge**

BOX 28.4

Cycle rickshaws and traffic congestion in Dhaka, Bangladesh

The teeming streets of Dhaka are being brought to a standstill by a boom in the number of cycle rickshaws, which offer the only means of travel through the business district that is marginally faster than walking. Seventy per cent of vehicles in the Bangladeshi capital are tricycle rickshaws, with their motorised cousins – 'baby taxis' in local parlance – claiming another 15 per cent. Cars inch through this *mêlée* at the mercy of rickshaw wallahs, whose aim is erratic. Buses and lorries move hardly at all.

The government is trying to impose controls, but nothing in Bangladesh works quite as it should. About 90,000 cycle rickshaws have licences, fitted to the backs of the vehicles like number plates; the other 300,000 or so use fake plates. Rickshaw pullers earn no more than £2 a day, but it is enough to attract the rural poor into town. By their mid-thirties most are too worn-out to work. The one blessing is that Dhaka is as flat as a table.

Rickshaws provide jobs directly and indirectly to about 1.5 million people in the capital, whose population is probably about 8.5 million – nobody knows any more. The side streets are full of cycle mechanics' shops, providing work for an army of young people.

Rickshaw artists, working in fetid slum workshops or from their own huts part-time, decorate the vehicles with paintings of gods, sporting heroes and film stars.

Rickshaws arrived in Dhaka sixty years ago from Calcutta, and both cities have since tried to get rid of them as symbols of backwardness and poverty. Calcutta retreated from an attempted clampdown in the face of a pullers' protest. Dhaka declared in 1986 that it would issue no more licences, in the hope of gradually forcing the vehicles off the road, but it never happened. Two-stroke motorised rickshaws were introduced around that time – a disastrous mistake, because they immediately covered the city in a cloud of smoke which has never left. Two years ago the government imposed limits on the number of baby taxis in the hope of curbing pollution, but the air remains eye-wateringly foul.

Opposition parties allege that the ruling Awami League is involved in supplying fake cycle rickshaw licences for a comparatively hefty £4. Such practices are condemned by the city's first elected mayor, but there has been no concerted effort to stop it. Police are generally happy to extort a few takas and send a suspect on his way.

Source: The Times 26 March 1998

Phnom Penh and the *becaks* of Indonesia have the passenger seat in front of the pedaller, while on the *trishaws* in Indian towns it is usually behind. Among the positive features of bicycle-based modes of transport are that over short distances in the rush hour, bicycles often provide the quickest form of transport and require relatively little parking space. Their use conserves energy resources and does not add to air pollution. Repairs are easily effected and the bicycle is suitable for a wide age range. In general terms of environmental impact, bicycles rate highly. On the other hand, bicycles and other non-motorised forms of transport are major contributors to urban road congestion (Box 28.4).

RAIL-BASED TRANSIT SYSTEMS

In contrast to urban growth in industrialised countries, the growth of few Third World cities was based on tracked transport. Nevertheless, mounting road-traffic problems in large Third World cities have encouraged transport planners to consider rail-based transit systems. These include conventional tram or streetcar systems, light rail units running along existing streets, rapid-rail units using a segregated track, and suburban rail or metro-type units, possibly sharing track with inter-city trains.

The capital and operating costs of different rail systems vary and will be influenced by local conditions (such as the existing urban fabric, terrain and route network) and the efficiency with which the system is managed. In general, however, because of their high costs, rail transits can provide only restricted spatial coverage of the city, and most have been provided to serve corridors of high demand. The Bangkok Mass Transit System (Skytrain) opened in 1999 at a cost of $1.2 billion, consists of two lines and covers a distance of 23 km. Common problems for developers of rail transit systems are difficulties of land acquisition, disputes over routes, changes in plans (in some cases related to frequent changes of government) and shortages of

Plate 28.3   Pedal power. Bicycles are the chief form of personal transport in China, with over 5 million in Beijing, where 30 per cent of work journeys are made by bicycle. Fortunately for this cyclist, the urban terrain is relatively flat

materials and finance. Calcutta's ten-mile (16 km) metro line opened in 1995 at a cost of £300 million and some twelve years behind schedule. Once railways are in operation, use has often been below forecast levels, and rail systems have found it difficult to compete successfully with buses. Although many metro systems are heavily subsidised, generally fares are still beyond the pockets of the poorest sections of society. The impact on traffic congestion has usually been minimal, and rail transits have made only limited improvements in the overall quality of urban transport.

According to Allport (1991),[15] the conditions necessary for a successful rail transit system include the following:

1. They should be confined to arterial corridors. Peak bus flows should approach 15,000 persons an hour before a metro is even considered.
2. Such corridors are unlikely to exist in cities of fewer than 5 million inhabitants, or where movement patterns are dispersed and not highly linear.
3. Cities with rail transit systems should have *per capita* incomes in excess of $1,800.
4. Large-scale immigration tends to reduce average income levels and is detrimental to a metro's success.

In view of the socio-spatial characteristics of most Third World cities, a rail-based transport strategy should, of necessity, be approached with caution. Evidence suggests that, in general, purpose-built bus rapid transit systems (with higher carrying capacity, lower fares, and wider geographical coverage), may be more appropriate than a rail-based system of mass public transport in Third World cities (Table 28.3). Bus rapid transit can offer high-speed, reliable and cheap transport along the major routes within cities. By using new efficient, comfortable, high capacity vehicles pollution levels are reduced at source and, when combined with flexible ticketing and real time route information, can increase ridership.[16]

TABLE 28.3 EXAMPLES OF MASS TRANSIT SYSTEMS IN THIRD WORLD CITIES

City	No. of Mass Transit Corridors	Passengers carried (million/year)	
		Total	Per Route Km
Metros			
Buenos Aires	6	242	5.1
Bangkok	2	90	3.8
Mexico DF	11	1433	7.1
Singapore	2	296	3.6
Light Rail Transit			
Kuala Lumpur	2	61	1.1
Manila	2	109	3.5
Medellin	3	105	3.2
Bus Rapid Transit			
Bogota	3	184	4.7
Curitiba	5	684	1.3
Sao Paulo	4	273	4.4

Source: Adapted from D. Banister (2005) Unsustainable Transport London: Routledge

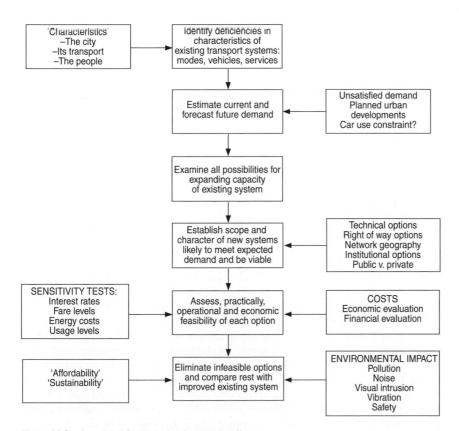

Figure 28.2   A protocol for assessing transport options
Source: D. Hilling (1996) *Transport and Developing Countries* London: Routledge

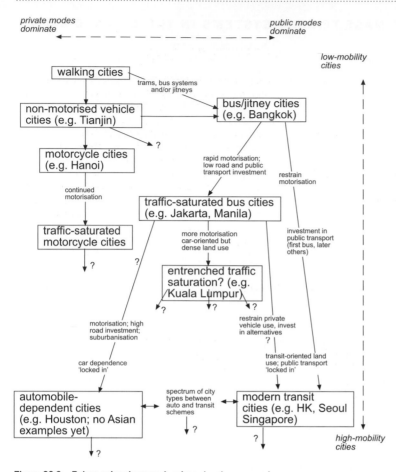

Figure 28.3 Future urban transport–urban structure scenarios

Source: Adapted from P. Barter (2004) Transport, urban structure and lock-in in the Kuala Lumpur metropolitan area *International Development Planning Review* 26(1), 18

URBAN TRANSPORT STRATEGIES IN THE THIRD WORLD

Capital-intensive high-technology transport strategies have provided few benefits to the mass of the people in Third World cities.[17] An alternative approach underlines the unique features of each urban setting and focuses on the relationship between the characteristics of the urban area, appropriate transport technology, and the social, economic and demographic structure of the population providing the demand.[18]

These principles are encapsulated in the protocol shown in Figure 28.2, which first establishes the characteristics of the city, its people and its transport, then identifies deficiencies in the existing transport systems. Once future demand for transport has been estimated, the current systems are examined for all possible ways of expanding capacity (e.g. changes in management, maintenance or operational procedures) before attention is given to new capacity options, each of which must be scrutinised against financial, technical and environmental factors. Since in most Third World cities scarcity of public finances is likely to be the major determinant of transport policy, there is little point in developing options that are unaffordable by either society or individual users. In view of this it would also be unwise to discount the likely continuing contribution of walking, non-motorised and motorised IPT to mobility and accessibility in the Third World city. Figure 28.3, developed with particular reference to Asian cities, indicates a range of future urban transport-urban structure scenarios that may apply more generally as levels of vehicle ownership rise in Third world cities.

FURTHER READING

BOOKS

A. Armstrong-Wright (1993) *Public Transport in Third World Cities* London: HMSO

M. Heraty (1991) *Developing World Transport* London: Grosvenor Press

D. Hilling (1996) *Transport and Developing Countries* London: Routledge

United Nations Centre for Human Settlements (1994) *Urban Public Transport in Developing Countries* Nairobi: UNCHS

E. Vasconcellos (2001) *Urban Transport, Environment and Equity: The Case for Developing Countries* London: Earthscan

JOURNAL ARTICLES

G. Adumo and R. Abadho (1992) Urban transport systems: a case of the *matuta* mode of transport in the city of Nairobi, Kenya *African Urban Quarterly* 7(1/2), 120–9

C. Bae and Y. Suthivanart (2003) Policy options towards a sustainable urban transportation system for Bangkok *International Development Planning Review* 25(1), 31–51

R. Cervero (1991) Paratransit in South East Asia *Review of Urban and Regional Development Studies* 3, 3–27

J. Pucher, N. Korattyswaropam, N. Mittal and N. Ittyerah (2005) Urban transport crisis in India *Transport Policy* 12, 185 98

J. Zacharias (2003) The search for sustainable transport in a developing city: the case of Tianjin *International development Planning Review* 25(3), 283–99

KEY CONCEPTS

- Walking cities
- Tracked cities
- Rubber cities
- Intermediate public transport

- Paratransit
- Non-motorised transport
- Rail-based transit
- Traffic congestion

STUDY QUESTIONS

1. Consider the relationship between transport and urban form in the Third World city.
2. With the aid of specific examples, assess the role of paratransit in Third World cities.
3. With reference to a particular Third World city, examine the problems of and prospects for urban traffic and transportation.

PROJECT

Write a consultant's report providing an overview of traffic problems in Third World cities. (Appropriate information sources in academic and professional publications will be found by searching relevant bibliographic databases available in your institutional library.) Your report should: (1) identify the nature and scale of urban traffic problems; (2) critically examine current traffic and transport strategies employed by urban authorities; and (3) offer a set of 'best practice' recommendations on how to address traffic and transport problems in Third World cities.

29

Poverty, Power and Politics in the Third World City

Preview: the social relations of power; partiality, inequality and the soft state; the bases of social power; clientelism; urban social movements; the role of women; community-based organisations; christian base communities; urban NGOs; community participation in urban governance; globalisation and social justice

INTRODUCTION

As the pace of urbanisation and urban growth has increased, the capacity of most Third World governments to manage the consequences of these trends has decreased. The social, economic and environmental impacts of this failure fall most heavily upon the poor, who are generally excluded from the benefits of capitalist urban development. As we have seen, the majority of the Third World urban population has a standard of living so low as to be inconceivable to the average citizen of an advanced country. More than 3 billion people, or two-thirds of the world's population, have incomes that are less than 10 per cent of the *per capita* income in the USA, and by 2025 this figure is expected to encompass over 70 per cent of the global population. In this chapter we consider the relationship between poverty, power and politics in the Third World city and examine the ways in which the urban masses of the Third World cope with their disadvantaged position. We begin by exploring the nature and distribution of power in the Third World city. We then examine particular mechanisms through which poor households seek to make gains within the urban power structure, ranging from the cultivation of individual patron–client relationships to collective urban social movements. We assess the role of non-governmental organisations (NGOs) in countering disadvantage and advancing the concept of community participation in Third World urban development. Finally, we consider the links between globalisation and social justice.

THE SOCIAL RELATIONS OF POWER

The principal relations of power within a society are indicated in Figure 29.1. This identifies four overlapping domains: the state, civil society, the corporate economy and the political economy. Each domain has an autonomous core of institutions that governs its respective sphere. The core of the state consists of the executive and judicial institutions; the core of civil society is the household; the core of the corporate economy is the corporation; and the core of the political economy is independent political organisations and social movements. For each domain a distinctive form of power can be identified: state power, social power, economic power and political power. This reflects the resources available to actors in the domain.

In practice, the integrity of each domain is not complete, and boundaries between domains may merge. The state, for example, has many centres of power, and they do not always act in a co-ordinated manner. Equally, civil society is divided along lines of social class, caste, ethnicity, race, religion and gender. The corporate economy consists of actors who are in competition and who combine only to further their collective interests, while the political economy is the quintessential terrain of conflict among different groups and factions involving the three other domains. Moreover, each domain exists at several different levels of territorial organisation – from households, neighbourhood and city, through district, region and nation, to supra-national groupings of states. This schema represents a useful framework for understanding power

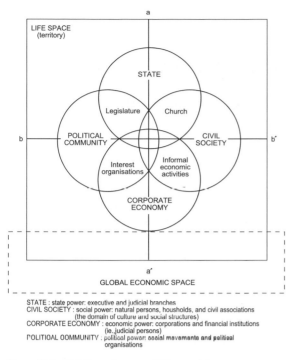

STATE : state power: executive and judicial branches
CIVIL SOCIETY : social power: natural persons, housholds, and civil associations
(the domain of culture and social structures)
CORPORATE ECONOMY : economic power: corporations and financial institutions
(ie.,judicial persons)
POLITICAL COMMUNITY : political power: social movements and political
organisations

Figure 29.1   Relations of power within a society
Source: J. Friedmann (1992) *Empowerment* Oxford: Blackwell

conflicts among different communities. As Friedmann (1992 p. 28) explains:

> it maps a socio-political space that allows us to locate specific institutions such as legislatures (between state and political community) or the Catholic Church in Latin America and Mediterranean Europe (between state and civil society) and permits us to gauge the relative importance of each domain in given national societies.[1]

For example, in many African countries a one-party state is firmly entrenched and a political community exists in only rudimentary form, whereas in several Latin American countries civil society and political community have generated urban social movements with the ability to exert pressure on the state and capital in pursuit of particular social and economic goals.

Over time, in capitalist societies, power has accumulated along the axis linking the state with the corporate economy, largely at the expense of power along the axis connecting civil society and the political community. This bifurcation of power is especially evident in the polarised socio-economic structure of Third World cities, where exclusion from political and economic

power is a fact of life for many urban dwellers. As Gilbert and Gugler (1992 p. 177) observe:

> government leaders are less than responsive to the needs of the masses. They are usually preoccupied with securing the support of the armed forces and promoting the investment of indigenous and foreign capital. They are constrained by multinational corporations and foreign governments which provide investments, buy exports, supply raw materials and spare parts, and wield the ultimate threat of subversion and military intervention. In this concept of the powerful, the voices of the mass of the population usually remain muted.[2]

The disadvantaged position of the Third World urban poor is accentuated by the practice of partiality inherent in the management of many Third World cities.

PARTIALITY, INEQUALITY AND THE SOFT STATE

In all countries, public officials have the power to allocate a variety of resources (ranging from infrastructure and services to public-sector employment) and enforce a range of sanctions (such as building permits and trade licences). Many Third World governments exercise weak control over the activities of public administration. In these **soft states**, laws and regulations are more prone to be flouted than in most Western countries. There is widespread disobedience by public officials at various levels of rules and directives handed down to them, and often they collude with powerful persons and groups whose conduct they should regulate. The corruption inherent in soft states reinforces patterns of inequality, since the mass of the population have little leverage in dealing with the elite and its representatives who control resources, whether it be managers in private firms or government officials. In many Third World countries, corruption is an accepted part of everyday life, manifested in nepotism in obtaining career positions, commission for providing government contracts, premiums to cover breaches of planning regulations, or additional administrative payments to speed up the bureaucratic process, or to obtain services.

In Jakarta a prevailing 'culture of corruption' has allowed private-sector developers to circumvent the land-use provisions of the city's master plan. Development permits for non-conforming land uses may be obtained for a price, while the lengthy official process of obtaining building permits can be accelerated by payment of fees to the various agents involved at each stage. For large projects, developers often employ middlemen to arrange all the necessary documentation. The involvement of these 'permit brokers' depends on their contacts with amenable government officials. Corruption is also rife in the collection of municipal taxes and user fees. Land-owners find it advantageous to pay a fee to have their property undervalued for tax purposes. Large consumers of water can have a lower use recorded by meter readers. In 1992 the city sanitation agency collected only 8 per cent of fees due for the garbage disposal service, while in 1990 city revenues from billboard advertising were only 50 per cent of the expected amount. A 'mafia' within the city parking agency enabled unofficial parking attendants to collect fees without issuing receipts, with the proceeds split between the two parties. This leakage of funds undermines the city's goal of cost recovery for its services and also prevents the extension of infrastructure and services into unserved areas for the benefit of the urban poor. The poor are also affected directly by corruption in urban administration, with informal-sector street traders obliged to pay unofficial levies and fees for security and other services which are collected by unauthorised public servants, including the forces of law and order.[3]

A principal factor underlying the widespread practice of corruption in many Third World cities is the generally low level of public-sector salaries, which provokes employees to augment their official incomes with illegal salary supplements. Throughout the Third World there is a tendency to view such action as an acceptable and inevitable part of urban life. However, the mass of the urban poor who exist outwith waged employment are unable to participate in this particular form of income generation. The in-built bias against the poor in Third World urban political economies raises the question of how they respond to their disadvantaged position. We address this by considering the bases of social power, then examining particular strategies for empowerment of the urban poor.

THE BASES OF SOCIAL POWER

Each form of power indicated in Figure 29.1 is based on certain resources that may be accessed. The state enjoys the support of the law, corporations have access to finance and the ability to move capital from one place to another, and the political community has the power to vote, demonstrate and lobby politicians. The power of civil society is gauged by the differential access of households to the bases of *social power*. These represent the principal means available to a household economy in the production of its life and livelihood.[4] As Figure 29.2 indicates, there are eight main bases of social power:

1. *Defensible life space*. The territorial base of the household economy includes the physical space of the home, in which household members eat, sleep and secure personal possessions, and the wider neighbourhood space, where socialising and other life-supporting activities take place, primarily in the context of the moral economy of non-market relations.

2. *Surplus time*. Time available to the household economy beyond that necessary for gaining a subsistence livelihood is a function of a range of factors, including time spent on the journey to work; ease of obtaining basic consumption items such as food, water and fuel; frequency of illness and access to medical services; time required for performance of essential domestic activities; and the gender division of labour.

3. *Knowledge and skills*. Education and technical training for at least some members of a household can enhance its long-term economic prospects.

4. *Appropriate information*. Access to relevant information on opportunities for wage-paying work, available public services, proven methods of health care, improved methods of housing production and changing political configurations can have a direct bearing on household survival.

5. *Social organisation*. This refers to formal and informal organisations to which household members may belong, including churches, mothers' clubs, credit circles, discussion groups, tenant associations and neighbourhood improvement associations. Organisations provide not only a means for a more convivial life, but also a source of relevant information, mutual support and collective action.

6. *Social networks*. These are essential for self-reliant action based on reciprocity. Households may have extensive horizontal networks among family, friends and neighbours that give them space to manoeuvre in times of economic crisis. Vertical networks up through the social hierarchy give households a chance to access other forms of power, but may lead to dependent patron–client relationships.

7. *Instruments of work and production*. The tools of household production include the physical strength produced by good health as well as tools used in the household's informal work (e.g. a bicycle or sewing machine) and in the domestic sphere (such as a stove, kitchen implements or toilet facilities).

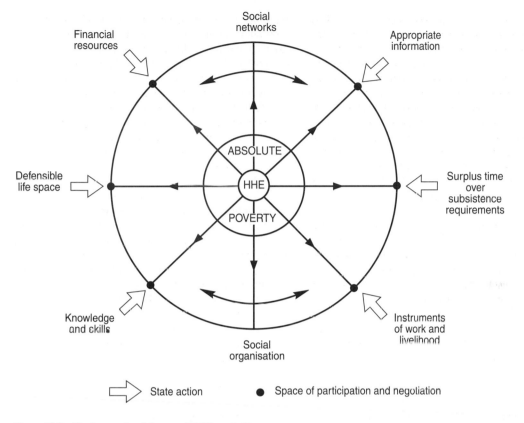

Figure 29.2   The bases of social power. *HHE* Household economy
Source: J. Friedmann (1992) *Empowerment* Oxford: Blackwell

8. *Financial resources*. These include the net monetary income of households as well as formal and informal credit arrangements.

Households may be regarded as having different levels of access to each of the bases of social power, with an absolute level of poverty indicating a position from which households may be unable to escape without external assistance (Figure 29.2). Individual households make their own decisions on how to use their resources to gain greater access to the different bases of social power. Most households, however, initially seek a secure territorial base for their activities, such as minimally adequate housing in the city. Surplus time is often a second priority, and both may be dependent on households' social networks and participation in popular organisations (see Chapter 24).

The struggle of the poor to improve their position in respect to one or more bases of social power may be based on either individual–familistic action or collective action. In Figure 29.2 six spaces of participation and negotiation are shown in which individual households may take action to resolve their problems with agents of the state. Such spaces are not depicted for either social organisation or social networks, which are power bases of civil society from which the state is excluded. By acting in collaboration with others through these avenues, households can improve their chances of gaining greater access to the other bases of power (as indicated by the lateral arrows in Figure 29.2). In the remainder of the discussion we examine specific mechanisms by which poor households seek to overcome their disadvantaged position in the urban power structure.

PATRON–CLIENT RELATIONS

Individuals and households may try to offset their disadvantage by cultivating as patron someone in a position to further their employment prospects or assist them in a crisis. The patron–client relationship is a reciprocal one between two individuals. In exchange for the patron's help the client gives political support and contributes to the patron's status. The exchange is not based on legal or contractual requirements but is an informal understanding. The relationship is also a highly unequal one. Not only does the patron have greater power, more economic resources and higher status, but he usually has numerous clients, and the leverage any one of them can exert on the patron is therefore narrowly circumscribed. The scope for **clientelism** within the Third World city is a function of the lack of impersonal rules for allocating resources. If, for example, job entry and tenure were institutionalised and purely a function of qualification, performance and/or seniority, it could not be a source of patronage. Unlike in Western countries, where most political activity takes place during the input (policy-making) stage, in the Third World often a large part of individual and collective lobbying, representation of interests and the rise and resolution of conflicts takes place during the output (execution) stage of the political process.[5]

In addition to the instrumental aspect of patron–client relationships, there are also usually affective elements. In some cases, kinship, common origin and shared experiences can be the source of genuine affective ties. Sometimes a fictive kinship relationship is established, as in the *compadrazgo* system in Latin America and the Philippines, where the client asks his actual or potential patron to act as godfather at his child's baptism, whereby the two men become *compadres*.[6] Some patrons, such as the owner of a small firm or manager of a municipality, have ultimate control over the resources provided to their clients. In contrast, many patrons are themselves clients for higher-placed patrons. Thus in a large city the poor man's patron frequently operates as a broker, obtaining many of the benefits for his client through a patron of his own to whom he, in return, provides specific resources, such as votes.

Clientelism, both by individuals and organised on behalf of neighbourhoods, has a long tradition in many Third World societies.[7] The logic of clientelist politics is evident in the *favela* community of Vila Brasil in Rio de Janeiro. The area has long been involved in the process of marketing its votes in return for promises of goods and services at election time, but prior to 1980 this had provided only limited benefits for the community, which remained a typically neglected squatter settlement. By the mid-1980s, however, each of the 500 dwellings in the *favela* was served by piped water and a sewage system, all streets were cleaned and well lit, and a community-association building and covered recreational area had been provided. This transformation was due almost entirely to the ability of the new president of the neighbourhood association to act as a 'power broker' and manipulate clientelist arrangements to the advantage of the community. Of particular importance was the president's relationship with a local politician who:

> was uninterested in political discussion and whose political discourse was the politics of favours. His only concern was to guarantee the necessary number of votes to maintain his position in office, and to this end, he employed local community leaders to work as his campaign officers.
>
> (Gay 1991 p. 106)[8]

In return for promised votes, one week before election day in 1982 a fleet of local-government trucks moved into the *favela* and paved all the community's roads and alleys. The politician was duly elected to office, living conditions in the *favela* improved, and the prestige and power of the neighbourhood-association president were enhanced. Politics in Vila Brasil, and similar communities elsewhere, is a simple matter in which allegiance is given exclusively to those candidates who bring tangible support to the community, irrespective of political ideology. This is reflected in the political philosophy of the president of the *favela* association, who regarded all politicians as corrupt and interested only in exploiting the poor as a cheap source of votes, and who viewed his own role as extracting as much as possible from them whenever the opportunity arose. Given the reality of politics in large parts of the Third World, clientelism clearly represents a rational strategy for many urban dwellers.

Clientelism may be interpreted either as a mechanism that responds to the demands of those in need or as a means of domination that excludes those who do not submit to it and that perpetuates the unequal status quo in society. Some communities have sought to make gains through collective action that eschews the electoral bargaining inherent in clientelism (Box 29.1). This form of struggle to improve the quality of personal and collective consumption within marginalised communities is highlighted by the emergence of urban social movements.

URBAN SOCIAL MOVEMENTS

There are many forms of grass-roots co-operative organisation by the urban poor of the Third World. These range from single-issue interest groups to more broadly based urban social movements. In the *barrios* of Santiago, Chile, women come together in communal

BOX 29.1

Clients and radicals in a Lima shanty town, Peru

Two very different strategic logics are visible among Lima's shanty-town residents. The strategy of radicals is to extract services and concessions by uniting the poor to exert maximum pressure on the state and other potential benefactors. Other residents prefer a clientelistic strategy of cultivating friendly relations with officials and others. These distinct strategies reflect divergent and competing views of the state. Radicalised residents see the state as a representative agent of capitalists which will respond to lower-class interests and needs only when subjected to pressure and threats. Clientalist residents view the state as embracing more universal interests and therefore open to negotiation with representatives of the poor.

These perceptions were translated into differing strategies. Radical activists employed marches, street demonstrations and invasions of public buildings as a means of underlining their demands of the government. Clientelistic leaders were more respectful of the established procedures, partly in the belief that such an approach yields rewards and partly because they viewed themselves as socially similar to the officials with whom they were negotiating.

The goals of political action were shared to the extent that both clientalists and radicals sought to obtain material improvements in their community. Radicals, however, also consider themselves as part of a broader social movement aimed at political transformation and social levelling.

Source: **S. Stokes (1991) Politics and Latin America's urban poor: reflections from a Lima shanty town** *Latin American Research Review* **26(2), 75–101**

kitchens (*ollas comunes*) to prepare one hot meal a day for members of the *olla*, using ingredients bought collectively or obtained from relief organisations. Labour is provided free of charge and meals are sold at cost price. In addition to offsetting a major problem of malnutrition, especially among children, participation in the *olla* also teaches the value of organisation, joint decision-making, leadership and collective effort.

As well as initiatives that seek to lower the cost of household subsistence, co-operative activities are also aimed at providing extra income. The artisanal workshops (*talleres*) in Santiago comprise groups of workers, often female, who manufacture goods for the market. An external agency, such as a Catholic action group, may

provide initial assistance in the form of start-up capital and training. The work (e.g. production of knitted garments) may be carried out together in a member's spare room or individually in each woman's house. Regular meetings of the *taller* allow for collective discussion and distribution of profits.[9]

In many Third World cities such single-issue community initiatives provide a basis for the development of an urban social movement[10] (Box 29.2). The Tierra y Libertad (Land and Liberty) urban social movement in the Mexican city of Monterrey illustrates how the organised poor managed to struggle successfully for living space in a repressive political environment; it

BOX 29.2

The definition of an urban social movement

Urban social movements have as their basic aim the improvement of the quality of individual and collective consumption within marginalised local spaces. For Castells (1983),[11] a grass-roots organisation is an urban social movement only if its aims include a fundamental change in 'urban meaning' or the political economic *status quo*. Urban social movements must mobilise around the goals of collective consumption, community culture and political self-management. According to Castells, urban groups that do not seek to effect structural change in society are merely interest groups rather than urban social movements. This perspective excludes many collective attempts to improve the quality of individual and collective consumption in the city, and does not reflect the daily practice of existing urban territorial organisations in the Third World.

In short, the Castellsian definition of an urban social movement is best seen as an 'ideal form' of such an organisation. A less restrictive definition of an urban social movement, which reflects the reality of grass-roots action within a dominant political economy, is 'a social organisation with a territorially based identity which strives for emancipation by way of collective action'. This does not deny that urban social movements can contribute to structural social reform or that, in many cases, structural reform is the only way to ensure the emancipation of the urban poor, but equally neither does it exclude the host of Third World social organisations engaged in a daily struggle with the state over issues of collective consumption.

Source: **F. Schuurman and T .van Naerssen (1989)** *Urban Social Movements in the Third World* **London: Routledge**

maintained its integrity and political identity in the face of official attempts at co-optation and repression.[12] The *colonia* Tierra y Libertad was formed in 1973 by a land invasion of 1,500 families, and in subsequent years became a centre from which the activities of the squatters (*posesionarios*) were co-ordinated. The solidarity of the social movement was fostered by shared poverty and a common employment base in the informal sector, a general struggle for a place to live against constant threat of eviction, a desire to secure access to transport, education, medical care and other facilities, and, in particular, shared origin in terms of village or region, family ties, *compadrazgo* and friendship.

The squatter movement was organised on a hierarchical basis designed to guarantee the participation of all members in decision-making and make rapid mobilisation possible in case of external threat. It also functioned as an effective means of social control in the *colonia*. At the head was the Asamblea General, the supreme decision-making forum, which comprised representatives from the neighbourhoods. Below the General Assembly was a series of committees with particular tasks such as internal order and defence (the maximum penalty for transgression of the rules being expulsion from the *colonia*), labour affairs (e.g. providing legal advice to workers), economic affairs (including setting up small co-operatives to produce daily necessities, as well as construction materials, transport facilities, a clothing plant and shoe factory), education, and medical affairs. Fundamentally the organisational structure was intended to create maximum possibilities for autonomous mobilisation. Ties with official society were rejected.

Urban services, such as water and electricity, were obtained collectively and illegally. Education and medical care were also provided without official aid, and police were not allowed to enter the *colonia*. The neighbourhoods developed the characteristics of enclaves in which the inhabitants organised their own lives, maintaining judicial exclusivity, addressed questions of collective consumption and promoted a community culture. In addition, through the 'conscientisation' of their members and solidarity action with the inhabitants of other *colonias*, the *posesionarios*, aided by left-wing university activists, sought to develop a base for a wider movement for radical socio-economic change. In short, the Tierra y Libertad urban social movement exhibited many of the characteristics of Castells's ideal type (see Box 29.2).

In 1976 the Tierra y Libertad united with thirty-one other *colonias* (comprising 50,000 *posesionarios*), sixteen tenants' associations, three workers' unions and three *ejido* (rural land) organisations to form the Frente Popular Tierra y Libertad. The objective was to define a common basis of interests among different sections of the urban poor that would permit the co-ordination of actions into a single struggle. In practice, however, after the formation of the Frente Popular, popular participation in decision-making decreased and the movement was riven by intra-leadership power struggles. Official politicians promoted destabilisation of the movement by entering into negotiations with only one of the parties, and by supporting competing left political movements seeking to establish a power base in the *colonias*. Despite its subsequent decline, for much of the 1970s the Tierra y Libertad urban social movement achieved significant social, economic and political gains on behalf of the urban poor and managed to maintain its autonomy in the face of a hostile state.

THE ROLE OF WOMEN IN THIRD WORLD URBAN SOCIAL MOVEMENTS

Gender relations – that is, relations of unequal power based on sexual difference – have shaped the forms of collective action undertaken in the poor neighbourhoods of Third World cities.[13] The number of women involved in the formation and organisation of popular urban social movements in Brazil has increased rapidly since the 1970s, with women constituting, on average, 80 per cent of participants.[14] This is in marked contrast with their position in political parties and trade unions, where they are heavily underrepresented. The high level of female participation in urban social movements has been linked with the sexual division of labour in industrialising countries and, in particular, with women's traditional role as wives and mothers. While few women among the Third World urban poor can afford to devote themselves exclusively to unpaid domestic work, the nature of their participation in the informal sector excludes them from formal political structures and restricts them to their homes and neighbourhoods for large amounts of their time. This socio-spatial confinement to poorly serviced communities heightens the women's sensitivity to the problems of the local area and makes them more willing to participate in collective action to overcome disadvantage. Health, housing and unemployment are typical of the kind of social concerns that have attracted the interest of low-income women seeking to improve their living conditions, and the active participation and influence of women in urban social movements have moulded their organisation and the forms of action taken in pursuit of social justice.

In São Paulo a period of intense industrialisation and population growth during the 1970s placed enormous pressure on the already limited capacity of public services and created the conditions for the growth of social movements in the unserviced squatter communities on the urban periphery. Women played a key role by using personal contacts and relationships within the community to network separate nuclei (such as church groups or mothers' clubs), publicise aims and objectives, and encourage people to participate. Strategies employed included the following:

1. *House meetings.* The principle of collective decision-making involved lengthy informal meetings in members' houses, with the times of meetings changed to accommodate the availability and commitments of the majority of participants. In addition to the practical advantages, house meetings provide a less intimidating environment and promote a non-hierarchical form of organisation in which each person has a chance to express their views. House meetings also strengthened personal relationships between members of the local community and fostered solidarity based on close familiar ties.
2. *Communication and publicity.* Most of those who participate in popular movements are invited to do so by a friend, relative or neighbour. Personal contact and verbal communication are the most effective means of mobilising people in the local community, and women are particularly adept at this form of political organisation.
3. *Spontaneous political action.* When more than a single popular movement exists in a *barrio*, they frequently draw support from each other to reinforce a political action. In São Paulo, when supplies of dried baby milk repeatedly failed to arrive at the local health centre, the *barrio* health movement enlisted members of the local unemployment committee and neighbourhood association to support a protest at the city hall. Other forms of direct political action include the sacking of supermarkets by the hungry poor (Box 29.3).
4. *Community research.* During negotiations following successful protests, social movements often found themselves outmanoeuvred by the bureaucracy, since the government could marshal statistics to support its arguments. In response, the women augmented qualitative information gained through informal discussions of shared experience with quantitative data based on simple questionnaire surveys. For example, regular surveys of

BOX 29.3

Direct action by the hungry poor

The sacking of food stores in Brazilian cities has been a common form of action by the organised poor in which women played a prominent role. In Rio de Janeiro a group of women from the Nova Hollanda *favela* removed US$3,125 worth of basic foodstuffs from a supermarket on 14 September 1983, and more than fifty similar incidents took place elsewhere in the city over the next ten days. Such events were rarely accompanied by violence or the destruction of property. All the sackings were organised and they involved a general code of practice whereby only basic foodstuffs such as beans, rice and oil were taken. In most cases a small group of women would get together and decide to ransack a supermarket or local store, pick the time and place, then inform friends, neighbours and relatives.

The snowball effect generated by the community network is described by a participant: 'I'd never participated in a sacking before until that night. Oh, I'd heard about women in the *barrio* who did, but no one knew who they were, and if they did they wouldn't say anything. Then one Thursday night I was looking out of my window when I saw a group of women filing past my house with empty shopping bags. I just knew there was going to be a sacking, so I grabbed my bag and a few paper sacks and followed them. No one said a word, everyone was very quiet, and as we went along the streets more women joined us. By the time we got to the local supermarket there must have been about fifty of us or more, and that was the first time anyone spoke. One of the women at the front, who I'd seen at the meetings of the unemployed in the vila, said, "Only take the basics, nobody touch the till," and that's what we did. When we'd filled our bags we all quietly returned home. That's the way it always is, everybody knows the rules . . . if you want to go, you go; if you don't, you don't . . . Nobody ever says anything. Round here they don't have to.'

Source: **Y. Corcoran-Nantes (1990) Women and popular urban social movements in São Paulo, Brazil *Bulletin of Latin American Research* 9(2), 249–64**

local medical posts by the health movement provided more up-to-date information than that held by the public health agency, which could be used to press the health movement's case.

Although urban social movements are not created specifically by women for women, nevertheless women have formed the motive force in many movements established to press for a more equitable distribution of the benefits as well as the costs of urbanisation.

COMMUNITY-BASED ORGANISATIONS

Community-based organisations (CBOs) are grass-roots locally based membership organisations that work for the development of their own communities. The classification includes many types of group, such as community leisure groups, residents' associations, savings and credit clubs, child-care groups and minority support groups. The range reflects the heterogeneous nature of slum populations and their interests and needs. They can exist informally, entirely outside the state, or they can be semi-official or have legal status (perhaps with some senior members receiving government salaries). The vast majority of CBOs are not profit-making organisations. The two most common types of CBO are local development associations, such as neighbourhood associations that represent an entire community, and interest associations, such as women's clubs, that represent particular groups within a community. A third type includes borrowers' groups and co-operatives that may make a profit yet can be distinguished from private businesses due to their community-development goals.

In 1998 there were over 200,000 CBOs functioning in Asia, Africa and Latin America alone.[15] Their rapid growth over recent decades reflects the promotion of 'structural adjustment programmes' by the World Bank from the late 1980s onwards that reduced already meagre state support for disadvantaged population groups in Third World cities. The diversity of residents' groups has resulted in a wide range of strategies for acquiring resources. While some CBOs depend entirely upon voluntary labour and financial contributions to sustain their activities, most interact with outside support organisations, including government, Churches or other CBOs or NGOs.

CHRISTIAN BASE COMMUNITIES

The efforts of the urban poor to improve their living conditions have been aided by external agencies sympathetic to the goal of social justice. A prominent force in Latin American cities is the basic ecclesiastical communities (CEBs) or Christian base communities. Rooted in the principles of liberation theology, these organisations seek to make religion of direct relevance to the demands of everyday life in poor communities.

The Catholic Church has been central to the growth of CEBs in Latin American cities.[16] CEBs are groups of ten to twenty people from the same area who come together to discuss concrete problems in the light of the Bible. The overwhelming majority are poor, and the initial stimulus is normally concern over lack of services (water, health care), cases of injustice (eviction from land) or natural disasters (earthquake, landslides) which impact most heavily on the poor. CEBs are fostered by the active participation of local priests, who seek to raise awareness of the structural causes of local problems and encourage local capacity for critical analysis as a basis for political action.[17] The CEB may be regarded as a particular form of NGO that seeks to actively promote the welfare of the urban poor.

URBAN NGOs IN THE THIRD WORLD

Third World NGOs attempt to fill the gap between the needs of vulnerable urban groups and the partial provision of services by the public sector. They operate primarily by:

1. Providing technical, legal and financial services to low-income households for constructing or improving housing, or by working with community organisations to provide basic infrastructure.
2. Advocating policies designed to reform state services, and lobbying directly for political change.

Although NGOs differ widely in their priorities, client groups and organisation, their strategies are conditioned by the political climate and prevailing bureaucratic procedures. Traditionally, most NGOs have operated in a climate of mutual suspicion with Third World governments, the state's natural tendency to centralise power being in contrast to the NGOs' emphasis on flexibility, local democracy and public participation. Increasingly, however, NGOs have acknowledged that, since government controls the political-economic framework within which they must operate, developing an interactive relationship may be necessary to effect changes in policy and practice for the benefit of the poor. Inevitably, however, this advocacy approach will be a long-term process, since public-service agencies are generally suspicious of change and lower-level officials are often discouraged from innovation or experimentation. In addition, NGOs are generally 'small players' compared with organisations such as the World Bank when it comes to influencing Third World governments. The decision to work with, but not for, government has to be based on an assessment of the 'reformability' of the political structures and of the costs and benefits of this strategy in relation to others. Clearly, it may be difficult for an NGO to work both within government and as an advocate for fundamental change in social and political structures.[18]

Rather than working through government, many NGOs seek to stimulate grass-roots organisations to participate in the political process by supporting 'local conscientisation', leadership and group formation, and by offering training in management skills. In practice it is often necessary to combine an empowerment approach with direct provision of services for the local community. Thus, in the *barrio* San Jorge on the periphery of Buenos Aires, only sixteen people were involved in community activities prior to the construction of a mother-and-child centre in 1987 by the International Institute for the Environment and Development – Latin America (IIED-LA). Three years later the community was electing representatives to a commission to develop a long-term improvement programme for the *barrio*, including transfer of public land to the squatters. The NGO played a key role in this transformation, maintaining the momentum of community action and providing advice and support to a community that had become disaffected by the ineffectual assistance offered by other external agents.[19] Provision of credit for housing construction is another important area for NGO activity aimed at promoting community self-help (Box 29.4).

COMMUNITY PARTICIPATION IN THIRD WORLD URBAN GOVERNANCE

Grass-roots community organisations and urban social movements arise because the urban poor are excluded from effective participation in the formal political decision-making process. The political ideology of the state can have a critical influence on the extent of popular participation in urban development. The election of progressive local governments in the municipalities of Moreno, Buenos Aires[20] and Vila El Salvador in Lima[21] was accompanied by decentralisation of decision-making to the community level, while in several Brazilian cities, participatory policies of various kinds were introduced following the electoral success of the Workers' Party in the late 1980s[22] (Box 29.5).

In practice, a vertical scale or 'ladder of community participation' may be identified for Third World urban society, which reflects varying levels of government support for community action.[23] This comprises:

BOX 29.4

Establishing autonomous housing co-operatives in Karachi, Pakistan

Catholic Social Services provides credit to residents in squatter settlements in Karachi through a revolving fund. The organisation has experimented with several different mechanisms and has developed a system that offers loans to groups of borrowers who collectively guarantee the repayment of individual loans. When people approach the society for a loan they are organised into a co-operative. Each member is asked to contribute Rs 50 (US$2.40) a month into a common fund, and for at least a year this amount cannot be touched. Thereafter the funds are available to members of the group requiring small loans with the consent of other members. Members of the co-operative who require a housing loan are asked to increase their contribution to the common fund to Rs 300 (US$14) a month for a six-month period. In addition, every prospective borrower is required to attend a basic loan and co-operative education course. Loans of up to Rs 10,000 (US$475) are granted. Recently Catholic Social Services has been encouraging several of the original co-operatives to become autonomous organisations. This will enable them to raise their own funds from new sources and allow the revolving fund and Catholic Social Services to concentrate on newly formed or potential co-operatives.

Korangi Christian Co-operative Society was the first group with which the society started work. The area is a resettlement site for those evicted from other parts of the city. The co-operative began with just twelve members in 1984. By June 1989 forty-five members had purchased plots and most had completed the construction of houses with permanent roofs. The group has also purchased land and built a community centre using the resources of the common fund. The Korangi Christian Co-operative Society was formally registered in 1989, and since then has begun to work autonomously.

Source: **M. Edwards and D. Hulme (1992)** *Making a Difference* **London: Earthscan**

---

1. *Empowerment*. Community members initiate and have control over a project or programme, possibly with the assistance of outside organisations and with a supportive municipal government. The low-income community of Jardin Celeste in São Paulo undertook a self-help project to provide

BOX 29.5

Participative budgeting in Belo Horizonte, Brazil

Belo Horizonte is the third largest city in Brazil, with 2 million inhabitants and 160 *favela* settlements. Prior to January 1993, when a progressive local government was elected, all decisions regarding public works were made centrally without popular input. The new city government introduced a participative budget system designed to:

1. Involve and give importance to the views of people's organisations.
2. Share information about the financial and administrative situation of the city, including revenue and expenditure.
3. Define investment priorities within each of the nine regions into which the city is divided, and also within sub-regions.
4. Guarantee citizens the right to be involved in defining government goals and strategies in order to meet social needs.

In 1994 US$15.6 million (40 per cent of the total city investment budget) was divided among the regions for the execution of identified projects. Understandably, in view of the sub-standard living conditions in the *favelas*, most of the public works approved by the participative budget were linked with sanitation (37 per cent), followed by road paving (28 per cent), education (13 per cent), health (8 per cent), housing (5 per cent), and the remainder for social services, transport, road infrastructure and sports facilities. By 2000 more than 40,000 people were taking part in the sixteen popular assemblies held in local neighbourhoods.

In addition to the physical improvements in the living conditions of the poor, which stem from socially progressive participative budgeting, the procedure also helps to strengthen democracy (often at the expense of clientelist neighbourhood leaders), broadens citizens' consciousness and increases the people's direct control over government. The participatory budgeting initiative has been copied in almost 200 municipalities

Source: R. Bretas (1996) Participative budgeting in Belo Horizonte: democratisation and citizenship *Environment and Urbanization* 8(1), 213–22; A. Novy and R. Leubolt (2005) Participatory budgeting in Porto Alegre *Urban Studies* 42(11), 2023–36

1,400 houses and neighbourhood facilities, founded by a government agency (FUNACON) and assisted by an independent technical assistance team contracted by the local government.[24]

2. *Partnership*. Members of a community, outside decision-makers and planners agree to share managerial responsibility for development projects. In Tegucigalpa, Honduras, the extension of a water supply to peripheral low-income communities requires the community's request to be approved by the national water authority to ensure that the applicants are able to construct and maintain the system. The central authority designs an appropriate system, covers many of the capital costs and provides technical assistance. The community forms a water association, then supplies the work force to construct the facilities, purchases some of the materials, takes responsibility for the administration and maintenance of the completed system and collects user fees.[25]

3. *Conciliation*. Conciliation occurs when government devises a development strategy for ratification by the people. Community representatives may be appointed to advisory groups or even decision-making bodies but are frequently forced to accept the views of a more powerful elite. This top-down paternalistic approach to urban management is evident in the master planning strategy for the city of Curitiba, Brazil.[26]

4. *Dissimulation*. People are appointed to rubber-stamp advisory committees in order to achieve a semblance of participation, as in the case of the advisory board to the Urban Planning Unit of Campo Grande, Brazil, on which one representative of the low-income peripheral areas sat with twenty members of professional associations and groups involved in speculative development.[27] From this level down the ladder of participation, government increasingly leaves communities to fend for themselves.

5. *Diplomacy*. Diplomacy is also a form of manipulation in which the government, owing to lack of interest, shortage of financial resources or incompetence, expects the community to undertake any necessary improvement projects, usually with the aid of an outside NGO. This approach may involve government in attitude surveys, consultation with residents and public hearings, but with no assurance that projects will be implemented or that support for any community effort will be forthcoming. Government may provide some limited aid, mainly for political reasons, if it appears that a community is making real progress. When the low-income community of Baldia in Karachi initiated a major sanitation project in partnership with a foreign NGO, the metropolitan government

subsequently surfaced roads, provided street lighting and improved the water supply.[28]

6. *Informing.* Informing comprises a top-down, one-way flow of information from public officials to the community of their rights, responsibilities and options, without an opportunity for feedback or negotiation. The lack of community participation may lead to problems of excessive cost and poor ongoing maintenance, as in the scheme to relocate squatters on an area of flood-prone land to the north of Dhaka.[29]

7. *Conspiracy.* In this case, participation of low-income communities in the formal decision-making process is not even considered. The poor appear little more than an embarrassment to government, the most vivid examples being the forced evictions of squatters from urban areas throughout the Third World.

8. *Self-management.* Self-management indicates a situation in which government does nothing to resolve local problems, and members of a community, possibly with the aid of an NGO, plan and implement improvements to their neighbourhood, though not always successfully. In contrast to empowerment, self-management emanates from lack of government interest, or even opposition to the demands of the poor. An example of a successful self-management initiative is the installation of a sanitation system in the Orangi neighbourhood of Karachi (Box 29.6).

BOX 29.6

The Orangi Pilot Project in Karachi, Pakistan

Orangi is an unauthorised settlement with around 1 million inhabitants extending over 8,000 ha (20,000 acres). Most inhabitants built their own houses and none received official help in doing so. There was no public provision for sanitation; most people used bucket latrines that were emptied every few days, usually on to the unpaved lanes running between houses. More affluent households constructed soakpits but these filled up after a few years. Some households living near creeks constructed sewage pipes that emptied into the creeks. The cost of persuading local government agencies to lay sewage pipes in Orangi was too much for local residents – who anyway felt that they should be provided free.

A local organisation called the Orangi Pilot Project (OPP), established in 1980, was sure that if local residents were fully involved a cheaper, more appropriate sanitation system could be installed. Research undertaken by OPP staff showed that the inhabitants were aware of the consequences of poor sanitation for their health and their property but they could not afford conventional systems, nor had they the technical or organisational skills to use alternative options. OPP organised meetings for those living in ten to fifteen adjacent houses each side of a lane and explained the benefits of improved sanitation and offered technical assistance. Where agreement was reached among the households of a lane they elected their own leader, who formally applied for technical help. Their site was surveyed, plans were drawn up and cost estimates prepared. Local leaders kept their group informed and collected money to pay for the work. Sewers were then installed, with maintenance organised by local groups.

The scope of the sewer construction programme grew as more local groups approached OPP for help, and the local authorities began to provide some financial support. By 1996 households in Orangi had constructed close to 69,000 sanitary pour-flush latrines in their homes plus 4,459 sewerage lines and 345 secondary drains – using their own funds and under their own management. One indication of the viability of this work is shown by the fact that some lanes have organised and undertaken lane sewerage investments independently of OPP; another is the households' willingness to make the investment needed in maintenance. Women were very active in local groups; many were elected group leaders and it was often women who found the funds to pay for the sewers out of household budgets.

By 2006 local people in Orangi had built and financed (to the extent of $1.4 million) sewers reaching almost 100,000 houses. It is estimated that to do the same work would have cost the local authority $10.5 million. The lesson of the Orangi Pilot Project – that, if organised and provided with technical support and managerial guidance, local communities can finance, manage and build neighbourhood sewerage systems – is being replicated in 250 other locations throughout Pakistan.

Sources: United Nations Centre for Human Settlements (1996) *An Urbanising World* Oxford: Oxford University Press; A. Hassan (2006) Orangi Pilot Project *Environment and Urbanization* 18(2), 451–80

GLOBALISATION AND SOCIAL JUSTICE

Globalisation has created new problems and exacerbated existing difficulties for many of the urban poor. However, globalisation may also enhance the potential of disadvantaged communities to address their problems. Advances in information and communications technologies facilitate the diffusion of values and norms that represent alternatives to capital accumulation goals of economic globalisation. Globalisation may foster demands for democratisation, human rights and social justice by promoting the formation of well-informed local civil social networks. Some low-income communities, using modern communications technologies, have broadened their perspective and have begun to reconstitute themselves as overlapping, and sometimes transnational, networks that share information, material resources, and solidarity. These developments signal new opportunities for civil society to engage government and the private sector in new forms of 'capacity building' co-operation that enable the poor to participate as empowered partners.

The activities of the Alliance in Mumbai, India, provides a working example of this form of political advocacy[30] (Box 29.7). In pursuit of an improved living environment for the urban poor, the Alliance employs a 'politics of patience'. This involves a strategy of accommodation, negotiation and long-term pressure rather than confrontation or threats of political reprisal. This 'realpolitick' is based on a view of 'politics without parties'. The Alliance does not view the poor as a vote bank for any political party or candidate but seeks to develop political affiliations with various levels of national and local government bureaucracy responsible for housing and infrastructure matters. It also maintains a working relationship with the police and an arm's-length relationship with the underworld that is involved in slum landlording and housing markets. The Alliance rejects the 'project model' of urban change in favour of a longer-term capacity building approach designed to empower the poor to 'own' as much as possible of the expertise that is necessary for them to claim their basic rights in urban housing.

The work of the Alliance in Mumbai may also be situated in the wider context of the emergence of transnational advocacy networks and the internationalisation of grass-roots CBOs that, for some, represents 'globalisation from below'. The Alliance is an active member of Slum/Shack Dwellers International (SDI), a transnational network with federations in fourteen

BOX 29.7

Structure and goals of the Alliance in Mumbai, India

The Alliance, formed in 1987, consists of three partner organisations:

1. Society for the Promotion of Area Resources Centres (SPARC), an NGO formed by social work professionals in 1984 to tackle problems of urban poverty, that provides technical knowledge and elite connections with state authorities and the private sector.
2. National Slum Dwellers' Federation (NSDF), a powerful grass-roots community-based organisation established in 1974.
3. Manila Milan (Women Together), an organisation of poor women set up in 1986 and now networked throughout India, with a particular interest in establishing local self-organising savings schemes among the poor.

The Alliance also has strong links with Mumbai's pavement dwellers and street children, whom it has organised into a group called Sadak Chaap (Street Impact) that has its own social and political agenda. The common goals of the Alliance concern gaining secure tenure in land, adequate and durable housing, and access to urban infrastructure. The politics of housing is at the heart of their activities. The power of the Alliance to influence political decisions that relate to urban management is exemplified in the 'people-managed' resettlement of 60,000 squatters from railway land in Mumbai. In contrast with most other resettlement schemes in Third World cities the affected low-income communities were involved in decision-making throughout the programme. In consequence the resettlement programme proceeded voluntarily and without coercion.[31]

countries concerned with mutual learning through shared experience.[32] Such transnational networks assist communities in 'thinking locally and acting globally' by disseminating information and invoking global norms that help to build new alliances, and by projecting local struggles into wider arenas to create extra local leverage.[33] Fundamentally, however, local organisation must precede global action and must persist to achieve its goals.

The political ideology and attitude of government are a key determinant of the success of initiatives to improve the living conditions of the Third World urban poor. Governments may support, manipulate, reject or neglect the demands of the poor. Strategies to improve the quality of life of the urban masses of the Third World must seek to reconstruct the relationship between the disadvantaged and the polity through enhanced participative democracy, meaningful dialogue and decentralised decision-making. It is also essential to adopt a long-term multi-sectoral approach to the problems of urban poverty. As the examples of 'empowerment' (in Jardin Celeste) and 'self-management' (in the Orangi project) indicate, individual and collective determination also play an important role in meeting basic needs, with or without government support. Nevertheless, in view of the disadvantaged position of the poor, in order to achieve long-term sustainable development, ongoing support from NGOs or government will be required at least to a point where community activities can become self-sustaining.

FURTHER READING

BOOKS

J. Friedmann (1992) *Empowerment* Oxford: Blackwell

J. Haynes (1997) *Democracy and Civil Society in the Third World: Politics and New Political Movements* Cambridge: Polity Press

F. Schurman and T. van Naerssen (1989) *Urban Social Movements in the Third World* London: Routledge

G. Shatkin (2007) *Collective Action and Urban Poverty Alleviation* Aldershot: Ashgate

B. Turner (1988) *Building Community: A Third World Case Book* London: Building Community Books

JOURNAL ARTICLES

M. Choguill (1996) A ladder of community participation for underdeveloped countries *Habitat International* 20(3), 431–44

Y. Corcoran-Nantes (1990) Women and popular urban social movements in São Paulo *Bulletin of Latin American Research* 9(2), 249–64

M. Hordijk (2005) Participatory governance in Peru *Environment and Urbanization* 17(1), 219–36

F. Miraftab (1997) Flirting with the enemy: challenges facing NGOs in development and empowerment *Habitat International* 21(4), 361–75

L. Peattie (1990) Participation: a case study of how invaders organise, negotiate and interact with government in Lima, Peru *Environment and Urbanization* 2(1), 19–30

O. Server (1996) Corruption: a major problem for urban management *Habitat International* 20(1), 23–41

L. Winayanti and H. Lang (2004) Provision of urban services in an informal settlement: a case study of Kampung Penas Tanggul, Jakarta *Habitat International* 28, 41–65

KEY CONCEPTS

- Partiality
- Soft states
- Culture of corruption
- Social power
- Clientelism
- *Compadrazgo* system

- Urban social movement
- Community-based organisations
- Christian-base community
- Non-governmental organisation
- Community participation
- Globalisation from below

STUDY QUESTIONS

1. Explain what is meant by a 'soft state'.
2. Illustrate how the urban poor respond to their disadvantaged position within the political power structure.
3. Critically examine the operation of clientelism in Third World cities.
4. Explain the role of urban NGOs in the power structure of Third World states.
5. With reference to relevant examples, examine the role of urban social movements in the Third World.
6. Consider the position of women in urban politics in the Third World and examine their contribution to actions aimed at promoting the interests of the urban poor.

PROJECT

Non-governmental organisations of various kinds operate within most cities of the Third World. Information on NGOs is available in academic journals (such as *Environment and Urbanization, Habitat International* and *Cities*), as well as on the World Wide Web. Identify the nature of Third World urban NGOs. Explain their aims, objectives and operating methods. Examine the relationship between NGOs and governments. Consider their potential role in advancing the welfare of disadvantaged groups. Illustrate your report with appropriate examples drawn from different Third World cities.

PART SIX

Prospective: The Future of the
City – Cities of the Future

30

The Future of the City –
Cities of the Future

Preview: an urban prospect; sustainable urban development; urban metabolism; waste management; energy consumption; the green city; the eco-city; the dispersed city; the car-free city; the compact city; the regional city; the network city; the informational city; the virtual city; the study of urban geography

INTRODUCTION

Cities have a long history, but the growth of very large cities and the transition towards a global urban society date from the advent of industrial urbanism in the early nineteenth century (see Chapter 3). Since then, two different population waves have affected Western Europe and North America. For much of this period the dominant direction of population movement was from rural to urban areas, reflecting the emergence of urban industrial society. More recently, since the Second World War, a reversal of this long-standing pattern has become apparent, with people on both sides of the Atlantic, and in Australia, reoccupying peri-urban areas. Despite evidence of a reurbanisation trend in some metropolitan regions, counter-urbanisation remains a major characteristic of contemporary Western societies (see Chapter 4). In contrast to the centrifugal pattern of urban population change in advanced societies, the centripetal processes of urbanisation and urban growth continue to dominate urban population dynamics in the Third World (see Chapters 21 and 23).

Continuation of these trends means that by 2025 65 per cent of the world's population will be urban dwellers. In the developed world, continued deconcentration of population at the national level and decentralisation at the local urban level are producing a 'rurban' settlement pattern in which urban lifestyles influence most of the country. In the Third World levels of urban development vary between countries. While a minority of states, such as Ethiopia, Uganda, Afghanistan and Cambodia, have yet to experience rapid urbanisation, the future social and settlement structure of most Third World countries will be dominated by a growing number of primate cities, many of which will be megacities. The early part of the twenty-first century will be characterised by continued large-scale urban development, heralding a future in which a growing majority of the world's population will live in urban places.

Urbanisation and urban growth on this unprecedented scale pose fundamental questions as to whether this magnitude of urban development can be sustained. How will the urban population be fed, housed, employed and cared for? In addition to meeting these basic needs, how can the increasing demand for mobility, recreation and satisfaction of higher-order needs such as self-esteem and human development be met? What effects will these concentrations of population have on local and global ecosystems? In this concluding chapter we adopt a prospective future-oriented viewpoint to consider the problems and prospects for cities and city life in the twenty-first century. We discuss the concept of sustainable urban development and examine the metabolism of cities, with particular attention to the basic problems of waste management and energy consumption. We evaluate a number of models of future urban form and consider the geographies of cities and city life in the third millennium.

SUSTAINABLE URBAN DEVELOPMENT

As we noted in Chapter 8, sustainable development is 'development which meets the needs of the present without compromising the ability of future generations to meet their own needs'.[1] This concept is based on the following three principles:

1. *Intergenerational equity*, which requires that natural capital assets of at least equal value to those of the present are passed on to future generations. This requires attention to the Earth's regenerative capacity and the ability of its systems to recuperate and maintain productivity. Ideally, the present generation should bequeath an improved environment in areas that are degraded or socially deprived.
2. *Social justice*, which requires that fair and equitable use is made of present resources in terms of meeting the basic needs of all and extending to all the opportunity to satisfy their aspiration to a better life. While for many people growing affluence has transformed luxuries into needs, the poorest are often unable to obtain basic necessities. On a global scale, the environmental costs of supporting the living standards of the rich while meeting the needs of the poor may prove impossible to sustain.
3. *Trans-frontier responsibility*, which requires recognition and control of cross-border pollution. Ideally the impacts of human activity should not involve an uncompensated geographical displacement of environmental problems. At a world level, rich nations should not overexploit the resources of other areas, thereby distorting regional economies and ecosystems. At the city scale the environmental costs of urban activities should not be displaced across metropolitan boundaries, in order to subsidise urban growth.[2]

The ideal world envisaged at the Rio Earth Summit in 1992[3] was one in which the objectives of sustainable development would be fulfilled at all levels of spatial organisation (Figure 30.1). Agenda 21 of the Earth Summit focused particular attention on the challenge of sustainable development at the urban scale. In this, concern for the sustainability of cities is expressed at two levels. The first is global and involves a range of issues concerning the long-term sustainability of the Earth's environment and the implications for urban life. The world's cities cannot continue to prosper if the aggregate impact of their economies' production and their inhabitants' consumption draws on global resources at unsustainable rates and deposits waste in global sinks at levels that lead to detrimental climatic change. The second is local and involves the possibility that urban life may be undermined from within

because of congestion, pollution and waste generation and their accompanying social and economic consequences (Box 30.1).

The concept of urban sustainability may be viewed as comprising five dimensions:

1. *Economic sustainability*, the ability of the local economy to sustain itself without causing irreversible damage to the natural resource base on which it depends. This implies maximising the productivity of a local (urban or regional) economy not in absolute terms (e.g. profit maximisation) but in relation to the sustainability of the other four dimensions. The difficulty of achieving economic sustainability in capitalist societies is compounded by economic globalisation that is promoting competition among cities, and between cities and their surrounding regions.

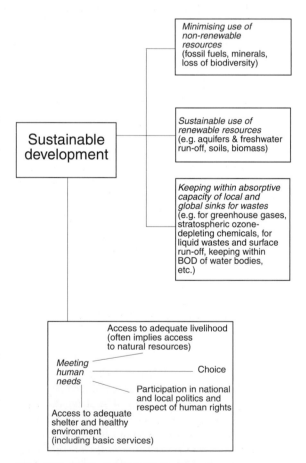

Figure 30.1   Goals of sustainable development
Source: J. Hardoy, D. Mitlin and D. Satterthwaite (1992) *Environmental Problems in Third World Cities* London: Earthscan.

BOX 30.1

Main urban dimensions of Agenda 21 of the Rio Earth Summit

Provision of adequate shelter for all:
- Adopt/strengthen national shelter strategies, including legal protection against unfair eviction from homes or land.
- Provide shelter for the homeless and the urban poor.
- Seek to reduce rural–urban drift by improving rural shelter.
- Introduce resettlement programmes for displaced persons.
- Develop multinational co-operation to support the efforts of developing countries.

Improve human settlement management
- Improve urban management.
- Strengthen urban data systems.
- Encourage intermediate city development.

Promote sustainable land-use planning and management
- Develop national land inventory and classification systems.
- Create efficient and accessible land markets, with land registers, etc.
- Encourage public–private partnerships in managing the land resource.
- Establish appropriate forms of land tenure.
- Develop fiscal and land-use planning solutions for a more rational and environmentally sound use of the land resource.
- Promote access to land for the urban poor.
- Adopt comprehensive land-use strategies.
- Encourage awareness of the problems of unplanned settlement in vulnerable areas.

Ensure integrated provision of environmental infrastructure: water, sanitation, drainage and solid waste management
- Introduce policies to minimise environmental damage.

- Undertake environmental impact assessments (EIAs).
- Promote policies to recover infrastructure costs, while extending services to all households.
- Seek joint solutions where issues cross localities.

Develop sustainable energy and transport systems in human settlements
- Develop and transfer technologies which are more energy-efficient and involve renewable resources.
- Improve urban transport systems.

Encourage human settlement planning and management in disaster-prone areas
- Promote a culture of safety.
- Develop pre-disaster planning.
- Initiate post-disaster reconstruction and rehabilitation planning.

Promote sustainable construction industry activities
- Encourage greater use of local natural materials and greater energy efficiency in design and materials.
- Strengthen land-use controls in sensitive areas.
- Encourage self-help schemes.

Meet the urban health challenge
- Develop municipal health plans.
- Promote awareness of primary health care.
- Strengthen environmental health services.
- Establish city collaboration networks.
- Improve training.
- Adopt health impact and EIA procedures.

2. *Social sustainability*, a set of actions and policies aimed at the improvement of quality of life and at fair access to and distribution of rights over the use and appropriation of the natural and built environment. This implies the improvement of local living conditions by reducing poverty and increasing satisfaction of basic needs.

3. *Natural sustainability*, the rational management of natural resources and of the pressures exerted by the waste produced by every society. Over-exploitation of natural capital and growing inequality in access to, and rights over, the natural resources of a city or region compromise the sustainability of natural capital.

4. *Physical sustainability*, the capacity of the urban built environment to support human life and productive activities. Crises of physical sustainability are evident particularly in metropolitan areas of the Third World as a result of the imbalance between in-migration of population and the 'carrying capacity' of the cities.

5. *Political sustainability*, the democratisation and participation of the local civil society in urban governance. Attainment of this goal may be undermined by the increasing influence of non-local and market forces in urban change.

Figure 30.2 illustrates the relationship among the five main dimensions of urban sustainability. In this, political sustainability is represented as the governance framework regulating the performance of the other four dimensions. The extent to which social, economic, natural and physical performance is sustainable depends on whether these activities can be kept within the ecological capacity of the urban regional ecosystem. In practice, conflicts may arise between the particular goals of each of the major dimensions of urban sustainable development.

Given the diversity of cities in terms of size, population growth rates and their economic, social, political, cultural and ecological settings, it is difficult to apply the concept of sustainable development generally. In most cities there are contradictions between the goals of sustainability and development. Most of the world's highly developed cities exhibit the highest *per capita*

use of environmental capital – in terms of consumption of non-renewable resources, pressure on watersheds, forests and agricultural systems, *per capita* emissions of greenhouse gases and stratospheric ozone depletion gases, and excess demand on ecosystems' waste-absorption capacities (Box 30.2). By contrast, most of the world's cities making least demand on environmental capital are 'underdeveloped', with high proportions of their population lacking safe and sufficient water, sanitation, adequate housing, access to health care, a secure livelihood and, often, basic civil and political rights. Under these circumstances the priorities for each city in relation to sustainability and development inevitably vary. It is unrealistic to expect poverty-stricken residents of Third World cities to attach as much importance to long-term environmental sustainability as the more comfortably placed proponents of green politics in advanced societies. This premise is enshrined in the concept of **urban environmental transition** that identifies changes in environmental challenges with urban development.[4] At an early stage of development the priority environmental challenges are those relating to the 'brown agenda' such as water supply, sewage and sanitation issues. As cities industrialise they are confronted by 'grey agenda' challenges such as those associated with industrial and auto-related pollution. Cities in post-industrial are facing 'green agenda' challenges such as greenhouse gas emissions, ozone-depletion substances, and increasing volumes of municipal waste. In considering the concept of sustainable urban development, therefore, it is

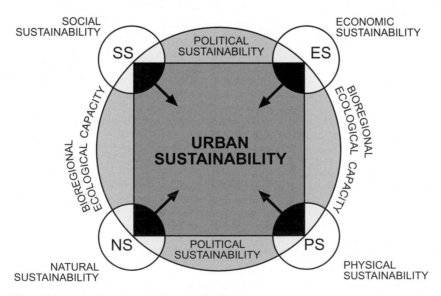

Figure 30.2   Major dimensions of urban sustainability

BOX 30.2

Cities and environmental issues

1. Of the world's population 42.6 per cent was urban in 1990: 72.6 per cent in developed countries, 33.6 per cent in developing countries.
2. Sixty per cent of the world's gross national product is produced in cities.
3. Sixty per cent of cities with over 2.5 million people are coastal, several already at or below sea level.
4. On average, each city of 1 million people daily consumes 625,000 tonnes of water, 2,000 tonnes of food and 9,500 tonnes of fuel, and generates 500,000 tonnes of wastewater, 2,000 tonnes of waste solids and 950 tonnes of air pollutants.
5. Urban populations in different countries vary hugely in their environmental demands. Urban residents in developed-country cities generate on average 0.7–1.8 kg (1.5–4.0 lb) of domestic waste daily, compared with 0.4–0.9 kg (0.9–2.0 lb) daily in developing countries.
6. In the USA almost a third of urban land is now devoted to the needs of the car.
7. Idling car engines in traffic jams in the USA in 1984 alone accounted for 4 per cent of petrol consumption.
8. Land loss has become a pressing problem. In Egypt, for example, more than 10 per cent of the most productive farmland has been lost to urban encroachment over the past three decades. In many Western European countries it is estimated that 2 per cent of agricultural land is being lost to cities each decade.

Source: **G. Haughton and C. Hunter (1994)** *Sustainable Cities* **London: Kingsley**

necessary to draw a distinction between the 'green agenda' for long-term environmental security and the 'brown agenda' of environmental issues associated with the immediate problems of survival and development in the cities of the Third World. As Nwaka (1996 p. 119) states:

for us in the developing world, the 'ecological debt' to future generations is not nearly as urgent as the 'social debt' for the future if today's young people lack the standard of health, education and skills to cope with tomorrow's world.[5]

Even in the West, sustainable development is not accepted universally as a key goal of urban growth, particularly if it involves constraints on personal patterns of consumption. Ideally, for richer cities with high levels of resource use, a priority should be the reduction of fossil fuel use and waste generation while maintaining a productive economy and achieving a more equitable distribution of the benefits of urban living. For poorer cities the priority is the attainment of basic social, economic and political goals within a context of seeking to minimise demands on environmental capital (Table 30.1). We do not, of course, live in an ideal world, and the goals of sustainable urban development are often difficult, if not impossible, to realise. We have considered each of these issues in the course of the book. Here we illustrate the difficulty of achieving sustainable urban development by considering the concept of urban metabolism.

URBAN METABOLISM

Cities occupy only 2 per cent of the world's land surface but use over 75 per cent of the world's resources.[6] As White and Whitney (1992) point out, most modern cities have spread far beyond their carrying capacity, and draw resources from far afield.[7] London's ecological footprint extends to 125 times its surface area. With 12 per cent of Britain's population, London requires the equivalent of Britain's entire productive land. In practice, this impact is global in extent, drawing on, among other resources, the wheat prairies of Kansas, the tea plantations of Assam, the forests of Amazonia and the copper mines of Zambia[8] (Box 30.3). In addition to providing resources for urban consumption, areas immediately around cities often receive much of the waste and pollution from urban growth, in the form of contaminated soil, water and air.

The fundamental cause of these problems is the essentially *linear* nature of the metabolism of modern cities. Resources flow through the urban system with little concern about their origin or about the destination of wastes; inputs and outputs are considered as largely unrelated. Raw materials are extracted, combined and processed into consumer goods that eventually end up as rubbish that is rarely reabsorbed into living nature. Fossil fuels are extracted, refined and burned and their fumes discharged into the atmosphere. Nutrients are taken from the land as food is harvested and not returned, as most urban sewage systems discharge into rivers and coastal waters (Table 30.2). This linear model of production, consumption and

TABLE 30.1 THE MULTIPLE GOALS OF SUSTAINABLE DEVELOPMENT AS APPLIED TO CITIES

Meeting the needs of the present . . .

Economic needs include access to an adequate livelihood or productive assets; also economic security when unemployed, ill, disabled or otherwise unable to secure a livelihood

Social, cultural and health needs include a shelter which is healthy, safe, affordable and secure, within a neighbourhood with provision for piped water, sanitation, drainage, transport, health care, education and child development. Also, a home, workplace and living environment protected from environmental hazards, including chemical pollution. Also important are needs related to people's choice and control, including homes and neighbourhoods that they value and where their social and cultural priorities are met. Shelters and services must meet the specific needs of children and of adults responsible for most child-rearing (usually women). Achieving this implies a more equitable distribution of income between nations and within nations

Political needs include freedom to participate in national and local politics and in decisions regarding management and development of one's home and neighbourhood – within a broader framework that ensures respect for civil and political rights an the implementation of environmental legislation

. . . without compromising the ability of future generations to meet their own needs

Minimising use or waste of non-renewable resources includes minimising the consumption of fossil fuels in housing, commerce, industry and transport plus substituting renewable sources where feasible. Also, minimising waste of scarce mineral resources (reduce use, reuse, recycle, reclaim). There are also cultural, historical and natural assets within cities that are irreplaceable and thus non-renewable – for instance, historic districts and parks and natural landscapes that provide space for play, recreation and access to nature

Sustainable use of finite renewable resources. Cities drawing on fresh-water resources at levels that can be sustained (with recycling and reuse promoted). Keeping to a sustainable ecological footprint in terms of land area on which city-based producers and consumers draw for agricultural and forest products and biomass fuels

Biodegradable wastes not overtaxing capacities of renewable sinks, e.g. capacity of a river to break down biodegradable wastes without ecological degradation

Non-biodegradable wastes/emissions not overtaxing (finite) capacity of local and global sinks to absorb or dilute them without adverse effects, e.g. persistent pesticides, greenhouse gases and stratospheric ozone-depleting chemicals

Source: **D. Satterthwaite (1997) Sustainable cities or cities that contribute to sustainable development?**
***Urban Studies* 34(10), 1667–91**

disposal differs markedly from nature's circular metabolism in which every output by an organism is also an input that renews and sustains the whole living environment. In its current form, urban metabolism is disruptive of natural cycles, promotes waste and undermines the goal of sustainable urban development. As urbanisation and urban growth continue apace, a critical question will be whether cities can move closer to being sustainable, self-regulating systems, not only in their internal functioning but also in their relationship with the outside world. We can examine this question by considering the particular issues of urban waste management and energy consumption.

URBAN WASTE MANAGEMENT

We have already examined a wide range of environmental issues in cities across the developed world and the Third World, including problems related to housing, health, transport, natural hazards, air and water pollution, noise, crowding, crime, poverty and stress. Here we focus on the particular problem of waste management to illustrate the prospects for improving the 'circularity' of urban metabolism and enhancing the potential for more sustainable cities.

Most city governments are confronted by mounting problems regarding the collection and disposal of solid wastes. In high-income countries the problems usually

BOX 30.3

London's ecological footprint

Number of people: 7,000,000

Surface area: 158,000 ha

Area required for food production (1.2 ha per person): 8,400,000 ha

Forest area required for wood products: 768,000 ha

Land area that would be required for carbon sequestration fuel production (1.5 ha per person): 10,500,000 ha

Total footprint (125 times London's surface area): 19,700,000 ha

Britain's productive land: 21,000,000 ha

Britain's surface area: 24,400,000 ha

Source: H. Girardet (1998) Sustainable cities: a contradiction in terms? *Journal of the Scottish Association of Geography Teachers* 27, 50–7

TABLE 30.2 THE METABOLISM OF GREATER LONDON (TONNES PER YEAR)

Inputs	
Total tonnes of fuel oil equivalent	20,000,000
Oxygen	40,000,000
Water	1,002,000,000
Food	2,400,000
Timber	1,200,000
Paper	2,200,000
Plastics	2,100,000
Glass	360,000
Cement	1,940,000
Bricks, blocks, sand and tarmac	6,000,000
Metals (total)	1,200,000
Wastes	
CO_2	60,000,000
SO_2	400,000
NO_x	280,000
Wet, digested sewage sludge	7,500,000
Industrial and demolition wastes	11,400,000
Household, civic and commercial wastes	3,900,000

Source: H. Girardet (1998) Sustainable cities: a contradiction in terms? *Journal of the Scottish Association of Geography Teachers* 27, 50–7

centre on the difficulties and high costs of disposing of the large volume of waste generated by households and businesses. In lower-income countries the main problems are related to collection, with between one-third and one-half of all solid wastes generated in Third World cities remaining uncollected (see Chapter 26). In many urban centres of the lowest-income countries only 10–20 per cent of solid waste is collected.

Urban waste may be viewed as a health and environmental hazard or as an economic resource from which marketable products can be derived. Despite evidence of a trend in favour of recycling, the dumping of urban waste in landfill sites remains the main means of disposal in Western cities. In London less than 10 per cent of household waste is recycled.[9] The level of waste recycling is generally higher in Third World cities, where 'recuperative production' is an integral component of the urban informal economy (Box 30.4). We must not, however, neglect the human dimensions of the ecological sustainability demonstrated by the recuperative economy of Third World cities. The low levels of resource use and waste generation and high levels of waste reuse or recycling are indicative of the inadequate incomes and poor living standards of a large proportion of the population. As we have seen, low levels of water use per person are usually the result of much of the city's population having no piped water supply to their dwelling. High levels of reclamation and recycling are the result of tens of thousands of people eking out a precarious living on city waste dumps, often with serious health risks.

Most city authorities in the Third World seem set on replicating the solid-waste management systems of advanced countries and give little consideration to the current or potential role of those who make a living picking saleable items from waste. However, some urban authorities have sought to combine social and environmental goals in their solid waste collection system by recognising that the people previously regarded as scavengers and waste-pickers are, in fact, recyclers and reclaimers who can be incorporated into city-wide waste-management schemes in ways that benefit them and the city environment.[10] In Buenos Aires the government has legalised the activities of informal garbage collectors (cartoneros) recognising their contribution to recycling and urban sanitation.[11] In Bogotá waste pickers have formed co-operatives that have bid successfully for some municipal waste-collection contracts.[12] The Indonesian city of Bandung has developed an innovative 'integrated resource recovery' strategy for waste management based on co-operation between the municipal authority, an NGO and a local community of scavengers.

BOX 30.4

The urban waste economy in Bangalore, India

In Bangalore, India's sixth largest city (4.1 million inhabitants within the municipality), between 40,000 and 50,000 people make a living by waste recovery or recycling. This represents between 1.6 per cent and 2.0 per cent of the work force. The waste economy also means that Bangalore recovers and recycles most of the solid wastes that are generated.

The recovery and trading network consists of some 25,000 waste pickers (predominantly women and children); 3,000–4,000 itinerant waste buyers of newspapers, plastics, glass, metals, clothes and other materials; some 800 small dealers, fifty medium-size dealers and fifty wholesalers in wastes; a great variety of enterprises using recycling materials, including two glass and four paper recycling plants, eight aluminium recyclers, 350–500 plastics factories using waste materials and an uncounted number of small recycling enterprises; householders, household servants and around 7,600 municipal street-sweeping and garbage collection workers, shop cleaners and office caretakers; piggery and poultry workers who collect food wastes from hotels and institutions; and farmers who collect compost from the garbage dumps or persuade garbage-truck drivers to deliver wastes direct to their farms.

Street pickers are estimated to receive about 15 per cent of waste put out on the streets and collected from over 12,000 bins – amounting to perhaps 300 tonnes of materials per day within the city. Municipal collectors and sweepers are estimated to take out thirty-seven tonnes per day, in addition to the waste removed by pickers. Itinerant buyers recover about 40 kg (88 lb) of material per day, and this implies the recovery of between 400,000 and 500,000 tonnes of materials per year. Middle and lower middle-income households are their main residential customers, although they also buy waste from offices and shops. High-income households with their own motor vehicles often store their own waste materials such as newspapers and plastics and take them direct to a dealer or even a wholesaler, although they will sell old clothes to itinerant buyers.

The fact that so much material is recovered from wastes by residents and shopkeepers or by waste pickers allows the remaining wastes, which are largely composed of organic matter, to be taken direct to farms and for natural composting to take place on garbage dumps. At one dump in 1990 about fifteen truckloads (each with around five tonnes of fresh waste) were delivered per day and about twelve farmers' truckloads of compost were removed. There was also a semi-mechanical compost plant that processed 50–100 tonnes of market wastes per day, producing about twenty tonnes of compost. About 210 tonnes of cow dung was collected each day from the roads for use as fuel by low-income people. A considerable amount of kitchen waste, leaves and tree trimmings is eaten by stray dogs, cows and pigs from street bins, amounting to perhaps 5 per cent by weight of garbage put in bins. Some citizen groups have been experimenting with decentralised composting and vermicomposting, so a small amount of further household organic waste is being recycled.

No study has been done of industrial wastes (metals, wiring, batteries, plastics, rubber, leather scraps, etc.) diverted by waste exchange or trading, or of bones sent to fertiliser factories and food wastes used by pig and poultry farms. None of the major industrial recyclables reaches the dumps. Food wastes generated by restaurants and hotels are traded. Construction wastes are used for filling low-lying land. Largely because of these varied activities of recovery and reuse, only about 335 tonnes of solid waste per day are handled by the corporation.

Although not all Indian cities have the capacity to recover and recycle as thoroughly as Bangalore, this study demonstrates that, where convenient markets exist, traditions of separation and informal waste trading thrive. It suggests that frugal habits are well established across a spectrum of household classes and that financial incentives reinforce these habits in lower-income groups, shops and factories. Such waste-reducing practices are found in other Third World countries, although the proportions of materials taken by itinerant waste buyers and waste pickers, and the patterns of control in the trade, may vary.

Source: **United Nations Centre for Human Settlements (1996)** *An Urbanising World* **Oxford: Oxford University Press**

Over a three-year period a pilot project enabled the community to achieve a number of significant economic and social advances on the basis of waste-recycling activity. These included shelter upgrading, health-care provision, toilet construction, and various economic activities such as composting of organic waste, seed farming using seeds collected from waste and rabbit farming. The longer-term (ten to fifteen-year) aim is to extend the system to cover the metropolitan area with a network of local composting and waste-recycling modules. This would be facilitated by government initiatives such as tax incentives for industries using recycled waste materials and direct purchase of compost – for example, for government reforestation schemes. Such schemes also underline the general conclusion that in seeking to achieve higher levels of sustainability in Third World cities it is essential to develop linkages between socio-economic and environmental goals.

ENERGY CONSUMPTION

Urban energy consumption and energy-related problems are likely to intensify in the course of the century. Already many city dwellers are exposed to unhealthy levels of energy-generated pollution. Globally, over 1,000 million people live in urban settlements where air-pollution levels exceed recommended health standards. In the USA 28 per cent of the urban population is exposed to harmful levels of particulates that cause the premature death of 40,000 people each year, while 46 per cent of the US urban population experience levels of ozone that exacerbate respiratory and cardio-vascular diseases. Conditions are even more extreme in Third World cities.[13] Urban emissions can also have regional impacts, reducing crop yields and forest integrity. In parts of the Third World the harvesting of wood for fuel by impoverished city dwellers has led to extreme deforestation around urban areas. Greenhouse gas emissions generated in the course of providing power to the world's cities are contributing to the problem of global climate change. Significantly, as the negative environmental impacts of urban energy consumption are manifesting themselves on local, regional and global scales, demand for energy is growing.

The World Commission on Environment and Development (1987)[14] forecast an increase in global energy consumption from the equivalent of 10 billion tons of coal per year in 1980 to 14 billion tons by 2025, allowing for a population increase from 4.5 billion to 8.2 billion and assuming the same *per capita* use and differences in use between the developed and less developed countries. The apparent unsustainability of such future scenarios led the World Commission to advocate a new, low-energy path based on energy efficiency and conservation, and accelerated development of renewable resources. The measures required to effect this strategy will have profound implications for urban development in both developed and Third World countries. Significantly, in the short term at least, the development needs of the latter imply additional energy consumption, even assuming major improvements in efficiency both in

the supply of energy and in its final use. The onus therefore falls on developed countries, which at present consume more than twice as much energy as the Third World, to reduce their energy demand and to assist others to develop benign energy strategies through technology transfer and financial aid.[15] A key question is how this may be achieved in an urban context.

Towns and cities are major energy consumers and there is a recursive relationship between energy systems and the structure of the urban environment. On one hand, falling real energy prices during the twentieth century have permitted increasing spatial separation of activities and the outward spread of urban areas at decreasing densities. On the other, the structure of urban areas is itself an important determinant of energy demand, especially for transport and space heating or cooling in buildings.[16] Interaction between energy systems and urban structure takes place at all scales from that of the individual building to the metropolitan region:

1. *Neighbourhood level*. Micro-scale measures to conserve energy include improved thermal insulation of buildings and changes in urban form. Provided it satisfies consumer demands, a systematic trend towards built forms like terraced housing or low-rise flats could lead to significant reductions in energy demand. Despite the fact that 85 per cent of new households in the UK in 2001 comprise single people,[17] the acceptability of this form of urban living is culturally contingent, with, in general, European urban dwellers more comfortable with the concept of apartment living than those committed to the goal of single-family detached residence.

   Energy savings can also be achieved by passive solar design in which siting, orientation, layout and landscaping are planned to make optimum use of solar gain and micro-climatic conditions to minimise the need for space heating or cooling of buildings from conventional sources. Measures include designing buildings to face south (in the northern hemisphere), with larger windows towards the sun and small windows to the rear, the use of particular building materials and the planting of shelter belts of trees. Another strategy to expand city-based energy production involves the use of modern biomass technologies to turn waste materials into sources of useful power. The huge volumes of solid and liquid waste produced by metropolitan areas are replete with combustible resources that can be fed into a variety of incineration systems, thereby simultaneously reducing the volume of waste while generating heat and electricity from an inexpensive, plentiful urban resource.

2. *District level.* The separate generation of power and heat from the combustion of fossil fuels is a wasteful process. Conventional power stations convert primary fuel into electricity with a maximum efficiency of about 38 per cent, with even the best plants dumping 60 per cent of the energy input in the form of steam from cooling towers and in flue gases. Combined heat and power (CHP) schemes, in which heat produced during the process of electricity generation is used for space and water heating, increase the efficiency of conversion of primary fuel to around 80 per cent and also reduce the environmental impact per unit of delivered energy. This advantage applies to small-scale CHP plant serving individual buildings or groups of buildings as well as large-scale schemes serving industry or providing heat for whole towns.[18]

Urban CHP schemes linked to district heating (DH) are well established in Scandinavia, with one of the most extensive schemes being in the Danish city of Odense, where 75,000, or 95 per cent, of homes are heated directly through the city's CHP plant, which also supplies schools, hospitals, industry, commerce and horticulture. In Sweden over one-third of local authorities operate district heating systems. These meet 30 per cent of the total heat demand and serve half the apartments in the country. In Helsinki district heating meets over 80 per cent of the annual heat demand of the city, with 81 per cent of the heat and 72 per cent of the electricity used in the city supplied by CHP plant. The development of urban and district-level CHP/DH schemes as a means of increasing energy efficiency has attracted international attention from countries as far afield as China, Australia and the UK, and over 300 micro-CHP/DH schemes are currently in operation in Britain.[19]

3. *Metropolitan level.* One of the most pressing challenges for future urban development is to achieve sustainable mobility. As we have seen, most cities have been transformed by the growth in the number of road vehicles (see Chapters 13 and 28) (Box 30.5), and the increasing reliance on private automobile use that has been built into much of the urban landscape is not easily changed. The convenience of private cars for those able to afford them and the extent to which many developments in retailing, in concentrations of job opportunities and in new residential developments encourage their

use suggest that it will be extremely difficult to shift car-users to other transport modes. One means to reduce energy consumption and environmental pollution is to use fuel cells to power automobiles, as well as in residential and commercial buildings. Pure water is the only by-product of hydrogen-powered fuel cells. Fuel cells may also be operated using methane gas collected from sewage systems and municipal landfills. While widespread use of fuel cells is likely to occur first in affluent cities, fuel-cell-powered buses, motor scooters and stationary power generators could also ameliorate the growing problems of Third World cities.

The need for, and extent of, private transport are closely related to the structure of the built environment. The physical separation of activities affects travel needs and therefore energy requirements for transport. The key variables at the urban scale are density and degree of mixing of different land uses, with, in general, lower energy use for transport with higher urban densities (Figure 30.3).

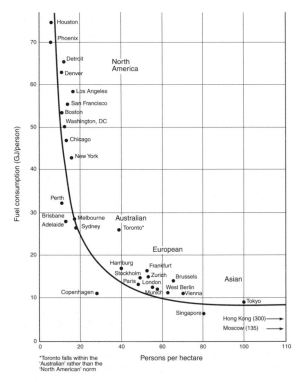

Figure 30.3 The relationship between urban population density and gasoline consumption

Source: UNCHS (eds) (1996) *An Urbanising World* Oxford: Oxford University Press

BOX 30.5

Car-related urban problems

1. In the UK road transport accounts for 18 per cent of all carbon dioxide emissions, 85 per cent of carbon monoxide emissions, 30 per cent of volatile organic compounds and 45 per cent of nitrous oxides, about 80 per cent of which are released in urban areas.
2. Only 15–20 per cent of urban nitrous oxide emissions are deposited in the city, with the rest transported outside.
3. In OECD countries, transport is responsible for 50 per cent of atmospheric lead emissions, 80 per cent of all benzene emissions and about 50 per cent of total hydrocarbons in urban areas.
4. In the heavily polluted Athens basin, and in Mexico City, motor vehicles account for 83 per cent of air pollution.
5. The manufacture of cars constitutes a major environmental impact in itself, in terms of materials and energy use.
6. Air conditioning for cars accounts for 10 per cent of global CFC-12 use.
7. Public transport is less land-intensive, with the passengers of 200 filled cars able to occupy one tramcar, on average.
8. Roads demand more land in construction than rail, and cost up to eighty times as much to build.
9. Material and energy use in road construction is high, as is disruption of nearby residents and ecosystems.
10. In cities in the USA and Australia road supply per capita tends to be three to four times as high as in European cities and seven to nine times higher than in Asian cities.
11. Cities in the USA provide 80 per cent more car parking spaces per thousand workers than European cities, and six times more than in Asian cities.
12. Road transport is associated with high accident and death tolls, and can deter cyclists and pedestrians.
13. Congestion costs are high: in Athens city centre, road traffic moves at an average of 7–8 km/hr (4–5 mph), in Paris 18 km/hr (11 mph) and in London 20 km/hr (12.5 mph).
14. The occupancy rate of cars is generally low: in Paris 1.25 passengers for every 4.5 seats.

Source: G. Haughton and C. Hunter (1994) *Sustainable Cities* London: Kingsley

One way to reduce travel needs would be to concentrate homes, jobs and services in a relatively compact urban centre to achieve high levels of accessibility with a reduced need for movement. A number of studies suggest this is an energy-efficient form of development[20] that also increases the viability of DH and CHP systems. Concentration of development in existing urban centres may also contribute to their rehabilitation and revitalisation. But there are also disadvantages. Above a certain size of urban centre, problems of congestion lead to inefficient energy use and loss of accessibility. Higher-density development may also give rise to concerns about 'town cramming' and loss of urban green space. As we shall see later, an alternative way to reduce physical separation of urban activities is through decentralised mixed development, or 'decentralised concentration' of jobs, services and housing to promote a significant degree of self-containment (Figure 30.4).

A key factor in determining the energy efficiency of centralisation or decentralised concentration is the way in which people value mobility and freedom of choice. This lifestyle perspective suggests that if rising energy costs or policy restraints effectively restrict personal mobility, a pattern of decentralised concentration will be energy-efficient because people will tend to use the jobs and services closest to home. However, if travel costs pose only a minimum determinant, such a pattern is likely to be more energy-intensive than centralisation because of the potentially large amount of cross-commuting and other non-work travel engaged in. To date, the latter situation has prevailed in many countries, including the UK, the USA and Australia. Furthermore, as well as seeking to reduce the distances people need to travel, planning for sustainable mobility in urban regions will also require a greatly enhanced role for public transport (Figure 30.5). A fundamental question is how these and other factors are likely to affect the form of cities in the future.

TOWARDS THE CITY OF THE FUTURE

The challenge of creating new and, one hopes, superior forms of settlement has occupied philosophers,

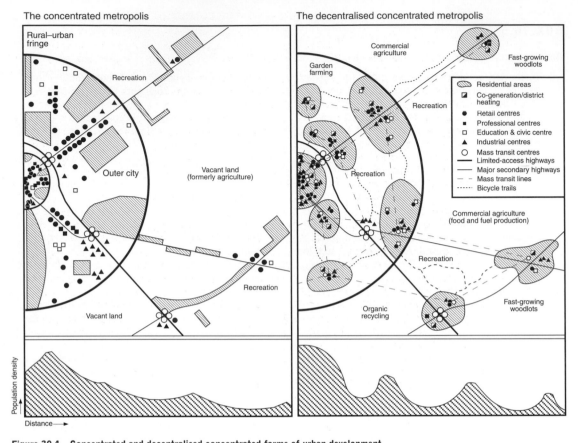

Figure 30.4   Concentrated and decentralised concentrated forms of urban development
Source: S. Brown and J. Williams (1983) *Cities of the World* New York: Harper & Row

architects, planners and urban theorists for centuries. The plethora of novel urban forms proposed includes idealistic designs such as Plato's Republic, More's Utopia and Skinner's Walden II, as well as actual built environments such as the Oneida community in the USA, the Israeli kibbutzim and the British new town.[21] The desire to improve urban quality of life and prognosticate on future urban form remains a powerful element in urban geography. The major contemporary models of future urban form include the following.

THE GREEN CITY

The desire to plan urban development within the context of the local natural environment was central to Howard's concept of the garden city (see Chapter 8). Geddes (1915)[22] also believed that urban planning should be based on a knowledge of natural regions and

their resources, and specifically that the river basin was a suitable natural unit for city development. Foreseeing the impact of the motor vehicle on urban expansion, Geddes proposed a stellar form of urban settlement that would allow axes of natural space to penetrate into the city. This approach of 'designing with nature' was developed subsequently by McHarg (1969),[23] who advocated detailed examination of the environmental condition of an area prior to urban development in order to identify those areas where urbanisation would least damage natural ecosystems. Unlike Howard, McHarg was less concerned with the human dimension of urban development than with its impact on nature. This tradition of planning in concert with the natural environment and the resource base is continued in the concept of the eco-city, which is based on the principles of appropriate technology, community economic development, social ecology, bio-regionalism, the green movement and sustainable development.[24]

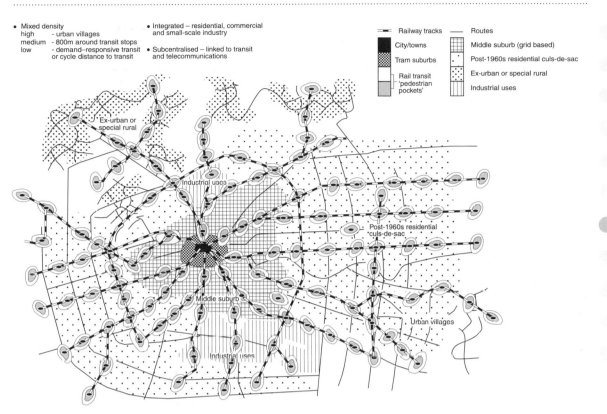

Figure 30.5   Remodelling the automobile-dependent city
Source: United Nations (1996) *An Urbanising World* Oxford; Oxford University Press

THE ECO-CITY

Making existing cities and new urban development more ecologically based and liveable is a central component of the quest for urban sustainability. The vision of an eco-city involves compact mixed-use urban form (see below), well defined higher-density, human-oriented centres that are focal points for population and employment growth and are linked by public transport, priority to the development of superior public transport systems and conditions for non-motorised modes of travel, with minimal road capacity increases to curtail automobile dependence, and protection of the city's natural areas and food-producing capacity. The eco-city aims to employ innovative 'closed loop' environmental technologies for water, energy and waste management, economic growth based on creativity and innovation and sensitive to local environmental and cultural contexts, sustainable urban design principles, and a high quality public realm both physically in terms of public space and socially in terms of social

capital and promoting good governance. The vision would be pursued via a decision-making process based on a 'debate and decide' rather than 'predict and provide' approach within a strong community-oriented, democratic framework.[25]

THE DISPERSED CITY

The promotion of dispersed or decentralised settlements as an alternative to large cities is part of the green city genre.[26] Key themes of the ideal model include the centrality of small-scale economic and political organisation, grass-roots political empowerment, an emphasis on collective action, local economic self-reliance, including farming and industry, the use of appropriate technologies, recycling and re-use of materials, and the value of 'natural' ecological or resource areas as potential political boundaries. The decentralisation theme reached an extreme in the low-density (one family per acre) high-technology 'exurbia' proposed by Wright

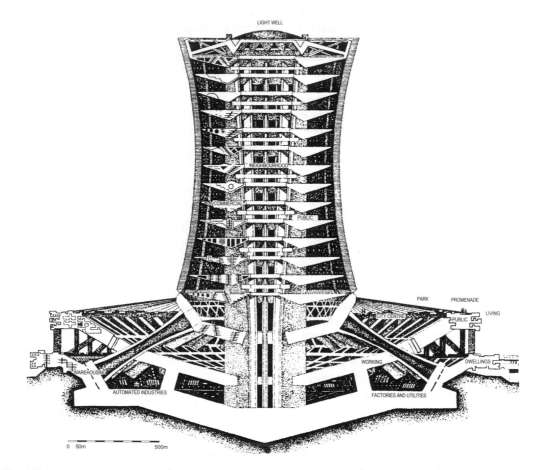

Figure 30.8   Soleri's three-dimensional city: Babel II D

the city centres available to all residents of an urban region by designing a constellation of thirty towns, each of 50,000 people and made up of smaller (1,900 person) neighbourhoods, around a central downtown area of 65,000 people – a model similar in concept to a scaled-up version of Howard's social city region (see Chapter 9). Advocates of New Urbanism and Smart Growth, such as Calthorpe and Fulton (2001)[38] conceptualise the regional city as the necessary scale at which to combat the twin problems of urban sprawl and inner city decline. Initiatives required to achieve this goal include a region-wide distribution of affordable housing, tax sharing between cities and suburbs, revived mass transit, and regional growth boundaries in order to redirect growth back to the core, framed within a vision of a coherent metropolitan region forming a single economic, cultural, environmental and civic entity.

THE NETWORK CITY

The transnational political-economic processes that have given rise to 'world cities' (see Chapter 14) have also influenced urban form and development at the regional level with the growth of 'corridor cities' linking knowledge-intensive centres with larger metropolises, as in the case of the London–Cambridge corridor. In these bicentric urban systems, close links have been forged between places of complementary function rather than simply on the basis of physical proximity (Figure 30.9). A small but growing number of modern urban agglomerations consist of an intricate web of corridor cities whose functional and locational relationships can provide them with holistic competitive advantages over some of their monocentric rivals. Examples of these network cities include Randstad Holland, Singapore and Malaysia's

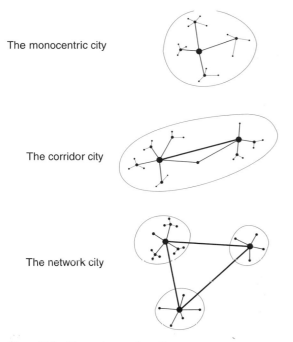

The monocentric city

The corridor city

The network city

Figure 30.9   Three urban configurations

Multimedia Super Corridor,[39] while the Kansai region of Japan represents a particularly innovative case (Box 30.6 Some observers claim that certain network cities may enjoy greater diversity and creativity, less congestion and more locational freedom than monocentric cities of comparable size,[40] and that the model may be of particular relevance for the still-growing world cities.

THE INFORMATIONAL CITY

In advanced societies, globalisation, deindustrialisation and the growth of a service-oriented economy based on manipulation of knowledge and information facilitated by advances in telematics have transformed many industrial cities into post-industrial or informational cities.[41] For example, in many North American and European cities some public services are delivered via telematics.[42] The replacement of the physical offices of certain government services with electronically mediated touch-screen kiosks in shopping malls can reduce costs and improve the quality and timeliness of information delivery to citizens. Nevertheless, the impact of telematics on urban form is ambiguous. The suggestion that telematics-based flows of electronic information can substitute for environmentally damaging and

energy-intensive physical flows overlooks the fact that telematics may also stimulate more travel, as cheaper and more widely accessible forms of communication generate new demands for the physical movement of people and goods. New services such as road information systems and auto route guidance can help drivers avoid congestion and improve the attractiveness of the road network. Furthermore, for many human activities, telecommunications is not an adequate substitute for face-to-face contact. In the **informational city**, although energy consumption might be reduced by partially substituting electronic for physical movements, equally it might be increased as the combined decentralising power of the automobile and telematics leads to a more dispersed metropolitan region.[43] Some see the annihilation of distance and time by telematics as heralding a trend towards an extreme form of urban decentralisation that will inevitably lead to the dissolution of the city.[44] Other theorists see modern communications technologies enhancing the power of the city by reinforcing the centrality of cities within nodal communications networks.[45] Even though some activities will disperse away from cities, despite the physical distance, they remain under the control of higher-order functions based in the city.

The social impact of telematics is also equivocal. Telematics may promote socially progressive urban change by overcoming the isolation of disabled, housebound and marginalised groups, but those excluded from the information society are unlikely to benefit, while some technologies (e.g. closed-circuit television) may even be used to exploit or monitor them more effectively.[46] Whereas affluent groups will be linked to the global urban system via the Internet, e-mail, teleshopping, telebanking and telecommuting, others will be confined to information ghettos where even a basic telephone service is not universally available. The informational city is likely to become more socially fragmented and polarised as physical and electronic space are used in new ways.[47] An insight into the form of the future informational city is offered by Japanese communications-based investment in new cities for the twenty-first century (Box 30.7), which includes science cities, as at Tsukuba and Kansai, a national technopole programme, and participation in a multi-functional polis planned for Adelaide in Australia.[48]

THE VIRTUAL CITY

The virtual city is an extension of the concept of the informational city. For some 'cyber-utopians' the

66. M. Smith (2001) *Transnational Urbanism* Oxford: Blackwell.

67. J. Robinson (2006) *Ordinary Cities: Between Modernity and Development* London: Routledge.

68. D. Harvey (1996) *Justice, Nature and the Geography of Difference* Oxford: Blackwell; M. Pacione (1999) Applied geography: in pursuit of useful knowledge *Applied Geography* 19, 1–12; M. Pacione (1999) The relevance of religion for a relevant human geography *Scottish Geographical Journal* 115(2), 117–31.

69. P. Jackson (1984) Social disorganisation and moral order in the city *Transactions of the Institute of British Geographers* 9, 168–80; F. Driver (1988) Moral geographies: social science and the urban environment in mid-nineteenth-century England *Transactions of the Institute of British Geographers* 13, 275–87.

70. D. Harvey (1973) *Social Justice and the City* Oxford: Blackwell; D. Smith (1977) *Human Geography: A Welfare Approach* London: Arnold.

71. A. Merrifield and E. Swyngedouw (1997) *The Urbanization of Injustice* New York: New York University Press; D. Smith (1994) Geography, community and morality *Environment and Planning A* 31, 19–35.

72. I. Young (1990) *Justice and the Politics of Difference* Princeton NJ: Princeton University Press.

73. S. White (1991) *Political Theory and Postmodernism* Cambridge: Cambridge University Press; D. Smith (1994) *Geography and Social Justice* Oxford: Blackwell.

74. D. Smith (1994) *Geography and Social Justice* Oxford: Blackwell.

75. R. Johnston (1980) On the nature of explanation in human geography? *Transactions of the Institute of British Geographers* 5, 402–12; L. Lees (2002) Rematerialising geography: the 'new' urban geography *Progress in Human Geography* 26(1), 101–12.

76. H. Wai-chung Yeung (1997) Critical realism and realist research in human geography: a method or a philosophy in search of a method? *Progress in Human Geography* 21(1), 51–74.

Chapter 3   THE ORIGINS AND GROWTH OF CITIES

1. P. Wheatley (1971) *The Pivot of the Four Quarters* Chicago: University of Chicago Press.

2. V. G. Childe (1950) The urban revolution *Town Planning Review* 21, 3–17.

3. O. Duncan (1961) From social system to ecosystem *Sociological Inquiry* 31, 140–9.

4. K. Wittfogel (1957) *Oriental Despotism: A Comparative Study of Tribal Power* New Haven CT: Yale University Press.

5. W. Sanders and B. Price (1968) *Mesoamerica: The Evolution of a Civilization* New York: Random House.

6. P. Wheatley (1971) *The Pivot of the Four Quarters* Chicago: University of Chicago Press.

7. G. Sjoberg (1960) *The Preindustrial City: Past and Present* New York: Free Press.

8. C. Redman (1978) *The Rise of Civilization* San Francisco: Freeman.

9. L. Wooley (1965) *Ur of the Chaldees* New York: Norton; A. Bossuyt, L. Broze and V. Ginsburgh (2001) On invisible trade relations between Mesopotamian cities during the third millennium BC *Professional Geographer* 53(3), 374–83.

10. K. Chang (1968) *The Archaeology of Ancient China* New Haven CT: Yale University Press.

11. G. Sjoberg (1960) *The Preindustrial City: Past and Present* New York: Free Press.

12. J. Vance (1971) Land assignment in pre-capitalist, capitalist and post-capitalist cities *Economic Geography* 47, 101–20.

13. J. Langton (1975) Residential patterns in pre-industrial cities: some case studies from seventeenth-century Britain *Transactions of the Institute of British Geographers* 65, 1–27.

14. J. Radford (1979) Testing the model of the pre-industrial city: the case of antebellum Charleston, South Carolina *Transactions of the Institute of British Geographers* 4(3), 392–410.

15. A. Rivet (1964) *Town and Country Planning in Roman Britain* London: Hutchinson.

16. R. Hilton (1976) *The Transition from Feudalism to Capitalism* London: Verso.

17. D. Goodman (1999) Medieval cities, in C. Chant and D. Goodman (eds) *Pre-industrial Cities and Technology* London: Routledge, 115–62; K. Lilley (2002) *Urban Life in the Middle Ages 1000–1450* London: Palgrave.

18. T. S. Ashton (1948) *The Industrial Revolution 1760–1830* London: Oxford University Press.

19. M. Weber (1958) *The Protestant Ethic and the Spirit of Capitalism*, translated by T. Parsons, New York: Scribner.

20. J. de Vries (1984) *European Urbanization 1500–1800* Cambridge MA: Harvard University Press.

21. P. Hohenberg and L. Lees (1995) *The Making of Urban Europe 1000–1994* Cambridge MA: Harvard University Press; A. Cowan (1998) *Urban Europe 1500–1700* London: Arnold.

22. R. Lawton (1972) An Age of Great Cities *Town Planning Review* 43, 199–224.

23. J. Johnston and C. Pooley (1982) *The Structure of Nineteenth Century Cities* London: Croom Helm.

24. C. Pooley (1979) Residential mobility in the British city *Transactions of the Institute of British Geographers* 4, 258–77.

25. R. Dennis (1977) Intercensal mobility in a Victorian city *Transactions of the Institute of British Geographers* 2, 349–63; P. Knights (1971) *The Plain People of Boston 1830–1860* New York: Oxford University Press; R. Barrows (1981) 'Hurryin' hoosiers' and the American pattern: geographical mobility in Indianapolis and urban North America *Social Science History* 5, 197–222.

26. R. Dennis (1987) People and housing in industrial society, in M. Pacione (ed.) *Historical Geography: Progress and Prospect* London: Croom Helm, 184–216.

27. Bournville Village Trust (1941) *When We Build Again* London: Allen & Unwin.

28. F. Engels (1958) *The Condition of the Working Class in England*, translated and edited by W. O. Henderson and W. H. Chaloner, Oxford: Blackwell.

29. M. Pacione (1995) *Glasgow: The Socio-spatial Development of the City* Chichester: Wiley.

30. J. Reps (1965) *The Making of Urban America* Princeton NJ: Princeton University Press.

31. J. Reps (1979) *Cities of the American West* Princeton NJ: Princeton University Press.

32. J. Reps (1965) *The Making of Urban America* Princeton NJ: Princeton University Press.

33. J. Borchert (1967) American metropolitan evolution *Geographical Review* 57, 301–22.

34. R. Bernard and B. Price (1983) *Sunbelt Cities: Politics and Growth since World War I* Austin TX: University of Texas Press; A. Markusen, P. Hall, S. Campbell and S. Deitrick (1991) *The Rise of the Sunbelt* Oxford: Oxford University Press; L. Suarez-Villa (2002) Regional inversion in the United States *Tijdschrift voor Economische en Sociale Geografie* 93(4), 424–42.

35. D. Bell (1973) *The Coming of Post-industrial Society* New York: Basic Books.

36. G. MacLeod, M. Raco and K. Ward (2003) Negotiating the contemporary city *Urban Studies* 40(9), 1655–71.

37. N. Ellin (1996) *Postmodern Urbanism* Oxford: Blackwell.

38. S. Flusty and M. Dear (1999) Invitation to a postmodern urbanism, in R. Beauregard and S. Body-Gendrot (eds) *The Urban Moment* Thousand Oaks CA: Sage, 25–50.

39. E. Soja (1995) Postmodern urbanism: the six restructurings of Los Angeles, in S. Watson and K. Gibson (eds) *Postmodern Cities and Spaces* Oxford: Blackwell, 125–37.

40. S. Lash and J. Urry (1983) *The End of Organised Capitalism* Oxford: Blackwell; E. Swyngedouw (1997) Neither global nor local: glocalization and the politics of scale, in K. Cox (ed.) *Spaces of Globalization* New York: Guilford Press, 137–66.

41. E. Soja (1995) Postmodern urbanism: the six restructurings of Los Angeles, in S. Watson and K. Gibson *Postmodern Cities and Spaces* Oxford: Blackwell, 125–37.

42. M. Davis (1990) *City of Quartz: Excavating the Future in Los Angeles* New York: Vintage.

43. J. Hannigan (1998) *Fantasy City* London: Routledge.

44. E. Soja (1996) *Thirdspace: Journeys to Los Angeles and other Real and Imagined Places* Oxford: Blackwell; A. Scott and E. Soja (1996) *The City: Los Angeles and Urban Theory at the End of the Twentieth Century* Los Angeles: University of California Press; D. Sudjic (1993) *The 100 Mile City* London: Flamingo.

Chapter 4   THE GLOBAL CONTEXT OF URBANISATION AND URBAN CHANGE

1. A. Weber (1899) *The Growth of Cities in the Nineteenth Century* New York: Macmillan.

2. S. Findley (1993) The Third World city: development policy and issues, in J. Kasarda and A. Parnell (eds) *Third World Cities* London: Sage, 1–33.

3. United Nations (1995) *World Urbanization Prospects: The 1994 Revision* New York: United Nations.

4. N. Harris (1991) *City, Class and Trade* London: I. B. Tauris.

5. H. Nagpaul (1988) *India's Great Cities* volume 1, Newbury Park CA: Sage.

6. K. Davies (1969) *World Urbanization* Los Angeles: University of California Press.

7. A. Fielding (1982) Counterurbanisation in Western Europe *Progress in Planning* 17(1), 1–52; A. Champion (1989) *Counterurbanisation: The Changing Pace and Nature of Population Decentralisation* London: Arnold; D. Cross (1990) *Counterurbanisation In England and Wales* Aldershot: Avebury; A. Nucci and L. Long (1996) Spatial and demographic dynamics of metropolitan and non-metropolitan territory in the United States *International Journal of Population Geography* 1(2), 165–81.

8. H. Richardson (1980) Polarisation reversal in developing countries *Papers of the Regional Science Association* 45, 67–85.

9. B. Berry (1976) *Urbanization and Counterurbanization* Beverly Hills CA: Sage.

10. H. Geyer and T. Kontuly (1993) A theoretical foundation for the concept of differential urbanization *International Regional Science Review* 15(12), 157–77.

11. M. Jefferson (1939) The law of the primate city *Geographical Review* 29, 226–32.

12. W. Frey and A. Speare (1988) *Regional and Metropolitan Growth and Decline in the United States* New York: Russell Sage Foundation; A. Fielding (1989) Migration and urbanisation in Western Europe since 1950 *Geographical Journal* 155, 60–9; N. Cook and T. Dyson (1982) Urbanisation in India *Population and Development Review* 8, 145–55; P. Townroe and D. Keen (1984) Polarisation reversal in the state of São Paulo, Brazil *Regional Studies* 18, 45–54; T. Gwebu (2006) Towards a theoretical exploration of the differential urbanisation model in sub-Saharan Africa: the Botswana case *Tijdschrift voor Economische en Sociale Geografie* 97(4), 418–33.

13. L. Klaassen, W. Molle and J. Paelinck (1981) *Dynamics of Urban Development* New York: St Martin's Press; L. van den Berg, R. Drewett, L. Klaassen, L. Rossi and C. Vijverberg (1982) *A Study of Growth and Decline* Oxford: Pegasus.

14. P. Cheshire (1995) A new phase of urban development in Western Europe? *Urban Studies* 32(7), 1045–63.

of America, in J. Short and R. Utt (eds) *A Guide to Smart Growth* Washington DC: Heritage Foundation.

35. T. Daniels (1999) *When City and Country Collide* Washington DC: Island Press; C. Dawkins and A. Nelson (2002) Urban containment policies and housing prices *Land Use Policy* 19, 1–12; E. Ben-Zadock (2005) Consistency, concurrency and compact development *urban Studies* 42(12), 2167–90; T. Carlson and Y. Dierwechter (2007) Effects of urban growth boundaries on residential development in Pierce County, Washington *Professional Geographer* 59(2), 209–20.

36. E. Gerber and J. Phillips (2004) Direct democracy and land use policy *Urban Studies* 41(2), 463–79.

37. R. Burchell, D. Listokin and C. Galley (2000) Smart growth *Housing Policy Debate* 11, 821–79; W. Weivel and K. Schaffer (2001) Learning to think as a region *European Planning Studies* 9(5), 593–611.

38. J. Cullingworth and R. Caves (2003) *Planning in the USA* London: Routledge.

39. R. French and F. Hamilton (1979) *The Socialist City* Chichester: Wiley.

40. R. French (1995) *Plans, Pragmatism and People* London: UCL Press.

41. G. Andrusz, M. Harloe and I. Szelenyi (1996) *Cities after Socialism* Oxford: Blackwell.

42. M. Pavlovskaya and S. Hanson (2001) Privatization of the urban fabric *Urban Geography* 22(1), 4–28.

43. O. Vendina (1997) Transformation processes in Moscow and intra-urban stratification of population *Geoforum* 42, 341–63.

44. J. Alden, S. Crow and Y. Beigulenko (1998) Moscow: planning for a world capital city toward 2000 *Cities* 15(5), 361–74; G. Kostinskiy (2001) Post-socialist cities in flux, in R. Paddison (ed.) *Handbook of Urban Studies* London: Sage, 451–65; R. Rudolph and I. Brade (2005) Moscow: processes of restructuring in the post-Soviet metropolitan periphery *Cities* 22(2), 135–50.

45. S. Lehmann and B. Ruble (1997) From Soviet to European Yaroslavl *Urban Studies* 34(7), 1085–107; R. Struyk (ed.) *Restructuring Russia's Housing Sector 1991–1997* Washington DC: Urban Institute.

46. D. Smith and J. Scarpaci (2000) Urbanization in transitional societies *Urban Geography* 21(8), 745–57.

47. Brundtland Commission (1987) *Our Common Future* Oxford: Oxford University Press.

48. European Commission Expert Group on the Urban Environment (1994) *European Sustainable Cities* Brussels: Commission of the European Communities.

Chapter 9   NEW TOWNS

1. Barlow Report (1940) *Report of the Royal Commission on the Distribution of the Industrial Population* Cmd 6153, London: HMSO; P. Abercrombie (1945) *The Greater London Plan 1944* London: HMSO.

2. F. Schaffer (1970) *The New Town Story* London: Paladin.

3. Reith Committee (1946) *Final Report of the New Towns Committee* Cmd 6876 London: HMSO.

4. T. Bendixson and J. Platt (1992) *Milton Keynes: Image and Reality* Cambridge: Granta.

5. Ministry of Housing and Local Government (1964) *The South East Study 1961–1981* London: HMSO.

6. R. Goodlad and S. Scott (1996) Housing and the Scottish new towns *Urban Studies* 33(2), 317–35.

7. J. Herington (1984) *The Outer City* London: Harper & Row.

8. P. Hall (1989) *London 2001* London: Unwin Hyman.

9. A. Stockdale and G. Lloyd (1998) Forgotten needs? The demographic and socio-economic impact of free-standing new settlements *Housing Studies* 13(1), 43–58.

10. F. Osborn and A. Whittick (1977) *New Towns: Their Origins, Achievements and Progress* London: International Textbook Co.; P. Merlin (1971) *New Towns* London: Methuen; G. Golany (1978) *International Urban Growth Policies: New Town Contributions* Chichester: Wiley; F. Atash and Y. Shirazi Beheshtika (1998) New towns and their practical challenges: the experience of Poulad Shahr in Iran *Habitat International* 22(1), 1–13; A. Jacquemin (1999) *Urban Development and New Towns in the Third World: Lessons from the New Bombay Experience* Aldershot: Ashgate.

11. F. Osborn and A. Whittick (1977) *New Towns* London: International Textbook Co.

12. L. van Grunsven (2000) Singapore, in P. Marcuse and R. van Kempen (eds) *Globalizing Cities* Oxford: Blackwell, 95–126; L. Sim, L. Malone-Lee and K. Chin (2001) Integrating land use and transport planning to reduce work-related travel *Habitat International* 25, 399–414.

13. C.-M. Lee and K.-H. Ahn (2005) Five new towns in the Seoul metropolitan area and their attractions in non-working trips *Habitat International* 29, 647–66.

14. A. Vedula (2007) Blueprint and reality: Navi Mumbai, the city of the twenty-first century *Habitat International* 31, 12–23; M. Pacione (2006) Mumbai *Cities* 23(3), 229–38.

15. C. Stein (1966) *Toward New Towns for America* Cambridge MA: MIT Press; C. Lee and B. Stabin-Nesmith (2001) The continuing value of a planned community *Journal of Urban Design* 6(2), 151–84.

16. H. Wright (1972) Radburn revisited *Ekistics* 33, 196–201.

17. J. Arnold (1971) *The New Deal in the Suburbs: A History of the Greenbelt Town Program* Columbus OH: Ohio State University Press; K. Parsons (1990) Clarence Stein and the greenbelt towns *Journal of the American Planning Association* 56(2), 161–83.

18. C. Corden (1977) *Planned Cities* Beverly Hills CA: Sage.

19. E. Eichler and B. Norwitch (1970) New towns, in D. Moynihan (ed.) *Toward a National Urban Policy* New York: Basic Books.

20. H. Gans (1967) *The Levittowners* New York: Random House.

21. N. Griffin (1974) *Irvine: The Genesis of a New Community* Washington DC: Urban Land Institute; A. Forsyth (2002) Planning lessons from three US new towns of the 1960s and 1970s: Irvine, Columbia and the Woodlands *Journal of the American Planning Association* 68(4), 387–415.

22. G. Breckenfield (1971) *Columbia and the New Cities* New York: Ives Washington.

23. K. Al-Hindi and C. Staddon (1997) The hidden histories and geographies of neo-traditional town planning: the case of Seaside, Florida *Environment and Planning D* 15, 349–72; M. Southworth (1997) Walkable suburbs? An evaluation of neo-traditional communities at the urban edge *Journal of the American Planning Association* 63(1), 28–44; C. Lee and K. Ahn (2003) Is Kentlands better than Radburn? The American garden city and New Urbanism paradigms *Journal of the American Planning Association* 69(1), 50–70.

24. K. Al-Hindi (201) The new urbanism: where and for whom? *Urban Geography* 22(3), 202–19; L. Ford (1999) Lynch revisited: new urbansim and theories of good city form *Cities* 16(4), 247–57.

Chapter 10   RESIDENTIAL MOBILITY AND NEIGHBOURHOOD CHANGE

1. P. Rossi (1955) *Why Families Move* New York: Free Press.

2. W. Clark and J. Onaka (1983) Life cycle and housing adjustment as explanations of residential mobility *Urban Studies* 20, 47–57.

3. D. Rowland (1982) Living arrangements and the later family life cycle in Australia *Australian Journal of Ageing* 1, 3–6.

4. H. Morrow-Jones and M. Wenning (2005) The housing ladder, the housing life-cycle and the housing life-course *Urban Studies* 42(10), 1739–54.

5. J. Huff (1986) Geographic regularities in residential search behavior *Annals of the Association of American Geographers* 76, 208–27.

6. R. Palm (1976) The role of real estate agents as mediators in two American cities *Geografiska Annaler* 58B, 28–41.

7. G. Weisbrod and A. Vidal (1981) Housing search barriers for low-income renters *Urban Affairs Quarterly* 16, 465–82.

8. M. Deurloo, W. Clark and F. Dieleman (1990) Choice of residential environment in the Randstad *Urban Studies* 27, 335–51.

9. J. Rex (1971) The concept of housing class and the sociology of race relations *Race* 12, 293–301.

10. P. Harrison (1983) *Inside the Inner City* Harmondsworth: Penguin.

11. O. Valins (2003) Stubborn identities and the construction of socio-spatial boundaries: ultra-orthodox Jews living in contemporary Britain *Transactions of the Institute of British Geographers* 28, 158–75.

12. I. Townsend (2002) Age-related and gated retirement communities in the third age *Environment and Planning B* 29, 371–96.

13. R. Kitchin (2002) Sexing the city *City* 6(2), 205–18.

14. E. Hoover and R. Vernon (1962) *Anatomy of a Metropolis* New York: Doubleday; K. Temkin and W. Rohr (1996) Neighbourhood change and urban policy *Journal of Planning Education and Research* 15, 159–70.

15. A. Downs (1981) *Neighborhoods and Urban Development* Washington DC: Brookings Institution.

16. W. Grigsby, M. Baratz, G. Galster and D. MacLennan (1987) The dynamics of neighbourhood change and decline *Progress in Planning* 28(1), 1–76.

17. D. Wilson, H. Margulis and J. Ketchum (1994) Spatial aspects of housing abandonment in the 1990s: the Cleveland experience *Housing Studies* 9(4), 493–510; T. Accordino and G. Johnson (2000) Addressing the vacant and abandoned property problem *Journal of Urban Affairs* 22(3), 301–15.

18. D. Bartelt (1995) Urban housing in an era of global capital *Annals of the American Academy of Political and Social Science* 551, 121–36.

19. N. Smith and P. Williams (eds) (1986) *Gentrification of the City* London: Allen & Unwin; N. Smith (1996) *The New Urban Frontier: Gentrification and the Revanchist City* London: Routledge; L. Lees (2000) A reappraisal of gentrification *Progress in Human Geography* 24(3), 389–408.

20. R. Beauregard (1986) The chaos and complexity of gentrification, in N. Smith and P. Williams (eds) *Gentrification of the City* London: Allen & Unwin, 35–55.

21. S. Zukin (1982) Loft living as historic compromise in the urban core *International Journal of Urban and Regional Research* 6, 256–67; C. Hamnett and D. Whitelegg (2007) Loft conversion and gentrification in London *Environment and Planning A* 39, 106–24.

22. A. Smith (1989) Gentrification and the spatial contribution of the state *Antipode* 21, 232–60; J. Carpenter and L. Lees (1995) Gentrification in New York, London and Paris: an international comparison *International Journal of Urban and Regional Research* 19(2), 286–303; T. Butler and G. Robson (2003) Negotiating their way *Urban Studies* 40(9), 1791–809.

23. G. Bentham and M. Moseley (1980) Socio-economic change and disparities within the Paris agglomeration *Regional Studies* 14, 55–70.

24. L. Lees (1994) Gentrification in London and New York: an Atlantic gap? *Housing Studies* 9(2), 199–217; J. Hackworth and N. Smith (2001) The changing state of gentrification *Tijdschrift voor Economische en Sociale Geografie* 92(4), 464–77.

Chapter 14   THE ECONOMY OF CITIES

1. I. Wallerstein (1974) *The Modern World System* New York: Academic Press.
2. M. Castells (1996) *The Rise of the Network Society* Oxford: Blackwell; S. Sassen (2002) *Global Networks, Linked Cities* London: Routledge.
3. J. Kurtzman (1993) *The Death of Money* New York: Simon & Schuster.
4. F. Hamilton (1984) Industrial restructuring: an international problem *Geoforum* 15, 349–64.
5. United Nations (1993) *World Investment Report* New York: United Nations; J. Dunning (1993) *Multinational Enterprises and the Global Economy* Wokingham: Addison-Wesley.
6. R. Pringle (1992) Financial markets versus governments, in T. Banuri and J. Schor (eds) *Financial Openness and National Autonomy* Oxford: Clarendon Press, 89–109.
7. C. Kossis (1992) A miracle without end? Japanese capitalism and the world economy *International Socialism* 54, 105–32; C. Johnson (1995) *Japan: Who Governs? The Rise of the Developmental State* New York: Norton.
8. M. Castells and P. Hall (1994) *Technopoles of the World* London: Routledge.
9. W. Stöhr and R. Ponighaus (1992) Towards a data-based evaluation of the Japanese technopolis policy *Regional Studies* 26(7), 605–23.
10. M. Castells (1996) *The Rise of the Network Society* Oxford: Blackwell.
11. A. Saxenian (1983) The genesis of Silicon Valley *Built Environment* 9, 7–17; A. Scott (1993) *Technopolis: High-technology Industry and Regional Development in Southern California* Berkeley CA: University of California Press.
12. R. Sternberg (1996) Reasons for the genesis of high-tech regions *Geoforum* 15, 349–64; N. Komninos (2002) *Intelligent Cities* London: Routledge.
13. I. Audirac (2003) Information Age landscapes outside the developed world *Journal of the American Planning Association* 69(1), 16–32.
14. S. Walcott (2002) Chinese industrial and science parks *Professional Geographer* 54(3), 349–64.
15. M. Indergaard (2003) The webs they weave *Urban Studies* 40(2), 379–401.
16. S. Tatsumo (1986) *The Technopolis Strategy: Japan, High Technology and the Control of the Twenty-first Century* Englewood Cliffs NJ: Prentice-Hall.
17. P. Daniels (1993) *Service Industries in the World Economy* Oxford: Blackwell.
18. T. Hutton (2004) Service industries, globalisation, and urban restructuring within the Asia-Pacific *Progress in Planning* 61, 1–74.
19. S. Sassen (1995) Urban impacts of economic globalisation, in J. Brotchie, M. Boddy, E. Blakeley, P. Hall and P. Newton (eds) *Cities in Competition* London: Longman, 36–57.
20. N. Buck, I. Gordon, P. Hall, M. Harlow and M. Kleinman (2002) *Working Capital: Life and Labour in Contemporary London* London: Routledge; C. Hamnett (2003) *Unequal City: London in the Global Arena* London: Routledge.
21. Organisation for Economic Co-operation and Development (1993) *Education at a Glance* Geneva: OECD.
22. M. Storper and A. Scott (1990) Work organisation and local labour markets in an era of flexible production *International Labour Review* 129(6), 573–91.
23. P. Drucker (1993) *Post-capitalist Society* London: Butterworth Heinemann.
24. R. Florida (2002) *The Rise of the Creative Class: and How it's Transforming Work, Leisure, Community and Everyday Life* New York: Basic Books; A. Scott (2006) Creative cities: conceptual issues and policy questions *Journal of Urban Affairs* 28(1), 1–17.
25. K. Newman (1999) *No Shame in my Game: The Working Poor in the Inner City* New York: Knopf; L. Gobillon *et al.* (2007) The mechanisms of spatial mismatch *Urban Studies* 44(2), 2401–27.
26. Office of National Statistics (2005) *Social Trends* London: Stationery Office.
27. A. Manning and B. Petrongolo (2004) *The Part Time Pay Penalty* London: DTI.
28. K. Ward *et al.* (2006) Living and working in urban working-class communities *Geoforum* 38, 312–25.
29. G. Haughton, S. Johnson, L. Murphy and K. Thomas (1993) *Local Geographies of Unemployment* Aldershot: Avebury.
30. M. Pacione (1995) *Glasgow: The Socio-spatial Development of the City* Chichester: Wiley.
31. R. Meegan (1989) Paradise postponed: the growth and decline of Merseyside's outer estates, in P. Cooke (ed.) *Localities* London: Unwin Hyman, 198–234.
32. S. Sassen (1994) *Cities in the World Economy* London: Pine Forge Press.
33. A. Portes, M. Castells and L. Benton (1989) *The Informal Economy: Studies in Advanced and Less Developed Countries* Baltimore MD: Johns Hopkins University Press; R. Waldinger and M. Lapp (1993) Back to the sweatshop or ahead to the informal sector? *International Journal of Urban and Regional Research* 17(1), 6–29; E. Bonacich and R. Applebaum (2000) The return of the sweatshop, in E. Bonacich and R. Applebaum (eds) *Behind the Label* Berkeley CA: University of California Press, 1–25.
34. J. Peck and N. Theodore (2001) Contingent Chicago *International Journal of Urban and Regional Research* 25(3), 471–96; N. Theodore (2003) Political economies of day labour *Urban Studies* 40(9), 1811–28.
35. P. Hall (1966) *The World Cities* London: Weidenfeld & Nicolson.
36. F. Braudel (1985) *Civilization and Capitalism* volume 3 *The Perspective of the World* London: Fontana.
37. R. Ross and K. Trachte (1983) Global cities and global classes *Review* 6(3), 393–431.

38. J. Feagin (1985) The global context of metropolitan growth *American Journal of Sociology* 90(6), 1204–30.

39. J. Friedmann (1986) The world city hypothesis *Development and Change* 17, 69–74.

40. N. Thrift (1989) The geography of international economic disorder, in R. Johnston and P. Taylor (eds) *A World in Crisis* Oxford: Blackwell, 16–79.

41. J. Beaverstock, R. Smith and P. Taylor (2000) World-city network *Annals of the Association of American Geographers* 90(1), 123–34.

42. S. Sassen (1994) *Cities in the World Economy* London: Pine Forge Press; D. Smith and M. Timberlake (1995) Conceptualising and mapping the structure of the world system's city system *Urban Studies* 32(2), 287–302; B. Warf (2000) New York: the Big Apple in the 1990s *Geoforum* 31, 487–99; J. Beaverstock, R. Smith and P. Taylor (2000) World-city network *Annals of the Association of American Geographers* 90(1), 123–34; J. Short (2004) *Global Metropolitan* London: Routledge.

43. J. Friedmann and G. Wolff (1982) World city formation *International Journal of Urban and Regional Research* 6(3), 309–44.

44. S. Sassen (1981) *The Global City: London, New York, Tokyo* Princeton NJ: Princeton University Press.

45. S. Sassen (1981) *The Global City: London, New York, Tokyo* Princeton NJ: Princeton University Press.

46. H. Wial and A. Friedhoff (2006) *Bearing the Brunt* Washington DC: Brookings Institution.

47. R. Martin (1988) Industrial capitalism in transition: the contemporary reorganisation of the British space-economy, in D. Massey and J. Allen (eds) *Uneven Redevelopment* London: Hodder & Stoughton, 202–31.

48. N. Flyn and A. Taylor (1986) Inside the rust belt: an analysis of the decline of the West Midlands economy *Environment and Planning A* 18, 865–900.

49. M. Healey and D. Clark (1985) Industrial decline in a local economy: the case of Coventry 1974–1982 *Environment and Planning A* 17, 1351–67.

50. R. Martin and B. Rowthorn (1986) *The Geography of Deindustrialisation* London: Macmillan.

51. D. Rehfeld (1995) Disintegration and reintegration of production clusters in the Ruhr area, in P. Cooke (ed.) *The Rise of the Rustbelt* London: UCL Press, 85–102.

52. M. Pacione (1995) *Glasgow: The Socio-spatial Development of the City* Chichester: Wiley.

53. A. Gamble (1981) *Britain in Decline: Economic Policy, Political Strategy and the British State* London: Macmillan.

54. G. Rees and J. Lambert (1985) *Cities in Crisis* London: Arnold.

55. B. Jessop (1980) The transformation of the state in post-war Britain, in R. Scase (ed.) *The State in Western Europe* London: Routledge, 23–93.

56. J. Mackintosh (1977) *The British Malaise: Political or Economic?* Southampton: Southampton University Press.

57. N. Gardner (1987) *Decade of Discontent: The Changing British Economy since 1973* Oxford: Blackwell.

58. R. Dennis (1978) The decline of manufacturing employment in Greater London *Urban Studies* 15, 69–73.

59. Community Development Project (1977) *The Cost of Industrial Change* London: CDP Editorial Team; J. May et al. (2007) Keeping London working *Transactions of the Institute of British Geographers* 32, 151–67.

60. B. Robson (1988) *Those Inner Cities* Oxford: Oxford University Press.

Chapter 15   POVERTY AND DEPRIVATION IN THE WESTERN CITY

1. S. Lansley and J. Mack (1991) *Breadline Britain in the 1990s* London: Harper Collins; M. O'Higgins and S. Jenkins (1990) Poverty in the EC, in R. Teehens and B. van Praag (eds) *Analysing Poverty in the European Community* Luxembourg: Eurostat.

2. P. Barclay (1995) *Inquiry into Income and Wealth* volume 1, York: Joseph Rowntree Foundation.

3. R. Silburn (1998) United Kingdom, in J. Dixon and D. Macorov (eds) *Poverty: A Persistent Global Reality* London: Routledge, 204–28.

4. US Bureau of the Census (1995) *Statistical Abstracts of the United States* Washington DC: Bureau of the Census.

5. J. Midgley and M. Livermore (1998) United States of America, in J. Dixon and D. Macarov (eds) *Poverty: A Persistent Global Reality* London: Routledge, 229–47.

6. United States Census Bureau (2000) *Poverty in the United States* Washington DC: US Census Bureau.

7. A. Power (1996) Area-based poverty and resident empowerment *Urban Studies* 33(9), 1535–64; A. Green (1996) Changing local concentrations of poverty and affluence in Britain 1981–1991 *Geography* 81(1), 15–25; M. Pacione (1997) Urban restructuring and the reproduction of inequality in Britain's cities: an overview, in M. Pacione (ed.) *Britain's Cities: Geographies of Division in Urban Britain* London: Routledge, 7–60.

8. Y. Shaw-Taylor (1998) Profile of social disadvantage in the 100 largest cities of the US 1980–1990/93 *Cities* 15(5), 317–26; J. Pack (1998) Poverty and urban public expenditures *Urban Studies* 35(11), 1995–2019.

9. O. Lewis (1966) The culture of poverty *Scientific American* 215, 19–25.

10. J. Rowntree (1901) *Poverty: A Study of Town Life* London: Macmillan.

11. W. Beveridge (1942) *Social Insurance and Allied Services* Cmnd 6404, London: HMSO.

12. S. Thake and R. Staubach (1993) *Investing in People* York: Joseph Rowntree Foundation.

13. M. Pacione (1986) Quality of life in Glasgow: an applied geographical analysis *Environment and Planning A* 18, 1499–1520.

18. P. Jones, D. Hillier and D. Comfort (2003) Business improvement districts *Town and Country Planning* 72(5), 158–61; M. Symes and M. Steel (2003) Lessons from America: the role of Business Improvement Districts as an agent of urban change *Town Planning Review* 74(3), 301–13.

19. J. Heinig (1985) *Public Policy and Federalism* New York: St Martin's Press.

20. J. McCarthy (2003) Regeneration and community involvement: the Chicago empowerment zone *City* 7(1), 95–105; D. Oakley and H. Tsao (2006) Socioeconomic gains and spillover effects of geographically targeted initiatives to combat economic distress *Cities* 24(1), 43–59.

21. S. McGreal, J. Berry, J. Lloyd and J. McCarthy (2002) Tax-based mechanisms in urban regeneration *Urban Studies* 39(10), 1819–31.

22. G. Peterson and C. Lewis (1986) *Reagan and the Cities* Washington DC: Urban Institute.

23. M. Levine (1989) The politics of partnership: urban redevelopment since 1945, in G. Squires (ed.) *Unequal Partnerships* New Brunswick NJ: Rutgers University Press, 12–34.

24. M. Levine (1989) The politics of partnership: urban redevelopment since 1945, in G. Squires (ed.) *Unequal Partnerships* New Brunswick NJ: Rutgers University Press, 12–34.

25. B. Berry (1985) Islands of renewal in seas of decay, in P. Peterson (ed.) *The New Urban Reality* Washington DC: Brookings Institution, 69–98.

26. M. Levine (1987) Downtown redevelopment as an urban growth strategy: a critical appraisal of the Baltimore renaissance *Journal of Urban Affairs* 9(2), 133–8.

27. P. Ambrose (1994) *Urban Process and Power* London: Routledge.

28. P. Malone (1996) *City, Capital and Water* London: Routledge.

29. I. Turok (1992) Property-led urban regeneration: panacea or placebo? *Environment and Planning A* 24, 361–79.

30. S. Fainstein (1994) *The City Builders: Property, Planning and Politics in London and New York* Oxford: Blackwell.

31. A. Merrifield (1993) The Canary Wharf debate *Environment and Planning A* 25, 1247–65; P. Daniels and J. Boke (1993) Extending the boundary of the City of London *Environment and Planning A* 25, 539–52.

32. A. Church and M. Frost (1995) The Thames Gateway *Geographical Journal* 161(2), 199–209.

33. A. Scott (1997) The cultural economy of cities *International Journal of Urban and Regional Research* 21(2), 323–39; A. Pratt (1997) The cultural industries production system *Environment and Planning A* 29, 1953–74.

34. F. Mort (1996) *Cultures of Consumption* London: Routledge; J. Montgomery (2003) Cultural quarters as mechanisms for urban regeneration *Planning Practice and research* 18(4), 293–306.

35. N. Oatley (1996) Sheffield's cultural industries quarter *Local Economy* 11(2), 172–9.

36. L. Crewe and J. Beaverstock (1998) Fashioning the city: cultures of consumption and contemporary urban spaces *Geoforum* 29(3), 287–308.

37. S. Zukin (1995) *The Cultures of Cities* Oxford: Blackwell.

38. C. Law (2002) *Urban Tourism* London: Continuum.

39. S. Shaw, S. Bagnall and J. Karmowska (2004) Ethnoscapes as spectacle *Urban Studies* 41(10), 1983–2000.

40. Audit Commission (1987) *The Management of London's Authorities* London: Audit Commission.

41. M. Bradford and B. Robson (1995) An evaluation of urban policy, in R. Hambleton and H. Thomas (eds) *Urban Policy Evaluation* London: Paul Chapman, 37–54.

42. E. Blakely (1989) *Planning Local Economic Development* Newbury Park CA: Sage.

43. J. Gold and S. Ward (1994) *Place Promotion* Chichester: Wiley; S. Ward (1998) *Selling Places* London: Spon.

44. A. Eisenschitz and J. Gough (1993) *The Politics of Local Economic Policy* London: Macmillan.

45. C. Hasluck (1987) *Urban Unemployment* London: Longman.

46. J. Guyford (1985) *The Politics of Local Socialism* London: Allen & Unwin; S. Quilley (2000) Manchester First: from municipal socialism to the entrepreneurial city *International Journal of Urban and Regional Research* 24(3), 601–15.

47. M. Boddy and C. Fudge (1984) *Local Socialism* London: Macmillan.

48. T. Chapin (2002) Beyond the entrepreneurial city *Journal of Urban Affairs* 24(5), 565–81.

49. N. Krumholz (1982) A retrospective view of equity planning: Cleveland 1969–1979 *Journal of the American Planning Association* 48, 163–83.

50. D. Keating, N. Krumholz and J. Metzger (1989) Cleveland: post-populist public–private partnership, in G. Squires (ed.) *Unequal Partnerships* New Brunswick NJ: Rutgers University Press, 121–41; N. Krumholz (1995) Equity and local economic development, in R. Caves (ed.) *Exploring Urban America* London: Sage, 170–83.

51. J. Dawson and C. Walker (1990) Mitigating the social costs of private development *Town Planning Review* 61(2), 157–70; M. Smith (1988) The uses of linked development policies in US cities, in M. Parkinson, B. Foley and D. Judd (eds) *Regenerating the Cities* Manchester: Manchester University Press, 93–109.

52. L. Sagalyn (1997) Negotiating for public benefits: the bargaining calculus of public–private development *Urban Studies* 34(12), 1955–70.

53. T. Brindley (2000) Community roles in urban regeneration *City* 4(3), 363–77.

54. P. Medoff and H. Sklar (1994) *Streets of Hope: The Fall and Rise of an Urban Neighbourhood* Boston MA: South End Press.

55. B. Colenutt (1991) The London Docklands Development Corporation: has the community benefited? in M. Keith and A. Rogers (eds) *Hollow Promises* London: Mansell, 31–41.

56. N. Wates (1976) *The Battle for Tolmer Square* London: Routledge.

57. A. Leyshon, R. Lee and C. Williams (2003) *Alternative Economic Spaces* London: Sage; A. Amin, A. Cameron and R. Hudson (2002) *Placing the Social Economy* London: Routledge; F. Moulaerts and O. Ailenei (2005) Social economy, third sector and solidarity relations *Urban Studies* 42(11), 2037–53.

58. K. Hayton (1996) A critical evaluation of the role of community business in urban regeneration *Town Planning Review* 67(1), 1–2.

59. A. McArthur (1994) Community business and urban regeneration, in R. Paddison, J. Money and B. Lever (eds) *International Perspectives in Urban Studies* volume 2, London: Jessica Kingsley, 251–77.

60. M. Tietze (1989) Neighbourhood economics: local communities and regional markets *Economic Development Quarterly* 3(2), 111–22; C. Walker (2002) *Community Development Corporations and their Changing Support Systems* Washington DC: Urban Institute.

61. S. Bryan (1992) A new direction for community development in the United States, in P. Ekins and M. Max-Neef (eds) *Real Life Economics* London: Routledge, 372–83; T. Robinson (1996) Inner-city innovation: the non-profit community development corporation *Urban Studies* 33(9), 1647–70; A. Vidal (1997) Can community development reinvent itself? *Journal of the American Planning Association* 63(4), 429–38.

62. A. Twelvetrees (1997) *Organising for Development* Aldershot: Avebury.

63. M. Pacione (1997) Local exchange trading systems as a response to the globalisation of capitalism *Urban Studies* 34(8), 1179–99; M. Pacione (1999) The other side of the coin: local currency as a response to the globalisation of capital *Regional Studies* 33(1), 63–72.

64. D. Wilson (2007) *Cities and Race: America's New Black Ghetto* London: Routledge.

65. C. Williams and J. Windebank (2001) *Revitalisaing Deprived Urban Neighbourhoods* Aldershot: Ashgate.

Chapter 17 COLLECTIVE CONSUMPTION AND SOCIAL JUSTICE IN THE CITY

1. M. Castells (1977) *The Urban Question* London: Arnold.
2. C. Pierson (1991) *Beyond the Welfare State?* Cambridge: Polity Press.
3. J. Wolch (1990) *The Shadow State* New York: Foundation Center.
4. S. Pinch (1997) *Worlds of Welfare* London: Routledge.
5. K. Walsh (1995) *Public Services and Market Mechanisms* Basingstoke: Macmillan.

6. P. Samuelson (1954) The pure theory of public expenditure *Review of Economics and Statistics* 36, 387–9.

7. R. Musgrave (1958) *The Theory of Public Finance* New York: McGraw-Hill.

8. J. Buchanan (1965) An economic theory of clubs *Economica* 32, 1–14.

9. C. Tiebout (1956) A pure theory of local expenditures *Journal of Political Economy* 64, 416–24.

10. S. Pinch (1985) *Cities and Services* London: Routledge.

11. C. Tiebout (1956) A pure theory of local expenditures *Journal of Political Economy* 64, 416–24.

12. V. Ostrom, C. Tiebout and R. Warren (1961) The organization of government in metropolitan areas *American Political Science Review* 55, 831–42.

13. J. O'Connor (1973) *The Fiscal Crisis of the State* New York: St Martin's Press.

14. A. Maslow (1954) *Motivation and Personality* New York: Harper.

15. D. Smith (1994) *Geography and Social Justice* Oxford: Blackwell.

16. A. Duncan and P. Smith (1996) On the use of statistical techniques to infer territorial spending needs, in G. Pola, G. France and R. Levaggi (eds) *Developments in Local Government Finance* Cheltenham: Edward Elgar, 24–39; M. Pacione (1999) The geography of poverty and deprivation, in M. Pacione (ed.) *Applied Geography: Principles and Practice* London: Routledge, 400–13.

17. P. Knox (1979) Medical deprivation and public policy *Social Science and Medicine* 13D, 111–21.

18. S. Bowlby (1979) Accessibility, mobility and shopping provision, in B. Goodall and A. Kirkby (eds) *Resources and Planning* Oxford: Pergamon Press, 293–324.

19. P. Bloch (1974) *Equality of Distribution of Police Services* Washington DC: Urban Institute.

20. P. Knox (1978) The intra-urban ecology of primary medical care *Environment and Planning A* 10, 415–35; D. Martin and H. Williams (1992) Market-area analysis and accessibility to primary health-care centres *Environment and Planning A* 24, 1009–19.

21. J. Bradley, A. Kirby and P. Taylor (1978) Distance decay and dental decay *Regional Studies* 12, 529–40.

22. E. Savas (1979) On equity in providing social services *Ekistics* 46, 144–8.

23. I. Robertson (1978) Planning the location of recreation centres in an urban area *Regional Studies* 12, 419–27; E. Talen and L. Anselin (1998) Assessing spatial equity: an evaluation of measures of accessibility to public playgrounds *Environment and Planning A* 30, 595–618.

24. A. Kirkby (1979) *Education, Health and Housing* Farnborough: Saxon House; M. Pacione (1989) Access to urban services: the case of secondary schools in Glasgow *Scottish Geographical Magazine* 105, 12–18; T. Smith and M. Noble (1995) *Education Divides* London: Child Poverty Action Group.

25. K. Cole and A. Gattrell (1986) Public libraries in Salford *Environment and Planning A* 18, 253–68.

26. T. Hagerstrand (1970) What about people in regional science? *Papers and Proceedings of the Regional Science Association* 24, 7–21.

27. S. Rowe and J. Wolch (1990) Social networks in time and space: homeless women on Skid Row, Los Angeles *Annals of the Association of American Geographers* 80(2), 184–204.

28. R. Hodgart (1978) Optimising access to public services *Progress in Human Geography* 2, 17–48; B. Massam (1993) *The Right Place: Shared Responsibility and the Location of Public Facilities* New York: Longman; K. Witten, D. Exeter and A. Field (2003) The quality of urban environments *Urban Studies* 40(1), 161–77.

29. R. Rich (1979) Neglected issues in the study of urban service distributions *Urban Studies* 16, 143–56.

30. R. Tawney (1952) *Equality* London: Allen & Unwin.

31. S. Pinch (1997) *Worlds of Welfare* London: Routledge; J. Wolch (1990) *The Shadow State* New York: Foundation Center.

32. R. Lineberry (1977) *Equality and Urban Policy* Beverly Hills CA: Sage.

33. K. Mladenka and K. Hill (1977) The distribution of benefits in an urban environment *Urban Affairs Quarterly* 13, 73–94.

34. F. Levy, A. Meltsner and A. Wildavsky (1974) *Urban Outcomes* Berkeley CA: University of California Press.

35. R. Lineberry (1977) *Equality and Urban Policy* Beverly Hills CA: Sage.

36. N. Kamel and A. Loukaitou-Sideris (2004) Residential assistance and recovery following the Northridge earthquake *Urban Studies* 41(3), 532–62.

37. S. Cutter (1995) Race, class and environmental justice *Progress in Human Geography* 19, 107–18.

38. L. Pulido, S. Sidawi and R. Voss (1996) An archaeology of environmental racism in Los Angeles *Urban Geography* 17(5), 419–39.

39. A. Hawley (1951) Metropolitan population and municipal government expenditures in central cities *Journal of Social Issues* 7(1), 100–8.

40. H. Brazer (1959) *City Expenditures in the United States* Occasional Paper 6, New York: National Bureau of Economic Research.

41. W. Neenan (1972) *Political Economy of Urban Areas* Chicago: Markham; H. Green (1974) *Fiscal Interactions in a Metropolitan Area* Lexington MA: Heath.

42. R. Warner (1989) Deinstitutionalisation: how did we get where we are? *Journal of Social Issues* 45, 17–30.

43. R. Kearns and S. Taylor (1989) Daily life experience of people with chronic mental disabilities in Hamilton, Ontario *Canada's Mental Health* 37, 1–4; J. Wolch (1980) Residential location of the service-dependent poor *Annals of the Association of American Geographers* 70, 330–41; H. Parr (1997) Mental health, public space, and the city: questions of individual and collective access *Environment and Planning D* 15, 435–54; M. Dear and J. Wolch (1987) *Landscapes of Despair: From Deinstitutionalisation to Homelessness* Princeton NJ: Princeton University Press.

44. M. Dear and S. Taylor (1982) *Not on our Street: Community Attitudes toward Mental Health Care* London: Pion; M. Dear (1992) Understanding and overcoming the NIMBY syndrome *Journal of the American Planning Association* 58(3), 288–300; L. Takahashi (1998) Community responses to human service delivery in US cities, in R. Fincher and J. Jacobs (eds) *Cities of Difference* New York: Guilford Press, 120–48.

45. S. Taylor (1989) Community exclusion of the mentally ill, in J. Wolch and M. Dear (eds) *The Power of Geography* Boston MA: Unwin Hyman, 316–30.

46. M. Pacione (1989) Access to urban services: the case of secondary schools in Glasgow *Scottish Geographical Magazine* 105, 12–18.

47. K. Roberts (1995) *Youth and Employment in Modern Britain* Oxford: Oxford University Press.

48. P. Hill (1996) Education, in G. Galster (ed.) *Reality and Research* Washington DC: Urban Institute Press, 131–55.

49. M. Pacione (1997) The geography of educational disadvantage in Glasgow *Applied Geography* 17(3), 169–92.

50. D. Dorling (1995) *A New Social Atlas of Britain* Chichester: Wiley.

51. D. Smith (1977) *Human Geography: A Welfare Approach* London: Arnold.

52. D. Smith (1973) *The Geography of Social Well-being in the United States* New York: McGraw-Hill; P. Knox (1975) *Social Well-being: A Spatial Perspective* Oxford: Oxford University Press; D. Harvey (1973) *Social Justice and the City* London: Arnold.

53. D. Harvey (1993) Class relations, social justice and the politics of difference, in M. Keith and S. Pile (eds) *Place and the Politics of Identity* London: Routledge, 41–66; D. Smith (1994) *Geography and Social Justice* Oxford: Blackwell; I. Young (1990) *Justice and the Politics of Difference* Princeton NJ: Princeton University Press.

54. A. Merrifield and E. Swyngedouw (1995) *The Urbanisation of Injustice* London: Lawrence & Wishart; D. Harvey (1996) *Justice, Nature and the Geography of Difference* Oxford: Blackwell.

55. K. Joseph and J. Sumption (1979) *Equality* London: John Murray.

56. Pope John Paul II (1988) *Solicitudo Rei Socialis* Vatican: Holy See.

Chapter 18 RESIDENTIAL DIFFERENTIATION AND COMMUNITIES IN THE CITY

1. F. Tonnies (1887) *Community and Society* New York: Harper & Row.

2. E. Durkheim (1893) *De la division du travail social* Paris: Alcan.

3. O. Spengler (1918) *The Decline of the West* New York: Knopf.

4. L. Wirth (1938) Urbanism as a way of life *American Journal of Sociology* 44, 1–24.

5. J. Jacobs (1961) *The Death and Life of Great American Cities* New York: Vintage.

6. H. Gans (1962) *The Urban Villagers* New York: Free Press; M. Young and P. Willmott (1957) *Family and Kinship in East London* London: Routledge; T. Jablonsky (1993) *Pride in the Jungle* Baltimore MD: Johns Hopkins University Press; M. O'Brien and D. Jones (1996) Social class: continuity and change: family life in Barking and Dagenham, in T. Butler and M. Rustin (eds) *Rising in the East* London: Lawrence & Wishart, 61–80.

7. C. Booth (1902) *Life and Labour of the People in London* London: Macmillan.

8. H. Zorbaugh (1961) The natural areas of the city, in G. Theodorson (ed.) *Studies in Human Ecology* New York: Harper & Row, 45–9.

9. H. Zorbaugh (1929) *The Gold Coast and the Slum* Chicago: University of Chicago Press.

10. P. Hatt (1946) The concept of the natural area *American Sociological Review* 11, 423–7.

11. L. Wirth (1928) *The Ghetto* Chicago: University of Chicago Press.

12. E. Shevky and W. Bell (1955) *Social Area Analysis* Stanford CA: Stanford University Press.

13. E. Shevky and W. Bell (1955) *Social Area Analysis* Stanford CA: Stanford University Press.

14. D. Timms (1971) *The Urban Mosaic* Cambridge: Cambridge University Press.

15. W. Davies (1984) *Factorial Ecology* Aldershot: Avebury.

16. D. Ley (1983) *A Social Geography of the City* New York: Harper & Row.

17. M. White (1987) *American Neighborhoods and Residential Differentiation* New York: Russell Sage Foundation.

18. D. Erwin (1984) Correlates of urban residential structure *Sociological Focus* 17, 59–75.

19. B. Berry and P. Rees (1969) The factorial ecology of Calcutta *American Journal of Sociology* 74, 445–91.

20. Y.-F. Tuan 1974 *Topophilia* Englewood Cliffs NJ: Prentice-Hall.

21. M. Fried (1963) Grieving for a lost home: psychological costs of relocation, in L. Duhl (ed.) *The Urban Condition* New York: Basic Books.

22. E. Soja (1971) *The Political Organisation of Space* Resource Paper 8, Commission on College Geography, Washington DC: Association of American Geographers.

23. J. Uexkull (1957) A stroll through the worlds of animals and men, in C. Schiller (ed.) *Instinctive Behaviour* London: Methuen, 5–80.

24. R. Ardrey (1972) *The Social Contract* London: Collins.

25. A. Rapaport (1977) *Human Aspects of Urban Form* Oxford: Pergamon.

26. E. Soja (1996) *Thirdspace: Journeys to Los Angeles and other Real-and-imagined Places* Oxford: Blackwell.

27. R. Glasser (1986) *Growing up in the Gorbals* London: Chatto & Windus.

28. R. Ford (1986) *The Sportswriter* London: Collins Harvill.

29. R. Downs and D. Stea (1973) *Image and Environment* London: Arnold.

30. T. Lee (1968) Urban neighbourhood as a socio-spatial schema *Human Relations* 21, 241–67.

31. D. Herbert and J. Raine (1976) Defining communities within urban areas *Town Planning Review* 47, 325–38.

32. F. Ladd (1970) Black youths view their environment: neighbourhood maps *Environment and Behavior* 22, 591–607.

33. M. Matthews (1984) Environmental cognition of young children *Transactions of the Institute of British Geographers* 9, 89–105.

34. D. Herbert and S. Peace (1980) The elderly in an urban environment, in D. Herbert and R. Johnston (eds) *Geography and the Urban Environment* volume 3, Chichester: Wiley, 223–55.

35. D. Ley and R. Cybriwsky (1974) Urban graffiti as territorial markers *Annals of the Association of American Geographers* 64, 491–505.

36. E. Relph (1976) *Place and Placelessness* London: Pion.

37. A. Schultz (1960) The social world and the theory of social action *Social Research* 27, 205–21.

38. D. Ley (1977) Social geography and the taken-for-granted world *Transactions of the Institute of British Geographers* 2, 498–512.

39. C. Geertz (1980) *Negara: The Theatre State in Nineteenth Century Bali* Princeton NJ: Princeton University Press.

40. P. Jackson (1985) Urban ethnography *Progress in Human Geography* 9, 157–76.

41. M. Castells (1997) *The Power of Identity* Oxford: Blackwell.

42. G. Hillery (1955) Definitions of community: areas of agreement *Rural Sociology* 20, 111–32.

43. G. Herbert (1963) The neighbourhood unit principle and organic theory *Sociological Review* 11, 165–213.

44. J. Gold (1980) *An Introduction to Behavioural Geography* Oxford: Oxford University Press.

45. A. Blowers (1973) The neighbourhood, in P. Sarre, H. Brown, A. Blowers, C. Hamnett and D. Boswell (eds) *The City as a Social System* Milton Keynes: Open University Press; R. Meegan and A. Mitchell (2001) It's not community round here, it's neighbourhood *Urban Studies* 38(12), 2167–94.

46. R. Park, E. Burgess and R. McKenzie (1925) *The City* Chicago: University of Chicago Press.

47. L. Wirth (1938) Urbanism as a way of life *American Journal of Sociology* 44, 1–24.

48. E. Shevky and W. Bell (1955) *Social Area Analysis* Stanford CA: Stanford University Press.

49. L. Wirth (1938) Urbanism as a way of life *American Journal of Sociology* 44, 1–24.

50. M. Webber (1963) Order in diversity: community without propinquity, in L. Wingo (ed.) *Cities and Space* Baltimore MD: Johns Hopkins University Press, 23–56.

51. S. Keller (1968) *The Urban Neighborhood* New York: Random House; R. Warren (1978) *The Community in America* Chicago: Rand McNally.

52. B. Wellman and B. Leighton (1979) Networks, neighbourhoods and communities *Urban Affairs Quarterly* 14(3), 363–90.

53. M. Janowitz (1952) *The Community Press in an Urban Setting* Chicago: University of Chicago Press.

54. K. Finsterbusch (1980) *Understanding Social Impacts* London: Sage.

55. J. Buckhardt (1971) Impact of highways on urban neighbourhoods *Highway Research Record* 356, 85–94; P. Rossi (1972) Community social indicators, in A. Campbell and P. Converse (eds) *The Human Meaning of Social Change* New York: Russell Sage Foundation.

56. R. Smith (1975) Measuring neighbourhood cohesion *Human Ecology* 3(3), 143–60.

57. M. Pacione (1984) Local areas in the city, in D. Herbert and R. Johnston (eds) *Geography and the Urban Environment* volume 6, Chichester: Wiley, 349–92.

58. C. Booth (1902) *Life and Labour of the People in London* London: Macmillan.

59. J. Allen and E. Turner (1988) *We the People: An Atlas of America's Ethnic Diversity* New York: Macmillan.

60. W. Clark (1986) Residential segregation in American cities *Population Research and Policy Review* 5, 95–127.

61. S. Wallace (1965) *Skid Row as a Way of Life* New York: Harper & Row.

62. G. Suttles (1968) *The Social Order of the Slum: Ethnicity and Territory in the Inner City* Chicago: University of Chicago Press.

63. M. Dear and J. Wolch (1987) *Landscapes of Despair* Oxford: Polity Press.

64 D. Massey (1980) Residential segregation and the spatial distribution of non-labour force population: the needy, elderly and disabled *Economic Geography* 56, 190–200.

65. M. Dear and S. Taylor (1982) *Not on our Street* London: Pion.

66. E. McKenzie (1994) *Privatopia* New Haven CT: Yale University Press.

67. D. Judd (1995) The rise of the new walled cities, in H. Liggett and D. Perry (eds) *Spatial Practices* London: Sage, 144–60; D. Luymens (1997) The fortification of suburbia *Landscape and Urban Planning* 39, 187–203.

68. M. Davis (1992) Fortress Los Angeles, in M. Sorkin (ed.) *Variations on a Theme Park* New York: Noonday Press, 154–80.

69. M. Davis (1992) *City of Quartz: Excavating the Future in Los Angeles* New York: Vintage.

70. W. Kowinski (1985) *The Malling of America* New York: Morrow.

71. D. Sibley (1992) Outsider in society and space, in K. Anderson and F. Gale (eds) *Inventing Place* Melbourne: Longman Cheshire, 107–22.

72. J. Binnie (1997) Coming out of geography *Environment and Planning D* 15, 223–37; A. Collins (2004) Sexual dissidence, enterprise and assimilation *Urban Studies* 41(9), 1789–806.

73. R. Kitchen (2002) Sexing the city *City* 6(2), 205–18.

74. L. Knopp (1990) Some theoretical implications of gay involvement in an urban land market *Political Geography Quarterly* 9, 337–52.

75. T. Rothenberg (1995) And she told two friends, in D. Bell and G. Valentine (eds) *Mapping Desire* London: Routledge, 165–81.

76. B. Forrest (1995) West Hollywood as a symbol: the significance of place in the construction of a gay identity *Environment and Planning D* 13, 133–57.

77. M. Gordon (1964) *Assimilation in American Life* New York: Oxford University Press; A. Khakee (1999) *Urban Renewal, Ethnicity and Social Exclusion in Europe* Aldershot: Ashgate.

78. F. Boal (1976) Ethnic residential segregation, in D. Herbert and R. Johnston (eds) *Social Areas in Cities* volume 1, Chichester: Wiley, 41–79.

79. P. Jackson and S. Smith (1981) *Social Interaction and Ethnic Segregation* London: Academic Press.

80. R. Ward and R. Jenkins (1984) *Ethnic Communities in Business* Cambridge: Cambridge University Press; G. Barrett, T. Jones and D. McEvoy (1997) Ethnic minority business, in R. Paddison and W. Lever (eds) *International Perspectives in Urban Studies* London: Jessica Kingsley, 188–217; J. Light, G. Sobough, M. Bozorgmehr and C. der Martirosian (1994) Beyond the ethnic enclave economy *Social Problems* 41, 601–16; C. Rhodes and N. Naki (1992) Brick Lane: a village economy in the shadow of the City, in L. Budd and S. Whimster (eds) *Global Finance and Urban Living* London: Routledge, 333–52; W. Li (1998) Anatomy of a new ethnic suburb: the Chinese ethnoburb in Los Angeles *Urban Studies* 35(3), 479–501; D. Lee (1995) Koreatown and Korean small firms in Los Angeles *Professional Geographer* 47(2), 184–95.

81. L. Wirth (1928) *The Ghetto* Chicago: University of Chicago Press; O. Valins (2003) Stubborn identities and the construction of socio-spatial boundaries *Transactions of the Institute of British Geographers* 28, 158–75.

82. J. Benyon and J. Solomos (1987) *The Roots of Urban Unrest* Oxford: Pergamon; H. Graham and T. Garr (1979) *Violence in America* London: Sage.

83. F. Boal (1987) Segregation, in M. Pacione (ed.) *Social Geography: Progress and Prospect* London: Croom Helm, 96–128; R. van Kempen and A. Ozuekren (1998) Ethnic segregation in cities *Urban Studies* 35(10), 1631–56.

84. D. Varady, S. Mantel, C. Hinitz-Washafsky and H. Halpern (1981) Suburbanisation and dispersion: a case study of Cincinnati's Jewish population *Geographical Research Forum* 3, 5–15.

85. S. Waterman and B. Kosmin (1988) Residential patterns and processes: a study of Jews in three London boroughs *Transactions of the Institute of British Geographers* 13, 79–95.

86. H. Rose (1970) The development of an urban subsystem: the case of the Negro ghetto *Annals of the*

Association of American Geographers 60, 1–17; W. Carter, M. Schill and S. Wachter (1998) Polarisation, public housing and racial minorities in US cities *Urban Studies* 35(10), 1889–911.

87. G. Kearsley and S. Srivastara (1974) The spatial evolution of Glasgow's Asian community *Scottish Geographical Magazine* 90, 110–24; P. Rees, D. Phillips and D. Medway (1995) The socio-economic geography of ethnic groups in two northern British cities *Environment and Planning A* 27, 557–91.

88. E. Hattman (1991) *Urban Housing Segregation of Minorities in Western Europe and the United States* Durham NC: Duke University Press.

89. M. Pacione (1996) Ethnic segregation in the European city: the case of Vienna *Geography* 81(2), 120–32.

90. D. Massey and N. Denton (1993) *Apartheid American Style* Cambridge MA: Harvard University Press; R. Farley and R. Wilger (1987) *Recent Changes in the Residential Segregation of Blacks from Whites: an Analysis of 203 Metropolises* Population Studies Center Report 15, Ann Arbor MI: Population Studies Center.

91. E. Skop and W. Li (2005) Asians in America's suburbs *Geographical Review* 95(2), 167–88.

92. H. Rose (1976) *Black Suburbanization* Cambridge MA: Ballinger

93. R. Warren and L. Lyon (1988) *New Perspectives on the American Community* Chicago: Dorsey Press; J. Palen (1995) *The Suburbs* New York: McGraw-Hill.

94. B. Lief and S. Goering (1987) The implementation of the federal mandate for fair housing, in G. Tobin (ed.) *Divided Neighborhoods* Beverly Hills CA: Sage, 227–67.

95. A. Shlay (1988) Not in that neighbourhood: the effects of population and housing on the distribution of mortgage finance within the Chicago SMSA *Social Science Research* 17, 137–63.

96. M. Winsberg (1985) Flight from the ghetto *American Journal of Economics and Sociology* 44, 411–21; M. Clapson (2003) *Suburban Century* Oxford: Berg; S. Meyer (2000) *As Long as they Don't Move Next Door* New York: Rowman & Littlefield; M. Patillo-McCoy (2000) The limits of out-migration for the black middle class *Journal of Urban Affairs* 22(3), 225–41.

97. N. Denton and D. Massey (1991) Patterns of neighbourhood transition in a multi-ethnic world: US metropolitan areas 1970–1980 *Demography* 28, 41–63.

98. R. Greene (1997) Chicago's new immigrants, indigenous poor and edge cities *Annals of the American Academy of Political and Social Science* 551, 178–90.

99. P. Marcuse (1996) Space and race in the post-Fordist city, in E. Mingione (ed.) *Urban Poverty and the Underclass* Oxford: Blackwell, 176–216.

100. R. Aponte (1991) Urban hispanic poverty, disaggregations and explanations *Social Problems* 38, 516–28; W. Wilson (1987) *The Truly Disadvantaged: The Inner City, the Underclass and Public Policy* Chicago: University of Chicago Press; C. Marks (1991) The urban underclass *Annual Review of Sociology* 17, 445–66.

101. D. Wilson (2007) *Cities and Race* London: Routledge.

102. D. Coleman and J. Salt (1992) *The British Population* Oxford: Oxford University Press.

103. D. Byrne (1998) Class and ethnicity in complex cities: the cases of Leicester and Bradford *Environment and Planning A* 30, 703–20; D. Shanahan (1997) Race, segregation and restructuring *Scottish Geographical Magazine* 113(3), 150–8; D. Phillips (1998) Black minority ethnic concentration, segregation and dispersal in Britain *Urban Studies* 35(10), 1681–702.

104. HMSO (1989) *Bangladeshis in Britain* London: HMSO.

105. D. Ley, C. Clark and C. Peach (1984) *Geography and Ethnic Pluralism* London: Allen & Unwin; E. Jones (1996) Social polarization in post-industrial London, in J. O'Laughlin and J. Friedrichs (eds) *Social Polarization in Post-industrial Metropolises* New York: de Gruyter, 19–44.

106. J. Henderson and V. Karn (1984) Race, class and the allocation of public housing in Britain *Urban Studies* 21, 115–28; Commission for Racial Equality (1984) *Race and Council Housing in Hackney* London: CRE; M. Harrison (1994) Housing empowerment, minority ethnic organisations, and public policy in the UK *Canadian Journal of Urban Research* 3(1), 29–39.

107. J. Morris and M. Winn (1990) *Housing and Social Inequality* London: Hilary Shipman.

108. J. Waheb (1989) *Muslims in Britain* London: Runnymede Trust.

Chapter 19   URBAN LIVEABILITY

1. M. Pacione (1990) Urban liveability: a review *Urban Geography* 11(1), 1–30.

2. P. Knox (1976) Fieldwork in urban geography *Scottish Geographical Magazine* 92(2), 101–7.

3. B. Liu (1982) Environmental quality indicators for large metropolitan areas *Journal of Environmental Management* 14, 127–38.

4. M. Pacione (1982) The use of objective and subjective indicators of quality of life in human geography *Progress in Human Geography* 6, 495–514.

5. M. Pacione (1990) A tale of two cities: the migration of the urban crisis in Glasgow *Cities* 7(4), 304–14.

6. J. Porteous (1977) *Environment and Behavior* Reading MA: Addison-Wesley; E. Krupat (1985) *People in Cities* Cambridge: Cambridge University Press; D. Walmsley (1988) *Urban Living* Harlow: Longman; D. Frick (1986) *The Quality of Urban Life* New York: de Gruyter.

7. B. Liu (1975) *Quality of Life Indicators in the United States Metropolitan Areas 1970* Washington DC: US Government Printing Office.

8. R. Boyer and D. Savageau (1981) *Places Rated Almanac* Chicago: Rand McNally; (1985) *Places Rated Almanac* Chicago: Rand McNally; (1989) *Places Rated Almanac* Englewood Cliffs NJ: Prentice-Hall.

87. P. Cozens, D. Hillier and G. Prescott (2002) Criminogenic associations and characteristic British housing designs *International Planning Studies* 7(2), 119–36; R. Scheider and R. Kitchen (2002) *Planning for Crime Prevention* London: Routledge; I. Colquhoun (2004) *Design out Crime* Oxford: Architectural Press.

88. J. English (1979) Access and deprivation in local authority housing, in C. Jones (ed.) *Urban Deprivation and the Inner City* London: Croom Helm, 113–35.

89. G. Francescato (1980) The quality of the residential environment, in M. Romanos (ed.) *Western European Cities in Crisis* Lexington MA: Lexington Books, 103–34.

90. G. Rent and C. Rent (1978) Low-income housing: factors related to residential satisfaction *Environment and Behavior* 10, 459–88; A. Campbell, P. Converge and W. Rodgers (1976) *The Quality of American Life* New York: Russell Sage Foundation; R. Marans and W. Rodgers (1974) Toward an understanding of community satisfaction, in A. Hawley and V. Rock (eds) *Metropolitan America* Washington DC: National Academy of Sciences, 111–17.

91. T. Duncan (1971) *Measuring Housing Quality* Occasional Paper 20, Birmingham: Centre for Urban and Regional Studies, University of Birmingham

92. M. Fried and P. Gleicher (1961) Some sources of residential satisfaction in an urban area *Journal of the American Institute of Planners* 27, 305–15; G. Tauber and J. Levin (1971) Public housing as neighbourhood *Social Science Quarterly* 52, 534–42; R. Merton (1968) *Social Theory and Social Structure* New York: Free Press; H. Gans (1967) Planning and city planning for mental health, in H. Eldredge (ed.) *Taming Metropolis* New York: Praeger; T. Duncan (1971) *Measuring Housing Quality*, Occasional Paper 20, Birmingham: Centre for Urban and Regional Studies, University of Birmingham; A. Campbell and P. Converse (1972) *The Human Meaning of Social Change* New York: Russell Sage Foundation.

93. A. Onibokun (1974) Evaluating consumers' satisfaction with housing *Journal of the American Institute of Planners* 40, 189–200; M. Pacione (1982) Evaluating the quality of the residential environment in a deprived council estate *Geoforum* 13(1), 45–55.

94. L. Susskind and M. Elliott (1981) Learning from citizen participation and citizen action in Western Europe *Journal of Applied Behavioural Science* 17, 497–517.

95. S. Langton (1978) *Citizen Participation in America* Lexington, MA: Lexington Books.

96. R. Alterman (1982) Planning for public participation *Environment and Planning B* 9, 295–313.

97. J. Zeisel (1977) *Low-rise Housing for Older People: Behavioral Criteria for Design* Washington DC: US Government Printing Office.

98. S. Milgram, J. Greenwald, S. Kessler, W. McKenna and J. Waters (1972) A psychological map of New York City *American Scientist* 60, 194–200.

99. M. Pacione (1993) Fear of crime in the city *Journal of the Scottish Association of Geography Teachers* 22, 31–40.

100. D. Ley (1974) *The Black Inner City as Frontier Outpost* Washington DC: American Association of Geographers.

101. F. Boal (1969) Territoriality on the Shankill–Falls divide *Irish Geography* 6, 30–50; F. Boal (2002) Belfast: walls within *Political Geography* 21, 687–94.

102. K. England (1993) Suburban pink-collar ghettos: the spatial entrapment of women? *Annals of the Association of American Geographers* 83(2), 225–42; E. Wyly (1998) Containment and mismatch: gender differences in commuting in metropolitan labour markets *Urban Geography* 19(5), 395–430; S. Hanson and G. Pratt (1988) Spatial dimensions of the gender division of labour in a local labour market *Urban Geography* 9, 180–202.

103. W. Markham (1987) Sex, relocation and occupational advancement, in L. Larwood and B. Gutek (eds) *Women and Work* Beverly Hills CA: Sage, 207–31.

104. L. Pickup (1988) Hard to get around, in J. Little, L. Peake and P. Richardson (eds) *Women in Cities* New York: New York University Press; S. Hanson and G. Pratt (1995) *Gender, Work and Space* London: Routledge, 83–92.

105. E. Jackson and K. Henderson (1995) Gender-based analysis of leisure constraints *Leisure Sciences* 17, 31–51.

106. R. Dixey (1988) A means to get out the house, in J. Little, L. Peake and P. Richardson (eds) *Women in Cities* New York: New York University Press.

107. J. Foord and N. Gregson (1986) Patriarchy *Antipode* 18(2), 180–211.

108. A. Warnes (1994) Cities and elderly people *Urban Studies* 31(4/5), 799–816.

109. S. Golant, G. Rowles and J. Meyer (1988) Aging and the aged, in G. Gaile and J. Willmott (eds) *Geography in America* Columbus OH: Merrill, 451–66.

110. R. Marans, M. Hunt and K. Vakalo (1984) Retirement communities, in I. Altman, M. Lawton and J. Wohlwill (eds) *Elderly People and the Environment* New York: Plenum Press, 57–93.

111. J. Piaget and B. Inhelder (1967) *The Child's Conception of Space* New York: Norton.

112. A. Siegel and S. White (1975) The development of spatial representations of large-scale environments, in W. Reese (ed.) *Advances in Child Development and Behavior* New York: Academic Press, 9–55.

113. M. Matthews (1992) *Making Sense of Place: Children's Understanding of Large-scale Environments* Hemel Hempstead: Harvester Wheatsheaf; P. Christensen and M. O'Brien (2003) *Children in the City* London: Routledge.

114. R. Imrie (1996) *Disability and the City* London: Paul Chapman; R. Butler and S. Bowlby (1997) Bodies and spaces: an exploration of disabled people's experiences of public space *Environment and Planning D* 15, 411–33.

115. C. Barnes (1991) *Disabled People in Britain and Discrimination* London: Hurst.

116. E. Howard (1902) *Garden Cities of Tomorrow* London: Faber.

117. Le Corbusier (1947) *City of Tomorrow and its Planning* London: Architectural Press.

118. P. Goodman and P. Goodman (1960) *Communitas* New York: Random House.

119. D. Zimmerman (1982) Small is beautiful but: an appraisal of the optimum city *Humboldt Journal of Social Relations* 9, 120–42.

120. A. Rapaport (1977) *Human Aspects of Urban Form* Elmsford NY: Pergamon; G. Moore and R. Golledge (1976) *Environmental Knowing* Stroudsburg PA: Dowden Hutchinson & Ross; K. Lynch (1981) *A Theory of Good City Form* Cambridge MA: MIT Press.

Chapter 20 POWER, POLITICS AND URBAN GOVERNANCE

1. H. Lasswell (1958) *Who Gets What When How* Cleveland OH: World Publishing.

2. T. Gurr and D. King (1987) *The State and the City* Chicago: University of Chicago Press.

3. F. Hunter (1953) *Community Power Structure* Chapel Hill NC: University of North Carolina Press.

4. R. Dahl (1961) *Who Governs?* New Haven CT: Yale University Press

5. J. Logan and H. Molotch (1987) *Urban Fortunes* Berkeley CA: University of California Press.

6. C. Stone (1976) *Economic Growth and Neighborhood Discontent* Chapel Hill NC: University of North Carolina Press; K. Ward (1998) Rereading urban regime theory: a sympathetic critique *Geoforum* 27(4), 427–38.

7. O. Williams and C. Adrian (1963) Community types and policy differences, in J. Wilson (ed.) *City Politics and Public Policy* New York: Wiley.

8. D. Yates (1977) *The Ungovernable City* Cambridge MA: MIT Press.

9. M. Castells (1976) *The Urban Question* Cambridge MA: MIT Press.

10. J. Bollens and H. Schmandt (1970) *The Metropolis* New York: Harper & Row.

11. H. Wolman and M. Goldsmith (1992) *Urban Politics and Policy* Oxford: Pergamon.

12. S. Leach (1994) The local government review *Regional Studies* 28, 537–49.

13. R. Johnston (1982) *The American Urban System* Harlow: Longman; D. Howe (1998) The shrinking central city amidst growing suburbs: case studies of Ohio's inelastic cities *Urban Geography* 19(8), 714–34.

14. T. Clark (1981) Race, class and suburban housing discrimination *Urban Geography* 2, 327–38.

15. T. Clark (1982) Federal incentives promoting the dispersal of low-income housing in suburbs *Professional Geographer* 34, 136–46.

16. J. Boudreau and R. Keil (2001) Seceding from responsibility? *Urban Studies* 38(10), 1701–31.

17. K. McHugh, P. Gober and D. Borough (2002) The Sun City wars *Urban Geography* 23(7), 627–48.

18. Advisory Commission on Intergovernmental Relations (1974) *Urban America and the Federal System* Washington DC: US Government Printing Office.

19. S. Nunn and M. Rosentraub (1997) Dimensions of interjurisdictional cooperation *Journal of the American Planning Association* 63(2), 205–19; F. Frisken (2001) The Toronto story *Journal of Urban Affairs* 23(5), 513–41.

20. L. Sharpe (1995) *The Government of World Cities: The Future of the Metro Model* Chichester: Wiley.

21. N. Brenner (2002) Decoding the newest metropolitan regionalism in the USA *Cities* 19(1), 3–21.

22. P. Dunleavy (1980) *Urban Political Analysis* London: Macmillan.

23. M. Pinto-Duschinsky (1977) Corruption in Britain *Political Studies* 25, 274–84.

24. E. Sharp (1990) *Urban Politics and Administration* Harlow: Longman.

25. J. Gyford, S. Leach and C. Game (1989) *The Changing Politics of Local Government* London: Unwin Hyman.

26. R. Wolfinger and S. Rozenstone (1980) *Who Votes?* New Haven CT: Yale University Press.

27. R. Johnston (1979) *Political, Electoral and Spatial Systems* Oxford: Oxford University Press.

28. M. Busteed (1975) *Geography and Voting Behaviour* Oxford: Oxford University Press.

29. G. Webster (1997) The potential impact of recent Supreme Court decisions on the use of race and ethnicity in the redistricting process *Cities* 14(1), 13–19.

30. G. Norcliffe (1977) Discretionary aspects of scientific districting *Area* 9, 240–6.

31. D. Thompson and T. Slocum (1982) A geographic information system for political redistricting in Maryland *Proceedings of the Applied Geography Conferences* 5, 21–35.

32. S. Arnstein (1969) A ladder of citizen participation *Journal of the American Institute of Planners* 35, 216–24; E. Rocha (1997) A ladder of empowerment *Journal of Planning Education and Research* 17, 31–44; P. Somerville (1998) Empowerment through residence *Housing Studies* 13(2), 233–57.

33. M. Fagence (1977) *Citizen Participation in Planning* Oxford: Pergamon.

34. P. Hambleton (1978) *Policy Planning and Local Government* London: Hutchinson.

35. G. Washnis (1972) *Municipal Decentralization and Neighborhood Resources* New York: Praeger; B. Webster (1982) Area management, in J. Stewart (ed.) *Public Policy and Local Government* London: Allen & Unwin; L. Susskind and M. Elliott (1981) Learning from citizen participation and citizen action in Western Europe *Journal of Applied Behavioural Science* 17(4), 497–517.

36. E. Nordlinger (1972) *Decentralizing the City: A Study of Boston's Little City Halls* Cambridge MA: MIT Press.

37. E. Nordlinger and J. Hardy (1972) Urban decentralisation: an evaluation of four models *Public Policy* 20(3), 359–96.

31. S. Patel, C. D'Cruz and S. Burra (2002) Beyond evictions in a global city *Environment and Urbanization* 14(1), 159–72.

32. S. Patel, S. Burra and C. D'Cruz (2001) Slum/Shack Dwellers International *Environment and Urbanization* 13(2), 45–59.

33. P. Evans (2000) Fighting marginalisation with transnational networks *Contemporary Sociology* 29(1), 230–41.

Chapter 30   THE FUTURE OF THE CITY – CITIES OF THE FUTURE

1. World Commission on Sustainable Development (1987) *Our Common Future* Oxford: Oxford University Press.

2. G. Haughton and C. Hunter (1994) *Sustainable Cities* London: Jessica Kingsley.

3. J. Quarrie (1992) *Earth Summit '92: The United Nations Conference on Environment and Development* London: Regency Press.

4. P. Marcotullio and Y. Lee (2003) Urban environmental transitions and urban transportation systems *International Development Planning Review* 25(4), 325–54.

5. G. Nwaka (1996) Planning sustainable cities in Africa *Canadian Journal of Urban Research* 5(1), 119–36.

6. H. Girardet (1992) *Cities: New Directions for Sustainable Urban Living* London: Gaia Books.

7. R. White and J. Whitney (1992) Cities and the environment: an overview, in R. Stren, R. White and J. Whitney (eds) *Sustainable Cities* Oxford: Westview Press, 8–51.

8. H. Girardet (1996) *Getting London in Shape for 2000* London: London First; Chartered Institution of Wastes Management (2002) *City Limits* London: IWM.

9. H. Girardet (1996) *Getting London in Shape for 2000* London: London First; A. Read (2001) Where there's muck there's brass: the cost of London's waste *Area* 33(1), 103–6.

10. C. Furedy (1992) Garbage: exploring non-conventional options in Asian cities *Environment and Urbanization* 4(2), 42–61; Vincentian Missionaries (1998) The Payatas environmental development programme: micro-enterprise promotion and involvement in solid waste management in Quezon City *Environment and Urbanization* 10(2), 55–68; W. Fahmi and K. Sutton (2006) Cairo's zabaleen garbage recyclers *Habitat International* 30, 809–37.

11. T. Chronopoulos (2006) Neo-liberal reform and urban space *City* 10(2), 167–82.

12. M. Pacheco (1992) Recycling in Bogotá: developing a culture for urban sustainability *Environment and Urbanization* 4(2), 74–9.

13. US Environmental Protection Agency (1997) *Public Health Effects of Ozone and Fine Particulate Pollution* Washington DC: USEPA; J. Leitmann (1999) *Sustaining Cities* New York: McGraw-Hill.

14. United Nations World Commission on Environment and Development (1987) *Our Common Future* Oxford: Oxford University Press.

15. A. Webb and C. Gossop (1993) Towards a sustainable energy policy, in A. Blowers (ed.) *Planning for a Sustainable Environment* London: Earthscan, 52–68.

16. S. Owens (1992) Energy, environmental sustainability and land use planning, in M. Breheny (ed.) *Sustainable Development and Urban Form* London: Pion, 79–105.

17. Department of the Environment (1988) *1985-based Estimates of Numbers of Households 1985–2001* London: Department of the Environment.

18. D. Hutchinson (1992) Towards sustainability: the combined production of heat and power, in M. Breheny (ed.) *Sustainable Development and Urban Form* London: Pion, 268–85.

19. D. Hutchinson (1992) Towards sustainability: the combined production of heat and power, in M. Breheny (ed.) *Sustainable Development and Urban Form* London: Pion, 268–85.

20. Commission of the European Communities (1990) *Green Paper on the Urban Environment* Brussels: CEC.

21. R. Moos and R. Brownstein (1977) *Environment and Utopia* New York: Plenum.

22. P. Geddes (1915) *Cities in Evolution* London: Williams & Norgate.

23. I. McHarg (1969) *Design with Nature* Philadelphia: Natural History Press.

24. M. Roseland (1997) Dimensions of the eco-city *Cities* 14(4), 197–202; D. Paterson and K. Connery (1997) Reconfiguring the edge city: the use of ecological design parameters in defining the form of community *Landsape and Urban Planning* 36, 327–46.

25. J. Kenworthy (2006) The eco-city *Environment and Urbanization* 18(1), 67–85.

26. E. Schumacher (1974) *Small is Beautiful* New York: Harper Colophon; E. Callenbach (1975) *Ecotopia* New York: Bantam.

27. F. Wright (1974) City of the future, in A. Blowers, C. Hammett and P. Saare (eds) *The Future of Cities* London: Hutchinson, 43–6.

28. J. Crawford (2000) *Carfree Cities* Utrecht: International Books.

29. Le Corbusier (1929) *The City of Tomorrow and its Planning* London: Rodher.

30. P. Soleri (1969) *Arcology* Cambridge MA: MIT Press.

31. Commission of the European Communities (1990) *Green Paper on the Urban Environment* Brussels: CEC; T. Elkin, D. McLaren and M. Hillman (1991) *Reviving the City* London: Friends of the Earth; P. Newman and J. Kenworthy (1989) *Cities and Automobile Dependence* Aldershot: Gower; E. Holden and I. Norland (2005) Three challenges for the compact city as a sustainable urban form *Urban Studies* 42(2), 2145–66.

32. M. Breheny (1995) The compact city and transport energy consumption *Transactions of the Institute of British Geographers* 20, 81–101; P. Gordon and

H. Richardson (1997) Are compact cities a desirable planning goal? *Journal of the American Planning Association* 63(1), 95–106.

33. M. Breheny (1997) Urban compaction: feasible and acceptable? *Cities* 14(4), 209–17.

34. S. Durmisevic (1999) The future of the underground space *Cities* 16(4), 233–45.

35. K. Lynch (1981) *Good City Form* Cambridge MA: MIT Press.

36. K. Lynch (1961) The pattern of metropolis *Daedalus* winter, 79–98.

37. V. Gruen (1973) *Centers for the Urban Environment* New York: Van Nostrand Reinhold.

38. P. Calthorpe and W. Fulton (2001) *The Regional City* London: Island Books.

39. K. Corey (2000) Intelligent corridors *Journal of Urban Technology* 7(2), 1–22.

40. W. Clark and M. Kuypers-Linde (1994) Commuting in restructured regions *Urban Studies* 31, 465–83; A. Townsend (2001) The Internet and the rise of the new network cities, 1969–1999 *Environment and Planning B* 28(1), 39–58.

41. M. Castells (1991) *The Informational City* Oxford: Blackwell.

42. R. Caves and M. Walshok (1999) Adopting innovations in information technology *Cities* 16(1), 3–12.

43. A. Gillespie (1992) Communications technologies and the future of the city, in M. Breheny (ed.) *Sustainable Development and Urban Form* London: Pion, 67–78.

44. M. McLuhan (1964) *Understanding Media* London: Sphere Books; J. Pelton (1992) *Future View: Communications Technology and Society in the Twenty-first Century* New York: Johnson Press.

45. M. Castells (1991) *The Informational City* Oxford: Blackwell.

46. S. Graham (1998) Spaces of surveillant simulation *Environment and Planning D* 16, 483–504.

47. S. Graham and S. Marvin (1996) *Telecommunications and the City* London: Routledge; R. Warren *et al.* (1998) The future of the future in planning: appropriating cyberpunk visions of the city *Journal of Planning Education and Research* 18, 49–60.

48. N. Carter and J. Brine (1995) MFP Australia: a version of sustainable development for a post-industrial society *Planning Practice and Research* 10(1), 25–43; P. Rimmer (1991) Exporting cities to the western Pacific rim: the art of the Japanese package, in J. Brotchie, M. Batty, P. Hall and P. Newton (eds) *Cities of the Twenty-first Century* Harlow: Longman, 243–61.

49. S. Graham (1999) Towards urban cyberspace planning, in J. Downey and J. McGuigan (eds) *Technocities* London: Sage, 9–33.

50. B. Ryan (2004) AlphaWorld *Journal of Urban Design* 9(3), 287–309.

51. http://secondlife.com.

commodification   The use of private markets rather than public sector allocation mechanisms to allocate goods and services.

common-interest development   A residential development planned and marketed as a spatially segregated environment. A synonym is *gated community*.

community   A social network or group of interacting individuals usually concentrated into a defined territory.

community business   An organisation owned and controlled by the local community, with membership open to all residents, which aims to create ultimately self-supporting jobs for local people, and to use profits made from business activities either to create more employment or to provide local services.

community development corporation   A community-based group organised to promote the economic and social development of a neighbourhood.

community/social economy   Not-for-profit activities geared towards meeting social needs, such as credit union, cooperatives, housing associations and community enterprises.

congestion   Occurs when the existing use of some facility exceeds its carrying capacity, often generating additional costs for individuals and the wider society, as evident in urban traffic congestion.

congregation   The residential clustering of an ethnic minority through choice, as opposed to involuntary segregation as a result of structural constraints and discrimination.

consumer services   Services usually supplied to individual consumers (e.g. retailing), in contrast to producer services that are supplied to businesses and government (for example, financial services).

contracting out   A situation whereby an organisation contracts with another external organisation for the provision of goods or a service. It may also be termed out-sourcing.

conurbation   A built-up area created by the coalescence of once-separate urban settlements, initially through ribbon development along major inter-urban routes.

core–periphery model   A model of the spatial organisation of human activity based on the unequal distribution of economic, social and political power between a dominant core (e.g. the capital city) and a subordinate and dependent periphery.

counter-urbanisation   A process of population deconcentration away from the large urban settlements.

crowding   A psychological and subjective experience that stems from recognition that one has less space than desired.

cultural industries   Activities (such as those involved in printing and publishing, radio, television and theatre, libraries, museums and art galleries and high fashion) that reflect the post-industrial concept of flexible specialisation, and which can give rise to a cultural industries quarter in a city.

cyberbia   A post-suburban human settlement in an information-based, electronic (cybernetic) society.

cycle of poverty   A self-perpetuating social process in which poverty and deprivation are transmitted intergenerationally.

daily hassles   Minor annoyances that, taken individually or as once-occurring events, would not tax the city dweller's coping mechanism. In the modern city, however, these events tend to be multiple, chronic and repetitive.

daily urban system   The area surrounding an urban centre with which there is a substantial amount of daily commuting traffic.

defensible space   An environment that inhibits crime by creating a physical expression of a social fabric that defends itself. The goal is to create an environment in which latent territoriality and a sense of community in the residents are translated into responsibility for ensuring a safe and well maintained living space.

deindustrialisation   A sustained decline in industrial, especially manufacturing, activity.

demographic transition   A general model describing the evolution of levels of fertility and mortality in a country over time.

dependence   A relationship between two or more societies which implies that the ability of a society to survive and reproduce itself derives, in large part, from its links with other dominant imperialist societies.

dependency theory   Relates the backwardness of Third World economies to the hegemony of Western nations over the world economic order and the consequent ability of these states and multinational corporations to exploit peripheral areas. Development and underdevelopment are viewed as different outcomes of the same process.

dialectic   A form of reasoning or analysis involving the use of and possible reconciliation of opposites (as, for example, in the opposition of global and local forces in urban change, resolved in the concept of glocalisation).

displacement   The process by which a social or economic change removes people involuntarily, such as from their homes or from employment.

distance decay   The attenuation of a pattern or process over distance (e.g. the number of consumers travelling to a shopping centre).

division of labour   The separation of tasks within the labour process and their allocation to different groups of workers. The division of labour may be based on product sectors (e.g. textile workers and shipbuilders), specialisation (e.g. managers and assembly workers), space (with the concentration of particular economic sectors or tasks in specific geographical areas) or gender.

ecological fallacy   Refers to the problem of inferring characteristics of individuals from aggregate data referring to a population. Since data based on spatial areas are used frequently in geography, researchers (and students) must be aware of the danger of inferring spurious correlations that may arise from the use of aggregate data.

ecological footprint   The land-area and natural-resource capital on which the city draws to sustain its population and production structure.

economic determinism   Theories that relate social changes directly to underlying economic changes in society and that minimise the ability of people to make decisions to affect their destinies.

economies of scale   The cost advantages gained by large-scale production as the average cost of production falls with increasing output.

economies of scope   The cost advantages that may arise when performing two or more activities together within a single firm rather than performing them separately.

ecumenopolis   The term used by Doxiodis to describe a continuously built-up environment covering the habitable area of the globe.

edge city   An office, entertainment and shopping node with 'more jobs than bedrooms' that has emerged in suburban locations to challenge the dominance of the metropolitan downtown (city centre).

embeddedness   The view that economic behaviour is not determined by universal values that are invariant, as in neo-classical economics, but is intimately related to cultural values that may be highly specific in time and space.

eminent domain   The right of a government to take private property for public use. In the UK this is normally exercised in the form of compulsory purchase under planning legislation.

empowerment zone   A US federal initiative to tackle urban regeneration by offering investment plus tax incentives in designated urban areas that were expected to secure a commitment of private-sector investment and to involve the community in the planning and implementation of the project.

enclave   A residential cluster of an ethnic minority that is a long-term phenomenon, although generally not as segregated a the ghetto.

enclave economy   A local economy in which a high percentage of workers are employed by members of their own ethnic group, usually within a few industries.

enterprise zone   An area within which special policies (such as tax concessions and simplified planning procedures) apply to encourage economic development through private investment.

entrepreneurialism   A perspective on urban development that views the city as a product that needs to be marketed. This marketing approach and emphasis on restructuring the city to appeal to global business, assign pre-eminence to economic interests in the decision-making process of urban planning.

environmental determinism   The belief that human activities are controlled by the environment. Specific offshoots at the urban scale include architectural determinism. A more general synonym is *environmentalism*.

epidemiological transition   A general model of the relationship between disease and development which proposes that, with development, pandemics of infectious diseases are replaced as the chief cause of death by degenerative diseases largely occasioned by lifestyle.

equity   Justice or fairness in the distribution of income and other aspects of human life chances.

equity planning   An approach to urban planning that emphasises providing additional choices for people with the fewest resources. A synonym is *pragmatic radicalism*.

ethclass   A term for the intersection of class and ethnicity in determining a person's identity and assimilation. Thus a middle-class Puerto Rican may feel more kinship with a middle-class Italian than with a lower-class Puerto Rican.

ethnicity   Both a way in which individuals define their personal identity (e.g. Italian) and a type of social stratification that emerges when people form groups based on real or perceived common cultural characteristics that mark them as different from others.

ethnoburb   Suburban ethnic clusters of residential areas and business districts in large metropolitan areas created by the deliberate efforts of an ethnic group.

ethnocentrism   A form of prejudice or stereotyping that assumes the superiority of one's own culture or ethnic group.

exchange value   The worth of an object or service based on the price it could bring if it were sold.

exclusionary zoning   Planning policies in the US that restrict certain types of activity and people from moving into a local government area.

exo-urbanisation   A pattern of foreign-investment-induced urbanisation in the Third World characterised by labour-intensive and assembly-manufacturing types of export-oriented industrialisation based on the low-cost input of large quantities of labour and land, which has in turn promoted rural–urban population migration.

extended metropolitan region   The spatial outcome of a process of population deconcentration from the core area of a metropolitan region and higher growth rates in the outer areas.

externalities   The usually unintended effects of one person's actions on another, over which the latter has no control. Externalities may be either positive (as when well kept gardens raise the value of all properties in a neighbourhood) or negative (as when noisy neighbours reduce the quality of life for others).

exurb   A low-density, low-population settlement beyond the suburbs on the far fringes of an urbanised area.

feminist geographies   Perspectives that draw on feminist politics and theories to explore how gender relations and geographies are mutually structured and transformed.

festival market place   A speciality centre that combines retailing and entertainment, often in a converted formerly non-retailing building.

feudalism   A form of social organisation characterised by the interrelationship of two main social groups: direct producers (peasants working the land) who were subject to politico-legal domination by social superiors (feudal lords), who formed a status hierarchy headed by a monarch.

filtering   A process of neighbourhood change whereby housing vacated by more affluent groups passes down the income scale to become accessible to lower-income groups.

financial exclusion   The withdrawal of retail financial services from poor urban areas, as part of a 'flight to quality' by banks and other financial institutions, leaving residents without access to credit and dependent upon alternative informal sources, including moneylenders or loan sharks.

fiscal crisis   A situation in which an urban government's expenditure increases more rapidly than its income from taxation and other sources, thereby undermining its ability to finance its operations.

food security   A situation in which people do not live in hunger or fear of starvation

Fordism   A system of economic and political organisation in which large-scale companies producing standardised goods dominate the economy. The economic success of the mass-production Fordist era from the end of the Second World War to the mid-1970s was based on a social compromise between workers and owners, manifested in institutions such as collective bargaining that linked wage increases with productivity measures. The net result was to distribute sufficient income to workers to support the consumption of industrial products on a mass scale.

free-rider   Someone who obtains benefits that they have not directly paid for.

fringe belt   A zone of mixed land uses at the edge of a built-up area.

garden city   A planned new settlement designed to provide a high-quality, low-density residential environment in a garden setting. The original garden cities of Howard stimulated new-town development in the UK, as well as smaller-scale intra-urban garden suburbs.

gated community   A residential area with defensive measures such as gates, fences and security guards to exclude social groups deemed undesirable.

gay village   A visible physical clustering of gay enterprises and community within a city

Gemeinschaft   A term used by Tonnies to describe a small, close-knit community in which tradition, family and religion govern social life.

gender   Social, psychological and cultural differences between men and women rather than biological differences of sex.

gentrification   The process of neighbourhood upgrading by relatively affluent incomers who move into a poorer neighbourhood in sufficient numbers to displace lower-income groups and transform its social identity.

geography   The study of the Earth as the home of people.

gerrymandering   The deliberate drawing of the boundaries of electoral constituencies to produce an advantage for an interested party.

Gesellschaft   A term used by Tonnies to describe a large, impersonal social setting in which formal institutions such as the law govern social relations.

ghetto   An urban residential district that is almost exclusively the preserve of one ethnic or cultural group.

globalisation   A complex of related economic, cultural and political processes that have served to increase the interconnectedness of social life in the contemporary world. The concept refers to both the 'space–time compression' of the world and the intensification of consciousness of the world as a whole. Significantly, there is a reflexive relationship between the global and the local. Globalisation does not lead automatically to the disintegration of local life. Individuals can either disembed themselves from a locality by operating within a global milieu or embed themselves by attachment to a particular locale. Neither condition is exclusive.

glocalisation   The process by which developments in particular places are the outcome of both global and local forces.

governance   A term used to indicate the shift away from direct government control of the economy and society via hierarchical bureaucracies (government) towards more indirect control via diverse non-governmental organisations. It represents a broader approach to urban management and is often associated with the declining power of local government.

green belt   An area of open, low-density land use around an existing city where further development is strictly controlled by planning policy.

growth coalition   Partnerships of mutual advantage that may involve both private and public-sector interests to promote and implement strategies that enhance the economic development of cities.

growth machine   A local pro-growth coalition of businesses, commercial landowners and *rentiers* (persons who profit from rental income) that dominates urban politics.

habitus   A term coined by Bourdieu to describe a distinctive set of values, ideas and practices – a collective life world that derives from its members' everyday experiences and operates at a subconscious level, through common daily practices, dress codes, use of language, comportment, and patterns of material consumption

heritage area   Historic elements of city structure that have been preserved through renovation or conversion to new uses.

hinterland   The spatial extent of the sphere of influence of a settlement; also referred to as the *catchment area* or *urban field*.

home rule   The power vested in a US city to change its charter and manage its own affairs with only limited state intervention.

housing association   An independent, non-profit-making organisation funded primarily by government to build, improve and manage affordable housing for sale or rent.

housing class   A concept introduced by Rex to denote different groups of people characterised by their access to particular types of housing, usually defined by tenure.

human ecology   The study of the relationships of human beings with their physical and social environments. As developed by the Chicago school, it represented an interactive perspective on urban social life which replaced notions of environmental determinism.

humanist geography   An approach to human geography distinguished by the central and active role that it gives to human awareness, consciousness, creativity and human agency in an attempt to understand the meaning, value and human significance of life events.

hypermarket   A superstore of at least 50,000 ft² (4,600 m²) of sales area.

imageability   A term used by Lynch to describe the degree to which a city is visibly legible or evokes a strong image in any observer's mind.

imagineering   The conscious creation of places with characteristics similar to other places, as in Disneyland.

import substitution model   A trade and economic policy based on the premise that a country should attempt (using protective tariffs or quotas) to substitute locally produced alternatives for products that it imports. Successful import substitution usually requires heavy expenditure on infrastructure and can involve economic risks (e.g. in the form of potential production inefficiencies and higher prices).

incorporation   The formation of a new city in the US from previously unincorporated territory, in accordance with state law.

index of dissimilarity   A quantitative measure of the extent to which a minority group is residentially segregated within a city.

industrial city   A city dominated by manufacturing activity.

industrial revolution   A transformation of the forces of production centring on, though not confined to, the circuit of industrial capital. The term is most usually applied to the series of changes in the British economy between 1750 and 1850.

informal sector   That part of an economy beyond official recognition and record which performs productive, useful and necessary labour without formal systems of control and remuneration. Informal-sector activities constitute a major part of the urban economy in the Third World. In the developed world the sector is often referred to as the black economy.

informational city   A city that acts as a focus for information flows, via high-technology media, and has a large proportion of the labour force employed in those service industries based on the manipulation of information (such as banking, insurance and legal services).

inner city   An area around the central business district usually associated with dilapidation, poor housing, and economic and social deprivation.

inverse care law   The concept proposed by Hart that those in greatest need of health care have the poorest access to it.

Islamic city   A model describing a city with a structure that reflects the religious principles of Islam. The concept is not uniformly accepted; some geographers maintain that many of the structural characteristics (such as narrow streets) are climate-related rather than religious-based, and may be found in many pre-industrial cities.

kampung   A largely unplanned and primarily low-income residential area, often a former traditional village, that is being absorbed by urban growth.

Kondratieff cycles   Long waves of economic development (over forty to sixty years) that involve fundamental qualitative transformations of economic systems. Long waves are distinguished by particular types of technological revolution (with, for example, the fifth Kondratieff characterised by innovations in information technology, telecommunications and biotechnology).

land invasion   A process, common in cities of Latin America in particular, whereby organised groups of squatters occupy an area of land in a planned operation to obtain space for the construction of shelter.

land sharing   A negotiated agreement to partition land occupied by a squatter settlement but required for development into two parts, one for the use of the landowner and the other for the use of present occupants of the site.

legibility   Building design or estate layout that allows the user to locate functions and navigate easily from the logical arrangement of elements.

liberal democracy   A form of representative democracy in which the ability of elected representatives to exercise decision-making power is subject to the rule of law and, in some states, moderated by a constitution that protects individual rights from government power.

linkage   A programme that requires the developers of new commercial construction projects to contribute financial or other resources for public projects in return for development approval.

locale   Distinct setting or context in which interactions between people take place.

local exchange trading system   A community attempt to develop a local alternative to the capitalist economy by creating a complementary form of social and economic organisation that uses a non-commodified local currency to generate trading activity among members.

local socialism   A policy followed by a number of Labour-controlled urban authorities in the UK in the early 1980s which sought to defend levels and standards of collective service provision, protect local employment and create alternative economic strategies to that of market capitalism.

local state   The set of institutions charged with the maintenance of social relations at the subnational level, including local government, the local judiciary as well as quangos and other local administrative agencies. The term is most commonly used with reference to metropolitan-level governance.

malapportionment   An electoral abuse in which a party promotes its own interests by defining constituencies of differing population or electorate sizes (for example, by creating small constituencies for one's own party to win and much larger ones for opposition party victories).

managerialism   An analytical approach that focuses upon the influence of managers on access to scarce resources and local services.

manipulated-city hypothesis   The argument that coalitions of private interests can operate through legal and institutional frameworks in cities to achieve favourable resource allocations.

Marxian geography   The study of geographical questions using the analytical insights, concepts and theoretical frameworks of historical materialism, an approach that emphasises the material bases of society and looks to the historical development of social relations to comprehend social change. From a Marxian or political economy perspective the urban geography of capitalism is the outcome of the relationship between political and economic forces within society. Urbanisation and urban development are seen as the most rational geographical means of furthering the capitalist goal of accumulation.

master-planned community New private-sector housing development designed for people with similar lifestyle and residential preferences.

megacity A giant metropolis with a population of at least 7 million by the year 2000.

megalopolis A term employed by Gottmann to describe the interconnected urban complex of the north-eastern seaboard of the US.

mental map A subjective or psychological representation of urban space, also commonly referred to as a *cognitive* map.

meta-city An urban agglomeration of over 20 million people.

metropolitan area A concentrated, dense settlement of people in a core city together with the city's suburban population who are economically and socially interdependent.

metropolitan village A dormitory settlement within commuting distance of an urban workplace and in which more than 20 per cent of the resident population are employed in towns or cities.

micro-finance Provision of small-scale financial services such as micro-credit or micro-insurance to poor people unable to access formal sector commercial financial services

migration The permanent or semi-permanent change of residence of an individual or group of people.

mobility The ability to move between different activity sites.

mobility transition A model proposed by Zelinski that relates the changing nature of migration to levels of development. While the universality of the model has been questioned, the basic assumption of systematic changes in the nature of mobility over time retains analytical value.

mode of production The structured social relations through which human societies organise productive activity, the extraction of surplus value and the reproduction of social life. Capitalism is one such mode of production.

modernisation A process of social change resulting from the diffusion of the characteristics of advanced societies and their adoption by less advanced societies.

modernisation theory A now discredited view of Third World development as a convergent and evolutionary process in which diffusion of economic and cultural innovation from the West would move less developed societies towards the kind of advanced economic, social and political structures that prevailed in Western Europe and North America.

modernism A mode of thinking characterised by a belief in universal progress through scientific analysis, together with the notion that social problems can be solved by the application of rational thought.

morphogenesis A process involving evolutionary or revolutionary change in form, as in the study of urban morphogenesis.

multinational corporation A large business organisation operating in a number of national economies; a transnational corporation operates in at least two countries.

natural area A term employed by the Chicago human ecologists to refer to a residential district characterised by physical individuality and the socio-cultural attributes of its inhabitants (e.g. a slum).

negative equity A situation in which the value represented by the market price of a house is insufficient to cover the cost of repayment of the mortgage taken on to purchase the property.

neighbourhood An urban district, in a strict sense defined as one in which there is an identifiable subculture to which the majority of residents conform.

neighbourhood effect A process of local influence whereby the characteristics of people's local social milieux exert an influence on how they think and act.

neighbourhood unit A relatively self-contained residential area, most often found in planned developments of either new suburban districts in existing towns or in new towns. The concept was popular in UK new towns planned in the 1950s and 1960s.

neoclassical economics Characterised by belief in the value of market mechanisms, the approach focuses on micro-level individual market problems rather than broader economic issues. It searches for universal principles of human economic behaviour and tends to ignore the social context of economic activity.

neo-liberalism An intellectual and political movement that espouses economic liberalism (involving, for example, free trade, privatisation, and limits on government intervention) as a means of promoting economic development and political liberty; and supports reform of centralised post-war economic institutions in favour of decentralised ones.

new federalism An approach to urban policy in the US based on reducing the level of direction and support from the federal government; its introduction in 1969 by President Richard Nixon brought to an end a previous era of creative federalism and its wide range of categorical urban programmes.

new industrial spaces The geographical concentration of firms involved in dense networks of subcontracting and collaboration, often related to innovative firms in sectors such as electronics and biotechnology.

new international division of labour A form of the division of labour associated with the internationalisation of production and the spread of industrialisation in a number of newly industrialising countries of the Third World as capital seeks to maintain its levels of profit.

New Right Those holding a set of ideas that share a common belief in the superiority of market mechanisms as the most efficient means of ensuring the production and distribution of goods and services.

new town A free-standing, self-contained and socially balanced settlement primarily planned to accommodate over-spill population and employment from nearby conurbations. The original British concept has been widely copied, but new communities based on other principles (such as private-sector development) have also emerged.

new urbanism A broad school of urban design that advocates a return to 'traditional' human-scale neighbourhood development, liveable communities, transit-oriented development and smart growth instead of low-density car-oriented urban development.

Nimbyism   NIMBY is an acronym for 'not in my back yard'. Nimbyism signifies an attitude typical of individuals who resist the siting of a source of negative externalities close to their homes.

normative theory   A theory that deals with what ought to be.

overurbanisation   The concept that, in Third World countries, economic growth is unable to keep pace with urban population growth, leading to major social and economic problems in most large cities.

parasitic economy   An exploitative local economic system that thrives on spatially immobile low-income customers in disadvantaged communities by charging exorbitant rates of interest for credit and heavily-inflated prices for commonplace goods and services.

paratransit   A term encompassing the full range of small vehicles available for public hire on a trip basis and which form a significant part of the total transport system in Third World cities. A synonym is *intermediate public transport*.

participatory democracy   Whereas traditional representative democracies tend to limit citizen participation to voting, leaving actual governance to politicians, participatory democracy seeks to create opportunities for all citizens to make a meaningful contribution to decision-making. For some, community-based activity is an important element of participatory democracy.

patriarchy   A broad system of social arrangements and institutional structures in which gender relations are characterised by the dominance of men over women and of masculinity over femininity.

peripheral urbanisation   A model that employs a political economy perspective to provide a generalised description of the impact of global capitalism on national urban systems in the Third World. The expansion of capitalism into peripheral areas is seen to generate a strong process of urbanisation.

personal space   Both the 'bubble' around each person and the processes by which people demarcate and personalise the spaces they inhabit.

pink-collar   A popular term to indicate female-concentrated occupations, such as clerical work, nursing, teaching and home-making.

pirate subdivision   The illegal subdivision of land for housing in Third World cities, as in the partitioning of an area into building plots for sale in direct contravention of planning regulations.

place   A unique and special location in space where the regular activities of human beings occur and which may furnish the basis of our sense of identity as human beings, as well as of our sense of community.

place marketing   Also known as place promotion, refers to the conscious use of publicity to transmit selective images of places (such as a city) to a target audience (e.g. business investors or tourists).

planning appeal   A provision in UK planning law whereby an applicant who is refused permission to develop land or property may appeal to the Secretary of State for a reconsideration of the local authority's verdict and possibly a reversal in the applicant's favour.

planning blight   The adverse effect of long-term planning proposals or the consequences of planning indecision on the value of property affected.

planning gain   The benefit (in money or in kind) provided to a local authority by a private developer in return for the granting of planning permission. The procedure is similar to the US concept of incentive zoning.

polarisation   A process of inequality that occurs when both the top and the bottom of the scale of income or wealth distribution grow faster than the middle, thus shrinking the middle and sharpening social differences between two extreme segments of the population.

political economy   An approach to the study of urban geography that emphasises the impact of political and economic institutions on the physical form and social life of cities.

pollution   The release or continued presence of substances that damage or degrade the environment.

popular planning   Planning by local communities in their own neighbourhoods that involves the formulation of planning proposals and their implementation by local community organisations. The success of the process depends on close collaboration between the community and the local planning authority which agrees to adopt the popular plan as official policy.

positive discrimination   A policy designed to favour disadvantaged groups within society in order to reduce or eliminate inequalities. If targeted at particular areas, these policies are referred to as area-based policies of positive discrimination.

positivism   A philosophy of science characterised by adherence to the 'scientific method' of investigation based on hypothesis testing, statistical inference and theory construction. The approach was central to the spatial analysis school of urban geography in the 1950s, but has since been superseded by approaches that accord greater importance to human agency and the prevailing social, economic and political structures in determining the nature of cities and urban life.

postcolonialism   An approach that rejects the notion, embedded in many Western representations of non-Western societies in the colonial and contemporary periods, that Western thought is superior. It attempts to expose the ethnocentricism of the dominant (white, male, Anglo-Saxon) culture.

post-Fordism   A set of workplace practices, modes of industrial organisation and institutional forms identified with the period since the mid-1970s characterised by the application of more flexible methods of production, including, for example, programmable machines, greater labour versatility, subcontracting, just-in-time production, and the closer integration of product development, marketing and production.

post-industrial city   A city with an employment profile that exhibits growth of the quaternary sector (i.e. the professions, management, administration and skilled technical areas) and a declining manufacturing work force. The dual labour market of a service economy contributes to income inequality and social polarisation in a city geared to middle-class consumption.

postmodernism   A broad-based movement in philosophy, the arts and social science characterised by scepticism towards the 'grand theory' of the modern era, stressing instead an openness to a wide range of views in social inquiry. In urban geography the postmodern city shares many of the characteristics of the post-industrial city, with an emphasis on new economic structures, social differentiation and variety of lifestyles.

post-suburbia   A relatively new spatial form pioneered in Los Angeles in the inter-war period and developed elsewhere after the Second World War, characterised by a complex, decentralised mix of urban, suburban and rural space and a mix of residents in terms of class and ethnicity.

poverty   An institutionally defined concept that refers to a level of resources below which it is not possible to achieve the standard of living considered to be the minimum norm in a given society at a given time.

power   The ability to make people do things they would not otherwise do through control over resources. Power is exercised at all scales from the individual to the world economy. Those with greatest power (economic, political, social) have most control over the organisation of society and of the allocation of society's benefits. This unequal power is reflected in the social relations between classes and is manifested in the capitalist city.

pre-industrial city   All cities prior to the industrial revolution and those in non-industrialised regions today. Initially used by Sjoberg, the concept argued that all pre-industrial cities shared similar reasons for their existence, social hierarchies and internal spatial structures, but this interpretation is no longer generally accepted.

primate city   A country's leading city which is disproportionately larger than any other in the system and dominant not only in population size but also in its role as the political, economic and social centre of the country.

privatisation   A diverse set of policies designed to introduce private ownership and/or private market allocation mechanisms to goods and services previously allocated and owned by the public sector.

producer service   Service that is supplied to businesses and government (e.g. R&D, financial services) rather than directly to individual users of consumer services.

pro-growth coalition   Groups with an interest in promoting urban growth.

proto-industrialisation   A phase in the development of modern industrial economies that preceded and paved the way for industrialisation proper. Proto-industrialisation was dominated by the putting-out system in which a merchant capitalist distributed raw materials to families working at home, took in the processed goods, paid piece rates for labour and arranged the finishing and sale of commodities.

public goods   Goods that are either freely available to all (such as air) or those provided equally to all citizens of a defined territory.

quality of life   The state of social well-being of an individual or group, either perceived or as identified by observable indicators.

queer theory   A theory inspired by the work of Foucault that emphasises the fluid, socially constructed character of sexual identities. Deliberate use of the pejorative term for homosexuals (queers) is intended to highlight (ironically) the repressive nature of dominant discourses on sexuality.

racism   An ideology of difference that assigns negative characteristics to culturally constructed categories of 'race'. It can lead to practices of racial discrimination (as, for example, in urban housing markets).

recuperative production   The recovery and recycling of waste, largely undertaken by workers in the informal sector, which forms an important element in the economy of Third World cities.

red-lining   The demarcation by financial institutions of residential areas of a city as being in decline and thus not suitable for investment.

regime theory   A variant of the urban-growth-machine model that develops the concept of fluid, overlapping alliances among local business and political leaders in order to achieve desired solutions to particular problems. A regime is defined as the informal arrangements by which public and private interests function together to make and carry out governing decisions.

regional shopping centre   A large free-standing shopping mall usually with more than 500,000 ft² (4,600 m²) of gross lettable area.

rent gap   The hypothesised gap between actual rent attracted by a piece of land or property and the rent that could be obtained under a higher and better use. As the rent gap enlarges, opportunities for profitable reinvestment increase, as in the process of gentrification.

representative democracy   A form of government founded on the principles of popular sovereignty by people's representatives who are responsible for acting in the people's best interests though not necessarily always according to their wishes.

retail warehouse park   An organised development of at least three retail warehouses (defined as single-storey retail units of at least 10,000 ft², or 930 m²) totalling at least 50,000 ft² (4,600 m²) of gross lettable area.

ribbon development   The process of urban sprawl along the main roads leading from a built-up area. Within urban areas the term refers to commercial strips along roads.

road pricing   A strategy to relieve traffic congestion which views the problem as a market failure that can be addressed by increasing the price motorists pay for the use of road space which is in short supply.

ruralopolis   A rural region, with a population density equal to or above the urban threshold, created by a process of *in situ* population growth or 'urbanisation by implosion', and that represents a hybrid settlement system that is spatially urban but economically, socially and institutionally agrarian.

rural–urban continuum   A continuous gradation of ways of life between the two poles of truly rural community and truly urban society. The concept has been used as a theory of social change which emphasises the transformations in ways of life from one pole to the other.

rural–urban fringe A transition zone between the continuously built-up urban and suburban areas of the city and the rural hinterland.

scanscape A term coined to describe the electronic surveillance strategies in Los Angeles to exclude groups regarded as undesirables from certain parts of the city.

search space The area within which a potential migrant searches for a new location.

segregation A concept that refers both to processes of social differentiation and to the spatial patterns that result from such processes, seen most visibly in the ghetto.

sense of place The attachment that people feel to a place.

sexuality Ideas and concepts about sex. Implicit in the term is the belief that human sexual activity is not biologically determined but primarily a learned form of behaviour shaped by cultural values.

shadow state The tendency for the voluntary sector to take over services that were previously allocated by the state.

shopping mall A shopping centre, usually comprising one or more anchor stores and several smaller units, in one building or an architecturally unified group of buildings, and usually with a single ground landlord.

site and services scheme A government initiative in Third World cities under which land is prepared and a service infrastructure provided, with lots then sold or leased for residents to organise the construction of a shelter.

Skid Row A decaying section of a city, usually close to the centre, which houses a concentration of transient population, notably those on the margins of economic survival.

slum An area of overcrowded and dilapidated, usually old, housing occupied by people who can afford only the cheapest dwellings available in the urban area, generally in or close to the inner city.

smart growth A set of planning techniques designed to achieve more sustainable development, that includes infill development, revitalisation of existing neighbourhoods, mixed-use developments, environmental preservation, and integrated regional transport and land-use planning

smokestack cities The cities created by the industrial revolution and characterised by heavy manufacturing industry.

social engineering An attempt to improve society by rational comprehensive planning based on scientific principles, as in the post-war comprehensive slum clearance programmes.

social movement A collective attempt to bring about or resist social change often through non-institutional means, as for example in a campaign against an urban motorway or to obtain basic infrastructure in a Third World squatter settlement.

social polarisation The trend towards increased inequality between rich and poor in society leading to a reduced middle class, and an 'hourglass' society.

socialist city A city that is planned and designed according to the principles of socialism and that operates as part of the command economy.

socio-spatial dialectic The mutually interacting process whereby people shape the structure of cities and, at the same time, are affected by the structure of those cities.

soft state Usually a Third World state in which governments exercise weak control over the activities of public administration and in which laws and regulations are more likely to be flouted than in most Western countries.

special district An independent unit of US local government established to provide one or more limited functions, such as fire protection. Special districts are usually created to meet problems that transcend local government boundaries, or to avoid taxation and debt restrictions imposed on local units of government by state law.

sprawl Unplanned suburban growth, referring to both continuing outward development on the periphery of metropolitan areas and to the specific form such development has taken in the construction of freeways, strip malls and other car-centred uses of space.

squatter settlement An illegal urban development, usually predominantly residential, on land neither owned nor rented by its occupants.

streetcar suburbs The first wave of US suburbs resulting from improved transport technology during the last quarter of the nineteenth century.

subculture A group of people who define themselves as different from the dominant culture in terms of some standards of behaviour and values but who do not constitute an entirely different culture. Occupation, ethnicity, age, sexuality, religion and social background can be bases for subcultures.

suburb An outer district lying within the commuting zone of an urban area.

subsidiarity The idea that national-level decision making should be devolved to the most appropriate level, usually downwards to local communities.

sunbelt/snowbelt (or rustbelt) Terms that characterise and contrast the main growing and declining regions of the US economy in recent decades.

supermarket A single-level self-service food store of between 5,000 ft^2 and 25,000 ft^2 (460 m^2 and 2,300 m^2) sales area.

super-regional shopping centre A free-standing shopping mall of at least 800,000 ft^2 (74,000 m^2) of gross lettable area.

surplus value A concept used by Marx in his analysis of capitalism to explain the source of profits. Thus, if a worker spends five hours in 'necessary labour' (necessary to subsist) and three hours in 'surplus labour' (producing exclusively for the employer), the surplus labour time is the source of surplus value.

sustainable development Development that meets the needs of the present without compromising the ability of future generations to meet their own needs.

sweatshop A factory or a home work operation that involves multiple violations of employment law typically relating to non-payment of minimum wages and violations of health and safety regulations. Sweatshops first appeared in the US apparel industry in the last decades of the nineteenth century with the mass production of garments in New York City by female immigrant workers.

taken-for-granted world   Also referred to as 'lifeworld', the concept focuses on the significance of everyday life and the personal geographies practised within it. The aim is to examine the everyday social practices through which people deal with the strategies of formal institutions of power.

technoburb   A perimeter city or zone that is functionally independent of the central city and can generate urban diversity without urban concentration, made possible by technologically advanced industries.

technopole   A concentrated area of development within which there is a deliberate strategy to plan and promote technologically innovative, industrially related production. A synonym is *science city*.

teleworking   The substitution of electronic communication for physical movement of employees to a central workplace. A synonym is *telecommuting*.

territorial justice   The allocation of resources across a set of areas in direct proportion to the needs of the areas.

territorial social indicator   A measure of social well-being in a defined territory, such as wards within a city.

territoriality   The propensity of people to define certain areas and defend them, underlain by the concept of a 'valued environment'.

text   A key concept in cultural studies which refers to any form that represents social meanings, including paintings, landscapes and buildings, as well as the written word.

tipping point   A situation when a new minority group migrating into a residential area becomes such a significant presence that they provoke a rapid exit of the remainder of the original population.

townscape   The combination of the town plan, its land use units and architectural forms, the study of which formed a central part of urban geography in its early development.

tragedy of the commons   A metaphor used to describe and account for situations in which the depletion of natural resources occurs because individual and collective interests do not coincide, and no institution has the power to ensure that they do. The classic example is that of individual graziers using common land and continually adding to their herds, even through it means that, in the absence of a regulatory authority, the collective resource is being exhausted.

transit village   A suburban community designed in neo-traditional terms with a mix of land uses, moderate residential densities, pedestrian circulation, and small offices and shops clustered around a transit station.

transnational corporation   A business corporation that operates in at least two countries.

trickle-down effect   The concept that, in the longer term, an expanded city revenue base created by central-area revitalisation provides funds that can be used to address social needs.

underclass   A term coined by Myrdal to describe the poor being forced to the margins or out of the labour market by post-industrial society. Collectively viewed as an unprivileged class of unemployed, unemployable and under-employed people at the bottom of the social scale and generally excluded from mainstream life of society.

underconsumption   A concept of particular importance in Marxian economics, where it refers to a persistent shortfall in demand, frequently explained in terms of inadequate purchasing power. The corollary is overproduction. These represent a crisis for capitalism, which seeks to maintain its perpetual goal of accumulation. The post-Second World War suburbanisation wave in the US was in part stimulated by government and financial institutions to overcome a crisis of underconsumption and overproduction.

underurbanisation   The achievement of high industrial growth without a parallel growth of urban population, typical of pre-reform socialist economies.

uneven development   A systematic process of economic and social development that is uneven in space and time, which is integral to the capitalist development process, as capitalists search for the point of maximum profit. Uneven development is evident at all geographic scales.

urban   Relating to towns and cities.

urban development corporations   Agencies established in the 1980s by the Conservative government in the UK as part of a strategy to privatise urban policy. The UDC was invested with wide powers over land use and development and charged with the primary task of creating an environment attractive to private investment.

urban ecology   A term applied by researchers of the Chicago school to their study of the social and spatial organisation of the city.

urban environmental transition theory   Contends that the priority environmental challenges change from those of a 'brown' agenda (relating to, for example, the provision of basic infrastructure) to those of a 'green' agenda (relating, for example to climate change), and that with development the environmental impact of cities shifts from being local, immediate and health-threatening to become globalised, delayed and ecosystem threatening.

urban liveability   A relative term whose precise meaning depends on the place, time and purpose of the assessment and on the value system of the assessor. In contrast to an objective definition of urban environmental quality, this view contends that quality is not an attribute inherent in the environment but a behaviour-related function of the interaction of environmental characteristics and person characteristics.

urban managers   Professionals and bureaucrats who make decisions that influence the internal spatial structure of urban areas through their control of resources such as access to council housing. Bureaucrats who work in parts of the state apparatus are normally termed urban managers, whereas professionals engaged in the private sector, such as real-estate agents, are termed gatekeepers.

urban place   Spatial concentration of human economic, social, cultural and political activities distinguished from non-urban/rural places by both physical aspects such as population density or administrative definition and lifestyle characteristics.

urban system   A set of interdependent urban places, either within a national territory or at the world scale.

urbanisation   The process by which an increasing proportion of a national population lives in towns and cities.

urbanism   A way of life associated with residence in an urban area.

use value   The value that derives from using an object, such as a house, apart from its monetary or exchange value.

vertical integration   A corporate structure in which the corporation controls all the processes involved in the manufacture and sale of a product.

welfare state   A set of institutions and social arrangements designed to assist people when they are in need as a result of factors such as illness, unemployment and dependence through youth and old age.

white-collar   A popular term used to denote a category of workers engaged primarily in non-manual occupations, such as professional and technical workers (e.g. teachers, lawyers) and office workers.

whole-economy model   A model of the Third World urban economy proposed by Friedman in which the household economy is seen as the basic economic unit or miniature political economy; it makes decisions on a continuing daily basis concerning the best use of available resources in response to changing internal (e.g. illness) and external (e.g. job loss) circumstances.

world city   A city in which a disproportionate part of the world's most important business is conducted; a focal point or command centre for the organisation of the global economy. A common synonym is *global city*.

world economy   The capitalist economic system that emerged between ad1400 and ad1700 and still exists today, linking different nations into a global economy in which different countries have different positions.

world-systems theory   Views the world as a single entity, the capitalist world economy, and considers that the study of social change cannot proceed country by country but must incorporate the whole world system.

Index

Note: Page numbers in *italic type* denote tables, page numbers in **bold type** denote figures and plates.

Questioning Cities Series

The Questioning Cities series brings together an unusual mix of urban scholars under the title. Rather than taking a broadly economic approach, planning approach or more socio-cultural approach, it aims to include titles from a multi-disciplinary field of those interested in critical urban analysis.

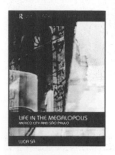

Life in the Megalopolis: Mexico City and Sao Paulo
Lucia Sa

Life in the Megalopolis discusses how contemporary literature, cinema, popular music and visual arts describe life in two of the largest cities in the world.

October 2007: 192pp / HB: 978-0-415-39271-6: £80.00 / PB: 978-0-415-39272-3: £23.99

Globalization, Violence and the Visual Culture of Cities
Christoph Lindner

This book is the first interdisciplinary volume to examine the complex relationship between globalization, violence, and the visual culture of cities

August 2009: 224pp / HB: 978-0-415-48214-1: £80.0

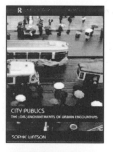

Title	ISBN	Binding	Date	Price	
Cities and Race	978-0-415-35806-4	PB	2006	£24.99	b
Cities in Globalization	978-0-415-40984-1	HB	2006	£95.00	
Cities, Nationalism and Democratization	978-0-415-41947-5	HB	2007	£80.00	
City Publics	978-0-415-31228-8	PB	2006	£24.99	b
Global Metropolitan	978-0-415-30542-6	PB	2004	£24.99	
In the Nature of Cities	978-0-415-36828-5	PB	2005	£25.99	
Ordinary Cities	978-0-415-30488-7	PB	2005	£26.99	
Reason in the City of Difference	978-0-415-28767-8	PB	2004	£25.99	
Small Cities	978-0-415-36658-8	PB	2006	£24.99	b

Routledge
Taylor & Francis Group
ROUTLEDGE

For more details, or to sign up to our **FREE** Geography newsletter please contact:

Gemma-Kate Hartley, Senior Marketing Co-ordinator
Gemma-kate.hartley@tandf.co.uk

The Urban and Regional Planning Reader

Eugénie Birch,
University of Pennsylvania, USA

The Urban and Regional Planning Reader draws together the very best of classic and contemporary writings to illuminate the planning of cities and metropolitan areas. Forty-seven generous selections include contributions from Lewis Mumford, Jane Jacobs, Ian McHarg, Paul Davidoff, Charles Harr, Susan Fainstein and Charles J. Hoch through to Timothy Beatley; Jonathan Barnett, Alex Garvin, Tom Daniels, Andres Duany and Barbara Faga. The variety and wide selection of readings offers one of the most innovative amalgamations of planning research and practice.

December 2008: / PB: 978-0-415-31998-0: **£28.99** / HB: 978-0-415-31997-3: **£95.00**

The Reader lays out the context, range of concerns, history, methods and key topics for 21st century urban and regional planning. Sections on the world of planning, history and theory, classic readings, practice and current issues include writings with a focus on the distribution of space and place, essays on housing, transportation design, environment, community development, the effects of cultural diversity and information technology on land use and other topics. It displays the techniques used to direct and control growth, including zoning, master planning, public budgeting and citizen participation. It explores different types of plans distinguished by their scale and reference type. It references analytical and presentation techniques and outlines ethical issues confronting planners. This *Urban and Regional Planning Reader* provides an essential resource, for students of planning, drawing together important but widely dispersed writings and the associated bibliography is a resource which enables deeper investigations. The synthesis is also valuable for lecturers and researches in the area and the pertinent editorial commentaries preceding each entry not only demonstrate its significance, but also outline the issue surrounding the topic.

Contents:
Introduction (EUGÉNIE L. BIRCH)
PART 1. THE WORLD OF URBAN AND REGIONAL PLANNING
PART 2. HISTORY AND THEORY OF URBAN AND REGIONAL PLANNING
PART 3. CLASSICS IN URBAN AND REGIONAL PLANNING
PART 4. THE PLAN: ITS ORIGINS AND CONTEMPORARY USES
PART 5. PLANNING PRACTICE AND METHODS
PART 6. KEY TOPICS IN URBAN AND REGIONAL PLANNING
PART 7. EMERGING ISSUES IN URBAN AND REGIONAL PLANNING

Visit www.routledge.com/9780415319980 for more details.

Routledge
Taylor & Francis Group

For more details, or to request a copy for review, please contact:

Gemma-Kate Hartley, Senior Marketing Co-ordinator
Gemma-kate.hartley@tandf.co.uk   020 7017 5911